Phosphor Handbook

Phosphor Handbook

Experimental Methods for Phosphor Evaluation and Characterization

Edited by

Ru-Shi Liu
Xiao-Jun Wang

Third Edition

CRC Press
Taylor & Francis Group
Boca Raton London New York

CRC Press is an imprint of the
Taylor & Francis Group, an **informa** business

Third edition published 2022
by CRC Press
6000 Broken Sound Parkway NW, Suite 300, Boca Raton, FL 33487-2742

and by CRC Press
2 Park Square, Milton Park, Abingdon, Oxon, OX14 4RN

© 2022 Taylor & Francis Group, LLC

First edition published by CRC Press 1998

CRC Press is an imprint of Taylor & Francis Group, LLC

ISBN: 978-0-367-55515-3 (hbk)
ISBN: 978-0-367-56622-7 (pbk)
ISBN: 978-1-003-09866-9 (ebk)

DOI: 10.1201/9781003098669

Typeset in Times LT Std
by KnowledgeWorks Global Ltd.

In Memoriam the Early Editors of the Handbook.

Shigeo Shionoya
Formerly of The
    University of Tokyo
Tokyo, Japan

Hajime Yamamoto
Formerly of Tokyo University
    of Technology
Tokyo, Japan

William M. Yen
Formerly of The University
    of Georgia
Athens, GA, USA

Contents

Foreword to the Third Edition of the *Phosphor Handbook*

The field of luminescence and phosphors has a long history, starting from early observations of light in the dark from afterglow materials. Centuries of extensive research followed aimed at providing insight into optical phenomena, now resulting in an increasing role of phosphors in our daily lives. Applications of luminescence grow more diverse and include, for example, phosphors in the color displays that our eyes seem to be glued to, energy-efficient LED lighting, data communication, luminescent probes in medical imaging and sensing, gadgets relying on afterglow phosphors and even luminescent lanthanides in our banknotes. It is interesting to note the central role that Asia has played in the discovery and development of new luminescent materials. Early applications involved afterglow paints in China, creating alternative images in the dark. While fundamental luminescence research was carried out in the 20th century on all continents, there has been a remarkably strong role of Japan and China in research, development and discovery of new luminescence processes and phosphors. It is therefore not surprising that the first edition of the *Phosphor Handbook* (*Keikotai Handobukku*) was initiated by the Phosphor Research Society in Japan in the 1980s.

The first *Phosphor Handbook* was a great book but with an impact limited to those speaking Japanese. Fortunately, about ten years later, the book was translated into English and edited by two giants in the field of luminescence: Shigeo Shionoya and William Yen. It is this version of the book that I acquired soon after it was released and it has been a source of information ever since. All aspects of luminescent materials were covered: phosphor synthesis, optical measuring techniques, fundamentals of luminescence processes, operation principles of light-emitting devices, light and color perception and of course an almost complete overview of all luminescent materials known, indexed by host material and activator ion. I cannot count how often I consulted this book, to quickly look up the optical properties of an ion-host combination, to find a suitable material with specific luminescence characteristics, to understand the operation principles of phosphors in various applications and to learn about careful measurements and analysis of phosphor properties. The authors, except for one, were Japanese, underpinning the central role of Japan in phosphor research.

As the field of luminescence continued to evolve and expand, it became clear that a second edition of the *Phosphor Handbook* was needed. Sadly Shigeo Shionoya has passed away and in 2006, William Yen together with Hajime Yamamoto edited the second edition of the *Phosphor Handbook*. The new edition was updated mostly by asking the original authors to adapt the various chapters to include recent developments. The *Phosphor Handbook* continued to play a prominent role in the luminescence community as a source of information on any topic related to phosphors. Almost 15 years later, it was again time to adapt the Handbook to cover important new developments in the rapidly changing phosphor field where new applications and new materials emerge and also measuring techniques have changed with the introduction of, for example, cheap (pulsed) diode lasers, fiber optics and compact CCD-based spectrometers. Our great colleagues William Yen and Hajime Yamamoto are unfortunately no longer with us and also many of the authors of the various chapters of the first and second editions of the *Phosphor Handbook* have passed on. This made it far from trivial to realize a third edition. We can be extremely grateful that Ru-Shi Liu and Xiao-Jun Wang have taken the initiative to edit and write this third edition of the *Phosphor Handbook*. It is very appropriate that the book is dedicated to the three founders, Shigeo Shionoya, William Yen and Hajime Yamamoto. At the same time, it is appropriate to sincerely thank Ru-Shi Liu and Xiao-Jun Wang for their strong commitment and time invested to organize, write and edit this third edition.

The third edition of the *Phosphor Handbook* is in some aspects different from the two previous editions. The authors are not the same and it is wonderful to see that so many highly respected colleagues in the field have taken the time to contribute their expertise and knowledge to this third edition. Interestingly, again almost all of the authors of this third edition are Asian (with well over 100 contributing authors you can count the non-Asian authors on the fingers of one hand). This illustrates the continued strong position of Asia in phosphor research. Just as in the previous editions, all aspects and the broad scope of phosphor research are covered which makes this Handbook a worthy successor of the previous editions. It will serve as a comprehensive resource describing a wide variety of topics that was also included in the previous editions. It will educate newcomers and help everyone in the field to quickly access all relevant knowledge in the exciting field of phosphor research. In addition to the "classic" topics that continue to be relevant (but sometimes forgotten), many new topics are included, in theory (e.g. first principle calculations), materials (e.g. recent developments in quantum dots and upconversion nanocrystals) and applications (e.g. LED phosphors for NIR sensing and agriculture). All this information no longer fits in a single volume and this third edition is therefore divided into three volumes.

At the time of writing this foreword, I have not read the new edition of the Handbook but did receive an overview of all the chapters and contributing authors. Based on this information, it is clear that the full phosphor community, from students to professors, can benefit from this new comprehensive source of everything you always wanted to know about phosphors – and more. The third edition of the *Phosphor Handbook* will be a classic and continues to promote progress and development of phosphors, in the spirit of the first edition. I look forward to reading it and hope that you as a reader will enjoy exploring this great book and be inspired by it in your research on luminescent materials.

Andries Meijerink
Utrecht, June 2021

Preface to the Third Edition

The last version of the *Phosphor Handbook* was well received by the phosphor research community since its publication in 2007. However, in the last 14 years, many notable advances have occurred. The success of the blue LED (Nobel Prize in Physics, 2014) and its phosphor-converted solid illumination greatly advanced the traditional phosphor research. New phosphorescent materials such as quantum dots, nanoparticles and efficient upconversion, quantum cutting phosphors and infrared broadband emission phosphors have been quickly developed to find themselves in ever-broader applications, from phototherapy to bioimaging, optics in agriculture to solar cell coating. These applications have all expanded beyond the traditional use in lighting and display. All of these developments should be included in the popular Handbook, making it necessary to publish a new version that reflects the most recent developments in phosphor research. Unfortunately, all the three well-respected editors of the previous version have passed away. As their former students and colleagues, we, the editors, feel a strong sense of responsibility to carry on the legacy of the Handbook and to update accordingly to continue serving the phosphor community. The aim of the third edition of the Handbook is to continue to provide an initial and comprehensive source of knowledge for researchers interested in synthesis, characterization, properties and applications of phosphor materials.

The third edition of the Handbook consists of three separate volumes. Volume 1 covers the theoretical background and fundamental properties of luminescence as applied to solid-state phosphor materials. New sections include the rapid developments in principal phosphors in nitrides, perovskite and silicon carbide. Volume 2 provides the descriptions of synthesis and optical properties of phosphors used in different applications, including the novel phosphors for some newly developed applications. New sections include Chapters 5 – Smart Phosphors, 6 – Quantum Dots for Display Applications, 7 – Colloidal Quantum Dots and Their Applications, 8 – Lanthanide-Doped Upconversion Nanoparticles for Super-Resolution Imaging, 9 – Upconversion Nanophosphors for Photonic Application, 16 – Single-Crystal Phosphors, 19 – Phosphors-Converting LEDs for Agriculture, 20 – AC-Driven LED Phosphors and 21 – Phosphors for Solar Cells. Volume 3 addresses the experimental methods for phosphor evaluation and characterization and the contents are widely expanded from the second edition, including the theoretical and experimental designs for new phosphors as well as the phosphor analysis through high pressure and synchrotron studies. Almost all the chapters in the third edition, except for some sections in the Fundamentals of Luminescence, have been prepared by the new faces who are actively and productively working in phosphor research and applications.

We commemorate the memory of the three mentors and editors of the previous editions – Professors Shigeo Shionoya, Hajime Yamamoto and William M. Yen. It was their efforts that completed the original Handbook that guided and inspired numerous graduate students and researchers in phosphor studies and applications. We wish to dedicate this new edition to them.

As the editors, we sincerely appreciate all the contributors from across the world who overcame various difficulties through such an unprecedented pandemic year to finish their chapters on time.

We are grateful to Professor Andries Meijerink of Utrecht University for writing the foreword to the Handbook. We also highly appreciate help from Nora Konopka, Prachi Mishra, and Jennifer Stair of CRC Press/Taylor & Francis Group and perfect editing work done by Garima Poddar of KGL. Finally, we hope that this third edition continues the legacy of the Handbook to serve as a robust reference for current and future researchers in this field.

Coeditors:
Ru-Shi Liu
Taipei, Taiwan

Xiao-Jun Wang
Statesboro, GA, USA
May, 2021

Preface to the Second Edition

We, the editors as well as the contributors, have been gratefully pleased by the reception accorded to the *Phosphor Handbook* by the technical community since its publication in 1998. This has resulted in the decision to reissue an updated version of the Handbook. As we had predicted, the development and the deployment of phosphor materials in an ever-increasing range of applications in lighting and display have continued its explosive growth in the last decade. It is our hope that an updated version of the Handbook will continue to serve as the initial and preferred reference source for all those interested in the properties and applications of phosphor materials.

For this new edition, we have asked all the authors we could contact to provide corrections and updates to their original contributions. The majority of these responded and their revisions have been properly incorporated in the present volume. It is fortunate that the great majority of the material appearing in the first edition, particularly those sections summarizing the fundamentals of luminescence and describing the principal classes of light-emitting solids, maintains its currency and hence its utility as a reference source.

Several notable advances have occurred in the last decade, which necessitated their inclusion in the second edition. For example, the wide dissemination of nitride-based LEDs opens the possibility of white light solid-state lighting sources that have economic advantages. New phosphors showing the property of "quantum cutting" have been intensively investigated in the last decade and the properties of nanophosphors have also attracted considerable attention. We have made an effort, in this new edition, to incorporate tutorial reviews in all of these emerging areas of phosphor development.

As noted in the preface of the first edition, the Handbook traces its origin to one first compiled by the Phosphor Research Society (Japan). The society membership supported the idea of translating the contents and provided considerable assistance in bringing the first edition to fruition. We continue to enjoy the cooperation of the Phosphor Research Society and value the advice and counsel of the membership in seeking improvements in this second edition.

We have been, however, permanently saddened by the demise of one of the principals of the society and the driving force behind the Handbook itself. Professor Shigeo Shionoya was a teacher, a mentor and a valued colleague who will be sorely missed. We wish then to dedicate this edition to his memory as a small and inadequate expression of our joint appreciation.

We also wish to express our thanks and appreciation for the editorial work carried out flawlessly by Helena Redshaw of Taylor & Francis.

William M. Yen
Athens, GA, USA

Hajime Yamamoto
Tokyo, Japan
December, 2006

Preface to the First Edition

This volume is the English version of a revised edition of the *Phosphor Handbook* (*Keikotai Handobukku*) that was first published in Japanese in December, 1987. The original Handbook was organized and edited under the auspices of the Phosphor Research Society (in Japan) and issued to celebrate the 200th Scientific Meeting of the Society that occurred in April, 1984.

The Phosphor Research Society is an organization of scientists and engineers engaged in the research and development of phosphors in Japan that was established in 1941. For more than half a century, the Society has promoted interaction between those interested in phosphor research and has served as a forum for discussion of the most recent developments. The Society sponsors five annual meetings; in each meeting, four or five papers are presented reflecting new cutting-edge developments in phosphor research in Japan and elsewhere. A technical digest with extended abstracts of the presentations is distributed during these meetings and serves as a record of the proceedings of these meetings.

This Handbook is designed to serve as a general reference for all those who might have an interest in the properties and/or applications of phosphors. This volume begins with a concise summary of the fundamentals of luminescence and then summarizes the principal classes of phosphors and their light-emitting properties. Detailed descriptions of the procedures for synthesis and manufacture of practical phosphors appear in later chapters and in the manner in which these materials are used in technical applications. The majority of the authors of the various chapters are important members of the Phosphor Research Society and they have all made significant contributions to the advancement of the phosphor field. Many of the contributors have played central roles in the evolution and remarkable development of lighting and display industries of Japan. The contributors to the original Japanese version of the Handbook have provided English translations of their articles; in addition, they have all updated their contributions by including the newest developments in their respective fields. A number of new sections have been added to this volume to reflect the most recent advances in phosphor technology.

As we approach the new millennium and the dawning of a radical new era of display and information exchange, we believe that the need for more efficient and targeted phosphors will continue to increase and that these materials will continue to play a central role in technological developments. We, the coeditors, are pleased to have engaged in this effort. It is our earnest hope that this Handbook becomes a useful tool to all scientists and engineers engaged in research in phosphors and related fields and that the community will use this volume as a daily and routine reference, so that the aims of the Phosphor Research Society in promoting progress and development in phosphors is fully attained.

Coeditors:
Shigeo Shionoya
Tokyo, Japan

William M. Yen
Athens, GA, USA
May, 1998

About the Authors

Ru-Shi Liu

Professor Ru-Shi Liu received his Bachelor's degree in Chemistry from Soochow University (Taiwan) in 1981. He got his Master's Degree in nuclear science from the National Tsing Hua University (Taiwan) in 1983. He obtained two PhD degrees in Chemistry from National Tsing Hua University in 1990 and the University of Cambridge in 1992. He joined Materials Research Laboratories at Industrial Technology Research Institute as an Associate Researcher, Research Scientist, Senior Research Scientist and Research Manager from 1983 to 1995. Then he became an Associate Professor at the Department of Chemistry of the National Taiwan University from 1995 to 1999. Then he was promoted to a Professor in 1999. In July 2016, he became the Distinguished Professor.

He got the Excellent Young Person Prize in 1989, Excellent Inventor Award (Argentine Medal) in 1995 and Excellent Young Chemist Award in 1998. He got the 9th Y. Z. Hsu scientific paper award due to the excellent energy-saving research in 2011. He received the Ministry of Science and Technology awards for distinguished research in 2013 and 2018. In 2015, he received the distinguished award for Novel and Synthesis by IUPAC & NMS. In 2017, he got the Chung-Shang Academic paper award. He got "Highly Cited Researchers" by Clarivate Analytics in 2018 and 2019. He got Hou Chin-Tui Award in 2018 due to the excellent research on basic science. He got the 17th Y. Z. Hsu Chair Professor award for the contribution to the excellent research on "Green Science & Technology" in 2019. He then got the 26th TECO award for the contribution to make the combination of the academic and practical application of materials chemistry in 2019. He got the Academic Award of the Ministry of Education and the Academic Achievement Award of the Chemical Society Located in Taipei in 2020.

His research is concerning with Materials Chemistry. He is the author and coauthor of more than 600 publications in international scientific journals. He has also granted more than 200 patents.

Xiao-Jun Wang

Professor Xiao-Jun Wang obtained his BS degree in physics from the Jilin University in 1982, his MS degrees in physics from the Chinese Academy of Sciences in 1985 and the Florida Institute of Technology in1987, and his PhD in physics from The University of Georgia in 1992 (supervisors: William M. Yen and William M. Dennis). He served as a Research Associate from 1992 to 1993 at the University Laser Center of Oklahoma State University, and then received a fellowship from the National Institutes of Health (NIH) as a Postdoctoral Fellow from 1993 to 1995 at the Beckman Laser Institute of the University of California, Irvine. In 1995, he received an NIH training grant and joined Georgia Southern University as an Assistant Professor. He was promoted to full professor in 2004 and continues to teach there today.

List of Contributors

Michele Back
Ca'Foscari University of Venice
Mestre-Venezia, Italy

Mikhail G. Brik
University of Tartu
Tartu, Estonia

Ming-Hsien Chan
Academia Sinica
Taipei, Taiwan

Lei Chen
Hefei University of Technology
Hefei, China

Xueyuan Chen
Fujian Institute of Research on the Structure
of Matter, CAS
Fuzhou, China

Mu-Huai Fang
National Taiwan University
Taipei, Taiwan

Chen Gao
University of Chinese Academy
of Sciences
Beijing, China

Michael Hsiao
Academia Sinica
Taipei, Taiwan

Decai Huang
Xiamen Institute of Rare Earth Materials,
Haixi Institutes, CAS
Xiamen, China

Wen-Tse Huang
National Taiwan University
Taipei, Taiwan

Won Bin Im
Hanyang University
Seoul, Republic of Korea

Yoon Hwa Kim
Korea Advanced Institute of Science and
Technology
Daejeon, Republic of Korea

Guogang Li
China University of Geosciences
Wuhan, China

Sisi Liang
Xiamen Institute of Rare Earth Materials,
Haixi Institutes, CAS
Xiamen, China

Chun Che Lin
National Taipei University of Technology
Taipei, Taiwan

Quanlin Liu
University of Science and Technology Beijing
Beijing, China

Ru-Shi Liu
National Taiwan University
Taipei, Taiwan

Chong-Geng Ma
Chongqing University of Posts and
Telecommunications
Chongqing, China

Sebastian Mahlik
University of Gdansk
Gdansk, Poland

Celina Matuszewska
Hong Kong Baptist University
Hong Kong, China

Zhiqiang Ming
University of Science and Technology Beijing
Beijing, China

Michal Piasecki
Jan Dlugosz University
Częstochowa, Poland

Anatoli I. Popov
University of Latvia
Riga, Latvia

Sarmad Ahmad Qamar
National Taipei University of Technology
Taipei, Taiwan

Zhen Song
University of Science and Technology Beijing
Beijing, China

Kengo Suzuki
Hamamatsu Photonics K.K.
Hamamatsu, Japan

Setsuhisa Tanabe
Kyoto University
Kyoto, Japan

Jumpei Ueda
Kyoto University
Kyoto, Japan

N. S. M. Viswanath
Chonnam National University
Gwangju, Republic of Korea

Yi Wei
China University of Geosciences
Wuhan, China

Ka-Leung Wong
Hong Kong Baptist University
Hong Kong, China

Zhiguo Xia
South China University of Technology
Guangzhou, China

Jian Xu
National Institute for Materials
   Science
Tsukuba, Japan

Tomoyuki Yamamoto
Waseda University
Tokyo, Japan

Shuai Zha
Hong Kong Baptist University
Hong Kong, China

Ming Zhao
University of Science and Technology Beijing
Beijing, China

Haomiao Zhu
Xiamen Institute of Rare Earth Materials,
   Haixi Institutes, CAS
Xiamen, China

1 First-Principles Methods as a Powerful Tool for Fundamental and Applied Research in the Field of Optical Materials

Mikhail G. Brik, Chong-Geng Ma, Tomoyuki Yamamoto, Michal Piasecki, and Anatoli I. Popov

CONTENTS

1.1 INTRODUCTION

Since the invention of lasers in 1960s the world witnesses a rapid development of numerous devices such as abovementioned lasers, light-emitting diodes, and displays, whose applications are entirely based on various optical materials, that make an important part of all of those. Very often, although not always, those optical materials are crystalline solids, in the form of single crystals, polycrystalline powders, ceramics, containing a small amount of different transition metal or rare-earth ions. These positively charged ions, which can be introduced artificially by different chemical methods, replace one of the host's cations forming a substitutional impurity center, that consists of that impurity ion and its nearest neighbors (called ligands) in the crystal lattice. A key feature of those impurity ions is that they have incomplete, or not filled, electron shells. Various interactions among the electrons in those unfilled shells give rise to a large number of energy levels, many of which are highly degenerated. In crystalline solids, these energy levels are split into a number of sublevels, and optical transitions between these states with emission or absorption of photons can take place. Those split energy levels of the transition metal and rare-earth ions cover a wide spectral range, from infrared through visible to ultraviolet, thus creating numerous opportunities for applications.

One of the most important consequences of introducing impurities into solids is that some of the electronic energy levels of impurity ions are located somewhere inside the host's band gap. The

DOI: 10.1201/9781003098669-1

relative position of those energy levels with respect to the valence and conduction bands of a host plays a decisive role in determining whether a given material can be suitable for a particular application. Therefore, the impurity ions and the potential hosts should not be considered independently of each other; the interplay of their electronic properties appears to be a crucial factor to determine whether the combination of the host's properties and impurity energy levels would ensure the best solution.

The understanding of electronic properties of already existing optical materials and efficient search for new ones can be facilitated by application of the modern first-principles methods based on the density functional theory (DFT) for an in-depth analysis of their structural, electronic, optical, elastic, thermodynamic properties and finding correlations – sometimes not obvious ones – between all those. Following smoothly or drastically changing trends in characteristics of optical materials that can be extracted from the results of the DFT-based calculations, it may be possible to avoid many unsuccessful experiments and narrow down the field of search for new materials with superior performance.

In view of these arguments, systematic and consistent applications of the same computational approach to large numbers of isostructural compounds with varying composition or impurity ions is a powerful tool for smart materials engineering. Even sometimes, the perfect reproduction of some key parameters of solids is not possible (e.g. the band gaps are always underestimated, if the common DFT is used), the trends across different families of compounds are extremely helpful in unveiling the "structure–property" and "property–property" relations, that are of paramount importance in practical materials science. Moreover, such comparative studies of large number of isostructural materials constitute a necessary prerequisite for the machine learning-based materials science [1, 2], and the first-principles methods are a very important tool for achieving rapid progress in that direction [3].

In this chapter, we consider in detail several examples of such calculations and describe the main conclusions that can be extracted from them. At first, the scrutinized studies of neat hosts that can be suitable for doping with many impurity ions are given. The spinel and elpasolite families of compounds were chosen for that purpose, and the dependences of their structural, electronic, optical, elastic, thermodynamic properties on the composition are described. An important view of chemical stability of solids and the way of its assessing from the first-principles point of view is also addressed. As the second step, the optical materials doped with different transition metal (Ti^{3+}, Mn^{4+}) and rare earth (Ce^{3+}) are analyzed in the frame of DFT-based techniques. Special attention is paid to the location of the impurity ions energy levels in the host's band gap and the practical implications of this information.

1.2 HIGH-THROUGHPUT FIRST-PRINCIPLES CALCULATIONS AS APPLIED TO ISOSTRUCTURAL COMPOUNDS

1.2.1 AB_2X_4 (A = Be, Mg, Ca, Sr, Ba; B = Al, Ga, In; X = O, S) Spinel Compounds

Spinel crystals have the general chemical formula AB_2X_4. In their cubic crystal structure described by the $Fd\overline{3}m$ space group (no. 227) there are different sites available for the A and B cations. As a rule, the divalent cations A and the trivalent B cations are four-fold and six-fold coordinated by the X anions, respectively. This is the so-called normal spinel structure. Sometimes, however, the position exchange between both types of cations can exist, and the "inverse" spinel structure can be formed, in which all the A cations occupy the half of the B cations octahedral sites, that can be shown by the $B(AB)X_4$ formula. Intermediate distributions can also be realized, covering the whole range between the normal and inverse spinels. Applications of the spinel crystals are numerous, and some of them can be mentioned such as lighting, lasing, optical sensing, displays manufacturing, and solar cell production [4–12]; various structural defects (especially caused by radiation) were studied in Refs. [13–15]. Many spinel compounds have wide band gap, and because of that they can easily be doped with different transition metal and rare-earth ions, to be used as the phosphor materials or solid-state laser media [16–19].

Since spinel compounds are so important, there exist considerable amount of publications that simultaneously consider a large number of these compounds. For example, the structural properties and stability of spinels were analyzed in Refs. [20, 21]; it was established that using the ionic radii and electronegativities of the constituting elements it is possible to perform empirical estimations of the lattice constant and formulate certain criteria of stability for existing and potential spinel compounds.

In this paragraph we consider 30 spinel compounds AB_2X_4 (A = Be, Mg, Ca, Sr, Ba; B = Al, Ga, In; X = O, S). Their structural, electronic, elastic, and thermodynamic properties were calculated using the periodic ab initio CRYSTAL14 code based on the LCAO (linear combination of atomic orbitals) method with local Gaussian-type basis sets (BSs) [22]. All relevant details of calculations can be found in Ref. [23].

As the first step of all calculations, the crystal structures were optimized for all 30 considered spinels. The initial structural inputs were built up based on the fact that the coordinates of the A^{2+} and B^{3+} cations occupy the tetrahedral (0, 0, 0) and octahedral (0.625, 0.625, 0.625) positions in terms of the cubic lattice constant a. The X^{2-} anions are located at the (u, u, u) position, the value of which is close to 0.25 and slightly varies from host to host depending on the chemical composition. The experimental structural data for only four spinels from the considered group are known: $MgAl_2O_4$, $MgGa_2O_4$, $MgIn_2O_4$, and $MgIn_2S_4$. The lattice constants of the remaining 26 spinels were estimated with the help of the empirical equation derived in Ref. [21] and used to create unit cells for optimization. All atoms were allowed to relax until the convergence criteria were reached. As a result, the maximum difference between the empirical and optimized lattice constants in this group of spinels was only about 3%. Figure 1.1 visualizes the trends in variation of the lattice constants among these 30 compounds when the chemical composition is changed.

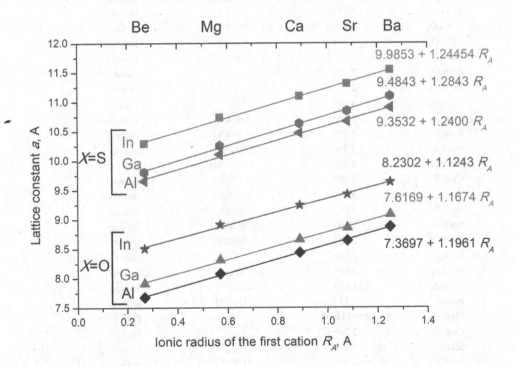

FIGURE 1.1 Variation of the calculated lattice constant versus ionic radius of the A cation in the group of the AB_2X_4 (A = Be, Mg, Ca, Sr, Ba; B = Al, Ga, In; X = O, S) spinels. The calculated values and the linear fits are shown by the symbols and solid lines, respectively. Linear fit equations are shown in the right part of the figure. Second cations B are indicated above the top horizontal line. (Adapted with permission from Ref. [23].)

It follows from Figure 1.1 that the lattice constant is a linear function of the first cation A (if the anion X is fixed), which gives an easy way of estimating the lattice constant through the linear fit equations given in the figure.

After the geometrical structures were optimized, the electronic properties of all spinels from the considered group were calculated. The obtained values of the band gaps are listed in Table 1.1. The character of the band gaps (direct/indirect) is also given.

An easy way to visualize the variation is to plot a two-dimensional color diagram, as shown in Figure 1.2. By comparing the numbers in the corresponding squares, one can easily see that the calculated band gaps for the oxide spinels are always wider than for their sulfide counterparts. This is in line with increased lattice constants of the sulfide spinels. An overall trend is that with increased

TABLE 1.1

Calculated Band Gaps E_g (eV) for Various Spinel Compounds AB_2X_4 (A = Be, Mg, Ca, Sr, Ba; B = Al, Ga, In; X = O, S) with Space Group $Fd3m$

Spinel	Band Gaps		Transition Types
	Γ–Γ	Γ–Another k Point	
$BeAl_2O_4$	8.439		Direct
$MgAl_2O_4$	7.160 (**7.8**[a])	7.147(SM)	Indirect
$CaAl_2O_4$	6.475	6.122(X)	Indirect
$SrAl_2O_4$	5.839	5.352(X)	Indirect
$BaAl_2O_4$	4.762	4.190(X)	Indirect
$BeGa_2O_4$	5.806	5.804(SM)	Indirect
$MgGa_2O_4$	5.132		Direct
$CaGa_2O_4$	4.746	4.637(X)	Indirect
$SrGa_2O_4$	4.288	3.904(X)	Indirect
$BaGa_2O_4$	3.407	2.807(X)	Indirect
$BeIn_2O_4$	3.352		Direct
$MgIn_2O_4$	3.529 (**3.4**[b])		Direct
$CaIn_2O_4$	3.843		Direct
$SrIn_2O_4$	3.630	3.604(L)	Indirect
$BaIn_2O_4$	3.040	2.864(X)	Indirect
$BeAl_2S_4$	3.554	3.508(SM)	Indirect
$MgAl_2S_4$	4.420	4.385(SM)	Indirect
$CaAl_2S_4$	4.457	4.278(X)	Indirect
$SrAl_2S_4$	3.913	3.581(X)	Indirect
$BaAl_2S_4$	3.289	2.812(X)	Indirect
$BeGa_2S_4$	2.160	2.013(SM)	Indirect
$MgGa_2S_4$	2.727	2.705(SM)	Indirect
$CaGa_2S_4$	2.610		Direct
$SrGa_2S_4$	2.152	1.982(X)	Indirect
$BaGa_2S_4$	1.659	1.262(X)	Indirect
$BeIn_2S_4$	2.350	2.313(SM)	Indirect
$MgIn_2S_4$	2.962 (**2.1**[c])		Direct
$CaIn_2S_4$	2.795		Direct
$SrIn_2S_4$	2.494		Direct
$BaIn_2S_4$	2.167	2.027(X)	Indirect

If available, the experimental data are given in bold. [a]Ref. [24], [b]Ref. [25], [c]Ref. [26].

Source: Reprinted with permission from Ref. [23].

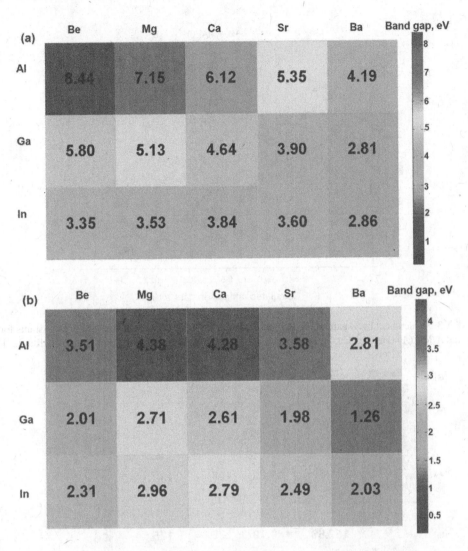

FIGURE 1.2 Relations between the calculated band gaps (eV) and composition for the AB_2O_4 (a) and AB_2S_4 (b) (A = Be, Mg, Ca, Sr, Ba; B = Al, Ga, In) spinels. (Reprinted with permission from Ref. [23].)

atomic numbers of both cation and anions the calculated band gap decreases in the considered group of spinels, as is depicted in Figure 1.3, where the direct band gaps at the Brillouin zone center are plotted versus calculated lattice constant.

Increase of the lattice constant and the interatomic separation for the spinels with greater atomic numbers of both cations and anions can be accompanied by the chemical bond softening, which further manifests itself in decreased values of the bulk modulus. Figure 1.4 confirms such an expectation by showing the calculated values of the bulk moduli (in GPa). $BeAl_2O_4$ appears to be the hardest material among all those considered, with the bulk modulus value of 228 GPa, whereas $BaIn_2S_4$ is the softest one, the bulk modulus of which is more than three times smaller (about 61 GPa only). A consequence of these results for the optical properties can be that the Stokes shift (the difference between the lowest in energy absorption and emission band) and associated with it thermal losses of the electron excitation energy may be anticipated to increase in the hosts with smaller elastic constants and easily deformable chemical bonds.

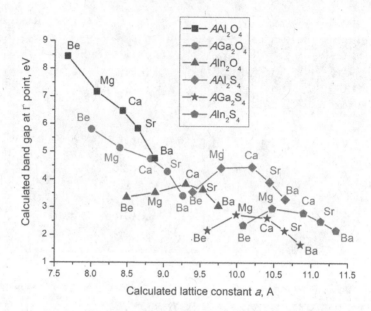

FIGURE 1.3   Calculated band gaps at the Brillouin zone center (eV) versus calculated lattice constants for the AB_2O_4 and AB_2S_4 (A = Be, Mg, Ca, Sr, Ba; B = Al, Ga, In) spinels. (Adapted with permission from Ref. [23].)

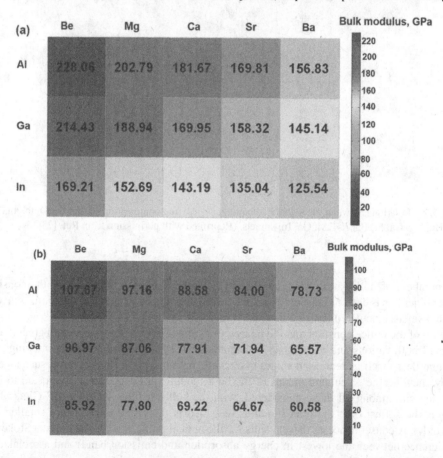

FIGURE 1.4   Relations between the calculated bulk moduli (GPa) and composition for the AB_2O_4 (a) and AB_2S_4 (b) (A = Be, Mg, Ca, Sr, Ba; B = Al, Ga, In) spinels. (Reprinted with permission from Ref. [23].)

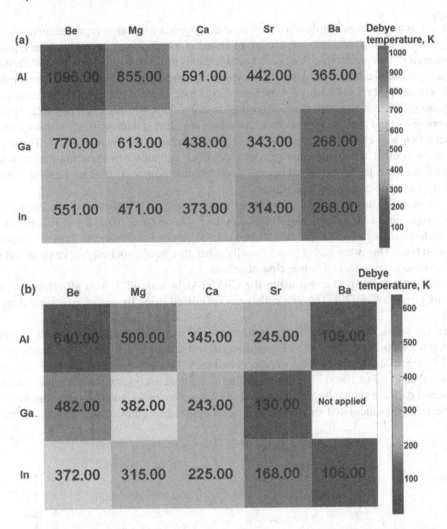

FIGURE 1.5 Relations between the calculated Debye temperature (K) and composition for the AB_2O_4 (a) and AB_2S_4 (b) (A = Be, Mg, Ca, Sr, Ba; B = Al, Ga, In) spinels. (Reprinted with permission from Ref. [23].)

As an additional illustration of the power of the first-principles calculations and their wide area of applicability, we show in Figure 1.5 calculated Debye temperatures for all these spinels. This parameter is proportional to the highest energy of the acoustic phonons in a given host, which is important to know when analyzing non-radiative transitions in solids. All necessary equations can be found in Refs. [23, 27]. The highest Debye temperature is again for the spinel composed of the lightest atoms in the considered group – $BeAl_2O_4$ (1096K), whereas $BaInl_2S_4$ has the lowest Debye temperature of 106K only. No value of Debye temperature for $BaGa_2S_4$ (in fact, an imaginary value was obtained, which is unphysical) indicates structural instability of this compound.

1.2.2 Cs_2NaLNX_6 (LN = La, ..., Lu, X = F, Cl, Br, I) Cubic Elpasolites

Another large group of materials considered in this chapter consists of 60 cubic elpasolites, the general chemical formula of which is M_2ALnX_6, where M and A are the monovalent cations of the alkali metals, Ln is any trivalent metal (lanthanide, actinide, transition metal, or aluminum), and X = F, Cl, Br, I is a halogen. Since many different chemical elements may be involved to make elpasolites, it is a very large family of isostructural compounds with many representative members. They crystallize in the $Fm\bar{3}m$ cubic space group or one of its subgroups, and only cubic elpasolites are considered in this paragraph. Both Ln^{3+} and A^+ cations are coordinated by six halide anions X^- making an ideal octahedron with O_h local site symmetry, whereas the M^+ cations are surrounded by 12 halide anions X^- at the sites with T_d symmetry [28]. Different trivalent cations can easily replace the Ln^{3+} cations without any need for charge compensation. This minimizes all possible crystal lattice deformations, and because of that cubic elpasolites are ideal systems to study impurity ions and their interaction with crystal lattice vibrations in the perfect octahedral crystal field. This circumstance, taken together with wide band gaps of elpasolites, explains why there are many examples of studies focused on the spectroscopic and crystallographic properties of the transition metal and rare-earth ions in cubic elpasolites [29–38].

Here we describe first-principles calculations of the structural, electronic, elastic, and thermodynamic properties of 60 halide elpasolites Cs_2NaLnX_6 (Ln = La, …, Lu; X = F, Cl, Br, I). Four groups, each consisting of 15 compounds, were selected among the considered crystals, depending on the anion type. They were called (conditionally, since this would not be a proper chemical term) fluoride, chloride, bromide, and iodide elpasolites.

All calculations were performed using the CRYSTAL14 code [22], with all relevant calculation details given in Ref. [39]. The unit cells were optimized using the initial structural data from Ref. [31].

Figure 1.6 shows how the calculated lattice constants in this group of materials depend on the lanthanide atomic number. The obtained numerical values of the lattice constants are represented by the filled symbols; they were approximated by the linear functions of the Ln ions atomic number Z (solid lines). The linear fitting equations are given for each line. Available experimental data are shown by the open symbols for some elpasolites; good agreement demonstrated between the calculated and measured lattice constants in these cases serves as a proof of

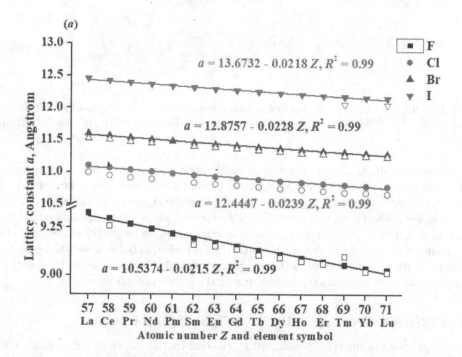

FIGURE 1.6 Dependence of the calculated lattice constants for the Cs_2NaLnX_6 (Ln = La, …, Lu; X = F, Cl, Br, I) elpasolites. The calculated values are shown by the filled symbols, whereas the empty symbols indicate the available experimental data [31]. (Reprinted with permission from Ref. [39].)

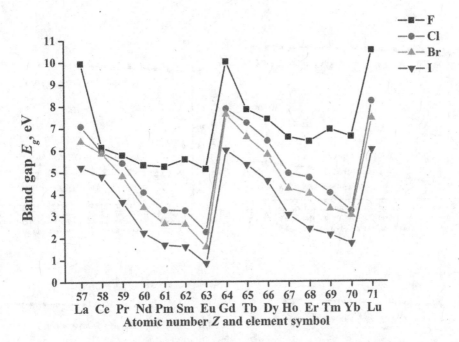

FIGURE 1.7 Variation of the calculated band gaps with the Ln atomic number for the Cs_2NaLnX_6 (Ln = La, ..., Lu; X = F, Cl, Br, I) elpasolites. (Reprinted with permission from Ref. [39].)

reliability of the predicted structural data for other (unexamined experimentally yet) considered elpasolites. The lattice constants decrease linearly with the lanthanide atomic number, which is due to the well-known lanthanide contraction (decrease of the ionic radii of rare-earth ions) with increasing atomic number. On the other hand, the lattice constants increase with increased anion atomic number, if keeping the same cations. The calculated band gaps of all 60 elpasolites in this group are plotted in Figure 1.7.

The calculated band gaps decrease with increasing anion atomic number (while keeping the same cations). The lanthanide fluoride elpasolites are characterized by the widest band gaps, about 10 eV for Cs_2NaLaF_6 and 10.5 eV for Cs_2NaLuF_6. The iodide elpasolite has more narrow band gaps, with the smallest band gap for Cs_2NaEuI_6 – just about 0.8 eV. If the variation of the calculated band gaps with varied Ln cations while keeping the same anion is considered, then the "zigzag" behavior can be observed, which is in line with the calculated band gaps of Ln^{3+}-doped $LaSi_3N_5$ (like in Figure 11 of Ref. [40]) and with the P. Dorenbos model describing location of the lowest 4f and 5d levels of the di- and trivalent lanthanides in inorganic solids [41].

After performing optimization of the optimized crystal structures, it is possible to analyze the elastic properties. For the cubic crystal structures, there are only three independent elastic tensor components: C_{11}, C_{12}, and C_{44}. Variation of these constants and bulk modulus B depicted in Figure 1.8 show the following trends across the considered groups of elpasolite crystals:

1. Values of all elastic parameters increase with growing Ln atomic number because of decreased chemical bond lengths and subsequent bonds stiffening when moving from La to Lu while keeping the same anion.
2. Values of all elastic parameters decrease with growing anion's atomic number keeping the same Ln element, because of increased anion size and atomic number thus leading to the increased chemical bond lengths and their weakening.

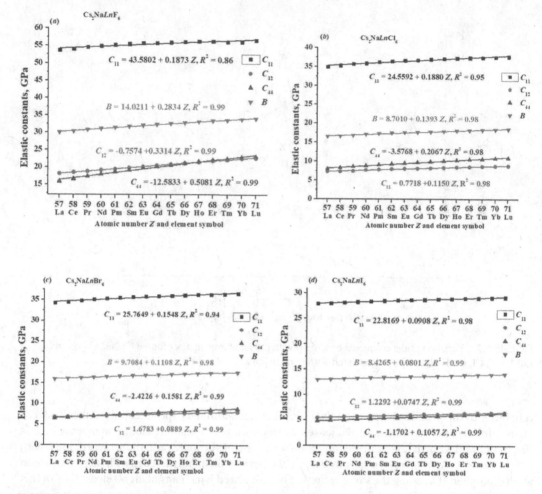

FIGURE 1.8 Dependence of the calculated elastic constants for the Cs_2NaLnX_6 (Ln = La, ..., Lu; X = F (a), Cl (b), Br (c), I (d)) elpasolites. The calculated values are shown by the symbols, whereas the linear fits are shown by the solid lines along with the linear fitting equations. (Reprinted with permission from Ref. [39].)

The knowledge of the elastic constants gives opportunity to estimate Debye temperature of solids. Figure 1.9 shows the calculated Debye temperatures for all 60 studied elpasolite systems. The variation follows the above-described trends in the elastic properties: a gradual increase of the Debye temperature toward the end of the lanthanide series within each of the four groups with different halide ions can be noticed. The Debye temperature of the considered elpasolites decreases with increased halogen atomic number in the series if the same Ln cations are kept in the chemical composition.

A set of important conclusions can be drawn from the analysis of the vibrational properties of solids. Elpasolite crystals have 27 normal vibrational modes and many of which are degenerate. The $k = 0$ vibrations of these systems may be classified in terms of the moiety modes [v_i ($i = 1$–6)] of the LnX_6^{3-} ionic group [42], or the [S_i ($i = 1$–10)] vibrations of the (Bravais) unit cell [43]. The S_6, S_7, S_8, and S_9 modes are infrared active, while the S_1, S_2, S_4, and S_5 modes are Raman active, as indicated by our calculation results. Figure 1.10 shows the variation of the calculated normal modes frequencies for 60 considered elpasolites. It can be found that the vibrational frequencies of those modes related to the LnX_6^{3-} complexes (i.e., six vibrational modes S_1, S_2, S_4, S_6,

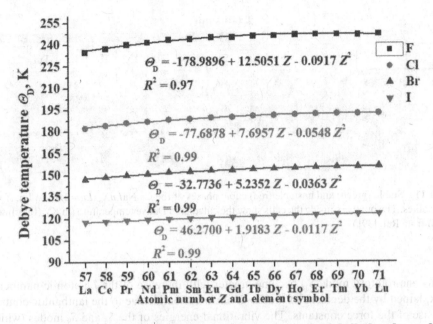

FIGURE 1.9 Dependence of the calculated Debye temperature for the Cs_2NaLnX_6 (Ln = La, ..., Lu; X = F, Cl, Br, I) elpasolites. The calculated values are shown by the symbols, whereas the quadratic fits are shown by the solid lines along with the quadratic fit equations. (Reprinted with permission from Ref. [39].)

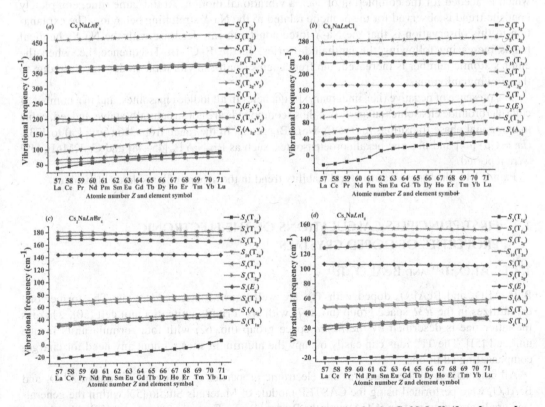

FIGURE 1.10 Calculated vibrational frequencies of the normal modes for the Cs_2NaLnX_6 (Ln = La, ..., Lu; X = F (a), Cl (b), Br (c), I (d)) elpasolites. The lines are the guides to the eye only. (Reprinted with permission from Ref. [39].)

Lanthanides

	La	Ce	Pr	Nd	Pm	Sm	Eu	Gd	Tb	Dy	Ho	Er	Tm	Yb	Lu
F	235	238	241	242	244	244	246	245	246	246	247	247	247	247	248
Cl	182	184	186	187	188	189	190	190	191	191	192	192	193	193	193
Br	147	149	150	151	152	153	153	154	154	155	155	155	156	156	157
I	118	118	119	119	120	120	121	121	121	122	122	123	123	123	124

Halides

FIGURE 1.11 Stable (green) and unstable (red) cubic phases of the Cs_2NaLnX_6 (Ln = La, ..., Lu; X = F, Cl, Br, I) elpasolites. The numbers inside the cells show the calculated Debye temperature (in K). (Reprinted with permission from Ref. [39].)

S_7, S_{10} plus one rotatory mode S_3) become greater with increase of the Ln atomic number. This can be explained by the decrease of the Ln–X chemical bonds due to the lanthanide contraction and increase of the force constants. The vibrational energies of the S_5 and S_9 modes (which are related to the Cs^+ translatory motion) also increase across the lanthanide series. This is because the increase of the host lattice rigidity with Ln atomic number makes vibrational displacement of the Cs^+ ion from its crystallographic site to be more difficult and greater amount of energy, which is needed for the completion of such a vibrational motion. At the same time, completely opposite trend is observed for the S_8 mode related to the Na-X stretching behavior. The explanation to this observation is that the Na-X force constants can decrease with the Na-X chemical bonds increasing following the lanthanide series. In the F–Cl–Br–I sequence, i.e. when the anion's atomic number is increased, the frequencies of all the vibration modes decrease due to the growing anion mass.

The presence of negative (i.e., imaginary) frequencies in all iodide elpasolites, and in a number of chloride/bromide elpasolites suggests that those compounds are not structurally stable. Indeed, cubic to tetragonal phase transitions for the $Cs_2NaLnBr_6$ group, in the range from 316 (Ln = La) to 173K (Ln = Gd) [44] and in some hexafluoroelpasolites, such as Rb_2NaYF_6 [45–48] and Cs_2KMnF_6 [49] were reported.

Figure 1.11 illustrates this structural stability trend in the groups of considered elpasolites.

1.3 FIRST-PRINCIPLES CALCULATIONS OF THE ELECTRONIC PROPERTIES OF DOPED CRYSTALS

1.3.1 Al_2O_3:Ti^{3+} AND $BeAl_2O_4$:Ti^{3+}

Both Al_2O_3 and $BeAl_2O_4$ doped with Ti^{3+} are well-known optical materials. The former one crystallizes in the $R\bar{3}c$ space group (no. 167) with six formula units in a unit cell [50], whereas the latter one is described by the $Pnma$ space group (no. 62) with four formula units in one unit cell [51]. The Ti^{3+} ions can easily occupy the aluminum sites without any need for charge-compensating additions.

All calculations of the structural and electronic properties of neat and Ti^{3+}-doped Al_2O_3 and $BeAl_2O_4$ were performed using the CASTEP module of Materials Studio [52] within the generalized gradient approximation (GGA) with the Perdew-Burke-Ernzerhof functional [53]. All relevant calculating details are given in Refs. [54, 55]. Optimization of crystal lattices returned good agreement with the experimental data (Table 1.2).

TABLE 1.2

Experimental and Calculated Lattice Constants for Al_2O_3 and $BeAl_2O_4$

	Al_2O_3				$BeAl_2O_4$	
	Exp.	Calc. (This Work)	Calc. [56]	Calc. [57]	Exp. [51]	Calc. (This Work)
a, Å	4.7606	4.8090	4.790	4.8211	9.4019	9.5184
b, Å	4.7606	4.8090	4.790	4.8211	5.4746	5.5425
c, Å	12.994	13.1264	13.076	13.1609	4.4259	4.4911

The calculated band structures of Al_2O_3 and $BeAl_2O_4$ are depicted in Figure 1.12, whereas Figure 1.13 shows the density of states (DOSs) diagrams for both compounds.

Both compounds have direct band gaps. For Al_2O_3 it was calculated to be 5.937 eV, which agrees with previously calculated results, e.g. 6.26 [58], 6.1 [59], and 6.5 eV [57]. All these calculated band gaps are underestimated by 2.4–2.7 eV when compared to the experimental value of 8.7 eV [60]. Such underestimation of the DFT-calculated band gaps is a usual phenomenon; it can be corrected by introducing the scissor operator, the action of which is merely restricted to the upward shift of the conduction band. In the case of $BeAl_2O_4$ the band gap was calculated to be 6.094 eV, which is consistent with another calculated result of 6.45 eV [61], and which is, of course, underestimated if compared to the experimental value of about 9 eV [62].

The composition of the calculated electronic bands (Figure 1.13) in the case of Al_2O_3 is as follows: the conduction band is made of the Al unoccupied 3s, 3p states. The valence band with the width of about 7 eV is composed of the O 2p states. A narrow deeply located band between −19 and −16 eV arises from the O 2s states; the 3s and 3p states of Al produce a minor contribution to the oxygen states because of mixing with the neighboring oxygen ions orbitals.

The electronic bands in $BeAl_2O_4$ have the following origin: the conduction band is made of the Be 2s, Al 3s, 3p states, whereas the oxygen 2p states are clearly dominating in the valence band.

To analyze the difference between the neat and Ti^{3+}-doped hosts, one Al^{3+} ion in the unit cells of these two compounds was replaced by one Ti^{3+} ion. The calculated electronic band structures are shown in Figure 1.14.

There are well-seen nearly dispersionless localized 3d states of Ti^{3+} ions in the band gap of both crystals. The Ti^{3+} ion has a single 3d electron and only one LS term ^2D. The 3d orbitals in an octahedral crystal field are split into the t_2 (lower in energy) and e (higher in energy) states. The separation between them is called the crystal field strength and is usually denoted as $10Dq$ [63].

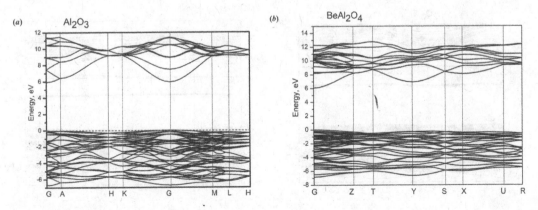

FIGURE 1.12 Calculated band structure of Al_2O_3 (a) and $BeAl_2O_4$ (b). (Reproduced with permission from Refs. [54, 55].)

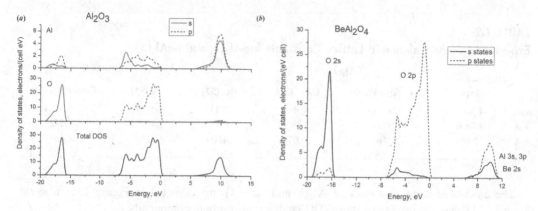

FIGURE 1.13   Calculated density of states (DOSs) diagrams for Al_2O_3 (a) and $BeAl_2O_4$ (b). (Reproduced with permission from Refs. [54, 55].)

The calculated DOS diagrams for both Al_2O_3:Ti^{3+} and $BeAl_2O_4$:Ti^{3+} are shown in Figure 1.15. It is another manifestation of the presence of the Ti 3d states in the band gap; they are clearly seen as two peaks in the band gap. The lowest in energy peak comes from the t_2 state, whereas the highest in energy 3d peak is from the e state. The maximum of the lowest 3d states is at about 4.78 eV above the top of the valence band in Al_2O_3 and at about 4.4 eV above the top of the valence band in $BeAl_2O_4$. The separation between the maxima of the t_2 and e states ($10Dq$) is about 2.198 eV or 17,730 cm^{-1}, which is somewhat underestimated if compared to the experimental values of about 19,130 cm^{-1} [64]. The same analysis of the $10Dq$ value for $BeAl_2O_4$ returns the value of 2.5 eV (~20,000 cm^{-1}), which agrees well with the experimental position of the maxima of the $BeAl_2O_4$:Ti^{3+} absorption spectra at about 19,980 cm^{-1} [65].

It is possible now to plot a complete energy diagram, which includes both the crystal field split 3d states of the Ti^{3+} ions and the host band structure. The energy levels of Ti^{3+} in Al_2O_3 were calculated to be (all in cm^{-1}): 0, 587 (doublet), 19,802 (doublet), and crystal field splitting of the Ti^{3+} states in $BeAl_2O_4$:Ti^{3+} was calculated to be 0, 1450, 1684 (t_2 states) and 19,568, 22,488 (e states) [66]. Due to a low symmetry of the Ti^{3+} position, all t_2 and e states are split completely in $BeAl_2O_4$, and degeneracy of the electronic states partially remains in Al_2O_3 due to the trigonal symmetry of the Al site in this host. Figure 1.16 combines together the results of the DFT-based calculations (electronic band structure) and crystal field theory (splitting of the d orbitals). The conduction bands in both

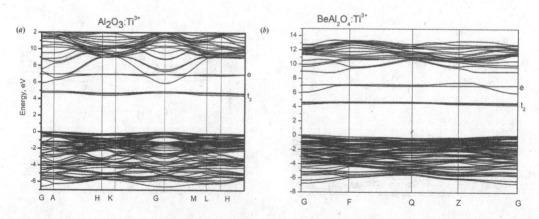

FIGURE 1.14   Calculated band structure of Al_2O_3:Ti^{3+} (a) and $BeAl_2O_4$:Ti^{3+} (b). (Reproduced with permission from Refs. [54, 55].)

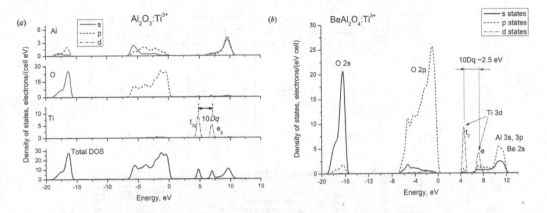

FIGURE 1.15 Calculated density of states diagrams for Al_2O_3:Ti^{3+} (a) and $BeAl_2O_4$:Ti^{3+} (b). (Reproduced with permission from Refs. [54, 55].)

cases were shifted up to overcome the underestimation of the DFT calculations and bring the band gap in accordance with the experimental results.

1.3.2 $YAlO_3$:Ce^{3+}

$YAlO_3$ (YAP) is an important material because of its scintillator applications [67], solid-state lasing [68], optical data storage [68–70], etc. It has a very wide band gap of YAP – about 8.8 eV [71], and because of it YAP can be doped with various impurity transition metal [72] and rare-earth ions. All rare-earth ions, if incorporated into the YAP lattice, occupy the Y^{3+} site, which are surrounded by 12 oxygen ions.

The experimental crystal structure data for YAP were taken from Ref. [73]. The calculations of the band structure and optical properties of YAP:Ce^{3+} were performed using the Materials Studio package and its CASTEP module [52]; all further details can be found in Ref. [74]. The calculated value of the band gap was 5.75 eV, which can be compared with the previously obtained values of 5.30 [75], 5.6 eV [76]. The scissor operator of 3 eV was applied to overcome the usual underestimation of the calculated band gap value. Figure 1.17 shows the calculated band structure of YAP:Ce^{3+}.

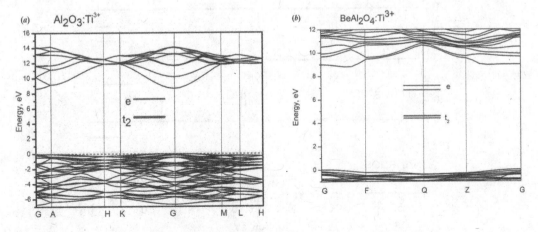

FIGURE 1.16 Calculated electronic band structure and Ti^{3+} energy levels in Al_2O_3 (a) and $BeAl_2O_4$ (b). (Reproduced with permission from Refs. [54, 55].)

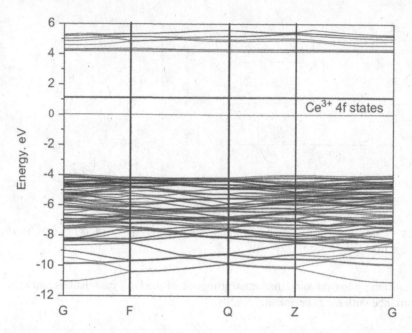

FIGURE 1.17   Calculated band structure of YAP:Ce³⁺. (Reprinted with permission from Ref. [74].)

The main difference from the pure YAP case is that the Ce 4f states appear in the band gap. They are located at 4 eV above the top of the valence band. The composition of the calculated bands becomes clear from the analysis of the partial DOS diagrams, given in Figure 1.18. It follows from Figure 1.18 that the band at about −45 eV is formed by the Y 4s states. The Y 4p states produce a narrow band at about −25 eV. The Y 4d states contribute to the conduction band at about 5 eV. The Al 3s and 3p states are located at about −20 eV and in the valence band between −10 and −4 eV.

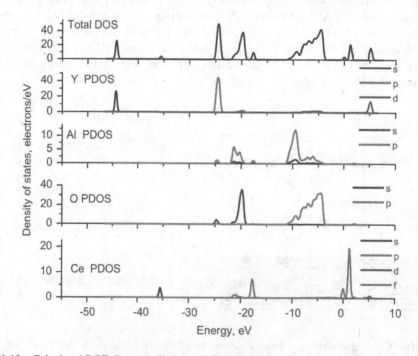

FIGURE 1.18   Calculated DOF diagrams for YAP:Ce³⁺. (Reprinted with permission from Ref. [74].)

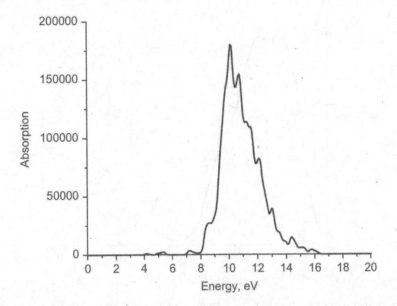

FIGURE 1.19 Calculated absorption spectrum of YAP:Ce^{3+}. (Reprinted with permission from Ref. [74].)

The oxygen 2s states are peaked at about −20 eV, and the 2p states make the valence band with a maximum at −5 eV. The Ce states are distributed as follows: the 5s states – at about −36 eV, the 5p states – from −21 to −18 eV, 6s states (which are mixed up with the 4f, 5p, 5d states) are at about −5 eV. The most interesting and important from the application point of view is that the 4f states are between 0 and 1 eV, and, finally, the 5d states are at about 5 eV.

Figure 1.19 shows the calculated YAP:Ce^{3+} absorption spectrum. A very wide intensive band between 7 and 16 eV corresponds to the band-to-band transitions, basically from the Al and O p states to unoccupied Y 4d states in the conduction band. A considerably weaker absorption at about 4–6 eV can be also noticed in the spectrum; it is caused by the Ce^{3+} ions. It is compared with the experimental spectrum and results of the crystal field calculations of the Ce^{3+} 5d energy levels in YAP [74] in Figure 1.20.

As seen from Figure 1.20, agreement between the calculated energy levels (shown by the vertical lines) and experimental spectrum is good. However, it should be emphasized that the highest 5d states of Ce^{3+} in YAP are very close to (or even located in) the conduction band of the host. Therefore, exact assignment of the Ce^{3+} highest 5d peaks can be difficult.

1.3.3 A_2SiF_6:Mn^{4+}, A = ALKALINE METAL

One of the recent hot topics in the field of phosphor materials is a rare-earth-free phosphor. Phosphor materials doped with rare-earth ions have been widely studied due to their excellent fluorescence properties. Considering a raw material cost and stable securing of rare-earths, the rare-earth-free materials are in high demand now not only for phosphor materials but also for other materials in industrial applications. Among the various phosphor materials, rare-earth-free red phosphors are important ones to enhance color rendering quality and lower the color temperature of the white LED, for which Mn^{4+} ions are good candidate for the emission center to fulfill the demanding properties. There are a lot of red phosphor materials doped with Mn^{4+} ions, which have been widely studied in detail for such purpose and some of them have been already commercialized. A_2SiF_6 doped with Mn^{4+}, where A is alkaline ions such as K^+, Rb^+, and Cs^+, are known to be efficient red phosphors [78, 79]. For further development of a series of the compounds, in the present study, the

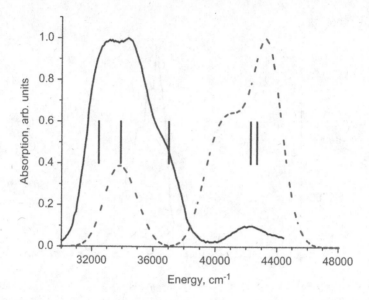

FIGURE 1.20   Calculated 5d energy levels (vertical lines) of Ce^{3+} in YAP in comparison with experimental excitation spectrum (solid line) [77]. Dashed line is the ab initio calculated absorption spectrum. (Reprinted with permission from Ref. [74].)

electronic structures of these A_2SiF_6:Mn^{4+} have been investigated by the first-principles calculations with a local spin-polarized density approximation (LSDA) within a DFT level.

The full-potential augmented plane wave plus local orbital package, WIEN2k [80], was employed here. Calculated electronic DOSs and band structure of K_2SiF_6 are shown in Figure 1.21(a) and (b), respectively. When Mn^{4+} ions are doped in A_2SiF_6, Mn^{4+} ions substitute at Si^{4+} site, which is a center of SiF_6 octahedron with the O_h symmetry. In the crystal field with O_h symmetry, three d-electrons in Mn^{4+} occupy up-spin t_{2g} state in the ground state, which is schematically shown in Figure 1.22. Within a one-electron approximation as performed in the current calculations, $10Dq$ (the crystal field strength) can be defined as the energy difference between t_{2g} and e_g states in the majority spin. Then the DOS of Mn^{4+}-doped K_2SiF_6 was calculated using $2 \times 1 \times 1$ supercell, which is shown in Figure 1.23. The energy of the highest occupied band is set to zero in the DOS, which suggests that up-spin t_{2g} level is occupied by three d electrons on Mn^{4+}. Overall structure of the calculated DOS shows similar profiles as a schematic one in Figure 1.22.

Using the calculated results of K_2SiF_6, Rb_2SiF_6, and Cs_2SiF_6 doped with Mn^{4+}, $10Dq$ was obtained by difference in energy between up-spin t_{2g} and e_g states, which are compared with earlier reports [81–83] in Figure 1.24. Present calculations within a one-electron approximation qualitatively reproduced a tendency of $10Dq$ values of A_2SiF_6:Mn^{4+}, though the calculated $10Dq$ is slightly lower than experimental ones by 0.12–0.14 eV. The red emission energy can be estimated also within current calculations. As shown in Figure 1.25, ground state of d^3 electron configuration is $^4A_2$, (up-t_{2g})^3, and first excited state is ^2E, (up-2_g)^2(down-t_{2g})^1. Then, within a one-electron approximation, it is possible to estimate the transition energy as a difference of the energy levels between up-2_g and down-t_{2g} states, which can be compared with the experimental ones [83–85] in Figure 1.26. The calculated transition energies overestimate the experimental ones by 0.28–0.29 eV, but a tendency of change in emission energies could be well explained by the current calculations even within a framework of one-electron approximation. It has been emphasized that the multi-electron effect is crucial to obtain $10Dq$ and transition energy such as the one from ^2E to $^4A_2$ states. However, it has been revealed in this study that a current approach within a one-electron approximation can qualitatively explain the important parameters in phosphor materials such as $10Dq$ and emission energies.

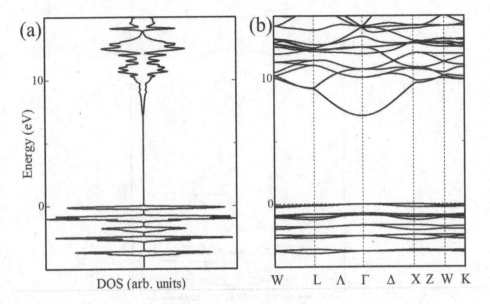

FIGURE 1.21 Calculated (a) electronic density of states and (b) band structure of K_2SiF_6.

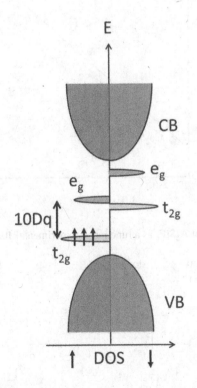

FIGURE 1.22 Schematic illustration of DOS with d^3-impurity levels. CB and VB stance for the conduction band and valence band, respectively.

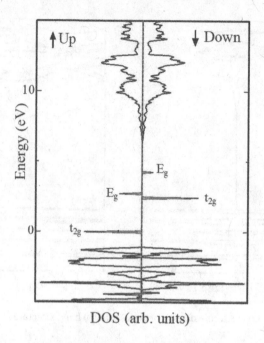

FIGURE 1.23 Calculated total DOS (black) and Mn-d partial DOS (red) of K_2SiF_6 doped with Mn^{4+}.

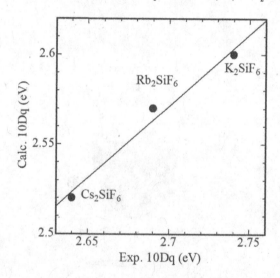

FIGURE 1.24 Calculated $10Dq$ of A_2SiF_6 as a function of experimental $10Dq$.

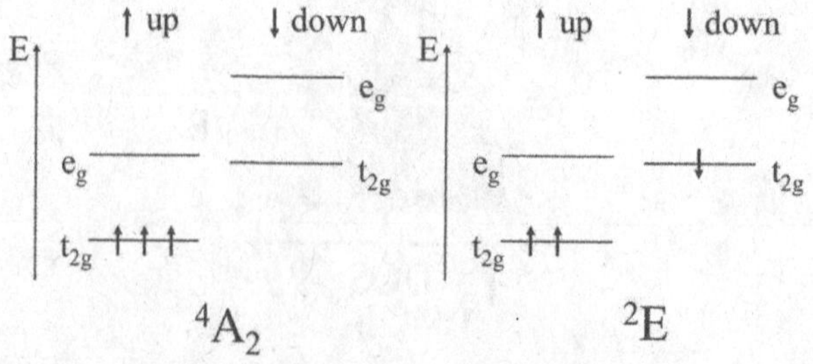

FIGURE 1.25 Electronic configuration of $^4A_2$ (left) and ^2E (right) states.

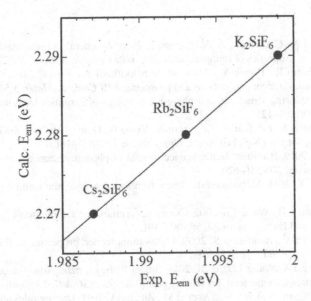

FIGURE 1.26 Calculated emission energies of A_2SiF_6 as a function of experimental ones.

1.4 CONCLUSIONS

Several examples of first-principles calculations of structural, electronic, optical, elastic, and thermodynamic properties for large groups of neat isostructural compounds, such as spinels and elpasolites, were described in detail in this chapter. Changing only one element in a chemical formula allows to find "composition-property" trends that are of paramount importance for a search for new and more efficient materials for applications. The chapter was continued with several examples of calculations of electronic and optical properties for doped optical materials, containing impurity ions in the host crystal lattice. Special attention was paid to the location of the ground state of impurity located inside the band gap of the host. Several examples of the most important and promising dopants – such as Ti^{3+}, Ce^{3+}, Mn^{4+} were considered. The proposed attempt to the analysis of the structural, electronic, optical properties, and opportunity to determine – important from application point of view – the thermal stability of doped materials can easily be extended to other compounds as well.

ACKNOWLEDGMENTS

M.G. Brik thanks the supports from the Chongqing Recruitment Program for 100 Overseas Innovative Talents (Grant No. 2015013), the Program for the Foreign Experts (Grant No. W2017011), and Wenfeng High-end Talents Project (Grant No. W2016-01) offered by Chongqing University of Posts and Telecommunications (CQUPT), Estonian Research Council grant PUT PRG111, European Regional Development Fund (TK141) and NCN project 2018/31/B/ST4/00924. C.-G. Ma acknowledges the supports of the Innovation and Entrepreneurship Program for Returned Overseas Chinese Scholars offered by Chinese Ministry and Chongqing Bureau of Human Resources and Social Security (Grant Nos. [2014] 167 and CX2019055). A.I. Popov is thankful for financial support from Latvian Research Council via Project LZP-2018/1-0214. This work was partly carried out at the Joint Research Center for Environmentally Conscious Technologies in Materials Science (Project No. 30009, 31017, and 02018) at ZAIKEN, Waseda University.

REFERENCES

1. Ward L., Agrawal A., Choudhary A., Wolverton C. 2016. A general-purpose machine learning framework for predicting properties of inorganic materials. *NPJ Comput. Mater.* 2:16028.
2. Ramprasad R., Batra R., Pilania G., Mannodi-Kanakkithodi A., Kim C. 2017. Machine learning in materials informatics: recent applications and prospects. *NPJ Comput. Mater.* 3:54.
3. Seko A. 2010. Exploring structures and phase relationships of ceramics from first principles. *J. Am. Ceram. Soc.* 93(5):1201–1214.
4. Bessiere A., Jacquart S., Priolkar K., Lecointre A., Viana B., Gourier D. 2011. $ZnGa_2O_4:Cr^{3+}$: a new red long-lasting phosphor with high brightness. *Opt. Express* 19:10131–10137.
5. Xu J., Tanabe S. 2019. Persistent luminescence instead of phosphorescence: history, mechanism, and perspective. *J. Lumin.* 205:581–620.
6. Liu Y.M., Zhang X. 2011. Metamaterials: a new frontier of science and technology. *Chem. Soc. Rev.* 40:2494–2507.
7. Yu M., Lin J., Zhou Y.H., Wang S.B. 2002. Citrate-gel synthesis and luminescent properties of $ZnGa_2O_4$ doped with Mn^{2+} and Eu^{3+}. *Mater. Lett.* 56:1007–1013.
8. Omkaram I., Rao B.V., Buddhudu S. 2009. Photoluminescence properties of $Eu^{3+}:MgAl_2O_4$ powder phosphor. *J. Alloys Compd.* 474:565–568.
9. Wang S.R., Zhang J.X., Yang J.D., et al. 2015. Spinel $ZnFe_2O_4$ nanoparticle-decorated rod-like ZnO nanoheterostructures for enhanced gas sensing performances. *RSC Adv.* 13:10048–10057.
10. Paudel T.R., Zakutayev A., Lany S., d'Avezac M., Zunger A. 2011. Doping rules and doping prototypes in A_2BO_4 spinel oxides. *Adv. Funct. Mater.* 21:4493–4501.
11. Zakutayev A. 2016. Design of nitride semiconductors for solar energy conversion. *J. Mater. Chem. A* 4:6742–6754.
12. Bajor A.L., Chmielewski M., Diduszko R., et al. 2014. Czochralski growth and characterization of $MgAl_2O_4$ single crystals. *J. Cryst. Growth* 401:844–848.
13. Seeman V., Feldbach E., Kärner T., et al. 2019. Fast-neutron-induced and as-grown structural defects in magnesium aluminate spinel crystals with different stoichiometry. *Opt. Mater.* 91:42–49.
14. Platonenko A., Gryaznov D., Kotomin E.A., Lushchik A., Seeman V., Popov A.I. 2020. Hybrid density functional calculations of hyperfine coupling tensor for hole-type defects in $MgAl_2O_4$. *Nucl. Instrum. Methods Phys. Res. B* 464:60–64.
15. Lushchik A., Dolgov S., Feldbach E., et al. 2018. Creation and thermal annealing of structural defects in neutron-irradiated $MgAl_2O_4$ single crystals. *Nucl. Instrum. Methods Phys. Res. B* 435:31–37.
16. Brik M.G., Papan J., Jovanović D.J., Dramićanin M.D. 2016. Luminescence of Cr^{3+} ions in $ZnAl_2O_4$ and $MgAl_2O_4$ spinels: correlation between experimental spectroscopic studies and crystal field calculations *J. Lumin.* 177:145–151.
17. Luo W., Ma P., Xie T.F., et al. 2017. Fabrication and spectroscopic properties of $Co:MgAl_2O_4$ transparent ceramics by the HIP post-treatment. *Opt. Mater.* 69:152–157.
18. Omkaram I., Rao B.V., Buddhudu S. 2009. Photoluminescence properties of $Eu^{3+}:MgAl_2O_4$ powder phosphor. *J. Alloys Compd.* 474:565–568.
19. Ladgaonkar B.P., Kolekar C.B., Vaingankar A.S. 2002. Infrared absorption spectroscopic study of Nd^{3+} substituted Zn-Mg ferrites. *Bull. Mater. Sci.* 25:351–354.
20. O'Neill H.S.C., Navrotsky A. 1983. Simple spinels; crystallographic parameters, cation radii, lattice energies, and cation distribution. *Am. Mineral.* 68:81–194.
21. Brik M.G., Suchocki A., Kaminska A. 2014. Lattice parameters and stability of the spinel compounds in relation to the ionic radii and electronegativities of constituting chemical elements. *Inorg. Chem.* 53:5088–5099.
22. Dovesi R., Saunders V.R., Roetti C., et al. 2014. *CRYSTAL14 User's Manual.* University of Torino, Torino.
23. Wang Y., Chen W.-B., Liu F.-Y., et al. 2019. High-throughput first-principles calculations as a powerful guiding tool for materials engineering: case study of the AB_2X_4 (A = Be, Mg, Ca, Sr, Ba; B = Al, Ga, In; X = O, S) spinel compounds. *Results Phys.* 13:102180.
24. Bortz M.L., French R.H., Jones D.J., Kasowski R.V., Ohuchi F.S. 1990. Temperature dependence of the electronic structure of oxides: MgO, $MgAl_2O_4$ and Al_2O_3. *Phys. Scripta* 41:537–541.
25. Ueda N., Omata T., Hikuma N., et al. 1992. New oxide phase with wide band gap and high electroconductivity, $MgIn_2O_4$. *Appl. Phys. Lett.* 61:1954–1955.
26. Sirimanne P.M., Sonoyama N., Sakata T. 2000. Semiconductor sensitization by microcrystals of $MgIn_2S_4$ on wide bandgap $MgIn_2O_4$. *J. Sol. State Chem.* 154:476–482.

27. Anderson O.L. 1963. A simplified method for calculating the Debye temperature from elastic constants. *J. Phys. Chem. Solids* 24(7):909–917.

28. Ning L.X., Tanner P.A., Xia S.D. 2003. Unit cell group analysis of rare earth elpasolites. *Vibrat. Spectrosc.* 31(1):51–61.

29. Zhu Y.W, Chen D.Q., Huang L., et al. 2018. Phase-transition-induced giant enhancement of red emission in Mn^{4+}-doped fluoride elpasolite phosphors. *J. Mater. Chem. C* 6(15):3951–3960.

30. Qin J.G., Xiao J., Zhu T.H., et al. 2018. Characteristic of a $Cs_2LiLaBr_6$:Ce scintillator detector and the responses for fast neutrons. *Nucl. Instrum. Meth. Phys. Res. A* 905:112–118.

31. Meyer G. 1982. The synthesis and structures of complex rare-earth halides. *Progr. Solid State Chem.* 14:141–219.

32. Gamelin D.R., Güdel H.U. 2001. Upconversion processes in transition metal and rare earth metal systems. *Top. Curr. Chem.* 214:1–56.

33. Tanner P.A. 2004. Spectra, energy levels and energy transfer in high symmetry lanthanide compounds. *Top. Curr. Chem.* 241:167–278.

34. Brik M.G., Krasnenko V., Tanner P.A. 2014. Density functional studies of cubic elpasolites Cs_2NaYX_6 (X = F, Cl, Br) at ambient and elevated hydrostatic pressure. *J. Lumin.* 152:49–53.

35. Stranger R., Moran G., Krausz E., Güdel H., Furer N. 1990. Octahedral monomeric molybdenum(III) – a magneto-optical study of Mo^{3+} doped in Cs_2NaYCl_6. *Mol. Phys.* 69(1):11–31.

36. Schwartz R.W., Hill N.J. 1974. Electron paramagnetic resonance study of Ce^{3+}, Dy^{3+} and Yb^{3+} in Cs_2NaYCl_6. A crystal with sites of perfect octahedral symmetry. *J. Chem. Soc., Faraday Trans. 2*(70):124–131.

37. Biswas K., Du M.H. 2012. Energy transport and scintillation of cerium-doped elpasolite Cs_2LiYCl_6: hybrid density functional calculations. *Phys. Rev. B* 86(1):014102.

38. Jiang C.Y., Brik M.G., Li L.H., et al. 2018. The electronic and optical properties of a narrow-band red-emitting nanophosphor K_2NaGaF_6:Mn^{4+} for warm white light-emitting diodes. *J. Mater. Chem. C* 6(12):3016–3025.

39. Wu B., Yang M.L., Yan Y.C., et al. 2021. Effects of chemical composition on the structural stability, elastic, vibrational, and electronic properties of Cs_2NaLnX_6 (*Ln* = La…Lu, X = F, Cl, Br, I) elpasolites. *J. Am. Ceram. Soc.* 104:1489–1500.

40. Ibrahim I.A.M., Lenčéš Z., Benco L., Šajgalík P. 2016. Lanthanide-doped $LaSi_3N_5$ based phosphors: ab initio study of electronic structures, band gaps, and energy level locations. *J. Lumin.* 172:83–91.

41. Dorenbos P. 2013. The electronic level structure of lanthanide impurities in $REPO_4$, $REBO_3$, $REAlO_3$, and RE_2O_3 (RE = La, Gd, Y, Lu, Sc) compounds. *J. Phys.: Condens. Matter.* 25(22):225501.

42. Tanner P.A., Shen M.Y. 1994. Superparameterization of low-temperature vibrational data for lanthanide hexahalide anions. *Spectrochim. Acta A* 50(5):997–1003.

43. Lentz A. 1974. Lattice vibration analysis of cryolite $A_3B'\,X_6$ and elpasolite $A_3BB'\,X_6$ compounds. A force constant calculation of Sr_2ZnTeO_6. *J. Phys. Chem. Solids* 35(7):827–832.

44. Meyer G., Gaebell H.C. 1978. On bromo-elpasolites, $Cs_2B^IM^{III}Br_6$ (B^I = Li, Na; M^{III} = Sc, Y, La-Lu, V, Cr). *Z. Naturforsch.* 33b(12):1476–1478.

45. Falin M.L., Gerasimov K.I., Latypov V.A., Leushin A.M., Khaidukov N.M. 2013. EPR and optical spectroscopy of structural phase transition in a Rb_2NaYF_6 crystal. *Phys. Rev. B* 87(11):115145.

46. Flerov I.N., Gorev M.V., Grannec J., Tressaud A. 2002. Role of metal fluoride octahedra in the mechanism of phase transitions in A_2BMF_6 elpasolites. *J. Fluorine Chem.* 116(1):9–14.

47. Falin M.L., Latypov V.A., Leushin A.M., Korableva S.L. 2016. EPR of Ce^{3+} in the Rb_2NaYF_6 single crystal. *J. Alloys Compd.* 688:295–300.

48. Krylov A.S., Vtyurin A.N., Oreshonkov A.S., Voronov V.N., Krylova S.N. 2013. Structural transformations in a single-crystal Rb_2NaYF_6: Raman scattering study. *J. Raman Spectr.* 44(5):763–769.

49. Xu Y., Carlson S., Sjödin A., Norrestam R. 2000. Phase transition in Cs_2KMnF_6: crystal structure of low and high-temperature modifications. *J. Solid State Chem.* 150(2):399–403.

50. Tsirel'son V.G., Antipin M.Y., Gerr R.G., Ozerov R.P., Struchkov Y.T. 1985. Ruby structure peculiarities derived from X-ray diffraction data localization of chromium atoms and electron deformation density. *Phys. Stat. Solidi A* 87:425–433.

51. Pilati T., Demartin F., Cariati F., Bruni S., Gramaccioli C.M. 1993. Atomic thermal parameters and thermodynamic functions for chrysoberyl ($BeAl_2O_4$) from vibrational spectra and transfer of empirical force fields. *Acta Cryst. B* 49:216–222.

52. Clark S.J., Segall M.D., Pickard C.J., et al. 2005. First principles methods using CASTEP. *Z. für Kristall.* 220:567–570.

53. Perdew J.P., Burke K., Ernzerhof M. 1996. Generalized gradient approximation made simple. *Phys. Rev. Lett.* 77:3865–3868.

54. Brik M.G. 2018. Ab-initio studies of the electronic and optical properties of Al_2O_3:Ti^{3+} laser crystals. *Physica B* 532:178–183.

55. Su P., Ma C.-G., Brik M.G., Srivastava A.M. 2018. A short review of theoretical and empirical models for characterization of optical materials doped with the transition metal and rare earth ions. *Opt. Mater.* 79:129–136.

56. d'Amour H., Schiferl D., Denner W, Schulz H., Holzapfer W.B. 1978. High-pressure single-crystal structure determinations for ruby up to 90 kbar using an automatic diffractometer. *J. Appl. Phys.* 49:4411–4416.

57. Lima A.F., Dantas J.M., Lalic M.V. 2012. An ab-initio study of electronic and optical properties of corundum Al_2O_3 doped with Sc, Y, Zr, and Nb. *J. Appl. Phys.* 112:093709.

58. Perevalov T.V., Shaposhnikov A.V., Gritsenko V.A., Wong H., Han J.H., Kim C.W. 2007. Electronic structure of α-Al_2O_3: ab initio simulations and comparison with experiment. *JETP Lett.* 85:165–168.

59. Samantaray C.B., Sim H., Hwang H. 2004. Electronic structures of transitional metal aluminates as high-k gate dielectrics: first principles study. *Appl. Surf. Sci.* 239:101–108.

60. Xu Y.-N., Ching W.Y. 1991. Self-consistent band structures, charge distributions, and optical-absorption spectra in MgO, α-Al_2O_3, and $MgAl_2O_4$. *Phys. Rev. B* 43:4461–4472.

61. Ching W.Y., Xu Y.-N., Brickeen B.K. 2001. Comparative study of the electronic structure of two laser crystals: $BeAl_2O_4$ and $LiYF_4$. *Phys. Rev. B* 63:115101.

62. Ivanov V.Y., Pustovarov V.A., Shlygin E.S., Korotaev A.V., Kruzhalov A.V. 2005. Electronic excitations in $BeAl_2O_4$, Be_2SiO_4, and $Be_3Al_2Si_6O_{18}$ crystals. *Phys. Solid State* 47:466–473.

63. Sugano S., Tanabe Y., Kamimura H. 1970. *Multiplets of Transition Metal Ions in Crystals*. Academic Press, New York, NY.

64. García-Revilla S., Rodríguez F., Valiente R., Pollnau M. 2002. Optical spectroscopy of Al_2O_3:Ti^{3+} single crystal under hydrostatic pressure. The influence on the Jahn-Teller coupling. *J. Phys.: Condens. Matter.* 14:447–459.

65. Sugimoto A., Segawa Y., Kim P.H., Namba S. 1989. Spectroscopic properties of Ti^{3+}-doped $BeAl_2O_4$. *J. Opt. Soc. Am. B* 6:2334–2337.

66. Brik M.G., Avram N.M., Avram C.N. 2013. Exchange charge model of crystal field for 3d ions. In *Optical Properties of 3d-Ions in Crystals: Spectroscopy and Crystal Field Analysis*, eds. N.M. Avram, M.G. Brik, 29–94. Tsinghua University Press, Beijing and Springer-Verlag, Berlin Heidelberg.

67. Nikl M. 2000. Wide band gap scintillation materials: progress in the technology and material understanding. *Phys. Stat. Sol. A* 178:595–620.

68. Zhydachevskii Y., Galanciak D., Kobyakov S., et al. 2006. Photoluminescence studies of Mn^{4+} ions in $YAlO_3$ crystals at ambient and high pressure. *J. Phys.: Condens. Matter.* 18:11385–11396.

69. Loutts G.B., Warren M., Taylor L., et al. 1998. Manganese-doped yttrium orthoaluminate: a potential material for holographic recording and data storage. *Phys. Rev. B* 57:3706–3709.

70. Kamenskikh I.A., Guerassimova N., Dujardin C., et al. 2003. Charge transfer fluorescence and f–f luminescence in ytterbium compounds. *Opt. Mater.* 24:267–274.

71. Lushchik C., Feldbach E., Frorip A., et al. 1994. Multiplication of electronic excitations in CaO and $YAlO_3$ crystals with free and self-trapped excitons. *J. Phys.: Condens. Matter.* 6:11177–11187.

72. Piskunov S., Isakoviča I., Putnina M, Popov A.I. 2020. Ab initio calculations of the electronic structure for Mn^{2+}-doped $YAlO_3$ crystals. *Fiz. Nizk. Temp.* 46:1365–1370.

73. Diehl R., Brandt G. 1975. Crystal structure refinement of $YAlO_3$, a promising laser material. *Mater. Res. Bull.* 10:85–90.

74. Brik M.G., Sildos I., Kiisk V. 2011. Calculations of physical properties of pure and doped crystals: ab initio and semi-empirical methods in application to $YAlO_3$:Ce^{3+} and TiO_2. *J. Lumin.* 131:396–403.

75. Bercha D.M., Rushchanskii K.Z., Sznajder M., Matkovskii A., Potera P. 2002. Elementary energy bands in ab initio calculations of the $YAlO_3$ and SbSI crystal band structure. *Phys. Rev. B* 66:195203.

76. Singh D.J. 2007. Antisite defects and traps in perovskite $YAlO_3$ and $LuAlO_3$: density functional calculations. *Phys. Rev. B* 76:214115.

77. Hao D., Zhao G., Dong Q., Chen J., Cheng Y., Ding Y. 2010. Spectral properties and energy transfer in $YAlO_3$ crystals doped with Ce ions and Mn ions. *Chin. Opt. Lett.* 8:303–305.

78. Setlur A.A., Radkov E.V., Henderson C.S., et al. 2010. Energy-efficient, high-color-rendering LED lamps using oxyfluoride and fluoride phosphors. *Chem. Mater.* 22:4076–4082.

79. Beers W.W., Smith D.J., Cohen W.E., Srivastava A.M. 2018. Temperature dependence (13–600 K) of Mn^{4+} lifetime in commercial $Mg_{28}Ge_{7.55}O_{32}F_{15.04}$ and K_2SiF_6 phosphors. *Opt. Mater.* 84:614–617.

80. Blaha P., Schwarz K., Madsen G.K.H., et al. 2001. *WIEN2k, an Augmented Plane Wave+Local Orbitals Program for Calculating Crystal Properties*. Karlheinz Schwarz Technische Universitat, Wien.
81. Paulusz A.G. 1973. Efficient Mn(IV) emission in fluorine coordination. *J. Electrochem. Soc.* 120:942–947.
82. Sakurai S., Nakamura T., Adachi S. 2016. Rb_2SiF_6:Mn^{4+} and Rb_2TiF_6:Mn^{4+} red-emitting phosphors ECS. *J. Solid State Sci. Technol.* 5:R206–R210.
83. Arai Y., Adachi S. 2011. Optical transitions and internal vibronic frequencies of MnF_6^{2-} ions in Cs_2SiF_6 and Cs_2GeF_6 red phosphors. *J. Electrochem. Soc.* 158:J179–J183.
84. Arai T., Adachi S. 2011. Excited states of $3d^3$ electrons in K_2SiF_6:Mn^{4+} red phosphor studied by photoluminescence excitation spectroscopy. *Jpn. J. Appl. Phys.* 50:092401.
85. Chodos S.L., Black A.M., Flint C.D. 1976. Vibronic spectra and lattice dynamics of Cs_2MnF_6 and $A^1_2M^{IV}F_6$:MnF^{2-}_6. *J. Chem. Phys.* 65:4816–4821.

2 Crystal Structure Design and Luminescence Performance of NIR Phosphors

Guogang Li and Yi Wei

CONTENTS

2.1 INTRODUCTION: BACKGROUND AND DRIVING FORCES

Near-infrared (NIR) light (700–1700 nm) has a widespread application in food composition and freshness analysis, night vision, agriculture, biological tissues, medical fields and so on [1–6]. Its high penetration ability ensures biosensing and in vivo bioimaging in human body for health monitoring with a harmless and convenient mode [7–10]. What's more, different absorption region of water, fat, carbohydrate, sugar, hemoglobin and organic elements can enable food composition analysis or blood distribution in human body [11–14]. To require multiple compositions inspection, broad-region NIR light source is highly expected. In recent decades, NIR light-emitting diodes (NIR-LEDs) gradually replace traditional tungsten halogen lamps due to unique advantages of excellent stability, low energy consumption and long lifetime. Typical NIR-LEDs device for food freshness and composition measurement is known as SFH 4735 IR LED that was developed by OSRAM in 2016 [15]. This commercial NIR-LEDs device exhibits a broadband NIR light in the region of 650–1050 nm and 16 mW power output by driving at 350 mA current. However, NIR emission intensity is relatively low. Generally, NIR-LEDs device is mainly composed of blue or near-ultraviolet (n-UV) LED chip and inorganic NIR phosphors. Hence, the photoluminescence performances of NIR phosphors have a significant influence on the lighting quality of NIR-LEDs devices.

Activators are the core of phosphors that determine intrinsic electron transition mechanism. Up to now, many researchers have devoted to exploring novel NIR phosphors and optimizing current NIR phosphors. Cr^{3+} ions are the most popular NIR-emitting activators as they can easily emit broadband NIR light ranging from 600 to 1300 nm in many oxide and halide matrixes, such as $ScBO_3:Cr^{3+}$, $La_3Ga_5GeO_{14}:Cr^{3+}$, $La_2MgZrO_6:Cr^{3+}$ and $Cs_2AgInCl_6:Cr^{3+}$ [16–22]. Recently reported Cr^{3+}-activated phosphors have acquired an achievement that the full width at half maximum (fwhm) value of emission band can reach as broad as 330 nm and internal quantum yield (IQY) is higher than 72.8% [15, 23]. However, the potential risk of Cr^{3+}-activated phosphors is the existence of $Cr^{3+} \rightarrow Cr^{6+}$ process [24, 25]. More importantly, NIR light of Cr^{3+} can only be realized in octahedral coordination configuration, and highly efficient NIR light is hard to achieve in other coordination environment [26, 27]. Except for Cr^{3+} ions, many activator ions such as $Eu^{2+}/Ce^{3+}/Bi^{3+}$ also exhibit NIR light in some inorganic phosphors and luminescent glasses [28–30]. As shown in Figure 2.1, NIR-emitting activators are mainly classified into four categories: 4f–4f electron transition (Pr^{3+}, Nd^{3+}, Ho^{3+}, Er^{3+}, Tm^{3+}, Sm^{3+} and Yb^{3+}) [31–33], 4f–5d electron transition (Eu^{2+}, Ce^{3+}) [34], 6s–6p electron transition ($Bi^{3+/2+/+}$) [35] and 3d–3d electron transition ($Mn^{5+/4+}$, Cr^{3+}) [36]. Compared with Cr^{3+} ions, the prominent superiority of other activator ions is that they can occupy abundant lattice sites except for octahedral coordination environment. Till now, many novel NIR phosphors have been reported with excellent NIR luminescence performance that originates from $Eu^{2+}/Ce^{3+}/Bi^{3+}/Mn^{4+}$ ions (non-Cr^{3+} ions), including $K_3LuSi_2O_7:Eu^{2+}$, $K_3ScSi_2O_7:Eu^{2+}$, $Ca_3Sc_2Si_3O_{12}:Eu^{2+}$, $Ba_3ScB_3O_9$: Eu^{2+} and $BaBPO_5$: Bi^{3+} [37–39]. These studies initiated a new insight to obtain NIR emission. This is because 4f–5d and 6s–6p electron transitions are sensitive to local coordination environment, NIR emission position and fwhm values can be easily adjusted, which is beneficial

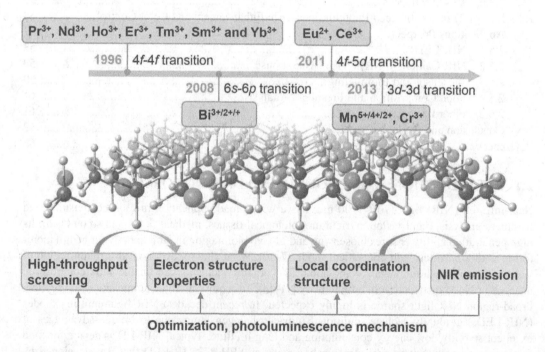

FIGURE 2.1   The overview of common NIR-emitting activator ions (Cr^{3+}, Eu^{2+}, Ce^{3+}, Bi^{3+}, $Mn^{2+/4+}$ and Ln^{3+}) (Ln = Pr, Nd, Ho, Er, Tm, Sm and Yb) with proposed study insight and photoluminescence optimization process for novel NIR-emitting inorganic phosphors.

to meet the requirements of different optical applications. Nevertheless, luminescence quantum efficiency of previously reported non-Cr^{3+}-activated NIR phosphors is relatively low in comparison to Cr^{3+}-doped phosphors. Furthermore, potential luminescence mechanism hasn't been systematically studied in non-Cr^{3+}-activated phosphors. Hence, a new insight are the key challenges in developing novel and highly efficient NIR phosphors.

Matrix compound can offer abundant crystallographic lattice sites to accommodate activator ions, so crystal structure and electron properties of matrix materials play a vital role in luminescence properties of activators. This is because electron transition of activator is sensitive to its circumambient crystal structure, luminescence properties can be tuned by modulating local coordination environment surrounding activators. According to crystal field theory, lattice symmetry, coordination number and bond length between central atom and ligands are main factors to tune luminescence properties [40–43]. Briefly, a long-wavelength luminescence is easily achieved in a strong crystal field lattice with short bond length and small coordination number. Many studies have systematically revealed the relationship between luminescence properties and local coordination structure of $Ce^{3+}/Eu^{2+}/Bi^{3+}$ ions. These strategies containing "covalency effect", "crystallographic sites engineering" and "component substitutions" have been widely approved to generate long-wavelength light [44–48]. Xia's group has reported plenty of Ce^{3+}/Eu^{2+}-activated oxide phosphors and revealed underlying luminescence mechanism based on "crystallographic sites engineering" and "component substitutions" approaches, such as $RbLi(Li_3SiO_4)_2:Eu^{2+}$, $Ca_{10}M(PO_4)_7:Eu^{2+}$, $K_3YSi_2O_7:Eu$ and $[(Sr,Ba)_3AlO_4F–Sr_3SiO_5]:Ce^{3+}$ phosphors [46, 49–56]. Liu's group successfully designed "component substitutions" strategies to obtain excellent nitride phosphors including $Al_{1-x}Si_xC_xN_{1-x}:Eu^{2+}$ and $SrLi(Al_{1-x}Ga_x)_3N_4:Eu^{2+}$ phosphors [57–62]. Besides, Li's group, Wang's group, Zhang's group and Im's group have reported many photoluminescence-tunable and highly efficient $Ce^{3+}/Eu^{2+}/Bi^{3+}$-activated phosphors based on crystal structure design and component substitution strategies [63–70]. These researches exploited many superior phosphors in visible light region, and corresponding "structure–luminescence" mechanism was systematically revealed. However, corresponding studies and crystal structure design strategies are still deficient in the exploitation of NIR phosphors.

Additionally, traditional "trial-and-error" experiment is commonly utilized in discovering novel phosphors [71]. However, long experiment period largely lowers operation and production efficiency. Recently, many strategies have been proposed to accelerate the discovery of phosphors, such as "high-throughput methodology" [72–74] "combinatorial chemistry screening" [75–77] and "single-particle diagnosis" [78, 79]. High-throughput methodology is usually known as machine learning based on an adequate database. To date, large batch of novel hybrid organic–inorganic perovskites and double perovskites have been screened for high-sensitivity X-ray detectors, solar cell and white LEDs applications with the acid of high-throughput density functional theory (DFT) prediction [80, 81]. Combinatorial chemistry technique is employed to multicompositional search, and Sohn's group developed many excellent nitride, oxide and fluoride phosphors by using combinatorial chemistry technique [82]. In 2014, Xie's group first initiated single-particle-diagnosis method with using single-crystal X-ray diffraction and single-particle fluorescence imaging and spectroscopy as microscale tools for rapidly accomplishing phase identification, photoluminescence properties assessment and structural determination of a single microcrystalline particle with a diameter down to 10 μm [83]. This method can rapidly obtain targeted phosphors. Overall, the preceding targeted-driven methodologies exhibit the promising applications in phosphors with specific properties.

To date, some reviews have summarized down-converted NIR phosphors involving quantum dots, lanthanide-doped nanomaterials and Cr^{3+}-activated phosphors [84–88]. However, these reviews mainly concentrated on synthesis approaches, luminescence properties and applications and rare reviews reported non-Cr^{3+}-activated inorganic phosphors. Although the "structure–luminescence" relationships have been systematically summarized in inorganic phosphors, it mainly focuses on revealing the luminescence properties of visible light, and it is still scarce to cover NIR "structure–luminescence" mechanism [89–92]. Herein, this chapter focuses on "crystal structure design-photoluminescence properties and mechanism" study of NIR phosphors that is originated from Eu^{2+}, Ce^{3+}, Bi^{3+}, $Mn^{2+/4+}$ and Ln^{3+} (Ln = Pr, Nd, Ho, Er, Tm, Sm and Yb) ions. First, intrinsic electron transitions of non-Cr^{3+} activator ions are summarized to identify the potential luminescence mechanism. This is because the photoluminescence performances of involved activator ions are sensitive to surrounding coordination environment and local electron structure. Crystal structure design strategies are subsequently proposed to generate superior NIR light. The proposed "structure–luminescence" mechanism mainly contains: (1) ligand covalency, (2) strong crystal field energy and distorted lattice, (3) selective sites occupation and (4) mixed valences. Such a structure design principle can be extended to achieve new NIR light and tune photoluminescence color in new inorganic phosphors. Importantly, this section proposes the concept of "high-throughput synthesis-DFT calculations prediction-structure design-photoluminescence performances optimization" based on the combination of theoretical analysis and experimental method to rapidly screen a series of novel NIR phosphors. The proposed insight largely shortens the experimental period and optimizes the lighting quality of NIR phosphors.

2.2 COMMON NIR ACTIVATORS WITH INTRINSIC TRANSITION MECHANISM

2.2.1 4F–5D TRANSITION

Inorganic phosphors are mainly composed of activators (or the codoping of sensitizers and activators) and matrix compound [93–96]. Activator ions mainly contribute as luminescence centers, so the selection and understanding of activators are very important. Cr^{3+} ions are usually considered as excellent NIR-emitting activator ions, and the luminescence properties of Cr^{3+}-activated phosphors can well meet the demands of various optical applications [97–99]. However, the improvement of photoluminescence quantum efficiency is still a crucial challenge for NIR phosphors. In the last five years, many rare-earth, bismuth and transitional metal ions also exhibit outstanding NIR emission, and some typical luminescence properties are summarized in Table 2.1. This result indicates that the emerging non-Cr^{3+}-activated phosphors can also be the promising NIR optical applications. Among non-Cr^{3+} activator ions, rare-earth Ce^{3+} and Eu^{2+} ions are the most popular activators and sensitizers in inorganic phosphors. Both of photoluminescence excitation and emission properties of rare-earth Ce^{3+}/Eu^{2+} ions are ascribed to typical 4f–5d electron transitions [46, 49, 78, 100–102]. Generally, the energy difference between $[Xe]4f^n$ ground state and the lowest $[Xe]4f^{n-1}5d$ excited state of Ce^{3+}/Eu^{2+} ions can be expressed with following equations [103–108]:

$$E_{abs}(n, Q, A) = E_{free}(n, Q) - D(Q, A) \tag{2.1}$$

$$E_{em}(n, Q, A) = E_{free}(n, Q) - D(Q, A) - E_{Stokes} \tag{2.2}$$

where n equals to 1 and 7 for Ce^{3+} and Eu^{2+}, respectively. Q represents ionic charge. $E_{free}(n, Q)$ is transition energy for free lanthanide ions in vacuum, and it is a constant for each lanthanide ions. $E_{abs}(n, Q, A)$ and $E_{em}(n, Q, A)$ are transition energy corresponding to absorption (excitation) spectra and emission spectra in compound A, respectively. $D(Q, A)$ and E_{Stokes} stand for redshift and Stokes shift energy, respectively. Redshift process is the movement of the lowest $4f^{n-1}5d$ energy level, related to the interaction with surrounding crystal field energy and the types of anion and cation. Owing to the sensitivity of 4f–5d transition to surrounding coordination environment, it is easy to achieve large-scale photoluminescence color tuning. After 4f–5d absorption (excitation) process, the lowest $4f^{n-1}5d$ energy level is further dropped down because of lattice relaxation effect, and such

TABLE 2.1

General Photoluminescence Properties (Optimal Excitation Position λ_{ex}, Maximal Emission Position λ_{em} and fwhm Values) of Recently Reported NIR Inorganic Phosphors and Some Luminescent Glasses

Compounds	λ_{ex} (nm)	λ_{em} (nm)	fwhm (nm)	Ref.
$K_3ScSi_2O_7$:Eu^{2+}	465	735	170	[37]
$Ca_3Mg[Li_2Si_2N_6]$:Eu^{2+}	440	734	124	[269]
$K_3LuSi_2O_7$:Eu^{2+}	460	740	160	[34]
$Ca_3Sc_2Si_3O_{12}$:Eu^{2+}	520	840	2050 cm^{-1}	[276]
$Ba_3ScB_3O_9$:Eu^{2+}	376	735	205	[38]
$Ca_3Sc_2Si_3O_{12}$:Eu^{2+}, Ce^{3+}, Yb^{3+}	520	840, 969	–	[313]
$Eu_xM_{1-x}H_2$ (M = Ca, Sr, Ba)	390	764, 728, 750	–	[314]
$Li_{38.7}Y_{3.3}Ca_{5.7}[Li_2Si_{30}N_{59}]O_2F$:Ce^{3+}	440	638	144	
$Li_{38.7}La_{3.3}Ca_{5.7}[Li_2Si_{30}N_{59}]O_2F$:Ce^{3+}	405	638	156	[29]
$Li_{38.7}Ce_{3.3}Ca_{5.7}[Li_2Si_{30}N_{59}]O_2F$	540	651	175	
$SrSnO_3$:5%Bi^{2+}	298	808	–	[315]
$BaBPO_5$:Bi^{3+}	476	1164	–	[39]
Bi-doped germanium–borate glasses	460	1235	250	[295]
$Ca_{14}Zn_6Ga_{10-x}O_{35}$:$xMn^{4+/5+}$	600–700	1152	–	[294]
$GdAlO_3$:0.1%Mn^{4+},0.9%Ge^{4+}	325	731	–	[280]
$(Mg_{1-x}Zn_x)_{2.97}(PO_4)_2$:0.03Mn^{2+}	417	730	–	[281]
$Li_2ZnGe_3O_8$:xMn^{2+}	475	832	–	[63]
$MgAl_2O_4$:Mn^{2+}	450	825	125	[164]
$Bi_5(AlCl_4)_3$	457.9, 785	1160	> 510	[317]
$MgAlSiN_3$:Mn^{2+}	470	754	150	[318]
Bi$^+$-doped sodalite	488	1146–1263		[319]
Bismuth-embedded zeolites	514.5	1145	160	[320]
60Leu40Dio5Bi glass	470	1188	–	[287]
$20KGaSi_2O_6$–$1Bi_2O_3$	470	1202	–	
Gd_2O_3:Bi^{3+}	490	1025	–	[321]
Bismuth-doped aluminosilicate NPs	514.5, 690	1200	–	[322]
$Bi_{8.5}Na_{26.6}[Al_{53.5}Si_{138.5}O_{384}]$ (Bi$^+$ embedded zeolites)	488, 786	1150	230	[288]
$XAl_{12}O_{19}$:Bi^{3+} (X = Ba, Sr and Ca)	330/280	770/808	82/30–35	[284]

a process is named as Stokes shift. The schematic connection between photoluminescence properties and energy-level transition is displayed in Figure 2.2.

Main factors on redshift index $D(Q, A)$ contain centroid shift (ε_c) and crystal field splitting (ε_{cfs}). The relationship among them can be calculated with following equation [42, 54, 104, 109–111].

$$D(Q, A) = \varepsilon_c + \frac{\varepsilon_{cfs}}{r(A)} - B \tag{2.3}$$

where $1/r(A)$ fraction is a constant and depends on the type of coordination polyhedron, which equals to 1.7 and 2.4 for cubic and octahedral coordination, respectively. B is a constant. The movement of $4f^{n-1}5d$ energy level is proportional to centroid shift and crystal field splitting energy. To achieve large redshift of photoluminescence excitation and emission spectra, the understanding and adjustment of centroid shift and crystal field splitting energy are the most effective methods. Centroid shift is defined as the movement of average $4f^{n-1}5d$ energy level in comparison with transition energy for free lanthanide ions in vacuum ($E_{Afree}(n, Q)$. This is because spherically symmetric

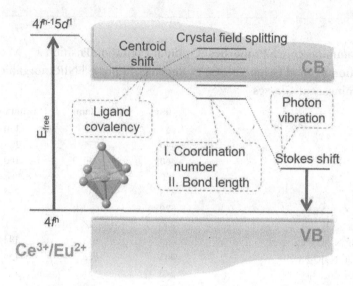

FIGURE 2.2 Schematic diagram illustrates three influence factors of the redshift of $4f^n5d$ energy levels in Ce^{3+}/Eu^{2+}-activated inorganic phosphors, including centroid shift, crystal field splitting and Stokes shift.

electron distribution function locates at the nucleus of Ce^{3+}/Eu^{2+} ions with summing wavefunctions of five 5d-orbitals. Therefore, local coordination environment can largely affect centroid shift value. The relationship between local coordination environment and centroid shift can be described as following equations [69, 112–116]:

$$\varepsilon_c = A \sum_{i=1}^{CN} \frac{\alpha_{sp}}{(R_i - 0.5\Delta R)^6} \tag{2.4}$$

$$A = \frac{e^2}{4\pi\varepsilon_0} \left(<r^2>_{6s6p} - <r^2>_{6s^2} \right) \tag{2.5}$$

where e is elementary charge, ε_0 is permittivity of vacuum, r is radial radius of an electron in $4f^{n-1}5d$ and $4f^n$ configuration for Ce^{3+}/Eu^{2+} ions, A is constant. R_i is distance between the ith ligand and central cation, CN is coordination numbers, ΔR is ionic radius difference between substituted matrix cation and activator ions. α_{sp} is spectroscopic polarizability, which is calculated with $\alpha_{sp} = 0.33 + 4.8/\chi_{av}^2$, and χ_{av} is electronegativity. Electronegativity plays an important role in the movement of centroid shift, of which a stronger electronegativity easily generates a larger centroid shift value. In addition, covalency between anion ligand and matrix cation also provides an important interaction, which is affected by nephelauxetic effect. According to previous studies, common anions can be subdivided in the following descending order: selenides > sulfides > nitrides > oxides > fluorides, namely, $Se^{2-} > S^{2-} > N^{3-} > O^{2-} > F^-$ [40, 117–120]. Generally, the stronger covalency also leads to the larger centroid shift value. For example, long-wavelength photoluminescence of Ce^{3+}/Eu^{2+} is easier to obtain in nitrides materials in comparison with oxide materials. As a result, the type of anion and the connection between ligand and cation are considered as an important factor to increase centroid shift value.

Within electrostatic crystal field approximation, d-wavefunction can be perturbed by ligands and point charges. Then d orbital can split into five-fold degenerate energy levels. Crystal field splitting energy is another vital index that is defined as energy difference between the lowest and highest degenerate energy level [121–124]. Crystal field splitting energy is closely associated with local coordination structure, and corresponding equation is described as [40, 125–128].

$$\varepsilon_{cfs} = \beta_{ploy} R^{-2} \qquad (2.6)$$

where β_{ploy} is a constant that relies on the type of coordination polyhedron. β_{ploy} value equals to 1, 0.89, 0.79 and 0.42 for octahedra, cube, dodecahedra, tricapped trigonal prism/cuboctahedra (with the unit of 1.35×10^9 pm²/cm), respectively. R represents average bond length between ligand and central cation. It can be observed that crystal field splitting energy is closely associated with polyhedral categorization and bond length between ligand and central cation. On the one hand, polyhedra with less ligands could reinforce crystal filed splitting energy. On the other hand, the shorter average bond length between central cation and ligands also induces a larger crystal field splitting energy. The large crystal field splitting energy is promotional to obtain the longer photoluminescence excitation and emission wavelength. Suitable photoluminescence excitation position in blue region is the first basic property for NIR phosphors. According to centroid shift and crystal field splitting energy analysis, it concludes that redshift of $4f^{n-1}5d$ energy level of Ce^{3+}/Eu^{2+} is significantly influenced by local lattice coordination environment, then NIR light could be designed by doping Ce^{3+}/Eu^{2+} ions in the lattice with highly covalent ligand, small coordination and short bond length.

When electron of Ce^{3+}/Eu^{2+} is excited from ground state to excited state, electron easily shifts to high vibronic states due to phonon energy interacts. This process might cause energy difference between excitation and emission position and a broadening process of photoluminescence spectra [129, 130]. Stokes shift (E_{Stokes}) describes energy movement to low-energy direction, which is caused by photon energy interaction and a shift of electron to high vibrational states [89, 131–134]. Stokes shift is intuitively reflected as the energy difference between photoluminescence excitation and emission maximum value. Intrinsically, Stokes shift and fwhm value of photoluminescence emission band is related to photon vibration model of matrix structure with following equation [135–137]:

$$E_{Stokes} = (2S - 1)hv \qquad (2.7)$$

$$fwhm = \sqrt{8 \ln 2} \times hv \times \sqrt{S} \times \sqrt{\coth\left(\frac{hv}{2kT}\right)} \qquad (2.8)$$

where E_{Stokes} represents Stokes shift energy, S stands for Huang–Rhys coupling factor, hv is phonon energy and T is measured temperature. Such an electron–lattice interaction indicates that Stokes shift and fwhm values will increase with increasing local photon vibration of matrix. Thus, the electron structure of matrix materials can also have a significant influence on the photoluminescence properties of Ce^{3+}/Eu^{2+} ions. Although most of previously reported Ce^{3+}/Eu^{2+}-activated inorganic phosphors exhibit visible emission from blue to red light, the abovementioned factors for electron transitions are also universal to NIR light. Hence, NIR light with suitable photoluminescence excitation range can be achieved by screening matrix materials with suitable crystal structure, chemical environment and photon vibration frequency. The understanding of intrinsic transition mechanism can offer a useful guidance to find suitable crystal structure and local coordination environment for NIR emission.

2.2.2 6s–6p Transition

In the past decades, bismuth-activated phosphors attracted much attention owing to unique photoluminescence properties and beneficial applications in telecommunication, biomedicine, white light illumination and lasers area [138–143]. Bismuth is known as "the wonder metal" with electronic configuration of $[Xe]4f^{14}5d^{10}6s^26p^3$. Under diverse synthesis conditions, bismuth ions can stably exist in different oxide states involving 0, +1, +2, +3, +5 [144–146]. Among them, Bi^{3+} is the most common activators in inorganic phosphors, and the unique photoluminescence behavior is attributed to $6s^2 \rightarrow 6s6p$ electron transition [147–152]. The ground state of free Bi^{3+} ion is $^1S_0$, whereas $6s6p$ excited states give rise to triplet levels [$^3P_0$, $^3P_1$, $^3P_2$] and $^1P_1$ singlet state [152–157]. $^1S_0 \rightarrow {}^3P_0$ is strongly forbidden, whereas $^1S_0 \rightarrow {}^3P_1$ transition (A-band) becomes allowed. This phenomenon is mainly ascribed to the generation of spin–orbit coupling between $^3P_1$ and $^1P_1$. $^1S_0 \rightarrow {}^3P_2$ transition (B-band) is spin-forbidden, which is allowed by coupling with unsymmetrical lattice vibrational modes. $^1S_0 \rightarrow {}^1P_1$ transition (C-band) is an allowed transition. The allowed higher electron transition of $^1S_0 \rightarrow {}^1P_1$ is appropriately 10,000 cm^{-1}, in comparison with $^1S_0 \rightarrow {}^3P_1$ transition (45,700 cm^{-1}). According to $^1S_0 \rightarrow {}^3P_0$ and $^1S_0 \rightarrow {}^3P_1$ electron transitions, Bi^{3+}-activated phosphors usually display broad absorption (excitation) bands in UV or n-UV region [158–160]. Figure 2.3 exhibits schematic configuration coordinate diagram and energy level transition model of Bi^{3+} ions. Due to naked $6s^2$ electron configuration, $6s^2 \rightarrow 6s6p$ electron transition is sensitive to surrounding coordination environment. This phenomenon is extremely similar with 4f–5d electron transition model. Thus, photoluminescence absorption (excitation) and emission properties (spectral shape, position and fwhm) can be adjusted in different lattice environment. In 2012, Wang et al. proposed environmental factor to illustrate corresponding relationship between $6s^2 \rightarrow 6s6p$ electron transition and chemical structure environment with following equation [157]:

$$h_e = \sum \left[f_c(i)\alpha(i)Q(i)^2 \right]^{\frac{1}{2}} \tag{2.9}$$

where $f_c(i)$ is fractional covalence of chemical bond from central ion to ith ligand; $\alpha(i)$ stands for polarizability of ith chemical bond volume; $Q(i)$ is the charge presented by ith neighboring anion in bond subformula. The sum equals to coordination number N. According to environmental factor calculation, $6s^2 \rightarrow 6s6p$ electron transition is mainly ascribed to four factors: ligand covalency, chemical bond volume polarizability, coordination number and the charge of binding ligands. Based on dielectric theory of chemical bond for complex crystals and refractive index, the corresponding

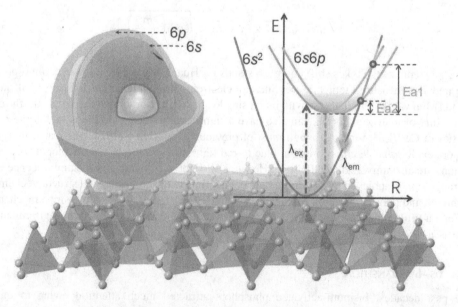

FIGURE 2.3 Schematic diagram illustrating the electron configuration of Bi^{3+} and configurational coordinate diagram of $6s^2 \rightarrow 6s6p$ energy level transition.

environment factors and coordination parameters of matrixes with known crystal structure can be calculated. Such a result can offer a valuable prediction for photoluminescence properties of Bi^{3+}-activated phosphors. A-band ($^1S_0 \rightarrow {}^3P_1$ transition) and C-band ($^1S_0 \rightarrow {}^1P_1$ transition) reflect absorption properties of Bi^{3+} ions. Both energy values of A-band and C-band gradually decrease with the increase of environment factors (f_c). The relationship between A-band and C-band energy and environment factors is expressed as [161, 162]:

$$E_A = 2.972 + 6.206 \exp\left(\frac{-h_e}{0.551}\right) \tag{2.10}$$

$$E_C = 3.236 + 10.924 \exp\left(\frac{-h_e}{0.644}\right) \tag{2.11}$$

Usually, photoluminescence excitation wavelength of Bi^{3+} is shorter than rare earth Ce^{3+}/Eu^{2+} ions and mainly locates in UV and n-UV region. Considering other types of $ns^2–nsp$ electron transitions including Pb^{2+}, Sn^{2+} and Sb^{3+}, fwhm value generated from $ns^2–nsp$ electron transition is relatively larger and broader than that of Ce^{3+}/Eu^{3+} ions [163–167]. Besides, Stokes shift value of $ns^2–nsp$ is usually larger than that of 4f–5d transition, and a larger Stokes shift effectively avoids spectral overlap between photoluminescence excitation and emission spectra [168]. Up to now, many previous studies reported broadband NIR light of Bi^{3+}-activated inorganic luminescent glasses, including bulk glasses, silica glasses, germanate glasses, fluoride glasses, chloride glasses and chalcogenide glasses [169–180]. These reported Bi^{3+}-doped luminescent glasses can successfully achieve broadband NIR light in the region of 800–1600 nm, and fwhm value reaches as high as 600 nm with IQY value of 32% [30, 181–184]. Due to broadband NIR light, Bi^{3+}-doped luminescent glasses can well apply in bioimaging and glass fibers laser and so on. Nevertheless, luminescent glasses still have some drawbacks, for example, low transmittance may largely decrease luminescence efficiency. Except for NIR luminescent glasses, some NIR-emitting Bi^{3+}-activated inorganic phosphors were also successfully obtained [39, 145]. These Bi^{3+}-activated inorganic phosphors have significant advantages over luminescent glasses. On the one hand, photoluminescence performance is easily tuned by modulating local lattice coordination environment. On the other hand, photoluminescence properties of Bi^{3+}-doped phosphors can be predicted by analyzing environment factors and structure parameters. Therefore, NIR light of Bi^{3+}-activated phosphors can be controllably achieved by modulating local lattice structure.

Except for common Bi^{3+} ions, Bi^{2+} ion can also enter in some inorganic matrix materials and present unique photoluminescence properties. Bi^{2+} ion is known as $[Xe]6s^26p$ electron configuration. Characteristic electron transition of Bi^{2+} is shown in Figure 2.4(a), ground state energy level of Bi^{2+} is $^2P_{1/2}$, and excited energy levels are $^2S_{1/2}$, $^2P_{3/2(2)}$ and $^2P_{3/2(1)}$. There are three kinds of typical excited transition pathways, including $^2P_{1/2} \rightarrow {}^2S_{1/2}$ (violet area, ~280 nm), $^2P_{1/2} \rightarrow {}^2P_{3/2(2)}$ (blue area, ~430 nm) and $^2P_{1/2} \rightarrow {}^2P_{3/2(1)}$ (red area, ~640 nm) [185–188]. $^2P_{1/2}$, $^2P_{3/2(2)}$, $^2P_{3/2(1)}$ ground and excited energy levels are split from ^2P ground state through spin–orbit splitting and crystal field splitting process. Usually, $^2P_{1/2} \rightarrow {}^2S_{1/2}$ transition (6p $\rightarrow$ 7s) dominates excitation spectra. After energy relaxation via a Stokes shift process, $^2P_{3/2(1)} \rightarrow {}^2P_{1/2}$ electron transition dominates photoluminescence emission process of Bi^{2+} [146]. Photoluminescence emission spectra of Bi^{2+} ions usually locate at 1000–1700 nm with a lifetime of hundreds of microseconds (lifetime value of Bi^{3+} ranges from hundreds of nanoseconds to dozens of microseconds) [188,189]. Although an accurate Bi^{2+} ionic radius has not been reported, photoluminescence behavior and mechanism of Bi^{2+} have been systematically studied. Intriguingly, Bi^{2+} usually emits long-wavelength light from orange to NIR light. Recent studies revealed that Bi^{2+} ion widely exists in crystalline matrixes such as strontium tetraborate, borophosphate and strontium phosphate. In 1886, Blasse et al. reported a red light for bismuth in alkaline-earth-metal sulfates [190]. Figure 2.4(b) exhibits typical photoluminescence excitation

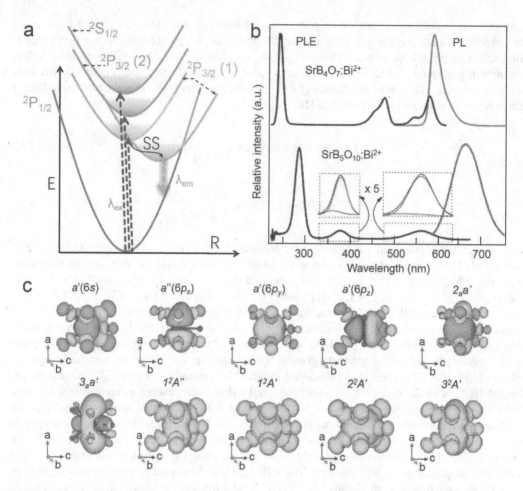

FIGURE 2.4 (a) Schematic energy level transition of Bi^{2+}, where SS represents Stokes shift process, λ_{ex} is excitation wavelength, λ_{em} stands for emission wavelength. (b) Typical photoluminescence excitation and emission wavelength of Bi^{2+} in SrB_4O_7 and SrB_5O_{10} phosphors [185]. (c) Active natural orbitals (rows 1 and 2) and electron densities (row 3) of spin–orbit free states (MS-CASPT2 calculations). There are small differences between NOs $a'(6s)$, $a''(6p_x)$, $a'(6p_y)$ and $a'(6p_z)$ of different states, which are not visible in the scale of figure [186]. (Modified from "The American Ceramic Society", Peng 2010, "The American Ceramic Society", Jong 2014, with permission.)

and emission spectra of Bi^{2+}-activated strontium tetraborate phosphors. Photoluminescence excitation spectra present three obvious excitation bands, in good agreement with electron transition theory. Upon UV light excitation (at around 245 or 286 nm), both compounds show NIR light from 600 to 800 nm. Besides, Jong et al. reported electronic states of spin–orbit coupling free Hamiltonian and full Hamiltonian to understand electron transition properties of Bi^{2+} (Figure 2.4(c)) [186]. This calculation can provide a theory to clarify the relationship between photoluminescence properties and electron transition mechanism. In conclusion, Bi^{2+} ions are also considered as wonderful activators to develop NIR light in suitable matrix materials.

2.2.3 3D–3D TRANSITION

Transition metal manganese (Mn) element also attracts much attention as activator ions, and the electron configuration of Mn is $[Ar]3d^54s^2$. This is because 3d electron shell is unfilled, and the oxidation states of Mn can easily be adjusted from Mn^+ to Mn^{7+}. Notably, Mn^{2+} and Mn^{4+} ions are considered as the most popular green/red-emitting activators in inorganic phosphors. The electron configuration of Mn^{2+} ion is $[Ar]3d^5$, and the photoluminescence properties of Mn^{2+} ions strongly depend on crystal field strength of surrounding coordination environment. Generally, ultra-narrow-band green light of Mn^{2+} ions is assigned to the occupation in tetrahedral sites that possesses weak crystal field strength. While Mn^{2+} ions locate in octahedral cation sites with strong crystal field strength, they usually emit broadband orange, red and NIR light. Different emission phenomena can be applied in different optical scenes, of which narrow-band green light is suitable in liquid crystal display applications (such as γ-AlON:Mn^{2+},Mg^{2+} and $Sr_2MgAl_{22}O_{36}$:Mn^{2+}) [191, 192]. Broad orange, red and NIR emissions have potential prospect in anti-counterfeiting, bioimaging, food analysis and so on. Schematic diagram of Mn^{2+} ion is exhibited in Figure 2.5(a). In 2018, Zhou et al. systematically reviewed electron properties, photoluminescence properties and corresponding relationship of Mn^{2+}-activated inorganic phosphors [193]. The ground state of Mn^{2+} is $^4A_1$, and excited states are $^4T_1(1)$ and $^4T_1(2)$. The simplified electron transition of Mn^{2+} is presented in Figure 2.5(b). Photoluminescence absorption (excitation) band is mainly attributed to electron transitions of $^6A_1 \rightarrow {}^4T_1(1)$ and $^6A_1 \rightarrow {}^4T_1(2)$, and absorption (excitation) band usually ranges from n-UV to blue region. While intrinsic electron transition of photoluminescence emission band belongs to $^4T_1(1) \rightarrow {}^6A_1$ and $^4T_1(2) \rightarrow {}^6A_1$ [194–198]. Tanabe–Sugano diagram can give the relationship between lattice symmetry and luminescence properties [199, 200]. The involved energy parameters include crystal field strength D_q, Racah parameter B and C. Crystal field strength D_q represents the symmetry of cation sites that are occupation by Mn^{2+} ions. Racah parameter B stands for covalency between activators and binding ligands. Racah parameter C corresponds to the interrelation of $C = 4B$. Tanabe–Sugano diagram for d^5 configuration is shown in Figure 2.5(c). When D_q/BB_3 in strong crystal fields, spin-doublet $^2T_2$ becomes ground state, which is a special feature. This result indicates

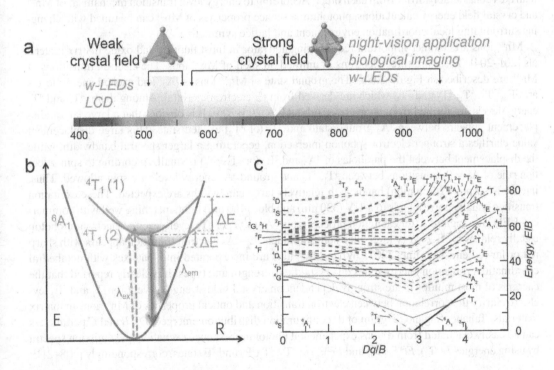

FIGURE 2.5 (a) Schematic photoluminescence properties of Mn^{2+} in response to local lattice environment and corresponding optical applications. (b) Configurational coordinate diagram of ground states and excited states of Mn^{2+} ions. (c) Tanabe–Sugano diagram for d^5 electron configuration in octahedral crystal field, $C/B = 4.5$. Spin-quartet and spin-doublet states are shown in solid and dashed lines, respectively [193]. (Modified from "The Royal Society of Chemistry", Zhou 2018, with permission.)

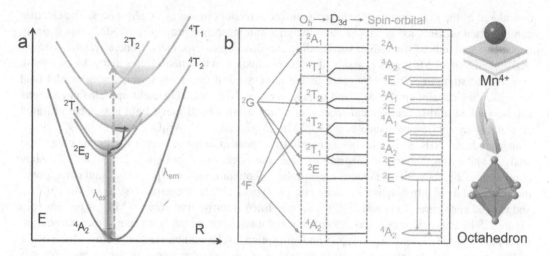

FIGURE 2.6 (a) The configurational coordinate diagram of ground states and excited states of Mn^{4+} ions. (b) Deduced energy level splitting of Mn^{4+} ions at a crystallographic site of D_{3d} symmetry in fluoride phosphors [207]. (Modified from "The Authors", Zhu 2014, with permission.)

that crystal field splitting strength D_q becomes smaller than spin-pairing energy. In this case, low-spin complex is stabilized in strong crystal field. The positions of lower excited energy levels of Mn^{2+} ions strongly depend on crystal field strength, demonstrating that emission (absorption) spectra in various matrixes considerably differ from each other. According to energy level transition mechanism of Mn^{2+} and crystal field energy calculations, photoluminescence properties of Mn^{2+} can be tuned with changing surrounding local coordination environment and lattice symmetry.

Mn^{4+} ions are another important red-emitting activator in most fluoride and oxide matrix materials [201–204]. There are three electrons in unfilled 3d shell of Mn^{4+} ions. The electron transitions of Mn^{4+} are described in Figure 2.6(a). The ground state of Mn^{4+} ions is $^4A_2$, and excited energy levels are ^2E_g, $^2T_2$, $^4T_1$, $^4T_2$ and $^2T_1$, which are derived from t^2_3 electronic orbital. Among them, $^4T_1$ and $^4T_2$ energy levels are formed from another t_2^2e orbital [96, 205, 206]. It is obvious that a large lateral displacement appears between $^4A_2$ ground state and $^4T_1$ (or $^4T_2$) excited state. This large displacement value clarifies a stronger electron–phonon interaction, generating a larger spectral bandwidth, while the displacement between the parabolas of $^4A_2$ and ^2E_g (or $^2T_1$, $^2T_2$) is small. According to spin selection rule of $\Delta S = 0$, transitions between $^4T_1$, $^4T_2$ and ground $^4A_2$ energy levels are spin-allowed. Thus, intense absorption (excitation) bands with relatively large bandwidths are expected. These excitation transition usually corresponds to n-UV (330 nm) and blue (460 nm) light, matching well with commercial LED chip region. After excitation process from $^4A_2$ to $^4T_1$ or $^4T_2$ levels, excited electron or photon usually relaxes to ^2E_g energy level, then spin-forbidden ^2E_g–$^4A_2$ electron transition appears with sharp red or far-red emission lines [207]. When Mn^{4+} ions are incorporated into matrixes with octahedral coordination, transition energy relies on crystal field strength and types. It is clearly reported that the energies of most multiplets are strongly dependent on crystal field strength except for $^2T_1$ and ^2E_g levels. To clarify the correlation between electron transition and optical properties of Mn^{4+} ions in matrix materials, Tanabe–Sugano diagram of d^3 configuration distribution emerged. D_q, B and C parameters can be easily evaluated from d^3 ions experimental photoluminescence excitation and emission spectra by using energies $E(^4T_2)$, $E(^4T_1(^4F))$ and $E(^2E_g)$ of $^4T_2$, $^4T_1(^4F)$ and ^2E states, correspondingly [208–211]:

$$E(^4T_2) = 10D_q \tag{2.12}$$

$$\frac{B}{D_q} = \frac{(\Delta E/D_q)^2 - 10(\Delta E/D_q)}{15(\Delta E/D_q - 8)} \tag{2.13}$$

$$\frac{E(^2E)}{B} = \frac{3.05C}{B} + 7.90 - \frac{1.80B}{D_q} \qquad (2.14)$$

where $\Delta E = E(^4T_1(^4F)) - E(^4T_2)$. Spin-doublet states locate between these spin-quartets. It is suggested that these spin-doublet states are in the form of additional structures and/or narrow peaks superimposed onto the broad bands of spin-allowed transitions.

Because photoluminescence properties of Mn^{4+} ions are sensitive to their local coordination environment, excited state position strongly dependents on the local lattice symmetry of cationic sites that are occupied by Mn^{4+} ions. Taken K_2TiF_6:Mn^{4+} as an example, Mn^{4+} ions probably substitute for Ti^{4+} sites with local site symmetry of D_{3d} that is descended from O_h. Based on group theory, Zhu et al. first analyzed the influence of symmetry descending on energy level splitting by combining with spin–orbit interaction. Detailed energy levels of Mn^{4+} at site symmetry of D_{3d} were deduced and plotted in Figure 2.6(b) [207]. As a result, the photoluminescence properties of Mn^{4+} can be influenced by crystal field strength and local lattice symmetry. Generally, Mn^{4+}-doped fluoride phosphors present shorter emission wavelength (630 nm) than Mn^{4+}-doped oxide materials (660–670 nm) [204, 212–214]. Hence, far-red even NIR emission are possibly achieved by choosing suitable matrix materials with strong crystal field strength and weak local lattice symmetry.

2.2.4 4F–4F TRANSITION

Lanthanide elements including Pr^{3+}, Nd^{3+}, Ho^{3+}, Er^{3+}, Tm^{3+}, Sm^{3+} and Yb^{3+} ions belong to typical 4f–4f transition, which exhibit multiple emission lines ranging from visible to NIR light region [215–218]. According to Pauli exclusion principle and Hund rule, energy difference between excited energy level and ground energy level is always fixed. Unlike 4f–5d and 6s–6p transition, the peak position of photoluminescence excitation and emission spectra cannot change in large-scale region by tuning local lattice structure. However, relative emission peak intensity might change, which can also effectively achieve photoluminescence color tuning. To clarify the generation of NIR light of 4f–4f transition, the understanding of intrinsic energy level transition is helpful to design these activators into suitable matrixes for some expected applications. The electron configuration of Pr^{3+} is $[Xe]4f^2$, and Pr^{3+} ion has unique features, one of which is broad light region in visible light to NIR regions depending on matrix structure and activators concentration [219, 220]. When doping into different matrixes, Pr^{3+} ion can be utilized in many applications including laser lighting, fiber optical communication, scintillator and LEDs lighting [221–224]. Figure 2.7(a) exhibits ground and excited energy levels of Pr^{3+} ions, under $^3H_4 \rightarrow {}^1D_2$ electron transition process (605 nm), Pr^{3+} ions mainly show linear NIR emission at around 988 nm, which is ascribed to $^1D_2 \rightarrow {}^3F_4$ electron transition. Er^{3+} ion is also a common NIR-emitting activator, whose electron configuration is $[Xe]4f^{11}$. The ground energy level is $^4I_{15/2}$, and excited energy levels mainly include $^4F_{7/2}$, $^2H_{11/2}$, $^4S_{3/2}$, $^4F_{9/2}$, $^4I_{11/2}$ and $^4I_{15/2}$ [225, 226]. When Er^{3+} ions are doped into proper matrix materials, main excitation transition is ascribed to $^4I_{15/2} \rightarrow {}^2F_{7/2}$ energy transition with 455 nm excitation wavelength. Then nonirradiative relaxation occurs across $^4S_{3/2}$, $^4F_{9/2}$, $^4I_{11/2}$ and $^4I_{15/2}$ energy levels, leading to large energy movement. Finally, excited photons may initiate energy transition from $^4I_{13/2}$ excited energy level to ground $^4I_{15/2}$ energy level, and thus, NIR emission at around 1540 nm occurs. Yb^{3+} ion attracts much attention as NIR-emitting activators in up-converted and down-converted applications [227, 228]. The electrons configuration of Yb^{3+} is $[Xe]4f^{13}$, Yb^{3+} ions are also a common NIR-emitting center in up-converted phosphors as well as down-converted phosphors. According to Pauli exclusion principle and Hund rule, Yb^{3+} electron spectrum term is ^2F that only possesses two spectral subbranches $^2F_{5/2}$ and $^2F_{7/2}$. Based on Hund second rule, $^2F_{5/2}$ level energy is excited state and $^2F_{7/2}$ is ground state, and energy difference between two energy levels is approximately 10,000 cm^{-1} [229, 230]. Usually,

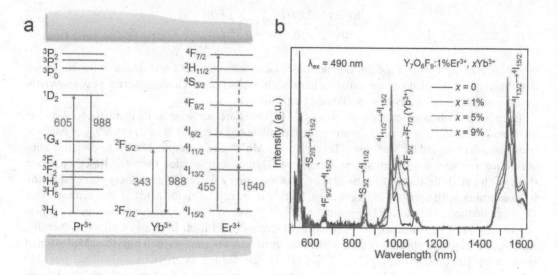

FIGURE 2.7 (a) Typical 4f–4f coordinate diagram of ground states and excited states of Mn^{2+} ion. (b) The linear emission band of 4f–4f transition, taken $Y_7O_6F_9$:1%Er^{3+}, xYb^{3+} as an example [241]. (Modified from "American Chemical Society", Wang 2018, with permission.)

the excitation peak position of Yb^{3+} ions in down-converted model locates at around 343 nm. After nonradiative relaxation and Stokes shift processes, $^2F_{5/2} \rightarrow {}^2F_{7/2}$ energy transition occurs, generating NIR emission (at around 988 nm) [231]. In addition, Nd^{3+}, Ho^{3+}, Tm^{3+} and Sm^{3+} ions also exhibit similarly linear NIR light with the electron transition of $^4F_{5/2} \rightarrow {}^4I_{9/2}$ ($^4F_{7/2} \rightarrow {}^4I_{9/2}$), $^5F_4 \rightarrow {}^5I_7$ ($^5I_6 \rightarrow {}^5I_8$), $^3F_{2,3} \rightarrow {}^3H_6$ ($^3H_4 \rightarrow {}^3H_6$) and $^4G_{5/2} \rightarrow {}^6F_{5/2}$, respectively [232–235]. These preceding activators can have wide applications due to their stable NIR light.

Judd–Ofelt theory offers intrinsic analysis of f–f transition and can clarify some underlying luminescence mechanism. Experimental oscillator strength (f_{exp}) of f–f transition is expressed in following equations [236, 237]:

$$f_{exp} = \frac{mc^2}{\pi e^2 \lambda^2 N} \times \frac{1}{0.43L} \int 0D(\lambda)\,dy \qquad (2.15)$$

where m ($0.91093897 \times 10^{-27}$ g) and e (4.803242×10^{-10} esu) are the charge of electron, c (2.99792458×1010 cm/s) is velocity of light in vacuum, N (ion/cm³) is the concentration of doped trivalent lanthanide ions, $\int 0D(\lambda)dy$ is integrated absorption coefficient that is calculated from the area of absorption peaks. Generally, experimental oscillator strength includes electric-dipole contribution and magnetic-dipole contribution that are defined theoretical oscillator strength (f_{cal}). According to Judd–Ofelt theory, theoretical oscillator of electric-dipole transition (f_{cal}^{ed}) from initial state $\langle (S, L)J \rangle$ to final state $\langle (S', L')J' \rangle$ can be calculated as [238]:

$$f_{cal}^{ed} = \frac{8\pi^2 mc}{3h(2J+1)\lambda} \frac{(n^2+2)^2}{9n} \sum_t$$
$$= _{2,4,6}\, \Omega_t \left| \left\langle 4f^N(S,L)J \left| U^{(t)} \right| 4f^N(S',L')J' \right\rangle \right|^2 \qquad (2.16)$$

where n is refractive index, h ($6.6260755 \times 10^{-27}$ erg·s) is Planck constant, λ is the wavelength of transition, J and J' is the total angular momentum of initial level and terminal level, respectively. $4f^N(S,L)J|U^{(t)}|4f^N(S',L')J'$ is reduced matrix element insensitive to matrix environment. While theoretical oscillator of magnetic-dipole transitions (f_{cal}^{md}) are restricted by selection rules: $\Delta S = \Delta L = 0$, $\Delta J = 0, \pm 1$, formula can be expressed as [239, 240]:

$$f_{cal}^{md} = \frac{2\pi^2 n}{3hmc\lambda(2J+1)} A$$

$$A = \left| \sum_{SL,S'L'} C(S,L)C(S'L')\langle 4f^N(S,L)J|L+2S|4f^N(S',L')J'\rangle \right|$$

(2.17)

where $C(S,L)C(S'L')$ is intermediary coupling coefficient. Matrix element $4f^N(S,L)J$ $|L+2S|4f^N(S',L')J'$ equals to $\hbar\{[(S+L+1)^2 - J^2][J^2 - (L-S)^2 / 4J]\}^{1/2}$. Using experimentally measured values of oscillator strength for different transitions, Ω_t ($t = 2, 4, 6$) parameters have been calculated by using a least squares fitting approach. Ω_t parameter follows the trend of $\Omega_2 > \Omega_4 > \Omega_6$. Ω_2 parameter is affected by ligand symmetry of a matrix material, while Ω_4 and Ω_6 parameters are related to the covalency of medium in which rare-earth ions are situated. On the basis of Judd–Ofelt theory, the different response of magnetic-dipole transitions and electric-dipole transition can sensitively tune relative emission peak intensity by adjusting local lattice symmetry. Combined with Judd–Ofelt theory and optical properties of trivalent Ln^{3+} ions, further understanding and underlying luminescence mechanism can also be clarified. The crucial challenge is that photoluminescence quantum efficiency of trivalent lanthanide ions is low. Previously, many studies designed energy transfer to enrich luminescence color and enhance luminescence intensity. For example, Figure 2.7(b) exhibits energy transfer between Er^{3+} and Yb^{3+} ions in $Y_7O_6F_9$ matrix, which presents multiplet emission peaks in the region of 600–1600 nm [241]. This phenomenon indicates that Pr^{3+}, Nd^{3+}, Ho^{3+}, Er^{3+}, Tm^{3+}, Sm^{3+} and Yb^{3+} ions have promising applications as NIR-emitting activators.

2.3 HIGH-THROUGHPUT CALCULATION AND CRYSTAL STRUCTURE DESIGN TO REALIZE NIR EMISSION

According to preceding discussion, it is obvious that many non-Cr^{3+} ions including Eu^{2+}/Ce^{3+}, Mn^{4+}, Ln^{3+} (Ln = Pr, Nd, Ho, Er, Tm, Sm and Yb) can act as the promising NIR-emitting centers in inorganic phosphors. Moreover, these novel phosphors have attracted more and more attention. Up to now, traditional "trial-and-error" method is commonly used in discovering novel phosphors. However, owing to a long experiment period, "trial-and-error" method cannot well meet the rapid development of basic theory researches. Recent years, the presence of artificial intelligence industry profoundly accelerates materials discovery process and then stimulates productivity over various fields. As one of pivotal methods, the combination of high-throughput experiments and machine learning offers a new insight to explore novel phosphors with excellent optical properties, and this emerging approach can even become popular tendency for the exploitation of future novel materials. Such a data-driving high-throughput approach can quickly and directionally screen suitable matrix materials and activator ions with intelligent data analysis.

The proposed screening process can be described as a "high-throughput screening approach-DFT calculations prediction-structure design-photoluminescence performance optimization" process, which is subdivided into following three sections (Figure 2.8): (I) High-throughput calculations, which is also called as massive data import process. Based on detailed physical and optical performance analysis on the periodic table of elements, possible elementals are picked out and then inputted into calculations system. This is because common NIR inorganic phosphors are composed of activators (C-*cation*) and matrixes. Matrixes materials are roughly divided into three parts, containing A-*cations* (alkaline-earth metals and alkaline metals), B-*cations* (IIIA–VA cations) and X-*anions* (VA–VIIA anions). Based on periodic table of elements, around 41 types of elementals

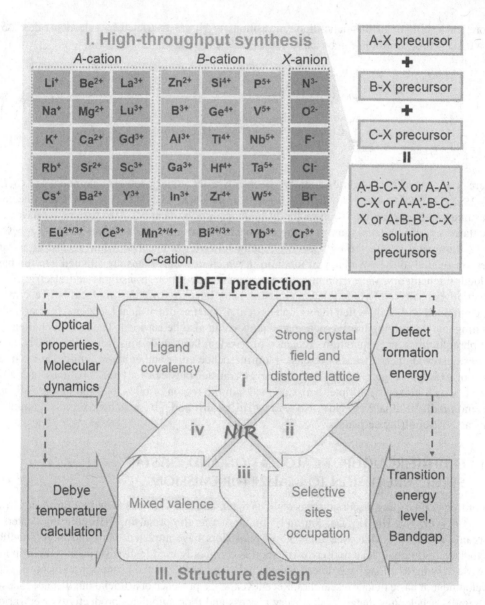

FIGURE 2.8 The overview of "high-throughput approach-DFT calculations prediction-structure design-photoluminescence performance optimization" process to obtain NIR light, process "I" represents high-throughput calculation, of which selected elements are the possible composition of NIR phosphors. Process "II" stands for DFT calculation that preliminarily screens matrixes and activator ions. DFT calculations mainly contain: Debye temperature, defect formation energy, optical properties, molecular dynamics, transition energy level and bandgap calculations. Process "III" represents matrix crystal structure design strategies, including i. ligand covalency, ii. strong crystal field energy and distorted lattice, iii. selective sites occupation and iv. mixed valences.

are screened, and thousands of solutions components can be synthesized in theory. Then, theoretical calculations are performed to further screen stable phosphors and give a prediction of optical properties. (II) Further physical and chemical properties of materials are obtained via theoretical calculations process. Generally, first-principle DFT calculation is one of the most popular approaches to offer detailed electron structure of phosphors, which also belongs to high-throughput strategies. Based on DFT calculations, Debye temperature, defect formation energy, optical properties, molecular dynamics, transition energy and optical bandgaps could be calculated in details. DFT calculation information can give a targeted matrix structure prediction and screen the combinations of activator and matrix. (III) Matrix crystal structure design based on local coordination environment modulation. This is because the photoluminescence properties of activators ions are closely associated with surrounding coordination environment and crystal structure. Crystal structure and local coordination environment are crucial to achieve highly efficient NIR light. Generally, the NIR emission of activators is strongly influenced by crystal field strength, local lattice symmetry, bond length and ligand covalency. In addition, the charge valence of activators also plays an important role in electron transition and photoluminescence color. When theoretical calculations, local crystal structure and electron transition of activator ions are taken into consideration, highly efficient NIR emission can be achieved and corresponding luminescence mechanism can be clearly identified.

2.3.1 HIGH-THROUGHPUT SYNTHESIS APPROACH

In the development of novel NIR phosphors, the urgent bottleneck of traditional "trial-and-error" method is slow experimental period. Notably, the emerging high-throughput approach can effectively shorten the growing gap between experiment and theory. What's more, high-throughput synthesis can also assist to discover and understand corresponding properties of phosphors [242, 243]. High-throughput approach is first proposed in 1982 by Alan P. Uthman et al. [244]. There are two classes of high-throughput pathways: (1) automatic, highly efficient machine learning method and high-performing calculation [245] and (2) single-particle diagnosis technique [79, 83]. The detailed illustration diagram is shown in Figure 2.9. Highly efficient machine learning approach mainly takes advantage of high-performance computer or professional software based on an adequate inorganic crystal data input. After efficiently, accurately screening activators and matrixes compounds based on theoretical analysis, material synthesis process becomes 10× faster than traditional "trial-and-error" approach, and device assembly process also becomes 10× faster, and even diagnostics process is 10–100 times faster. Conversely, materials synthesis, device assembly and diagnostics processes can further give a guidance to optimize high-throughput screening theory. Through enormous data analysis, machine learning is very accurate and can give reliable prediction. To date, high-throughput synthesis with machine learning prediction has been widely utilized in the development of novel functional materials for various photoelectric area. Sun's group utilized a combination of traditional and machine-learning-aided approach to design and realize a high-throughput experiment platform. Around 75 unique compounds of perovskite inspired materials were investigated in two months with 90% accuracy [246]. Yin's group proposed a strategy by combining machine learning and DFT calculations to engineer stable halide perovskites materials, and the potential relationship between structure and chemistry feature are clarified based on 354 types of halide perovskite candidates [80]. Wang's group developed a target-driven method to quickly predict and screen undiscovered hybrid organic–inorganic perovskites for photovoltaics through high-throughput synthesis [247]. Zhang's group exploited lanthanide-doped NIR-II nanoparticles with engineered luminescence lifetimes [248], which was also considered as one of high-throughput methods. Based on high-throughput studies, automatic, highly efficient machine learning approach can quickly screen targeted energy materials or phosphors. These preceding examples demonstrate that the development of high-throughput synthesis is mature and can promote various materials discovery process.

Up to now, these studies about the assistance of high-throughput synthesis are still very rare in the exploitation of inorganic phosphors, especially for NIR phosphors [81, 82, 249–251]. Therefore, the development of high-throughput approach is very important. Single-particle diagnosis technique was first developed by Xie's group in 2014 [83], and the main technique contains single-particle fluorescence imaging and spectroscopy techniques. With the help of these techniques, researchers can evaluate the photoluminescence of a phosphor particle distinguished from a complex powder mixture. Then, the picked single particle was performed on a high-resolution single-crystal

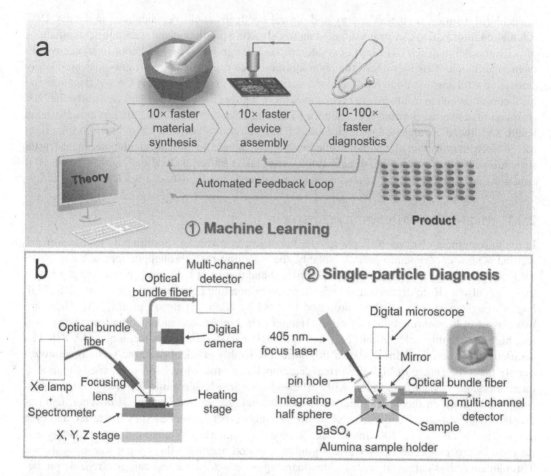

FIGURE 2.9 (a) The automated feedback loop, driven by machine learning, drives process improvement. Theory, synthesis and device processes take advantage of high-performance computing and materials databases. For many materials systems, an ~10 times multiplier is a minimum necessary to bridge the mismatch in actual and aspirational materials development timelines are referred [253]. Note that the cycle of learning is limited by the slowest step (bottleneck). (b) Schematics of home-built single-particle fluorescence microspectroscopy setups for measuring photoluminescence spectra, and quantum efficiency [83]. (Modified from "Elsevier Inc", Correa-Baena 2018, "American Chemical Society", Hirosaki 2014, with permission.)

X-ray diffractometer for determining detailed crystal structure. This method can accurately screen expected luminescence color and exploit unparalleled crystal structure phase. In 2015 and 2019, Xie's group subsequently reported a series of nitride and oxynitride phosphors with excellent photoluminescence properties based on the combinational approach of single-particle diagnosis technology and DFT calculations [78, 79, 252]. The preceding results demonstrate that high-throughput synthesis has been a popular trend to accurately acquire targeted materials. To develop high-performance NIR phosphors, the development of high-throughput calculation or single-particle diagnosis pathways can precisely screen novel inorganic phosphors.

2.3.2 DFT CALCULATIONS

First-principle DFT is considered one of the most popular high-throughput strategies. As presented in Figure 2.10, crystal and electron structure can be first optimized by Perdew–Burke–Ernzerhof (PBE) generalized gradient approximation (GGA) or exchange-correlation function and

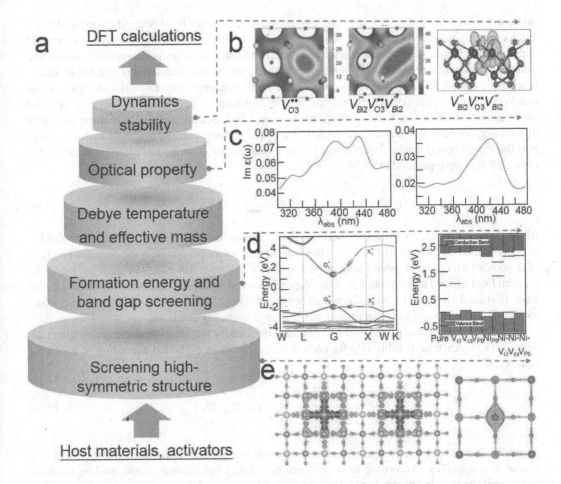

FIGURE 2.10 (a) The overview of DFT calculations. (b) Positron density in $Bi_{2.14}Sr_{0.75}Ta_2O_9$ along (0 0 1) plane for trapped positrons in $V_{O3}^{\bullet\bullet}$ and $V_{Bi2}^{'''}V_{O3}^{\bullet\bullet}V_{Bi2}^{'''}$ vacancies associates and mid-gap localized level. Bi, blue; Sr, green; Ta, gray; and O, red [258]. (c) Absorption spectrum for $Sr_2LiAlO_4{:}0.0625Eu^{2+}$ (left) and $Sr_2LiAlO_4{:}0.125Ce^{3+}$ (right) [81]. (d) GW band structure of $Cs_2AgInCl_6$. The orbital characters and free-exciton wavefunction are plotted as a fat-band structure. Green, blue, cyan and red colors denote Cl 3p, Ag 4d, In 5s and Ag 5s orbitals, respectively. Magenta circles indicate the lowest free-exciton amplitude $|Avck|$, where $|S\rangle = \Sigma vck\, Avck|vc\rangle$ is exciton wavefunction, v and c denote valence and conduction states and k is wavevector. $|S\rangle$ is derived from electron and hole states with same parity (labels at zone center G and X) along GX, implying a dark transition; and defect levels for both pure and doped $CsPbCl_3$. Red lines represent the highest defect level under Fermi level, whereas blue lines represent lowest defect level above Fermi level. VBM, CBM and defect levels derive from corresponding band structures [130, 259]. (e) Configuration showing strengthened self-trapped excitons confinement by surrounding $[NaCl_6]$ octahedra. Self-trapped excitons are confined within two lattice parameters surrounded by $[NaCl_6]$ octahedra [130]. (Modified from "Wiley-VCH Verlag GmbH & Co. KGaA", Li 2019, "Elsevier Inc", Wang 2018, "Springer Nature Limited", Luo 2018, "American Chemical Society", Yong 2018, with permission.)

Hartree-Fock hybrid functional (HSE) methods. Based on this calculation, detailed structure information can be predicted, such as structure symmetry, atoms distribution and cation–anion binding types. Hence, matrix and activators can be initially screened to obtain NIR light. Second, further calculations of optical bandgap and formation energy can sift out suitable matrix structure. Optical bandgap value usually needs to be large enough to accommodate typical electron transition of activator ions. If energy difference between conduction band and valence band is too small, ground or excited states might be embedded in valence or conduction band, easily generating electron migration from ground state or excited state of activators to valence band or conduction band. Such an electron migration process is not beneficial for highly efficient NIR light because of elevated emission loss. Besides, formation energy is a useful pathway to predict prior sites occupation and even gives an evident prediction of photoluminescence performance of activators and defect formation position and defect type. Nowadays, many studies have reported sites formation energy calculation to clarify prior sites occupation in some inorganic phosphors. For example, Qiao et al. clarified sites occupation of Eu^{2+} in $K_2BaCa(PO_4)_2$ oxide matrix by calculating defect formation energy in different cation sites and occurrence probability P_i [254]. Such calculations well assist and support the experimental results. For a defect (D) in a charge state q, formation enthalpy is calculated through following equation [255, 256]:

$$\Delta H_{D,q}(\mu, E_F) = (E_{D,q} - E_H) - \sum n_\alpha \mu_\alpha + q(E_F + E_V) \tag{2.18}$$

where $(E_{D,q} - E_H)$ is energy difference between defect and matrix supercells, n_α indicates the number of α atoms added ($n_\alpha < 0$) or removed ($n_\alpha > 0$) when a defect is formed, and μ_α is the chemical potential of α atom, which can be expressed with respect to that of elemental phase (μ_α^{el}) by $\mu_\alpha = \mu_\alpha^{el} + \Delta\mu_\alpha$. E_F is Fermi level referred to VBM (E_V). μ_α value depends on experimental conditions. Equilibrium Fermi levels (E_F) are determined by using calculated density of states (DOS), and it could solve semiconductor statistic equations self-consistently and satisfy charge neutrality condition.

In addition, occurrence probability (P_i) can further quantify the relative preference of site occupations. Occurrence probability (P_i) for each defect complex is evaluated according to following equation [254]:

$$P_i = \frac{1}{Z_{tot}} \exp\left(-\frac{E_i}{kT}\right), (i = 1,\ldots,70) \tag{2.19}$$

where Z_{tot} is partition function, E_i is defect formation energy, k is Boltzmann constant and T is synthesis temperature. According to occurrence probability calculations, some unclear luminescence performances in phosphors are demonstrated. The representative studies are sites occupation analysis of red $Rb_3YSi_2O_7$:Eu [46] and photoluminescence tuning of $Na_{(1-x)}Li_{(x)}Al_{(1-x)}Si_{(1+x)}O_4$:$Eu^{2+}$ [47]. The preceding researches clearly clarify red emission mechanism by occupying into small lattice. Besides, photoluminescence tuning can be achieved with designing composition substitution under the assistance of occurrence probability calculations. Based on formation energy and occurrence probability analysis, state density distribution and electron configuration of activator-doped luminescence in different cation sites can be further obtained. For example, Wei et al. give detailed DOS distribution analysis of Bi^{3+}-activated La_4GeO_8 in different La^{3+} sites [257]. Therefore, luminescence behavior can be predicted and give a guidance for the synthesis of NIR phosphors.

Third, Debye temperature calculation usually identifies matrix crystal structure rigidity, and Debye temperature was usually determined with quasiharmonic model [260, 261]. To screen matrix materials with stable physicochemical properties and excellent environmental stability, Debye temperature and effective mass analysis take effect. Finally, optical properties and dynamics calculations can give optical properties prediction for the involved phosphors. For example, it can give a prediction of corresponding absorption (excitation) and the emission position. The relationship between luminescence position and energy level transition of activators can also be calculated. Besides, the interaction between activators and binding anions can be revealed. The preceding calculation results can give punctilious filtrate for NIR phosphors. Combining high-throughput synthesis and DFT calculations, targeted data-driven structure prediction and screening process become much faster and more accurate.

Owing to the sensibility of Bi^{3+}/Eu^{2+}/Ce^{3+} ions to surrounding lattice environment and multiple valences of Bi and Mn ions, NIR light can be designed by modulating microscopic crystal structure.

According to crystal field theory, photoluminescence properties of activator ions are closely associated with local crystal field environment. Corresponding equation can be described in following equations [262, 263].

$$D_q = \frac{1}{6} Z e^2 \frac{r^4}{R^5} \tag{2.20}$$

where D_q is the magnitude of excited energy level separation of activators ions, Z is anion charge or valence, e is electron charge, r is the radius of d wavefunction and R is average bond length. Although this equation is derived by using a point charge model, it can be approximately described crystal field splitting trends with bond distance. Usually, the stronger crystal field splitting energy might be generated by a shorter bond length and, thus, resulting in a longer excitation and emission wavelength.

Up to now, many studies have proposed the relationship between crystal structure and photoluminescence properties in visible light region. For instance, Fang et al. reviewed the control of narrow-band emission by modulating local lattice structure [89]. They also reviewed that the photoluminescence color of Eu-doped alkali–lithosilicate phosphors can be adjusted by controlling the cuboid-size of local lattice in UCr_4C_4 structure [92]. These summaries could provide useful guidance to design highly efficient phosphors in response to practical requirements. These previous studies can clearly clarify the photoluminescence control of phosphors among the whole visible light area. Nevertheless, the relationship between NIR light and local crystal structure is unclear. Herein, we summarize four types of structure design criteria to obtain NIR light, which can help to screen suitable matrix materials and lattice sites. As shown in Figure 2.11, crystal structure design

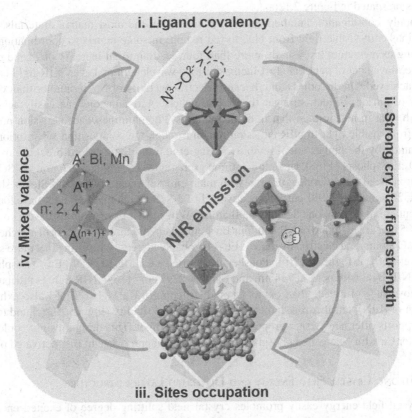

FIGURE 2.11   The overview of structure design strategies for the generation of NIR emission. Crystal structure design mainly contains: i. ligand covalency, ii. strong crystal field energy and distorted lattice environment, iii. selective sites occupation and iv. mixed valences.

criteria mainly include: i. ligand covalency, ii. strong crystal field energy and distorted lattice environment, iii. selective sites occupation and iv. mixed valences charge. The influence of four strategies on the generation of NIR emission are discussed in details.

2.3.3 LIGAND COVALENCY

For $Ce^{3+}/Eu^{2+}/Bi^{3+}$ ions, the decrease of excited energy levels (5d, 6p) relative to free ion is known as centroid shift. It could lead to centroid shift to lower energy due to a reduction of interelectron repulsion, which depends on the type of polyhedra. The size of nephelauxetic effect can be characterized by nephelauxetic ratio β as follows [40]:

$$1 - \beta = hk \tag{2.21}$$

where h characterizes anion ligands and k is central cations. When anions or ionic complexes are put in the order of increasing h-parameter, the well-known nephelauxetic effect series is obtained, i.e., no ligands $< F^- < H_2O < NH_3 < Cl^- < Br^- < N^{3-} < I^{2-} < O^{2-} < S^{2-} < Se^{2-}$ and $H_2O < SO4^{2-} < CO_3^{2-} < PO_4^{3-} < BO_3^{3-} < SiO_4^{4-} < AlO_4^{5-} < O^{2-}$. The reduction of interelectron repulsion and the increase of h-parameter are often ascribed to the sharing of electrons between central cation and surrounding ligands. Therefore, the movement of centroid shift is also related to covalency between activator ions and surrounding anion ligands. The larger centroid shift will happen in the higher covalency (detailed discussion is exhibited in Section 2.2.1). On the one hand, coordination numbers of ligand anion can affect covalency owing to interaction between central cations and ligands. Schematic diagram is presented in Figure 2.12(a).

Commonly, coordination numbers are less than four in the most matrix materials, and thus, activator ions emit visible light from blue to red region. In some matrixes, coordination numbers surrounding oxygen atom are usually more than four, leading to an increase of ligand covalency. Hence, the movement of centroid shift increases, long-wavelength and even NIR light can be successfully achieved. On the other hand, because electronegativity of N^{3-} is smaller than O^{2-}, ligand covalency of nitride is much stronger than oxide matrix. Activator ions can easily achieve long-wavelength light in nitride than that in oxide matrixes. Photoluminescence emission position can be tuned from visible light to NIR emission region through the composition substitution of lowly covalent anion by the higher one. Previously, Tang et al. reported a record-breaking oxide-based $Ba_3ScB_3O_9:Eu^{2+}$ phosphor [38]. Under an optimal excitation wavelength at 376 nm, $Ba_3ScB_3O_9:Eu^{2+}$ shows a broadband long-wavelength light with main emission peak, fwhm and Stokes shift values of 735 nm, 205 nm and 12,991 cm^{-1}, respectively. In $Ba_3ScB_3O_9$, atoms O_3 and O_6 have four binding ligands containing two Ba, one Sc and one B atoms. While other oxygen atoms have a coordination number of 5 that are assign to one B and four Ba atoms. More Ba atoms could form a higher electron density on O atoms, and electron overlap of Eu–O probably increases, resulting in a more significant nephelauxetic effect (Figure 2.12(b)). Poesl et al. reported a $Ca_3Mg[Li_2Si_2N_6]:Eu^{2+}$ phosphor, which exhibited NIR light with peak at 734 nm and fwhm $\approx$ 2293 cm^{-1} [264]. NIR light is attributed to the stronger covalency of nitride than oxide. Crystal structure of $Ca_3Mg[Li_2Si_2N_6]$ also exhibits high connection of $[LiN_4]$ tetrahedron, indicating high structure rigidity (Figure 2.12(c) and (d)). These previous reports offer new design proposal to obtain NIR light. This design insight could not only be useful for Ce^{3+}/Eu^{2+}-activated phosphors but also can be appropriate to Bi^{3+}-activated phosphors.

2.3.4 STRONG CRYSTAL FIELD ENERGY AND DISTORTED LATTICE STRUCTURE

Strong crystal field energy easily promotes crystal field splitting degree of excited energy level, thus leading to a large redshift of excitation and emission position. To the best of our knowledge, typical garnet structure possesses high structure rigidity and lattice symmetry [265, 266]. When activators are introduced into garnet structure, highly efficient long-wavelength emission were

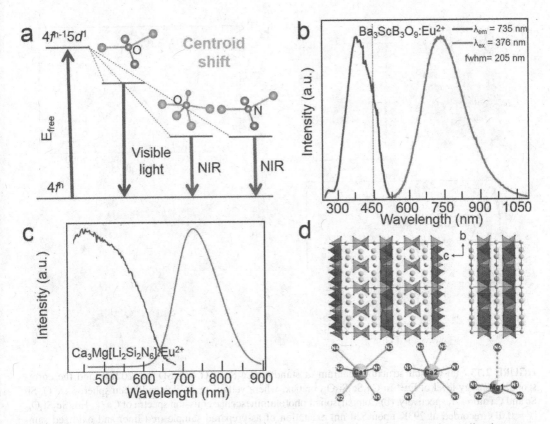

FIGURE 2.12 (a) The overview schematic diagram of the relationship between local coordination structure of ligands and centroid shift variation. (b) Photoluminescence excitation and emission spectra for NIR Ba$_3$ScB$_3$O$_9$:1%Eu^{2+} phosphor [38]. (c) Photoluminescence excitation (blue line) and emission (red line) spectra of deep-red Ca$_3$Mg[Li$_2$Si$_2$N$_6$]:Eu^{2+} [264]. (d) Detailed crystal structure information of layers in Ca$_3$Mg[Li$_2$Si$_2$N$_6$]:Eu^{2+}. Left: [Si$_2$N$_6$]$^{10-}$ bow-tie units (turquoise) and achter rings of [LiN$_4$] tetrahedra (violet). Right: Layer made up of bow-tie units and achter rings forming two sublayers of condensed dreier rings [264]. (Modified from "Elsevier B.V.", Berezovskaya 2013, "American Chemical Society", Poesl 2017, with permission.)

usually achieved based on strong crystal field energy. Commercially available green LuAG:Ce, yellow YAG:Ce phosphor materials belong to typical garnet structure [267–270]. Surprisingly, Berezovskaya et al. reported a NIR Eu^{2+}-activated Ca$_3$Sc$_2$Si$_3$O$_{12}$ garnet phosphors [271]. As shown in Figure 2.13(a), Ca$_3$Sc$_2$Si$_3$O$_{12}$ matrix possesses two types of polyhedra for Eu^{2+} ions doping. Ca atom is coordinated by eight oxygen atoms, forming a slightly distorted cubic polyhedron. Sc atom is coordinated with six oxygen atoms, forming a symmetric octahedron. Octahedra and dodecahedra connect with each other by sharing common faces and edges. High connection among polyhedra may be a main reason for the generation of strong crystal field energy. In addition, polyhedra distortion degree also has an influence on crystal field strength. Lattice distortion degree can be described with following equation [272, 273].

$$D = \frac{1}{n}\sum_{i=1}^{n}\frac{|d_t - d_{av}|}{d_{av}} \qquad (2.22)$$

where d_i is the distance from central atom to ith bonding atom, d_{av} is average bond length and D is the distortion index of activator-anion polyhedron. A large lattice distortion degree easily results in a large crystal field splitting energy of 5d orbitals of activator and causes a redshift phenomenon

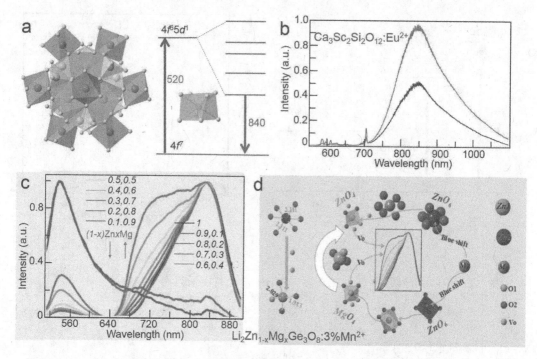

FIGURE 2.13 (a) Crystal structure diagram of standard $Ca_3Sc_2Si_3O_{12}$ (ICSD No. 20214), and the corresponding energy level of Eu^{2+} in $Ca_3Sc_2Si_3O_{12}$ matrix, where yellow, gray, blue and green spheres are O, Si, Sc and Ca atoms, respectively. (b) Comparison of photoluminescence emission spectra of $Ca_{3(1-x)}Eu_{3x}Sc_2Si_3O_{12}$ ($x = 0.01$), recorded at 293K upon 520 nm excitation of as-prepared sample (red line) and oxidized sample (black line). Inset shows photoluminescence emission spectrum of as-prepared sample at 77K [271]. (c) Normalized photoluminescence emission spectra of $Li_2Zn_{1-x}Mg_xGe_3O_8$:3%Mn^{2+} ($0 \leq x \leq 0.9$) [274]. (d) Normalized photoluminescence emission spectra of $(Mg_{1-x}Zn_x)_{2.97}(PO_4)_2$:0.03$Mn^{2+}$ ($x = 0, 0.2, 0.4, 0.5, 0.6$ and 0.8) under 417 nm excitation wavelength [276]. (Modified from "Elsevier B.V.", Berezovskaya 2013, "The Royal Society of Chemistry", Cheng 2017, "The Royal Society of Chemistry", Ma 2015, with permission.)

in photoluminescence excitation and emission spectrum. Accordingly, when Eu^{2+} ions embed into distorted Ca^{2+} sites of $Ca_3Sc_2Si_3O_{12}$ matrix, $4f^65d^1$ excited energy level generates a large crystal field splitting energy. Thus, $Ca_3Sc_2Si_3O_{12}$:Eu^{2+} presents broadband NIR light with peak at 840 nm under 520 nm excitation (Figure 2.13(b)). In addition, Cheng et al. reported NIR phosphors with different spectral widths by adjusting Zn/Mg ratio in $Li_2Zn_{1-x}Mg_xGe_3O_8$:xMn^{2+} [274]. Photoluminescence excitation spectrum is characterized by a broad band from 240 to 640 nm, centered at 475 nm. Emission spectrum presents a broad band from 650 to 900 nm with main peak locating at 832 nm (Figure 2.13(c)). Corresponding mechanism is described in Figure 2.13(d). Newly formed Mg2–O bonds would become looser and strain of Mg2–O bonds might decrease with Mg^{2+} ions occupying in Zn_2 sites in contrast to Zn_1 sites. Thus, Mg2–O bonds are easily broken. As a result, it is easy for $[ZnO_6]$ unit to transfer into $[MgO_4]$ or $[ZnO_4]$ in $Li_2Zn_{1-x}Mg_xGe_3O_8$ matrix, which generates some defects and leads to an obvious broadening of emission spectra. In conclusion, NIR light can be obtained by incorporating in crystal structure with strong crystal field energy and distorted lattice environment. Combined with this design proposal, Li et al. fabricated a series of novel Mn^{4+}-doped $MAlO_3$ (M = La, Gd) persistent phosphors with emission maximum at around 730 nm [275]. Ma et al. reported that a series of novel red- to NIR-emitting Mn^{2+}-activated $(Mg_{1-x}Zn_x)_{2.97}(PO_4)_2$ ($x = 0$, 0.2, 0.4, 0.5, 0.6 and 0.8) [276]. Five- and six-coordinated structure for Mn^{2+} enables the generation of NIR light (peaking at 730 nm, lower energy) and red light (peaking at 630 nm, higher energy). The main reason is that Mn^{2+} ions are distributed at distinct crystal field environment with a different coordination. To increase crystal field splitting energy of activators, local lattice coordination structure plays a vital role. Local lattice structure mainly contains following two aspects: short bond length and low lattice symmetry (distorted lattice structure).

2.3.5 Selective Sites Occupation

Typically, transition-metal and main-group elements are usually coordinated by 2–6 ligand atoms, while coordination numbers of rare-earth Ce^{3+}/Eu^{2+} ions or Bi^{3+} ions are usually more than 6. Since photoluminescence properties of activators are sensitive to surrounding coordination environment, the coordination number of central cation has an important influence on crystal field splitting energy. According to Van Uitert equation, the relationship between emission position and local coordination structure can be simply estimated with following empirical equation [277, 278]:

$$E = Q\left[1 - \left(\frac{V}{4}\right)^{1/V} 10^{-\frac{near}{80}}\right] \tag{2.23}$$

where E is energy location of the lower d-band edge of activator ions. Q is dependent on the types of activators and remains unchanged with specific activator ions. V is the valence of active cation. n refers to coordination number. r is the radius of matrix cation that is replaced by activator. ea value is a constant that is determined by the classification of matrixes. Hence, E is proportional to n and r value, demonstrating that the smaller coordination numbers and smaller ionic radii easily contribute to a long-wavelength emission. In multiple-cations matrix materials, when designing activators in small lattice (small coordination number and short bond length), novel NIR light can be successfully achieved. Based on sites occupation design proposal, a set of matrixes with less than six-fold lattice structures are picked out to accommodate $Ce^{3+}/Eu^{2+}/Bi^{3+}$ ions. Intriguingly, novel NIR light has been successfully achieved. Qiao et al. reported an unprecedented phosphors $K_3LuSi_2O_7:Eu^{2+}$ that gave a NIR emission band centered at 740 nm with an fwhm value of 160 nm upon 460 nm blue light excitation (Figure 2.14(a)) [34]. The inset of Figure 2.14(a) exhibits that annealed phosphors show dark red under natural light. The interesting NIR light is mainly attributed to prior site occupation of Eu^{2+} in smaller $[LuO_6]$ octahedron, which obeys well with proposed design principle. Subsequently, Zhang et al. reported an isomorphic $K_3ScSi_2O_7:Eu^{2+}$ NIR phosphor, which emits broadband NIR light in the range of 600–1100 nm, and main peak locates at 735 nm with an fwhm of 170 nm (Figure 2.14(b)) [37]. According to Gaussian fitting of NIR emission band, spectrum can be well decomposed into two sub-bands with a maximum at 13,997 cm^{-1} (714 nm) and 13,112 cm^{-1} (763 nm), demonstrating that Eu^{2+} ions substitute for two different sites in $K_3ScSi_2O_7$ matrix. As shown in Figure 2.14(d), three cationic sites are suitable for Eu^{2+} ions occupying in $K_3Lu(Sc)Si_2O_7$ matrixes. Average bond length of K1–O ranges from 2.846 to 2.86 Å, and K1 is coordinated with nine oxygen atoms. K2 and Sc are coordinated with six oxygen atoms, and the average bond length of K2–O ranges from 2.759 to 2.77 Å. The average bond length of Sc–O is in the region of 2.147–2.22 Å. Due to the smaller coordination number and shorter bond length, NIR light of Eu^{2+} can be ascribed to sites occupation in K2 and Sc sites. Besides, Liu et al. showed that low-temperature topotactic reduction by using Al metal powder as an oxygen getter can be adopted as a powerful technique for the conversion of bismuth-doped red-emitting systems into their NIR cousins [39]. For the first time, authors identified that the major occupation of Bi^{3+} in the P^{5+} and/or B^{3+} sites of $BaBPO_4$ phosphors based on detailed analysis of X-ray absorption data, which were the main reason for unparalleled NIR light (Figure 2.14(c)). This work anticipated that low-temperature topotactic reduction strategy can be applied in the development of more novel Bi-doped phosphors in various functional optical applications. As shown in Figure 2.14(d), B/P atoms are coordinated with four oxygen atoms and Ba atom is coordinated by ten oxygen atoms. The average bond length values of B–O, P–O and Ba–O are 1.482, 1.538 and 2.845 Å, respectively. NIR emission of Bi^{3+} is attributed to selective site occupation in B and P sites with shorter average bond length and smaller

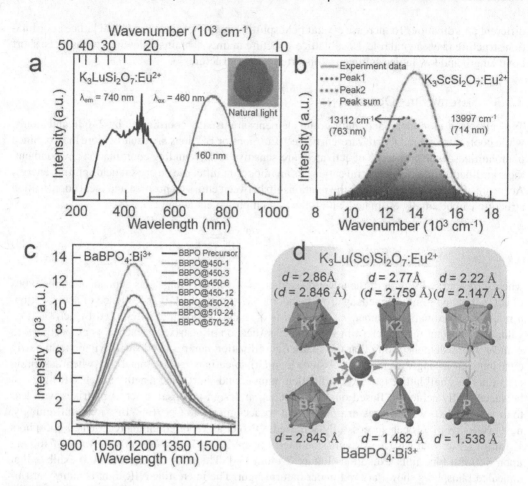

FIGURE 2.14 (a) Photoluminescence excitation and emission spectra of $K_3LuSi_2O_7$:$0.01Eu^{2+}$ measured/monitored at 460 and 740 nm, and inset shows photographs of $K_3LuSi_2O_7$:$0.01Eu^{2+}$ with deep-red body color under natural light [34]. (b) Gaussian fitting curves of photoluminescence emission spectrum for $K_3ScSi_2O_7$:$0.03Eu^{2+}$ upon 465 nm excitation wavelength [37]. (c) Photoluminescence emission spectra of $BaBPO_4$:Bi^{3+} phosphors in NIR region under 476 nm excitation [39]. (d) A schematic of selective sites occupation for Bi^{3+}/Eu^{2+} activators to generate efficient NIR light. (Modified from "The authors", Qiao 2019, "The Royal Society of Chemistry", Zhang 2020, "The Royal Society of Chemistry", Liu 2016, with permission.)

coordination numbers. Recently, our group also reported new NIR emission of Bi^{3+} in highly symmetric $XAl_{12}O_{19}$ (X = Ba, Sr and Ca) ranging from 700 to 1000 nm upon n-UV excitation. We also clarify that abnormal NIR emission is attributed to selective site occupation in small $[AlO_6]$ and $[AlO_4]$ polyhedra [279]. Combined with the preceding discussion, designing activators into small lattice can effectively achieve NIR emission. It is very vital to exploit the approaches to control sites occupation of activators. The strategies mainly contain (1) composition substitution [61], (2) defect control [280, 281], and (3) synthesis condition optimization (such as topotactic reduction) [39]. Through these selective site occupation strategies, emission can be even adjusted from visible to NIR light region, which can well meet the demands for various optical applications.

2.3.6 MIXED VALENCES

As discussed earlier, some activators including Bi, Eu, Mn have multiple charge valences, various luminescence performances can be obtained owing to different characteristic electron transition from different charge valences. For instance, europium ions have two common charge valences, including Eu^{2+} and Eu^{3+} ions. Generally, for Eu^{2+}-doped phosphors, the trace of Eu^{3+} can usually be observed. Owing to the coexistence of Eu^{2+} and Eu^{3+} ions, photoluminescence color can be largely modulated from blue to red light and even white light. Hence, mixed valence is also a common approach to achieve large photoluminescence tuning in single-component matrix materials. Apart from Eu, bismuth and manganese are also known as activators with multiple charge valences. Bismuth can be formed of "+5", "+3", "+2", "+1", "0" in response to different coordination environment and synthesis condition [148]. Among them, Bi^{3+} ions are most popular activators that emit light from blue to red region. Recent decades, Bi^{2+}-doped phosphors usually show orange-red light [142, 185, 187]. In addition, Bi^{+}-doped phosphors or luminescent glasses generally emit broadband NIR light [35, 282–284]. Manganese ions may also present multiple charge valences, and common manganese ions are Mn^{2+}, Mn^{4+} and Mn^{5+} [285, 286]. In various synthesis conditions and coordination structures, Mn in different valence states can emit green, orangish-red, red and even NIR light. Owing to unique photoluminescence properties of activator ions, controllable photoluminescence behavior can be achieved by controlling the valence charge. On the one hand, when changing preparation condition of precursor and raw materials, mixed valences are successfully obtained. As shown in Figure 2.15(a), two adjacent valences of activators ions simultaneously exist in matrixes

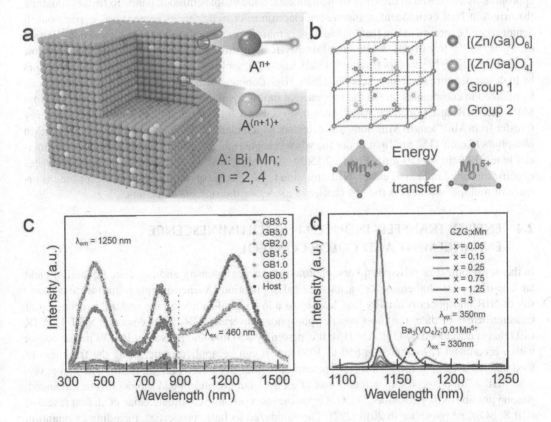

FIGURE 2.15 (a) A schematic mixed valence distribution diagram in oxide matrixes. (b) The interaction relationship of mixed valence, taken Mn^{4+}/Mn^{5+} in $Ca_{14}Zn_6Ga_{10}O_{35}$ matrix as an example [294]. (c) Photoluminescence excitation ($\lambda_{em} = 1235$ nm) and emission ($\lambda_{ex} = 460$ nm) spectra of GB_x ($0.5 \leq x \leq 3.5$) luminescence glasses. Inset shows NIR luminescence decay curves corresponding to GB3.5 and GB3.0 samples [290]. (d) PL spectra of $Ca_{14}Zn_6Ga_{10}O_{35}$:Mn under 350 nm excitation, and photoluminescence emission spectrum of $Ba_2(VO_4)_3$:0.01Mn samples under 330 nm excitation is measured as reference [289, 291]. (Modified from "Elsevier Ltd.", Firmani 2019, "The American Ceramic Society", Liu 2019, "Elsevier Ltd.", Liao 2020, "Elsevier B.V.", Brik 2009, with permission.)

materials. Under excitation light irradiation, both charge valences can exhibit themselves' characteristic luminescence performances. In comparison with single-valence activator, mixed valence obviously enriches the photoluminescence color. On the other hand, activators ions with different charge valences are sensitive to surrounding coordination environment. When activators are successfully doped in matrix materials, local lattice structure may induce self-oxidation/self-reduction process to keep charge balance and energy optimization. Taken $Ca_{14}Zn_6Ga_{10}O_{35}$:Mn as an example, there are two groups of lattices can be occupied by Mn ions, Mn^{4+} prefers to replace $[(Zn/Ga)O_6]$ octahedra, while Mn^{5+} ions prefer to occupy in $[(Zn/Ga)O_4]$ tetrahedra (Figure 2.15(b)).[12] [294]. In this phosphors system, Mn^{4+} exhibits bright red emission and Mn^{5+} ions emit NIR light. Due to selective site occupation and self-oxidation process, Mn^{4+} and Mn^{5+} ions simultaneously exist in $Ca_{14}Zn_6Ga_{10}O_{35}$. Energy transfer from Mn^{4+} to Mn^{5+} also occurs. Based on the preceding design criterions, many studies obtained NIR light with the coexistence of multiple charge valences. Liu et al. reported a series of bismuth-doped germanium-borate glasses with composition of $40GeO_2$–$25B_2O_3$–$25Gd_2O_3$–$10La_2O_3$–xBi_2O_3 via a melt-quenching method, in which multiple Bi centers were simultaneously stabilized. Dual-modulating modes of visible (380–750 nm) and NIR (1000–1600 nm) broadband photoluminescence emissions were effectively controlled under flexible excitation scheme[183, 287, 288]. Photoluminescence emission spectra at low temperature of 10–298K were appropriately employed to interpret such an unusual wide visible emission band. To further illustrate the origin of NIR component, transmission electron microscopy measurement was carried out. It demonstrated experimentally that visible light is mainly originated from the collective contribution of $^3P_1/^3P_0 \rightarrow {}^1S_0$ transitions, while broadband NIR light should be related to the formation of low valent Bi^+ and (or) Bi^0 centers (Figure 2.15(c)). Liao et al. reported $Ca_{14}Zn_6Ga_{10-x}O_{35}$:$x$Mn phosphors by high-temperature solid-state reaction [289]. High doping concentration 30 mol% for Mn ions can be realized in compounds due to the existence of multiple valances of Mn ions. Both red light from Mn^{4+} and NIR emission from Mn^{5+} were observed in $Ca_{14}Zn_6Ga_{10-x}O_{35}$:$x$Mn phosphors. Energy transfer from Mn^{4+} ions to Mn^{5+} ions was detected, which made it possible for $Ca_{14}Zn_6Ga_{10-x}O_{35}$:$x$Mn phosphors to emit 1152 nm light under the whole visible light excitation, especially under commercial blue LED chips (450 nm) (Figure 2.15(d)). Accordingly, the coexistence of multiple valances of activator ions can help to enrich NIR emission region. Corresponding mechanism for the occurrence of multiple valences is the key challenge to choose the suitable activators.

2.4   ENERGY TRANSFER–INDUCED PHOTOLUMINESCENCE ENHANCEMENT AND COLOR CONTROL

In the development of NIR phosphors, photoluminescence intensity and associated quantum yield are one of crucial challenges for various optical applications. Generally, photoluminescent intensity of NIR phosphors is usually low, leading to a low signal-to-noise ratio and a low accuracy of measurements. To date, the most popular phosphor-converted NIR-LEDs device is SFH 4735 IR LED that is developed by OSRAM [15]. It is driven by 350 mA and possesses 16 mW power output with a broadband emission from 650 to 1050 nm. It can be applied in detecting the freshness of food and the calorie of food. However, the emission intensity of NIR region is low. The newer version, SHF 4776, possesses an output power of 24 mW, but quantum yield should also be enhanced. Among possible NIR phosphors, $ScBO_3$:Cr^{3+} is the most possible candidate. Shao et al. first revealed NIR $ScBO_3$:Cr^{3+} phosphor in 2018 [292]. They analyzed its basic properties, including its quantum yield (65%), NIR-LEDs package (~26 mW) and photoluminescence properties. Recently, Fang et al. reported $ScBO_3$:Cr^{3+} NIR phosphors with remarkably higher quantum yield (72.8%) and stronger output power (39.11 mW) [15]. Such a highly efficient NIR emission can well meet the requirement of optical applications. However, the luminescence quantum yield of non-Cr^{3+} NIR phosphors is much lower, which limits their wide applications. Many strategies have been reported to enhance NIR-emission intensity of non-Cr^{3+}-incorporated phosphors. Energy transfer strategy is proposed as one of the most efficient methods, which can not only increase photoluminescence intensity and luminous quantum yield, but also broaden photoluminescence emission spectra. As a result, energy transfer has been widely used to improve luminous quantum yield. To date, many kinds of energy transfer models have been proposed, and detailed schematic diagram is shown in Figure 2.16(a).

Main sensitizer and activator ions are composed of Ln^{n+} (Ln^{n+}: Eu^{2+}, Ce^{3+}, Tb^{3+}, Nd^{3+}, Ho^{3+}, Yb^{3+}, Er^{3+}, Pr^{3+}, Sm^{3+}, Tm^{3+}), Bi^{3+} and Cr^{3+} and $Mn^{2+/4+}$ ions. There are mainly seven types of energy transfer models, including (1) $Ln^{n+} \rightarrow Ln^{n+}$ ($Eu^{2+} \rightarrow Nd^{3+}$, $Nd^{3+} \rightarrow Yb^{3+}$, $Ce^{3+} \rightarrow Yb^{3+}$, $Ho^{3+} \rightarrow Yb^{3+}$, $Tb^{3+} \rightarrow Yb^{3+}$, $Er^{3+} \rightarrow Yb^{3+}$, $Ce^{3+} \rightarrow Pr^{3+}$ and so on), (2) $Ln^{n+} \rightarrow Cr^{3+}$ ($Yb^{3+} \rightarrow Cr^{3+}$), (3) $Bi^{3+} \rightarrow Ln^{n+}$ ($Bi^{3+} \rightarrow Yb^{3+}$, $Bi^{3+} \rightarrow Er^{3+}$, $Bi^{3+} \rightarrow Tb^{3+}$), (4) $Bi^{3+} \rightarrow Cr^{3+}$, (5) $Cr^{3+} \rightarrow Ln^{n+}$ ($Cr^{3+} \rightarrow Yb^{3+}$ and $Cr^{3+} \rightarrow Pr^{3+}$), (6) $Mn^{2+/4+} \rightarrow Ln^{n+}$ and (7) at least two Ln^{3+} ions as multiple sensitizers in phosphors. Through the seven strategies, photoluminescence properties of NIR light can be largely tuned, including (1) large photoluminescence intensity or luminous quantum yield enhancement, (2) absorption region optimization, especially in n-UV and blue region, and (3) spectral position and emission band width tuning. The common energy transfer of NIR emission and corresponding energy transfer efficiency or optimization is summarized in Table 2.2. For example, Yao et al. reported a novel NIR-emitting $LiScP_2O_7$:Cr^{3+} phosphor that was developed completely in NIR spectral range (750–1100 nm) (Figure 2.16(b)) [293]. Under 470 nm excitation, $LiScP_2O_7$:$0.06Cr^{3+}$ shows broadband

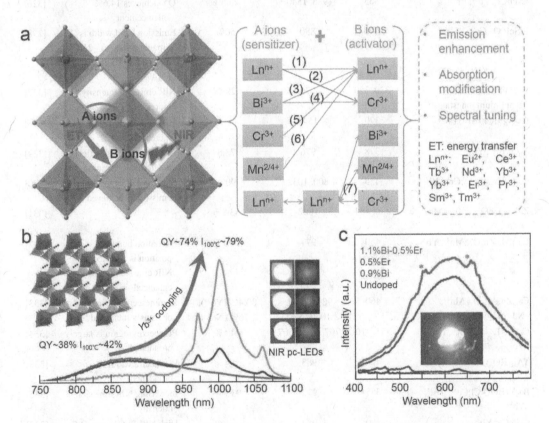

FIGURE 2.16 (a) Schematic energy transfer diagram (left) and energy transfer models for NIR light (right), where energy transfer can be divided into seven models, where A ions and B ions are sensitizers and activators, respectively. "ET" represents energy transfer process, "NIR" stands for near-infrared light. (b) Photoluminescence emission spectra with NIR-LEDs illuminations and crystal structure diagram in NIR $LiScP_2O_7$:Yb^{3+}, Cr^{3+} phosphors based on energy transfer process [293]. (c) Photoluminescence emission spectra in the visible region of Bi^{3+}–Er^{3+} codoped $Cs_2AgInCl_6$. Inset exhibits the digital photograph of fabricated white-light-emitting diodes device from Bi^{3+}–Er^{3+} codoped $Cs_2AgInCl_6$ coated on a commercial n-UV LED chip [294]. (Modified from "American Chemical Society", Yao 2020, "Wiley-VCH Verlag GmbH & Co. KGaA, Weinheim", Arfin 2020, with permission.)

TABLE 2.2

The Optimization of Photoluminescence Properties (Excitation Peak Position and Emission Peak Position) Via Energy Transfer Process in NIR Phosphors

Phosphors	λ_{ex} (nm)	λ_{em} (nm)	Energy Transfer Efficiency	Optimization	Ref.
$Sr_2Si_5N_8:Eu^{2+}$, Nd^{3+}	450	1100	–	High quantum yield (QY = 22%)	[323]
$CaSc_2O_4:Ce^{3+}$, Yb^{3+}	470	967	58%	NIR emission is enhanced 2.4 times	[324]
$Lu_2O_3:Ho^{3+}$, Yb^{3+}	446	1208	62%	Both Ho^{3+} and Yb^{3+} contribute to NIR emission, and emission band width is broadened	[325]
$Y_7O_6F_9:Er^{3+}$, Yb^{3+}	490	1000, 1540	58%	NIR emission intensity is obviously enhanced	[243]
$CsPbCl_3:Yb^{3+}$, Er^{3+}	365	986, 1540	23.5%	QY achieves 14.6% enhancement	[326]
$LiScP_2O_7:Cr^{3+}$, Yb^{3+}	470	880	50%	Emission band width is broadened (up to ~210 nm) and IQY is increased (~74%)	[298]
Bi^{3+}/Yb^{3+}-codoped gadolinium tungstate	330, 350	976	29%	NIR emission intensity increases by 3 times	[327]
$Cs_2AgInCl_6:Bi^{3+}$, Er^{3+}	370	1540	–	NIR emission of Er^{3+} increase 45 times	[328]
$Y_2O_3:Bi^{3+}$, Yb^{3+}	378	976	73.8	Excitation peak position is tuned to n-UV region	[329]
$Ca_2LuZr_2Al_3O_{12}:Cr^{3+}$, Yb^{3+}	455	800, 1032	90%	Super broadband NIR emission is achieved	[330]
$Zn_{1.96-x}GeO_{4+1/2x}:0.04Mn$, xYb.	330, 424	1002	63.8%	–	[331]
$Ca_{14}Al_{10}Zn_6O_{35}:Mn^{4+}$, Yb^{3+}	460	977	26.1%	Excitation band and position is adjusted and NIR emission intensity is enhanced	[332]
$Ca_{14}Zn_6Al_{10}O_{35}:Mn^{4+}$, Nd^{3+}/Yb^{3+}	460	980 (Yb^{3+}), 1060 (Nd^{3+})	80.4% (Yb^{3+}), 76% (Nd^{3+})	Luminescence efficiency is obviously increased	[333]
$Ca_3La_2Te_2O_{12}:Mn^{4+}$, Nd^{3+}, Yb^{3+}	355, 410	707, 978, 1067	44.5%	PL spectra region is largely broadened	[334]
$YAl_3(BO_3)_4:Cr^{3+}$, Yb^{3+}/Nd^{3+}	420	983	–	PL spectra width is broadened	[335]
$Ba_2Y(BO_3)_2Cl:Ce^{3+}$, Tb^{3+}, Yb^{3+}	355	972	93.2%	Increasing NIR emission intensity	[336]
$CsPbCl_3:Yb^{3+}$	300	984	–	High NIR PLQY, over 60%	[337]

NIR light peaking at ~880 nm, with a fwhm value of ~170 nm and an IQY of ~38%. Moreover, photoluminescence intensity improvements of $LiScP_2O_7:Cr^{3+}$ are achieved by codoping Yb^{3+}, leading to broadened fwhm (up to around 210 nm), increased IQY (η_{max} = ~74%), and reduced thermal quenching performances. Energy transfer processes in $LiScP_2O_7:Cr^{3+}$, Yb^{3+} are quantitatively analyzed on the basis of photoluminescence decay lifetime and QY measurements. Photoluminescence emission intensity improvements are principally originated from energy transfer from Cr^{3+} to more efficient and thermally stable Yb^{3+} emitters. Finally, NIR-light-emitting diodes (NIR-LEDs) are fabricated by combining $LiScP_2O_7:Cr^{3+}$, Yb^{3+} phosphors with blue LED chips, giving a maximum NIR output power of ~36 mW and photoelectric efficiency of ~12% at 100 mA drive current. These results suggest that $LiScP_2O_7:Cr^{3+}$, Yb^{3+} phosphors could be the promising luminescence converters for broadband NIR-LEDs applications. In addition, Bi^{3+} and trivalent lanthanide ions have been codoped in metal oxides as optical sensitizers and emitters (Figure 2.16(c)). Codoping of Bi^{3+} and Ln^{3+} (Ln = Er and Yb) in $Cs_2AgInCl_6$ double perovskite is presented, of which metal halide perovskite with coordination number 6 provided an opportunity to codope Bi^{3+} and lanthanide ions [294]. Bi^{3+}–Er^{3+}-codoped $Cs_2AgInCl_6$ phosphors showed Er^{3+} f-electron transition at 1540 nm (suitable for optical communication). The incorporation of Bi^{3+} obviously decreased the excitation (absorption) energy due to 6s–6p electron transition, enabling that NIR phosphors can be efficiently excited by 370 nm light (typical n-UV LED chip). What's more, Bi^{3+}–Er^{3+}-codoped $Cs_2AgInCl_6$ showed 45 times higher emission intensity compared to Er^{3+}-doped $Cs_2AgInCl_6$. Similar results were also observed in Bi^{3+}–Yb^{3+} codoped $Cs_2AgInCl_6$ with maximal emission at around 994 nm. In consequence, energy transfer strategy is an efficient tool to enhance photoluminescence intensity and to optimize photoluminescence excitation (absorption) and emission position. Energy transfer strategies have the promising applications in down-converted as well as many up-converted phosphors. For long period time in future, energy transfer strategy can also play an important role in the enhancement of NIR emission performances.

2.5　APPLICATIONS PROSPECT

Because NIR light has advantages of high penetration and low energy consumption, it has many promising applications as a real-time monitoring analytical and nondestructive tool (Figure 2.17). Traditional NIR light source usually contains tungsten halogen lamp, laser diodes and supercontinuum lasers [295]. The abovementioned light sources have several disadvantages, including poor

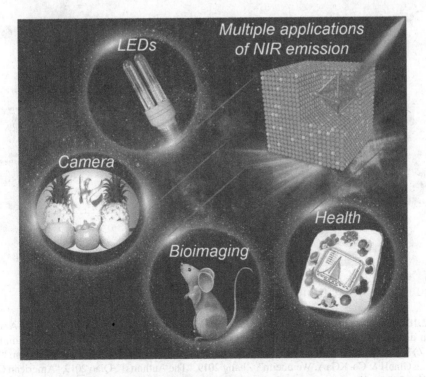

FIGURE 2.17　An overview of multiple applications of NIR emission.

spectra stability, narrow emission spectrum, high electrical consumption and shorter lifetime. To overcome these disadvantages, alternative NIR-LEDs device emerges and gives efficient photo-electric conversion. Owing to high penetration, NIR light can be utilized as NIR camera or video camera in many night-vision situations, which can clearly capture black-and-white images [34]. Based on the principle of characteristic light reflection, transmission and scattering for complex bonds, NIR light can be applied in food processing industry for freshness detection and composition analysis [296]. NIR light can harmlessly penetrate into human biological tissues at a high attenuation property. Therefore, NIR light attracts much attention in medical fields.

2.5.1 NIR-LEDs

Compared with traditional tungsten halogen lamps, the prominent virtues of NIR-LEDs devices present excellent stability, low energy consumption, long lifetime and so on. Hence, many research-ers have devoted to the lighting-quality optimization of NIR-LEDs devices. Usually, NIR-LEDs devices are fabricated by n-UV/blue InGaN LED chips and NIR phosphors with the assignment of silica gel. When mixtures are adequately dried in an oven, NIR-LEDs device can be used for fur-ther electroluminescence performance measurements. The photos of fabricated NIR-LEDs devices are presented in Figure 2.18(a). Figure 2.18(b) exhibits detailed distribution of NIR phosphors on

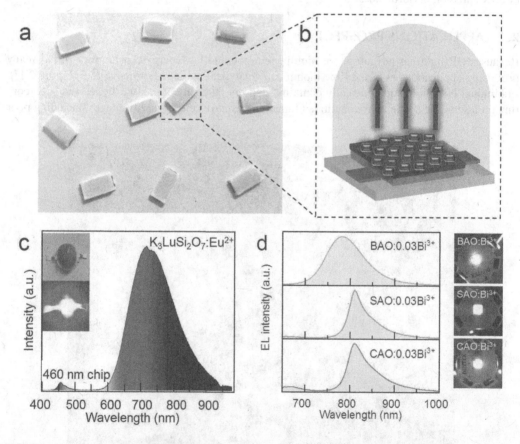

FIGURE 2.18 (a) Photos of NIR-LEDs device fabricated by $Ca_2LuHf_2Al_3O_{12}:0.08Cr^{3+}$ [297]. (b) A schematic fabrication diagram of NIR-LEDs devices. Electroluminescence spectra and corresponding luminescence photos of (c) $K_3LuSi_2O_7:Eu^{2+}$ [34] and (d) $MAl_{12}O_{19}:Bi^{3+}$ (M = Ba, Sr and Ca) [279]. (Modified from "WILEY-VCH Verlag GmbH & Co. KGaA, Weinheim", Zhang 2019, "The Author(s)", Qiao 2019, "American Chemical Society", Wei 2020, with permission.)

an InGaN LED chip, and then it emits NIR light under blue light excitation. Figure 2.18(c) and (d) show representative electroluminescence properties of $K_3LuSi_2O_7:Eu^{2+}$, $BaAl_{12}O_{19}:Bi^{3+}$ (BAO:Bi^{3+}), $SrAl_{12}O_{19}:Bi^{3+}$ (SAO:Bi^{3+}) and $CaAl_{12}O_{19}:Bi^{3+}$ (CAO:Bi^{3+}). All electroluminescence results exhibit stable NIR light, which are similar with photoluminescence results. Such NIR-LEDs light can have promising potentials in multiple optical applications. Accordingly, luminescence intensity and position optimization play a vital role in lighting-quality enhancement of NIR-LEDs devices. However, the serious issues for NIR phosphors are low photoelectric conversion efficiency. In future, the exploitation of new NIR phosphors is still urgent to promote the development of NIR-LEDs lighting industry.

2.5.2 NIR CAMERA IN NIGHT-VISION SITUATIONS

Owing to high penetrance properties, NIR emission light source has potential application in night-vision area. As presented in Figure 2.19(a), when an object is placed in the middle of NIR light source and NIR detectors, the profile and shape of object can be clearly observed. NIR camera is originated from the preceding design principle. Many reports have confirmed the excellent performances of NIR camera based on Cr^{3+}/Eu^{2+}-activated phosphors. Zhang et al. fabricated a broadband NIR-LEDs by coating $Ca_2LuHf_2Al_3O_{12}:0.08Cr^{3+}$ phosphor on a commercial 460 nm LED chip [297]. The fabricated NIR-LEDs was used as light source to light up and brighten the logo printed on an A4 paper sheet in a completely dark situation. The result shows an excellent lighting effect (Figure 2.19(b)). Then, the photos of NIR light were set to pass through palm and finger (Figure 2.19(c)). Blood vessels in the finger and palm can be clearly observed by chiaroscuro due to absorption of NIR light by chromophores in blood (Figure 2.19(c)). Qiao et al. also reported NIR emission of $K_3LuSi_2O_7:Eu^{2+}$ phosphors. Figure 2.19(d)–(f) show photographs obtained by different

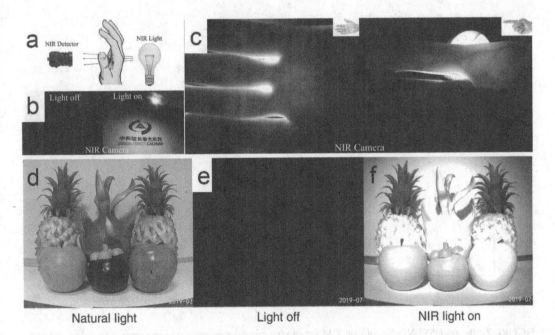

Natural light   Light off   NIR light on

FIGURE 2.19 (a) A schematic illumination of NIR-LEDs devices in dark environment. (b) The photos of A4 paper with and without NIR emission light. (c) Photos of NIR light transilluminating palm and finger, respectively. NIR-LEDs works at 100 mA current [297]. (d)–(f) Photographs under natural light and NIR-LEDs light captured by corresponding visible camera and a NIR camera [34]. (Modified from "WILEY-VCH Verlag GmbH & Co. KGaA, Weinheim", Zhang 2019, "The Author(s)", Qiao 2019, with permission.)

cameras under natural and NIR-LEDs light, respectively. Nothing can be detected by NIR camera once NIR-LEDs is off. In contrast, NIR camera can capture black-and-white images while NIR-LEDs lamp is lighted. The preceding results demonstrate that NIR emission can act as phosphors in night-vision situations [34].

2.5.3 BIOSENSOR AND BIOIMAGING

NIR light is harmless to human biological tissues owing to a high attenuation property and penetrability properties. The NIR emission attracts much attention in medical area. Up to now, many bioimaging and medical applications use the up-converted phosphors. Such up-conversion phosphors usually require NIR emission light as external light source. Then, up-converted phosphors can emit visible or NIR emission light to detector. Such a process is vital tool to disease diagnosis and treatment. Li et al. reported a biomimetic persistent luminescent nanoplatform ($Zn_{1.25}Ga_{1.5}Ge_{0.25}O_4$:0.5%$Cr^{3+}$, 2.5%$Yb^{3+}$, 0.25%$Er^{3+}$) for noninvasive high-sensitive diagnosis (Figure 2.20(a)) [298]. Sun et al. reported $NaYF_4$:Yb,Nd@CaF_2 core/shell nanoparticles by designing $Nd^{3+} \rightarrow Yb^{3+}$ energy transfer [299]. The liver of nude mouse can be observed clearly after injection within 1 minute, which presents clear in vivo photoluminescence profile and image (Figure 2.20(b)). In conclusion, no matter

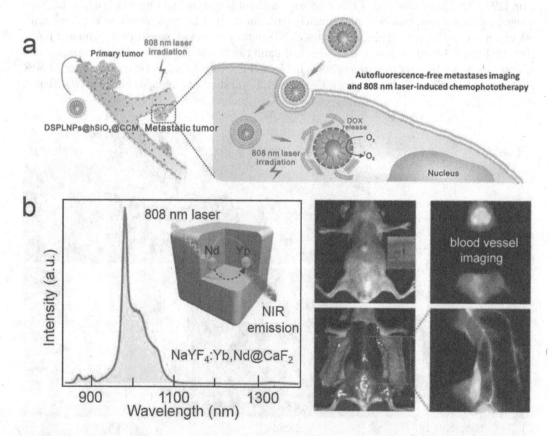

FIGURE 2.20 (a) DSPLNPs@hSiO$_2$@CCM for in vivo autofluorescence-free metastases imaging and chemophototherapy in vivo autofluorescence-free imaging and chemophototherapy of metastases [298, 300]. (b) Photoluminescence emission spectra of $NaYF_4$:Yb,Nd@CaF_2 nanoparticles (left), inset plots the illustration of $NaYF_4$:7%Yb, 60%Nd@CaF_2 core/shell structures under 808 nm excitation (left). In vivo imaging of bright field is displayed (right) [299]. (Modified from "American Chemical Society", Li 2017, "American Chemical Society", Cao 2018, "American Chemical Society", Sun 2018, with permission.)

for which bioimaging applications, NIR light plays an important role as external excitation source. Hence, the exploitation of high-power NIR emission is necessary.

2.5.4 FOOD COMPOSITION AND FRESHNESS ANALYSIS

As shown in Figure 2.21(a), the NIR spectra of foods mainly comprise a broad band, which arises from overlapping absorptions corresponding to overtones and combinations of vibrational modes of C–H, O–H and N–H chemical bonds [301]. The concentrations of constituents such as water, protein, fat and carbohydrate can be determined in principle by using classical absorption spectroscopy. The major advantage of NIR emission light is that usually no sample preparation is necessary, and the analysis approach is very simple and very swift (between 15 and 90 seconds) [301]. Due to a broad absorption region of foods, the requirement of NIR emission is its band width. Nowadays, many researchers are devoted to exploit ultra-broadband NIR emission light. $Mg_3Ga_2GeO_8:Cr^{3+}$ exhibits ultra-broadband NIR emission in the range of 650–1200 nm, which matches well with the overtones of molecular vibrations (e.g., O–H, C–H and N–H) presented in food composition (Figure 2.21(b)–(d)) [302]. In addition, previously reported $La_3Ga_{5(1-x)}GeO_{14}:5xCr^{3+}$ and $La_3GaGe_5O_{16}:Cr^{3+}$ materials also exhibit broadband NIR emission (650–1050 nm), which exhibits promising applications in food freshness and composition analysis area [24,25]. However, the urgent issue of some recent NIR-emitting phosphors is to extend emission band width. Therefore, ultra-broad NIR emission is the focus in food analysis field.

2.5.5 FIBER LASER

Fiber lasers are part of the family of solid-state lasers. In the fiber laser, the lasing medium is an optical fiber doped with low levels of a rare earth element, such as yttrium, erbium or thulium, which determine the wavelength of output light [303, 304]. Diode lasers are used to couple infrared light into the cladding of doped fiber and act as pump source. This pumping action excites dopant atoms, stimulating them to emit photons at a specific wavelength. Diffraction gratings are used

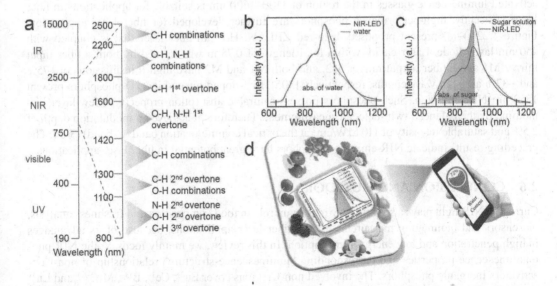

FIGURE 2.21 (a) Principal types of NIR absorption bands and their corresponding locations [301]. (b) and (c) Normalized transmission spectra of NIR light after penetrating water and sugar solutions, respectively. (d) A schematic food analysis diagram [302]. (Modified from "John Wiley & Sons, Ltd", Osborne 2006, "American Chemical Society", Wang 2019, with permission.)

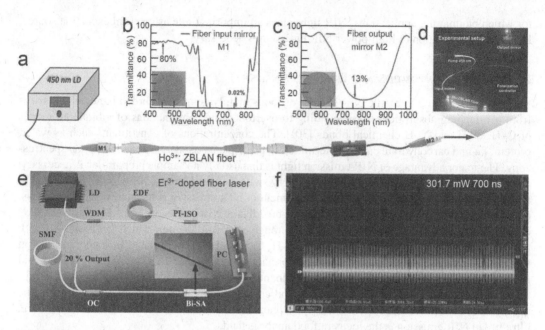

FIGURE 2.22 (a) Experimental schematic of the proposed deep-red ZBLAN: Ho³⁺ (fluorozirconate glass) all-fiber laser. (b) and (c) Optical transmission spectra of fiber input mirror M1 and fiber output mirror M2, respectively. Insets plot corresponding microscopy views of M1 and M2. (d) Experimental setup of the proposed green ZBLAN: Ho³⁺ all-fiber laser [305]. (e) Experimental setup of fundamentally mode-locked Er³⁺-doped fiber laser. (f) Oscilloscope pulse trace of harmonic and dual-wavelength operation [306]. (Modified from "The Royal Society of Chemistry", Li 2018, "American Chemical Society", Guo 2020, with permission.)

as rear mirror and output coupler, which form resonator. Fiber lasers have excellent beam quality and highest efficiency among solid-state lasers. Furthermore, in the last ten years, bismuth-activated luminescence glasses in the region of 1100–1600 nm is suitable for applications in fiber lasers. Recently, high-performance phosphors are further developed for fiber laser application. Figure 2.22(a)–(d) present proposed deep-red ZBLAN: Ho³⁺ all-fiber laser that is coupled with 450 nm laser-diode. Laser cavity with a total length of 0.75 m was formed by both a fiber input mirror M1 and a fiber output mirror M2, and both M1 and M2 have high transmissions of ~75% and ~87% at green wavelengths, respectively [305]. Er³⁺-doped bismuthine 2D phosphors present 1550 nm NIR emission. Such phosphors possess saturable absorption property of few-layer bismuthine by the balanced twin-detector measurement. Therefore, it achieved a modulation depth of 2.5% and saturable intensity of $110 \, \text{MW/cm}^2$ at the optical communication band (C-band) [306]. The preceding results indicate NIR-emitting phosphors have great potential in fiber laser applications.

2.6 CONCLUSION AND OUTLOOK

Currently, NIR light plays a significantly important role in food composition and freshness analysis, biosensors and bioimaging measurements and fiber laser applications, because of its advantages in high penetration and low energy consumption. In this review, we mainly focus on the NIR photoluminescence properties and corresponding "luminescence-structure" relationship of non-Cr³⁺-activators inorganic phosphors. The involved non-Cr³⁺ ions cover Eu²⁺, Ce³⁺, Bi³⁺, Mn²⁺/⁴⁺ and Ln³⁺ (Ln = Pr, Nd, Ho, Er, Tm, Sm and Yb) ions. First, intrinsic electron transitions of non-Cr³⁺ ions are summarized in details. The deep understanding of these activators can help to find proper matrix materials to achieve highly efficient NIR emission. Importantly, this review proposes the concept of "high-throughput synthesis-DFT calculations prediction-structure design-luminescence performance optimization" based on the combination theoretical calculation and the experimental method. This proposal is helpful to rapidly screen a series of new NIR phosphors. This is because high-throughput methodology and DFT calculation are very accurate and can give reliable prediction based on an adequate inorganic crystal data input. High-throughput methodology can effectively screen targeted activators and matrix materials. Due to the sensibility of activator ions to surrounding coordination environment, the focus is placed on the local crystal structure design of matrix to obtain highly efficient and tunable NIR light. These structure design strategies mainly include (1) ligand covalency, (2) strong crystal field and distorted lattice, (3) selective sites occupation and (4) mixed valences. Despite of considerable progresses over the past decades, there are also more challenges as well as opportunities to explore novel NIR phosphors for NIR camera, food analysis, biosensors/bioimaging and fiber laser applications. Many research areas need to be further studied, include but not limited to:

1. The optimization of photoluminescence quantum efficiency, stability of NIR phosphors, especially for $Eu^{2+}/Ce^{3+}/Bi^{3+}$-activated phosphors.
2. For non-Cr^{3+}-doped NIR phosphors, position and band width of photoluminescence excitation spectra should be further tuned to blue region. The fwhm values should be further optimized.
3. High-throughput synthesis and DFT calculation methods are still not very common to obtain novel NIR phosphors. Theoretical system needs to be further added and completed. Besides, the underlying mechanism of some abnormal luminescence behavior need be further revealed.
4. The small-sized NIR phosphors (microparticle or nanoparticle) are desired to be fabricated into micro-LEDs (μLEDs) via the optimization of synthesis conditions. Such a NIR-μLEDs technique can be helpful and convenient to the optical application and measurements by just using wearable smartwatch or other portable equipment. Besides, the NIR light can also be used in the imperceptible artificial skin and masks based on temperature sensibility and bifunctional nature [307].

REFERENCES

1. Vankayala R, and Hwang KC (2018) Near-Infrared-Light-Activatable Nanomaterial-Mediated Phototheranostic Nanomedicines: An Emerging Paradigm for Cancer Treatment. Adv. Mater. 30: e1706320.
2. Li Y, Yao L, Yin Z, Cheng Z, Yang S, and Zhang Y (2020) Defect-Induced Abnormal Enhanced Upconversion Luminescence in BiOBr:Yb^{3+}/Er^{3+} Ultrathin Nanosheets and its Influence on Visible-NIR light Photocatalysis. Inorg. Chem. Front. 7: 519.
3. Liang L, Chen N, Jia Y, Ma Q, Wang J, Yuan Q, and Tan W (2019) Recent Progress in Engineering Near-Infrared Persistent Luminescence Nanoprobes for Time-Resolved Biosensing/Bioimaging. Nano Res. 12: 1279.
4. Baianu I, Guo J, Nelson R, You T, and Costescu D (2011) NIR Calibrations for Soybean Seeds and Soy Food Composition Analysis: Total Carbohydrates, Oil, Proteins and Water Contents [v.2]. Nat. Prec. https://doi.org/10.1038/npre.2011.6611.1.
5. Wang J, Zhao J, Li Y, Yang M, Chang YQ, Zhang JP, Sun Z, and Wang Y (2015) Enhanced Light Absorption in Porous Particles for Ultra-NIR-Sensitive Biomaterials. ACS Macro Lett. 4: 392.
6. Ning Y, Zhu M, and Zhang JL (2019) Near-Infrared (NIR) Lanthanide Molecular Probes for Bioimaging and Biosensing. Coordin. Chem. Rev. 399: 213028.
7. Abdukayum A, Chen JT, Zhao Q, and Yan XP (2013) Functional Near Infrared-Emitting Cr^{3+}/Pr^{3+} Co-Doped Zinc Gallogermanate Persistent Luminescent Nanoparticles with Superlong Afterglow for In Vivo Targeted Bioimaging. J. Am. Chem. Soc. 135: 14125.
8. Jia Z, Yuan C, Liu Y, Wang XJ, Sun P, Wang L, Jiang H, and Jiang J (2020) Strategies to Approach High Performance in Cr^{3+}-Doped Phosphors for High-Power NIR-LED Light Sources. Light Sci. Appl. 9: 86.
9. Xing P, Niu Y, Mu R, Wang Z, Xie D, Li H, Dong L, and Wang C (2020) A Pocket-Escaping Design to Prevent the Common Interference with Near-Infrared Fluorescent Probes In Vivo. Nat. Commun. 11: 1573.

10. Gu Y, Guo Z, Yuan W, Kong M, Liu Y, Liu Y, Gao Y, Feng W, Wang F, Zhou J, Jin D, and Li F (2019) High-Sensitivity Imaging of Time-Domain Near-Infrared Light Transducer. Nat. Photonics 13: 525.

11. Porep JU, Kammerer DR, and Carle R (2015) On-Line Application of Near Infrared (NIR) Spectroscopy in Food Production. Trends Food Sci. Technol. 46: 211.

12. Firmani P, Nardecchia A, Nocente F, Gazza L, Marini F, and Biancolillo A (2020) Multi-Block Classification of Italian Semolina Based on Near Infrared Spectroscopy (NIR) Analysis and Alveographic Indices. Food Chem. 309: 125677.

13. Grassi S, and Alamprese C (2018) Advances in NIR Spectroscopy Applied to Process Analytical Technology in Food Industries. Curr. Opin. Food Sci. 22: 17.

14. Richter B, Rurik M, Gurk S, Kohlbacher O, and Fischer M (2019) Food Monitoring: Screening of the Geographical Origin of White Asparagus Using FT-NIR and Machine Learning. Food Control 104: 318.

15. Fang MH, Huang PY, Bao Z, Majewska N, Leśniewski T, Mahlik S, Grinberg M, Leniec G, Kaczmarek SM, Yang CW, Lu KM, Sheu HS, and Liu RS (2020) Penetrating Biological Tissue Using Light-Emitting Diodes with a Highly Efficient Near-Infrared $ScBO_3$:Cr^{3+} Phosphor. Chem. Mater. 32: 2166.

16. De Vos A, Lejaeghere K, Vanpoucke DE, Joos JJ, Smet PF, and Hemelsoet K (2016) First-Principles Study of Antisite Defect Configurations in $ZnGa_2O_4$:Cr Persistent Phosphors. Inorg. Chem. 55: 2402.

17. Zhu Q, Xiahou J, Guo Y, Li H, Ding C, Wang J, Li X, Sun X, and Li JG (2018) $Zn_3Ga_2Ge_2O_{10}$:Cr^{3+} Uniform Microspheres: Template-Free Synthesis, Tunable Bandgap/Trap Depth, and In Vivo Echargeable Near-Infrared-Persistent Luminescence. ACS Appl. Bio. Mater. 2: 577.

18. Mao M, Zhou T, Zeng H, Wang L, Huang F, Tang X, and Xie RJ (2020) Broadband Near-Infrared (NIR) Emission Realized by the Crystal-Field Engineering of $Y_{3-x}Ca_xAl_{5-x}Si_xO_{12}$:$Cr^{3+}$ (x = 0–2.0) Garnet Phosphors. J. Mater. Chem. C 8: 1981.

19. Zhou X, Geng W, Li J, Wang Y, Ding J, and Wang Y (2020) An Ultraviolet–Visible and Near-Infrared-Responded Broadband NIR Phosphor and Its NIR Spectroscopy Application. Adv. Opt. Mater. 8: 1902003.

20. Liu G, Molokeev MS, Lei B, and Xia Z (2020) Two-Site Cr^{3+} Occupation in the $MgTa_2O_6$:Cr^{3+} Phosphor Toward Broad-Band Near-Infrared Emission for Vessel Visualization. J. Mater. Chem. C 8: 9322.

21. De Clercq OQ, Martin LIDJ, Korthout K, Kusakovskij J, Vrielinck H, and Poelman D (2017) Probing the Local Structure of the Near-Infrared Emitting Persistent Phosphor $LiGa_5O_8$:Cr^{3+}. J. Mater. Chem. C 5: 10861.

22. Zhao F, Song Z, Zhao J, and Liu Q (2019) Double Perovskite $Cs_2AgInCl_6$:Cr^{3+}: Broadband and Near-Infrared Luminescent Materials. Inorg. Chem. Front. 6: 3621.

23. Zeng H, Zhou T, Wang L, and Xie RJ (2019) Two-Site Occupation for Exploring Ultra-Broadband Near-Infrared Phosphor-Double-Perovskite La_2MgZrO_6:Cr^{3+}. Chem. Mater. 31: 5245.

24. Rajendran V, Fang MH, Guzman GND, Lesniewski T, Mahlik S, Grinberg M, Leniec G, Kaczmarek SM, Lin YS, Lu KM, Lin CM, Chang H, Hu SF, and Liu RS (2018) Super Broadband Near-Infrared Phosphors with High Radiant Flux as Future Light Sources for Spectroscopy Applications. ACS Energy Lett. 3: 2679.

25. Rajendran V, Lesniewski T, Mahlik S, Grinberg M, Leniec G, Kaczmarek SM, Pang WK, Lin YS, Lu KM, Lin CM, Chang H, Hu SF, and Liu RS (2019) Ultra-Broadband Phosphors Converted Near-Infrared Light Emitting Diode with Efficient Radiant Power for Spectroscopy Applications. ACS Photonics 6: 3215.

26. Xu X, Shao Q, Yao L, Dong Y, and Jiang J (2020) Highly Efficient and Thermally Stable Cr^{3+}-Activated Silicate Phosphors for Broadband Near-Infrared LED Applications. Chem. Eng. J. 383: 123108.

27. Dai D, Wang Z, Liu C, Li X, Zhang L, Xing Z, Yang Z, and Li P (2019) Lithium Substitution Endowing Cr^{3+}-Doped Gallium Germanate Phosphors with Super-Broad-Band and Long Persistent Near-Infrared Luminescence. ACS Appl. Electron. Mater. 1: 2551.

28. Joos JJ, Van der Heggen D, Martin L, Amidani L, Smet PF, Barandiaran Z, and Seijo L (2020) Broadband Infrared LEDs Based on Europium-to-Terbium Charge Transfer Luminescence. Nat. Commun. 11: 3647.

29. Maak C, Strobel P, Weiler V, Schmidt PJ, and Schnick W (2018) Unprecedented Deep-red Ce^{3+} Luminescence of the Nitridolithosilicates $Li_{38.7}RE_{3.3}Ca_{5.7}[Li_2Si_{30}N_{59}]O_2F$ (RE = La, Ce, Y). Chem. Mater. 30: 5500.

30. Xu W, Peng M, Ma Z, Dong G, and Qiu J (2012) A New Study on Bismuth Doped Oxide Glasses. Opt. Exp. 20: 15692.

31. Zhao S, Zhang Y, and Zang Z (2020) Room-Temperature Doping of Ytterbium into Efficient Near-Infrared Emission $CsPbBr_{1.5}Cl_{1.5}$ Perovskite Quantum Dots. Chem. Commun. 56: 5811.

32. Chen J, Guo C, Yang Z, Li T, Zhao J, and McKittrick J (2016) Li$_2$SrSiO$_4$:Ce^{3+}, Pr^{3+} Phosphor with Blue, Red, and Near-Infrared Emissions Used for Plant Growth LED. J. Am. Ceram. Soc. 99: 218.
33. Qi Y, Li S, Min Q, Lu W, Xu X, Zhou D, Qiu J, Wang L, and Yu X (2018) Optical Temperature Sensing Properties of KLu$_2$F$_7$: Yb^{3+}/Er^{3+}/Nd^{3+} Nanoparticles under NIR Excitation. J. Alloys Compd. 742: 497.
34. Qiao J, Zhou G, Zhou Y, Zhang Q, and Xia Z (2019) Divalent Europium-Doped Near-Infrared-Emitting Phosphor for Light-Emitting Diodes. Nat. Commun. 10: 5267.
35. Romanov AN, Haula EV, Boldyrev KN, Shashkin DP, and Korchak VN (2019) Broadband Near-IR Photoluminescence of Bismuth-Doped Pollucite-Related Phase CsGaGe$_2$O$_6$. J. Lumin. 216: 116741.
36. Huang WT, Cheng CL, Bao Z, Yang CW, Lu KM, Kang CY, Lin CM, and Liu RS (2019) Broadband Cr^{3+}, Sn^{4+}-Doped Oxide Nanophosphors for Infrared Mini Light-Emitting Diodes. Angew. Chem. Int. Ed. 58: 2069.
37. Lai J, Shen W, Qiu J, Zhou D, Long Z, Yang Y, Zhang K, Khan I, and Wang Q (2020) Broadband Near-Infrared Emission Enhancement in K$_2$Ga$_2$Sn$_6$O$_{16}$:Cr^{3+} Phosphor by Electron-Lattice Coupling Regulation. J. Am. Ceram. Soc. 103: 5067.
38. Tang Z, Zhang Q, Cao Y, Li Y, and Wang Y (2020) Eu^{2+}-Doped Ultra-Broadband VIS-NIR Emitting Phosphor. Chem. Eng. J. 388: 124231.
39. Liu B-M, Yong Z-J, Zhou Y, Zhou D-D, Zheng L-R, Li L-N, Yu H-M, and Sun H-T (2016) Creation of Near-Infrared Luminescent Phosphors Enabled by Topotactic Reduction of Bismuth-Activated Red-Emitting Crystals. J. Mater. Chem. C 4: 9489.
40. Li G, Tian Y, Zhao Y, and Lin J (2015) Recent Progress in Luminescence Tuning of Ce^{3+} and Eu^{2+}-Activated Phosphors for pc-WLEDs. Chem. Soc. Rev. 44: 8688.
41. Zhao M, Zhang Q, and Xia Z (2020) Narrow-Band Emitters in LED Backlights for Liquid-Crystal Displays. Mater. Today 40: 246.
42. Wang L, Xie RJ, Suehiro T, Takeda T, and Hirosaki N (2018) Down-Conversion Nitride Materials for Solid State Lighting: Recent Advances and Perspectives. Chem. Rev. 118: 1951.
43. Lin H, Hu T, Cheng Y, Chen M, and Wang Y (2018) Glass Ceramic Phosphors: Towards Long-Lifetime High-Power White Light-Emitting-Diode Applications—A Review. Laser Photonics Rev. 12: 1700344.
44. Brik MG, Camardello SJ, and Srivastava AM (2015) Influence of Covalency on the Mn4+ ^{2}Eg→^{4}A$_{2g}$ Emission Energy in Crystals. ECS J. Solid State Sci. Technol. 4: R39.
45. Li J, Ding J, Cao Y, Zhou X, Ma B, Zhao Z, and Wang Y (2018) Color-Tunable Phosphor [Mg$_{1.25}$Si$_{1.25}$Al$_{2.5}$]O$_3$N$_3$:Eu^{2+}—A New Modified Polymorph of AlON with Double Sites Related Luminescence and Low Thermal Quenching. ACS Appl. Mater. Interfaces 10: 37307.
46. Qiao J, Ning L, Molokeev MS, Chuang YC, Zhang Q, Poeppelmeier KR, and Xia Z (2019) Site-Selective Occupancy of Eu^{2+} Toward Blue-Light-Excited Red Emission in a Rb$_3$YSi$_2$O$_7$:Eu Phosphor. Angew. Chem. Int. Ed. 58: 11521.
47. Zhao M, Xia Z, Huang X, Ning L, Gautier R, Molokeev MS, Zhou Y, Chuang Y-C, Zhang Q, Liu Q, and Poeppelmeier KR (2019) Li Substituent Tuning of LED Phosphors with Enhanced Efficiency, Tunable Photoluminescence, and Improved Thermal Stability. Sci. Adv. 5: eaav0363.
48. Shang M, Liang S, Qu N, Lian H, and Lin J (2017) Influence of Anion/Cation Substitution (Sr^{2+} → Ba^{2+}, Al^{3+} → Si^{4+}, N^{3-} → O^{2-}) on Phase Transformation and Luminescence Properties of Ba$_3$Si$_6$O$_{15}$:Eu^{2+} Phosphors. Chem. Mater. 29: 1813.
49. Zhao M, Liao H, Ning L, Zhang Q, Liu Q, and Xia Z (2018) Next-Generation Narrow-Band Green-Emitting RbLi(Li$_3$SiO$_4$)$_2$:Eu^{2+} Phosphor for Backlight Display Application. Adv. Mater. 30: e1802489.
50. Zhang Z, Ma C, Gautier R, Molokeev MS, Liu Q, and Xia Z (2018) Structural Confinement toward Giant Enhancement of Red Emission in Mn^{2+}-Based Phosphors. Adv. Funct. Mater. 28: 1804150.
51. Yu Z, Xia Z, Chen M, Xiang Q, and Liu Q (2017) Insight into the Preparation and Luminescence Properties of Yellow-Green-Emitting [(Sr,Ba)$_3$AlO$_4$F–Sr$_3$SiO$_5$]:Ce$_{3+}$ Solid Solution Phosphors. J. Mater. Chem. C 5: 3176.
52. Qiao J, Amachraa M, Molokeev M, Chuang Y-C, Ong SP, Zhang Q, and Xia Z (2019) Engineering of K$_3$YSi$_2$O$_7$ To Tune Photoluminescence with Selected Activators and Site Occupancy. Chem. Mater. 31: 7770.
53. Liao H, Zhao M, Molokeev MS, Liu Q, and Xia Z (2018) Learning from a Mineral Structure toward an Ultra-Narrow-Band Blue-Emitting Silicate Phosphor RbNa$_3$(Li$_3$SiO$_4$)$_4$:Eu^{2+}. Angew. Chem. Int. Ed. 57: 11728.
54. He L, Song Z, Jia X, Xia Z, and Liu Q (2018) Control of Luminescence in Eu^{2+}-Doped Orthosilicate-Orthophosphate Phosphors by Chainlike Polyhedra and Electronic Structures. Inorg. Chem. 57: 609.
55. Chen Y, He J, Zhang X, Rong M, Xia Z, Wang J, and Liu ZQ (2020) Dual-Mode Optical Thermometry Design in Lu$_3$Al$_5$O$_{12}$:Ce^{3+}/Mn^{4+} Phosphor. Inorg. Chem. 59: 1383.

56. Chen M, Xia Z, Molokeev MS, Lin CC, Su C, Chuang Y-C, and Liu Q (2017) Probing Eu^{2+} Luminescence from Different Crystallographic Sites in $Ca_{10}M(PO_4)7:Eu^{2+}$ (M = Li, Na, and K) with β-$Ca_3(PO_4)_2$-Type Structure. Chem. Mater. 29: 7563.

57. Wang L, Xie RJ, Li Y, Wang X, Ma CG, Luo D, Takeda T, Tsai YT, Liu RS, and Hirosaki N (2016) $Ca_{1-x}Li_xAl_{1-x}Si_{1+x}N_3:Eu^{2+}$ Solid Solutions as Broadband, Color-Tunable and Thermally Robust Red Phosphors for Superior Color Rendition White Light-Emitting Diodes. Light. Sci. Appl. 5: e16155.

58. Wang L, Wang X, Takeda T, Hirosaki N, Tsai YT, Liu RS, and Xie RJ (2015) Structure, Luminescence, and Application of a Robust Carbidonitride Blue Phosphor $(Al_{1-x}Si_xC_xN_{1-x}:Eu^{2+})$ for Near UV-LED Driven Solid State Lighting. Chem. Mater. 27: 8457.

59. Tsai YT, Nguyen HD, Lazarowska A, Mahlik S, Grinberg M, and Liu RS (2016) Improvement of the Water Resistance of a Narrow-Band Red-Emitting $SrLiAl_3N_4:Eu^{2+}$ Phosphor Synthesized under High Isostatic Pressure through Coating with an Organosilica Layer. Angew. Chem. Int. Ed. 55: 9652.

60. Lin CC, Meijerink A, and Liu RS (2016) Critical Red Components for Next-Generation White LEDs. J. Phys. Chem. Lett. 7: 495.

61. Fang M-H, Meng S-Y, Majewska N, Leśniewski T, Mahlik S, Grinberg M, Sheu HS, and Liu RS (2019) Chemical Control of $SrLi(Al_{1-x}Ga_x)_3N_4:Eu^{2+}$ Red Phosphors at Extreme Conditions for Application in Light-Emitting Diodes. Chem. Mater. 31: 4614.

62. Fang MH, Chen PY, Bao Z, Majewska N, Lesniewski T, Mahlik S, Grinberg M, Sheu HS, Lee JF, and Liu RS (2020) Broadband $NaK_2Li[Li_3SiO_4]_4:Ce$ Alkali Lithosilicate Blue Phosphors. J. Phys. Chem. Lett. 11: 6621.

63. Cheng J, Li P, Wang Z, Li Z, Tian M, Wang C, and Yang Z (2018) Color Selective Manipulation in $Li_2ZnGe_3O_8:Mn^{2+}$ by Multiple-Cation Substitution on Different Crystal-Sites. Dalton Trans. 47: 4293.

64. Xing Z, Li P, Dai D, Li X, Liu C, Zhang L, and Wang Z (2019) Self-Luminescence of Perovskite-Like $LaSrGaO_4$ via Intrinsic Defects and Anomalous Luminescence Analysis of $LaSrGaO_4:Mn^{2+}$. Inorg. Chem. 58: 4869.

65. Wang C, Wang Z, Li P, Cheng J, Li Z, Tian M, Sun Y, and Yang Z (2017) Relationships between Luminescence Properties and Polyhedron Distortion in $Ca_{9-x-y-z}Mg_xSr_yBa_zCe(PO_4)_7:Eu^{2+},Mn^{2+}$. J. Mater. Chem. C 5: 10839.

66. Zhao M, Liao H, Molokeev MS, Zhou Y, Zhang Q, Liu Q, and Xia Z (2019) Emerging Ultra-Narrow-Band Cyan-Emitting Phosphor for White LEDs with Enhanced Color Rendition. Light: Sci. Appl. 8: 1.

67. Zhang X, Fang MH, Tsai YT, Lazarowska A, Mahlik S, Lesniewski T, Grinberg M, Pang WK, Pan F, Liang C, Zhou W, Wang J, Lee J-F, Cheng B-M, Hung T-L, Chen Y-Y, and Liu R-S (2017) Controlling of Structural Ordering and Rigidity of β-SiAlON:Eu through Chemical Cosubstitution to Approach Narrow-Band-Emission for Light-Emitting Diodes Application. Chem. Mater. 29: 6781.

68. Li H, Pang R, Liu G, Sun W, Li D, Jiang L, Zhang S, Li C, Feng J, and Zhang H (2018) Synthesis and Luminescence Properties of Bi^{3+}-Activated K_2MgGeO_4: A Promising High-Brightness Orange-Emitting Phosphor for WLEDs Conversion. Inorg. Chem. 57: 12303.

69. Kim YH, Kim HJ, Ong SP, Wang Z, and Im WB (2020) Cation-Size Mismatch as a Design Principle for Enhancing the Efficiency of Garnet Phosphors. Chem. Mater. 32: 3097.

70. Ha J, Kim YH, Novitskaya E, Wang Z, Sanchez M, Graeve OA, Ong SP, Im WB, and McKittrick J (2019) Color Tunable Single-Phase Eu^{2+} and Ce^{3+} Co-Activated Sr_2LiAlO_4 Phosphors. J. Mater. Chem. C 7: 7734.

71. Lai S, Zhao M, Qiao J, Molokeev MS, and Xia Z (2020) Data-Driven Photoluminescence Tuning in Eu^{2+}-Doped Phosphors. J. Phys. Chem. Lett. 11: 5680.

72. Huber W, Carey VJ, Gentleman R, Anders S, Carlson M, Carvalho BS, Bravo HC, Davis S, Gatto L, Girke T, Gottardo R, Hahne F, Hansen KD, Irizarry RA, Lawrence M, Love MI, MacDonald J, Obenchain V, Oles AK, Pages H, Reyes A, Shannon P, Smyth GK, Tenenbaum D, Waldron L, and Morgan M (2015) Orchestrating High-Throughput Genomic Analysis with Bioconductor. Nat. Methods 12: 115.

73. Shen ZH, Wang JJ, Lin Y, Nan CW, Chen LQ, and Shen Y (2018) High-Throughput Phase-Field Design of High-Energy-Density Polymer Nanocomposites. Adv. Mater. 30: 1704380.

74. Singh AK, Montoya JH, Gregoire JM, and Persson KA (2019) Robust and Synthesizable Photocatalysts for CO_2 Reduction: A Data-Driven Materials Discovery. Nat. Commun. 10: 443.

75. Park WB, Singh SP, Kim M, and Sohn KS (2015) Combinatorial Screening of Luminescent and Structural Properties in a Ce^{3+}-Doped Ln–Al–Si–O–N (Ln = Y, La, Gd, Lu) System: The Discovery of a Novel $Gd_3Al_{(3+x)}Si_{(3-x)}O_{(12+x)}N_{(2-x)}:Ce^{3+}$ Phosphor. Inorg. Chem. 54: 1829.

76. Meyer J, and Tappe F (2015) Photoluminescent Materials for Solid-State Lighting: State of the Art and Future Challenges. Adv. Opt. Mater. 3: 424.
77. Liu Y, Hu Z, Suo Z, Hu L, Feng L, Gong X, Liu Y, and Zhang J (2019) High-Throughput Experiments Facilitate Materials Innovation: A Review. Sci. China Technol. Sci. 62: 521.
78. Takeda T, Hirosaki N, Funahashi S, and Xie RJ (2015) Narrow-Band Green-Emitting Phosphor $Ba_2LiSi_7AlN_{12}$:Eu^{2+} with High Thermal Stability Discovered by a Single Particle Diagnosis Approach. Chem. Mater. 27: 5892.
79. Li S, Xia Y, Amachraa M, Hung NT, Wang Z, Ong SP, and Xie R-J (2019) Data-Driven Discovery of Full-Visible-Spectrum Phosphor. Chem. Mater. 31: 6286.
80. Li Z, Xu Q, Sun Q, Hou Z, and Yin WJ (2019) Thermodynamic Stability Landscape of Halide Double Perovskites via High-Throughput Computing and Machine Learning. Adv. Funct. Mater. 29: 1807280.
81. Wang Z, Ha J, Kim YH, Im WB, McKittrick J, and Ong SP (2018) Mining Unexplored Chemistries for Phosphors for High-Color-Quality White-Light-Emitting Diodes. Joule 2: 914.
82. Park WB, Shin N, Hong KP, Pyo M, and Sohn KS (2012) A New Paradigm for Materials Discovery: Heuristics-Assisted Combinatorial Chemistry Involving Parameterization of Material Novelty. Adv. Funct. Mater. 22: 2258.
83. Hirosaki N, Takeda T, Funahashi S, and Xie RJ (2014) Discovery of New Nitridosilicate Phosphors for Solid State Lighting by the Single-Particle-Diagnosis Approach. Chem. Mater. 26: 4280.
84. Zhong Y, and Dai H (2020) A Mini-Review on Rare-Earth Down-Conversion Nanoparticles for NIR-II Imaging of Biological Systems. Nano Res. 13: 1281.
85. Wu J, Zheng G, Liu X, and Qiu J (2020) Near-Infrared Laser Driven White Light Continuum Generation: Materials, Photophysical Behaviours and Applications. Chem. Soc. Rev. 49: 3461.
86. Ma Y, Zhang Y, and Yu WW (2019) Near Infrared Emitting Quantum Dots: Synthesis, Luminescence Properties and Applications. J. Mater. Chem. C 7: 13662.
87. Fan X, Liu Z, Yang X, Chen W, Zeng W, Tian S, Yu X, Qiu J, and Xu X (2019) Recent Developments and Progress of Inorganic Photo-Stimulated Phosphors. J. Rare Earths 37: 679.
88. De Guzman GNA, Fang MH, Liang CH, Bao Z, Hu SF, and Liu RS (2020) Near-Infrared Phosphors and Their Full Potential: A Review on Practical Applications and Future Perspectives. J. Lumin. 219: 116944.
89. Fang MH, Leaño JL, and Liu RS (2018) Control of Narrow-Band Emission in Phosphor Materials for Application in Light-Emitting Diodes. ACS Energy Lett. 3: 2573.
90. Xia Z, and Liu Q (2016) Progress in Discovery and Structural Design of Color Conversion Phosphors for LEDs. Prog. Mater. Sci. 84: 59.
91. Qin X, Liu X, Huang W, Bettinelli M, and Liu X (2017) Lanthanide-Activated Phosphors Based on 4f–5d Optical Transitions: Theoretical and Experimental Aspects. Chem. Rev. 117: 4488.
92. Fang MH, Mariano COM, Chen PY, Hu SF, and Liu RS (2020) Cuboid-Size-Controlled Color-Tunable Eu-Doped Alkali–Lithosilicate Phosphors. Chem. Mater. 32: 1748.
93. Tsai YT, Chiang CY, Zhou W, Lee JF, Sheu HS, and Liu RS (2015) Structural Ordering and Charge Variation Induced by Cation Substitution in $(Sr,Ca)AlSiN_3$:Eu Phosphor. J. Am. Chem. Soc. 137: 8936.
94. Elzer E, Strobel P, Weiler V, Schmidt PJ, and Schnick W (2020) Illuminating Nitridoberylloaluminates: The Highly Efficient Red-Emitting Phosphor $Sr_2[BeAl_3N_5]$:Eu^{2+}. Chem. Mater. 32: 6611.
95. Nitta M, Nagao N, Nomura Y, Hirasawa T, Sakai Y, Ogata T, Azuma M, Torii S, Ishigaki T, and Inada Y (2020) High-Brightness Red-Emitting Phosphor $La_3(Si,Al)_6(O,N)_{11}$:Ce^{3+} for Next-Generation Solid-State Light Sources. ACS Appl. Mater. Interfaces 12: 31652.
96. Huang D, Zhu H, Deng Z, Zou Q, Lu H, Yi X, Guo W, Lu C, and Chen X (2019) Moisture-Resistant Mn^{4+}-Doped Core-Shell-Structured Fluoride Red Phosphor Exhibiting High Luminous Efficacy for Warm White Light-Emitting Diodes. Angew. Chem. Int. Ed. 58: 3843.
97. Jin Y, Hu Y, Yuan L, Chen L, Wu H, Ju G, Duan H, and Mu Z (2016) Multifunctional Near-Infrared Emitting Cr^{3+}-Doped $Mg_4Ga_8Ge_2O_{20}$ Particles with Long Persistent and Photostimulated Persistent Luminescence, and Photochromic Properties. J. Mater. Chem. C 4: 6614.
98. Liu S, Wang Z, Cai H, Song Z, and Liu Q (2020) Highly Efficient Near-Infrared Phosphor $LaMgGa_{11}O_{19}$:Cr^{3+}. Inorg. Chem. Front. 7: 1467.
99. Ding S, Guo H, Feng P, Ye Q, and Wang Y (2020) A New Near-Infrared Long Persistent Luminescence Material with Its Outstanding Persistent Luminescence Performance and Promising Multifunctional Application Prospects. Adv. Opt. Mater. 8: 2000097.
100. Pust P, Weiler V, Hecht C, Tucks A, Wochnik AS, Henss AK, Wiechert D, Scheu C, Schmidt PJ, and Schnick W (2014) Narrow-Band Red-Emitting $Sr[LiAl_3N_4]$:Eu^{2+} as A Next-Generation LED-Phosphor Material. Nat. Mater. 13: 891.

101. Zhao Y, Yin L, ten Kate OM, Dierre B, Abellon R, Xie R-J, van Ommen JR, and Hintzen HT (2019) Enhanced Thermal Degradation Stability of the $Sr_2Si_5N_8:Eu^{2+}$ Phosphor by Ultra-Thin Al_2O_3 Coating through the Atomic Layer Deposition Technique in A Fluidized Bed Reactor. J. Mater. Chem. C 7: 5772.

102. Liu Y, Zhang J, Zhang C, Xu J, Liu G, Jiang J, and Jiang H (2015) $Ba_9Lu_2Si_6O_{24}:Ce^{3+}$: An Efficient Green Phosphor with High Thermal and Radiation Stability for Solid-State Lighting. Adv. Opt. Mater. 3: 1096.

103. Dorenbos P (2003) Relation between Eu^{2+} and Ce^{3+} f↔d-Transition Energies in Inorganic Compounds. J. Phys.: Condens. Matter 15: 4797.

104. Dorenbos P (2013) Ce^{3+} 5d-Centroid Shift and Vacuum Referred 4f-Electron Binding Energies of All Lanthanide Impurities in 150 Different Compounds. J. Lumin. 135: 93.

105. Ning L, Zhou C, Chen W, Huang Y, Duan C, Dorenbos P, Tao Y, and Liang H (2015) Electronic Properties of Ce^{3+}-Doped $Sr_3Al_2O_5Cl_2$: A Combined Spectroscopic and Theoretical Study. J. Phys. Chem. C 119: 6785.

106. Zhou L, Zhou W, Pan F, Shi R, Huang L, Liang H, Tanner PA, Du X, Huang Y, Tao Y, and Zheng L (2016) Spectral Properties and Energy Transfer of a Potential Solar Energy Converter. Chem. Mater. 28: 2834.

107. Strobel P, Weiler V, Hecht C, Schmidt PJ, and Schnick W (2017) Luminescence of the Narrow-Band Red Emitting Nitridomagnesosilicate $Li_2(Ca_{1-x}Sr_x)_2[Mg_2Si_2N_6]:Eu^{2+}$ (x = 0–0.06). Chem. Mater. 29: 1377.

108. Dorenbos P (2013) Lanthanide 4f-Electron Binding Energies and the Nephelauxetic Effect in Wide Band Gap Compounds. J. Lumin. 136: 122.

109. MacKenzie KJD, Gainsford GJ, and Ryan MJ (1996) Rietveld Refinement of the Crystal Structures of the Yttrium Silicon Oxynitrides $Y_2Si_3N_4O_3$ (N-Melilite) and $Y_4Si_2O_7N_2$ (J-Phase). J. Eur. Ceram. Soc. 16: 553.

110. Dorenbos P (2000) 5d-Level Energies of Ce^{3+} and the Crystalline Environment. I. Fluoride Compounds. Phys. Rev. B 62: 15640.

111. Dorenbos P (2005) Thermal Quenching of Eu^{2+} 5d–4f Luminescence in Inorganic Compounds. J. Phys. Condens. Matter 17: 8103.

112. Li X, Li P, Wang Z, Liu S, Bao Q, Meng X, Qiu K, Li Y, Li Z, and Yang Z (2017) Color-Tunable Luminescence Properties of Bi^{3+} in $Ca_5(BO_3)_3F$ via Changing Site Occupation and Energy Transfer. Chem. Mater. 29: 8792.

113. ten Kate OM, Zhang Z, van Ommen JR, and Hintzen HT (2018) Dependence of the Photoluminescence Properties of Eu^{2+} Doped M–Si–N (M = Alkali, Alkaline Earth or Rare Earth Metal) Nitridosilicates on Their Structure and Composition. J. Mater. Chem. C 6: 5671.

114. Luo H, Bos AJ, Dobrowolska A, and Dorenbos P (2015) Low-Temperature VUV Photoluminescence and Thermoluminescence of UV Excited Afterglow Phosphor $Sr_3Al_xSi_{1-x}O_5:Ce^{3+},Ln^{3+}$ (Ln = Er, Nd, Sm, Dy and Tm). Phys. Chem. Chem. Phys. 17: 15419.

115. Zhou Y, Zhuang W, Hu Y, Liu R, Xu H, Chen M, Liu Y, Li Y, Zheng Y, and Chen G (2019) Cyan-Green Phosphor $(Lu_2M)(Al_4Si)O_{12}:Ce^{3+}$ for High-Quality LED Lamp: Tunable Photoluminescence Properties and Enhanced Thermal Stability. Inorg. Chem. 58: 1492.

116. Leng Z, Li R, Li L, Xue D, Zhang D, Li G, Chen X, and Zhang Y (2018) Preferential Neighboring Substitution-Triggered Full Visible Spectrum Emission in Single-Phased $Ca_{10.5-x}Mg_x(PO_4)_7:Eu^{2+}$ Phosphors for High Color-Rendering White LEDs. ACS Appl. Mater. Interfaces 10: 33322.

117. Chen SN, and Chen TM (2013) Luminescence and Spectroscopic Studies of $(Y_{0.99}Ce_{0.01})_3OF_3S_2$: A New Orange Yellow-Emitting Oxyfluorosulfide Phosphor for LEDs. J. Chin. Chem. Soc. 60: 961.

118. Wu YC, Chen YC, Chen TM, Lee CS, Chen KJ, and Kuo HC (2012) Crystal Structure Characterization, Optical and Photoluminescent Properties of Tunable Yellow- to Orange-Emitting $Y_2(Ca,Sr)F_4S_2:Ce^{3+}$ Phosphors for Solid-State Lighting. J. Mater. Chem. 22: 8048.

119. Lee SP, Huang CH, Chan TS, and Chen TM (2014) New Ce^{3+}-Activated Thiosilicate Phosphor for LED Lighting-Synthesis, Luminescence Studies, and Applications. ACS Appl. Mater. Interfaces 6: 7260.

120. Lee SP, Chan TS, and Chen TM (2015) Novel Reddish-Orange-Emitting $BaLa_2Si_2S_8:Eu^{2+}$ Thiosilicate Phosphor for LED Lighting. ACS Appl. Mater. Interfaces 7: 40.

121. Wang QY, Yuan P, Wang TW, Yin ZQ, and Lu FC (2020) Effect of Sr and Ca Substitution of Ba on the Photoluminescence Properties of the Eu^{2+} Activated $Ba_2MgSi_2O_7$ Phosphor. Ceram. Int. 46: 1374.

122. Xiao W, Wu D, Zhang L, Zhang X, Hao Z, Pan GH, Zhang L, Ba X, and Zhang J (2017) The Inductive Effect of Neighboring Cations in Tuning Luminescence Properties of the Solid Solution Phosphors. Inorg. Chem. 56: 9938.

123. Chen L, Fei M, Zhang Z, Jiang Y, Chen S, Dong Y, Sun Z, Zhao Z, Fu Y, He J, Li C, and Jiang Z (2016) Understanding the Local and Electronic Structures toward Enhanced Thermal Stable Luminescence of CaAlSiN3:Eu2+. Chem. Mater. 28: 5505.

124. Companion AL, and Komarynsky MA (1964) Crystal Field Splitting Diagrams. J. Chem. Educ. 41: 257.

125. Song Z, Xia Z, and Liu Q (2018) Insight into the Relationship between Crystal Structure and Crystal-Field Splitting of Ce^{3+} Doped Garnet Compounds. J. Phys. Chem. C 122: 3567.

126. Nazarov M, Brik MG, Spassky D, and Tsukerblat B (2017) Crystal Field Splitting of 5d States and Luminescence Mechanism in $SrAl_2O_4$:Eu^{2+} Phosphor. J. Lumin. 182: 79.

127. Song Z, and Liu Q (2020) Crystal-Field Splitting of Ce^{3+} in Narrow-Band Phosphor $SrLiAl_3N_4$. J. Rare Earths 39: 386.

128. George NC, Pell AJ, Dantelle G, Page K, Llobet A, Balasubramanian M, Pintacuda G, Chmelka BF, and Seshadri R (2013) Local Environments of Dilute Activator Ions in the Solid-State Lighting Phosphor $Y_{3-x}Ce_xAl_5O_{12}$. Chem. Mater. 25: 3979.

129. Jing Y, Liu Y, Jiang X, Molokeev MS, Lin Z, and Xia Z (2020) Sb^{3+} Dopant and Halogen Substitution Triggered Highly Efficient and Tunable Emission in Lead-Free Metal Halide Single Crystals. Chem. Mater. 32: 5327.

130. Luo J, Wang X, Li S, Liu J, Guo Y, Niu G, Yao L, Fu Y, Gao L, Dong Q, Zhao C, Leng M, Ma F, Liang W, Wang L, Jin S, Han J, Zhang L, Etheridge J, Wang J, Yan Y, Sargent EH, and Tang J (2018) Efficient and Stable Emission of Warm-White Light from Lead-Free Halide Double Perovskites. Nature 563: 541.

131. Brennan MC, Herr JE, Nguyen-Beck TS, Zinna J, Draguta S, Rouvimov S, Parkhill J, and Kuno M (2017) Origin of the Size-Dependent Stokes Shift in $CsPbBr_3$ Perovskite Nanocrystals. J. Am. Chem. Soc. 139: 12201.

132. de Jong M, Seijo L, Meijerink A, and Rabouw FT (2015) Resolving the Ambiguity in the Relation between Stokes Shift and Huang-Rhys Parameter. Phys. Chem. Chem. Phys. 17: 16959.

133. Sadeghi S, Bahmani Jalali H, Melikov R, Ganesh Kumar B, Mohammadi Aria M, Ow-Yang CW, and Nizamoglu S (2018) Stokes-Shift-Engineered Indium Phosphide Quantum Dots for Efficient Luminescent Solar Concentrators. ACS Appl. Mater. Interfaces 10: 12975.

134. Daicho H, Shinomiya Y, Enomoto K, Nakano A, Sawa H, Matsuishi S, and Hosono H (2018) A Novel Red-Emitting $K_2Ca(PO_4)F$:Eu^{2+} Phosphor with a Large Stokes Shift. Chem. Commun. 54: 884.

135. McCall KM, Stoumpos CC, Kostina SS, Kanatzidis MG, and Wessels BW (2017) Strong Electron–Phonon Coupling and Self-Trapped Excitons in the Defect Halide Perovskites $A_3M_2I_9$ (A = Cs, Rb; M = Bi, Sb). Chem. Mater. 29: 4129.

136. Yang B, Yin L, Niu G, Yuan JH, Xue KH, Tan Z, Miao XS, Niu M, Du X, Song H, Lifshitz E, and Tang J (2019) Lead-Free Halide Rb_2CuBr_3 as Sensitive X-Ray Scintillator. Adv. Mater. 31: e1904711.

137. Stadler W, Hofmann DM, Alt HC, Muschik T, Meyer BK, Weigel E, Muller-Vogt G, Salk M, Rupp E, and Benz KW (1995) Optical Investigations of Defects in $Cd_{1-x}Zn_xTe$. Phys. Rev. B Condens. Matter 51: 10619.

138. Dang P, Liang S, Li G, Wei Y, Cheng Z, Lian H, Shang M, Al Kheraif AA, and Lin J (2018) Full Color Luminescence Tuning in Bi^{3+}/Eu^{3+}-Doped $LiCa_3MgV_3O_{12}$ Garnet Phosphors Based on Local Lattice Distortion and Multiple Energy Transfers. Inorg. Chem. 57: 9251.

139. Liu D, Yun X, Dang P, Lian H, Shang M, Li G, and Lin J (2020) Yellow/Orange-Emitting $ABZn_2Ga_2O_7$:Bi^{3+}(A = Ca, Sr; B = Ba, Sr) Phosphors: Optical Temperature Sensing and White Light-Emitting Diode Applications. Chem. Mater. 32: 3065.

140. Lozhkina OA, Murashkina AA, Shilovskikh VV, Kapitonov YV, Ryabchuk VK, Emeline AV, and Miyasaka T (2018) Invalidity of Band-Gap Engineering Concept for Bi^{3+} Heterovalent Doping in $CsPbBr_3$ Halide Perovskite. J. Phys. Chem. Lett. 9: 5408.

141. Manna D, Das TK, and Yella A (2019) Tunable and Stable White Light Emission in Bi^{3+}-Alloyed $Cs_2AgInCl_6$ Double Perovskite Nanocrystals. Chem. Mater. 31: 10063.

142. Awater RHP, and Dorenbos P (2016) X-Ray Induced Valence Change and Vacuum Referred Binding Energies of Bi^{3+} and Bi^{2+} in $Li_2BaP_2O_7$. J. Phys. Chem. C 120: 15114.

143. Wang DY, Tang ZB, Khan WU, and Wang Y (2017) Photoluminescence Study of a Broad Yellow-Emitting Phosphor $K_2ZrSi_2O_7$:Bi^{3+}. Chem. Eng. J. 313: 1082.

144. Sun HT, Zhou J, and Qiu J (2014) Recent Advances in Bismuth Activated Photonic Materials. Prog. Mater. Sci. 64: 1.

145. Puchalska M, Bolek P, Kot K, and Zych E (2020) Luminescence of Bi^{3+} and Bi^{2+} Ions in Novel Bi-Doped $SrAl_4O_7$ Phosphor. Opt. Mater. 107: 109999.

146. Zheng J, Li L, and Peng M (2017) A New Red Aluminate Phosphor $CaAl_{12}O_{19}$ Activated by Bi^{2+} for White LEDs. Sci. Adv. Mater. 9: 485.

147. Kang F, Zhang Y, and Peng M (2015) Controlling the Energy Transfer via Multi Luminescent Centers to Achieve White Light/Tunable Emissions in a Single-Phased X2-type Y_2SiO_5:Eu^{3+},Bi^{3+} Phosphor for Ultraviolet Converted LEDs. Inorg. Chem. 54: 1462.

148. Dang P, Liu D, Li G, Al Kheraif AA, and Lin J (2020) Recent Advances in Bismuth Ion-Doped Phosphor Materials: Structure Design, Tunable Photoluminescence Properties, and Application in White LEDs. Adv. Opt. Mater. 8: 1901993.

149. Han J, Li L, Peng M, Huang B, Pan F, Kang F, Li L, Wang J, and Lei B (2017) Toward Bi^{3+} Red Luminescence with No Visible Reabsorption through Manageable Energy Interaction and Crystal Defect Modulation in Single Bi^{3+}-Doped $ZnWO_4$ Crystal. Chem. Mater. 29: 8412.

150. Kang F, Zhang H, Wondraczek L, Yang X, Zhang Y, Lei DY, and Peng M (2016) Band-Gap Modulation in Single Bi^{3+}-Doped Yttrium–Scandium–Niobium Vanadates for Color Tuning over the Whole Visible Spectrum. Chem. Mater. 28: 2692.

151. Kang F, Peng M, Lei DY, and Zhang Q (2016) Recoverable and Unrecoverable Bi^{3+}-Related Photoemissions Induced by Thermal Expansion and Contraction in $LuVO_4:Bi^{3+}$ and $ScVO_4:Bi^{3+}$ Compounds. Chem. Mater. 28: 7807.

152. Kang F, Peng M, Yang X, Dong G, Nie G, Liang W, Xu S, and Qiu J (2014) Broadly Tuning Bi^{3+} Emission via Crystal Field Modulation in Solid Solution Compounds $(Y,Lu,Sc)VO_4:Bi$ for Ultraviolet Converted White LEDs. J. Mater. Chem. C 2: 6068.

153. Kang F, Yang X, Peng M, Wondraczek L, Ma Z, Zhang Q, and Qiu J (2014) Red Photoluminescence from Bi^{3+} and the Influence of the Oxygen-Vacancy Perturbation in $ScVO_4$: A Combined Experimental and Theoretical Study. J. Phys. Chem. C 118: 7515.

154. Han J, Pan F, Molokeev MS, Dai J, Peng M, Zhou W, and Wang J (2018) Redefinition of Crystal Structure and Bi^{3+} Yellow Luminescence with Strong Near-Ultraviolet Excitation in $La_3BWO_9:Bi^{3+}$ Phosphor for White Light-Emitting Diodes. ACS Appl. Mater. Interfaces 10: 13660.

155. Peng M, Lei J, Li L, Wondraczek L, Zhang Q, and Qiu J (2013) Site-Specific Reduction of Bi^{3+} to Bi^{2+} in Bismuth-Doped Over-Stoichiometric Barium Phosphates. J. Mater. Chem. C 1: 5303.

156. Gao G, Peng M, and Wondraczek L (2014) Spectral Shifting and NIR Down-Conversion in Bi^{3+}/Yb^{3+} Co-Doped Zn_2GeO_4. J. Mater. Chem. C 2: 8083.

157. Wang L, Sun Q, Liu Q, and Shi J (2012) Investigation and Application of Quantitative Relationship between sp Energy Levels of Bi^{3+} Ion and Host Lattice. J. Solid State Chem. 191: 142.

158. Xue J, Wang X, Jeong JH, and Yan X (2018) Spectral and Energy Transfer in Bi^{3+}-Re^{n+} (n = 2, 3, 4) Co-Doped Phosphors: Extended Optical Applications. Phys. Chem. Chem. Phys. 20: 11516.

159. Boutinaud P (2013) Revisiting the Spectroscopy of the Bi^{3+} Ion in Oxide Compounds. Inorg. Chem. 52: 6028.

160. Setlur AA, and Srivastava AM (2006) The Nature of Bi^{3+} Luminescence in Garnet Hosts. Opt. Mater. 29: 410.

161. Lyu T, and Dorenbos P (2020) Vacuum-Referred Binding Energies of Bismuth and Lanthanide Levels in $ARE(Si,Ge)O_4$ (A = Li, Na; RE = Y, Lu): Toward Designing Charge-Carrier-Trapping Processes for Energy Storage. Chem. Mater. 32: 1192.

162. Awater RHP, and Dorenbos P (2017) The Bi^{3+} 6s and 6p Electron Binding Energies in Relation to the Chemical Environment of Inorganic Compounds. J. Lumin. 184: 221.

163. Song E, Jiang X, Zhou Y, Lin Z, Ye S, Xia Z, and Zhang Q (2019) Heavy Mn^{2+} Doped $MgAl_2O_4$ Phosphor for High-Efficient Near-Infrared Light-Emitting Diode and The Night-Vision Application. Adv. Opt. Mater. 7: 1901105.

164. Zeng R, Zhang L, Xue Y, Ke B, Zhao Z, Huang D, Wei Q, Zhou W, and Zou B (2020) Highly Efficient Blue Emission from Self-Trapped Excitons in Stable Sb^{3+}-Doped $Cs_2NaInCl_6$ Double Perovskites. J. Phys. Chem. Lett. 11: 2053.

165. Jiménez JA (2014) Luminescent Properties of Cu^+/Sn^{2+}-Activated Aluminophosphate Glass. Opt. Mater. 37: 347.

166. Masai H, Suzuki Y, Yanagida T, and Mibu K (2015) Luminescence of Sn^{2+} Center in $ZnO–B_2O_3$ Glasses Melted in Air and Ar Conditions. Bull. Chem. Soc. Jan. 88: 1047.

167. Torimoto A, Masai H, Okada G, and Yanagida T (2017) X-Ray Induced Luminescence of Sn^{2+}-Centers in Zinc Phosphate Glasses. Radiat. Meas. 106: 175.

168. Xing G, Gao Z, Tao M, Wei Y, Liu Y, Dang P, Li G, and Lin J (2019) Novel Orange-Yellow-Green Color-Tunable Bi^{3+}-Doped $Ba_3Y_{4-w}Lu_wO_9$ (0≤w≤4) Luminescent Materials: Site Migration and Photoluminescence Control. Inorg. Chem. Front. 6: 3598.

169. Wang M, Fan X, and Xiong G (1995) Luminescence of Bi^{3+} Ions and Energy Transfer from Bi^{3+} Ions to Eu^{3+} Ions in Silica Glasses Prepared by the Sol-gel Process. J. Phys. Chem. Solids 56: 859.

170. Romanov AN, Veber AA, Fattakhova ZT, Usovich OV, Haula EV, Trusov LA, Kazin PE, Korchak VN, Tsvetkov VB, and Sulimov VB (2013) Subvalent Bismuth Monocation Bi^+ Photoluminescence in Ternary Halide Crystals $KAlCl_4$ and $KMgCl_3$. J. Lumin. 134: 180.

171. Peng M, Wu B, Da N, Wang C, Chen D, Zhu C, and Qiu J (2008) Bismuth-Activated Luminescent Materials for Broadband Optical Amplifier in WDM System. J. Non Cryst. Solids 354: 1221.

172. Ren J, Yang L, Qiu J, Chen D, Jiang X, and Zhu C (2006) Effect of Various Alkaline-Earth Metal Oxides on the Broadband Infrared Luminescence from Bismuth-Doped Silicate Glasses. Solid State Commun. 140: 38.

173. Peng M, Chen D, Qiu J, Jiang X, and Zhu C (2007) Bismuth-Doped Zinc Aluminosilicate Glasses and Glass-Ceramics with Ultra-Broadband Infrared Luminescence. Opt. Mater. 29: 556.

174. Schweizer T, Samson BN, Hector JR, Brocklesby WS, Hewak DW, and Payne DN (1999) Infrared Emission from Holmium Doped Gallium Lanthanum Sulphide Glass. Infrared Phys. Tech. 40: 329.

175. Arai Y, Suzuki T, Ohishi Y, Morimoto S, and Khonthon S (2007) Ultrabroadband Near-Infrared Emission from a Colorless Bismuth-Doped Glass. Appl. Phys. Lett. 90: 261110.

176. Zhou S, Lei W, Jiang N, Hao J, Wu E, Zeng H, and Qiu J (2009) Space-Selective Control of Luminescence Inside the Bi-Doped Mesoporous Silica Glass by A Femtosecond Laser. J. Mater. Chem. 19: 4603.

177. Sontakke AD, Tarafder A, Biswas K, and Annapurna K (2009) Sensitized Red Luminescence from Bi^{3+} Co-Doped Eu^{3+}: $ZnO-B_2O_3$ Glasses. Physica B 404: 3525.

178. Zhang N, Qiu J, Dong G, Yang Z, Zhang Q, and Peng M (2012) Broadband Tunable Near-Infrared Emission of Bi-Doped Composite Germanosilicate Glasses. J. Mater. Chem. 22: 3154.

179. Aly KA, Abdel Rahim FM, and Dahshan A (2014) Thermal Analysis and Physical Properties of Bi–Se–Te Chalcogenide Glasses. J. Alloys Compd. 593: 283.

180. Denker B, Galagan B, Osiko V, Shulman I, Sverchkov S, and Dianov E (2009) Absorption and Emission Properties of Bi-Doped Mg–Al–Si Oxide Glass System. Appl. Phys. B 95: 801.

181. Jiang S, Denker BI, Digonnet MJF, Galagan BI, Shulman IL, Glesener JW, Dries JC, Sverchkov SE, and Dianov EM (2010) Spectral and Luminescent Properties of Bi-Doped Bulk Glasses and Factors Acting on Them. Opt. Compon. Mater. VII 7598: 759805.

182. Denker BI, Galagan BI, Osiko VV, Shulman IL, Sverchkov SE, and Dianov EM (2009) Factors Affecting the Formation of Near Infrared-Emitting Optical Centers in Bi-Doped Glasses. Appl. Phys. B 98: 455.

183. Peng M, Zhang N, Wondraczek L, Qiu J, Yang Z, and Zhang Q (2011) Ultrabroad NIR Luminescence and Energy Transfer in Bi and Er/Bi Co-Doped Germanate Glasses. Opt. Exp. 19: 20799.

184. Xu B, Zhou S, Guan M, Tan D, Teng Y, Zhou J, Ma Z, Hong Z, and Qiu J (2011) Unusual Luminescence Quenching and Reviving Behavior of Bi-Doped Germanate Glasses. Opt. Exp. 19: 23436.

185. Peng M, and Wondraczek L (2010) Orange-to-Red Emission from Bi^{2+} and Alkaline Earth Codoped Strontium Borate Phosphors for White Light Emitting Diodes. J. Am. Ceram. Soc. 93: 1437.

186. de Jong M, Meijerink A, Gordon RA, Barandiarán Z, and Seijo L (2014) Is Bi^{2+} Responsible for the Red-Orange Emission of Bismuth-Doped SrB_4O_7? J. Phys. Chem. C 118: 9696.

187. Li L, Peng M, Viana B, Wang J, Lei B, Liu Y, Zhang Q, and Qiu J (2015) Unusual Concentration Induced Antithermal Quenching of the Bi^{2+} Emission from $Sr_2P_2O_7$:Bi^{2+}. Inorg. Chem. 54: 6028.

188. Li L, Cao J, Viana B, Xu S, and Peng M (2017) Site Occupancy Preference and Antithermal Quenching of the Bi^{2+} Deep Red Emission in Beta-$Ca_2P_2O_7$:Bi^{2+}. Inorg. Chem. 56: 6499.

189. Wang L, Tan L, Yue Y, Peng M, Qiu J, and Mauro J (2016) Efficient Enhancement of Bismuth NIR Luminescence by Aluminum and Its Mechanism in Bismuth-Doped Germanate Laser Glass. J. Am. Ceram. Soc. 99: 2071.

190. Blasse G (1997) Classical Phosphors: A Pandora's Box. J. Lumin. 72: 129.

191. Dong Q, Yang F, Cui J, Tian Y, Liu S, Du F, Peng J, and Ye X (2019) Enhanced Narrow Green Emission and Thermal Stability in γ-AlON: Mn^{2+}, Mg^{2+} Phosphor via Charge Compensation. Ceram. Int. 45: 11868.

192. Zhu Y, Liang Y, Liu S, Li H, and Chen J (2019) Narrow-Band Green-Emitting $Sr_2MgAl_{22}O_{36}$:Mn^{2+} Phosphors with Superior Thermal Stability and Wide Color Gamut for Backlighting Display Applications. Adv. Opt. Mater. 7: 1801419.

193. Zhou Q, Dolgov L, Srivastava AM, Zhou L, Wang Z, Shi J, Dramićanin MD, Brik MG, and Wu M (2018) Mn^{2+} and Mn^{4+} Red Phosphors: Synthesis, Luminescence and Applications in WLEDs. A Review. J. Mater. Chem. C 6: 2652.

194. Liu W, Lin Q, Li H, Wu K, Robel I, Pietryga JM, and Klimov VI (2016) Mn^{2+}-Doped Lead Halide Perovskite Nanocrystals with Dual-Color Emission Controlled by Halide Content. J. Am. Chem. Soc. 138: 14954.

195. Bai D, Zhang J, Jin Z, Bian H, Wang K, Wang H, Liang L, Wang Q, and Liu SF (2018) Interstitial Mn^{2+}-Driven High-Aspect-Ratio Grain Growth for Low-Trap-Density Microcrystalline Films for Record Efficiency $CsPbI_2Br$ Solar Cells. ACS Energy Lett. 3: 970.

196. Majher JD, Gray MB, Strom TA, and Woodward PM (2019) $Cs_2NaBiCl_6$:Mn^{2+}—A New Orange-Red Halide Double Perovskite Phosphor. Chem. Mater. 31: 1738.

197. Yang X, Pu C, Qin H, Liu S, Xu Z, and Peng X (2019) Temperature- and Mn^{2+} Concentration-Dependent Emission Properties of Mn^{2+}-Doped ZnSe Nanocrystals. J. Am. Chem. Soc. 141: 2288.

198. Ji S, Yuan X, Li J, Hua J, Wang Y, Zeng R, Li H, and Zhao J (2018) Photoluminescence Lifetimes and Thermal Degradation of Mn^{2+}-Doped $CsPbCl_3$ Perovskite Nanocrystals. J. Phys. Chem. C 122: 23217.

199. Cao R, Shi Z, Quan G, Hu Z, Zheng G, Chen T, Guo S, and Ao H (2018) Rare-Earth Free Broadband $Ca_3Mg_3P_4O_{16}$:Mn^{2+} Red Phosphor: Synthesis and Luminescence Properties. J. Lumin. 194: 542.

200. Cao R, Peng D, Xu H, Jiang S, Luo Z, Ao H, and Liu P (2016) Synthesis and Luminescence Properties of $NaAl_{11}O_{17}$:Mn^{2+} Green Phosphor for White LEDs. J. Lumin. 178: 388.

201. Song E, Zhou Y, Yang XB, Liao Z, Zhao W, Deng T, Wang L, Ma Y, Ye S, and Zhang Q (2017) Highly Efficient and Stable Narrow-Band Red Phosphor Cs_2SiF_6:Mn^{4+} for High-Power Warm White LED Applications. ACS Photonics 4: 2556.

202. Lu W, Lv W, Zhao Q, Jiao M, Shao B, and You H (2014) A Novel Efficient Mn^{4+} Activated $Ca_{14}Al_{10}Zn_6O_{35}$ Phosphor: Application in Red-Emitting and White LEDs. Inorg. Chem. 53: 11985.

203. Huang L, Liu Y, Yu J, Zhu Y, Pan F, Xuan T, Brik MG, Wang C, and Wang J (2018) Highly Stable K_2SiF_6:Mn^{4+}@K_2SiF_6 Composite Phosphor with Narrow Red Emission for White LEDs. ACS Appl. Mater. Interfaces 10: 18082.

204. Lin H, Hu T, Huang Q, Cheng Y, Wang B, Xu J, Wang J, and Wang Y (2017) Non-Rare-Earth K_2XF_7:Mn^{4+} (X = Ta, Nb): A Highly-Efficient Narrow-Band Red Phosphor Enabling the Application in Wide-Color-Gamut LCD. Laser Photonics Rev. 11: 1700148.

205. Arai Y, and Adachi S (2011) Optical Properties of Mn^{4+}-Activated Na_2SnF_6 and Cs_2SnF_6 Red Phosphors. J. Lumin. 131: 2652.

206. Jiang C, Peng M, Srivastava AM, Li L, and M GB (2018) Mn^{4+}-Doped Heterodialkaline Fluorogermanate Red Phosphor with High Quantum Yield and Spectral Luminous Efficacy for Warm-White-Light-Emitting Device Application. Inorg. Chem. 57: 14705.

207. Zhu H, Lin CC, Luo W, Shu S, Liu Z, Liu Y, Kong J, Ma E, Cao Y, Liu RS, and Chen X (2014) Highly Efficient Non-Rare-Earth Red Emitting Phosphor for Warm White Light-Emitting Diodes. Nat. Commun. 5: 4312.

208. Brik MG, and Srivastava AM (2018) Review—A Review of the Electronic Structure and Optical Properties of Ions with d3 Electron Configuration (V^{2+}, Cr^{3+}, Mn^{4+}, Fe^{5+}) and Main Related Misconceptions. ECS J. Solid State Sci. Technol. 7: R3079.

209. Adachi S (2019) Review—Tanabe–Sugano Energy-Level Diagram and Racah Parameters in Mn^{4+}-Activated Red and Deep Red-Emitting Phosphors. ECS J. Solid State Sci. Technol. 8: R183.

210. Lu Z, Huang T, Deng R, Wang H, Wen L, Huang M, Zhou L, and Yao C (2018) Double Perovskite Ca_2GdNbO_6:Mn^{4+} Deep Red Phosphor: Potential Application for Warm W-LEDs. Superlattices Microstruct. 117: 476.

211. Zhang S, Hu Y, Duan H, Fu Y, and He M (2017) An Efficient, Broad-Band Red-Emitting $Li_2MgTi_3O_8$:Mn^{4+} Phosphor for Blue-Converted White LEDs. J. Alloys Compd. 693: 315.

212. Wang B, Lin H, Xu J, Chen H, and Wang Y (2014) $CaMg_2Al_{16}O_{27}$:Mn^{4+}-Based Red Phosphor: A Potential Color Converter for High-Powered Warm W-LED. ACS Appl. Mater. Interfaces 6: 22905.

213. Zhou Y, Song E, Deng T, Wang Y, Xia Z, and Zhang Q (2019) Surface Passivation toward Highly Stable Mn^{4+}-Activated Red-Emitting Fluoride Phosphors and Enhanced Photostability for White LEDs. Adv. Mater. Interfaces 6: 1802006.

214. Amarasinghe DK, and Rabuffetti FA (2019) Bandshift Luminescence Thermometry Using Mn^{4+}:$Na_4Mg(WO_4)_3$ Phosphors. Chem. Mater. 31: 10197.

215. Xiong P, Peng M, Qin K, Xu F, and Xu X (2019) Visible to Near-Infrared Persistent Luminescence and Mechanoluminescence from Pr^{3+}-Doped $LiGa_5O_8$ for Energy Storage and Bioimaging. Adv. Opt. Mater. 7: 1901107.

216. Tu D, Xu CN, Yoshida A, Fujihala M, Hirotsu J, and Zheng XG (2017) $LiNbO_3$:Pr^{3+}: A Multipiezo Material with Simultaneous Piezoelectricity and Sensitive Piezoluminescence. Adv. Mater. 29: 1606914.

217. Liu Y, Fan L, Cai Y, Zhang W, Wang B, and Zhu B (2017) Superionic Conductivity of Sm^{3+}, Pr^{3+}, and Nd^{3+} Triple-Doped Ceria through Bulk and Surface Two-Step Doping Approach. ACS Appl. Mater. Interfaces 9: 23614.

218. Guo Q, Zhao C, Liao L, Lis S, Liu H, Mei L, and Jiang Z (2017) Luminescence Investigations of Novel Orange-Red Fluorapatite $KLaSr_3(PO_4)_3F$: Sm^{3+} Phosphors with High Thermal Stability. J. Am. Ceram. Soc. 100: 2221.

219. Zhang H, Gao Z, Li G, Zhu Y, Liu S, Li K, and Liang Y (2020) A Ratiometric Optical Thermometer with Multi-Color Emission and High Sensitivity Based on Double Perovskite $LaMg_{0.402}Nb_{0.598}O_3$:$Pr^{3+}$ Thermochromic Phosphors. Chem. Eng. J. 380: 122491.

220. Morassuti CY, Andrade LHC, Silva JR, Baesso ML, Guimarães FB, Rohling JH, Nunes LAO, Boulon G, Guyot Y, and Lima SM (2019) Spectroscopic Investigation and Interest of Pr^{3+}-Doped Calcium Aluminosilicate Glass. J. Lumin. 210: 376.

221. Mhlongo GH, Ntwaeaborwa OM, Swart HC, Kroon RE, Solarz P, Ryba-Romanowski W, and Hillie KT (2011) Luminescence Dependence of Pr^{3+} Activated SiO_2 Nanophosphor on Pr^{3+} Concentration, Temperature, and ZnO Incorporation. J. Phys. Chem. C 115: 17625.

222. Wang S, Ma S, Zhang G, Ye Z, and Cheng X (2019) High-Performance Pr^{3+}-Doped Scandate Optical Thermometry: 200K of Sensing Range with Relative Temperature Sensitivity above 2%.K^{-1}. ACS Appl. Mater. Interfaces 11: 42330.

223. Wu D, Cai Z, Zhong Y, Peng J, Cheng Y, Weng J, Luo Z, and Xu H (2017) Compact Passive Q-Switching Pr^{3+}-Doped ZBLAN Fiber Laser With Black Phosphorus-Based Saturable Absorber. IEEE J. Sel. Top. Quant. Electron. 23: 7.

224. Wang B, Li X, Chen Y, Chen Y, Zhou J, and Zeng Q (2018) Long Persistent and Photo-Stimulated Luminescence in Pr^{3+}-Doped Layered Perovskite Phosphor for Optical Data Storage. J. Am. Ceram. Soc. 101: 4598.

225. Chen Q, Xie X, Huang B, Liang L, Han S, Yi Z, Wang Y, Li Y, Fan D, Huang L, and Liu X (2017) Confining Excitation Energy in Er^{3+}-Sensitized Upconversion Nanocrystals through Tm^{3+}-Mediated Transient Energy Trapping. Angew. Chem. Int. Ed. 56: 7605.

226. Liu L, Wang S, Zhao B, Pei P, Fan Y, Li X, and Zhang F (2018) Er^{3+} Sensitized 1530 nm to 1180 nm Second Near-Infrared Window Upconversion Nanocrystals for In Vivo Biosensing. Angew. Chem. Int. Ed. 57: 7518.

227. Rabouw FT, Prins PT, Villanueva-Delgado P, Castelijns M, Geitenbeek RG, and Meijerink A (2018) Quenching Pathways in $NaYF_4$:Er^{3+},Yb^{3+} Upconversion Nanocrystals. ACS Nano 12: 4812.

228. Suo H, Guo C, and Li T (2016) Broad-Scope Thermometry Based on Dual-Color Modulation Up-Conversion Phosphor $Ba_5Gd_8Zn_4O_{21}$:Er^{3+}/Yb^{3+}. J. Phys. Chem. C 120: 2914.

229. Zhang Y, Xin J, Zheng H, Zhao Z, Zhang J, Yang Y, Zheng S, Zhao R, and Ding B (2018) Yb3+ Doping Monoclinic Lu_2WO_6: Near-Infrared Emission and Energy-Transfer Luminescence Mechanism. J. Phys. Chem. C 122: 21607.

230. Tian Y, Tian Y, Huang P, Wang L, Shi Q, and Cui Ce (2016) Effect of Yb^{3+} Concentration on Upconversion Luminescence and Temperature Sensing Behavior in Yb^{3+}/Er^{3+} Co-Doped $YNbO_4$ Nanoparticles Prepared via Molten Salt Route. Chem. Eng. J. 297: 26.

231. Ueda J, Miyano S, and Tanabe S (2018) Formation of Deep Electron Traps by Yb^{3+} Codoping Leads to Super-Long Persistent Luminescence in Ce^{3+}-Doped Yttrium Aluminum Gallium Garnet Phosphors. ACS Appl. Mater. Interfaces 10: 20652.

232. Suo H, Zhao X, Zhang Z, and Guo C (2020) Ultra-Sensitive Optical Nano-Thermometer $LaPO_4$:Yb^{3+}/Nd^{3+} Based on Thermo-Enhanced NIR-to-NIR Emissions. Chem. Eng. J. 389: 124506.

233. Talewar RA, Gaikwad VM, Tawalare PK, and Moharil SV (2019) Sensitization of Er^{3+}/Ho^{3+} Visible and NIR Emission in $NaYMoO_4)_2$ phosphors. Opt. Laser Technol. 115: 215.

234. Zhang H, Fan Y, Pei P, Sun C, Lu L, and Zhang F (2019) Tm^{3+}-Sensitized NIR-II Fluorescent Nanocrystals for In Vivo Information Storage and Decoding. Angew. Chem. Int. Ed. 58: 10153.

235. Herrera A, Fernandes RG, de Camargo ASS, Hernandes AC, Buchner S, Jacinto C, and Balzaretti NM (2016) Visible–NIR Emission and Structural Properties of Sm^{3+} Doped Heavy-Metal Oxide Glass with Composition B_2O_3–PbO–Bi_2O_3–GeO_2. J. Lumin. 171: 106.

236. Wilson KE, Zhang W, Rong L, Ren J, Jia Y, Qian S, Lin J, Ma J, Liu L, Jiang H, and Ke X (2013) Judd-Ofelt Analysis and Mid-Infrared Emission Properties of Ho^{3+}-Yb^{3+} Co-Doped Tellurite Oxy-Halide Glasses. Proc. SPIE 8906: 89060P.

237. Sui G, Chen B, Zhang J, Li X, Xu S, Sun J, Zhang Y, Tong L, Luo X, and Xia H (2018) Examination of Judd-Ofelt Calculation and Temperature Self-Reading for Tm^{3+} and Tm^{3+}/Yb^{3+} Doped $LiYF_4$ Single Crystals. J. Lumin. 198: 77.

238. Fan H, Chen Y, Yan T, Lin J, Peng G, Wang J, Boughton RI, and Ye N (2018) Crystal Growth, Spectral Properties and Judd-Ofelt Analysis of Pr^{3+}:$LaMgAl_{11}O_{19}$. J. Alloys Compd. 767: 938.

239. Ferhi M, Bouzidi C, Horchani-Naifer K, Elhouichet H, and Ferid M (2015) Judd–Ofelt Analysis of Spectroscopic Properties of Eu^{3+} Doped $KLa(PO_3)_4$. J. Lumin. 157: 21.
240. Shivakumara C, Saraf R, Behera S, Dhananjaya N, and Nagabhushana H (2015) Synthesis of Eu^{3+}-Activated $BaMoO_4$ Phosphors and Their Judd-Ofelt Analysis: Applications in Lasers and White LEDs. Spectrochim. Acta, A: Mol. Biomol. Spectrosc. 151: 141.
241. Wang DY, Ma PC, Zhang JC, and Wang YH (2018) Efficient Down- and Up-Conversion Luminescence in Er^{3+}–Yb^{3+} Co-Doped $Y_7O_6F_9$ for Photovoltaics. ACS Appl. Energy Mater. 1: 447.
242. Mounet N, Gibertini M, Schwaller P, Campi D, Merkys A, Marrazzo A, Sohier T, Castelli IE, Cepellotti A, Pizzi G, and Marzari N (2018) Two-Dimensional Materials from High-Throughput Computational Exfoliation of Experimentally Known Compounds. Nat. Nanotechnol. 13: 246.
243. Dickinson ME, Flenniken AM, Ji X, Teboul L, Wong MD, White JK, Meehan TF, Weninger WJ, Westerberg H, Adissu H, Baker CN, Bower L, Brown JM, Caddle LB, Chiani F, Clary D, Cleak J, Daly MJ, Denegre JM, Doe B, Dolan ME, Edie SM, Fuchs H, Gailus-Durner V, Galli A, Gambadoro A, Gallegos J, Guo S, Horner NR, Hsu CW, Johnson SJ, Kalaga S, Keith LC, Lanoue L, Lawson TN, Lek M, Mark M, Marschall S, Mason J, McElwee ML, Newbigging S, Nutter LM, Peterson KA, Ramirez-Solis R, Rowland DJ, Ryder E, Samocha KE, Seavitt JR, Selloum M, Szoke-Kovacs Z, Tamura M, Trainor AG, Tudose I, Wakana S, Warren J, Wendling O, West DB, Wong L, Yoshiki A, Harwell MRC, MacArthur DG, Tocchini-Valentini GP, Gao X, Flicek P, Bradley A, Skarnes WC, Justice MJ, Parkinson HE, Moore M, Wells S, Braun RE, Svenson KL, de Angelis MH, Herault Y, Mohun T, Mallon AM, Henkelman RM, Brown SD, Adams DJ, Lloyd KC, McKerlie C, Beaudet AL, Bucan M, and Murray SA (2016) High-Throughput Discovery of Novel Developmental Phenotypes. Nature 537: 508.
244. Uthman AP, Koontz JP, Hlnderllter-Smlth J, Woodward WS, and Reilley CN (1982) High-Throughput Microcomputer-Based Binary-Coded Search Systems for Infrared, Carbon-13 Nuclear Magnetic Resonance, and Mass Spectral Data. Anal. Chem. 54: 1772.
245. Reuter JA, Spacek DV, and Snyder MP (2015) High-Throughput Sequencing Technologies. Mol. Cell 58: 586.
246. Sun S, Hartono NTP, Ren ZD, Oviedo F, Buscemi AM, Layurova M, Chen DX, Ogunfunmi T, Thapa J, Ramasamy S, Settens C, DeCost BL, Kusne AG, Liu Z, Tian SIP, Peters IM, Correa-Baena J-P, and Buonassisi T (2019) Accelerated Development of Perovskite-Inspired Materials via High-Throughput Synthesis and Machine-Learning Diagnosis. Joule 3: 1437.
247. Lu S, Zhou Q, Ouyang Y, Guo Y, Li Q, and Wang J (2018) Accelerated Discovery of Stable Lead-Free Hybrid Organic-Inorganic Perovskites via Machine Learning. Nat. Commun. 9: 3405.
248. Fan Y, Wang P, Lu Y, Wang R, Zhou L, Zheng X, Li X, Piper JA, and Zhang F (2018) Lifetime-Engineered NIR-II Nanoparticles Unlock Multiplexed In Vivo Imaging. Nat. Nanotechnol. 13: 941.
249. Wang Z, Chu I-H, Zhou F, and Ong SP (2016) Electronic Structure Descriptor for the Discovery of Narrow-Band Red-Emitting Phosphors. Chem. Mater. 28: 4024.
250. Ye W, Chen C, Wang Z, Chu IH, and Ong SP (2018) Deep Neural Networks for Accurate Predictions of Crystal Stability. Nat. Commun. 9: 3800.
251. Park WB, Son KH, Singh SP, and Sohn KS (2012) Solid-State Combinatorial Screening of $ARSi_4N_7$:Eu^{2+} (A = Sr, Ba, Ca; R = Y, La, Lu) Phosphors. ACS Comb. Sci. 14: 537.
252. Wang XJ, Wang L, Takeda T, Funahashi S, Suehiro T, Hirosaki N, and Xie RJ (2015) Blue-Emitting $Sr_3Si_{8-x}Al_xO_7+xN_{8-x}$:$Eu^{2+}$ Discovered by a Single-Particle-Diagnosis Approach: Crystal Structure, Luminescence, Scale-Up Synthesis, and Its Abnormal Thermal Quenching Behavior. Chem. Mater. 27: 7689.
253. Correa-Baena JP, Hippalgaonkar K, van Duren J, Jaffer S, Chandrasekhar VR, Stevanovic V, Wadia C, Guha S, and Buonassisi T (2018) Accelerating Materials Development via Automation, Machine Learning, and High-Performance Computing. Joule 2: 1410.
254. Qiao J, Ning L, Molokeev MS, Chuang YC, Liu Q, and Xia Z (2018) Eu^{2+} Site Preferences in the Mixed Cation $K_2BaCa(PO_4)_2$ and Thermally Stable Luminescence. J. Am. Chem. Soc. 140: 9730.
255. Xiao Z, Meng W, Wang J, and Yan Y (2016) Defect Properties of the Two-Dimensional $(CH_3NH_3)_2Pb(SCN)_2I_2$ Perovskite: A Density-Functional Theory Study. Phys. Chem. Chem. Phys. 18: 25786.
256. Cheng C, Ning L, Ke X, Molokeev MS, Wang Z, Zhou G, Chuang YC, and Xia Z (2019) Designing High-Performance LED Phosphors by Controlling the Phase Stability via a Heterovalent Substitution Strategy. Adv. Opt. Mater. 8: 1901608.
257. Wei Y, Xing G, Liu K, Li G, Dang P, Liang S, Liu M, Cheng Z, Jin D, and Lin J (2019) New Strategy for Designing Orangish-Red-Emitting Phosphor via Oxygen-Vacancy-Induced Electronic Localization. Light Sci. Appl. 8: 15.

258. Day J, Senthilarasu S, and Mallick TK (2019) Improving Spectral Modification for Applications in Solar Cells: A Review. Renew. Energy 132: 186.

259. Yong ZJ, Guo SQ, Ma JP, Zhang JY, Li ZY, Chen YM, Zhang BB, Zhou Y, Shu J, Gu JL, Zheng LR, Bakr OM, and Sun HT (2018) Doping-Enhanced Short-Range Order of Perovskite Nanocrystals for Near-Unity Violet Luminescence Quantum Yield. J. Am. Chem. Soc. 140: 9942.

260. Hariyani S, and Brgoch J (2020) Local Structure Distortion Induced Broad Band Emission in the All-Inorganic $BaScO_2F:Eu^{2+}$ Perovskite. Chem. Mater. 32: 6640.

261. Lin L, Ning L, Zhou R, Jiang C, Peng M, Huang Y, Chen J, Huang Y, Tao Y, and Liang H (2018) Site Occupation of Eu^{2+} in $Ba_{2-x}Sr_xSiO_4$ ($x = 0$–1.9) and Origin of Improved Luminescence Thermal Stability in the Intermediate Composition. Inorg. Chem. 57: 7090.

262. Sun L, Devakumar B, Liang J, Wang S, Sun Q, and Huang X (2020) A Broadband Cyan-Emitting $Ca_2LuZr_2(AlO_4)_3:Ce^{3+}$ Garnet Phosphor for Near-Ultraviolet-Pumped Warm-White Light-Emitting Diodes with an Improved Color Rendering Index. J. Mater. Chem. C 8: 1095.

263. Wei Q, Ding J, Zhou X, Wang X, and Wang Y (2020) New Strategy of Designing a Novel Yellow-Emitting Phosphor $Na_4Hf_2Si_3O_{12}:Eu^{2+}$ for Multifunctional Applications. J. Alloys Compd. 817: 152762.

264. Poesl C, and Schnick W (2017) Crystal Structure and Nontypical Deep-Red Luminescence of $Ca_3Mg[Li_2Si_2N_6]:Eu^{2+}$. Chem. Mater. 29: 3778.

265. Lin YC, Bettinelli M, and Karlsson M (2019) Unraveling the Mechanisms of Thermal Quenching of Luminescence in Ce^{3+}-Doped Garnet Phosphors. Chem. Mater. 31: 3851.

266. Dong B, Yeandel SR, Goddard P, and Slater PR (2019) Combined Experimental and Computational Study of Ce-Doped $La_3Zr_2Li_7O_{12}$ Garnet Solid-State Electrolyte. Chem. Mater. 32: 215.

267. Chewpraditkul W, Swiderski L, Moszynski M, Szczesniak T, Syntfeld-Kazuch A, Wanarak C, and Limsuwan P (2009) Scintillation Properties of LuAG:Ce, YAG:Ce and LYSO:Ce Crystals for Gamma-Ray Detection. IEEE Trans. Nucl. Sci. 56: 3800.

268. Liu S, Feng X, Zhou Z, Nikl M, Shi Y, and Pan Y (2014) Effect of Mg^{2+} Co-Doping on the Scintillation Performance of LuAG:Ce Ceramics. Phys. Status Solidi RRL 8: 105.

269. Hu C, Liu S, Fasoli M, Vedda A, Nikl M, Feng X, and Pan Y (2015) O-Centers in LuAG:Ce,Mg Ceramics. Phys. Status Solidi RRL 9: 245.

270. Aboulaich A, Michalska M, Schneider R, Potdevin A, Deschamps J, Deloncle R, Chadeyron G, and Mahiou R (2014) Ce-Doped YAG Nanophosphor and Red Emitting $CuInS_2/ZnS$ Core/Shell Quantum Dots for Warm White Light-Emitting Diode with High Color Rendering Index. ACS Appl. Mater. Interfaces 6: 252.

271. Berezovskaya IV, Dotsenko VP, Voloshinovskii AS, and Smola SS (2013) Near Infrared Emission of Eu^{2+} Ions in $Ca_3Sc_2Si_3O_{12}$. Chem. Phys. Lett. 585: 11.

272. Dai D, Wang Z, Xing Z, Li X, Liu C, Zhang L, Yang Z, and Li P (2019) Broad Band Emission Near-Infrared Material $Mg_3Ga_2GeO_8:Cr^{3+}$: Substitution of Ga-In, Structural Modification, Luminescence Property and Application for High Efficiency LED. J. Alloys Compd. 806: 926.

273. Liu D, Dang P, Yun X, Li G, Lian H, and Lin J (2019) Luminescence Color Tuning and Energy Transfer Properties in $(Sr,Ba)_2LaGaO_5:Bi^{3+},Eu^{3+}$ Solid Solution Phosphors: Realization of Single-Phased White Emission for WLEDs. J. Mater. Chem. C 7: 13536.

274. Cheng J, Li P, Wang Z, Sun Y, Bai Q, Li Z, Tian M, Wang C, and Yang Z (2017) Synthesis, Structure and Luminescence Properties of Novel NIR Luminescent Materials $Li_2ZnGe_3O_8:xMn^{2+}$. J. Mater. Chem. C 5: 127.

275. Li Y, Li YY, Sharafudeen K, Dong GP, Zhou SF, Ma ZJ, Peng MY, and Qiu JR (2014) A Strategy for Developing Near Infrared Long-Persistent Phosphors: Taking $MAlO_3:Mn^{4+},Ge^{4+}$ (M = La, Gd) as An Example. J. Mater. Chem. C 2: 2019.

276. Ma YY, Hu JQ, Song EH, Ye S, and Zhang QY (2015) Regulation of Red to Near-Infrared Emission in Mn^{2+} Single Doped Magnesium Zinc Phosphate Solid-Solution Phosphors by Modification of the Crystal Field. J. Mater. Chem. C 3: 12443.

277. Guo Y, Moon BK, Choi BC, Jeong JH, Kim JH, and Choi H (2016) Fluorescence Properties with Red-Shift of Eu^{2+} Emission in Novel Phosphor-Silicate Apatite $Sr_3LaNa(PO_4)_2SiO_4$ Phosphors. Ceram. Int. 42: 18324.

278. Geng W, Zhou X, Ding J, Li G, and Wang Y (2017) $K_7Ca_9[Si_2O_7]_4F:Ce^{3+}$: A Novel Blue-Emitting Phosphor with Good Thermal Stability for Ultraviolet-Excited Light Emitting Diodes. J. Mater. Chem. C 5: 11605.

279. Wei Y, Gao Z, Yun X, Yang H, Liu Y, and Li G (2020) Abnormal Bi^{3+}-Activated NIR Emission in Highly Symmetric $XAl_{12}O_{19}$ (X = Ba, Sr, Ca) by Selective Sites Occupation. Chem. Mater. 32: 8747.

280. Kim YH, Arunkumar P, Kim BY, Unithrattil S, Kim E, Moon SH, Hyun JY, Kim KH, Lee D, Lee JS, and Im WB (2017) A Zero-Thermal-Quenching Phosphor. Nat. Mater. 16: 543.
281. Wei Y, Yang H, Gao Z, Liu Y, Xing G, Dang P, Kheraif AAA, Li G, Lin J, and Liu RS (2020) Strategies for Designing Antithermal-Quenching Red Phosphors. Adv. Sci. 7: 1903060.
282. Romanov AN, Veber AA, Vtyurina DN, Fattakhova ZT, Haula EV, Shashkin DP, Sulimov VB, Tsvetkov VB, and Korchak VN (2015) Near Infrared Photoluminescence of the Univalent Bismuth Impurity Center in Leucite and Pollucite Crystal Hosts. J. Mater. Chem. C 3: 3592.
283. Sun HT, Matsushita Y, Sakka Y, Shirahata N, Tanaka M, Katsuya Y, Gao H, and Kobayashi K (2012) Synchrotron X-Ray, Photoluminescence, and Quantum Chemistry Studies of Bismuth-Embedded Dehydrated Zeolite Y. J. Am. Chem. Soc. 134: 2918.
284. Wang L, Long NJ, Li L, Lu Y, Li M, Cao J, Zhang Y, Zhang Q, Xu S, Yang Z, Mao C, and Peng M (2018) Multi-Functional Bismuth-Doped Bioglasses: Combining Bioactivity and Photothermal Response for Bone Tumor Treatment and Tissue Repair. Light Sci. Appl. 7: 1.
285. Pan J, Quan LN, Zhao Y, Peng W, Murali B, Sarmah SP, Yuan M, Sinatra L, Alyami NM, Liu J, Yassitepe E, Yang Z, Voznyy O, Comin R, Hedhili MN, Mohammed OF, Lu ZH, Kim DH, Sargent EH, and Bakr OM (2016) Highly Efficient Perovskite-Quantum-Dot Light-Emitting Diodes by Surface Engineering. Adv. Mater. 28: 8718.
286. Song EH, Zhou YY, Wei Y, Han XX, Tao ZR, Qiu RL, Xia ZG, and Zhang QY (2019) A Thermally Stable Narrow-Band Green-Emitting Phosphor MgAl$_2$O$_4$:Mn^{2+} for Wide Color Gamut Backlight Display Application. J. Mater. Chem. C 7: 8192.
287. Hamstra MA, Folkerts HF, and Blasse G (1994) Red Bismuth Emission in Alkaline-Earth-Metal Sulfates. J. Mater. Chem. 4: 1349.
288. Romanov AN, Haula EV, Fattakhova ZT, Veber AA, Tsvetkov VB, Zhigunov DM, Korchak VN, and Sulimov VB (2011) Near-IR Luminescence from Subvalent Bismuth Species in Fluoride Glass. Opt. Mater. 34: 155.
289. Liao Z, Xu H, Zhao W, Yang H, Zhong J, Zhang H, Nie Z, and Zhou ZK (2020) Energy Transfer from Mn^{4+} to Mn^{5+} and Near Infrared Emission with Wide Excitation Band in Ca$_{14}$Zn$_6$Ga$_{10}$O$_{35}$:Mn Phosphors. Chem. Eng. J. 395: 125060.
290. Liu X, Cheng C, Li X, Jiao Q, and Dai S (2019) Controllable Ultra-Broadband Visible and Near-Infrared Photoemissions in Bi-Doped Germanium-Borate Glasses. J. Am. Ceram. Soc. 103: 183.
291. Brik MG, Cavalli E, Borromei R, and Bettinelli M (2009) Crystal Field Parameters and Energy Level Structure of the MnO$_4^{3-}$ Tetroxo Anion in Li$_3$PO$_4$, Ca$_2$PO$_4$Cl and Sr$_5$(PO$_4$)$_3$Cl Crystals. J. Lumin. 129: 801.
292. Shao Q, Ding H, Yao L, Xu J, Liang C, and Jiang J (2018) Photoluminescence Properties of a ScBO$_3$:Cr^{3+} Phosphor and Its Applications for Broadband Near-Infrared LEDs. RSC Adv. 8: 12035.
293. Yao L, Shao Q, Han S, Liang C, He J, and Jiang J (2020) Enhancing Near-Infrared Photoluminescence Intensity and Spectral Properties in Yb^{3+} Codoped LiScP$_2$O$_7$:Cr^{3+}. Chem. Mater. 32: 2430.
294. Arfin H, Kaur J, Sheikh T, Chakraborty S, and Nag A (2020) Bi^{3+}–Er^{3+} and Bi^{3+}–Yb^{3+} Codoped Cs$_2$AgInC$_{16}$ Double Perovskite Near Infrared Emitters. Angew. Chem. Int. Ed. 59: 1.
295. Li N, Lau YS, Xiao Z, Ding L, and Zhu F (2018) NIR to Visible Light Upconversion Devices Comprising an NIR Charge Generation Layer and a Perovskite Emitter. Adv. Opt. Mater. 6: 1801084.
296. Li X, Zhang L, Zhang Y, Wang D, Wang X, Yu L, Zhang W, and Li P (2020) Review of NIR Spectroscopy Methods for Nondestructive Quality Analysis of Oilseeds and Edible Oils. Trends Food Sci. Technol. 101: 172.
297. Zhang L, Wang D, Hao Z, Zhang X, Pan Gh, Wu H, and Zhang J (2019) Cr^{3+}-Doped Broadband NIR Garnet Phosphor with Enhanced Luminescence and its Application in NIR Spectroscopy. Adv. Opt. Mater. 7: 1900185.
298. Li YJ, Yang CX, and Yan XP (2018) Biomimetic Persistent Luminescent Nanoplatform for Autofluorescence-Free Metastasis Tracking and Chemophotodynamic Therapy. Anal. Chem. 90: 4188.
299. Sun SK, Wang HF, and Yan XP (2018) Engineering Persistent Luminescence Nanoparticles for Biological Applications: From Biosensing/Bioimaging to Theranostics. Acc. Chem. Res. 51: 1131.
300. Cao C, Xue M, Zhu X, Yang P, Feng W, and Li F (2017) Energy Transfer Highway in Nd^{3+}-Sensitized Nanoparticles for Efficient Near-Infrared Bioimaging. ACS Appl. Mater. Interfaces 9: 18540.
301. Osborne BG. 2006. Near-Infrared Spectroscopy in Food Analysis, Encyclopedia of Analytical Chemistry: John Wiley & Sons, Ltd, North Ryde.
302. Wang C, Wang X, Zhou Y, Zhang S, Li C, Hu D, Xu L, and Jiao H (2019) An Ultra-Broadband Near-Infrared Cr^{3+}-Activated Gallogermanate Mg$_3$Ga$_2$GeO$_8$ Phosphor as Light Sources for Food Analysis. ACS Appl. Electron. Mater. 1: 1046.

303. Zervas MN, and Codemard CA (2014) High Power Fiber Lasers: A Review. IEEE J. Sel. Top. Quant. Electtron. 20: 219.

304. Wright LG, Christodoulides DN, and Wise FW (2017) Spatiotemporal Mode-Locking in Multimode Fiber Lasers. Science 358: 94.

305. Li W, Wu J, Guan X, Zhou Z, Xu H, Luo Z, and Cai Z (2018) Efficient Continuous-Wave and Short-Pulse Ho^{3+}-Doped Fluorozirconate Glass All-Fiber Lasers Operating in the Visible Spectral Range. Nanoscale 10: 5272.

306. Guo P, Li X, Feng T, Zhang Y, and Xu W (2020) Few-Layer Bismuthene for Coexistence of Harmonic and Dual Wavelength in a Mode-Locked Fiber Laser. ACS Appl. Mater. Interfaces 12: 31757.

307. Lee J, Sul H, Jung Y, Kim H, Han S, Choi J, Shin J, Kim D, Jung J, Hong S, and Ko SH (2020) Thermally Controlled, Active Imperceptible Artificial Skin in Visible-to-Infrared Range. Adv. Funct. Mater. 30: 2003328.

308. Zhou L, Tanner PA, Zhou W, Ai Y, Ning L, Wu MM, and Liang H (2017) Unique Spectral Overlap and Resonant Energy Transfer between Europium(II) and Ytterbium(III) Cations: No Quantum Cutting. Angew. Chem. Int. Ed. 56: 10357.

309. Kunkel N, Kohlmann H, Sayede A, and Springborg M (2011) Alkaline-Earth Metal Hydrides as Novel Host Lattices for Eu(II) Luminescence. Inorg. Chem. 50: 5873.

310. Qin X, Li Y, Wu D, Wu Y, Chen R, Ma Z, Liu S, and Qiu J (2015) A Novel NIR Long Phosphorescent Phosphor:SrSnO$_3$:Bi^{2+}. RSC Adv. 5: 101347.

311. Sun HT, Sakka Y, Gao H, Miwa Y, Fujii M, Shirahata N, Bai Z, and Li JG (2011) Ultrabroad Near-Infrared Photoluminescence from Bi$_5$(AlCl$_4$)$_3$ Crystal. J. Mater. Chem. 21: 4060.

312. Liu S, Zhang S, Mao N, Song Z, and Liu Q (2020) Broadband Deep-Red-to-Near-Infrared Emission from Mn^{2+} in Strong Crystal-Field of Nitride MgAlSiN$_3$. J. Am. Ceram. Soc. 103: 6793.

313. Sun HT, Fujii M, Sakka Y, Bai Z, Shirahata N, Zhang L, Miwa Y, and Gao H (2010) Near-Infrared Photoluminescence and Raman Characterization of Bismuth-Embedded Sodalite Nanocrystals. Optics Lett. 35: 1743.

314. Sun HT, Hosokawa A, Miwa Y, Shimaoka F, Fujii M, Mizuhata M, Hayashi S, and Deki S (2009) Strong Ultra-Broadband Near-Infrared Photoluminescence from Bismuth-Embedded Zeolites and Their Derivatives. Adv. Mater. 21: 3694.

315. Zhang K, Hou JS, Liu BM, Zhou Y, Yong ZJ, Li LN, Sun HT, and Fang YZ (2016) Superbroad Near-Infrared Photoluminescence Covering the Second Biological Window Achieved by Bismuth-Doped Oxygen-Deficient Gadolinium Oxide. RSC Adv. 6: 78396.

316. Sun HT, Yang J, Fujii M, Sakka Y, Zhu Y, Asahara T, Shirahata N, Ii M, Bai Z, Li JG, and Gao H (2011) Highly Fluorescent Silica-Coated Bismuth-Doped Aluminosilicate Nanoparticles for Near-Infrared Bioimaging. Small 7: 199.

317. Möller S, Katelnikovas A, Haase M, and Jüstel T (2016) New NIR Emitting Phosphor for Blue LEDs with Stable Light Output up to 180°C. J. Lumin. 172: 185.

318. Li J, Chen L, Hao Z, Zhang X, Zhang L, Luo Y, and Zhang J (2015) Efficient Near-Infrared Downconversion and Energy Transfer Mechanism of Ce^{3+}/Yb^{3+} Codoped Calcium Scandate Phosphor. Inorg. Chem. 54: 4806.

319. Xiang G, Ma Y, Zhou X, Jiang S, Li L, Luo X, Hao Z, Zhang X, Pan GH, Luo Y, and Zhang J (2017) Investigation of the Energy-Transfer Mechanism in Ho^{3+}- and Yb^{3+}-Codoped Lu$_2$O$_3$ Phosphor with Efficient Near-Infrared Downconversion. Inorg. Chem. 56: 1498.

320. Zhu Y, Pan G, Shao L, Yang G, Xu X, Zhao J, and Mao Y (2020) Effective Infrared Emission of Erbium Ions Doped Inorganic Lead Halide Perovskite Quantum Dots by Sensitization of Ytterbium Ions. J. Alloys Compd. 835: 155390.

321. Yadav RV, Yadav RS, Bahadur A, Singh AK, and Rai SB (2016) Enhanced Quantum Cutting via Li$^+$ Doping from a Bi^{3+}/Yb^{3+}-Codoped Gadolinium Tungstate Phosphor. Inorg. Chem. 55: 10928.

322. Arfin H, Kaur J, Sheikh T, Chakraborty S, and Nag A (2020) Bi^{3+}–Er^{3+} and Bi^{3+}-Yb^{3+} Codoped Cs$_2$AgInCl$_6$ Double Perovskite Near-Infrared Emitters. Angew. Chem. Int. Ed. 59: 11307.

323. Huang XY, Ji XH, and Zhang QY (2011) Broadband Downconversion of Ultraviolet Light to Near-Infrared Emission in Bi^{3+}–Yb^{3+}-Codoped Y$_2$O$_3$ Phosphors. J. Am. Ceram. Soc. 94: 833.

324. He S, Zhang L, Wu H, Wu H, Pan G, Hao Z, Zhang X, Zhang L, Zhang H, and Zhang J (2020) Efficient Super Broadband NIR Ca$_2$LuZr$_2$Al$_3$O$_{12}$:Cr^{3+},Yb^{3+} Garnet Phosphor for pc-LED Light Source toward NIR Spectroscopy Applications. Adv. Opt. Mater. 8: 1901684.

325. Gao G, and Wondraczek L (2013) Near-Infrared Down-Conversion in Mn^{2+}–Yb^{3+} Co-Doped Zn$_2$GeO$_4$. J. Mater. Chem. C 1: 1952.

326. Lu W, Jiao M, Shao B, Zhao L, Feng Y, and You H (2016) An Intense NIR Emission from $Ca_{14}Al_{10}Zn_6O_{35}$:Mn^{4+},Yb^{3+} via Energy Transfer for Solar Spectral Converters. Dalton Trans. 45: 466.
327. Gao X, Xia W, Chen T, Yang X, Jin X, and Xiao S (2016) Conversion of Broadband UV-Visible Light to Near Infrared Emission by $Ca_{14}Zn_6Al_{10}O_{35}$: Mn^{4+}, Nd^{3+}/Yb^{3+}. RSC Adv. 6: 7544.
328. Sun L, Devakumar B, Liang J, Wang S, Sun Q, and Huang X (2019) Novel High-Efficiency Violet-Red Dual-Emitting Lu_2GeO_5:Bi^{3+}, Eu^{3+} Phosphors for Indoor Plant Growth Lighting. J. Lumin. 214: 116544.
329. Liu P, Liu J, Zheng X, Luo H, Li X, Yao Z, Yu X, Shi X, Hou B, and Xia Y (2014) An Efficient Light Converter YAB:Cr^{3+},Yb^{3+}/Nd^{3+} with Broadband Excitation and Strong NIR Emission for Harvesting c-Si-Based Solar Cells. J. Mater. Chem. C 2: 5769.
330. Zhao J, Guo C, and Li T (2015) Enhanced Near-Infrared Emission by Co-Doping Ce^{3+} in $Ba_2Y(BO_3)2Cl$:Tb_{3+}, Yb_{3+} Phosphor. RSC Adv. 5: 28299.
331. Ishii A, and Miyasaka T (2020) Sensitized Yb3+ Luminescence in $CsPbCl_3$ Film for Highly Efficient Near-Infrared Light-Emitting Diodes. Adv Sci. 7: 1903142.

3 Solid Solution Design toward New Phosphor Discovery and Photoluminescence Tuning

Zhiqiang Ming, Ming Zhao, and Zhiguo Xia

CONTENTS

3.1 THE SOLID SOLUTION DESIGN PRINCIPLE OF PHOSPHOR DISCOVERY

3.1.1 THE BASIC KNOWLEDGE OF SOLID SOLUTION

Generally, a solid solution is formed by two or more solutes coexisting in a solvent. Such a multi-component system is considered a solution rather than a compound when the crystal structure of the solvent remains unchanged by the addition of the solutes [1, 2]. The word "solution" is used to describe the intimate mixing of components at the atomic level and distinguishes them from physical mixtures of components. The chemical components of a solid solution invariably maintain in a single homogeneous phase. As shown in Figure 3.1(a), the phase diagram exhibits a solid solution at all relative concentrations of the two solutes, which have a range of compositions namely $\alpha_x\beta_{1-x}$. Many examples can be found in metallurgy, geology, and solid-state chemistry studies. Depending on the crystal structure and chemical properties of solvent and solutes, there are mainly two types of solid solutions: substitutional and interstitial solid solution (Figure 3.1(b)). The most common type is the substitutional solid solution, in which the atoms of one crystal replace those of the other.

DOI: 10.1201/9781003098669-3

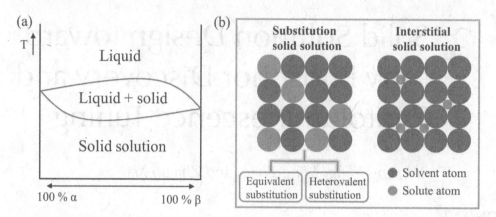

FIGURE 3.1 (a) A binary phase diagram exhibiting solid solutions over the full range of relative concentrations. (b) Schematic diagram of the different types of solid solutions.

Similar to isomorphism, they can be divided into equivalent substitution and heterovalent substitution according to the valence of the ions substituted for each other in the lattice, such as $K^+ \rightarrow Na^+$ in $(Na_{1-x}K_x)AlSi_3O_8$, $Ca^{2+} \rightarrow Na^+$, and $Al^{3+} \rightarrow Si^{4+}$ in $(1-x)NaAlSi_3O_8-xCaAl_2Si_2O_8$. Therefore, the substitutional solid solution can also be divided into single site's substitution and co-substitution at two sites. For the interstitial solid solution, the atoms usually occupy vacant positions or interstitial sites in the lattice. The formation of a solid solution has two approaches. One is the combination of two or more solids as they have been melted into liquids at high temperatures and then cooling the product to form the new solid. The other is to form thin films by depositing vapors of the starting materials onto the substrate. Many factors affect the formation of solid solutions, such as ionic size and valence, structural flexibility, and synthesis temperature [3]. The substances may be soluble in a partial or even complete range of relative concentrations, thereby producing a crystal the properties of which vary continuously within the range. Therefore, the electrical, thermal, magnetic, and optical properties of the solid solution may change continuously with the variation of chemical composition.

3.1.2   The Utilization of Solid Solution Design toward Phosphor Discovery

As is known to all, different solid solutions have been previously found in natural minerals with variable chemical compositions. And thus, the solid solution design has been extensively studied in the exploration of the solid-state functional materials, which is specially regarded as an efficient strategy to stimulate photoluminescence tuning of phosphor materials and discover new luminescence materials through changing the local environment around luminescent centers. Generally, the solid solution design is a method to continuously vary the chemical composition of the host in terms of chemical substitution, which involves constituent elements of the host replaced by other elements with a similar ionic radius. Hence, the solid solution can be formed via the chemical substitution. When the structure of the host remains invariant after element replacement, the chemical substitution is equated with the solid solution design.

As for the phosphor discovery and photoluminescence tuning therein, chemical substitutions can commonly be classified into three types: cationic substitution, anionic substitution, and chemical unit co-substitution [4, 5]. For cationic substitution and anionic substitution, a constituent ion of the host is usually replaced by another ion with an equivalent charge such as the replacement of Ca^{2+} for Sr^{2+} and Cl^- for Br^- in the hosts [6, 7]. However, anionic substitution is more difficult to achieve compared to the cationic substitution. The chemical unit co-substitution is defined as the simultaneous replacement of two or more cations, anions, complex anions, fundamental building units, or vacancies. For the part of substitution, the overall sum of the oxidation states is constant, but each component is not necessarily equivalent, for instance, the co-substitution of $[Al^{3+}-F^-]$ with $[Si^{4+}-O^{2-}]$ or $[Ca^{2+}-Si^{4+}]$ with $[Lu^{3+}-Al^{3+}]$ to build up the correlation of the two compounds [8, 9]. Based on the chemical unit co-substitution, the mineral-inspired strategy is proposed to discover the potential host lattices for phosphors via the mineral-inspired prototype evolution and new phase construction [10]. In other words, some typical mineral compounds with an especial formula can be deemed as the primary prototype compounds, which enlighten people on the discovery of new hosts via multiple substitutions.

Therefore, both the cation substitution and chemical unit co-substitution strategies are more common to discover the new materials or tune the photoluminescence of the phosphors. Apart from equivalent substitution, a new approach to the formation of solid solution is the single ion heterovalent substitution, which can more unexpectedly regulate the local coordination environment and maintain the total crystal structure unchanged in some special hosts. For example, the heterovalent substitution of Si^{4+} by Al^{3+} for some phosphors can realize surprising tuning in luminescence properties [11]. Thus, the formation of the solid solution in the phosphors will be important, and they can be widely used in different phosphors systems. However, the effects of solid solution design on different electronic transitions are different, and we should first understand the basic knowledge therein, as discussed later.

3.2 ELECTRONIC TRANSITIONS IN SOLID SOLUTION PHOSPHORS

3.2.1 Basic Knowledge of 4f–5d Transition

The $4f-5d$ transition in phosphors is often characterized by broadband emission, short decay times, and large oscillator strengths since its transition is parity-allowed electrical dipole transition [12]. The most widely used activators with $4f-5d$ transition are Eu^{2+} and Ce^{3+} ions, respectively. The photoluminescence properties of Eu^{2+}- and Ce^{3+}-doped phosphors are strongly dependent on the host lattice because their $5d$ electrons are unshielded from their environment by the $5s^2$ and $5p^6$ shells as they are in the excited state [13]. Therefore, $4f-5d$ transition plays an important role in the solid solution phosphors and the photoluminescence tuning therein.

The $4f^n$ energy-level splitting is mainly affected by the Coulomb field, spin-orbit coupling, and crystal field, and $5d$ energy-level splitting is mainly influenced by different matrix materials [14]. Compared with the $5d$ energy level, the $4f^n$ energy level is extremely insensitive to lattice and the energy of the $4f^n$ level can be regarded a constant in a host. Therefore, when studying the $4f-5d$ transitions of Eu^{2+} and Ce^{3+} ions, relevant factors affecting the $5d$ energy level need to be mainly considered. Figure 3.2 shows the effects of the host lattice on the $5d$ energy levels of free Eu^{2+} and Ce^{3+} ions. Based on the semiempirical Dorenbos model, the energy gap is strongly related to a given host lattice (A), which is dependent on the spectroscopic redshift $D(A)$ and Stokes shift $\Delta S(A)$ [15]. If the first $5d$ energy level is known for a specific lanthanide ion in host A, one may determine the redshift $D(A)$. From that, the positions of the unrelaxed and relaxed lowest $5d$ energy levels for all other lanthanides in the same compound can be predicted [16]. The $D(A)$ is mainly determined by the centroid shift (ε_c) and crystal field splitting (ε_{cfs}). Several electrons of the anionic ligands move into the bare orbitals of the central metal ion and reduce the cationic valency, which would bring about the reduction of the interaction between metal ions and ligands. This effect is named the nephelauxetic effect (also known as the covalency effect). An increase in the degree of covalency effect can lead to a larger centroid shift, which means that the $5d$ levels of the free ion move to lower energy due to the decrease in the interelectron repulsion [17]. Besides the covalency of the chemical bonds between the luminescent center and its ligands, the polarizability of the neighboring anion ligands plays an essential role in the covalency effect. If anions or ionic complexes are put in, then the magnitude of centroid shift can be obtained according to the degree of the covalency effect. The extent of centroid

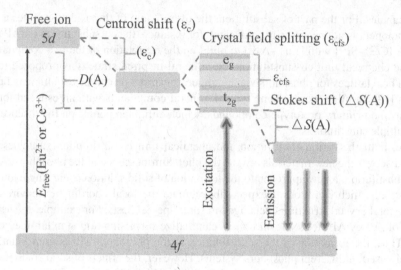

FIGURE 3.2 Schematic of the effects of the host lattice on the energy levels of free Eu^{2+} and Ce^{3+} ions.

shift is given in the following order: no ligands < F^- < Cl^- < Br^- < I^- < O^{2-} < N^{3-} < S^{2-} < Se^{2-} and SO_4^{2-} < CO_3^{2-} < PO_4^{3-} < BO_3^{3-} < SiO_4^{4-} < AlO_4^{5-}. In addition, the higher the degree of covalency, the larger centroid downshift will be generated, which reduces the emission energy of the $5d$–$4f$ transition.

Crystal field splitting effect also contributes to change the value of the redshift, thereby leading to tunable photoluminescence in all cover spectra ranges from near-ultraviolet (NUV) to near-infrared (NIR). The crystal field splitting effect describes the difference in energy between the highest and the lowest $5d$ energy levels [18]. A series of factors can affect the crystal field splitting, such as the bond lengths between the activator ion and its coordinating anions, the degree of covalency between the activator ion and its coordinating ligands, the local environment, and the symmetry of the activator site [19]. According to the previous studies of Dorenbos, when the charge valence of the center metal ion is determined, the crystal field strength is arranged in the order that follows the sequence of octahedron > cubic > dodecahedron > tricapped trigonal prism > cuboctahedron [17]. Therefore, the polyhedral distortion depending on solid solution design plays a remarkable role in the crystal field strength and the lattice symmetry, which can be characterized by the polyhedral distortion indexes D [20],

$$D = \frac{1}{n} \sum_{i=1}^{n} \frac{|l_i - l_{av}|}{l_{av}} \tag{3.1}$$

where l_i represents the distance between the luminescence center and the ith coordinating atom and l_{av} is the average bond length of the luminescence center and its first ligand. An increase in D can promote an enhancement of the crystal field strength and the decrease of site symmetry, thereby reducing the energy of the first excited level of $5d$ orbital and broadening the emission band. Any adjustments in the host composition and microstructure environment via solid solution would influence the distance between the luminescence center ion and its ligand, which may affect the luminescence properties of the Eu^{2+}- and Ce^{3+}-doped phosphors by adjusting the crystal field splitting strength. Therefore, as a method to change the host composition and microstructure, the solid solution method can effectively tune the energy of $4f$–$5d$ transition, which can achieve spectral adjustment and discover new luminescent materials for Eu^{2+}- and Ce^{3+}-doped phosphors.

3.2.2 Basic Knowledge of 4ف–4ف Transition

Compared with the $4f$–$5d$ transition, the electric dipole transition between $4f$ energy levels is parity-forbidden, and its luminescence features generally show narrow or sharp emission and excitation lines, long lifetimes, and excellent coherence properties, because the $4f$ electrons are well shielded by the outer $5s^2$ and $5p^6$ electrons [21–23]. The $4f$ orbitals have a smaller radial extension than the outer $5s$ and $5p$ orbitals, which brings about a low sensitivity to the crystal field, small electron-phonon coupling strengths, and low exchange perturbations. Normally, the luminescence with the $4f$–$5d$ transition has little variations in the designed solid solution phosphors. However, the chemical compositions still affect the site occupation and further modify the luminescence.

Trivalent lanthanide ions such as Eu^{3+} and Tb^{3+} ions are typical activators with the $4f$–$4f$ transition, and the luminescence color and the excited energy are almost fixed because they are hardly affected by the chemical composition of the host and the surrounding environment [24]. However, the emission intensity of some $4f$–$4f$ transitions like Eu^{3+} ion in a special case can depend strongly on the site symmetry in a host crystal [25]. It is well known that only when Eu^{3+} ions occupy the crystallographic sites without the inversion symmetry, the electric dipole transition is allowed. If the Eu^{3+} ions occupy the high symmetry sites, the electric dipole emission is weak, and the magnetic dipole transition becomes relatively stronger and dominates. For the hypersensitive electronic dipole transition, the parity-forbidden $4f$–$4f$ transition would further break if the site symmetry of Eu^{3+} decreases [26]. Hence, when the Eu^{3+} doping sites in the lattice lack inversion symmetry, the solid solution strategy can adjust the Eu^{3+} emission intensity by changing the symmetry of the Eu^{3+} site.

3.2.3 Basic Knowledge of 3D–3D Transition

The electron transition between $3d$ and $3d$ energy level in transition metal ions is common with very small probability owing to parity-forbidden [27–29]. Transition metal ions with $3d$–$3d$ transition are utilized widely in phosphors, which mainly include Mn^{2+}, Mn^{4+}, and Cr^{3+} ions. For Mn^{2+} ion, it has five d electrons in the unfilled $3d$ shell. The Mn^{2+}-doped phosphors exhibit a broad emission band in the visible range because of the $^4T_1 \rightarrow {}^6A_1$ transition, and the emission color of Mn^{2+} can be adjusted from green to red depending on the crystal field around the Mn^{2+} nearest neighbor coordination polyhedron [30, 31]. The Tanabe–Sugano diagram can clearly describe the relationship between the crystal field strength and the energy levels of the transition metal ions [32–34]. The corresponding Tanabe–Sugano diagram about Mn^{2+} ion with d^5 electron configuration in octahedral symmetry is shown in Figure 3.3(a) [28]. All transitions from the ground sextet state to the excited state are spin-forbidden because each excited level is either a spin doublet or a quartet, which implies that the quantum efficiency of Mn^{2+} is universally low. When the crystal field strength around the Mn^{2+} ion locating at coordination polyhedron is strong (the value of D_q/B is above about 2.8), the ground state energy level converts from spin-sextet $^6A_1$ to spin-doublet $^2T_2$. Obviously, the positions of the lowest excited state energy levels are strongly sensitive to the crystal field strength, so the emission color of Mn^{2+}-doped phosphors can be tuned via the variation of the host compositions. Most Mn^{2+} ion substitutions normally occur in the octahedron and tetrahedron crystallographic sites [35], and the $3d$–$3d$ transition of Mn^{2+} is spin-forbidden in both environments. Generally, the crystal field strength of an octahedron is much stronger than that of a tetrahedron, resulting in the larger splitting energy of the excited state in orbitals and the lower energy of Mn^{2+} emission in the octahedral crystal field. Accordingly, Mn^{2+} ion with the octahedral coordination usually exhibits an orange-to-red emission, whereas Mn^{2+} ion with the tetrahedral coordination gives a green emission.

For Mn^{4+} ion and Cr^{3+} ion, they have the same electronic configuration with three electrons occupying the outermost $3d$ orbitals. Thus, their spectroscopic properties nearly exhibit similar performance with the emission peaks located in the red to NIR region [36]. The Tanabe–Sugano diagram about Mn^{4+} and Cr^{3+} ions with d^3 electron configuration in octahedral symmetry is shown in Figure 3.3(b) [28]. Owing to the $^4A_2$ O$^4T_1(^4F)$ and $^4A_2$ a$^4T_2$ spin-allowed transitions, two broadbands

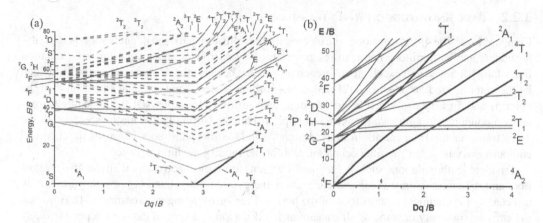

FIGURE 3.3 Tanabe–Sugano diagram for (a) Tanabe–Sugano diagram for the d^5 electron configuration in the octahedral crystal field. $C/B = 4.5$. The spin-quartet and spin-doublet states are shown by the solid and dashed lines, respectively. (b) The d^3 electron configuration in the octahedral crystal field. B and C are Racah parameters and $C/B = 4.7$. The spin-quartet and spin-doublet states are shown by the thick and thin lines, respectively. (Zhou, Q., Dolgov, L., Srivastava, A.M., Zhou, L., Wang, Z.L., Shi, J.X., Dramićanin, M.D., Brik, M.G., and Wu, M.M., *J. Mater. Chem. C*, 6, 2652, 2018. With permission)

are usually observed in the photoluminescence excitation spectra of Mn^{4+}- and Cr^{3+}-ions-doped phosphors, but another spin-allowed transition between $^4A_2$ and $^4T_1(^4P)$ is usually masked by the charge-transfer and the host lattice absorption band [36]. The ground state energy level is always $^4A_2$ state, whereas the first excited state energy level varies with the crystal field strength around Mn^{4+} ion or Cr^{3+} ion. When the crystal field strength is low (the value of D_q/B is less than about 2.2), the lowest excited state energy level is $^4T_2$ and the energy of $^4T_2$ state becomes higher with the crystal field strength increasing. For the emission spectra, on the one hand, the broadband emission stems from the spin-allowed transition between $^4A_2$ and $^4T_2$ energy levels. When the crystal field strength is high (the value of D_q/B is more than about 2.2), the ^2E state becomes the lowest excited state energy level, and the energy of the ^2E state maintains almost invariant as the crystal field changes. On the other hand, the sharp emission line origins from the spin-forbidden transition between $^4A_2$ and ^2E energy level, and its position keeps almost unchanged under different crystal field strengths [37]. Compared with Cr^{3+} ion doped in the host, Mn^{4+} ion is always located in a strong crystal field environment because of its highly effective positive charge [38, 39]. It means that Mn^{4+}-doped phosphors always show the sharp emission lines originating from the spin-forbidden $^2E \rightarrow {}^4A_2$ transition. Besides, all the excited energy levels are influenced by the Racah parameter B, which is closely related to the chemical bond between the $3d^3$ configuration ion and the ligand [40]. Normally, the Racah parameter B is small when the nephelauxetic effect is weak, so the ^2E energy level is located at a lower energy position. Therefore, the position of the emission peak can be tuned by changing the composition of the host to modify the crystal field strength and covalence of Mn^{4+} ion or Cr^{3+} ion. In summary, the photoluminescence from the $3d$–$3d$ transition of Mn^{2+}-, Mn^{4+}-, and Cr^{3+}-doped phosphors can be precisely tuned by the solid solution strategy.

3.2.4 BASIC KNOWLEDGE OF 6s–6p TRANSITION

The $6s$–$6p$ transition is a parity-allowed electrical dipole transition, which is regularly featured by much broader emission and excitation bands, compared with that of $4f$–$4f$ transition. As an excellent activator doped into the host, the Bi^{3+} ion exhibits a typical broadband emission from the $6s$–$6p$ transition. The electron configuration of Bi^{3+} ion is $[Xe]4f^{14}5d^{10}6s^2$, so the Bi^{3+} ion has two exposed electrons in the $6s$ orbital that are sensitive to the variation of the surrounding crystal field [41, 42].

The ground state level of Bi^{3+} free ion stemming from the $6s$ orbital is $^1S_0$ singlet state, and the excited state levels originating from the $6s6p$ orbital are ^3P_j ($j = 0, 1, 2$) triplet states and $^1P_1$ singlet level [43]. According to the selection rule, only the electronic transition from $^1S_0$ level to $^1P_1$ level is allowed, but the spin-orbit coupling that occurs between $^3P_1$ level and $^1P_1$ level would increase the transition probability of $^1S_0 \rightarrow ^3P_1$. For most hosts, the $^1S_0 \rightarrow ^3P_1$ transition is located in the UV region, and the $^1S_0 \rightarrow ^1P_1$ transition is located in the NUV region [44]. Hence, Bi^{3+} ions-activated phosphors often exhibit broad excitation spectra in the UV or NUV region. The emission transitions of Bi^{3+} ion include allowed $^3P_1 \rightarrow ^1S_0$ and forbidden $^3P_0 \rightarrow ^1S_0$ optical transitions [45]. Similar to the Eu^{2+} ion and Ce^{3+} ion, the centroid shift and the crystal field splitting play a vital role in the energy of the lowest excited level [46]. Thus, the diversity of the crystal field and covalence around the Bi^{3+} ion in the host lattice will lead to the shift of the spectra, and the Bi^{3+} ions doped in diverse host materials can show different emission colors ranging from UV to red. Therefore, the solid solution strategy can also accurately achieve controllable photoluminescence in Bi^{3+}-doped phosphors through efficiently modifying the local coordinated environment around Bi^{3+} ions.

3.3 DOPED SOLID SOLUTION PHOSPHORS WITH DIVERSE ELECTRON TRANSITIONS

3.3.1 Doped Solid Solution Phosphors with 4f–5d Transition

3.3.1.1 Eu^{2+}-Doped Solid Solution Phosphors

In the past few decades, Eu^{2+}-doped phosphors have attracted numerous scholars to conduct extensive research owing to the diversity of luminescence, and many commercial phosphors have been discovered, such as $(Sr,Ba)_2SiO_4:Eu^{2+}$ solid solution phosphor. Both Ba_2SiO_4 and Sr_2SiO_4 have the same orthorhombic β-K_2SO_4 structure, and they can maintain the structure after the addition of Eu^{2+} ions in the host lattice. Denault et al. utilized the cationic substitution method to modify the luminescence properties of Eu^{2+}-doped Ba_2SiO_4 phosphors owing to the variation of local structure [47]. As is shown in Figure 3.4(a) and (b), both the excitation and emission spectra exhibit different degrees of redshift with the enhancement of Sr content in $Sr_xBa_{2-x}SiO_4:Eu^{2+}$ ($x = 0$–2) phosphors. Moreover, Ba_2SiO_4 and Sr_2SiO_4 have formed the solid solution and the [Sr/BaO_9] and [Sr/BaO_{10}] polyhedral volumes continuously decrease with the increase of Sr^{2+} concentration in $Sr_xBa_{2-x}SiO_4:Eu^{2+}$ phosphors (Figure 3.4(c)). Since the Eu^{2+} ions enter into Sr and Ba site in $Sr_xBa_{2-x}SiO_4:Eu^{2+}$, the average distance between Eu^{2+} ion and O^{2-} ion would constantly reduce, resulting in an increase in the crystal field splitting strength. Furthermore, the decrease of Ba content would lead to an increased distortion of the [Sr/BaO_9] and [Sr/BaO_{10}] polyhedral and thereby enhance the crystal field splitting (Figure 3.4(d)). In addition, with the increase Sr content, more Eu^{2+} ions would occupy the Sr/Ba1 site, which leads to a larger centroid shift for the Eu^{2+} ion. Therefore, the reasons for the emission peak moving toward a longer wavelength in solid solution $Sr_xBa_{2-x}SiO_4:Eu^{2+}$ with increasing x are the above combination.

The same substitution strategy has different effects on tuning the luminescence for various phosphor systems. For example, Fang et al. observed that the substitution of Ba^{2+} for Sr^{2+} would bring about an unexpected adjustment in the emission spectra of $Sr_{4.7-x}Ba_xEu_{0.3}(PO_4)_3Cl$ phosphors, as shown in Figure 3.4(e) [48]. There exist two crystallography sites suitable for Eu^{2+} ions occupation in $Sr_5(PO_4)_3Cl$ host, in which Sr1 site is coordinated by nine O^{2-} ions and Sr2 is coordinated by six O^{2-} and two Cl^- ions. For $Sr_5(PO_4)_3Cl:Eu^{2+}$, the emission spectrum is symmetrical because Eu^{2+} ions only occupy the Sr2 sites before the cationic substitution. After introducing Ba^{2+} ions into the host lattice ($0.5 \leq x \leq 1.5$), larger Ba^{2+} ions would be inclined to locate at the Sr2 sites on account of the smaller space of the Sr1 site than that of the Sr2 site (Figure 3.4(f)). A part of Eu^{2+} ions deriving from the Sr2 sites begin to occupy the Sr1 sites due to the limited space of the Sr2 sites, inducing an abnormal redshift in the emission spectra. The increase in the full width at half maxima (FWHM) of the emission band can be mainly attributed to the appearance of a new luminescence center.

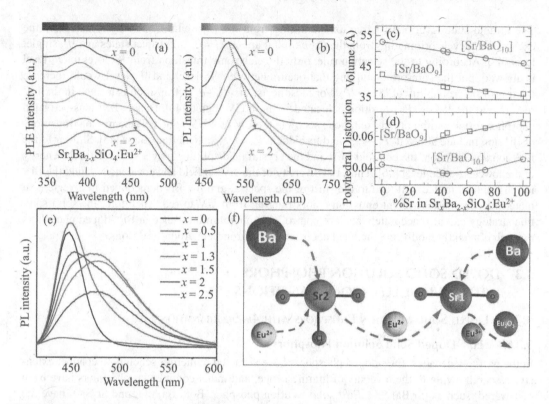

FIGURE 3.4 (a) The excitation and (b) emission spectra for the isostructural $Sr_xBa_{2-x}SiO_4:Eu^{2+}$ solid solution phosphors under the maximum emission/excitation wavelength. The variation of the $[Sr/BaO_9]$ and $[Sr/BaO_{10}]$ polyhedral (c) volume and (b) distortion with Sr concentration increasing in $Sr_xBa_{2-x}SiO4:Eu^{2+}$ samples. (e) The emission spectra of $Sr_{4.7-x}Ba_xEu_{0.3}(PO_4)_3Cl$ phosphors with different x values. (f) The mechanism of cation size-control Eu^{2+} ions redistribution. (Denault, K.A., Brgoch, J., Gaultois, M.W., Mikhailovsky, A., Petry, R., Winkler, H., DenBaars, S.P., and Seshadri, R., *Chem. Mater.*, 26, 2275, 2014. With permission; Fang, M.H., Ni, C., Zhang, X., Tsai, Y.T., Mahlik, S., Lazarowska, A., Grinberg, M., Sheu, H.S., Lee, J.F., Cheng, B.M., and Liu, R.S., *ACS Appl. Mater. Interfaces*, 8, 30677, 2016. With permission)

The space of the Sr2 sites achieves the limitation on Ba^{2+} ions occupation as the value of x exceeds 1.3, and then Ba^{2+} ions begin to occupy the Sr1 sites along with the further increasing content of Ba^{2+} ions. Moreover, more Ba^{2+} ions occupying the Sr1 sites would further compress the space of Eu^{2+} in the Sr1 site, which promotes the oxidation of Eu^{2+} with a larger ionic radius to Eu^{3+} with a smaller ionic radius. The abovementioned results indicate that the cation substitution can guide Eu^{2+} ions to selectively occupy various cation sites, thereby tuning the luminescence.

Except for cationic substitution, anionic substitution can also influence the position of the lowest excited energy level to achieve PL tuning via varying the centroid shift and the crystal field splitting in Eu^{2+}-doped phosphors. For instance, in the $Ba_5SiO_4F_xCl_{6-x}:Eu^{2+}$ solid solution phosphors, Xia's group reported an anion substitution of F^- for Cl^- modified the emission peak position [7]. As shown in Figure 3.5(a), the emission band shifts from 442 to 503 nm along with the introduction of F^- into the $Ba_5SiO_4F_xCl_{6-x}:Eu^{2+}$ system. The sample with $x = 0$ presents an asymmetric blue emission band centered at 442 nm, indicating that there are two different emission centers: Eu1 and Eu2. The luminescence from Eu1 dominants in the sample with $x = 0$ and then the luminescence from Eu2 becomes dominant with the addition of F^-. Meanwhile, there is a significant redshift in the excitation spectra as the F^- content increases (Figure 3.5(b)). The variation of spectra can be explained that the substitution of F^- for Cl^- induces a stronger polarization effect and then boosts Eu^{2+} ions preferentially occupy the Eu2 sites with the loose environment, resulting in the green

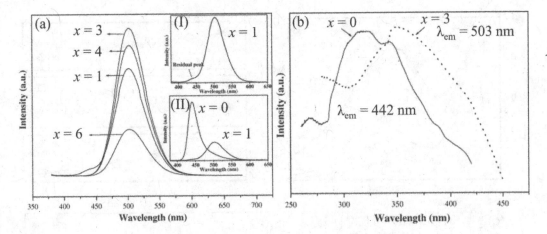

FIGURE 3.5 (a) The PL spectra of Eu^{2+}-doped $Ba_5SiO_4F_xCl_{6-x}$ phosphors under $\lambda_{ex} = 365$ nm. Inset I: the enlarged emission spectrum of $x = 1$ sample; and II: the emission spectra of $x = 1$ and 0 sample. (b) The PLE spectra of $x = 0$ and 3 sample under $\lambda_{em} = 442$ and 503 nm, respectively. (Xia, Z.G., Li, Q., and Sun, J.Y., *Mater. Lett.*, 61, 1885, 2007. With permission)

emission centered at 503 nm. In addition, when the F^- ions are introduced into the host lattice, it would cause a stronger crystal field and centroid shift of Eu^{2+}, because the distance between F^- and Eu^{2+} is shorter than that between Cl^- and Eu^{2+} and the electronegativity of F^- is larger than that of Cl^-. On the other hand, a similar spectral adjustment can also be achieved via anion group substitution such as $[BO_3]^{3-}$ replacing $[PO_4]^{3-}$. Dai et al. reported a continuous regulatable emission in Eu^{2+}-activated $Sr_5(PO_4)_{3-x}(BO_3)_xCl$ solid solution by adjusting the content of $[BO_3]^{3-}$ [49]. As shown in Figure 3.6(a), a new emission peak centered about 550 nm emerges in $Sr_5(PO_4)_{3-x}(BO_3)_xCl:Eu^{2+}$ phosphors with x increasing, indicating that a distinct luminescence center has been produced. With the increase of $[BO_3]^{3-}$ group concentration, the lattice parameters a, b, and cell volume V all reduce linearly, while the lattice parameters c increases linearly (Figure 3.6(b)), suggesting a continuous distortion in host lattice and a successive decline in the distance between Eu^{2+} and O^{2-}. The introduction of the anionic group $[BO_3]^{3-}$ into lattice cause the stronger crystal field around Eu^{2+}, because the Eu–O bond length decreases, and the polyhedron distortion enlarge. At the same time, owing to the structural variation, the Eu^{2+} ions selectively occupy different sites as the content of B^{3+} increases, which alters the amount of Eu^{2+} ions in different luminescence centers. Accordingly, the PL tuning of $Sr_5(PO_4)_{3-x}(BO_3)_xCl:Eu^{2+}$ phosphors is achieved by changing the relative emission peak intensity of 450 and 550 nm.

Furthermore, the chemical unit co-substitution is reported recently and work as an effective strategy to tune the luminescence properties of Eu^{2+}-doped phosphors. Based on the chemical unit co-substitution strategy, Xia's group studied the Eu^{2+}-doped $Ca_2(Al_{1-x}Mg_x)(Al_{1-x}Si_{1+x})O_7$ ($x = 0$–1) phosphors, which are solid solution between gehlenite $Ca_2Al(AlSiO_7)$ and akermanite $Ca_2Mg(Si_2O_7)$ [50]. $A_2B(T_2O_7)$ structure family is composed of pairs of connected tetrahedra in bow-tie form joined together into sheets by the B cations and isolated by the A cations. Figure 3.6(c) shows the schematic diagram about the chemical unit co-substitution of $[Mg^{2+}–Si^{4+}]$ for $[Al^{3+}–Al^{3+}]$, where the co-substitution only occurs at tetrahedral sites. Figure 3.6(d) gives the normalized photoluminescence spectra of Eu^{2+}-doped $Ca_2(Al_{1-x}Mg_x)(Al_{1-x}Si_{1+x})O_7$ phosphors for all values of x. The emission bands exhibit a blueshift and become narrow with the increasing content of $[Mg^{2+}–Si^{4+}]$, which is ascribed to the variation of the local coordination environment of the Eu^{2+} ion.

Another type of chemical unit co-substitution is related with the evolution of mineral models [51]. For example, a mineral compound UCr_4C_4 with high rigid structure is regarded as the initial prototype compound, and a series of new phosphors with narrow band emission have been reported

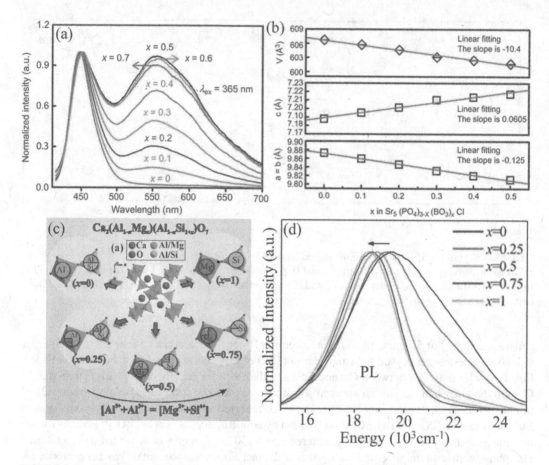

FIGURE 3.6   (a) The normalized PL spectra of the $Sr_5(PO_4)_{3-x}(BO_3)_xCl:Eu^{2+}$ ($x = 0$–0.7) solid solution under $\lambda_{ex} = 365$ nm. (b) The variation of unit cell parameters with x increasing in $Sr_5(PO_4)_{3-x}(BO_3)_xCl:Eu^{2+}$ phosphors. (c) Chemical unit co-substitution of $Ca_2(Al_{1-x}Mg_x)(Al_{1-x}Si_{1+x})O_7$: from $x = 0$ to 1 via the co-substitution of $[Mg^{2+}$–$Si^{4+}]$ for $[Al^{3+}$–$Al^{3+}]$. (d) The normalized PL spectra of the phosphors $Ca_2(Al_{1-x}Mg_x)(Al_{1-x}Si_{1+x})O_7:Eu^{2+}$ ($x = 0$–1). (Dai, P.P., Li, C., Zhang, X.T., Xu, J., Chen, X., Wang, X.L., Jia, Y., Wang, X.J., and Liu, Y.C., *Light Sci. Appl.*, 5, e16024, 2016. With permission; Xia, Z.G., Ma, C.G., Molokeev, M.S., Liu, Q.L., Rickert, K., and Poeppelmeier, K.R., *J. Am. Chem. Soc.*, 137, 12494, 2015. With permission)

by co-substitution (Figure 3.7(a)) [52], such as $NaLi_3SiO_4:Eu^{2+}$ blue phosphor ($\lambda_{em} = 469$ nm, FWHM = 32 nm), $RbLi(Li_3SiO_4)_2:Eu^{2+}$ green phosphor ($\lambda_{em} = 530$ nm, FWHM = 42 nm), and $Sr[LiAl_3N_4]:Eu^{2+}$ red phosphor ($\lambda_{em} = 654$ nm, FWHM = 50 nm) [53–55]. The mineral-inspired strategy has been widely used to find new hosts for activators doping and discover new phosphors. As shown in Figure 3.7(b), Xia's group designed a new phosphor $K_2BaCa(PO_4)_2$ with β-K_2SO_4 type structure through co-substitution of $K(M1)O_6 \rightarrow BaO_6$, $K(M2)O_7 \rightarrow (K/Ca)O_7$, $K(M3)O_6 \rightarrow KO_6$, and $SO_4 \rightarrow PO_4$ [56]. $K_2BaCa(PO_4)_2:Eu^{2+}$ exhibits a blue emission at 460 nm with zero thermal quenching owing to the existence of crystal defect levels. In addition, many new phosphate phosphors for white LEDs have been discovered derived from the β-$Ca_3(PO_4)_2$-type mineral structure, such as green-emitting $Sr_9Sc(PO_4)_7:Eu^{2+}$, yellow-emitting $Sr_9MgLi(PO_4)_7:Eu^{2+}$ and red-emitting $Sr_9MnLi(PO_4)_7:Eu^{2+}$ [57–59].

Chemical unit co-substitution can also control the valence of Eu activator in some specific hosts to achieve tunable emission. Zhang et al. observed an apparent variation in emission color of $Ca_{1+x}Y_{1-x}Al_{1-x}Si_xO_4:Eu$ ($x = 0$–0.3) solid solution phosphors owing to the co-substitution of $[Ca^{2+}$–$Si^{4+}]$ for $[Y^{3+}$–$Al^{3+}]$ [60]. The emission intensity of Eu^{2+} gradually increases, while that of Eu^{3+} has

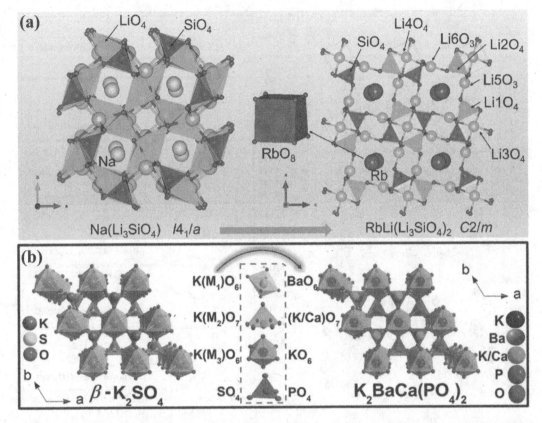

FIGURE 3.7 (a) Mineral prototype-inspired materials design from $NaLi_3SiO_4$ to $RbLi(Li_3SiO_4)_2$. (b) Structural phase transformation model from β-K_2SO_4 mineral prototype to the new phase of $K_2BaCa(PO_4)_2$ through the different substitution construction on different cation sites. (Meijerink, A., *Sci. China Mater.*, 62, 146, 2018. With permission; Qiao, J.W., Ning, L.X., Molokeev, M.S., Chuang, Y.C., Liu, Q.L., and Xia, Z.G., *J. Am. Chem. Soc.*, 140, 9730, 2018. With permission)

irregular variation as the content of [Ca^{2+}–Si^{4+}] increases, as shown in Figure 3.8(a). The crystal structure of this solid solution consists of an extremely condensed framework formed by vertex-sharing [$(Ca/Y)O_9$] polyhedrons and [$(Al/Si)O_6$] octahedrons. There is only one crystallographic site Ca/Y suitable for Eu ions occupation because of the difference of cation radius. The polyhedral volume of [$(Ca/Y)O_9$] expands while that of [$(Al/Si)O_6$] simultaneously reduces with increasing x in $Ca_{1+x}Y_{1-x}Al_{1-x}Si_xO_4$:Eu phosphors (Figure 3.8(b)), which demonstrates that the larger Ca^{2+} ions successfully replace the smaller Y^{3+} ions and the smaller Si^{4+} ions substitute the larger Al^{3+} ions. The decrement of the (Al/Si)–O bond lengths and increment of (Ca/Y)–O bond lengths with increasing x, which is in accordance with the variation of the [$(Ca/Y)O_9$] and [$(Al/Si)O_6$] polyhedron volumes. As shown in Figure 3.8(c), in $CaYAlO_4$ lattice, the Ca/Y site is coordinated by nine O^{2-} ions and connect with [AlO_6] octahedron by sharing O^{2-} vertexes. Before [Ca^{2+}–Si^{4+}] are introduced to into the crystal lattice, Eu^{3+} ions easily occupy the Ca/Y sites while Eu^{2+} ions are tough to locate in the Ca/Y sites because the space of [$(Ca/Y)O_9$] polyhedron is too small to tolerate Eu^{2+} ions. As the concentration of [Ca^{2+}–Si^{4+}] increases, the space of [$(Ca/Y)O_9$] polyhedron expands, which promotes the reduction of Eu^{3+} to Eu^{2+} in the Y/Ca sites (Figure 3.8(d)). Eventually, this co-substitution contributes to the increase in green emission intensity. As for the Eu^{3+} emission, they proposed two reasons to interpret the irregular variation. One is the content of Eu^{3+} decreases because of the reduction process. The other one is that the symmetry of Eu^{3+} cuts down as a result of the increase in the distortion of [$(Ca/Y)O_9$] polyhedron (Figure 3.8(b)).

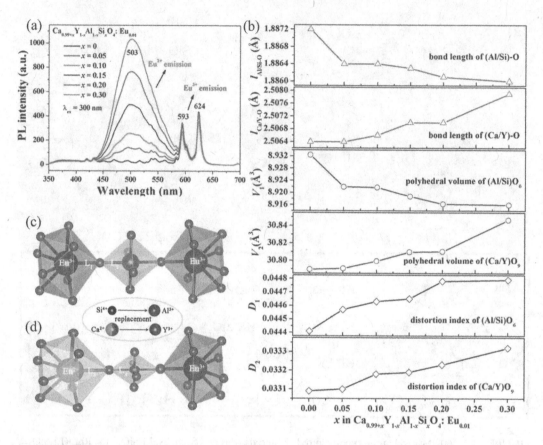

FIGURE 3.8 (a) Emission spectra of $Ca_{1+x}Y_{1-x}Al_{1-x}Si_xO_4$:Eu ($x = 0$–0.3) phosphors under $\lambda_{ex} = 300$ nm. (b) The variation of average bond length of (Al/Si)–O, (Ca/Y)–O, the polyhedral volumes of [(Al/Si)O$_6$], [(Ca/Y)O$_9$], and the related distortion index in $Ca_{1+x}Y_{1-x}Al_{1-x}Si_xO_4$:Eu samples with increasing x. Local structural coordination of Eu ions in the lattices of (c) $CaYAlO_4$:Eu^{3+}, and (d) $Ca_{0.99+x}Y_{1-x}Al_{1-x}Si_xO_4$:Eu samples. (Zhang, Y., Li, X.J., Li, K., Lian, H.Z., Shang, M.M., and Lin, J., *ACS Appl. Mater. Interfaces*, 7, 2715, 2015. With permission)

Although some isostructural compounds are quite different in chemical compositions, it can also design a solid solution to regulate the luminescence behaviors. Both $Sr_2Ca(PO_4)_2$ and $Ca_{10}Li(PO_4)_7$ host belong to β-$Ca_3(PO_4)_2$-type structure. Based on cation substitution occurring at multiple sites, Xia's group reported a controllable luminescence behavior in solid solution $xSr_2Ca(PO_4)_2$–$(1-x)$ $Ca_{10}Li(PO_4)_7$:Eu^{2+} [61]. As shown in Figure 3.9(a), except for an intrinsic excitation peak at 276 nm, an extra excitation band centered around 363 nm gradually appears in $xSr_2Ca(PO_4)_2$–$(1-x)$ $Ca_{10}Li(PO_4)_7$:Eu^{2+} phosphors with increasing x. It can be observed that a new emission peak at about 498 nm appears while the blue-violet emission peak gradually disappears with the content of $Sr_2Ca(PO_4)_2$ increasing (Figure 3.9(b)), which certifies the existence of different emission centers. Both the lattice parameter and volume present a nonlinear increase with the increase of the $Sr_2Ca(PO_4)_2$ component in solid solution (Figure 3.9(c)), which is attributed to the emergence of vacancy. There are five independent cation sites in $Ca_{10}Li(PO_4)_7$ host: Ca1, Ca2, Ca3, Li4, and Ca5 site. They thought that the partial Sr^{2+} ions would replace Li$^+$ ions and occupy Li4 sites accompanied by generating vacancies at Li4 sites when $x < 0.5$ (Figure 3.9(d)). After $x \geq 0.5$, partial Ca^{2+} ions would begin to replace Li$^+$ ions and create the same vacancy. In view of the space size of cation sites, Eu^{2+} ions will occupy four diverse sites except for Ca5 site in the initial host. Considering the difference in charge and radius between Eu^{2+} ion and Li$^+$ ion, Eu^{2+} ions barely occupy Li4 sites as

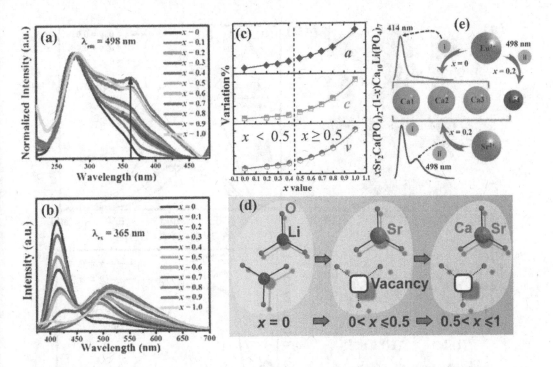

FIGURE 3.9 (a) The excitation and (b) emission spectra of solid solution $x\mathrm{Sr_2Ca(PO_4)_2}$–$(1-x)$ $\mathrm{Ca_{10}Li(PO_4)_7}$:$0.03\mathrm{Eu^{2+}}$ ($x = 0$–1). (c) The variation of unit cell parameters a and c and the unit cell volume v with x increasing. (d) Mechanism of generating vacancies at Li4 site via the substitution of $\mathrm{Li^+}$ ions for $\mathrm{Sr^{2+}}/\mathrm{Ca^{2+}}$ ions in $\mathrm{Ca_{10}Li(PO_4)_7}$ matrix. (e) Specific local environment of $\mathrm{Eu^{2+}}$ ions in different sites and the relationship between the distribution and luminescent behavior. (Chen, M.Y., Xia, Z.G., Molokeev, M.S., Wang, T., and Liu, Q.L., *Chem. Mater.*, 29, 1430, 2017. With permission)

$x = 0$. Although most $\mathrm{Eu^{2+}}$ ions are located at Ca1, Ca2, and Ca3 sites, the sample with $x = 0$ exhibits a blue-violet emission because the three sites have the same coordination number and nearly identical Ca–O average bond length. With the introduction of $\mathrm{Sr_2Ca(PO_4)_2}$, $\mathrm{Eu^{2+}}$ ions begin to occupy Li4 sites, resulting in the appearance of a new emission peak. Thus, the variation of the site environment would promote the redistribution of $\mathrm{Eu^{2+}}$ ions to tune the luminescence performance.

In addition, an alteration in the local structure around $\mathrm{Eu^{2+}}$ can be achieved in some specific hosts by a heterovalent substitution. Recently, Xia's group designed an interesting solid solution phosphor, $\mathrm{Eu^{2+}}$-doped $\mathrm{K_2Hf_{1-x}Lu_xSi_3O_9}$, on the basis of $\mathrm{Lu^{3+}}$ replacing $\mathrm{Hf^{4+}}$ [62]. As shown in Figure 3.10(a), the $\mathrm{K_2HfSi_3O_9}$ host owns a typical wadeite-layered structure, in which K atoms are coordinated by nine O atoms and Hf atoms are coordinated by six O atoms. A conspicuous emitting color modification from blue to cyan and a dramatic improvement on emission intensity are observed with the introduction of $\mathrm{Lu^{3+}}$ ions into $\mathrm{K_2HfSi_3O_9}$:$\mathrm{Eu^{2+}}$ phosphor (Figure 3.10(b)). The similar color variation can also be induced by increasing the content of $\mathrm{Eu^{2+}}$, which is ascribed to the appearance of a new luminescence center. Meanwhile, Figure 3.10(c) displays an obvious redshift in excitation spectra by increasing the value of x. Owing to the big difference between the radius of $\mathrm{Eu^{2+}}$ and $\mathrm{Hf^{4+}}$, $\mathrm{Eu^{2+}}$ ions tend to replace $\mathrm{K^+}$ ions, which accompanies with the generation of the $\mathrm{K^+}$ vacancies or the substitution of $\mathrm{Hf^{4+}}$ with $\mathrm{Eu^{3+}}$ to balance the charge. Thus, there are two types of the local coordination environment around $\mathrm{Eu^{2+}}$ sites in $\mathrm{K_2HfSi_3O_9}$ (Figure 3.10(d)) and the related emission spectrum presents two emission bands. Then, by adding $\mathrm{Lu^{3+}}$ ions into the $\mathrm{Eu^{2+}}$-doped $\mathrm{K_2HfSi_3O_9}$ phosphor, the replacement of $\mathrm{Lu^{3+}}$ for $\mathrm{Hf^{4+}}$ would become the dominant approach to balance the charge due to the lower defect formation energy. It means that $\mathrm{Eu^{2+}}$ sites possess only one kind of local coordination environment in $\mathrm{K_2Hf_{1-x}Lu_xSi_3O_9}$ as $x > 0$. Hence, the

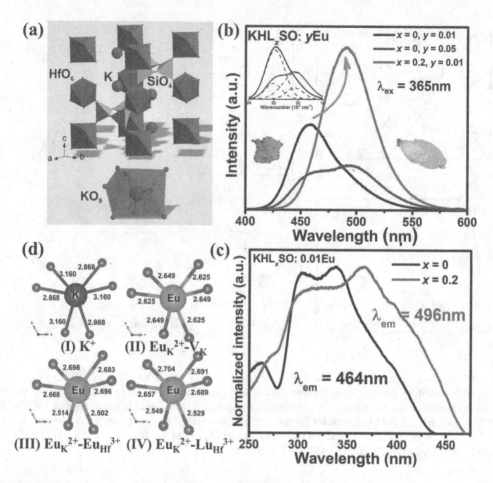

FIGURE 3.10 (a) The unit cell structure of $K_2HfSi_3O_9$ with representative coordination polyhedral. The (b) emission and (c) excitation spectra of $K_2Hf_{1-x}Lu_xSi_3O_9$:yEu^{2+} phosphors. (d) Optimized local structures of (I) K^+, (II, III) Eu^{2+} in $K_2HfSi_3O_9$ and (IV) in $K_2Hf_{1-x}Lu_xSi_3O_9$. (Cheng, C., Ning, L.X., Ke, X.X., Molokeev, M.S., Wang, Z.L., Zhou, G.J., Chuang, Y.C., and Xia, Z.G., *Adv. Opt. Mater.*, 8, 1901608, 2019. With permission)

heterovalent substitution of Hf^{4+} with Lu^{3+} can achieve the spectral tuning in $K_2Hf_{1-x}Lu_xSi_3O_9$:Eu^{2+} phosphors. Similarly, Xia's group found that the substitution of Li^+ for Al^{3+} besides Na^+ in nepheline $NaAlSiO_4$:Eu phosphor could achieve a tunable emission via controlling the distribution of Eu in different sites, along with an increased emission intensity owing to the enhanced reduction reaction of Eu^{3+} to Eu^{2+} [63].

3.3.1.2 Ce³⁺-Doped Solid Solution Phosphors

Similar to Eu^{2+}-activated solid solution phosphors, the luminescence properties of Ce^{3+}-doped phosphors are also extremely sensitive to the local coordination environments of Ce^{3+}. There are numerous reports about spectral tuning in Ce^{3+}-doped phosphors via the adjustment of host composition in the solid solution compounds. For instance, Ji et al. reported $Ba_{1.8-x}Sr_xSiO_4$:$0.1Ce^{3+}$, $0.1Na^+$ solid solution phosphors through varying the content of Sr^{2+} [64]. On the whole, an increase in x brings about a continuous redshift in the emission and excitation peak, accompanied by the variation in emission intensity, as displayed in Figure 3.11(a). The above mentioned results about redshift can be explained as that the Ce–O bond length decreases when the larger Ba^{2+} ions are replaced by Sr^{2+} ions, contributing to the increase in the effect of the centroid shift and the crystal field splitting. The

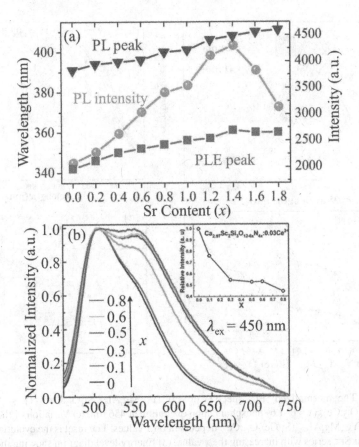

FIGURE 3.11 (a) Dependence of PLE and PL peak, PL intensity on content of Sr^{2+} for $Ba_{1.8-x}Sr_xSiO_4$:0.1Ce^{3+}, 0.1Na^+ phosphors. (b) The normalized emission spectra and relative intensity (inset) of $Ca_3Sc_2Si_3O_{12-6x}N_{4x}$:$Ce^{3+}$ ($x = 0-0.8$) phosphors upon excitation at 450 nm. (Ji, X.U., Zhang, J.L., Li, Y., Liao, S.Z., Zhang, X.G., Yang, Z.Y., Wang, Z.L., Qiu, Z.X., Zhou, W.L., Yu, L.P., and Lian, S.X., *Chem. Mater.*, 30, 5137, 2018. With permission; Liu, Y.F., Zhang, X., Hao, Z.D., Wang, X.J., and Zhang, J.H., *J. Mater. Chem.*, 21, 6354, 2011. With permission)

addition of Sr^{2+} ions improves the emission intensity of $Ba_{1.8-x}Sr_xSiO_4$:0.1Ce^{3+}, 0.1Na^+ phosphors as the value of x does not exceed 1.4. The anion substitution also has a strong effect on centroid shift, which can significantly modify the position of $5d$ energy level. Liu et al. utilized the variation of N/O in $Ca_3Sc_2Si_3O_{12-6x}N_{4x}$:$Ce^{3+}$ phosphors to obtain the controlled emission, as shown in Figure 3.11(b) [65]. When the content of N^{3-} ions increases in $Ca_3Sc_2Si_3O_{12-6x}N_{4x}$:$Ce^{3+}$, an additional emission peak centered around 550 nm gradually emerges, which causes a yellow-orange emission color. This is because the replacement of O^{2-} ion with N^{3-} ions leads to the variation in the local coordination environment around Ce^{3+} ions. The electronegativity of N^{3-} ion is stronger than that of O^{2-} ion; therefore, adding N^{3-} ion to the $Ca_3Sc_2Si_3O_{12}$:Ce^{3+} phosphor can make the $5d$ energy level possess a larger centroid shift, which boosts the emission peak shift toward longer wavelength. Nevertheless, the emission intensity gradually decreases in Ce^{3+}-doped $Ca_3Sc_2Si_3O_{12-6x}N_{4x}$ phosphors as the increase of x, because the substitution of N^{3-} ions for O^{2-} ions would produce oxygen vacancies, thereby increasing the probability of non-radiative transitions.

As mentioned earlier, the chemical unit co-substitution has a significant influence on the local crystal structure to modify the $4f-5d$ transition. Therefore, it is also a common method to tune the luminescence properties of Ce^{3+}-doped solid solution phosphors. Shang et al. investigated [Mg^{2+}–Si^{4+}/Ge^{4+}] substitution into $Y_3Al_5O_{12}$:Ce^{3+} garnet phosphors and the effect on the crystal structure

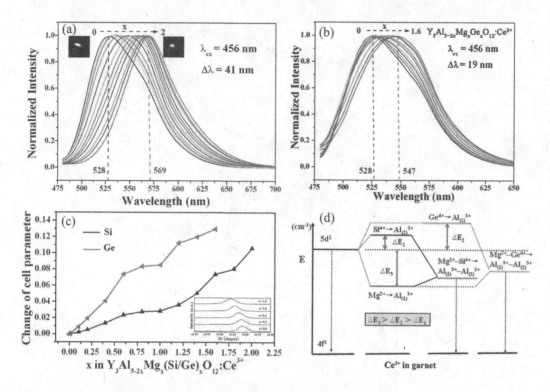

FIGURE 3.12 The normalized PL spectra of (a) $Y_{3-x}Mg_xAl_{5-2x}Si_xO_{12}:Ce^{3+}$ ($x = 0$–2) and (b) $Y_{3-x}Mg_xAl_{5-2x}Ge_xO_{12}:Ce^{3+}$ ($x = 0$–1.6) phosphors upon excitation at 456 nm. (c) Variation of the lattice parameter a ($= b = c$) in $Y_{3-x}Mg_xAl_{5-2x}(Si/Ge)_xO_{12}:Ce^{3+}$ dependent on x values. The inset is the evolution of reflections near $2\theta = 33.4°$ for Si^{4+} series with increasing the x value. (d) Energy-level diagram showing the lowest energy ground state and the first excited $5d$ state of Ce^{3+} in garnet. The energy change for the $5d$ state is indicated by ΔE. (Shang, M.M., Fan, J., Lian, H.Z., Zhang, Y., Geng, D.L., and Lin, J., *Inorg. Chem.*, 53, 7748, 2014. With permission)

and luminescence properties [66]. Figure 3.12(a) clearly shows the evolution of normalized emission spectra under 456 nm excitation as the ratio of $[Mg^{2+}-Si^{4+}]$ increases. The emission peak obviously moves toward a longer wavelength with $[Mg^{2+}-Si^{4+}]$ co-substitution for $[Al^{3+}-Al^{3+}]$ in $Y_{3-x}Mg_xAl_{5-2x}Si_xO_{12}:Ce^{3+}$ ($x = 0$–2) phosphors. Similarly, an obvious redshift of the emission bands occurs in Ce^{3+}-doped $Y_{3-x}Mg_xAl_{5-2x}Ge_xO_{12}$ phosphors with increasing the value of x (Figure 3.12(b)). The radius of Mg^{2+} in octahedral site is larger than that of Al^{3+}, so when the Mg^{2+} ions enter into the lattice and occupy the Al2 sites, the bond length of Mg–O would expand, resulting in a decrease in the distance between the Ce^{3+} and O^{2-}. Eventually, the crystal field strength of $[CeO_8]$ coordination polyhedron increases, which would lead to the reduction of the energy of the lowest $5d$ level. However, the smaller Si^{4+} or Ge^{4+} replacing Al^{3+} would contribute to a longer bond length of Ce–O and a weaker crystal field of $[CeO_8]$ polyhedron; hence, the energy of the lowest $5d$ level would increase. Figure 3.12(c) shows the enhancement of the cell parameter with increasing x. When the co-substitution of $[Mg^{2+}-Si^{4+}/Ge^{4+}]$ for $[Al^{3+}-Al^{3+}]$ occurs in the $Y_3Al_5O_{12}:Ce^{3+}$ sample, the crystal field strength around Ce^{3+} would increase. In other words, Mg^{2+} substitution has a stronger impact on the crystal field splitting than Si^{4+} or Ge^{4+} substitution, leading to a lower position of the lowest $5d$ level (Figure 3.12(d)). In addition, when the value of x is fixed in $Y_{3-x}Mg_xAl_{5-2x}(Si/Ge)_xO_{12}:Ce^{3+}$ phosphors, the $[Mg^{2+}-Si^{4+}]$ co-substitution system has a larger cell parameter and a stronger crystal field around Ce^{3+} than the $[Mg^{2+}-Ge^{4+}]$ co-substitution system. Therefore, there is a larger redshift in the emission peak of $Y_{3-x}Mg_xAl_{5-2x}Si_xO_{12}:Ce^{3+}$ phosphor systems.

The chemical unit co-substitution including anion can also efficiently regulate the spectral properties of Ce^{3+}-activated phosphors through the variation in crystal field strength and centroid shift. A continuous redshift in emission spectra of Ce^{3+}-doped $La_5(Si_{2+x}B_{1-x})(O_{13-x}N_x)$ ($x = 0–1$) solid solution phosphors is realized by the replacement of $[Si^{4+}–N^{3-}]$ for $[B^{3+}–O^{2-}]$ [67]. As displayed in Figure 3.13(a), the emission peak gradually shifts from 421 nm toward 463 nm with the value of x increases. Furthermore, the emission band gradually widens with increasing x when the value of x is less than 0.8, and then the FWHM declines slightly as x continues to increase. The lattice parameter a, c values and cell volume V show a conspicuous linear relationship with the value of x (Figure 3.13(b)), certifying the formation of solid solution between $La_5Si_2BO_{13}$ and $La_5Si_3O_{12}N$. However, the variation in the lattice parameters a and c present a converse tendency, leading to an increase in lattice distortion and disorder, which may result in the broadening emission band. There are two reasons for the redshift of the emission bands. On the one hand, the crystal field strength around Ce^{3+} is enhanced by increasing the polyhedral distortion of Ce sites. On the other hand, since the N^{3-} ion has stronger covalence compared to O^{2-} ion, adding N^{3-} ions in $La_5(Si_{2+x}B_{1-x})(O_{13-x}N_x):Ce^{3+}$ would enhance the centroid shift of the $5d$ level. As another example, Im et al. investigated the co-substituted of $[Si^{4+}–O^{2-}]$ for $[Al^{3+}–F^-]$ effect on luminescence properties of Ce^{3+}-doped $Sr_3Al_{1-x}Si_xO_{4+x}F_{1-x}$ ($x = 0–1$) solid solution phosphors [8]. As shown in Figure 3.13(c), a nonmonotonic

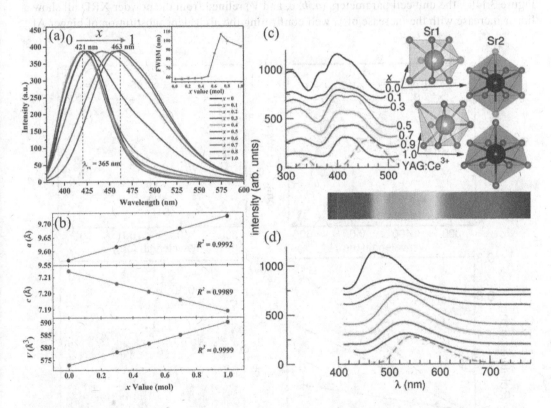

FIGURE 3.13 (a) Normalized emission spectra and relative FWHM (inset) of Ce^{3+}-doped $La_5(Si_{2+x}B_{1-x})(O_{13-x}N_x)$ ($x = 0–1$) solid solution phosphors upon excitation at 365 nm. (b) Variation in unit cell parameters (a, c, and V) with x values of $La_5(Si_{2+x}B_{1-x})(O_{13-x}N_x):Ce^{3+}$ solid solution series. (c) Excitation and (d) emission spectra of Ce^{3+}-doped $Sr_3Al_{1-x}Si_xO_{4+x}F_{1-x}$ ($x = 0–1$) solid solution phosphors. Sr1 and Sr2 polyhedra looking down the [1 0 0] direction are shown to the right of (c); the upper two polyhedra are from the Sr_3AlO_4F structure and the lower two are from the Sr_3SiO_5 structure. Light gray, black, orange, and green spheres represent Sr1, Sr2, O, and F atoms, respectively. (Xia, Z.G., Molokeev, M.S., Im, W.B., Unithrattil, S., and Liu, Q.L., *J. Phys. Chem. C*, 119, 9488, 2015. With permission; Im, W.B., George, N., Kurzman, J., Brinkley, S., Mikhailovsky, A., Hu, J., Chmelka, B.F., DenBaars, S.P., and Seshadri, R., *Adv. Mater.*, 23, 2300, 2011. With permission)

redshift of the excitation peak and an increasingly anisotropic of the coordination around both Sr sites are observed as the value of x increases in $Sr_3Al_{1-x}Si_xO_{4+x}F_{1-x}$:$Ce^{3+}$, while the emission bands monotonically shift toward longer wavelengths with the addition of [Si^{4+}–F^-] (Figure 3.13(d)). The above-mentioned redshift of emission spectra is partly because of an increased polyhedral distortion of the Ce^{3+} sites, which brings a stronger crystal field splitting. The replacement of F atoms with lower electronegative O atoms would generate a larger centroid shift in the $5d$ level of Ce^{3+}, which also causes a redshift in the excitation and emission bands.

As mentioned earlier, the heterovalent substitution is a new strategy in some particular hosts to fabricate the solid solution phosphor and discover new luminescence behaviors via modifying the local coordination environment around the activator sites. Recently, You et al. reported the Si^{4+} substitution with Al^{3+} producing an abnormal red emission in $La_3Al_xSi_{6-x}N_{11-x/3}$:$Ce^{3+}$ phosphors [11]. As displayed in Figure 3.14(a), the excitation bands around 300–400 and 400–500 nm gradually disappear, while that around 500–600 nm gradually generate as the Al^{3+} ions add into $La_3Si_6N_{11}$:Ce^{3+} phosphors. The variation of the excitation spectra might be caused by the change of the local structure around Ce^{3+}. Apart from the green-yellow emission band, a new red emission band gradually appears with the content of Al^{3+} ions increasing in $La_3Al_xSi_{6-x}N_{11-x/3}$:$Ce^{3+}$ (Figure 3.14(b)), implying the emergence of a new emission center. As exhibited in Figure 3.14(c), the unit cell parameters (a, b, c, and V) refined from the powder XRD all show a linear increase with the increase of x, well confirming the aliovalent substitution of bigger Al^{3+}

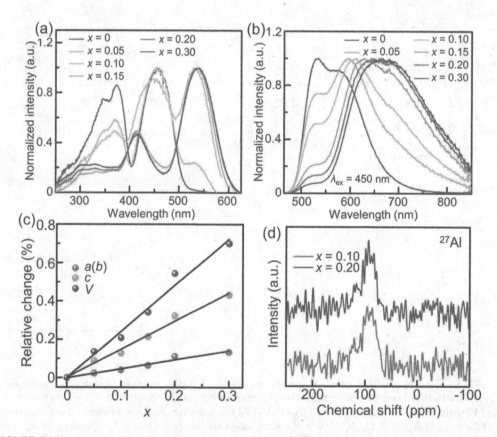

FIGURE 3.14 Normalized (a) excitation spectra monitored at 580, 597, 612, 627, 643, and 664 nm, respectively, and (b) emission spectra monitored at 450 nm of $La_{2.9}Al_xSi_{6-x}N_{11-x/3}$:$0.1Ce^{3+}$ ($x = 0$–0.3) phosphors. (c) Unit cell parameters (a, b, c, and V) as a function of Al content in $La_{2.9}Al_xSi_{6-x}N_{11-x/3}$:$0.1Ce^{3+}$. (d) ^{27}Al solid-state NMR spectra of $x = 0.1$, 0.2 samples. (You, S.H., Li, S.X., Jia, Y.C., and Xie, R.J., *Chem. Mater.*, 32, 3631, 2020. With permission)

for smaller Si^{4+}. Moreover, there are obvious ^{27}Al nuclear magnetic resonance (NMR) peaks around 80–100 ppm in $x = 0.1, 0.2$ samples (Figure 3.14(d)), which indicates that Al^{3+} ions enter host lattice to form $[AlN_4]$ tetrahedra. They proposed that the nonequivalent replacement of Si^{4+} by Al^{3+} would generate an interstitial site, which is suitable for Ce^{3+} occupation owing to the stimulation of charge compensation. Therefore, a special red emission band is formed in $La_3Al_xSi_{6-x}N_{11-x/3}:Ce^{3+}$. Besides, heterovalent substitution can lead to a tunable spectrum by efficiently affecting the crystal field splitting and centroid shift. For instance, Hu et al. successfully incorporated Be^{2+} ions in $Lu_2SrAl_4SiO_{12}:Ce^{3+}$ phosphor [68]. The replacement of Al^{3+}/Si^{4+} ions with Be^{2+} ions leads to an obvious redshift in emission and excitation spectra due to an enhanced crystal field strength and nephelauxetic effect.

3.3.2 Eu^{3+}-Doped Solid Solution Phosphors with 4f–4f Transition

Normally, Eu^{3+}-doped phosphors nearly maintain stationary emission peak position in different hosts because the $4f$–$4f$ transition is independent of the local crystal environment. Nevertheless, there are some special Eu^{3+}-doped phosphors, whose luminescence intensity can be adjusted via the solid solution strategy [69]. Liu et al. thoroughly investigated the effect of Sr^{2+} replacing Ba^{2+} on enhancing the luminescence intensity of Eu^{3+}-doped $BaLaMgSbO_6$ phosphors [70]. $BaLaMgSbO_6$ host belongs to a double perovskite structure with Ba^{2+} and La^{3+} ions locating at the 12-fold polyhedral site, and Mg^{2+} and Sb^{3+} ions locating at the 6-fold polyhedral site. Eu^{3+} ion occupies the LaO_{12} polyhedral sites with low crystal symmetry, leading to a stronger emission intensity of $^5D_0$–$^7F_2$ transition than that of $^5D_0$–$^7F_1$ transition. As shown in Figure 3.15(a) and (b), with the increase of the concentration of Sr^{2+} in $Ba_{1-x}Sr_xLaSbO_6:Eu^{3+}$, all the luminescence intensity gradually enhances under the different excitation wavelength, while the peak position of the emission spectra is almost unchanged. The emission intensity ratio of $^5D_0$–$^7F_2$ (615 nm)/$^5D_0$–$^7F_1$ (590 nm) transition increases with increasing the content of Sr^{2+}. As is known to all, the $^5D_0$–$^7F_1$ magnetic dipole transition is dominant when the Eu^{3+} ion occupies an inversion symmetry site. Whereas the $^5D_0$–$^7F_2$ electric dipole transition becomes quite strong and dominates if the position of the Eu^{3+} ion lacks the inversion symmetry. Since the luminescence intensity ratios are between 6 and 8 for different excitations in $Ba_{1-x}Sr_xLaSbO_6:Eu^{3+}$ phosphors, the Eu^{3+} ions in the lattice occupy the low symmetry sites. Generally, the intensity of the electronic dipole transition is hypersensitive to the site symmetry of Eu^{3+}. The $^5D_0$–$^7F_2$ transition can be promoted by the gradually decreased site symmetry with the replacement of Sr^{2+} for Ba^{2+}. Consequently, the intensity ratios of $^5D_0$–$^7F_2/^5D_0$–$^7F_1$ increase as Sr^{2+} ions are added into the host.

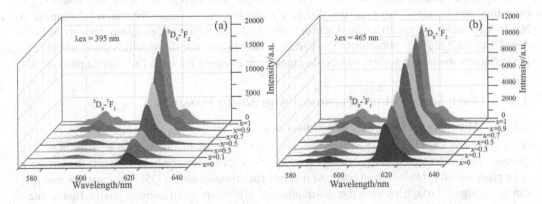

FIGURE 3.15 The photoluminescence emission spectra of $Ba_{1-x}Sr_xLa_{0.95}SbO_6:0.05Eu^{3+}$ ($x = 0$–1) phosphors upon excitation at (a) 395 nm and (b) 465 nm. (Liu, Q., Wang, L.X., Huang, W.T., Zhang, L., Yu, M.X., and Zhang, Q.T., *J. Alloys Compd.*, 717, 156, 2017. With permission)

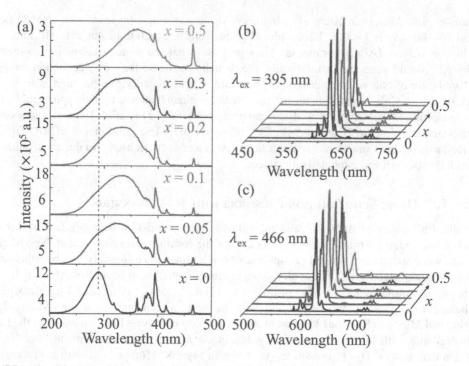

FIGURE 3.16 The excitation and emission spectra of LiYMgW$_{1-x}$Mo$_x$O$_6$:Eu^{3+} (x = 0–0.5) phosphors. The excitation spectra in (a) were obtained via monitoring 618 nm emission. The emission spectra were acquired when excited at (b) 395 nm, and (c) 466 nm, respectively. (Liang, Y.J., Noh, H.M., Ran, W.G., Park, S.H., Choi, B.C., Jeong, J.H., and Kim, K.H., *J. Alloys Compd.*, 716, 56, 2017. With permission)

In addition, the excitation peak derived from the charge transfer between the high positive ion and O^{2-} ion can be tuned by cation substitution. Figure 3.16(a) displays the excitation spectra of Eu^{3+}-doped LiYMgW$_{1-x}$Mo$_x$O$_6$ at different Mo^{6+} concentration [71]. The excitation peak caused by charge transfer obviously moves toward a longer wavelength with increasing the Mo^{6+} concentration, while the other excitation peak positions retain invariant. The above-mentioned redshift can be explained as follows, the 4d orbitals of Mo^{6+} have lower energy than that of 5d orbitals of W^{6+}; therefore, the electronegativity of Mo^{6+} is larger than that of W^{6+}, so Mo^{6+} substituting for W^{6+} would result in the reduction of the energy gap. As shown in Figure 3.16(b) and (c), the emission intensities of LiYMg W$_{1-x}$Mo$_x$O$_6$:Eu^{3+} samples first increase and then decrease with increasing x upon 395- and 466 nm excitation. Nevertheless, there are also different examples. For instance, the cation substitution of Na$^+$ for Li$^+$ would give rise to a remarkable decrease of the excitation and emission intensity in Eu^{3+}-doped Li$_{1-y}$Na$_y$YMgWO$_6$ phosphors. The earlier results imply that the specific cation substitution is an effective method to achieve tunable luminescence intensity for some Eu^{3+}-doped phosphors.

3.3.3 DOPED SOLID SOLUTION PHOSPHORS WITH 3D–3D TRANSITION

3.3.3.1 Mn^{2+}-Doped Solid Solution Phosphors

As is known to all, Mn^{2+} ion is extremely susceptible to the local coordination environment. The emission peak is located in the green region as Mn^{2+} occupies a tetrahedral site, while the emission color presents red if Mn^{2+} is in an octahedral coordination environment [35]. The emission energies can be changed through varying the distribution of Mn^{2+} ions at different crystallographic sites, and then the tunable luminescence would be realized in the Mn^{2+}-doped phosphors. Recently, Ma et al. studied the site-selective occupation of Mn^{2+} ions for the adjustment of luminescence property via Zn^{2+} replacing Mg^{2+} [72]. Under different excitation wavelengths, a manifest evolution of

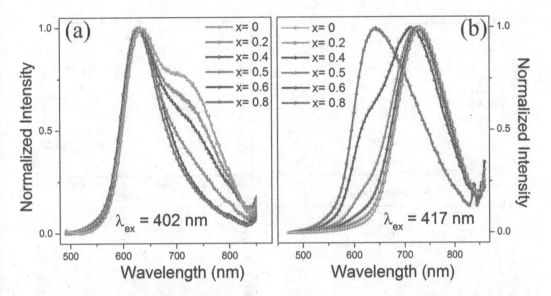

FIGURE 3.17 The normalized photoluminescence emission spectra of $(Mg_{1-x}Zn_x)_{2.97}(PO_4)_2:0.03Mn^{2+}$ ($x = 0$–0.8) phosphors under excitation of (a) 402 nm and (b) 417 nm. (Ma, Y.Y., Hu, J.Q., Song, E.H., Ye, S., and Zhang, Q.Y., *J. Mater. Chem. C*, 3, 12443, 2015. With permission)

luminescence behavior of $(Mg_{1-x}Zn_x)_{2.97}(PO_4)_2:0.03Mn^{2+}$ samples are observed by introducing Zn into the lattice to form a solid solution, as shown in Figure 3.17(a) and (b). In the γ-$Mg_3(PO_4)_2$ structure, there are two distinct Mg^{2+} sites. One is a sixfold coordinated polyhedral site and the other one is a fivefold coordinated polyhedral site, which can provide two diverse crystal field environments for Mn^{2+} ions. Considering the crystal structure and emission spectra, Mn^{2+}-doped $Mg_3(PO_4)_2$ phosphor exhibits two different luminescent centers. The high-energy emission originates from Mn^{2+} locating at $[MgO_6]$ octahedral site and the other emission derives from Mn^{2+} locating at $[MgO_5]$ polyhedral site. The increase of Zn/Mg ratio would promote more Mn^{2+} ions occupying $[MgO_6]$ octahedral site in $(Mg_{1-x}Zn_x)_{2.97}(PO_4)_2:Mn^{2+}$ samples, and then the longer wavelength emission peak gradually disappears under 402 nm excitation and the shorter wavelength emission peak gradually generates upon 417 nm excitation.

Furthermore, a series of solid solution methods that modify the local coordination environment of Mn^{2+} can realize the photoluminescence tuning in Mn^{2+}-doped phosphors. Cheng et al. controlled the distribution of Mn^{2+} ions in various lattice sites via multiple-cation substitution, and the emitting colors of Mn^{2+}-doped $Li_2ZnGe_3O_8$ phosphors could be regulated from green to NIR light [73]. For the $Li_2ZnGe_3O_8$ host lattice, there are different cationic sites, including 4-coordinated Li1/Zn1 site, 6-coordinated Li2/Zn2 site, and 6-coordinated Ge site. Considering the difference in charge and ionic radius, Mn^{2+} ions tend to occupy the ZnO_6 site in the $Li_2ZnGe_3O_8$ host and exhibit NIR emission. Based on the cation substitution of Ca^{2+} for Zn^{2+}, the emission peak centered around 832 nm gradually disappears with increasing the concentration of Ca^{2+} ions in $Li_2Zn_{1-x}Ca_xGe_3O_8:Mn^{2+}$ phosphors (Figure 3.18(a)). Nevertheless, as shown in Figure 3.18(b), with the increase of x, the emission intensity of peak centered around 548 nm gradually increases and then diminishes after x exceeds 0.4. The variation of luminescence in $Li_2Zn_{1-x}Ca_xGe_3O_8:Mn^{2+}$ phosphors can be explained that the increase in Ca content would stimulate the redistribution of Mn^{2+} ions in distinct sites and then cause intensity changes at different emission centers. At the same time, they also investigated separately the effect of the cation substitution of Zn^{2+} with Sr^{2+} and Ba^{2+} in $Li_2ZnGe_3O_8:Mn^{2+}$ phosphor on the luminescence properties. As displayed in Figure 3.18(c) and (d), the obvious intensity change of three different luminescence centers can be observed when Sr^{2+} and Ba^{2+} are introduced into the $Li_2ZnGe_3O_8:Mn^{2+}$ phosphor. Before the substitution ions reach the optimum content, the

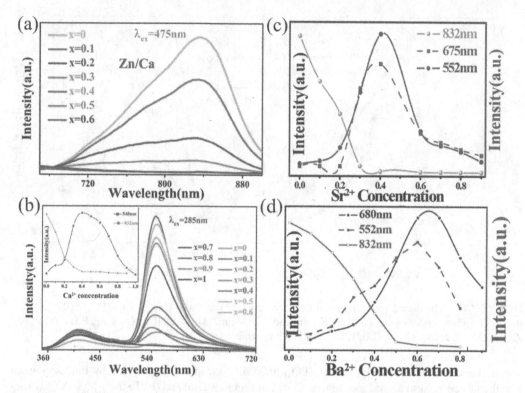

FIGURE 3.18 The emission spectra of $Li_2Zn_{1-x}Ca_xGe_3O_8$:Mn^{2+} phosphors under excitation of (a) 475 nm and (b) 285 nm. The inset is the intensity variation of different emission peaks of 548 and 832 nm in $Li_2Zn_{1-x}Ca_xGe_3O_8$:Mn^{2+}. (c) The intensity variation of different emission peaks of 552, 675, and 832 nm in $Li_2Zn_{1-y}Sr_yGe_3O_8$:Mn^{2+}. (d) The intensity variation of different emission peaks of 552, 680, and 832 nm in $Li_2Zn_{1-z}Ba_zGe_3O_8$:Mn^{2+}. (Cheng, J.G., Li, P.L., Wang, Z.J., Li, Z.L., Tian, M.M., Wang, C., and Yang, Z.P., *Dalton Trans.*, 47, 4293, 2018. With permission)

intensity of the emission peak centered 832 nm constantly reduces while the peaks centered 675 and 552 nm continuously enhances with the increase of Sr^{2+} and Ba^{2+} doping content. Since the cation substitution of Zn^{2+} with Sr^{2+} or Ba^{2+} would change the local environment of Mn^{2+} at ZnO_6, resulting in the ZnO_4 and GeO_6 sites being occupied by Mn^{2+}. Thus, the introduction of Sr^{2+} and Ba^{2+} into the Mn^{2+}-doped $Li_2ZnGe_3O_8$ sample would promote Mn^{2+} to exhibit additional green and red emissions.

Interestingly, the solid solution can adjust the spectra variation in some special Mn-doped phosphors by controlling the valence state of Mn. Since the crystal structures of $Ba_{0.75}Al_{11}O_{17.25}$ and $BaMgAl_{10}O_{17}$ are both evolved from $NaAl_{11}O_{17}$, one can design a series of solid solution compounds based on the chemical unit co-substitution strategy. Recently, Hu et al. reported a continuously tunable emission in Mn-doped $(1-x)Ba_{0.75}Al_{11}O_{17.25-x}BaMgAl_{10}O_{17}$ solid solution phosphors through varying the local crystal structure [74]. As shown in Figure 3.19(a), the green emission from Mn^{2+} diminishes constantly, while the red emission from Mn^{4+} gradually enhances with increasing the content of $BaMgAl_{10}O_{17}$. When Mn is doped in $Ba_{0.75}Al_{11}O_{17.25}$ host ($x = 0$), it nearly exhibits a prominent green emission since its crystal structure can provide suitable sites for the occupation of Mn^{2+} ions and Mn prefers to exist in the form of divalent manganese. However, the local crystal environment changes with the increase of x. The characteristic emission of Mn^{4+} and Mn^{2+} are observed in Mn-doped $(1-x)Ba_{0.75}Al_{11}O_{17.25-x}BaMgAl_{10}O_{17}$ solid solution phosphors and finally the emission of Mn^{4+} becomes dominant in the sample with $x = 1$. In the electron paramagnetic resonance (EPR) spectra of $(1-x)Ba_{0.75}Al_{11}O_{17.25-x}BaMgAl_{10}O_{17}$ solid solution, the characteristic signal

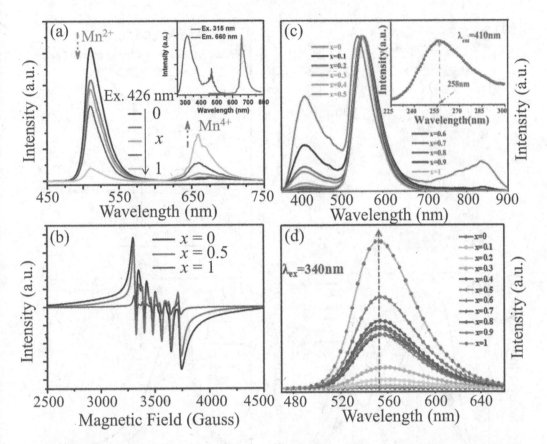

FIGURE 3.19   (a) The PL spectra ($\lambda_{ex}=426$ nm) and (b) electron paramagnetic resonance spectra of Mn-doped $(1-x)\mathrm{Ba_{0.75}Al_{11}O_{17.25-x}BaMgAl_{10}O_{17}}$ solid solution phosphors. The PL spectra of $\mathrm{Li_{2(1-x)}Zn_{1+x}Ge_3O_8:Mn^{2+}}$ phosphors upon excitation of (c) 258 nm and (d) 340 nm. (Hu, J.Q., Song, E.H., Zhou, Y.Y., Zhang, S.L., Ye, S., Xia, Z.G., and Zhang, Q.Y., *J. Mater. Chem. C*, 7, 5716, 2019. With permission; Cheng, J.G., Li, P.L., Wang, Z.J., Li, Z.L., Tian, M.M., Wang, C., and Yang, Z.P., *Dalton Trans.*, 47, 4293, 2018. With permission)

of $\mathrm{Mn^{2+}}$ gradually becomes weaker with increasing x (Figure 3.19(b)), which indicates that the addition of $\mathrm{BaMgAl_{10}O_{17}}$ obviously retards the reduction of $\mathrm{Mn^{4+}}$ for $\mathrm{Mn^{2+}}$. Hence, the increase in the proportion of $\mathrm{BaMgAl_{10}O_{17}}$ would precisely regulate the luminescence behaviors of Mn-doped $(1-x)\mathrm{Ba_{0.75}Al_{11}O_{17.25-x}BaMgAl_{10}O_{17}}$ phosphors because of controlling the progress of the reduction of $\mathrm{Mn^{4+}}$ to $\mathrm{Mn^{2+}}$.

Furthermore, heterovalent substitution can also successfully achieve spectral regulation by influencing the local environment surrounding $\mathrm{Mn^{2+}}$ in some solid solution phosphors [73]. As shown in Figure 3.19(c), a continuous decrease in the emission intensity of the peaks centered 414 and 832 nm is observed with an enhanced process in the heterovalent substitution of $\mathrm{Zn^{2+}}$ for $\mathrm{Li^+}$ under 258 nm excitation. However, the green emission gradually appears under 340 nm excitation with increasing the content of $\mathrm{Zn^{2+}}$ in $\mathrm{Li_{2(1-x)}Zn_{1+x}Ge_3O_8:Mn^{2+}}$ phosphors (Figure 3.19(d)). Since the addition of extra $\mathrm{Zn^{2+}}$ in the host would bring about a variation in the local environment of cationic sites, more $\mathrm{Mn^{2+}}$ ions selectively occupy $[\mathrm{ZnO_4}]$ sites, and the green emission from the $[\mathrm{ZnO_4}]$ site gradually enhances.

3.3.3.2 Mn⁴⁺-Doped Solid Solution Phosphors

Although the energy levels of $\mathrm{Mn^{4+}}$ ion are highly sensitive to the local crystal field environment theoretically, it is hard to tune the peak position owing to the specific crystal field in most $\mathrm{Mn^{4+}}$-doped phosphors. Considering the crystal field stabilization energy, the $\mathrm{Mn^{4+}}$ ion is more

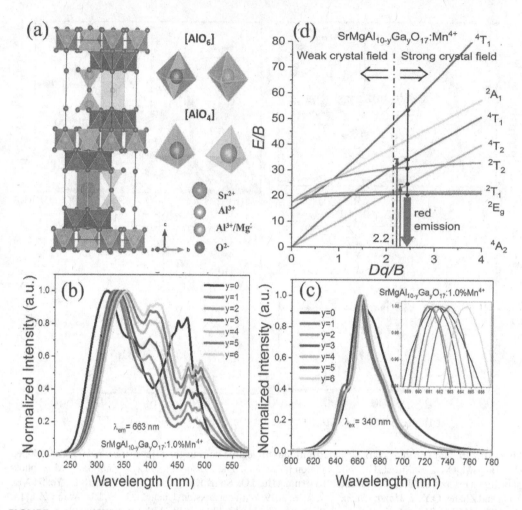

FIGURE 3.20   (a) The crystal structure of $SrMgAl_{10}O_{17}$ host, the coordination environment of the octahedron $[AlO_6]$, and tetrahedron $[AlO_4]$. The normalized photoluminescence (b) excitation and (c) emission spectra of $SrMgAl_{10-y}Ga_yO_{17}:Mn^{4+}$ ($y = 0$–6) samples, inset is the corresponding amplification emission spectra ranging from 670 to 700 nm. The inset is the corresponding amplification emission spectra ranging from 658 to 668 nm. (d) The Tanabe–Sugano energy level diagram of Mn^{4+} in $SrMgAl_{10-y}Ga_yO_{17}:Mn^{4+}$ phosphors. (Gu, S.M., Xia, M., Zhou, C., Kong, Z.H., Molokeev, M.S., Liu, L., Wong, W.-Y., and Zhou, Z., *Chem. Eng. J.*, 396, 125208, 2020. With permission)

stable in the octahedral environment than in other environments [75]. Figure 3.20(a) displays that $SrMgAl_{10}O_{17}$ host lattice is *P6₃/mmc* space group with a hexagonal structure, only $[AlO_6]$ octahedron is suitable for Mn^{4+} occupying because of the comparable ionic radius between Al^{3+} and Mn^{4+}. Gu et al. successfully achieved adjustable luminescence properties with the redshift of spectra in Mn^{4+}-doped $SrMgAl_{10-y}Ga_yO_{17}$ phosphors by introducing Ga^{3+} ion into the host lattice, as shown in Figure 3.20(b) and (c) [76]. The spectral shape of the samples without Ga^{3+} substitution is obviously distinguished from other samples with Ga^{3+} substituting Al^{3+}, which can be explained that a new luminescent center in Ga^{3+} site is generated in $SrMgAl_{10-y}Ga_yO_{17}$ phosphors after Ga^{3+} replacing Al^{3+}. The calculated consequence of D_q/B is greater than 2.2, suggesting that Mn^{4+} ion possesses a strong crystal field environment when Ga^{3+} substitutes Al^{3+}. It means that the emission originates from $^2E \rightarrow {}^4A_2$ parity-forbidden transition as shown in Figure 3.20(d). The nephelauxetic effect plays the dominant role in the $^2E \rightarrow {}^4A_2$ emission transition of Mn^{4+} in most oxide phosphors. Thus, they concluded that the redshift phenomenon in excitation and emission spectra with the

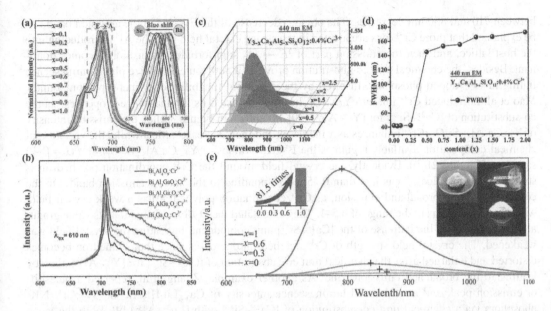

FIGURE 3.21 (a) Normalized photoluminescence emission spectra of the as-prepared $(Ba_{1-x}Sr_x)_2YSbO_6$:Mn^{4+} ($x = 0$–1.0) phosphors. The inset is the enlarged part of emission spectra ranging from 670 to 700 nm. (b) The emission spectra of $Bi_2Ga_{4-x}Al_xO_9$:Cr^{3+} ($x = 0$–4) under 610 nm excitation. (c) The emission spectra and (d) FWHM values of $Y_{3-x}Ca_xAl_{5-x}Si_xO_{12}$:0.4% Cr^{3+} measured at room temperature under 440 nm excitation. (e) The emission spectra of $Ca_{3-x}Lu_xHf_2Al_{2-x}Si_{1-x}O_{12}$:$Cr^{3+}$ ($x = 0$–1) phosphors. (Zhong, J.S., Chen, D.Q., Yuan, S., Liu, M.J., Yuan, Y.J., Zhu, Y.W., Li, X.Y., and Ji, Z.G., *Inorg. Chem.*, 57, 8978, 2018. With permission; Liu, C.Y., Xia, Z.G., Chen, M.Y., Molokeev, M.S., and Liu, Q.L., *Inorg. Chem.*, 54, 1876, 2015. With permission; Mao, M.Q., Zhou, T.L., Zeng, H.T., Wang, L., Huang, F., Tang, X.Y., and Xie, R.J., *J. Mater. Chem. C*, 8, 1981, 2020. With permission; Zhang, L.L., Wang, D.D., Hao, Z.D., Zhang, X., Pan, G.H., Wu, H.J., and Zhang, J.H., *Adv. Opt. Mater.*, 7, 1900185, 2019. With permission)

increase of Ga^{3+} in the host lattice is caused by the nephelauxetic effect. The nephelauxetic effect becomes weakened and the Racah parameter B decreases with increasing Ga^{3+} content. A smaller value of the Racah parameter B results in lower transition energy from the lowest excited state ^2E to the ground state $^4A_2$ level, corresponding to the redshift in the emission spectra of $SrMgAl_{10-y}Ga_yO_{17}$:Mn^{4+} phosphors.

Additionally, a slight blueshift in $(Ba_{1-x}Sr_x)_2YSbO_6$:Mn^{4+} red phosphors via the cationic substitution of larger Sr^{2+} for smaller Ba^{2+} was reported by Zhong et al., as shown in Figure 3.21(a) [77]. The slight blueshift of the emission spectra is primarily derived from the higher crystal field strength of Mn^{4+} ion in Sr_2YSbO_6 host than that in Ba_2YSbO_6 host. If Mn^{4+} is located in a stronger crystal field, the energy of the lowest level will become higher bringing about a larger energy difference between ^2E energy level and $^4A_2$ energy level and then inducing a blueshift of the emission band in $(Ba_{1-x}Sr_x)_2YSbO_6$:Mn^{4+} red phosphors.

3.3.3.3 Cr^{3+}-Doped Solid Solution Phosphors

Cr^{3+} ion is frequently considered to be a significant activator for NIR emission, and its luminescence properties can be more easily controlled in the solid solution compounds by modifying its local crystal field environment. Xia's group achieved the tunable spectra of $Bi_2Ga_{4-x}Al_xO_9$:Cr^{3+} ($x = 0$–4) phosphors by the solid solution strategy [78]. Figure 3.21(b) describes the variation in the emission spectra of Cr^{3+}-doped $Bi_2Ga_{4-x}Al_xO_9$ samples during the chemical composition modification of Al^{3+} replacing Ga^{3+}. Cr^{3+} ions occupy the GaO_6 polyhedron sites, which has a weak crystal field than that in AlO_6 octahedron owing to the smaller volume. Increasing the ratio of Al/Ga would lead to the enhancement of site symmetry around Cr^{3+} ions, and thus the emission intensity of broadband between 710 and 850 nm decreases. The obvious decrease of the FWHM in the emission band can be explained that more Cr^{3+} ions are located in the strong crystal field with the Al^{3+} introducing into the host lattice, and then the emission part of $^2E \rightarrow {}^4A_2$ spin-forbidden transition becomes dominant. Besides, the chemical unit co-substitution provides a new route to realize photoluminescence tuning in Cr^{3+}-doped phosphors. Inspired by the structural tunability of garnet-type compounds, Mao et al. introduced Cr^{3+} ion into $Y_3Al_5O_{12}$ host and adjusted its luminescence properties via the co-substitution of $[Ca^{2+}-Si^{4+}]$ for $[Y^{3+}-Al^{3+}]$ [79]. As shown in Figure 3.21(c), the emission intensity of $Y_{3-x}Ca_xAl_{5-x}Si_xO_{12}:Cr^{3+}$ enhances as x increases and then decreases. Meanwhile, the variation of chemical composition availably regulates the FWHM values of $Y_{3-x}Ca_xAl_{5-x}Si_xO_{12}:Cr^{3+}$ ($x = 0-2$) samples (Figure 3.21(d)). Evidently, the crystal field around the Cr^{3+} coordination polyhedron is strong when the value of x is less than 0.25, corresponding to the narrow emission band. On the contrary, it shows broadband emission, as Cr^{3+} coordination polyhedral has a weak crystal field, when the value of x is in the range of 0.5–1. They calculated the D_q/B and found that its value gradually decreased with the increase of the $[Ca^{2+}-Si^{4+}]$ unit component, indicating the crystal field was weakened. The crystal field strength of Cr^{3+} octahedron decreases since the octahedron becomes distorted and enlarged after the chemical unit co-substitution of $[Ca^{2+}-Si^{4+}]$ for $[Y^{3+}-Al^{3+}]$, thereby leading to the broadened emission band. As another example, Zhang et al. reported a blueshift of emission peak and an enhanced luminescence intensity of $Ca_{3-x}Lu_xHf_2Al_{2-x}Si_{1-x}O_{12}:Cr^{3+}$ NIR phosphors via a chemical unit co-substitution of $[Ca^{2+}-Si^{4+}]$ with $[Lu^{3+}-Al^{3+}]$ [9]. With the value of x increasing from 0 to 1, a tremendous enhancement in emission intensity along with a conspicuous blueshift of the emission peak from 855 to 785 nm occurs in $Ca_{3-x}Lu_xHf_2Al_{2-x}Si_{1-x}O_{12}:Cr^{3+}$ (Figure 3.21(e)). There are two distinct emission centers, including Cr1 with a shorter emission peak and Cr2 with a longer emission peak. Before the variation in the host component, the luminescence from Cr2 is dominant due to the large amount of Cr2. The introduction of $[Lu^{3+}-Al^{3+}]$ in $Ca_3Hf_2Al_2SiO_{12}:Cr^{3+}$ would boost an increase in the amount of Cr1, which brings about a remarkable blueshift in the emission spectra. And the $[Lu^{3+}-Al^{3+}]$ co-substitution for $[Ca^{2+}-Si^{4+}]$ induces the reduction of Cr^{3+} from Cr^{4+}, resulting in an enhanced total amount of the two Cr^{3+} centers and a great enhancement in emission intensity.

3.3.4 Bi^{3+}-Doped Solid Solution Phosphors with 6s–6p Transition

Generally, Bi^{3+}-doped phosphors have strong absorption in the UV and NUV region and their emission peaks are located in the region between UV and red. Solid solution design is a useful approach to adjust the spectra of Bi^{3+}-doped phosphors by affecting the coordination environment of Bi^{3+} since the 6s–6p transition is dependent on the local crystal field environment of the luminescence center [80]. Kang et al. reported the tunable luminescence property of $(Lu_{1-x-y}Sc_xY_y)VO_4:Bi^{3+}$ phosphors through the cation substitution of Sc^{3+} and Y^{3+} for Lu^{3+} [81]. Figure 3.22(a) describes the effect of Sc^{3+} and Y^{3+} concentration on the excitation spectra of Bi^{3+}-doped $(Lu_{1-x-y}Sc_xY_y)VO_4$ phosphors. The tail of the excitation peak of Bi^{3+} continuously moves to a longer wavelength with the increase of Sc^{3+} content, and the excitation band exhibits a blueshift with the replacement of Y^{3+} for Lu^{3+}. The emission peak has a similar variation. As shown in Figure 3.22(b), the larger Y^{3+} ion replacing smaller Lu^{3+} ion results in a blueshift of the emission peak, while the smaller Sc^{3+} introducing into lattice induces a redshift. Furthermore, the phosphor color varies from yellow toward the red region under a 254 nm UV lamp as the radius of lanthanide ions decreases. If the smaller rare-earth ions substitute the larger rare-earth ions in $(Lu_{1-x}Sc_x)VO_4:Bi^{3+}$ samples, the $Bi^{3+}-O^{2-}$ bond length decreases, and then the crystal field strength of Bi^{3+} coordination polyhedron increases, which brings about the emission peak regularly moving toward longer wavelength. Contrary to the introduction of Sc^{3+}, the increase of the Y/Lu ratio would weaken the crystal field owing to the reduction of the distance between Bi^{3+} and O^{2-}, which is the reason for the blueshift of the emission peak in $(Lu_{1-y}Y_y)VO_4:Bi^{3+}$ phosphors.

Similarly, a successively tunable luminescence in Bi^{3+}-doped $Sc(V_xP_{1-x})O_4$ ($x = 0-1$) solid solution phosphors was designed by Kang et al. [82]. As shown in Figure 3.23(a), the substitution of V^{5+} ions

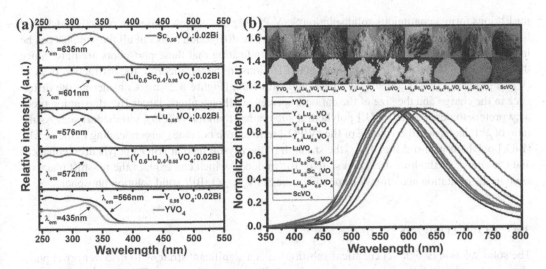

FIGURE 3.22 (a) The photoluminescence excitation spectra of solid solution phosphors $(Lu_{1-x-y}Sc_xY_y)$ VO$_4$:Bi^{3+} (x = 0–1). (b) The normalized emission spectra of $(Lu_{1-x-y}Sc_xY_y)$VO$_4$:Bi^{3+} upon excitation at 265 nm. Inset is the digital photographs of corresponding samples exposed to natural light (upper) and a 254 nm UV lamp. (Kang, F.W., Sun, G.H., Boutinaud, P., Gao, F., Wang, Z.H., Lu, J., Li, Y.Y., and Xiao, S.S., *J. Mater. Chem. C*, 7, 9865, 2019. With permission)

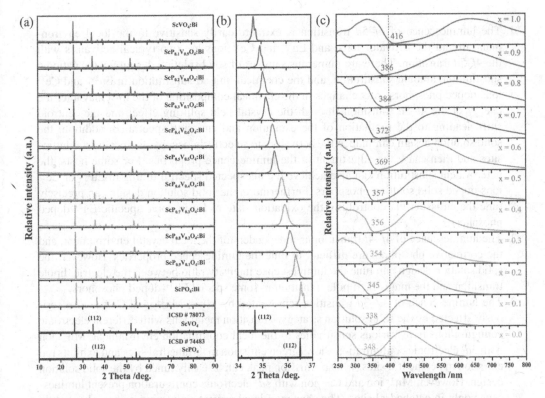

FIGURE 3.23 (a) XRD patterns of the Sc(V$_x$P$_{1-x}$)O$_4$:Bi^{3+} (x = 0–1) solid solution samples and (b) the corresponding enlarged diffraction patterns within the range of 34–37°. (c) The normalized photoluminescence excitation and emission spectra of the Sc(V$_x$P$_{1-x}$)O$_4$:Bi^{3+} solid solution phosphors. (Kang, F.W., Sun, G.H., Boutinaud, P., Gao, F., Wang, Z.H., Lu, J., Li, Y.Y., and Xiao, S.S., *J. Mater. Chem. C*, 7, 9865, 2019. With permission)

for P^{5+} ions form a continuous solid solution $Sc(V_xP_{1-x})O_4:Bi^{3+}$, which causes an obvious redshift of the excitation and emission spectra (Figure 3.23(c)). The diffraction peaks of all the samples can be indexed to the standard card of $ScPO_4$ and $ScVO_4$, indicating that these phosphors are pure. Since the ionic radius of the V^{5+} is larger than that of P^{5+} ions in the tetrahedron, the diffraction peak of (1 1 2) gradually moves toward the lower angle direction (Figure 3.23(b)). Considering the difference in the charge and the size of the cationic species of all phosphors, larger Bi^{3+} dopant ion probably prefers to occupy the $[ScO_8]$ polyhedral sites rather than the $[V/PO_4]$ tetrahedral site. As the ratio of V/P increases, the volume of the $[V/PO_4]$ tetrahedron becomes larger, leading to the nearby $[BiO_8]$ polyhedron shrinking. Thus, the strength of the crystal field at $[BiO_8]$ polyhedron enhances and the energy of the lowest excited state of Bi^{3+} reduces, which can explain the redshift phenomenon in the excitation and emission spectra of the $Sc(V_xP_{1-x})O_4:Bi^{3+}$ solid solution phosphors.

3.4   CONCLUSIONS

The solid solution design via chemical substitution is a significant strategy to discover novel phosphors and modify the luminescence properties of the existing phosphors. The formation of a solid solution would involve the change of local lattice environment and influence the electron transitions of $4f–5d$ transition, $4f–4f$ transition, $3d–3d$ transition, and $6s–6p$ transition in different aspects. In this chapter, based on the different electronic transition modes, we mainly summarize the solid solution design for tuning luminescence and seeking the new hosts in diverse forms of phosphors. The summary is as follows:

1. The luminescence of $4f–5d$ transition is extraordinarily sensitive to the local environment around the activator ions, and Eu^{2+} and Ce^{3+} ions are two typical activators with the $4f–5d$ transition. There are numerous samples of solid solution design via the cationic substitution, anionic substitution, and the chemical unit co-substitution in Eu^{2+}- and Ce^{3+} ions-doped phosphors. Any changes in the chemical composition of the host may alter the energy of the $4f–5d$ transition through the crystal field splitting effect and the centroid shift, leading to the regulation of the excitation and emission spectra. In addition, the solid solution design can induce the activators to selectively occupy distinct crystal lattice sites and then cause the adjustment of the luminescence properties. For some hosts, the heterovalent substitution can also achieve the unexpected spectral tuning in Eu^{2+}- or Ce^{3+} ions-doped solid solution phosphors. Furthermore, the solid solution design can precisely tune the emission via controlling the oxidation state of Eu in some specific Eu^{2+}-doped phosphors.
2. The luminescence of $4f–4f$ transition is independent of the local crystal environment, and the correlative phosphors are difficult to tune the luminescence properties. However, the solid solution design can tune the luminescence intensity ratio between the electric dipole transition and the magnetic dipole transition in some special Eu^{3+}-doped phosphors.
3. The luminescence of $3d–3d$ transition represented by Mn^{2+}, Mn^{4+}, and Cr^{3+} is also seriously affected by the solid solution strategy. Transition metal ions with different electronic configurations have various sensitivities to the local coordination environment. Mn^{2+} ion with $3d^5$ electronic configuration shows green emission in tetrahedral sites and exhibits red emission in octahedral sites, whose emission peak can be easily tuned by the solid solution design. However, Mn^{4+} ion and Cr^{3+} ion with $3d^3$ electronic configuration present luminescence only in octahedral sites. The change of local crystal environment can only slightly adjust the emission peak and regulate the luminescence intensity and FWHM of Mn^{4+}- and Cr^{3+}-doped phosphors. Nevertheless, the solid solution design is still an effective method to optimize the luminescence properties of existing phosphors and to develop new hosts in phosphors referring to the $3d–3d$ transition.

4. The $6s–6p$ transition is also highly sensitive to the local crystal environment around the activator. A number of Bi^{3+}-doped phosphors with various luminescence have been discovered by the solid solution strategy. The change of chemical components is usually accompanied by the variation of the crystal field splitting effect and the centroid shift that contribute to tuning the energy of the $6s–6p$ transition in Bi^{3+}-doped phosphors. Besides, the solid solution design can conduct Bi^{3+} ions selectively occupy the crystallography sites in some special hosts, resulting in the controllable spectral properties. In summary, abundant novel phosphors with excellent luminescence performance will be found via the solid solution strategy in the future.

ACKNOWLEDGMENTS

The project was supported by the National Natural Science Foundations of China (Grant No. 51972118, 51961145101, and 51722202), International Cooperation Project of National Key Research and Development Program of China (2021YFE0105700), Fundamental Research Funds for the Central Universities (FRFTP-18-002C1), Guangzhou Science & Technology Project (202007020005), and the Local Innovative and Research Teams Project of Guangdong Pearl River Talents Program (2017BT01X137).

REFERENCES

1. Chelikowsky JR (1979) Solid solubilities in divalent alloys. *Phys. Rev. B* 19: 686.
2. Hume-Rothery W, and Coles BR (1954) The transition metals and their alloys. *Adv. Phys.* 3: 149.
3. Zhang Y, Zhou YJ, Lin JP, Chen GL, and Liaw PK (2008) Solid-solution phase formation rules for multi-component alloys. *Adv. Eng. Mater.* 10: 534.
4. Park J, Lee SJ, and Kim YJ (2013) Evolution of luminescence of $Sr_{2-y-z}Ca_zSi(O_{1-x}N_x)_4$:$yEu^{2+}$ with N^{3-}, Eu^{2+}, and Ca^{2+} substitutions. *Cryst. Growth Des.* 13: 5204.
5. Wang L, Xie RJ, Li YQ, Wang XJ, Ma C-G, Luo D, Takeda T, Tsai YT, Liu RS, and Hirosaki N (2016) $Ca_{1-x}Li_xAl_{1-x}Si_{1+x}N_3$:$Eu^{2+}$ solid solutions as broadband, color-tunable and thermally robust red phosphors for superior color rendition white light-emitting diodes. *Light Sci. Appl.* 5: e16155.
6. Liu Q, Wang LX, Huang WT, Li XB, Yu MX, and Zhang QT (2017) Red-emitting double perovskite phosphors $Sr_{1-x}Ca_xLaMgSbO_6$:Eu^{3+}: luminescence improvement based on composition modulation. *Ceram. Int.* 43: 16292.
7. Xia ZG, Li Q, and Sun JY (2007) Luminescence properties of $Ba_5SiO_4(F, Cl)_6$:Eu^{2+} phosphor. *Mater. Lett.* 61: 1885.
8. Im WB, George N, Kurzman J, Brinkley S, Mikhailovsky A, Hu J, Chmelka BF, DenBaars SP, and Seshadri R (2011) Efficient and color-tunable oxyfluoride solid solution phosphors for solid-state white lighting. *Adv. Mater.* 23: 2300.
9. Zhang LL, Wang DD, Hao ZD, Zhang X, Pan GH, Wu HJ, and Zhang JH (2019) Cr^{3+}-doped broadband NIR garnet phosphor with enhanced luminescence and its application in NIR spectroscopy. *Adv. Opt. Mater.* 7: 1900185.
10. Xia ZG, and Poeppelmeier KR (2017) Chemistry-inspired adaptable framework structures. *Acc. Chem. Res.* 50: 1222.
11. You SH, Li SX, Jia YC, and Xie RJ (2020) Interstitial site engineering for creating unusual red emission in $La_3Si_6N_{11}$:Ce^{3+}. *Chem. Mater.* 32: 3631.
12. Xia ZG, and Liu QL (2016) Progress in discovery and structural design of color conversion phosphors for LEDs. *Prog. Mater. Sci.* 84: 59.
13. Wang L, Xie RJ, Suehiro T, Takeda T, and Hirosaki N (2018) Down-conversion nitride materials for solid state lighting: recent advances and perspectives. *Chem. Rev.* 118: 1951.
14. Qin X, Liu XW, Huang W, Bettinelli M, and Liu XG (2017) Lanthanide-activated phosphors based on 4f-5d optical transitions: theoretical and experimental aspects. *Chem. Rev.* 117: 4488.
15. Dorenbos P (2003) Energy of the first $4f^7 \rightarrow 4f^65d$ transition of Eu^{2+} in inorganic compounds. *J. Lumin.* 104: 239.
16. Dorenbos P (2002) 5d-level energies of Ce^{3+} and the crystalline environment. IV. Aluminates and "simple" oxides. *J. Lumin.* 99: 283.

17. Li GG, Tian Y, Zhao Y, and Lin J (2015) Recent progress in luminescence tuning of Ce^{3+} and Eu^{2+}-activated phosphors for pc-WLEDs. *Chem. Soc. Rev.* 44: 8688.

18. Ye S, Xiao F, Pan YX, Ma YY, and Zhang QY (2010) Phosphors in phosphor-converted white light-emitting diodes: recent advances in materials, techniques and properties. *Mater. Sci. Eng. R Rep.* 71: 1.

19. Xia ZG, and Meijerink A (2017) Ce^{3+}-doped garnet phosphors: composition modification, luminescence properties and applications. *Chem. Soc. Rev.* 46: 275.

20. Baur WH (1974) The geometry of polyhedral distortions. Predictive relationships for the phosphate group. *Acta Cryst.* 30: 1195.

21. Ronda CR, Justel T, and Nikol H (1998) Rare earth phosphors: fundamentals and applications. *J. Alloys Compd.* 275–277: 669.

22. Shao BQ, Huo JS, and You HP (2019) Prevailing strategies to tune emission color of lanthanide-activated phosphors for WLED applications. *Adv. Optic. Mater.* 7: 1900319.

23. Smet PF, Parmentier AB, and Poelman D (2011) Selecting conversion phosphors for white light-emitting diodes. *J. Electron. Mater.* 158: R37.

24. Li JH, Yan J, Wen DW, Khan WU, Shi JX, Wu MM, Su Q, and Tanner PA (2016) Advanced red phosphors for white light-emitting diodes. *J. Mater. Chem. C* 4: 8611.

25. Lee GH, and Kang S (2011) Solid-solution red phosphors for white LEDs. *J. Lumin.* 131: 2582.

26. Ye S, Wang CH, Liu ZS, Lu J, and Jing XP (2008) Photoluminescence and energy transfer of phosphor series $Ba_{2-z}Sr_zCaMo_{1-y}W_yO_6$:Eu,Li for white light UVLED applications. *Appl. Phys. B* 91: 551.

27. Li YM, Qi S, Li PL, and Wang ZJ (2017) Research progress of Mn doped phosphors. *RSC Adv.* 7: 38318.

28. Zhou Q, Dolgov L, Srivastava AM, Zhou L, Wang ZL, Shi JX, Dramićanin MD, Brik MG, and Wu MM (2018) Mn^{2+} and Mn^{4+} red phosphors: synthesis, luminescence and applications in WLEDs. A review. *J. Mater. Chem. C* 6: 2652.

29. Avram NM, and Brik MG. 2013. *Optical Properties of 3d-Ions in Crystals: Spectroscopy and Crystal Field Analysis.* Springer-Verlag Berlin Heidelberg.

30. Duan CJ, Delsing ACA, and Hintzen HT (2009) Photoluminescence properties of novel red-emitting Mn^{2+}-activated MZnOS (M = Ca, Ba) phosphors. *Chem. Mater.* 21: 1010.

31. Zhang JC, Zhao LZ, Long YZ, Zhang HD, Sun B, Han WP, Yan X, and Wang XS (2015) Color manipulation of intense multiluminescence from $CaZnOS:Mn^{2+}$ by Mn^{2+} concentration effect. *Chem. Mater.* 27: 7481.

32. Tanabe Y, and Sugano S (1954) On the absorption spectra of complex ions. I. *J. Phys. Soc. Jpn.* 9: 753.

33. Tanabe Y, and Sugano S (1954) On the absorption spectra of complex ions II. *J. Phys. Soc. Jpn.* 9: 766.

34. Tanabe Y, and Sugano S (1956) On the absorption spectra of complex ions, III. The calculation of the crystalline filed strength. *J. Phys. Soc. Jpn.* 11: 864.

35. Palumbo DT, and Brown JJ (1970) Electronic states of Mn^{2+}-activated phosphors. I. Green-emitting phosphors. *J. Electrochem. Soc.* 117: 1184.

36. Grinberg M, Lesniewski T, Mahlik S, and Liu RS (2017) $3d^3$ system – comparison of Mn^{4+} and Cr^{3+} in different lattices. *Opt. Mater.* 74: 93.

37. Peng MY, Yin XW, Tanner PA, Brik MG, and Li PF (2015) Site occupancy preference, enhancement mechanism, and thermal resistance of Mn^{4+} red luminescence in $Sr_4Al_{14}O_{25}:Mn^{4+}$ for warm WLEDs. *Chem. Mater.* 27: 2938.

38. Zhou Z, Zhou N, Xia M, Yokoyama M, and Hintzen HT (2016) Research progress and application prospects of transition metal Mn^{4+}-activated luminescent materials. *J. Mater. Chem. C* 4: 9143.

39. Chen D, Zhou Y, and Zhong J (2016) A review on Mn^{4+} activators in solids for warm white light-emitting diodes. *RSC Adv.* 6: 86285.

40. Brik MG, Camardello SJ, and Srivastava AM (2014) Influence of covalency on the Mn^{4+} $^2E_g \rightarrow {}^4A_{2g}$ emission energy in crystals. *ECS J. Solid State Sci. Technol.* 4: R39.

41. Blasse G, and Bril A (1968) Investigations on Bi^{3+}-activated phosphors. *J. Chem. Phys.* 48: 217.

42. Li X, Li PL, Wang ZJ, Liu SM, Bao Q, Meng XY, Qiu KL, Li YB, Li ZQ, and Yang ZP (2017) Color-tunable luminescence properties of Bi^{3+} in $Ca_5(BO_3)_3F$ via changing site occupation and energy transfer. *Chem. Mater.* 29: 8792.

43. Awater RHP, and Dorenbos P (2017) The Bi^{3+} 6s and 6p electron binding energies in relation to the chemical environment of inorganic compounds. *J. Lumin.* 184: 221.

44. Dang PP, Liu DJ, Li GG, Al Kheraif AA, and Lin J (2020) Recent advances in bismuth ion-doped phosphor materials: structure design, tunable photoluminescence properties, and application in white LEDs. *Adv. Opt. Mater.* 8: 1901993.

45. Sun HT, Zhou JJ, and Qiu JR (2014) Recent advances in bismuth activated photonic materials. *Prog. Mater. Sci.* 64: 1.

46. Liu DJ, Yun XH, Dang PP, Lian HZ, Shang MM, Li GG, and Lin J (2020) Yellow/orange-emitting $ABZn_2Ga_2O_7:Bi^{3+}$ (A = Ca, Sr; B = Ba, Sr) phosphors: optical temperature sensing and white light-emitting diode applications. *Chem. Mater.* 32: 3065.

47. Denault KA, Brgoch J, Gaultois MW, Mikhailovsky A, Petry R, Winkler H, DenBaars SP, and Seshadri R (2014) Consequences of optimal bond valence on structural rigidity and improved luminescence properties in $Sr_xBa_{2-x}SiO_4:Eu^{2+}$ orthosilicate phosphors. *Chem. Mater.* 26: 2275.

48. Fang MH, Ni C, Zhang X, Tsai YT, Mahlik S, Lazarowska A, Grinberg M, Sheu HS, Lee JF, Cheng BM, and Liu RS (2016) Enhance color rendering index via full spectrum employing the important key of cyan phosphor. *ACS Appl. Mater. Interfaces* 8: 30677.

49. Dai PP, Li C, Zhang XT, Xu J, Chen X, Wang XL, Jia Y, Wang XJ, and Liu YC (2016) A single Eu^{2+}-activated high-color-rendering oxychloride white-light phosphor for white-light-emitting diodes. *Light Sci. Appl.* 5: e16024.

50. Xia ZG, Ma CG, Molokeev MS, Liu QL, Rickert K, and Poeppelmeier KR (2015) Chemical unit cosubstitution and tuning of photoluminescence in the $Ca_2(Al_{1-x}Mg_x)(Al_{1-x}Si_{1+x})O_7:Eu^{2+}$ phosphor. *J. Am. Chem. Soc.* 137: 12494.

51. Xia ZG, Xu ZH, Chen MY, and Liu QL (2016) Recent developments in the new inorganic solid-state LED phosphors. *Dalton Trans.* 45: 11214.

52. Meijerink A (2018) Emerging substance class with narrow-band blue/green-emitting rare earth phosphors for backlight display application. *Sci. China Mater.* 62: 146.

53. Dutzler D, Seibald M, Baumann D, and Huppertz H (2018) Alkali lithosilicates: renaissance of a reputable substance class with surprising luminescence properties. *Angew. Chem. Int. Ed. Engl.* 57: 13676.

54. Zhao M, Liao HX, Ning LX, Zhang QY, Liu QL, and Xia ZG (2018) Next-generation narrow-band green-emitting $RbLi(Li_3SiO_4)_2:Eu^{2+}$ phosphor for backlight display application. *Adv. Mater.* 30: 1802489.

55. Pust P, Weiler V, Hecht C, Tucks A, Wochnik AS, Henss AK, Wiechert D, Scheu C, Schmidt PJ, and Schnick W (2014) Narrow-band red-emitting $Sr[LiAl_3N_4]:Eu^{2+}$ as a next-generation LED-phosphor material. *Nat. Mater.* 13: 891.

56. Qiao JW, Ning LX, Molokeev MS, Chuang YC, Liu QL, and Xia ZG (2018) Eu^{2+} site preferences in the mixed cation $K_2BaCa(PO_4)_2$ and thermally stable luminescence. *J. Am. Chem. Soc.* 140: 9730.

57. Dong XL, Zhang JH, Zhang LL, Zhang X, Hao ZD, and Luo YS (2014) Yellow-emitting $Sr_9Sc(PO_4)_7:Eu^{2+}$, Mn^{2+} phosphor with energy transfer for potential application in white light-emitting diodes. *Eur. J. Inorg. Chem.* 2014: 870.

58. Qiao JW, Xia ZG, Zhang ZC, Hu BT, and Liu QL (2018) Near UV-pumped yellow-emitting $Sr_9MgLi(PO_4)_7:Eu^{2+}$ phosphor for white-light LEDs. *Sci. China Mater.* 61: 985.

59. Zhang ZC, Ma CG, Gautier R, Molokeev MS, Liu QL, and Xia ZG (2018) Structural confinement toward giant enhancement of red emission in Mn^{2+}-based phosphors. *Adv. Funct. Mater.* 28: 1804150.

60. Zhang Y, Li XJ, Li K, Lian HZ, Shang MM, and Lin J (2015) Crystal-site engineering control for the reduction of Eu^{3+} to Eu^{2+} in $CaYAlO_4$: structure refinement and tunable emission properties. *ACS Appl. Mater. Interfaces* 7: 2715.

61. Chen MY, Xia ZG, Molokeev MS, Wang T, and Liu QL (2017) Tuning of photoluminescence and local structures of substituted cations in $xSr_2Ca(PO_4)_2-(1-x)Ca_{10}Li(PO_4)_7:Eu^{2+}$ phosphors. *Chem. Mater.* 29: 1430.

62. Cheng C, Ning LX, Ke XX, Molokeev MS, Wang ZL, Zhou GJ, Chuang YC, and Xia ZG (2019) Designing high-performance LED phosphors by controlling the phase stability via a heterovalent substitution strategy. *Adv. Opt. Mater.* 8: 1901608.

63. Zhao M, Xia ZG, Huang XX, Ning LX, Gautier R, Molokeev MS, Zhou YY, Chun CY, Zhang QY, Liu QL, and Poeppelmeier KR (2019) Li substituent tuning of LED phosphors with enhanced efficiency, tunable photoluminescence, and improved thermal stability. *Sci. Adv.* 5: eaav0363.

64. Ji XU, Zhang JL, Li Y, Liao SZ, Zhang XG, Yang ZY, Wang ZL, Qiu ZX, Zhou WL, Yu LP, and Lian SX (2018) Improving quantum efficiency and thermal stability in blue-emitting $Ba_{2-x}Sr_xSiO_4:Ce^{3+}$ phosphor via solid solution. *Chem. Mater.* 30: 5137.

65. Liu YF, Zhang X, Hao ZD, Wang XJ, and Zhang JH (2011) Generation of broadband emission by incorporating N^{3-} into $Ca_3Sc_2Si_3O_{12}:Ce^{3+}$ garnet for high rendering white LEDs. *J. Mater. Chem.* 21: 6354.

66. Shang MM, Fan J, Lian HZ, Zhang Y, Geng DL, and Lin J (2014) A double substitution of Mg^{2+}-Si^{4+}/Ge^{4+} for $Al_{(1)}^{3+}$-$Al_{(2)}^{3+}$ in Ce^{3+}-doped garnet phosphor for white LEDs. *Inorg. Chem.* 53: 7748.

67. Xia ZG, Molokeev MS, Im WB, Unithrattil S, and Liu QL (2015) Crystal structure and photoluminescence evolution of $La_5(Si_{2+x}B_{1-x})(O_{13-x}N_x):Ce^{3+}$ solid solution phosphors. *J. Phys. Chem. C* 119: 9488.

68. Hu T, Molokeev MS, Xia ZG, and Zhang QY (2019) Aliovalent substitution toward reinforced structural rigidity in Ce^{3+}-doped garnet phosphors featuring improved performance. *J. Mater. Chem. C* 7: 14594.

69. Li S, Wei XT, Deng KM, Tian XN, Qin YG, Chen YH, and Yin M (2013) A new red-emitting phosphor of Eu^{3+}-doped $Sr_2MgMo_xW_{1-x}O_6$ for solid state lighting. *Curr. Appl. Phys.* 13: 1288.

70. Liu Q, Wang LX, Huang WT, Zhang L, Yu MX, and Zhang QT (2017) Enhanced luminescence properties of double perovskite (Ba, Sr)LaMgSbO$_6$:Eu^{3+} phosphors based on composition modulation. *J. Alloys Compd.* 717: 156.

71. Liang YJ, Noh HM, Ran WG, Park SH, Choi BC, Jeong JH, and Kim KH (2017) The design and synthesis of new double perovskite (Na,Li)YMg(W,Mo)O$_6$: Eu^{3+} red phosphors for white light-emitting diodes. *J. Alloys Compd.* 716: 56.

72. Ma YY, Hu JQ, Song EH, Ye S, and Zhang QY (2015) Regulation of red to near-infrared emission in Mn^{2+} single doped magnesium zinc phosphate solid-solution phosphors by modification of the crystal field. *J. Mater. Chem. C* 3: 12443.

73. Cheng JG, Li PL, Wang ZJ, Li ZL, Tian MM, Wang C, and Yang ZP (2018) Color selective manipulation in $Li_2ZnGe_3O_8$:Mn^{2+} by multiple-cation substitution on different crystal-sites. *Dalton Trans.* 47: 4293.

74. Hu JQ, Song EH, Zhou YY, Zhang SL, Ye S, Xia ZG, and Zhang QY (2019) Non-stoichiometric defect-controlled reduction toward mixed-valence Mn-doped hexaaluminates and their optical applications. *J. Mater. Chem. C* 7: 5716.

75. Zhu HM, Lin CC, Luo WQ, Shu ST, Liu ZG, Liu YS, Kong JT, Ma E, Cao YG, Liu RS, and Chen XY (2014) Highly efficient non-rare-earth red emitting phosphor for warm white light-emitting diodes. *Nat. Commun.* 5: 1.

76. Gu SM, Xia M, Zhou C, Kong ZH, Molokeev MS, Liu L, Wong W-Y, and Zhou Z (2020) Red shift properties, crystal field theory and nephelauxetic effect on Mn^{4+}-doped $SrMgAl_{10-y}Ga_yO_{17}$ red phosphor for plant growth LED light. *Chem. Eng. J.* 396: 125208.

77. Zhong JS, Chen DQ, Yuan S, Liu MJ, Yuan YJ, Zhu YW, Li XY, and Ji ZG (2018) Tunable optical properties and enhanced thermal quenching of non-rare-earth double-perovskite $(Ba_{1-x}Sr_x)_2YSbO_6$:Mn^{4+} red phosphors based on composition modulation. *Inorg. Chem.* 57: 8978.

78. Liu CY, Xia ZG, Chen MY, Molokeev MS, and Liu QL (2015) Near-infrared luminescence and color tunable chromophores based on Cr^{3+}-doped mullite-type $Bi_2(Ga,Al)_4O_9$ solid solutions. *Inorg. Chem.* 54: 1876.

79. Mao MQ, Zhou TL, Zeng HT, Wang L, Huang F, Tang XY, and Xie R-J (2020) Broadband near-infrared (NIR) emission realized by the crystal-field engineering of $Y_{3-x}Ca_xAl_{5-x}Si_xO_{12}$:$Cr^{3+}$ ($x = 0$–2.0) garnet phosphors. *J. Mater. Chem. C* 8: 1981.

80. Dang PP, Liu DJ, Yun XH, Li GG, Huang DY, Lian HZ, Shang MM, and Lin J (2020) Ultra-broadband cyan-to-orange emitting $Ba_{1+x}Sr_{1-x}Ga_4O_8$:Bi^{3+} phosphors: luminescence control and optical temperature sensing. *J. Mater. Chem. C* 8: 1598.

81. Kang FW, Peng MY, Yang XB, Dong GP, Nie GC, Liang WJ, Xu SH, and Qiu JR (2014) Broadly tuning Bi^{3+} emission via crystal field modulation in solid solution compounds (Y,Lu,Sc)VO$_4$:Bi for ultraviolet converted white LEDs. *J. Mater. Chem. C* 2: 6068.

82. Kang FW, Sun GH, Boutinaud P, Gao F, Wang ZH, Lu J, Li YY, and Xiao SS (2019) Tuning the Bi^{3+}-photoemission color over the entire visible region by manipulating secondary cations modulation in the $ScV_xP_{1-x}O_4$:Bi^{3+} ($0 \leq x \leq 1$) solid solution. *J. Mater. Chem. C* 7: 9865.

4 Local Structure and Luminescence Tuning in Phosphors

Zhen Song and Quanlin Liu

CONTENTS

DOI: 10.1201/9781003098669-4

4.1   INTRODUCTION

Phosphors play an important role in many fields of modern society, such as solid-state lighting, liquid crystal display (LCD), scintillator, counterfeit-proofing and biomedical imaging. Its function mainly relies on the ability of light conversion from short to long wavelength, originating from the transitions between ground and excited energy levels of the luminescent centers embedded in the phosphor host. As a result, the central aim of phosphor research is to tune the luminescence via modulating those energy levels, on which crystal structure of the host has significant effects. Generally, luminescence is determined by the crystal structure and local structure of phosphor. Commonly used structure characterization technique such as X-ray diffraction (XRD), however, could only detect average structure, from which the local structure around the luminescence center may deviate. Identification of local bonding could be performed by a series of techniques, such as XRD combined with spectroscopic analysis, extended X-ray absorption fine structure spectroscopy [1], solid-state nuclear magnetic resonance [2], X-ray pair distribution function analysis [3] and transmission electron microscopy (TEM). Unlike band-to-band optical transitions, electronic transitions within the isolated luminescent activators are strongly influenced by the coordinated crystalline environment. Under the local structure symmetry, energy levels of central ions may be split and shifted, leading to peak shifts in both photoluminescence (PL) and PL excitation (PLE) spectra. Therefore, it is necessary and important to build connections between local structure and luminescence.

This chapter is devoted to review the local structure and luminescence tuning in phosphors. We first introduce the basic theory and concepts of local structure and luminescence from d–f transitions of lanthanide ions and then discuss the luminescence tuning caused by local structure regarding some representative phosphors, such as $Y_3Al_5O_{12}$:Ce^{3+} (YAG:Ce), (Ba,Sr,Ca)$_2$SiO$_4$:Eu^{2+} and (Ca,Sr)AlSiN$_3$:Eu^{2+}. We summarize the relationships among the local structure, crystal-field splitting, preferential occupancy, the degree of ordering, split-atom-site and optimal bonding, local structure evolution, remote control of neighboring cation, cation-size-mismatch, chemical substitution/co-substitution and luminescence tuning including thermal quenching properties. We demonstrate how to analyze the local structure mainly on the basis of XRD data and TEM technique and to further understand luminescent properties from the viewpoint of local structure. On the other hand, we also show how to speculate or analyze the local structure from luminescence spectra. We hope the detailed composition-structure-property relationships of these representative phosphors can help the readers to research on luminescent materials.

4.2   BASIC THEORY AND CONCEPTS

4.2.1   F–D TRANSITIONS AND CRYSTAL-FIELD THEORY

In recent years, much attention has been paid to rare earth–doped phosphors employing f–d electronic transitions. The most obvious advantage lies in the intensified oscillator strength of the parity-allowed f–d transitions compared to those of f–f or d–d transitions. Consequently, lanthanide ions Ce^{3+} and Eu^{2+} are widely used as luminescent center dopants. Meanwhile, f-d transitions are sensitive to the subtle change in the ligand environment. Unlike 4f electrons that are screened by outer-shell 5s^{2}5p^6 electrons, 5d electron is directly exposed to crystalline environment. Therefore, the surrounding ligands have significant impact on 5d energy levels, and the diversity in host structure produces various phosphors covering the whole spectral region from ultraviolet to far red.

It is, therefore, necessary to understand the effect of crystalline ligand environment on luminescence tuning.

Crystal-field theory is commonly acknowledged as a successful theory in explaining the relationship between ligand environment and energy levels, which states that the fivefold degenerated energy levels of d electron undergo splitting by Stark effect due to the electrostatic interaction with ligand ions. Its detailed analytical description has been developed for the ideal octahedron under O_h symmetry with six coordination [4, 5]. Due to the larger ionic size, however, rare-earth ions prefer to be accommodated in polyhedron having high coordination number. Although an ideal eight-coordinated cube also possesses the point symmetry of O_h, the splitting and magnitude of energy levels differ from that in octahedron. Suppose the center-ligand length is a for both octahedron and cube, then the crystal-field potentials are

$$V(\text{Octahedron}) = 6\frac{Ze^2}{a} + D\left(x^4 + y^4 + z^4 - \frac{3}{5}r^4\right) \tag{4.1}$$

and

$$V(\text{cube}) = 8\frac{Ze^2}{a} + D\left[\frac{4}{9}\left(x^4 + y^4 + z^4 + 6x^2y^2 + 6x^2z^2 + 6y^2z^2\right) - \frac{4}{5}r^4\right] \tag{4.2}$$

With the one-electron wavefunction $|2,m_l\rangle$ and substitutions of $D = 35Ze^2/4a^5$ and $q = (2/105)\int_0^\infty R_{nl}^2 r^4 r^2 dr = (2/105)\overline{r^4}$, the perturbation matrix elements could be obtained according to $\int \psi^* V \psi d\tau \equiv \langle \psi|V|\psi\rangle$ after a series of angular and radial integrations. The secular matrices are

$$\begin{vmatrix} Dq - E & 0 & 0 & 0 & 5Dq \\ 0 & -4Dq - E & 0 & 0 & 0 \\ 0 & 0 & 6Dq - E & 0 & 0 \\ 0 & 0 & 0 & -4Dq - E & 0 \\ 5Dq & 0 & 0 & 0 & Dq - E \end{vmatrix} \tag{4.3}$$

for octahedron and

$$\begin{vmatrix} -\frac{8}{9}Dq - E & 0 & 0 & 0 & -\frac{40}{9}Dq \\ 0 & \frac{32}{9}Dq - E & 0 & 0 & 0 \\ 0 & 0 & -\frac{16}{3}Dq - E & 0 & 0 \\ 0 & 0 & 0 & \frac{32}{9}Dq - E & 0 \\ -\frac{40}{9}Dq & 0 & 0 & 0 & -\frac{8}{9}Dq - E \end{vmatrix} \tag{4.4}$$

for cube. Finally, the roots are $-4Dq$ (triple), $6Dq$ (double) for octahedron and $-16/3Dq$(double) and $32/9Dq$(triple) for the cube. Although the total crystal-field splitting has shrunk from $90/9Dq$ to $80/9Dq$, it still preserves the reciprocal 5-power relationship. In addition, in the case of ideal cube, two orbitals are positioned in lower energy scale in comparison to the three degenerated orbitals in octahedron. Further symmetry degradation will cause splitting of the two energy levels, and in the excitation spectra of Ce^{3+}-doped phosphors, we could always observe two PLE peaks. It seems that the relationship between structure and luminescence becomes quite clear under the guidance of the reciprocal 5-power law, and in practice, phosphor researchers only need to consider the bond length magnitudes. This statement, however, meets many contradictory cases, and the relationship between local structure and luminescence tuning is complex and far beyond the mere bond length comparison, as will be discussed later.

4.2.2   Average Structure and Local Structure

Luminescent centers, such as transition-metal or rare-earth ions, usually take up a certain type of crystallographic site, i.e., Wyckoff position, in a crystallized inorganic host. Generally, from the viewpoint of average structure, cations or anions occupying a specific Wyckoff position are equivalent to each other, on the condition that they belong to the same chemical element. On the other hand, when it comes to local structure, those ions may become locally inequivalent. Take the dodecahedron in lanthanide aluminum garnets for example [6]. The anionic oxygen atoms are accommodated in only one kind of crystallographic site, i.e., the general position 96 h. Nevertheless, the eight coordinated ligands of the dodecahedral site could be categorized into two groups, as colored in red and blue in Figure 4.1(a). Although they are crystallographic equivalent, ligands 1, 2, 6, 7 share the same bond length, while ligands 3, 4, 5, 8 share the other bond length. The dodecahedron could be thought as a distorted cube, as shown in Figure 4.1(b), with ligand 1-4-7-8 constituting the bottom face, and ligand 2-3-6-5 the top face. Under the point symmetry D_2, some of the edges share the same distances, which are $d_{23} = d_{56} = d_{18} = d_{47}$, $d_{14} = d_{78} = d_{25} = d_{36}$, $d_{21} = d_{67}$ and $d_{58} = d_{34}$. In addition, the equivalent edges have the same connectivity with polyhedra, as illustrated in Figure 4.1(c). This example shows the subtle difference between local and average structures encountered frequently in phosphor host and suggests that it is necessary to focus more on the local structure around the dopant in luminescence tuning analysis.

4.2.2.1   Tolerance Factor of Crystal Structure

The chemical compositions of inorganic phosphors vary from binary (oxide, sulfide and nitride, for example) to ternary, quaternary or even more components. Different compositional combinations provide a huge number of alternatives for novel compounds. However, not all the combinations could result to stable phase. From the viewpoint of local-average structure, this means that the chemical-species–dependent polyhedra have to cooperate with each other to form a rational average structure. This cooperation could be characterized by tolerance factor.

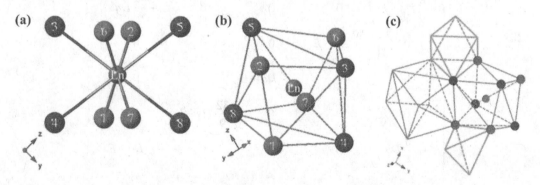

FIGURE 4.1   (a) Coordinated ligands of dodecahedral (A) site. The crystallographic equivalent eight oxygen atoms could be grouped into two categories according to the bond lengths. They are rendered by different colors; (b) Perspective view of the distorted cube of the dodecahedral ligands; (c) Schematic of the adjacent dodecahedron, octahedron and tetrahedron with respect to the oxygen ligands; (Song Z., Xia Z.G. and Liu Q.L., *J. Phys. Chem. C.* 122, 3567-3574, 2018. With permission.)

The concept of tolerance factor was first introduced to characterize the phase stability of perovskite structure ABX_3. The ideal cubic structure could be deconstructed into two kinds of cationic polyhedra, i.e., 12-coordinated cuboctahedron and 6-coordinated octahedron. Goldschmidt took notice of the geometrical relationship between edge lengths of the abovementioned polyhedra and set up the expression of tolerance factor by use of ionic radius R_A and R_B, which is $\tau = (R_A + R_X) / (\sqrt{2}(R_B + R_X))$ [7]. For ideal cubic perovskite structure, τ has the value of 1, but it fluctuates among the range 0.81–1.0 for real distorted perovskite-type structures. Similar methods could be applied to other cubic structures, such as garnet [8], spinel [9] and pyrochlore [10]. In the following part, we will present a detailed description of tolerance factor in garnet structure.

The garnet-type phosphors involve a great number of compositional derivatives [11]. The garnet structure belongs to space group of I a–3 d [12], and it could be thought as constituted by multi-type polyhedra, i.e., dodecahedron (A), octahedron {B} and tetrahedron [C], corresponding to Wyckoff sites 24 c, 16 a and 24 d, respectively; thus, the crystal chemical formula has the form of $(A)_3\{B\}_2[C]_3 <D>_{12}$. The ionic site preference of chemical species is mainly determined by the relative ionic size due to the huge difference in polyhedral sizes. For example, in $(Y)_3\{Al\}_2[Al]_3<O>_{12}$ ($Y_3Al_5O_{12}$, yttrium aluminum garnet, often abbreviated as YAG), dodecahedron has a volume of about 20 Å^3, while volumes of octahedron and tetrahedron only approximate 10 and 3 Å^3, respectively [6]. Generally, large ions are expected to reside in large polyhedral-volume site. Rare-earth and alkaline-earth elements generally occupy (A) sites, while {B} sites mainly accommodate smaller rare-earth and transition metal elements. In [C] sites, multivalence smaller cations have been reported, including V^{5+}/As^{5+}, Si^{4+}/Ge^{4+}, $Al^{3+}/Ga^{3+}/Fe^{3+}$ and Li^+. However, in the garnet structure exists the antisite defect, i.e., large cations could enter small polyhedra, such as Y replacing Al [13, 14] and the preferential occupancy of Ga to tetrahedron than octahedron [15]. In these cases, the electronic configuration may play an important role [15, 16]. For the anion type in <D> site, fluoride garnets also have been reported besides oxide garnets [17]. In addition, nitrogen atoms could partially substitute oxygen atoms in <D> site by joint substitution of $Si^{4+}–N^{3-}$ for $Al^{3+}–O^{2-}$ [18, 19].

A detailed investigation of the polyhedral connectivity of the garnet structure shows a special triangle unit of three dodecahedra combined by sharing edges. Those units are docked with two counterpart faces of octahedron, as shown in Figure 4.2(a). The packing of dodecahedra plays the main role in constituting the garnet structure, while octahedra and tetrahedra connect different parts of dodecahedra. From this point of view, the ionic radius of cations and anions from the chemical compositions has to fulfill the garnet structure compliance. To derive the geometrical relationship between polyhedral edges, the distorted dodecahedron is converted to regular cube, as shown in Figure 4.2(b). After this simplification, the octahedron still connects the interlaced two triangle units. The role of tetrahedron could be visualized in Figure 4.2(c), in which it connects the extension of the upper unit and the lower unit. Consequently, the distances between two neighboring triangle units and the tetrahedral height are selected to serve as the tolerance factor as follows. Let R_A, R_B, R_C and R_D stand for the radius of ions occupying dodecahedron (A), octahedron{B}, tetrahedron [C] and general position <D>, respectively. When it comes to the simplified model, (A) corresponds to the regular cube, and the edge length L has a relationship with the body diagonal length as $\sqrt{3}L = 2R_A + 2R_D$. The distance between the neighboring triangle units is related to L. The upper cap of the octahedron is a regular triangle with edge length L. Its center, {B} ion and <D> ion constitute a right-angled triangle. Therefore, the distance is calculated as $L' = 2\sqrt{(R_B + R_D)^2 - ((\sqrt{3}/3)L)^2}$. Meanwhile, by simple solid geometrical calculations, it is known that the height H of the tetrahedron with edge length L'' is $H = \sqrt{(2/3)}L''$, while $L'' = \sqrt{(8/3)}(R_C + R_D)$. Finally, the tolerance factor, τ, is defined as the ratio between the distance L' and the tetrahedral height H,

$$\tau = \frac{L'}{H} = \frac{2\sqrt{(R_B + R_D)^2 - ((\sqrt{3}/3)L)^2}}{\sqrt{(2/3)} \times \sqrt{(8/3)}(R_C + R_D)} = \frac{3\sqrt{(R_B + R_D)^2 - (4/9)(R_A + R_D)^2}}{2(R_C + R_D)} \quad (4.5)$$

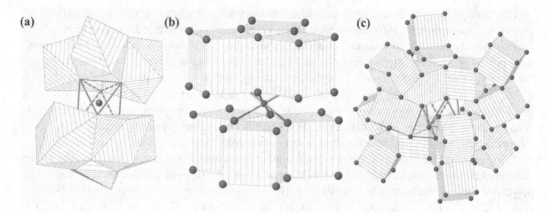

FIGURE 4.2 (a) Connectivity between octahedron/tetrahedron and dodecahedron with sharing edges in the garnet structure. (b) Modeling of the garnet structure. The distorted cubes are resembled by regular ones to demonstrate how the polyhedra are packed and connected. (c) The tetrahedron connects the extension of the upper and lower units. (Song Z., Zhou D.D. and Liu Q.L., *Acta Cryst.* C75, 1353–1358, 2019. With permission.)

From the analytical expression of tolerance factor for the garnet structure, it is clear that the ionic radii of all the chemical species are included.

For real garnet-based compounds, the value of τ is expected to fluctuate around 1, and the numerical limits could be identified by nonexistent garnet phase. More than 130 real or hypothetical garnet-relevant compounds are used to test the validity of the tolerance factor proposed. During the calculation, the Shannon ionic radii [20] are selected in accordance with the coordination number, i.e., 8 for (A), 6 for {B},4 for [C] and 4 for <D>. The results could be found from the original Ref. [8]. The data distribution for τ is then plotted in Figure 4.3, with the non-garnet phases denoted by crossed red circles. It is clear that the tolerance factor proposed is reliable in indicating the garnet phase formation from the given chemical compositions. The τ values of the oxide garnets are scattered in the range 0.748–1.333. Those data points could be coarsely categorized with regard to the chemical composition in [C] site. The data spread range seems related to the valence state of [C], since silicate and germanium garnets have much wider range than that of aluminum, gallium and iron garnets. The possible explanation lies in the element choice under valence constraint. For [C]$^{4+}$, the valence scheme is (A)$^{2+}$+{B}$^{3+}$, whereas for [C]$^{3+}$, only cations with +3 valence state are available for (A) and {B}. Consequently, more compounds exist in [C]$^{4+}$-series garnets. Besides, the τ values for fluoride garnets are distributed in lower positions than those of oxide garnets, which results from the smaller ionic radius of F$^-$ compared to O^{2-}. It is known from the definition that the tolerance factor is defined on the basis of close packing of hard spheres, which assumes larger anion volumetric occupation than that of cation. As a result, the polyhedral shrinkage of {B} is more severe than that of [C], which leads to smaller τ values.

In silicate-aluminum garnets containing alkaline-earth metal elements, $(M)_3\{Al\}_2[Si]_3<O>_{12}$, M = Ca, Sr, Ba, the Ca-garnet have been widely reported [21–25]. It has a τ value of 0.863, whereas Sr- and Ba-garnet have much smaller τ values, 0.690 and 0.391, respectively. The trend stems from the increasing ionic radius, R_A, in the sequence $R_{Ca} < R_{Sr} < R_{Ba}$. This implies that the [Si] tetrahedra are unable to connect the neighboring dodecahedral triangle unit. Therefore, from the viewpoint of tolerance factor, $(Sr)_3\{Al\}_2[Si]_3<O>_{12}$ and $(Ba)_3\{Al\}_2[Si]_3<O>_{12}$ can hardly form the garnet phase. This is confirmed by the failure in synthesizing garnets of those compositions reported by Gentile and Roy [26]. Meanwhile, Novak and Gibbs also gave explanations based on the anomalous features in the structural stability of the hypothetical silicate garnets structures [27]. Nevertheless, Sr-garnets are discovered in germanium garnets, such as $(Sr)_3\{Y\}_2[Ge]_3<O>_{12}$ [28, 29] with $\tau = 1.228$.

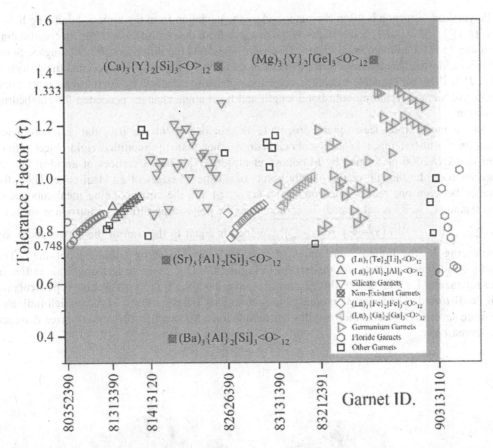

FIGURE 4.3 Tolerance factor (τ) of garnet-type compounds. The crossed red circles denote non-garnet phases. (Song Z., Zhou D.D. and Liu Q.L., *Acta Cryst.* C75, 1353-1358, 2019. With permission.)

Silicate-yttrium garnets containing metal elements with valence state +3 share the formula of $(Ca)_3\{M\}_2[Si]_3<O>_{12}$, in which M stands for metal elements with valence state +3. Transition metal elements from the 4th period contribute most to {M}. On increasing the ionic radius of {M}, $(Ca)_3\{In\}_2[Si]_3<O>_{12}$ is reported to form in the garnet phase ($\tau = 1.285$) [27, 30], but $(Ca)_3\{Y\}_2[Si]_3<O>_{12}$ is crystallized in an orthorhombic compound ($\tau = 1.423$) [31, 32]. On the contrary, when [Si] is replaced by larger [Ge], $(Ca)_3\{Y\}_2[Ge]_3<O>_{12}$ has the garnet phase with a smaller τ value of 1.319. The reason may lie in that larger [Ge] increases the tetrahedral volumetric size to connect the neighboring dodecahedral triangle units.

4.2.2.2 Polyhedral Distortion of Local Structure and Crystal-Field Splitting

For inorganic phosphors, the real coordination polyhedron could be thought as under certain deformation from a perfect model [33], even for those host with cubic structure. In the last section, all the cationic polyhedra in the garnet structure lose O_h point symmetry. For example, the eight-coordinated polyhedron could be considered as distorted from a perfect cube [34]. In lanthanide aluminum perovskite (LnAlO₃, abbreviated as LnAP), it is the 12-coordinated polyhedron deformed from a perfect cuboctahedron [35]. Polyhedron deformation will degrade the local symmetry and further remove energy level degeneracy. Consequently, the two lowest energy 5d levels are often intuitively assigned as sublevels from doublet e_g [36]. In this part, we will discuss the effect of polyhedral deformation on luminescence tuning by the determination of crystal-field splittings of both ideal cuboctahedron and real 12-coordinated polyhedron.

Polyhedron distortion is often characterized by the deviation from the average bond length, i.e., $D = (1/n) \sum_{i=1}^{n} \left(\left(\left| l_i - l_{av} \right| \right) / l_{av} \right)$, in which l_i is the distance from the central atom to the ith coordinating atom and l_{av} is the average bond length. However, it neglects the influences of bond angles. Some cases exist in which the ligands of the polyhedron are moved with equal distance, and the distortion index D will be zero, unable to characterize the polyhedron distortion. Therefore, a more comprehensive method including both bond length and bond angle changes is needed for polyhedron distortion.

Here a more direct least-square procedure is provided to fit the irregular 12-coordinated polyhedron with reference to ideal cuboctahedron other than the modified rigid-object refinement in JANA2006 performed by Mihóková et al. [36, 37]. The 12 vertices of an ideal cuboctahedron could be thought as the middle points of all the 12 edges of an ideal cube. Then, the distance between one real-polyhedron ligand (x_i, y_i, z_i) and the corresponding ideal-cuboctahedron ligand (x_r, y_r, z_r) is calculated. To characterize the polyhedron fitting, we introduce an index $M = \sum_{i=1}^{12} \sqrt{\left[\left(x_i - x_r \right)^2 + \left(y_i - y_r \right)^2 + \left(z_i - z_r \right)^2 \right]}$, which is equal to the sum of the 12 distances. By minimizing M, the ligand coordinates of the ideal cuboctahedron could be obtained. The results of polyhedron fitting are visualized in Figure 4.4. Due to the decreasing ionic radius in the sequence of La, Gd, Y, Lu, the 12-coordinated cuboctahedron in perovskite-type structure will be distorted. The index M increases monotonically from LaAP to LuAP, which indicates the more severe deformation for smaller lanthanide ions. Meanwhile, the center-vertex distance (R) decreases.

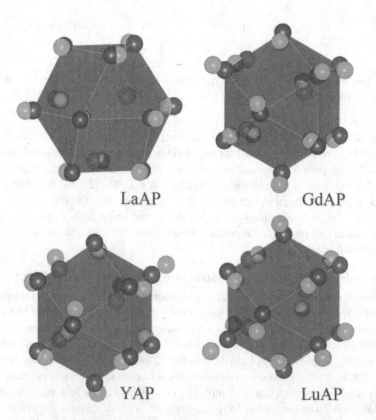

LaAP   GdAP

YAP   LuAP

FIGURE 4.4 The fitted ideal-cuboctahedron (purple) and the real polyhedron (green). (Song Z. and Liu Q.L., *Phys. Chem. Chem. Phys.* 21, 2372–2377, 2019. With permission.)

We perform crystal-field calculations to elucidate the relationship between the crystal-field splitting of Ce^{3+} and its local structure. For YAP, GdAP and LuAP, the local symmetry of the lanthanide sites is C_s, with the crystal-field potential expression for d electrons

$$V(C_s) = B_0^2 C_0^2 + B_2^2 (C_{-2}^2 + C_2^2) + B_2'^2 i (C_{-2}^2 - C_2^2)$$
$$+ B_0^4 C_0^4 + B_2^4 (C_{-2}^4 + C_2^4) + B_2'^4 i (C_{-2}^4 - C_2^4) + B_4^4 (C_{-4}^4 + C_4^4) + B_4'^4 i (C_{-4}^4 - C_4^4) \quad (4.6)$$

whereas the local symmetry D_3 has a crystal-field potential of

$$V(D_3) = B_0^2 C_0^2 + B_0^4 C_0^4 + B_3^4 (C_{-3}^4 - C_3^4) \quad (4.7)$$

in which B_q^k stands for the k-rank crystal-field parameter. Then, the crystal-field matrix element could be calculated as:

$$\langle \alpha SLJM_J | V | \alpha' S'L'J'M_J' \rangle = \sum_{kq} (-1)^{2J-M_J+S+L'+k} \times \begin{pmatrix} J & k & J' \\ -M_J & q & M_J' \end{pmatrix} \begin{Bmatrix} J' & k & J \\ L & S & L' \end{Bmatrix}$$
$$\times [J,J']^{1/2} \langle l \| C^k \| l \rangle \times \langle l^N \alpha SL \| U^k \| l^N \alpha' S'L' \rangle B_{kq} \delta_{SS'} \quad (4.8)$$

where (), {} are 3-j and 6-j symbols [38] and <> represent reduced matrix elements [39]. The effect of spin-orbit (SO) coupling is included with the SO coupling constant 991 cm^{-1} for the 5d electron of Ce^{3+} [40]. Thus, an SO matrix is constructed and added to the crystal-field matrix, with the expression

$$\langle l^N \alpha SLJM_J | H_{SO} | l^N \alpha' S'L'J'M_J' \rangle = \xi_{nl} \sqrt{l(l+1)(2l+1)} (-1)^{L+S'+J}$$
$$\times \begin{Bmatrix} L & S & J \\ S' & L' & 1 \end{Bmatrix} \times \langle l^N \alpha SL \| V^{11} \| l^N \alpha' S'L' \rangle \delta_{JM_J,J'M_J'} \quad (4.9)$$

Furthermore, B_q^k could be calculated by the coordinates of the local polyhedral ligands according to the point charge electrostatic model (PCEM) through

$$B_q^k = \sum_{L=1}^{12} Z e^2 \frac{\langle r^k \rangle}{R_L^{k+1}} \sqrt{\frac{4\pi}{2k+1}} (-1)^q \mathrm{Re} Y_k^q (\theta_L, \varphi_L) \quad (4.10)$$

in which Z is the net charge of the ligand, Y_k^q is the k-rank spherical harmonic function, and $(R_L, \theta_L, \varphi_L)$ are the spherical coordinates of ligand L.

Until this stage, B_q^k is expressed according to the real polyhedron. To investigate the effect of polyhedron deformation on 5d energy levels of Ce^{3+}, it is necessary to know the energy levels of Ce^{3+} in the corresponding artificial structure, i.e., the ideal cuboctahedron. We make the assumption that the ideal cuboctahedron and the real 12-coordinated polyhedron share the same radial integral of atomic wavefunctions. In this way, the perturbation eigenvalues of the ideal cuboctahedron could be obtained directly by diagonalizing the perturbation matrix. Diagonalization of the crystal-field matrix will give eigenfunctions showing the mix of five real basis functions of d electron, i.e., d_{xy}, d_{xz}, d_{yz}, $d_{x^2-y^2}$ and d_{z^2}.

With the geometrical configurations of the real and virtual polyhedra, it is possible to discuss the energy level change caused by deformation from ideal polyhedron by the crystal-field calculations for both ideal cuboctahedron and real 12-coordinated polyhedron as shown in Figure 4.5. Generally, the ideal cuboctahedron has local symmetry O_h, in which the original fivefold degenerated energy

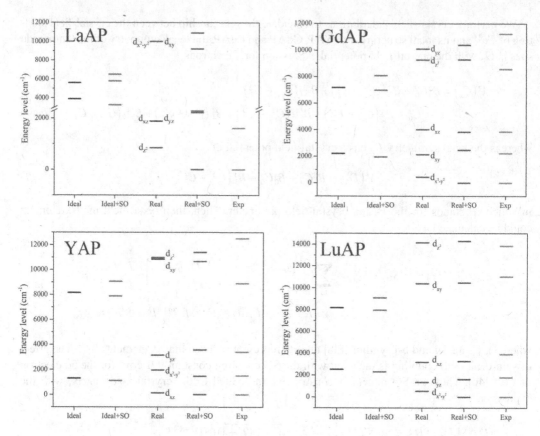

FIGURE 4.5 Energy levels for the 5d energy levels of Ce^{3+} in LnAP. They are subtracted from the lowest level of the corresponding experimental data. The leading orbital term appearing is denoted for each level. (Song Z. and Liu Q.L., *Phys. Chem. Chem. Phys.* 21, 2372–2377, 2019. With permission.)

levels will be split into low-energy doublet e_g and high-energy triplet t_{2g}. Deformation of ideal cuboctahedron will lower the symmetry and remove energy-level degeneration. For LaAP with local symmetry D_3, the reduction scheme from O_h is $e_g \rightarrow E$ and $e_{2g} \rightarrow A_1 + E$, which means one singlet and two doublets. For GdAP, YAP and LuAP, due to the common space group (*Pnma*), they follow almost the same energy level pattern. The local symmetry of the 12-coordinated polyhedron is C_s, under which the fivefold degeneration will be totally removed, and then the SO coupling further shifts those levels slightly.

The leading real orbital is also denoted for every energy level in Figure 4.5. The orbital population is rather versatile, which suggests that it is also dependent on the crystal-structure details. From the results, it is clear that the orbitals are mixed by polyhedral deformation. For ideal cuboctahedron, the low-energy doublet e_g consists of orbitals $d_{x^2-y^2}$ and d_{z^2}. For LaAP, the lowest level is pure d_{z^2}, while the next-lowest one has a leading term of d_{xz} other than $d_{x^2-y^2}$. For GdAP, YAP and LuAP, the lowest level is the combination of d_{xz}, $d_{x^2-y^2}$ and d_{z^2} with different leading terms. Therefore, it can be inferred that, on symmetry degradation ($O_h \rightarrow D_3, C_s$), the sublevels of e_g and t_{2g} get crossed at certain degree of deformation, and thus, the e_g sublevels will take up higher energy level in some cases.

4.2.3 Inductive Effect from Second-Nearest Cation and Charge Transfer Transition

Inductive effect exists widely in inorganic solids. Generally, a chemical bond is formed by a cation and an anion, and the arrangement of the two nuclei and their valence electrons differs from those in the separate atoms. If the other less electronegative cation is connected to the same anion, the

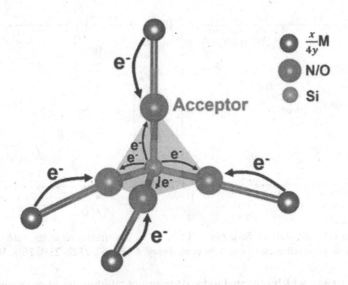

FIGURE 4.6 Schematic diagram of the inductive effect of M. The thickness of arrow represents the degree of ability of donating electron. (Kong Y.W. et al., *Inorg. Chem.* 57, 2320–2331, 2018. With permission.)

anion prefers to seize electron from the cation, as conceptually drawn in Figure 4.6. Consequently, the original bond will be loosened due to the "electron pressure" exerted by the new bond, which is named as the inductive effect by Noll [41] who uses this notation to explain the subtle Si–O bond length changes in silicates. It could be used to account for electric, magnetic and optical properties of inorganic solid states, as well as Fe^{3+}/Fe^{2+} redox potential position [42]. Recently, the inductive effect could be characterized quantitatively from the viewpoints of chemical bond and structural chemistry and used to reveal the effect of local structure on luminescence [43, 44].

The structures of nitridosilicates and oxysilicates usually consist of $[SiN_4]$ or $[SiO_4]$ tetrahedra. We introduce the inductive effect factor $-\mu\Delta\chi$, which characterizes the electronegativity difference between constituent metal elements and silicon. Through statistics over 100 compounds, we set up a linear relationship between average Si–N/Si–O bond length and the inductive factor [43]. It reveals the relevance between chemical composition and the local structure. Besides that, there also exists a positive correlation between the inductive factor and the centroid shift of 5d-levels of Ce^{3+}, as shown in Figure 4.7. With the increase of inductive factor, the anionic N/O atoms tend to seize electron from M instead of Si, leading to extended Si–N/O bonds but shortened M–N bonds. Therefore, it means the $[SiN_4]$ or $[SiO_4]$ tetrahedra framework takes up more space than that of M. After Ce enters the host, the binding between Ce and its ligands is changed due to the inductive effect, i.e., the covalency increases, which increases the centroid shift of 5d-levels in Ce^{3+}.

Charge transfer (CT) transition ($4f^n \leftrightarrow 4f^{n+1}L^{-1}$, L = ligand) occurs with one electron transferred from ligands to the luminescence center. It has important applications in rare earth elements–doped phosphors, such as the CT luminescence in Yb^{3+}-containing compounds [45–47], and the CT sensitized luminescence in Eu^{3+}-doped Y_2O_3 [45]. Since the CT process involves electron redistribution, the change of CT energy E^{CT} could be studied from the view point of inductive effect due to the second-nearest cations (SNCs). We introduce electronegativity factor $\sum_i \chi_i(A_i)/N - \chi(M)$ and ionic radius factor $\sum_i r_i(A_i)/N$ to investigate the inductive effect of the SNCs of the luminescence center on CT process and find positive correlation for the electronegativity factor and negative correlation for ionic radius factor with regard to the CT energy [44]. The results are selectively shown in Figure 4.8. As for $LnPO_4:Eu^{3+}$, Ln site is ninefold coordinated, surrounded by 6 Ln and 7 P when Ln = La and Gd, and eightfold coordinated, surrounded by 4 Ln and 6 P for Ln = Y, Lu and Sc. In Figure 4.8(b), the radius factor shows normal negative relationship with E^{CT}, but the electronegativity

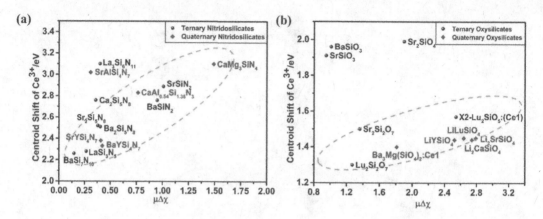

FIGURE 4.7 The centroid shift of 5d-levels of Ce^{3+}, ε_c, vs. the inductive factor $-\mu\Delta\chi$ in (qua)ternary (a) nitridosilcates and (b) oxysilcates. (Kong Y.W. et al., *Inorg. Chem.* 57, 2320–2331, 2018. With permission.)

factor has mixed trends, which is related to local structures. Within the ellipse regions of the same coordination number in Figure 4.8(b)(ii), the normal positive correlation between the electronegativity factor and the CT energy remains. This means local structure plays an important role in CT process, besides the electronegativity and the ionic radius.

4.2.4 Electronic Structure

For most phosphors, the ground and excited states of luminescence dopant are embedded in the forbidden gap, and therefore, electronic structure of the host plays an important role in determining

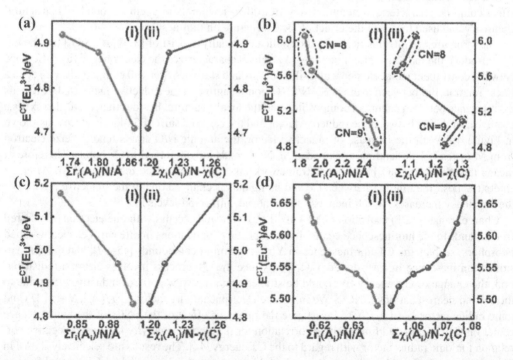

FIGURE 4.8 The CT energy, E^{CT} vs two inductive factor values in some compounds. (a) $Ln_2O_3:Eu^{3+}$ (Ln = Y, Lu, Gd); (b) $LnPO_4:Eu^{3+}$ (Ln = La, Gd, Lu, Y, Sc); (c) $La_{0.95-x}Gd_xPO_4:0.05Eu^{3+}$; (d) $Lu_xY_{1-x}PO_4:Eu^{3+}$; (Kong Y.W. et al., *Inorg. Chem.* 57, 12376–12383, 2018. With permission.)

the luminescent properties. The bandgap energy and relative positions of conduction/valence bands are related to the local structure. For example, the Sn–O–Sn bonds in $Re_2Sn_2O_7$ (Re = Y, La, Lu) are highly distorted from linear geometry in pyrochlore, leading to a relatively narrow conduction band and a wide bandgap [48]. Blasse and Corsmit found from the reflection spectra of A_2BWO_6 compounds that the absorption edge is mainly dependent on A element and weakly influenced by B element [49]. With the increase in the ionic radius of A or B, the absorption edge is continuously shifting to longer wavelength. To elucidate the CT mechanism, the local coordination and the relevant orbitals are redrawn in Figure 4.9(a). The absorption corresponds to an electronic transition from the occupied oxygen 2p orbitals to the empty tungsten t_{2g} orbitals. Since the spectral position of the charge-transfer band depends on the charge and radius of the coordinated cations, it is reasonable that the oxygen valence electron feels weaker field as A or B ion increases the ionic radius, and therefore, less energy is needed to initiate the electron transfer to the highly charged tungsten ion. Consequently, the CT is more sensitive to the four coordinated A ions other than only one coordinated B ion, although B ion is located closer to the tungstate group than A ion.

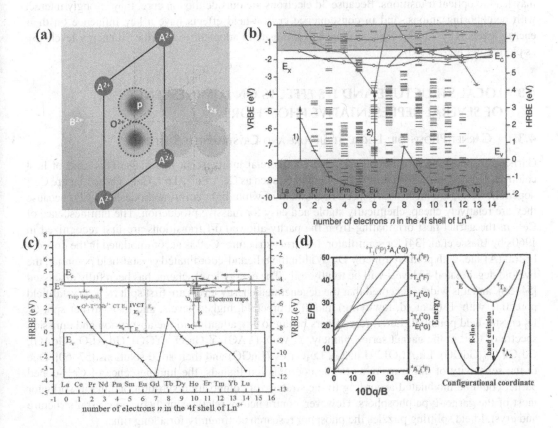

FIGURE 4.9 (a) Local anion coordination and schematic shape of molecular orbitals involved in the charge transfer transition in A_2BWO_6. (Song Z. et al., *Inorg. Chem. Front.*, 6, 2969–3011, 2019. With permission.) (b) VRBE scheme for $LaAlO_3$. Arrow (1) indicates the energy of the CT-band maximum for Ce^{4+} and arrow (2) for Eu^{3+}. (Dorenbos, *Phys. Condens. Matter.*, 25, 225501, 2013. With permission.) (c), HRBE and VRBE schemes of Ln^{3+}-doped $(Ca_{1-x}Na_x)[Ti_{1-x}Nb_x]O_3$. The 4f ground states ($Ln^{3+}$:4f) are labeled by the black inverted triangle and connected by solid curves. (E1: energy of electron trap depth, E2: energy of O2–Ti^{4+}/Nb^{5+} CT, E3: energy of IVCT, E4: electron transition energy from the top of valence band to $^3H_4$, E5: the band energy. (Zhang R.J. et al., *J. Alloy Compd.*, 729, 663–670, 2017. With permission.) (d), Left: Tanabe-Sugano energy level diagram for Cr^{3+} in the octahedral symmetry. Right: Configurational coordinate diagram illustrating different emission channels. The solid and dashed arrows represent emissions from Cr^{3+} ions located in the strong and weak field sites, respectively. (Zhao F.Y. et al., *Inorg. Chem. Front.*, 6, 3621–3628, 2019. With permission.)

For lanthanide, since the 4f shell is shielded from outer 5s and 5p shells, the excitation/emission wavelength from the f-f transition is relatively insensitive to the host. The energy levels of trivalent lanthanide ions in $LaCl_3$ were determined by Dieke et al. [50], which is called Dieke diagram. Dorenbos have systematically studied how lanthanide ion levels change with chemical composition and structure of inorganic compounds [51]. On the basis of the CT model and the chemical shift, the host referred binding energy schemes (HRBE) and vacuum referred binding energy schemes (VRBE) can be constructed. These two schemes are often called Dorenbos diagram, on the basis of which 4f and 5d energy level electronic structures of 14 lanthanide ions can be predicted according to the spectral data of one lanthanide ion (e.g. Eu^{2+} and Eu^{3+}). Furthermore, one can illustrate and predict phosphor properties, such as excitation/emission wavelength, luminescence thermal stability, persistent luminescence properties and scintillation performance. Figure 4.9(b) and (c) shows the electronic structures of lanthanide-doped $LaAlO_3$ and Pr^{3+}-doped $(Ca,Ti)_{1-x}[Na,Nb]_xO_3$ perovskite compounds [52, 53]. Besides, transition metal ions have also played an important role in luminescent materials. They have incomplete 3d shells and, therefore, have low-lying energy levels which may lead to optical transitions. Because 3d electrons are outside the ion core, they strongly interact with neighboring anions, and in consequence crystal-field effects have a key influence on their energy levels. The Tanabe-Sugano diagrams are usually adopted to describe 3d energy levels [5], as plotted in Figure 4.9(d).

4.3   LOCAL STRUCTURE AND ITS EFFECTS ON LUMINESCENCE OF SEVERAL REPRESENTATIVE PHOSPHORS

4.3.1   GARNET PHOSPHOR: LOCAL STRUCTURE AND CRYSTAL-FIELD SPLITTING

Garnet phosphor covers a large variety of commercial luminescent materials. This kind of host could accommodate various luminescent dopants, such as Ce^{3+}, Eu^{3+}, Tb^{3+}, Cr^{3+}. Those that are Ce^{3+} doped, however, are most widely used in the field of white light emitting diodes (WLED) because they are relatively cheap, chemically stable and easy for massive production. The luminescence of Ce^{3+} in the garnet host originating from the parity-allowed d-f transitions are first recognized in 1960s by Blasse et al. [34] for scintillator. In garnet structure, Ce^{3+} is accommodated in the dodeca-hedral (A) site, with point symmetry D_2. Within the 8 ligand-coordinated crystal-field potential, the fivefold degenerated 5d levels will be totally split. The energy level scheme has been fully discussed [54–56], and it is widely accepted that the degenerated energy levels are first split by T_d crystal-field potential, with ^2E (twofold, low-lying) and $^2T_2$ (threefold, higher) levels, and then further split by D_2 crystal-field potential. Later, researchers began to pay attention to the absorption and emission spectra of Ce^{3+} in the garnet series, namely, $Y_3Al_5O_{12}$(YAG), $Y_3Ga_5O_{12}$(YGG), $Gd_3Al_5O_{12}$(GdAG), $Gd_3Ga_5O_{12}$(GdGG), Lu_3Al_5O12 (LuAG), $Lu_3Ga_5O_{12}$(LuGG) and their solid solutions [57–60]. Due to the sensitivity of d–f transitions to the coordination ligands, the luminescence of Ce^{3+}-doped garnet could be modulated according to the strategy of composition substitution, and it works for most of the garnet-type phosphors. However, contradictory relationship between crystal structure and crystal-field splitting puzzles the phosphor research community for a long time.

4.3.1.1   The Reverse Garnet Effect

Composition substitution could effectively alter both average and local structures for the series of garnet-type compounds. The cell parameters, volumes of dodecahedron (A) and mean bond-lengths are consistent in trend with the ionic radius of substituted elements for different compositions. However, the change in crystal-field splitting demonstrates contradictory trends. On one hand, the cell is expanding from YAG to YGG due to the larger Ga, and the crystal-field splitting is decreasing [61], which is in accordance with expectation. On the other hand, the shrinkage of cell is observed from YAG to LuAG, while the crystal-field splitting is surprisingly reduced [62, 63]. Some researchers name this abnormal phenomenon as the reverse garnet effect [33, 64]. What confuses most is that

TABLE 4.1

Crystal Structure Information of Garnet Compounds

Compounds			YAG	YGG	GdAG	GdGG	LuAG	LuGG
Lattice parameter (Å)			12.002	12.280	12.114	12.379	11.906	12.188
Dodeca-hedron (A)	Bond lengths (Å)		2.439	2.428	2.522	2.527	2.383	2.393
			2.306	2.338	2.353	2.354	2.276	2.303
	Mean bond lengths (Å)		2.372	2.383	2.437	2.440	2.330	2.348
	Polyhedron volume (Å³)		22.895	23.052	24.895	25.026	21.621	22.027
ICSD #			41.145	23.852	192.184	192.181	23.846	23.850

Source: Song Z., Xia Z.G. and Liu Q.L., *J. Phys. Chem. C.* 122, 3567–3574, 2018. With permission.

the normal and reverse phenomena could be observed simultaneously in the same compound, just as the case of YAG. This self-contradictory phenomenon puts a great challenge to the reciprocal d^5 relationship between bond lengths and crystal-field splitting.

Composition change of the abovementioned garnets has no effect on the space group. Therefore, the structure change induced by dopant incorporation could be examined by checking lattice parameters and bond lengths. The detailed structural data is listed in Table 4.1. For the (A) site, two kinds of bond lengths exist; therefore, the mean bond lengths are considered. The relationship between composition, lattice parameter, bond length and volume of dodecahedra is displayed in Figure 4.10. It can be easily seen that the trends are in accordance with the ionic radius under eight-coordinated ligands, which is 0.977 Å for Lu^{3+}, 1.019 Å for Y^{3+} and 0.938 Å for Gd^{3+} [20]. The solid lines connecting the whole Al or Ga series show that the cell expands together with the bond length and dodecahedra as the sequence Lu < Y < Gd. Meanwhile, those garnets belonging to the same

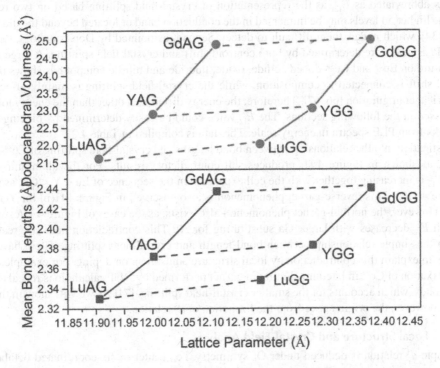

FIGURE 4.10 Mean bond lengths and dodecahedral volumes of garnet compounds as a function of lattice parameter (Song Z., Xia Z.G. and Liu Q.L., *J. Phys. Chem. C.* 122, 3567–3574, 2018. With permission.)

TABLE 4.2

Data of Crystal-Field Splitting of Ce^{3+}-Doped Garnets

	Wavenumber (cm^{-1})		Crystal-field splitting (cm^{-1})	Ref
YAG	29,239	21,739	7500	Bachmann et al. [68]
	29,239	21,929	7310	Chen et al. [69]
			8360	Ogieglo et al. [70]
	29,400	22,000	7400	Blasse et al. [34]
YGG	28,571	23,810	4762	Ueda [71]
GdAG	29,762	21,253	8509	Chen et al. [69]
GdGG	28,735	23,364	5371	Ogieglo et al. [72]
	28,700	23,500	5200	Kaminska et al. [73]
LuAG	28,233	21,138	7100	Ogieglo et al. [70]

Source: Song Z., Xia Z.G. and Liu Q.L., *J. Phys. Chem. C.* 122, 3567-3574, 2018. With permission.

rare-earth element are connected by dashed lines, which also give increasing trend as the composition varies from Al to Ga. The abovementioned analysis shows that it is the ionic size of different elements constituting garnets that determines the bond lengths between (A) site and ligands. In the next part, we will check whether this statement holds for crystal-field splitting.

Researchers are particularly interested in the two lowest 5d levels, because absorption transitions from the 4f ground levels to the two lowest 5d levels result in two PLE (photoluminescent excitation) peaks located in the visible-blue range. We treat the energy difference between those two peaks, which is abbreviated as E_{12}, as the representation of crystal-field splitting based on two reasons. First, the higher 5d levels may be immersed in the conduction band or located beyond the ultraviolet region [34], which makes them difficult to detect. Second, as pointed by Dorenbos [65], the location of PLE peaks are determined by both centroid shift and crystal-field splitting. A large survey of literature on Eu^{2+} and Ce^{3+}-doped sulfide, oxide, fluoride and nitride compounds shows that the centroid shift is connected to composition, while the crystal-field splitting is mainly affected by geometrical configuration [66, 67]. Therefore, the energy difference, other than the energy location, is discussed in the following sections. The E_{12} value could be easily determined by picking up the two peaks from PLE spectra in energy scale. The data is compiled in Table 4.2.

Investigation into the relationship between bond lengths and crystal-field splitting of Ce^{3+}-doped garnets, as shown in Figure 4.11, produces self-contradictory results. For the aluminum-garnet series, E_{12} is increasing together with the cell expansion in the sequence of Lu < Y < Gd, as shown in Figure 4.11(a). This reverse-garnet phenomenon is also observed in Figure 4.11(b) for YGG and GdGG. However, the normal-garnet phenomenon also exists, as the cases of Figure 4.11(c) and (d), in which E_{12} decreases with larger Ga substituting for Al. This contradiction puts up great challenge to the simple relationship between bond length and crystal-field splitting. There have been attempts to explain this contradiction by local structure adjustment on doping. For example, structural relaxation of Ce^{3+} in lutetium aluminum garnet performed by DFT calculations reveals longer Ce–O bonds, which accounts for the smaller crystal-field splitting. [74] As we will show in the next section, however, the origin arises from the limitation of d^5 relation.

4.3.1.2   Local Structure and Crystal-Field Analysis

The simple d^5 relation is deduced under O$_h$ symmetry, i.e., center of six-coordinated octahedron. The symmetry elements in O$_h$ point group will eliminate all the 2nd-rank crystal-field parameters in the crystal-field potential.[75, 76] As a result, for d electron the O$_h$ crystal-field potential could be

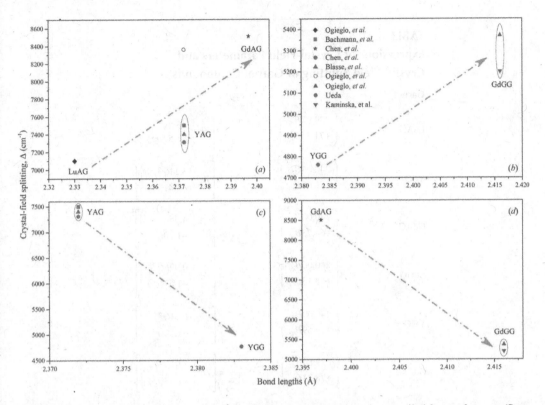

FIGURE 4.11 Relationship between crystal-field splitting and bond lengths compiled from references (Song Z., Xia Z.G. and Liu Q.L., *J. Phys. Chem. C.* 122, 3567–3574, 2018. With permission.)

expressed using only one 4th-rank crystal-field parameter, i.e., B_{40}. If expressed by the point-charge electrostatic model (PCEM), B_{40} has the form of d^5 in the denominator. However, when the site symmetry degrades, the symmetry limitations on the 2nd crystal-field parameters are broken, and they become nonzero, which will introduce d^3 together with d^5. As mentioned earlier, for garnet structure, the dopant Ce^{3+} is accommodated in the eight-coordinated (A) site, with point symmetry D_2. The crystal-field potential is expressed as [75]

$$V(D_2) = \left[B_{20}U_0^2 + B_{22}\left(U_2^2 + U_{-2}^2\right) \right]\langle l\|C^2\|l\rangle$$
$$+ \left[B_{40}U_0^4 + B_{42}\left(U_2^4 + U_{-2}^4\right) + B_{44}\left(U_4^4 + U_{-4}^4\right) \right]\langle l\|C^4C^4\|l\rangle \tag{4.11}$$

Therefore, it is expected that the 2nd-rank crystal-field parameters will be included in the analytical expression of split energy levels. In this case, the simple d^5 relationship takes no more effect.

Crystal-field analysis could be performed to calculate the energy level splitting of Ce^{3+}-doped garnets on the basis of local structure data. [61] In brief, the perturbation matrix is constructed according to the D_2 crystal-field potential, and then the eigenvalues are obtained by matrix diagonalization. The important point lies in how to treat crystal-field potential and SO coupling. Unlike transition-metal elements, for rare-earth elements the SO coupling is significant. If SO coupling is treated prior to crystal-field potential, the energy levels of $5d^1$ electron are first split into $^2D_{5/2}$ and $^2D_{3/2}$ terms, which will be further split under the crystal-field potential. However, this treatment is inconsistent with the real case, because the energy difference between the two ^2D terms is $5\xi/2$ (ξ, the SO coupling constant, has a value of 991 cm^{-1} for 5d electron of Ce^{3+}), [40] much smaller than the energy level separation. If the energy levels are first split by crystal-field potential and after

TABLE 4.3

Expressions of Crystal-Field Parameters and Crystal-Field Splitting of Garnet Compounds

Garnet	$(E_{12})^2$	
LuAG	$\approx \left(\begin{array}{c} 239.97B_{20} \\ +71.34B_{40} \end{array} \right)^2$	$= \left[Ze^2 \left(\begin{array}{c} 2.64\langle r^2 \rangle \\ -1.57\langle r^4 \rangle \end{array} \right) \right]^2$
YAG	$\approx \left(\begin{array}{c} 240.04B_{20} \\ +70.68B_{40} \end{array} \right)^2$	$= \left[Ze^2 \left(\begin{array}{c} 3.11\langle r^2 \rangle \\ -1.41\langle r^4 \rangle \end{array} \right) \right]^2$
GdAG	$\approx \left(\begin{array}{c} 240.00B_{20} \\ +64.73B_{40} \end{array} \right)^2$	$= \left[Ze^2 \left(\begin{array}{c} 4.38\langle r^2 \rangle \\ -1.20\langle r^4 \rangle \end{array} \right) \right]^2$
LuGG	$\approx \left(\begin{array}{c} 240.00B_{20} \\ +78.13B_{40} \end{array} \right)^2$	$= \left[Ze^2 \left(\begin{array}{c} 0.996\langle r^2 \rangle \\ -1.67\langle r^4 \rangle \end{array} \right) \right]^2$
YGG	$\approx \left(\begin{array}{c} 240.00B_{20} \\ +74.59B_{40} \end{array} \right)^2$	$= \left[Ze^2 \left(\begin{array}{c} 1.24\langle r^2 \rangle \\ -1.50\langle r^4 \rangle \end{array} \right) \right]^2$
GdGG	$\approx \left(\begin{array}{c} 239.86B_{20} \\ +78.23B_{40} \end{array} \right)^2$	$= \left[Ze^2 \left(\begin{array}{c} 3.01\langle r^2 \rangle \\ -1.33\langle r^4 \rangle \end{array} \right) \right]^2$

Source: Song Z., Xia Z.G. and Liu Q.L., *J. Phys. Chem. C.* 122, 3567-3574, 2018. With permission.

that SO coupling is considered, the final split energy will be only related to ξ. The SO coupling and crystal-field potential should be considered simultaneously, but this treatment will lead to solving a 5-power secular equation during the diagonalization process, which seldom gives analytical results. A more realistic routine is to assume cubic field $\gg$ the SO interaction $\gg$ D2 crystal-field, [56] and by this method, we have shown the decrease of crystal-field splitting from YAG to YGG. [61] Interpretation of these results in Table 4.3 provide convincing explanation for the abovementioned self-contradictory phenomena. For the aluminum and gallium series, it is clear that E_{12} increases as Lu < Y < Gd. For LnAG/LnGG paris, i.e., LuAG/LuGG, YAG/YGG and GdAG/GdGG, E_{12} drops as Ga substitutes for Al. Those results are in accordance with Figure 4.11 and the contradiction disappears. It should be noted that the results of crystal-field contributions of every ligand depend on the axis choice. Under different coordinate systems, the values may change, but the overall result remains the same due to physical equality. The normal and reverse phenomena on crystal-field splitting phenomena of garnet structures could be understood from the abovementioned treatment. [6]

However, it should be noted that PCEM is a coarse model, because the ligands are regarded as negative point-charges in this model, which deviates from the real case. Moreover, the bond lengths are extracted from pure garnet structures without dopant Ce^{3+}, and it is known that incorporation of larger Ce^{3+} will cause cell expansion, which means the real bond lengths should be larger than those used in this work. Nevertheless, PCEM could be regarded as the first approximation, and it could be applied to explain the crystal-field splitting in Ce^{3+}-doped garnets.

It is known from the preceding section that both incorporation of larger size lanthanides and substituting Al for Ga have the effect of expanding the cell and dodecahedron volume. However, subtle differences exist between how the oxygen ligands respond to different site substitutions. A detailed study of the dodecahedral ligand coordinates of garnet compounds reveals that the ligand movement has two distinct patterns with regards to site substitutions. In Figure 4.12(a) and (b), the evolvements from LuAG to YAG and LuAG to LuGG are selected to elucidate the two ligand movement patterns. The arrows indicate the ligand movement between different garnet compounds. For the (A) site substitution, the ligand movement shown in Figure 4.12(a) indicates a much larger elongation of distance between O2 and O3 than that between O1 and O2. This pattern is in accordance with the dodecahedron inflation from LuAG to YAG, considering the fact that edge 2-1 is shared by tetrahedron, while edge 2-3 is shared by another dodecahedron (Figure 4.1(b)). Therefore, d_{23}/d_{21} is increasing through Lu-Y-Gd among Al and Ga garnet series. However, for the {B} and [C] site substitutions, which are referred to Ga substitution for Al from LuAG to LuGG in Figure 4.12(a), they share a quite different mode. The ligand movement pattern indicates a further separation between O1 and O2, but an approach between O2 and O3. This means inflation of octahedron and tetrahedron simultaneously squeezes dodecahedron. As a result, we expect a shorter d_{23} and longer d_{12}, then finally a smaller d_{23}/d_{21}. From this point of view, it is known that the ligand oxygen atoms respond in different ways regarding to different site substitutions. Although the whole effect is the same, i.e., larger cell and longer bond lengths, the two ligand-movement patterns lead to the reverse trend of crystal-field splitting and cube distortions.

The key point to understand the reverse garnet phenomenon lies in changes of local structure, i.e., the dodecahedral deformation on multi-site substitutions. In the case of an isolated dodecahedron, isotropic expansion or shrinkage is expected on lanthanide substitution. However, in the garnet structure, polyhedra are linked by shared edges, and polyhedron deformations induced by compositional substitution behave in different manners with or without neighboring polyhedra. It is observed that the dodecahedral deformation is far beyond isotropic. The displacements of dodecahedron ligands exhibit distinct moving patterns depending on the site-substitution type. [64] Therefore, it is believed that competitions exist between dodecahedron and neighboring polyhedra. The dodecahedral deformation caused by lanthanide substitution is expected to cause suppression by the octahedron/tetrahedron grid. Meanwhile, the octahedron/tetrahedron deformation caused by Ga/Al substitution will also induce dodecahedral deformation, which further tunes the Ce^{3+} luminescence. Due to polyhedron competitions, the application of the simple negative relationship

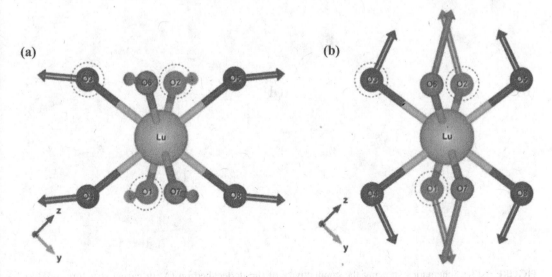

FIGURE 4.12 (a) Ligand movement pattern for LuAG to YAG; (b) Ligand movement pattern for LuAG to LuGG. The arrows indicate the moving direction of ligands when the garnet compound change. The magnitude of the arrow has been magnified 30 times for a better visualization. Plotted by VESTA[77] (Song Z., Xia Z.G. and Liu Q.L., *J. Phys. Chem. C.* 122, 3567–3574, 2018. With permission.)

between crystal-field splitting and polyhedron size is strongly limited in garnet structure. This may be the origin of the reverse garnet phenomenon.

Basically, the main character of dodecahedral deformation belongs to tetragonal distortion. Wu et al. [64] reported a simple relationship between excitation properties of Ce^{3+} in garnets and the polyhedron edge ratio. With an increase in the edge ratio, an approximately linear increase of the excitation maxima exists. The polyhedron edge-ratio could be regarded as an indicator of tetragonal distortion. Seijo and Barandiarán perform quantum chemistry calculations of Ce^{3+}-doped garnets based on AIMPs (Ab Initio Model Potentials). [78] They establish the eight-coordinated distorted cube by transforming a reference perfect cube through a series of SBT (stretching, bending and twisting) operations. Only the symmetric stretching (S1) and symmetric bending (S3) modes have important impacts on the lowest 4f-5d transition. In addition, the symmetric bending (S3) also leads to tetragonal distortion of a perfect cube, which induces elongation or compression of the cube, resulting in a cuboid. We will show that the relationship between crystal-field splitting and the cuboid geometry is a key factor to understand the anomalous reverse garnet effect.

The dodecahedron could be thought as distorted cube, as shown in Figure 4.13(a). The two caps, which are composed of ligand 2-3-6-5 and 1-4-7-8, are connected by two tetrahedrons, as shown in Figure 4.13(b). The two height edges d_{12} and d_{67} are shared between dodecahedron and tetrahedron, while the other two height edges d_{58} and d_{34} have no sharing counterparts. The two caps have the same polyhedron connectivity. Take the upper cap for example. Two edges d_{23} and d_{56} are

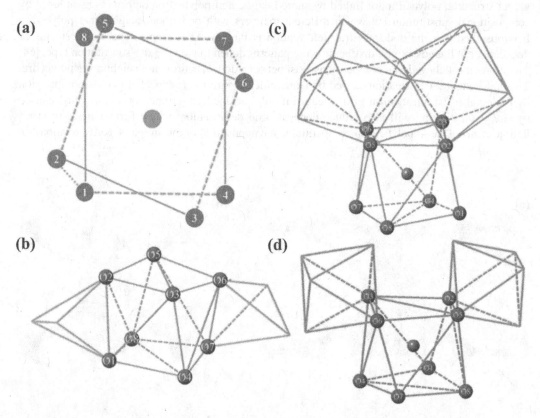

FIGURE 4.13 Schematics showing the connectivity of the dodecahedron (A) in garnet structure. (a) Top view of the distorted cube. Solid lines indicate shared edges with another dodecahedron, dashed lines octahedron and dotted lines tetrahedron. (b) Two height edges shared by tetrahedrons; (c) Two cap edges shared by two adjacent dodecahedra; (d) The other two cap edges shared by octahedra. (Song Z. and Liu Q.L., *J. Phys. Chem. C.* 123, 8656–8662, 2019. With permission.)

shared by another two adjacent dodecahedra (Figure 4.13(c)), while d_{25} and d_{36} by two octahedra (Figure 4.13(d)).

The three polyhedra have a general trend of volume expansion and edge lengthening on larger cations' substitution (Figure 4.14). However, due to the neighboring polyhedron competition, the edges may behave in different ways to compromise the volumetric change of all the polyhedra. For example, the d_{23} edge is shortened when Ga is substituting. Wu et al. ascribe the polyhedron deformation to the length ratio between the edges shared by an adjacent dodecahedron and tetrahedron. In this case, it is d_{23}/d_{21}, as shown in Figure 4.15. [64] It is obvious that changes of d_{23}/d_{21}, d_{36}/d_{21} and the lowest 5d level almost have the same pattern among the garnet series. In addition to the report by Wu et al. [64] it is understood that a significant correlation exists between polyhedron deformation and the lowest 5d level of Ce^{3+} in garnet compounds. In the following part, we establish a simplified cube deformation model to find the explanation for the correlation. [79]

The coordination environment of the dodecahedral sites could be evolved from a perfect cube by a series of SBT operations, as pointed out by Seijo and Barandiarán. [78] Among those operations, tetragonal distortion corresponds to elongation or compression along edge direction. Therefore, in Figure 4.16, we set up a simplified cuboid model to characterize the deformation from perfect cube. The Cartesian coordinates vertices are (a, a, b), $(a, -a, b)$, $(-a, a, b)$, $(-a, -a, b)$, $(a, a, -b)$, $(a, -a, -b)$, $(-a, a, -b)$ and $(-a, -a, -b)$, with Ce^{3+} accommodated in the origin $(0, 0, 0)$. The case of $a = b$ stands for a perfect cube, while $a/b < 1$ for compression and $a/b > 1$ for elongation of the cube. In the following section, we will focus on the effect of crystal-field potential on one d electron ($l = 2$) embedded in the cuboid.

It is widely known that the fivefold degenerated energy levels for d electron will be split into e_g(doublet) and t_{2g}(triplet) levels in the crystal-field of a perfect octahedron, with e_g levels higher in energy. When it comes to a perfect cube, the split results remain the same but with e_g levels lower in energy. The cuboid model for tetragonal distortion will further degrade the point symmetry from

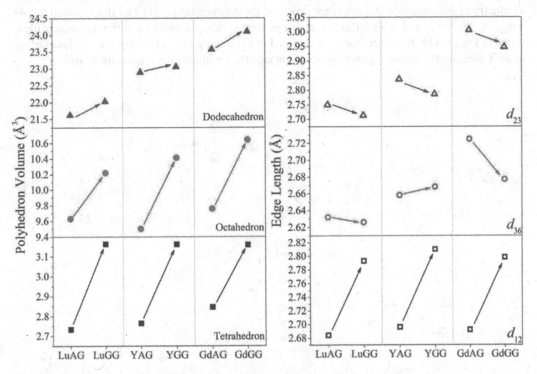

FIGURE 4.14 Volume and length change of polyhedra and sharing edges in a series of garnets. (Song Z. and Liu Q.L., *J. Phys. Chem. C.* 123, 8656–8662, 2019. With permission.)

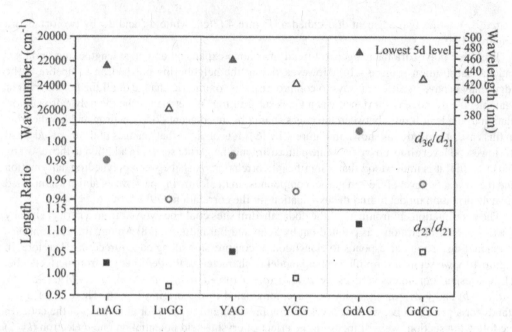

FIGURE 4.15 Correspondence between the lowest 5d level of Ce^{3+} and polyhedron distortion. (Song Z. and Liu Q.L., *J. Phys. Chem. C.* 123, 8656–8662, 2019. With permission.)

O_h to D_{4h}. According to group theory, the subduction scheme is $e_g \to a_{1g} \oplus b_{1g}$ and $t_{2g} \to b_{2g} \oplus e_g$, and therefore, it is known that under D_{4h}, the split levels will be three onefold and one twofold degenerated levels. In addition, if the local symmetry continues to descend to D_2, which is the real point symmetry of dodecahedral site in garnet structure, the degeneration will be totally removed ($a_{1g} \to a$, $b_{1g} \to a$, $b_{2g} \to b_1$ and $e_g \to b_2 \oplus b_3$). The energy level splitting scheme for different cases is illustrated in Figure 4.16. However, it should be noted that the energy levels are not arranged according to their magnitudes, because group theory is not capable for quantitative determinations.

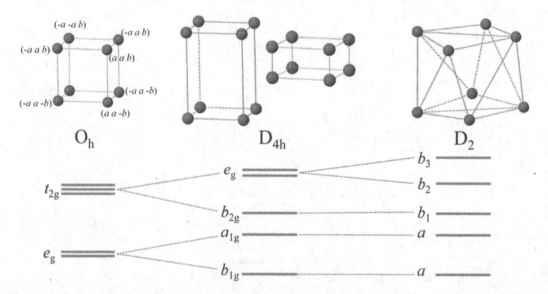

FIGURE 4.16 Cuboid model with Cartesian coordinates of vertex and energy level splitting scheme. (Song Z. and Liu Q.L., *J. Phys. Chem. C.* 123, 8656–8662, 2019. With permission.)

To obtain the quantitative description of the split energy levels in the crystal-field of elongated or compressed cuboids, we combine the perturbation calculation together with point charge electrostatic model (PCEM). For D_{4h}, the crystal-field potential expressed by crystal-field parameters (B^k) is

$$V(D_{4h}) = B_{20}U_0^2 \langle l\|C^2\|l\rangle + \left[B_{40}U_0^4 + B_{44}\left(U_4^4 + U_{-4}^4\right)\right]\langle l\|C^4\|l\rangle \tag{4.12}$$

Following the similar procedure in the former sections, we could obtain the eigenvalues expressed by a and b, which stand for the geometrical parameters of the cuboids. However, the complex analytical expressions for eigenvalues fail to demonstrate the magnitude order intuitively. To visualize the relative energy level positions, we set values to a, b in the purpose of investigating the eigenvalue/e^2Z coefficients numerically. The height of the cuboid, which approximates $2b$, is set in the range 2.0–3.0 Å while a/b ratio lies in the range 0.5–2. Accordingly, the case $a/b = 1$ stands for a perfect cube, $a/b > 1$ for compressed cuboid and $a/b < 1$ for elongated one. The numerical results are explored for fixed a/b ratio and cuboid height. In Figure 4.17(a), we selectively plot the eigenvalue coefficients against various cuboid heights at a/b ratio equal to 0.5, 1, 1.5 and 2.

Analysis of the numerical simulation in Figure 4.17 presents new relationship between edge ratio and the lowest crystal-field level. In Figure 4.17(a), the cuboid height is varying at fixed edge ratio (a/b). In the case of perfect cube ($a/b = 1$), a_{1g} and b_{1g} levels constitute the low-energy twofold degenerated level. Although the elongated and compressed cuboids belong to the same point symmetry D_{4h} and consequently have the same energy level splitting scheme, the magnitude order of those

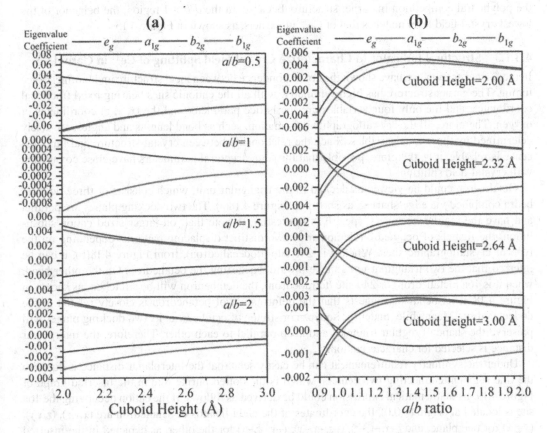

FIGURE 4.17 Numerical results of the eigenvalue coefficients at varying cuboid heights (a) and a/b ratios (b). (Song Z. and Liu Q.L., *J. Phys. Chem. C.* 123, 8656–8662, 2019. With permission.)

levels is quite different. The lowest energy level in the elongated ($a/b < 1$) and compressed ($a/b > 1$) cuboids change to b_{1g} and a_{1g} levels, respectively. As the cuboid height increases, the lowest energy level is shifting towards higher energy. In fact, for each a/b ratio, the cuboid width is increasing simultaneously with height, and the size enlargement is isotropic, which belongs to the deformation of isolated polyhedron. On the other hand, when cuboid heights are fixed for different values as shown in Figure 4.17(b), the trends of eigenvalue coefficients against a/b ratios exhibit non-monotonic behavior. Generally, they share the same pattern. In the $a/b < 1$ region, b_{1g} serves as the lowest level, and the total crystal-field splitting is shrinking as a/b ratio approaches 1. However, when it comes to the $a/b > 1$ region, the lowest level turns to a_{1g}. It is interesting to note that the lowest level first drops down and then increases at larger a/b ratios. In other words, the tetragonal distortion induces strengthening of the crystal-field splitting, and consequently, the lowest crystal-field level is continuously lowering in the beginning part of $a/b > 1$ region. The numerical simulation results just reproduce the dependence between d_{23}/d_{21} ratio and the lowest $5d$ level of Ce^{3+} in Figure 4.15. In this case, the cuboid size enlargement is anisotropic with increasing a/b ratio because of the fixed height, which resembles the competition between dodecahedron and neighboring polyhedrons in garnet structure.

The preceding results shed light on understanding the coincidence between polyhedron deformation and the lowest 5d level of Ce^{3+} in garnet compounds. We ascribe the origin of reverse garnet effect to the anisotropic tetragonal distortion caused by polyhedral competition. It should be noted that the a/b ratio is just an approximation to the edge-length ratio of dodecahedron in garnet structure. In the real case, it is difficult to define precisely the height or the width for the inclined and non-perpendicular edges. Nevertheless, we believe that the compressed cuboid model accounts for the polyhedral competition in garnet structure, because, in the $a/b > 1$ region, the behavior of the lowest crystal-field level matches that of Ce^{3+} in garnets, as shown in Figure 4.15.

4.3.1.3 Structural Indicator to Characterize Crystal-Field Splitting of Ce³⁺ in Garnet

In the former section, we have shown the correspondence between local structure and luminescence tuning. The garnet structure has high symmetry, with all the cationic sites bearing fixed fractional coordinates, and the only four variables are the lattice parameter and the (x, y, z) coordinates of oxygen. Therefore, all the crystallographic information, such as bond lengths and angles, are solely determined by those four variables. Since the relationship between crystal structure and luminescence is reliable, it is, therefore, possible that the four structural parameters have close connection with crystal-field splitting.

Octahedron could be viewed as docked to the triangular unit, which consists of three dodecahedra combined via edge-sharing, as shown in Figure 4.18(a). The two docking planes are parallel and have the regular triangle shapes. It is interesting to note that, on larger sized cation substitution, the ligands of octahedron demonstrate two distinct displacing patterns, depending on the type of crystallographic sites. When it is 24c site (dodecahedron), from Figure 4.18(b), it can be inferred that the two triangular units move inward, squeezing the octahedron. On the other hand, when it is 16a (octahedron) or 24d site (tetrahedron), the octahedron will be stretched, as shown in Figure 4.18. This behavior suggests that the deformation of octahedron is closely related to the dodecahedron. Meanwhile, under the squeeze or stretch of octahedron, its two docking planes will preserve the shape of regular triangles and keep parallel to each other. Therefore, the interplanar distance is selected for characterization.

Under the symmetry requirement, it can be easily seen that the interplanar distance is equal to the line segment connecting two centers, which is the dotted purple line in the inserted graph of Figure 4.19. Its mathematical expression could be derived as follows. If the cation occupying the 16a site is located at origin (0,0,0), the coordinates of the eight ligands of octahedron are (x,y,z), (z,x,y), (y,z,x) for one plane, and ($-x,-y,-z$), ($-z,-x,-y$), ($-y,-z,-x$) for the other, as depicted in the inserted graph of Figure 4.19. Their centers are located at $\left(\left((x+y+z)/3\right),\left((x+y+z)/3\right),\left((x+y+z)/3\right)\right)$

(a)

(b)

(c)

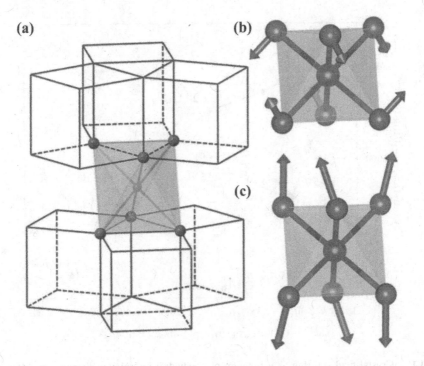

FIGURE 4.18 Geometrical relationship between polyhedra and ligands moving pattern. (a), Connectivity between dodecahedron and octahedron. The distorted dodecahedra in the triangular units are replaced by regular cubes for a better view. Octahedral ligand displacement of larger sized cation substitution on 24c site (dodecahedron) for (b), while 16a and 24d sites (octahedron and tetrahedron) for (c). (Song Z. and Liu Q.L., *J. Phys. Chem. C.* 124, 870–873, 2020. With permission.)

and $\left(-\left((x+y+z)/3\right),-\left((x+y+z)/3\right),-\left((x+y+z)/3\right)\right)$, with distance between them expressed as $\left(2/\sqrt{3}\right)|x+y+z|$. After multiplying by the cell parameter, the interplanar distance has the final form of $\left(2/\sqrt{3}\right)|x+y+z| \cdot L$, and it is picked as the structural indicator.[80] It is noticeable that the oxygen coordinates should resemble the form of (−0.0306, 0.0506, 0.1493), or else it could be picked out by examining the general positions of oxygen.

To check its relevance with crystal-field splitting, the relationship between indicator values and the energy difference between the lowest and the next lowest 5d levels is investigated in Figure 4.19. At the same time, the structural indicator could reproduce the reverse garnet effect. Therefore, this indicator proves to be effective in characterizing the crystal-field splitting, and owing to the simplicity in calculation, it is very promising to be used in predicting the structure-property relationship for the garnet-type structures. However, because the 5d–4f emission is determined not only by crystal-field splitting but also centroid-shift and Stokes-shift, the correlation between the interplanar distance and emission property is not obvious.

Since the tetragonal distortion of dodecahedron is the key factor of crystal-field splitting, we select two cases, YAG/GdAG and YAG/YGG, to show how the structural indicator is correlated to the tetragonal distortion. It has been reported that, on cation substitution, the ligand oxygen atoms exhibit two moving patterns, in which oxygen atoms are displacing along perpendicular directions. [6] The Gd substitution for Y requires volumetric increase of dodecahedron, and the octahedron gets squeezed by the two docking triangular units, just as Figure 4.18(b). In this case, the interplanar distance declines, characterizing a more tetragonally distorted dodecahedron and a larger crystal-field splitting. On the other hand, when larger Ga substitutes for Al, the octahedron expands and compresses the triangular unit, resulting in longer interplanar distance. In this case, the tetragonal

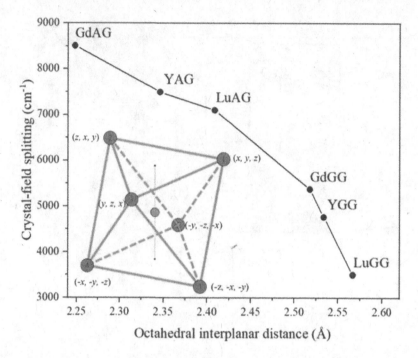

FIGURE 4.19   Crystal-field splitting as a function of octahedral interplanar distance for different garnet series. The inserted graph shows octahedron and its ligands numbered from 1 to 6, with the fractional coordinates aside. The centers of the two regular triangles are denoted as small purple dots. A dotted purple line connecting the two centers represents the interplanar distance. (Song Z. and Liu Q.L., *J. Phys. Chem. C.* 124, 870–873, 2020. With permission.)

distortion is released and gives a smaller crystal-field splitting. These two modes also give different slopes in Figure 4.19. For gallium garnets, the crystal-field splitting changes more steeply versus the interplanar distance than that of aluminum garnets. The reason lies in the structural feature that the interplanar distance is more sensitive to the cation substitution in the 16a site.

4.3.2   (Ba,Sr)$_2$SiO$_4$:Eu^{2+}: Local Structure, the Degree of Ordering and Thermal Quenching

The application of wLED devices requires good resistance to thermal luminescent quenching of phosphor, because the high working temperature up to 400K could cause intensity degradation of phosphor. It is, thus, important to understand the mechanisms of thermal quenching in phosphors and design new compositions with brilliant thermal quenching resistance. Generally, two mechanisms of thermal luminescence quenching exist in f–d transitions, i.e., the thermal-stimulated ionization from the 5d^1 level to the conduction band and the large displacement between the ground and excited states in the configurational coordinate diagram. [81–83] Meanwhile, it is widely recognized that a more rigid structure is favorable for achieving stable temperature-dependent luminescence, as demonstrated in the case of CaAlSiN$_3$:Eu^{2+} and YAG:Ce^{3+}. [68, 84] This section will show that how the local structure in the orthosilicate solid-solution phosphor (Ba$_{1-x}$Sr$_x$)$_2$SiO$_4$:Eu^{2+} promotes the thermal luminescence stability by noticing the coincidence in composition dependence of quenching temperatures and Sr/Ba atomic ordering degree in the host lattice. [85]

Alkaline-earth metal orthosilicate phosphors M$_2$SiO$_4$:Eu^{2+} (M = Sr, Ba) have aroused wide research interests in wLED applications. Its luminescence was first reported by Barry in 1968. [86] Among this silicate series, Ba$_2$SiO$_4$ and high-temperature phase α-Sr$_2$SiO$_4$ belong to the

β-K_2SO_4-type orthorhombic crystal system with space group *Pmnb*. [87,88] Two cation sites exist: ten-coordinated M1 site and nine-coordinated M2 site. Denault et al. reported the dependence of thermal luminescent stability on the composition in $(Ba_{1-x}Sr_x)_2SiO_4:Eu^{2+}$ phosphors, drawing the conclusion that the intermediate composition with 46% Sr has the highest resistance to the thermal quenching of luminescence and ascribed this to optimal bonding that creates a more rigid crystal structure compared to the end-member compositions. [3]

We find that different slopes of the fitting curves of the cell parameters occur at $x = 0.5$, indicating that the distribution of Sr and Ba on M1/M2 sites is ordered, as shown in Figure 4.20. The smaller Sr^{2+} ions prefer the smaller nine-coordinated M2 site, while the larger Ba^{2+} ions prefer the larger ten-coordinated M1 site. Atomic occupancies extracted from the Rietveld refinement results further confirm this trend. When x is less than 0.5, occupancy of Sr in M1 site is always higher than that in M2 site. The result shows that Sr^{2+} ions preferentially occupy M2 site when they are doped. The preferential substitution behavior of Sr^{2+} may arise from the shorter average bond length of M2–O and the smaller polyhedral volumetric size of M2 site.

Denault et al. reported the dependence of thermal luminescent stability on the composition in $(Ba_{1-x}Sr_x)_2SiO_4:Eu^{2+}$ phosphors and found that the intermediate composition of 46% Sr has the highest resistance to the thermal quenching of luminescence.[3] They ascribed this to optimal bonding that creates a more rigid crystal structure compared to the end-member compositions. Usually, the thermal luminescence quenching property is characterized by $T_{1/2}$, which means the temperature at which the emission intensity drops to half of that at room temperature. Figure 4.21 illustrates the normalized integrated intensity of $(Ba_{1-x}Sr_x)_2SiO_4:Eu^{2+}$ with different Sr doping content in temperature range 25–250°C and the inset is $T_{1/2}$ as a function of x. As can be seen in the figure, intermediate compositions($x = 0.4$–0.6) possess better thermal stability with $T_{1/2}$ over 200°C. We find that the composition dependence of quenching temperatures coincides with that of Sr/Ba atomic ordering degree in the host lattice.[85] Therefore, the improved thermal luminescence quenching resistance of the intermediate compositions in $(Ba_{1-x}Sr_x)_2SiO_4:Eu^{2+}$ phosphors originates from the ordered cationic distribution.

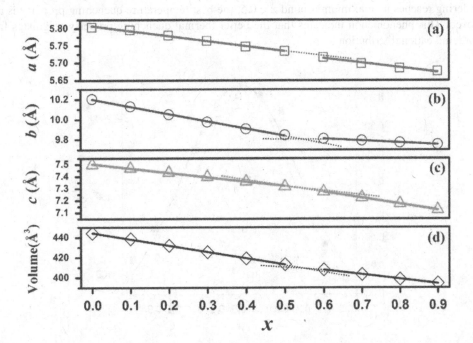

FIGURE 4.20 Cell parameters (a, b, c) and cell volume (d) of $(Ba_{1-x}Sr_x)SiO_4:Eu^{2+}$ with different x values. (He L.Z. et al., *J. Lumin.* 180, 163–168, 2016. With permission.)

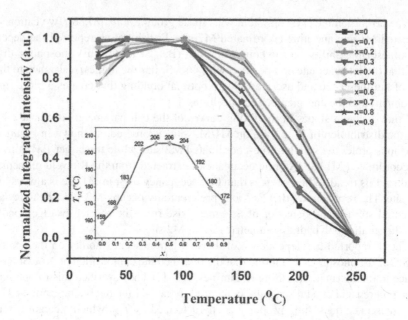

FIGURE 4.21   The normalized integrated intensity of $(Ba_{1-x}Sr_x)_2SiO_4{:}Eu^{2+}$ under 450 nm excitation at different temperatures and $T_{1/2}$ as a function of x. (He L.Z. et al., *J. Lumin.* 180, 163–168, 2016. With permission.)

For the $(Ba_{1-x}Sr_x)_2SiO_4$, the degree of ordering η can be defined as: $\eta = Occ_{Sr2} - Occ_{Sr1} = Occ_{Ba1} - Occ_{Ba2}$. [89] Then, the degree of Sr/Ba ordering and $T_{1/2}$ are depicted together in Figure 4.22. As can be seen in the figure, the fitting curve of the two parameters have the similar variation tendency, reaching their maxima at $x = 0.5$. It demonstrates that the temperature quenching behavior of $(Ba_{1-x}Sr_x)_2SiO_4{:}Eu^{2+}$ samples is closely related to the degree of cation ordering. When the degree of ordering reaches its maximum around $x = 0.5$, the best temperature quenching property is also achieved. This phenomenon indicates that the better thermal quenching property originates from the ordered cation distribution.

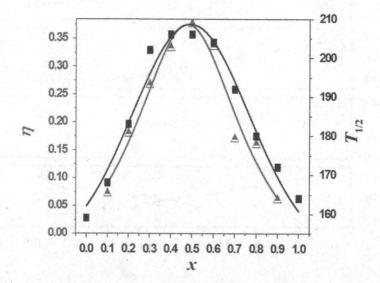

FIGURE 4.22   The comparison of degree of ordering η and $T_{1/2}$ of $(Ba_{1-x}Sr_x)_2SiO_4{:}Eu^{2+}$ samples against x. (He L.Z. et al., *J. Lumin.* 180, 163–168, 2016. With permission.)

Cation ordering could be thought as the local structure induced by size-mismatch. There are two crystallographic sites in the structure of M_2SiO_4: ten-coordinated M1 site and nine-coordinated M2 site. The volume of the M1 site is larger than that of the M2 site. The space volumes of the M1 and M2 site are suitable for Ba^{2+} and Sr^{2+} ions, respectively. For end-member Ba_2SiO_4, there is no cation ordering. Half of the Ba ions occupy M1 sites, while half of them have to occupy the smaller M2 sites, the rooms of which are not suitable (too small) for them. Similarly, for pure Sr_2SiO_4, half of the Sr ions have to occupy the larger M1 sites, which are not appropriate for them. This mismatching between size of atom and volume of crystallographic site will degrade structural rigidity and deteriorate the thermal luminescence quenching resistance. In the viewpoint of crystal chemistry, we can consider $BaSrSiO_4$ as end-member compound since there are one large and the other small crystallographic site for Ba and Sr, respectively, and so we can consider Ba_2SiO_4 as the intermediate composition where Ba completely substitutes for Sr and also Sr_2SiO_4 as the other intermediate composition where Sr fully replaces Ba.

4.3.3 $(Ba,Ca)_2SiO_4:Eu^{2+}$: ATOM SITE SPLITTING, OPTIMAL BONDING AND THERMAL QUENCHING

In the $(Ba,Ca)_2SiO_4$ system, six phases from the Ba-end to the Ca-end exist, which are denoted as Ba_2SiO_4, T-phase, X-phase, α-, β- and γ-Ca_2SiO_4, respectively. [90] Among them, the Eu^{2+}-doped T-phase has attracted much attention because of the unique crystal structure and excellent luminescent thermal quenching features. [91] Fukuda et al. [92] found that its crystal structure is disordered, and many crystallographic sites possess splitting phenomenon. [92] According to the research of Park et al. for $(Ba_{1.20}Ca_{0.7}Eu_{0.1})SiO_4$, the thermal quenching temperature $T_{1/2}$ (the temperature when the integrated emission intensity drops to half the initial intensity) can reach 175°C, which is higher than that of the commercial $(Ba,Sr)_2SiO_4$ phosphor. [91] Based on the crystallographic data of the representative T-phase $(Ba_{0.65}Ca_{0.35})_2SiO_4$, we carefully evaluate the coordination environments of all the cation sites to understand the origin of the atom site splitting. [93] By analyzing the cation site coordination environments using split- and unsplit-atom-site models, we find that the three cation sites in the split-atom-site structure are optimally bonded with ligand O atoms, leading to a more rigid structure and better luminescent thermal stability compared to end-members Ba_2SiO_4 and Ca_2SiO_4 crystals.

The representative crystal structure of the T-phase in $(Ba,Ca)_2SiO_4$ solid solutions is illustrated in Figure 4.23. It is trigonal with space group P-$3m1$. There are five crystallographic cationic sites for Ba and Ca, denoted as M1, M2, M3, M4 and M5. The M1–M3 sites are occupied by Ba^{2+}/Ca^{2+} ions. On the contrary, the M4 site is almost taken place by Ca^{2+} ions, and the M5 site is solely occupied by Ba^{2+} ions. Si atoms take two sites, and the SiO_4 tetrahedra are isolated without sharing any O atoms. O atoms occupy four sites, denoted as O1, O2, O3 and O4, where O1, O2 and O3 sites have site splitting. It is interesting that crystallographic site splitting exists in the crystal structure. As depicted in Figure 4.23(a) and (b), the M1 site splits from Wyckoff position 2d into the 6i position. One M2 splits from 1a into six 6g sites. M3, M4, M5 and two Si sites do not split. One O1 site (6i) splits into two sites (12j). One O2 site (2d) splits into three sites (6i). One O3 site splits into two sites with Wyckoff positions unchanged, which is 6i, but with different atomic fractional coordinates, denoted as O3a and O3b, respectively. O4 site does not split. For the split-atom-site model, the atom occupancies of M1, M2, O1, O2, O3a and O3b are correspondingly adjusted to be 1/3, 1/6, 1/2, 1/3, 1/2 and 1/2, respectively, due to atom site splitting.

Because many sites in the crystal structure of the T-phase have splitting and occupancy for split-sites decreases correspondingly, it is very difficult to determine the coordination polyhedron and coordination number. To analyze the local structure, we depict two-dimensional maps parallel to (0 0 1) at different z values. Figure 4.24(a) depicts the two-dimensional map at $z = 0$ containing M2 sites, which clearly shows that M2 sites are split into six positions, displacing from the threefold

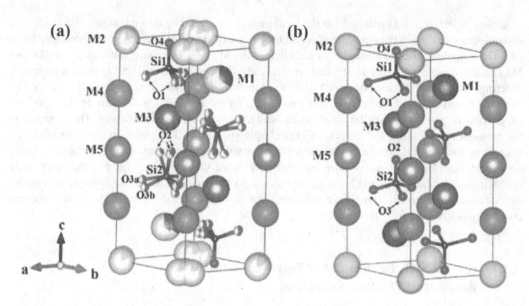

FIGURE 4.23   Crystal structures of T-phase $(Ba,Ca)_2SiO_4$ for the spilt-atom-site (a) and unsplit-atom-site models (b). (He L.Z. et al., *Inorg. Chem.* 57, 4146–4154, 2018. With permission.)

roto-inversion axis, to form hexatomic-rings. Each of the split positions is on the twofold rotation axes, which are parallel to [1 0 0] and [1 1 0]. Figure 4.24(b) shows the two-dimensional map at $z = 0.16$ including M1 and O1 sites. The M1 sites are split into three positions, displacing from the threefold roto-inversion axis. The O1 site departs from the mirror plane, which produces doublet atomic splitting. As the splitting phenomenon takes place only on the sites which accommodate Ba^{2+}/Ca^{2+} ions, it is highly possible that the splitting is due to the significant difference in ionic radii between Ba^{2+} and Ca^{2+} ions. Figure 4.24(c) gives the two-dimensional map at $z = 0.344$ including

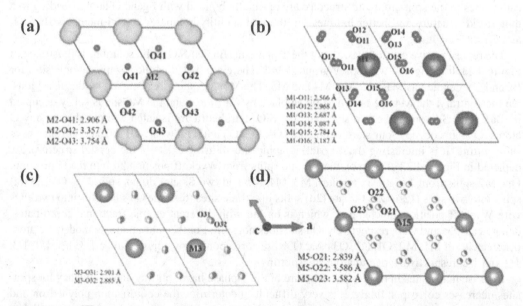

FIGURE 4.24   Two-dimensional map parallel to (0 0 1) at (a) $z = 0$, (b) $z = 0.1$, (c) $z = 0.344$ and (d) $z = 0.5$ with the corresponding bond lengths from the structure of T-phase $(Ba,Ca)_2SiO_4$. (He L.Z. et al., *Inorg. Chem.* 57, 4146–4154, 2018. With permission.)

M1, O3a and O3b sites. Figure 4.24(d) plots the two-dimensional map at $z = 0.5$ including M5 and O2 sites. It can be observed intuitively that, compared to the unsplit-atom-site model, the O2 site displaces from the threefold roto-inversion axis and splits into three sites.

To clearly address the origin of atom site splitting, the following notations are adopted. Since the multiplicity of the O1 site coordinating to the M1 site is six for the unsplit-atom-site model, these positions are denoted as O11, O12, O13, O14, O15 and O16. For the split-atom-site model, the six O1 sites displace from the mirror plane with double splitting to produce 12 O1 positions. After splitting, a and b are used for further classification, such as O11a and O11b (circled by dotted lines in Figure 4.25(b) and (d)), which are from the split of the same O1 site. The phenomenon of atom

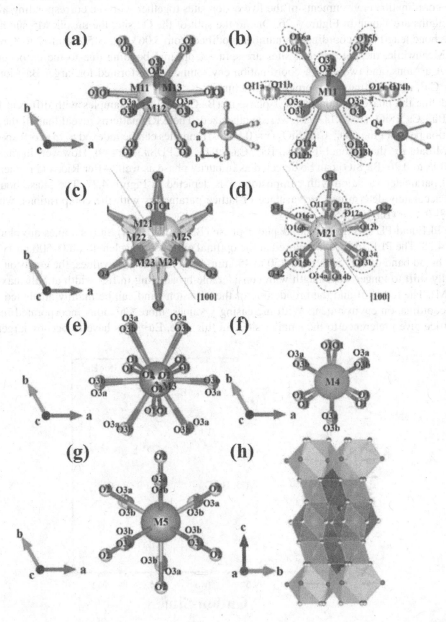

FIGURE 4.25 (a)–(g) Coordination environments of M1, M2, M3, M4 and M5 sites and (h) the form of cationic coordination polyhedra in the crystal structure of the T-phase. (He L.Z. et al., *Inorg. Chem.* 57, 4146–4154, 2018. With permission.)

site splitting arises from bond optimization during the co-occupation of Ba^{2+} and Ca^{2+} ions. Take M1 site for example. Due to the significant difference in the ionic radii of Ba^{2+} and Ca^{2+} ions, it is more energetically favorable to provide larger space for Ba^{2+} and smaller space for Ca^{2+} ions. The bond length data for M1–O1 can be classified into two groups, with one group having average bond length of 2.679 Å, which approximates that for Ca–O in Ca_2SiO_4 (2.53 Å), and the other 3.081 Å, in agreement with that for Ba–O in Ba_2SiO_4 (2.98 Å). In the unsplit-atom-site model, however, the bond length for M1–O1 is 2.878 Å, which is too long for Ca atoms. Therefore, the origin of atom site splitting for M1 and O1 is to optimize the bonding between Ca/Ba and O atoms. Similar analysis could be applied to the remaining splitting sites.

The coordination environments of the five cation sites together with the corresponding average bond lengths are shown in Figure 4.26. Due to the split of the O2 site, the unsplit M5 site has the average bond length and coordination number modified from 3.063 to 2.815 Å and 12 to 8, respectively. Meanwhile, the M1 and M2 sites are split for optimal bonding due to the co-occupation of Ba^{2+}/Ca^{2+} ions, and two sets of coordination environments are formed for larger Ba^{2+} ions and smaller Ca^{2+} ions, leading to a more rigid structure.

To define the phase boundaries of T-phase in $(Ba,Ca)_2SiO_4$, the samples with different Ba/Ca ratio, $(Ba_{1-x}Ca_x)_2SiO_4$ ($x = 0.18$–0.5), were synthesized. The XRD patterns reveal that all the major diffraction peaks of the $(Ba_{1-x}Ca_x)_2SiO_4$ ($x = 0.18$–0.5) samples can be indexed to the corresponding standard data for the trigonal T-phase $Ba_{1.31}Ca_{0.69}SiO_4$ (JCPDS#36–1449). However, in the range of $x = 0.18$ to 0.26, Ba_2SiO_4 can be detected as impurity phase as well. After Rietveld refinements, the cell parameter variation with composition x is depicted in Figure 4.27. The phase boundary can be accurately determined by variance of lattice parameters with the compositions, which is $x = 0.28$–0.4 for $(Ba_{1-x}Ca_x)_2SiO_4$.

The PLE and PL spectra of Eu^{2+}-doped T-phase $(Ba_{0.99-y}Ca_yEu_{0.01})_2SiO_4$ samples are plotted in Figure 4.28. The PLE spectra monitored at the optimal emission wavelength (460–500 nm) exhibit typical broad band of Eu^{2+} ions from 250 to 450 nm. With increasing y values, the emission peaks gradually shift to longer wavelength with considerable broadening in full width at half-maximum (FWHM). The red-shift and the broadening of the emission band can be mainly attributed to the looser coordination environment. With increasing y values, more Ca^{2+} ions incorporated into the host lattice give preference to the smaller sites. In this case, Eu^{2+} ions have to occupy larger sites

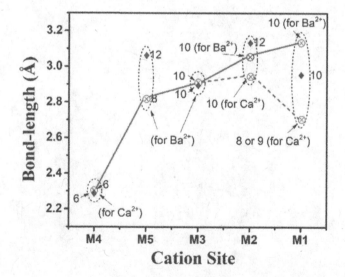

FIGURE 4.26 Average bond lengths and the coordination numbers of the different Ca^{2+}/Ba^{2+} sites for the split-atom-site model (⊗) and unsplit-atom-site model (♦) for T-phase $(Ba,Ca)SiO_4$, respectively. (He L.Z. et al., *Inorg. Chem.* 57, 4146-4154, 2018. With permission.)

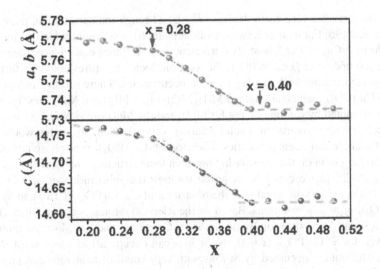

FIGURE 4.27 Refined cell parameters of $(Ba_{1-x}Ca_x)_2SiO_4$ ($x = 0.18-0.5$) samples with varying x values. (He L.Z. et al., *Inorg. Chem.* 57, 4146–4154, 2018. With permission.)

which are normally occupied by Ba^{2+} ions. Figure 4.28(c) shows the temperature dependence of the integrated intensities of $(Ba_{0.99-y}Ca_yEu_{0.01})SiO_4$ phosphors under 365 nm excitation. The inset is $T_{1/2}$ against Ca^{2+} contents (y). It is notable that the thermal stability of all T-phase $(Ba_{0.99-y}Ca_y Eu_{0.01})SiO_4$ phosphors is superior to that of $Ba_2SiO_4:Eu^{2+}$ ($y = 0$) and $Ca_2SiO_4:Eu^{2+}$ ($y = 0.99$). Their integrated luminescence intensities maintain about 90% of the initials while that of Ba_2SiO_4 starts to drop dramatically from 375K. It is inspiring to find that the thermal luminescence stabilities of $y = 0.28$ and 0.32 samples are even superior to that of the $BaSrSiO_4:Eu^{2+}$ sample, which possesses the best thermal quenching stability in the $(Ba,Sr)_2SiO_4:Eu^{2+}$ series. For T-phase $(Ba_{0.99-y}Ca_yEu_{0.01})$ SiO_4 phosphors, Ba^{2+} ions preferentially take over larger space while Ca^{2+} ions smaller space. Thus, a more rigid structure is formed due to a rational distribution of the Ba/Sr/Eu cations over the crystallographic sites and the optimized bonding appearing between them and the ligands due to atom site splitting, leading to the enhanced thermal stability.

4.3.4 β-Ca$_3$(PO$_4$)$_2$-Type Phosphors: Co-Substitution, Local Structure, Vacancy and Luminescence Tuning

Phosphate phosphors with β-Ca$_3$(PO$_4$)$_2$-type structure have aroused extensive interest because of their versatile structural types and the substitution-induced different local structures, as well as the

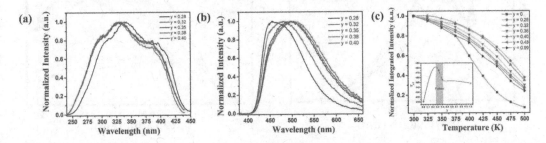

FIGURE 4.28 (a) Normalized PLE spectra monitored at the optimal emission wavelength and (b) normalized PL spectra at 340 nm excitation of T-phase $(Ba_{0.99-y}Ca_yEu_{0.01})_2SiO_4$ samples as varying y values. (c) Dependence of the integrated intensity of $(Ba_{0.99-y}Ca_yEu_{0.01})SiO_4$ phosphors with increasing y values (the inset is the $T_{1/2}$ vs y) on temperature. (He L.Z. et al., *Inorg. Chem.* 57, 4146–4154, 2018. With permission.)

tunable PL behaviors after rare-earth doping. [94, 95] Through ion substitution, many new phases can be constructed. [96] For example, two univalent metal M^+ ions could substitute for one divalent Ca^{2+} ion and form β-$Ca_{10}M(PO_4)_7$ host. Two trivalent metal R^{3+} ions can replace three divalent Ca^{2+} ions with final composition β-$Ca_9R(PO_4)_7$. In addition, local structures could be formed through different substituted cations and varying cationic occupancies. There are five independent cation sites (M) in β-$Ca_3(PO_4)_2$, which are named M(1), M(2), M(3), M(4) and M(5) sites, respectively. All sites are 100% occupied by Ca ions except for M(4), and the M(4) site is only 50% occupied by the Ca ions. By constructing isostructural solid solution, versatile local structures would emerge and give rise to different luminescent properties. Therefore, β-$Ca_3(PO_4)_2$-type phosphate phosphors are favorable for investigation on the relationship between local structure and luminescence tuning.

$Sr_2Ca(PO_4)_2$ (SCP) phase could be acquired through the substitution of Sr^{2+} for Ca^{2+} in the $Ca_3(PO_4)_2$ host. It could be combined with the isostructural $Ca_{10}Li(PO_4)_7$ (CLP) and $Sr_2Ca(PO_4)_2$ to form β-$Ca_3(PO_4)_2$-type solid solutions. Based on the Rietveld refinements, $Ca_{10}Li(PO_4)_7$ ($x = 0$) has M(1), M(2), M(3) and M(5) sites totally occupied by Ca^{2+} ions, and the M(4) site totally occupied by Li^+. However, for $Sr_2Ca(PO_4)_2$ ($x = 1$), the Sr ions can occupy all M sites except for M(5). It is found that the M(5) site is occupied by Sr ions with very small or near-zero concentration due to very short M(5)–O bond length (2.26 Å). Furthermore, the M(4) site is only 50% occupied by Ca^{2+} and Sr^{2+} ions, and the other 50% is vacancy. Figure 4.29 gives the emission spectra of solid solution xSCP–$(1 - x)$CLP:0.03Eu^{2+} ($x = 0$–1.0) monitored at 365 nm UV excitation as a function of x. The

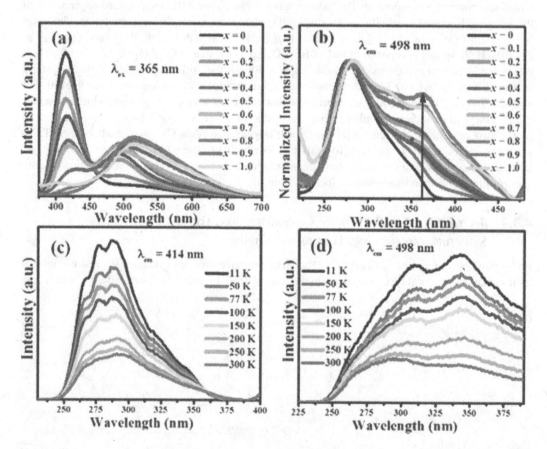

FIGURE 4.29 (a) Emission spectra of x(SCP)–$(1 - x)$ (CLP):0.03Eu^{2+} ($0 \leq x \leq 1$) samples for $x = 0$–1.0. (b) Normalized excitation spectra of x(SCP)–$(1 - x)$ (CLP):0.03Eu^{2+} ($0 \leq x \leq 1$) samples for $x = 0$–1.0. The temperature-dependent excitation spectra of 0.4(SCP)–0.6(CLP):0.03Eu^{2+} for different monitored wavelengths: (c) 414 nm and (d) 498 nm. (Chen M.Y. et al., *Chem. Mater.* 29, 1430–1438, 2017. With permission.)

emission band from 380 to 480 nm centered at 414 nm with a narrow bandwidth corresponding to CLP:0.03Eu^{2+} ($x = 0$). Nevertheless, another emission band located at 498 nm appears with the introduction of SCP into CLP:0.03Eu^{2+}. During this cation's replacement process, Sr^{2+} ions substitute Ca^{2+} and Li$^+$ ions. With the increasing of x, the emission intensity of the shorter wavelength side at 414 nm decreases rapidly together with the gradual increase of the longer wavelength side at about 498 nm. When $x = 1.0$, the emission spectra of SCP:0.03Eu^{2+} only exhibit a broad band from 420 to 680 nm centered at 515 nm. Based on such a construction of the β-Ca$_3$(PO$_4$)$_2$-type solid solution, the PL tuning has been realized. [97]

The two-band emission profile in Figure 4.29(a) could be attributed to the presence of different luminescence centers, originating from Eu^{2+} occupying different cation sites. As evidenced by the normalized PLE spectra of xSCP–(1 – x)CLP:0.03Eu^{2+} ($x = 0 - 1.0$) at room temperature in Figure 4.29(b), the dominant excitation band of Eu^{2+} in the CLP:0.03Eu^{2+} is located at about 276 nm in the high energy side, and an additional excitation band centered at 363 nm in the low energy side emerges accompanied by increasing x. The appearance of both excitation bands has verified the presence of different luminescence centers. Among the five independent cation sites in xSCP–(1 – x)CLP solid solution, Eu^{2+} ions are potentially distributed over M(1), M(2) and M(3) sites due to no restrictions for them, and the single emission band around 414 nm is originated from the three sites. With the introduction of SCP, the Sr ions can occupy all M sites except for the M(5) site due to very short M(5)–O bond length (2.26 Å). Therefore, the emission band around 498 nm is attributed to Eu^{2+} ions entering the M(4) site. The low-temperature PLE spectra of 0.4SCP–0.6CLP:0.03Eu^{2+} monitored at 414 and 498 nm are provided in Figure 4.29(c) and (d) for confirmation. The PLE spectra monitored at 414 nm recorded at 14K consist of a group of compact bands in the range of 250–350 nm, indicating that the emission band around 414 nm is originated from the Eu^{2+} ions mainly distributed over M(1), M(2) and M(3) sites because the average bond length and coordination number of the three sites are similar. However, the PLE spectra monitored at 498 nm present a new excitation peak at long wavelength (above 350 nm), which should be assigned to the Eu^{2+} ions entering the M(4) site.

The substitution of different alkaline-metal ions (Li, Na and K) also affects the luminescence of Ca$_{10}$M(PO$_4$)$_7$:Eu^{2+}.[98] As shown in Figure 4.30(a), the PL spectra of the Ca$_{10}$M(PO$_4$)$_7$:Eu^{2+} share the same band at 410 nm, but the second band on the long wavelength side only can be observed for M = Na and K. The average bond lengths of five sites in Ca$_{10}$M(PO$_4$)$_7$ as a function of ionic radii of M are exhibited in Figure 4.30(b). The differences of the M–O bond lengths, d(Li–O) < d(Na–O) < d(K–O), are remarkable when varying the M ions. However, the average bond lengths of Ca1–O, Ca2–O and Ca3–O almost remain the same. Therefore, the variation from Li to Na to K ions only changes the local environment of Eu^{2+} ions in the Na(4) and K(4) sites, which leads to different Eu^{2+} emissions. Meanwhile, Eu^{2+} has difficulty in entering the Li(4) site because of the tiny crystallographic site. The local environment of Eu^{2+} ions in the Ca(1), Ca(2) or Ca(3) sites remains nearly invariable, which yields nearly the same Eu^{2+} emission.

In Eu-doped apatite-type phosphate Ca$_2$La$_8$(SiO$_4$)$_6$O$_2$, Eu^{3+} ions could be gradually transformed to Eu^{2+} by co-substituting [Ca^{2+}–P^{5+}] for [La^{3+}–Si^{4+}] in the host, with the emission light consecutively varying from red to blue/green. It is caused by the preferential site occupancy and charge neutralization. The apatite structure contains two kinds of cation sites, namely, the nine-coordinated 4f sites with C$_3$ point symmetry and the seven-coordinated 6h sites with C$_s$ point symmetry [99]. The co-substitution of [Ca^{2+}–P^{5+}] for [La^{3+}–Si^{4+}] simultaneously occurs at 4f/6h cation sites and tetrahedral sites. As shown in Figure 4.31(a) and (b), both PLE and PL spectra show a mixture of sharp-line and broad-band profiles for most of the samples, suggesting the coexistence of Eu^{3+} and Eu^{2+}. However, all the samples are prepared under reducing atmosphere, which suggests that Eu enters La^{3+} site to gain the +3 valence state. The normalized Eu L$_3$-edge XANES spectra in Figure 4.31(c) display two evident peaks at 6975 and 6984 eV, which are attributed to the electron transitions of 2p$_{3/2}$→5d in Eu^{2+} and Eu^{3+}, correspondingly [100]. The transformation between Eu^{3+} and Eu^{2+} due to the co-substitution of [Ca^{2+}–P^{5+}] for [La^{3+}–Si^{4+}] could be understood from Figure 4.31(d).

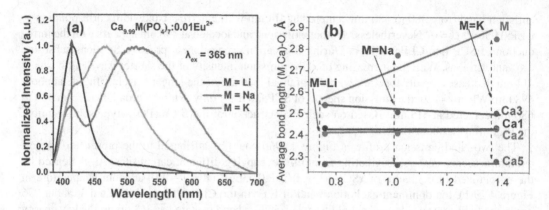

FIGURE 4.30 (a) Comparison of the normalized emission spectra of $Ca_{9.99}M(PO)_7:0.01Eu^{2+}$ (M = Li, Na and K) samples. (b) Main average bond lengths (Å) of five sites in $Ca_{9.9}M(PO_4)_7:0.1Eu^{2+}$ as a function of ionic radii of M. (Chen M.Y. et al., *Chem. Mater.* 29, 7563–7570, 2017. With permission.)

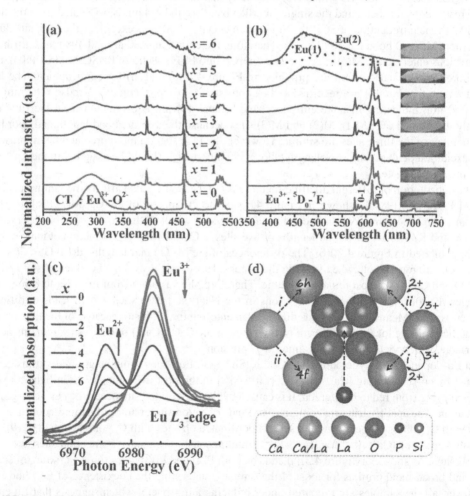

FIGURE 4.31 PLE (a) and PL (b) spectra of $Ca_{0.98(2+x)}La_{0.98(8-x)}Eu_{0.2}(SiO_4)_{6-x}(PO_4)_xO_2$ samples. The insets in (b) are the luminescence photos taken under a 390 nm UV lamp. (c) Normalized Eu L_3-edge X-ray absorption near edge structure (XANES) spectra. (d) Schematic explanation of the transformation between Eu^{3+} and Eu^{2+} due to the co-substitution of $[Ca^{2+}–P^{5+}]$ for $[La^{3+}–Si^{4+}]$. (Li G.G. et al., *Chem. Commun.*, 52, 7376–7379, 2016. With permission.)

When P^{5+} gradually substitutes Si^{4+}, the simultaneous substitution of Ca^{2+} for La^{3+}/Eu^{3+} at 6h sites will occur to maintain a charge balance. In this way, Eu^{3+} ions are continuously transformed to Eu^{2+} with broad-band emission [101].

4.3.5 CaAlSiN₃-Type Phosphors: Preferential Occupation, Local Structure Evolution, Remote Control and Luminescence Tuning

4.3.5.1 Preferential Substitution of O for N in CaAlSiN₃:Eu and Luminescence Tuning

$CaAlSiN_3$:Eu-type nitride phosphors have widely been commercialized in phosphor-converted light-emitting diode (pc-LEDs). Due to the strong nephelauxetic effect and large crystal-field splitting, $CaAlSiN_3$:Eu-type nitride phosphors could have visible light excitation accompanied by orange-red emissions. On the other hand, the rigid host framework is favorable for excellent resistance to thermal luminescence quenching. Compared to oxide phosphors, the luminescence of nitride phosphors could be tuned by both cationic and anionic compositional modifications. In the following section, we will show the luminescence tuning in $CaAlSiN_3$:Eu^{2+} by oxygen preferential substitutions for nitrogen located at one of two crystallographic sites.

It is well known that the broad emission bands in $CaAlSiN_3$:Eu are attributed to the allowed transition between $4f^7$ and $4f^65d$ state of Eu^{2+} ions, which could be affected by two factors: centroid of 5d orbital energy level and the effect of crystal-field splitting. In the crystal structure, Eu ions are doped to substitute for the Ca ions and coordinated with five N atoms [84]. In the view of the effects of covalence and crystal-field strength, it is necessary to focus on anion substitution, which will introduce O to replace N. Considering that oxygen atoms only occupy one of two N-occupied crystallographic site in the lattice, the crystal chemical formula of $CaAlSiN_3$ doped with O could be written more subtly as $Ca(Al/Si)_2N^I{}_2(N^{II}{}_{1-x}O_x)$, because the Al^{3+} and Si^{4+} ions randomly occupy 8b equivalent crystallographic site [102], Ca atoms take up 4a site and N atoms are located at two crystallographic site, i.e., 8b/4a represented as N^I and N^{II}.

Figure 4.32(a) shows the XRD patterns of $Ca_{0.875-0.5x}(Al_{0.75}Si_{1.25})N_2(N_{1-x}O_x)$:0.02Eu compositions. Samples exhibit different phase compositions for different oxygen contents x. In the range of $0 \leq x \leq 0.2$, compositions are almost single phase with $CaAlSiN_3$-type structure. When $x = 0.25$, a second phase, Ca–SiAlON, appears as the minor impurity of main $CaAlSiN_3$ phase. Sample with $x = 0.3$ is mixtures of $CaAlSiN_3$ and Ca–SiAlON, while sample with $x = 0.35$ is almost single Ca–SiAlON phase. These show that the maximum oxygen content for $CaAlSiN_3$ phase stability is $x \approx 0.25$, so we subsequently focus on samples with $x = 0-0.25$. For the XRD patterns of samples with $x = 0-0.25$, three pairs of peaks around 33°, 37° and 49°, are indexed as (0 2 0) and (3 1 0), (0 2 1) and (3 1 1),

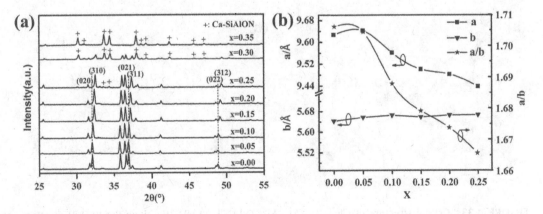

FIGURE 4.32 (a). XRD patterns of $Ca_{0.875-0.5x}(Al_{0.75}Si_{1.25})N_2(N_{1-x}O_x)$:0.02Eu. (b) Lattice parameters of $Ca_{0.875-0.5x}[Al_{0.75}Si_{1.25}]N_2(N_{1-x}O_x)$:0.02Eu. (Wang T. et al., *J. Lumin.* 137, 173–179, 2013. With permission.)

and (0 2 2) and (3 1 2) according to the JCPDS card 39-0747 [103], respectively. Each pair of peaks is separated with increasing O content x. This is ascribed to the fact that the lattice parameters of $CaAlSiN_3$ phase, a and b, have different change tendencies with varying x. Figure 4.32(b) performed various tendencies of cell parameters a and b, as well as a/b ratio with the increasing O content x. The lattice parameter a of these compositions shows a clear decrease with higher oxygen concentrations, as expected by the smaller ionic radii of O^{2-} vs. N^{3-}. However, the lattice parameter b does not follow a clear decrease with O^2 concentration as expected by Vegard's law. Thus, a/b ratio has a large change with x, resulting in separation of each pair of peaks.

The space group of $CaAlSiN_3$, $Cmc2_1$, is the maximum non-isomorphic subgroup of wurtzite-type structure with space group $P6_3mc$. The ideal structure of $CaAlSiN_3$ should present the wurtzite-type structure with $a/b = 1.732$. The observation that the crystal structure has a larger shrinkage along the [1 0 0] direction than the [0 1 0] direction with oxygen concentration increasing (the parameter a decreases and b almost keeps constant) indicates that the structure is substantially distorted from the ideal structure of $CaAlSiN_3$. In the crystal structure of $CaAlSiN_3$, Ca atoms occupy 4a site, Al/Si atoms take up 8b site randomly, while N atoms locate at two crystallographic site, i.e., 8b and 4a represented as N^I and N^{II}, respectively (in Figure 4.33(a), N^I green ball and N^{II} red ball). The multiplicity of N^I and N^{II} are 8 and 4, respectively, so the amount of N^I is twice as that of N^{II}. Due to the difference of the point symmetry of N^I and N^{II}, the crystal chemical formula of $CaAlSiN_3$ can be represented as $Ca(Al/Si)_2N^I_2N^{II}$. Substitutions of O atoms for N atoms generally give rise to the lattice contraction since the mean bond lengths Si–N, Si–O, Al–N and Al–O are 1.74, 1.64, 1.87 and 1.75 Å, respectively. The large difference in shrinkage between [1 0 0] and [0 1 0] directions suggests that O atoms only replace the N atoms in one of two crystallographic sites. In the crystal structure (Figure 4.33(a)), [SiN4] and [AlN4] tetrahedra connect with N atoms to form six-membered rings. N^I atoms are linked with three Si/Al atoms, while N^{II} atoms are chain-linked with two Si/Al atoms. In fact, a large amount of structural data in oxygen-containing silicates or alumina silicates demonstrate that divalent O atoms can either occupy terminal sites of Si/Al–O tetrahedra or bridge two Si/Al–O tetrahedra. O atoms that take up the sites linking three Si/Al–O tetrahedra are not observed, which is consistent with silicate chemistry and confirmed by the Pauling's second rule for anion distributions, as the divalent O anion is less coordinated by higher valence silicon. Therefore, only N^{II} atoms can be substituted by O atoms in $CaAlSiN_3$-based crystal structure and the crystal-chemical formula can be represented as $Ca(Al/Si)_2N^I_2(N^{II}_{1-x}O_x)$ [102].

We prepared samples according to nominal compositions $Ca_{0.875-0.5x}(Al_{0.75}Si_{1.25})N^I_2(N^{II}_{1-x}O_x)$: $0.02Eu$ ($x = 0, 0.05, 0.1, 0.15, 0.2, 0.25, 0.3$ and 0.35). Herein, the ratio of Al/Si was fixed as $0.75/1.25$ according to our previous reports that the samples with this ratio showed better PL [104]. The charge

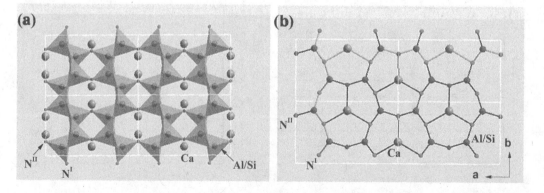

FIGURE 4.33 Crystal structure of $Ca_{0.875-0.5x}[Al_{0.75}Si_{1.25}]N^I_2(N^{II}_{1-x}O_x)$:$0.02Eu$ along the [0 0 1] direction. (Part (a) is a polyhedral view of the structure and (b) is a single layer.) (Wang T. et al., *J. Lumin.* 137, 173–179, 2013. With permission.)

imbalance caused by the substitutional O to N is compensated by forming the Ca ion vacancy [105], for which the occupancy of Ca is less than 1 to meet the requirements. Along the [0 1 0] direction (the orange chain in Figure 4.33(b)), the combination of $-N^{I}-Si/Al-N^{I}-$ without substituting of O contributed dominantly to the rigidness of the structure, which caused the b-axis length almost constant. However, along the [1 0 0] direction, the replacement of O for N^{II} formed the $-N^{I}-Si/Al-N^{II}/O-$ chains leading to a rapid decrease of the lattice parameter of this direction (a axis, the cyan chain in Figure 4.33(b)), which agreed well with the results of XRD analysis.

Figure 4.34(a) shows the emission spectra of the series Eu-doped $Ca(Al/Si)_2N_2(N_{1-x}O_x)$ with normalized intensity, excited by 450 nm. For the pure nitrogen sample, the emission band has a broad range from 500 to 800 nm peaking at 650 nm. An obvious blue shift of emission band is observed with the increasing O concentration. The emission peak moves to around 610 nm for sample with $x = 0.25$. Figure 4.34(b) shows the linear dependence of peak position for emission spectra on O content x. The broad emission bands for all the samples are attributed to the allowed transition between $4f^7$ and $4f^65d$ states of Eu^{2+} ions, which could be affected by two factors: centroid of 5d orbital energy level and crystal-field splitting. Incorporating more O to substitute for N atoms reduces the

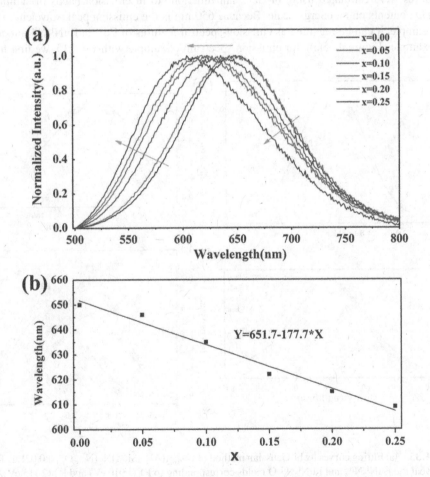

FIGURE 4.34 Photoluminescence of $Ca_{0.875-0.5x}[Al_{0.75}Si_{1.25}]N_2(N_{1-x}O_x):0.02Eu$ with various O content x. (a) Normalized emission spectra of $Ca_{0.875-0.5x}[Al_{0.75}Si_{1.25}]N_2(N_{1-x}O_x):0.02Eu$ samples excited by 450 nm. The figures x indicates O concentration. The arrows illustrate the band shift and change in the spectra with increasing x. (b) Graph showing the linear dependence of peak position for photoluminescence spectra on O content x. The inset equation is the linear fit of the peak position vs. O concentration. (Wang T. et al., *J. Lumin.* 137, 173–179, 2013. With permission.)

nephelauxetic effect around the Eu^{2+} and elevates the lowest 5d energy-level of Eu^{2+} in the crystal, since O^{2-} has higher electronegativity than N^{3-} and Eu–O has lower covalency than Eu–N. These result in the blue-shift of emission bands ultimately.

Interestingly, the shapes of the emission spectra for the samples containing O are not single band by Gaussian fitting, suggesting that there may be more than one luminescence center in the crystal. Piao et al. [106] reported that the emission spectrum was derived from the overlap of two types of Eu^{2+} ions in different environments:EuN4 tetrahedra and EuN3O tetrahedra. However, the Ca^{2+} ions are coordinated with five N atoms: one N^{II} atom at a bond length of 2.342 Å, two N^{I} atoms at about 2.540 Å, another N^{II} atom at about 2.529 Å. The fifth coordinated N^{II} atom located at about 2.644 Å has to be taken into account. This viewpoint has been confirmed by Uheda's group in their first publication about $CaAlSiN_3$ [84]. In this case, five coordination mode is chosen to explain the two environments of Eu^{2+} ions: One, denoted as $EuN^{I}_2N^{II}_3$, represents that the five coordinated N atoms contain two N^{I} and three N^{II}, and the other, $EuN^{I}_2N^{II}_2O$, means that Eu coordinates to two N^{I} atoms, two N^{II} atoms and one O atom substituting for one N^{II} atom.

To further clarify the luminescent feature of Eu^{2+} in the two coordinating environments, two fitting curves were estimated using bi-Gaussian functions to fit emission bands of samples with different O contents on an energy scale. Because 650 nm is the emission peak wavelength of pure nitrogen sample, it can be regarded as emission-spectral features of Eu^{2+} in $EuN^{I}_2N^{II}_3$ model. As a typical example (Figure 4.35(a)), for emission spectrum of sample with $x = 0.15$, we first fixed the

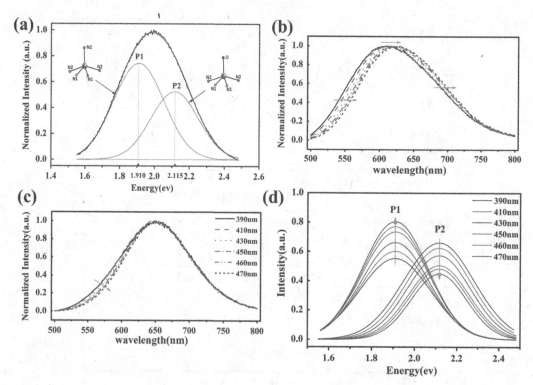

FIGURE 4.35 (a) Fitting curves by bi-Gaussian method of $Ca_{0.625}[Al_{0.75}Si_{1.25}]N_2$ $(N_{0.85}O_{0.15})$:0.02Eu. The inset figures reveal the $EuN^{I}_2N^{II}_3$ and $EuN^{I}_2N^{II}_2O$ molds corresponding to P1 (1.910 eV) and P2 (2.115 eV), respectively. (b) Emission spectra of $Ca(Al/Si)_2N_2N$:0.02Eu (a) and $Ca(Al/Si)_2N_2(N_{0.8}O_{0.2})$:0.02Eu (c). The figures indicate different excitation wavelengths. The arrows illustrate the band shift and change in the spectra with increasing excitation wavelength. (d) Fitting curves by bi-Gaussian method of $Ca(Al/Si)_2N_2(N_{0.8}O_{0.2})$:0.02Eu. The figures indicate different excitation wavelengths. The arrows show the relative intensity of fitting curves change with increasing excitation wavelength. With longer wavelength excitation, P1 intensities increase and P2 intensities decrease. (Wang T. et al., *J. Lumin.* 137, 173–179, 2013. With permission.)

fitting curve peaking at 1.910 eV (650 nm–P1) and its full width half maximum (FWHM) based on the experimental emission data of pure nitrogen sample (i.e., $x = 0$), and then fitted the whole emission band curve of sample with $x = 0.15$ by bi-Gaussian fitting. Consequently, another fitting curve was achieved, peaking at 2.115 eV (587 nm–P2), which can be attributed to emission band of Eu^{2+} in $EuN^I_2N^{II}_2O$, i.e., another luminescent center. Next, P1 and P2 were used as two type fitting curves for short, respectively. Using the method described earlier, all the emission bands of samples with $x = 0.0$-0.25 could be fitted well by two Gaussian curves P1 and P2 with relatively fixed peak positions, suggesting that there only exist two Eu-coordinated modes, i.e., $EuN^I_2N^{II}_3$ and $EuN^I_2N^{II}_2O$. In other words, Eu cannot coordinate more than one oxygen atoms.

However, with increasing O concentration, P1 intensities decrease while P2 intensities increase, indicating that more $EuN^I_2N^{II}_2O$ models form with more $EuN^I_2N^{II}_3$ models dismissing as O concentration increases. Summarily, in $CaAlSiN_3$ structure, oxygen atoms only substitute for one kind of N atoms, i.e., N^{II} at 4a crystallographic site and the substitution amount is severely limited to only form $EuN^I_2N^{II}_2O$ models resulting in a new luminescent band peaking at 587 nm. With increasing oxygen concentration, the increase of the $EuN^I_2N^{II}_2O$ models amount induces a blue shift from 650 to 610 nm and the emission peak position for PL spectra shows the linear dependence on O content x.

The normalized emission spectra excited by different wavelengths of the sample $Ca(Al/Si)_2N_2N$:0.02Eu and $Ca(Al/Si)_2N_2(N_{0.8}O_{0.2})$:0.02Eu at room temperature were presented in Figure 4.35(b) and (c). The emission peak positions of $Ca(Al/Si)_2N_2N$:0.02Eu almost keep constant, while those of $Ca(Al/Si)_2N_2(N_{0.8}O_{0.2})$:0.02Eu show a near-linear change with different excitation wavelengths. This can be interpreted well by the two-luminescence center models. According to the models, $Ca(Al/Si)_2N_2N$:0.02Eu contains only one luminescence center $EuN^I_2N^{II}_3$ and its emission spectra keep 650 nm peak position with different excited wavelengths. Relatively, $Ca(Al/Si)_2N_2(N_{0.8}O_{0.2})$:0.02Eu owns two luminescence centers and its emission bands are overlap of two specific emission wavelength–P1 and P2, which can be obtained by bi-Gaussian fitting, as shown in Figure 4.35(d). P1 and P2 have different relative intensities under various excited wavelengths. With longer wavelength excitation, P1 intensities increase and P2 intensities decrease. Since the emission bands of samples are the overlap of P1 and P2, it can be well-understood that emission peak position shifts to longer wavelength with excitation light moving to longer wavelength. These also indicate that P1 band can be excited more easily by longer wavelength light compared to P2 band.

Time-resolved PL analysis was employed to describe the energy transfer in the $Ca(Al/Si)_2N_2(N_{1-x}O_x)$ phosphors [107]. Two-peak emission behavior and variation in decay behavior were observed through two experiments. Due to the different environments of P1 and P2, the fitted peaks and their corresponding environments around Eu^{2+} ions are designated in time-resolved emission spectra. The peak positions of P1 and P2 change with the different Eu concentrations. Figure 4.36(a) and b show that the relative P1/P2 ratio remains almost constant for $Ca(Al/Si)_2N_2(N_{1-x}O_x)$: $0.0005Eu^{2+}$, which suggests that P1 keeps almost the same time evolvement with P2 in $Ca(Al/Si)_2N_2(N_{1-x}O_x)$: $0.0005Eu^{2+}$. Figure 4.36(c) and (d) show the relative P1/P2 ratio decreases dramatically as the decay time increases in $Ca(Al/Si)_2N_2(N_{1-x}O_x)$: $0.02Eu^{2+}$, resulting from the relatively declined P1 and rising P2 intensity. This is ascribed to the fact that P1 has a faster time evolvement than P2 for $Ca(Al/Si)_2N_2(N_{1-x}O_x)$: $0.02Eu^{2+}$ and the brisk energy transfer from Eu^{2+} ions at EuN4O environments to those at pure N environments. According to various time evolvements of P1 and P2 in two different $Ca(Al/Si)_2N_2(N_{1-x}O_x)$ hosts, the energy transfer barely occurs in $Ca(Al/Si)_2N_2(N_{1-x}O_x)$:$0.0005Eu^{2+}$ phosphor, while it occurs strongly in $Ca(Al/Si)_2N_2(N_{1-x}O_x)$:$0.02Eu^{2+}$ because the higher Eu^{2+} concentration shortens the distance for energy transmission and allows more chances for the energy transfer. The energy transfer appears sequentially from the high-energy Eu^{2+} ions at the EuN4O environments to the low energy Eu^{2+} ions at the EuN5 environments and leads to the decline of P1 emission and the simultaneous increasing of P2 emission in $Ca(Al/Si)_2N_2(N_{1-x}O_x)$:$0.02Eu^{2+}$. The active energy transfer eventually controls the spectral distribution. Therefore, the peak positions of emission spectra scarcely change with the delay time for $Ca(Al/Si)_2N_2(N_{1-x}O_x)$:$0.0005Eu^{2+}$, while

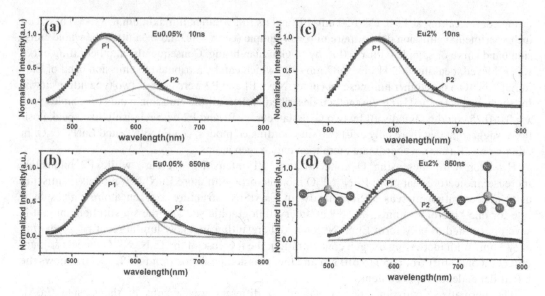

FIGURE 4.36 Time-resolved emission spectra of $Ca(Al/Si)_2N_2(N_{1-x}O_x)$: $0.0005Eu^{2+}$ (a and b) and $Ca(Al/Si)_2N_2(N_{1-x}O_x)$: $0.02Eu^{2+}$ (c and d). Fitted Gaussian emission bands are shown and their corresponding environments around Eu^{2+} ions are designated in the inset. (Wang T. et al., *J. Electrochem. Soc.* 161, H25–H28, 2014. With permission.)

the peak position shifts dramatically to the low-energy side for $Ca(Al/Si)_2N_2(N_{1-x}O_x)$:$0.02Eu^{2+}$ as evidenced in Figure 4.37.

4.3.5.2   Local Structure Evolution and Luminescence Tuning by Cation Co-Substitution in Eu^{2+}-Doped $(Ca_{1-x}Li_x)(Al_{1-x}Si_{1+x})N_3$ Solid Solutions

With a wurzite-type structure (space group $Cmc2_1$), it is also interesting to find that $LiSi_2N_3$ has isotypic crystal structure with $CaAlSiN_3$. Therefore, $LiSi_2N_3$ can be regarded as a complete substitution of the $(CaAl)^{5+}$ pair by a $(LiSi)^{5+}$ pair in $CaAlSiN_3$. In other words, Li^+ and Si^{4+} ions can be completely incorporated into the host of $CaAlSiN_3$, and the charge is kept balanced by a cooperative substitution, i.e., Li replacing Ca with Al substituting for Si. Thus, the solid solutions of $(Ca_{1-x}Li_x)_{0.98}(Al_{1-x}Si_{1+x})N_3$:$0.02Eu^{2+}$ could be synthesized following this co-substitution strategy [108]. Depending on the variation of the chemical composition, the emission band of Eu^{2+} shows a very interesting variation trend: first, a slight blue-shift from 669 to 663 nm, then an obvious red-shift from 663 to 738 nm, and finally an abrupt blue-shift to 600 nm when x = 1. By analyzing the changes of luminescence in detail, we reveal an evolution process of local structure around Eu^{2+} in $(Ca_{1-x}Li_x)_{0.98}(Al_{1-x}Si_{1+x})N_3$:$0.02Eu^{2+}$ which accounts for the interesting three-stage change of emission band.

Co-substitution of the $(LiSi)^{5+}$ pair for the $(CaAl)^{5+}$ pair in the $CaAlSiN_3$ host forms a completely miscible solid solution $(Ca_{1-x}Li_x)_{0.98}(Al_{1-x}Si_{1+x})N_3$:$0.02Eu^{2+}$ as shown in Figure 4.38(a). Meanwhile, the Bragg diffraction peaks of $(Ca_{1-x}Li_x)_{0.98}(Al_{1-x}Si_{1+x})N_3$:$0.02Eu^{2+}$ shift to higher angles with respect to those of $CaAlSiN_3$:Eu^{2+} (Figure 4.38(b)) with increasing x values. The lattice parameters (*a, b* and *c*) and cell volume (V) decrease with x monotonically (Figure 4.38(c)) due to the fact that the radius of Li^+(0.76 Å, CN = 6) is smaller than that of Ca^{2+}(1.00 Å, CN = 6).

Figure 4.39(a) shows the normalized PLE spectra with x = 0–1. It can be observed that the dominant excitation bands of Eu^{2+} in the series of $(Ca_{1-x}Li_x)_{0.98}(Al_{1-x}Si_{1+x})N_3$:$0.02Eu^{2+}$ (x = 0–1) compounds are located at about 310 nm in the high energy side, which is probably associated with their small crystallographic site. With the increasing x values, the excitation spectra on the high energy side can be considered to stay invariable, while a blue-shift of the PLE spectra on the low energy

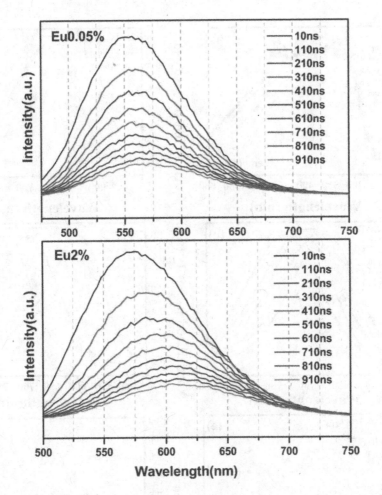

FIGURE 4.37 Time-resolved emission spectra of Ca(Al/Si)$_2$N$_2$(N$_{1-x}$O$_x$): 0.0005Eu^{2+} and Ca(Al/Si)$_2$N$_2$(N$_{1-x}$O$_x$): 0.02Eu^{2+}. The different colors represent different decay times. (Wang T. et al., *J. Electrochem. Soc.* 161, H25–H28, 2014. With permission.)

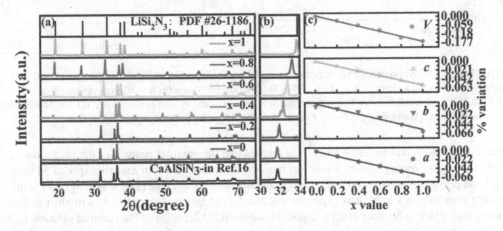

FIGURE 4.38 (a) Full range (15–75°) XRD patterns and (b) the selected diffraction peaks near 32° of (Ca$_{1-x}$Li$_x$)$_{0.98}$(Al$_{1-x}$Si$_{1+x}$)N$_3$:0.02Eu^{2+} (x = 0–1) samples. (c) The lattice parameters a, b and c and the unit cell volume V of (Ca$_{1-x}$Li$_x$)$_{0.98}$(Al$_{1-x}$Si$_{1+x}$)N$_3$:0.02Eu^{2+} (x = 0–1) samples depending on different x values. (Wang T. et al., *Inorg. Chem.* 55, 2929–2933, 2016. With permission.)

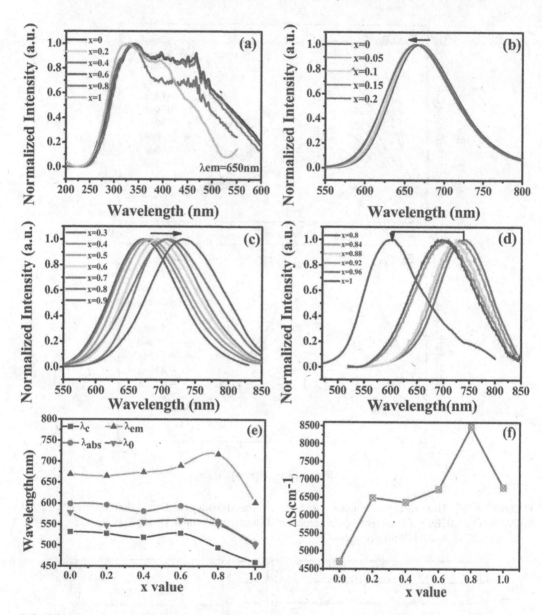

FIGURE 4.39   (a) Normalized PLE spectra of $(Ca_{1-x}Li_x)_{0.98}(Al_{1-x}Si_{1+x})N_3{:}0.02Eu^{2+}$ samples with $x = 0{-}1$. The normalized emission spectra of $(Ca_{1-x}Li_x)_{0.98}(Al_{1-x}Si_{1+x})N_3{:}0.02Eu^{2+}$ samples for different ranges: (b) $x = 0{-}0.2$, (c) $x = 0.3{-}0.9$, (d) $x = 0.8{-}1$. (e) The wavelength of λ_c, λ_{abs}, λ_{em} and λ_0 as a function of x. (f) The Stokes shift as a function of x. (Wang T. et al., *Inorg. Chem.* 55, 2929–2933, 2016. With permission.)

side can be observed in Figure 4.39(e). This implies that the centroid position of the excitation band shifts to a higher energy region. Moreover, more replacement results in a smaller split of 5d energy level and brings about narrower excitation bands. Because the average bond length of Li–N is much smaller than that of Ca–N, it is expected that the 5d excitation states of Eu^{2+} will shift to lower energy and split to a broader energy region. However, this blue-shift of the centroid position of the excitation bands is opposite to the expected behavior that smaller bond length generally results in a broad absorption band and the red-shift of the excitation peak due to stronger nephelauxetic effect and larger crystal-field splitting. This unusual behavior may be caused by two effects. First, the Eu^{2+} coordinate environment in the $(Ca_{1-x}Li_x)(Al_{1-x}Si_{1+x})N_3$ structure may stay unchanged when a small part of $(CaAl)^{5+}$ is replaced by $(LiSi)^{5+}$. To accommodate the smaller Li^+ cations, the distances between Eu^{2+} and the anions may not decrease or even become slightly larger. Thus, the crystal-field splitting of the d-manifold is smaller for Eu^{2+} in the solid solution with an increasing fraction of $(LiSi)^{5+}$, causing a blue-shift of the excitation band and a more narrower excitation spectrum. The second origin of the blue-shift is the decreased centroid shift caused by the replacement of Si^{4+} with Al^{3+}. Based on the model suggested by Morrison [109] and its development by Dorenbos [110], the centroid shift ε_c (the lowering of the average of the five 5d levels) can be affected by the so-called spectroscopic polarizability α^i_{sp}. With the substitution of $(LiSi)^{5+}$ for $(CaAl)^{5+}$, viz. $Li^+ \rightarrow Ca^{2+}$ and $Si^{4+} \rightarrow Al^{3+}$ simultaneously, $Si^{4+} \rightarrow Al^{3+}$ may be the dominant reason to modify the luminescence properties of $(Ca_{1-x}Li_x)_{0.98}(Al_{1-x}Si_{1+x})N_3:0.02Eu^{2+}$ ($x = 0–1$). With the increasing x, the average electronegativity of cations increases. This leads to the decrease of the spectroscopic polarizability α_{sp}^i. Therefore, the centroid of the 5d excitation band stays in a higher energy region with a larger x, which corresponds to the blue-shift of the excitation band.

The normalized emission spectra of $(Ca_{1-x}Li_x)_{0.98}(Al_{1-x}Si_{1+x})N_3:0.02Eu^{2+}$ in different composition ranges are demonstrated in Figure 4.39(b)–(d). These samples exhibit dark red emission with a broad band peaking at about 650–700 nm apart from the compound with $x = 1$, i.e., $LiSi_2N_3:Eu^{2+}$. From the normalized emission spectra, the spectral evolution depending on chemical compositions can be clearly observed. Three obvious changing stages can be found here: Stage I is a slight blue-shift of the emission bands in the compositions with $x = 0–0.2$ (Figure 4.39(b)). Stage II is a large red-shift in the compositions with x from 0.2 to less than 1 (Figure 4.39(c)). Stage III is an abrupt blue-shift when x is 1 (Figure 4.39(d)). The main emission band of $LiSi_2N_3:Eu^{2+}$ is observed at about 600 nm, and the same result has been mentioned in several reports [111, 112]. These changes are very interesting and cannot be interpreted by the conventional isostructural solid solution model. Here, we proposed a coordination model to explain the three-stage changes.

The model is proposed on the basis of local structural evolution of the $(Ca_{1-x}Li_x)_{0.98}(Al_{1-x}Si_{1+x})N_3:0.02Eu^{2+}$, from $CaAlSiN_3$ ($x = 0$) to $LiSi_2N_3$ ($x = 1$) via the $(LiSi)^{5+}$ substitution for $(CaAl)^{5+}$ couple as shown in Figure 4.40. When x is 0, $CaAlSiN_3$ has a relatively stable and rigid crystal structure, and the coordinated tetrahedra in the six-ring of Ca^{2+} ions contain three $[AlN_4]$ tetrahedra and three $[SiN_4]$ tetrahedra (labeled as M1). With the increasing fraction of $(LiSi)^{5+}$, the local coordinated structures of the luminescence centers appear as the following three models: two $[AlN_4]$ tetrahedra and four $[SiN_4]$ tetrahedra (M2), one $[AlN_4]$ tetrahedron and five $[SiN_4]$ tetrahedra (M3) and six $[SiN_4]$ tetrahedra (M4) (when $x = 1$). M1, M2, M3 and M4 are all the possible local coordination environments for the luminescence center Eu^{2+}. Although the compositions of $(Ca_{1-x}Li_x)_{0.98}(Al_{1-x}Si_{1+x})N_3:0.02Eu^{2+}$ can be tuned continuously and correspondingly and the lattice parameters show continuous changes, the local environments of Eu^{2+} have only four types as mentioned earlier and are shown in Figure 4.40. In $CaAlSiN_3$, only M1 can be found in the coordination environments of the central atoms. Due to the alternatively arranged three $[AlN_4]$ tetrahedra and three $[SiN_4]$ tetrahedra, the structure of M1 is relatively rigid and stable. With the increase of concentration values, a $[SiN_4]$ tetrahedron will replace an $[AlN_4]$ tetrahedron, and M2, M3 and M4 appear successively.

In three stages, different spectral shift behaviors could be related to the local structures of Eu^{2+} in the solid solutions. For stage I with $x = 0–0.2$, the slight blue-shift of the emission bands corresponds to the blue-shift of the 5d centroid (the average energy) of the excitation bands. The Stokes shift has an acute increase as can be seen in Figure 4.39(f). As depicted in Figure 4.40, in the beginning of the structural evolution, the coexistence of M1 and M2 breaks the relatively stable crystal structure and creates a less rigid crystal structure compared to the end-member compositions. Therefore, the appearance of the inhomogeneous ligand environments (M1 and M2) of the luminescence centers may account for the increase of Stokes shift and the FWHM of the emission bands. However, because the cavity in M1 is looser than that in M2, the Eu^{2+} ions with a much larger size than Ca^{2+} and Li^+ will reside preferentially in M1 environments. Therefore, the emission bands show a very slight blue-shift which is consistent with the excitation bands.

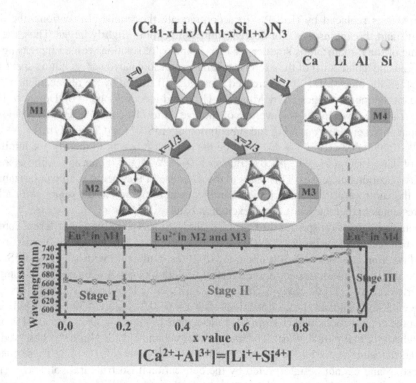

FIGURE 4.40 Proposed model on the chemical unit co-substitution strategy represented by structural evolution of $(Ca_{1-x}Li_x)_{0.98}(Al_{1-x}Si_{1+x})N_3:0.02Eu^{2+}$. From $CaAlSiN_3$ ($x = 0$) to $LiSi_2N_3$ ($x = 1$) via the $(LiSi)^{5+}$ substitution for $(CaAl)^{5+}$ couple. It corresponds to three stages in the evolution of photoluminescence. (Wang T. et al., *Inorg. Chem.* 55, 2929–2933, 2016. With permission.)

For the compounds with x from 0.2 to less than 1 in stage II, the red-shift of the emission band can be mainly attributed to the increasing Stokes shift (Figure 4.39(f)) arising from the large structural relaxation. With the increase of x, the six-ring ligands of Eu^{2+} ions have two appearances as shown in Figure 4.40: M2 and M3. The coexistence of M1, M2 and M3 is one of the dominant reasons for the larger Stokes shift. In isostructural alkaline-earth compounds, the observation of the larger Stokes shift can be explained with the configurational coordinate diagram. In the configurational coordinate diagram, the equilibrium distance in the excited state of lanthanide ions can be changed by comparing with the equilibrium distance in the ground state. The change of the equilibrium distance can be affected significantly by the rigidness of the host lattice. As a result of the less rigid lattice structure caused by the presence of M2 and M3, the change of the equilibrium distance (due to the contraction) in the $4f^65d^1$ excited state is increased. The increased change of the equilibrium distance brings about a larger shift of the parabolas in the configurational coordinate diagram, which results in a larger Stokes shift in the luminescence spectra.

The luminescence behavior of $LiSi_2N_3:Eu^{2+}$ is quite surprising, which shows an unexpected emission wavelength peaking at about 600 nm. If it follows the red-shift tendency of the emission in the series of solid solutions, $LiSi_2N_3:Eu^{2+}$ should have an emission band with a peak beyond 700 nm. Figure 4.39(d) shows more detailed work to study the abnormal luminescence behavior of $LiSi_2N_3:Eu^{2+}$. Apart from the emission band of Eu^{2+} in $LiSi_2N_3$, a continuous red-shift of the emission peaks is observed with the increasing x. The jump point can be found exactly at $x = 1$. As can be seen in Figure 4.39(a), a higher location of the lowest energy level of Eu^{2+} can be observed in the PLE spectrum of Eu^{2+} in $LiSi_2N_3$. The smaller FWHM of the excitation spectra also indicates that the crystal-field splitting of $LiSi_2N_3:Eu^{2+}$ is much smaller than that of $CaAlSiN_3:Eu^{2+}$ which could be the dominant reason for such a sudden blue-shift. Compared to the sample with $x = 0.8$, a smaller Stokes shift of $LiSi_2N_3:Eu^{2+}$ may be another reason for the sudden blue-shift. In the $LiSi_2N_3$ lattice, as the absence of Al^{3+}, the skeleton structure is built only by $[SiN_4]$ tetrahedra, and the only coordination environment type is M4. This results in the most homogeneous environment of Eu^{2+} and a more rigid lattice, which accounts for the smaller Stokes shift. The slightly larger value of the Stokes shift of $LiSi_2N_3:Eu^{2+}$ than that of $(Ca_{1-x}Li_x)_{0.98}(Al_{1-x}Si_{1+x})N_3:0.02Eu^{2+}$ ($x = 0–0.6$) may be attributed to the serious mismatching size and valence of Eu^{2+} and Li^+.

4.3.5.3 Luminescence Tuning through Remote Control of Neighboring-Cation Substitution

Besides Li^+, $CaAlSiN_3:Eu^{2+}$ could also accommodate La^{3+} in Ca^{2+} sites, and the substituted cations with different valence states have remote control of Eu^{2+} luminescence [1]. As shown in Figure 4.41, the trend of thermal quenching behavior progressively improves across the overall samples from the La series ($x = 0.15–0.03$) to the intermediate $x = 0.00$ sample, and then to the Li series ($x = 0.03–0.15$). The difference lies in the local structure around Eu^{2+} formed by neighboring cations. For La^{3+} series, L_3-edge extended X-ray absorption fine structure spectroscopy reveals shorter average La–N distances with regard to Ca–N bond due to the large valence state of La^{3+}. As a result, La–N bond

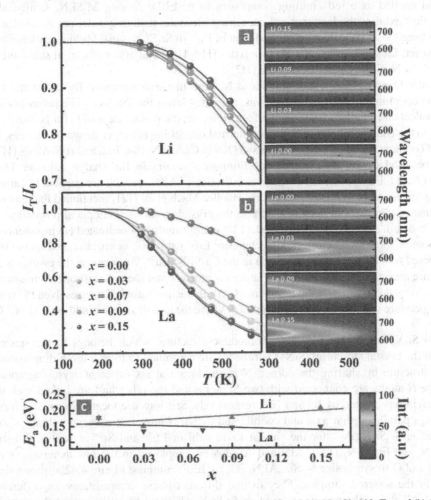

FIGURE 4.41 Thermal luminescence quenching behavior for (a) $(Ca_{1-x}Li_x)(Al_{1-x}Si_{1+x})N_3:Eu$ and (b) $(Ca_{1-x}La_x)(Al_{1+x}Si_{1-x})N_3:Eu$ samples. (c) Plot of activation energies with variable x for La and Li series. (Wang S.S. et al., *J. Am. Chem. Soc.* 135, 12504–12507, 2013. With permission.)

could be thought to have more covalency than Ca-N bond. Similarly, Li–N bond would be weaker in binding strength. Considering the local structure of Ca^{2+}, it is surrounded by the nearest ligand nitrogen anions, which are shared by the neighboring cations Si^{4+}, Al^{3+} and outer Ca^{2+}. Since La-N is more covalent than the outer Ca–N bond, the bond length of outer Ca–N will be stretched due to the inductive effect. The dopant Eu^{2+} prefers to occupy the looser Ca site to minimize the lattice strain. Thermal stability of La series materials worsens because the covalence of Eu–N herein decreases as x increases, which results in a low activation barrier and significant thermal quenching. On the contrary, a slightly tighter Eu–N bond in the Li series than that in the parent $CaAlSiN_3$:Eu material leads to a higher quenching activation barrier.

4.3.6 $Sr_2Si_5N_8$-Type Phosphors: Anion-Preferential Occupation, Cation-Size-Mismatch and Luminescence Tuning

4.3.6.1 Effect of Al–O Co-Substitution for Si–N on Local Structure and Luminescence of $Sr_2Si_5N_8$:Eu^{2+}

Alkaline-earth silicate nitride phosphor $M_2Si_5N_8$:Eu^{2+} (M=Ca, Sr, Ba) has been successfully patented and applied as a red emitting component in pc-LED. Among $M_2Si_5N_8$ family, although $Sr_2Si_5N_8$:Eu^{2+} exhibits the best thermal quenching behavior, it slightly inferiors to $CaAlSiN_3$:Eu^{2+} resulting from looser coordination environment of Eu^{2+} in $Sr_2Si_5N_8$ host. Multiple cationic substitution, however, fails to improve the PL properties [113–116]. The effect of partial substitution of O for N in $Sr_2Si_5N_8$:Eu^{2+} is also investigated [117].

Generally, O atoms occupy the same site as N atoms in nitride structure. By O substituting for N atoms and coordinating with rare-earth ions, they may affect the luminescence behaviors for their lower covalent and smaller crystal-field effect. However, the replacement of O for N could possibly give rise to the charge imbalance and poor thermal quenching effect as shown in silicates. By revelation of common occupied sites of Al/Si and O/N in $CaAlSiN_3$ [84, 102] and α-SiAlON [118–120], Al will be good co-dopants with O substitutions to maintain the charge balance. Therefore, $Sr_2Si_{5-x}Al_xN_{8-x}O_x$:$Eu^{2+}$ phosphors made by substituting of AlO^+ for SiN^+ are studied. In early papers regarding $Sr_2Si_{5-x}Al_xN_{8-x}O_x$:$Eu^{2+}$ phosphors, Mueller-Mach et al. [121] mentioned the reason of lattice parameter changes, but deep research on the crystal structure and optical properties was not reported in detail. Additionally, Chen et al. [122] mainly studied coordinated environments of Eu^{2+} and proposed a local anion clustering mechanism: Eu^{2+} gains excess nitride coordination in the Sr and Ba-based host lattice, but excess oxide in the $Ca_2Si_5N_8$:Eu^{2+}. We focus on the evolution of crystal structure according to the changes of lattice parameters and the red-shift of the emission bands based on changed coordination environment of Eu^{2+} in the host lattice. Time-resolved PL analysis at room temperature and below was employed to describe the energy transfer in $Sr_2Si_{5-x}Al_xN_{8-x}O_x$:$Eu^{2+}$ samples.

The $Sr_2Si_5N_8$ compound has an orthorhombic structure, which belongs to the space group $Pmn2_1$. In the crystal structure, [SiN4] tetrahedra are combined to form a three-dimensional rigid network structure by sharing the corner. N atoms locate at six different crystallographic sites: half of the N atoms are connected with two Si atoms and the other half are linked with three Si atoms, herein represented as N^{II} and N^{III}, respectively. Sr^{2+} ions are located in cavities formed by Si6N6 rings parallel with a axis and coordinated by ten N atoms. It is easy to obtain a complete solid-solution $Sr_2Si_5N_8$:Eu^{2+} for the similar ionic radii of Eu^{2+} and Sr^{2+} as well as iso-structure of $Sr_2Si_5N_8$ and $Eu_2Si_5N_8$ [123]. AlN and Al_2O_3 were employed in the raw materials as the origin of Al and O to synthesize $Sr_2Si_{5-x}Al_xN_{8-x}O_x$:0.02$Eu^{2+}$ samples. Figure 4.42(a) shows the XRD patterns of the sintered samples. They exhibit different phase compositions for different Al/O content x. All the peaks can be indexed as $Sr_2Si_5N_8$ (JCPDS: 85-0101) except the sample with $x = 0.4$. A small amount of Sr_2SiO_4 appeared as an impurity phase with more O atoms, doped. Figure 4.42(b) performs various tendencies of cell parameters a, b and c with increasing Al/O

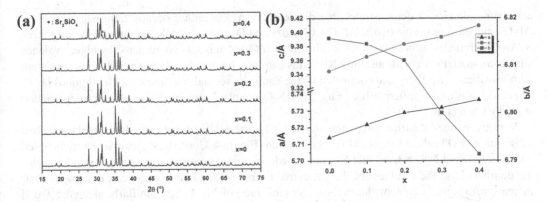

FIGURE 4.42 (a) XRD patterns of $Sr_2Si_{5-x}Al_xN_{8-x}O_x$:$Eu^{2+}$. (b) Lattice parameters of $Sr_2Si_{5-x}Al_xN_{8-x}O_x$:$Eu^{2+}$. (Wang T. et al., *J. Lumin.* 147, 173–178, 2014. With permission.)

content x. It is worth noting that both lattice parameters a and c of these compositions show clear up trend with higher Al/O concentrations, while the lattice parameter b follows a clear decrease with increasing Al/O concentration rather than increase trend as expected by Vegard's law.

Substitutions of Al atoms for Si atoms generally give rise to the lattice expansion, while substitutions of O atoms for N atoms cause the lattice contraction, since the mean bond lengths of Si–N, Si–O, Al–N and Al–O are 1.74, 1.64, 1.87 and 1.75 Å, respectively. Obviously, the extension of a and c is mainly attributed to the substitution of Al for Si atoms and the shrinkage along b axis is predominantly ascribed to the preferential substitutions of O for N atoms. Accordingly, the large different transformations among three directions suggest that preferential occupation occurred in the evolution of crystal structure.

The crystal structure of $Sr_2Si_5N_8$ along [0 0 1] and [1 0 0] directions are shown in Figure 4.43. The atom labels were marked according to their own atomic sites, notably N1, N2 and N5 atoms bonding two Si atoms are represented as N^{II}, while N3, N4 and N6 atoms connecting three Si atoms belong to N^{III}. On the base of a large amount of structural data in oxygen-containing silicates and

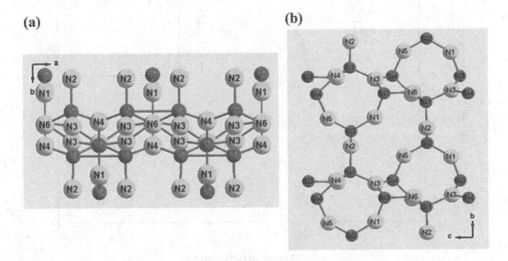

FIGURE 4.43 Crystal structure of $Sr_2Si_{5-x}Al_xN_{8-x}O_x$:$Eu^{2+}$. (a) and (b) is along [0 0 1] and [1 0 0] direction, respectively, wherein Sr atoms were removed for clear expression of the skeleton pattern. (Wang T. et al., *J. Lumin.* 147, 173–178, 2014. With permission.)

aluminum silicates, it demonstrates that divalent O ions can either occupy terminal sites of Si/
Al–O tetrahedra or bridge two Si/Al–O tetrahedra and O atoms that take up the sites linking three
Si/Al–O tetrahedral is not observed. As the divalent O anion is less coordinated by higher valence
silicon compared with N atoms, only N^{II} atoms can be substituted by O atoms, which is consistent
with oxosilicate chemistry and confirmed by the Pauling's second rule for anion distributions [124].
This order was also testified in $Ba_{1.95}Eu_{0.05}Si_4AlN_7O$ sample determined from a neutron diffraction
study by Chen et al. [122].

Various changes of lattice parameters showed in Figure 4.42(b) are also attributed to the selective
substitution of O for N. Along the [1 0 0] orientation (Figure 4.43(a)), the skeleton structure is based
on the condensed N^{III} (N3, N4 and N6) atoms, which cannot be replaced by O atoms and contribute dominantly to the rigidness of the structure. Therefore, the lattice parameter *a* shows a clear
increase with higher x content, due to the larger ionic radii of Al^{3+} vs Si^{4+}. Similarly, along the [0 0 1]
direction (Figure 4.43(b)), the $–N^{III}–Si–N^{III}–$ chains protect themselves against the effects of substitution of O for N. The substitution of Al^{3+} for Si^{4+} makes the principal contribution to the lattice
expansion. However, along the [0 1 0] direction, the framework is connected by N^{II} atoms, i.e., N1,
N2 and N5 atoms. The substitution of O for N^{II} leads to a rapid decrease of the lattice parameter on
this direction, matching well with the results of lattice parameters analysis shown in Figure 4.42(b).

In Figure 4.44, the emission spectra of $Sr_2Si_{5-x}Al_xN_{8-x}O_x:Eu^{2+}$ reveal small red shifts from 620
to 626 nm by increasing x value. This behavior is in contradiction with the expansion of crystal structure along *a* and *c* axis which should shift the emission of Eu^{2+} to the high energy side
by decreasing the crystal-field strength. However, an important factor of the shrinkage along
b axis demonstrated earlier, which changes local symmetry around the Eu^{2+} ions and increases the
crystal-field strength, can cause the red shift of emission spectra. With the increasing of x value,
more Al/O atoms were incorporated into lattice resulting in severe changes of the local structure
of Eu^{2+}. According to Dorenbos' work [110], spectroscopic polarizability α_{sp} should be regarded as
a phenomenological parameter representing the effects of ligand polarization, the effects of covalence and possible charge cloud expansion effects, which are all potentially responsible for centroid
shift. By calculating the α_{sp} value of various cations present in some compounds, it was found that

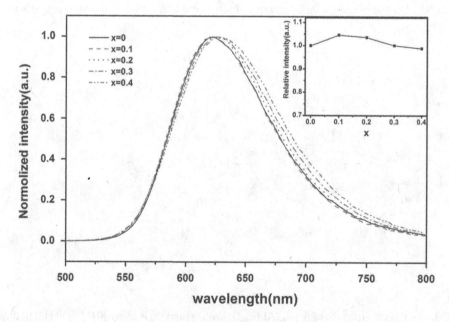

FIGURE 4.44 Photoluminescence spectra of $Sr_2Si_{5-x}Al_xN_{8-x}O_x:Eu^{2+}$ with various *x* values excited by
400 nm. (Wang T. et al., *J. Lumin.* 147, 173–178, 2014. With permission.)

the α_{sp} value of Si was smaller than that of Al. The substitution of Al for Si increases the average α_{sp} value, resulting in the larger centroid shift.(ε_c) and lower energy level of 5d configuration with the x value increasing, and eventually leading to the red-shifts of the emission wavelength. Jung et al. [125] reported the same results that the Eu^{2+} activators sited in Al-rich environment gave rise to the lower energy emission spectrum, while those in Si-rich environment were responsible to the higher energy emission band. However, on the other hand, incorporating more O to substitute for N atoms reduces the nephelauxetic effect around the Eu^{2+} and elevates the lowest 5d energy level of Eu^{2+} in the crystal, since O^{2-} has higher electronegativity than N^{3-} and Eu–O has lower covalency than Eu–N. These result in a blue-shift of emission band ultimately. The resultant red-shifts can be attributed to the subtle competition between red-shift effects from Al substitutions for Si and blue-shifts from Al replacement by O. The resultant red-shifts demonstrate that the Eu^{2+} ions are more sensitive to the change of local structure caused by Al.

4.3.6.2 Cation-Size-Mismatch and Luminescence Tuning in $M_2Si_{5-x}Al_xN_{8-x}O_x(M = Ca, Sr, Ba)$

The cationic compositional change of alkaline-earth elements can also modify local structure and tune the luminescence. The PL peak wavelengths in Figure 4.45(a) are arranged in the order Sr > Ba > Ca, which reflects a complex balance of lattice and local coordination effects caused by the structural change in M = Sr, Ca materials. In addition, with more AlO^+ substituted, PL peak wavelength has a red shift for M = Sr, Ba but blue shift for M = Ca. Figure 4.45(b) plots the energy shifts against the size difference $\Delta r = r(M^{2+}) - r(Eu^{2+})$ for each set of three samples with the same x. A series of lines cross near $\Delta r = 0$, showing an approximately linear dependence on size-mismatch. Meanwhile, it also has influence on thermal luminescence quenching. As shown in Figure 4.45(c), the M = Ca sample has onset temperature as low as 150K, and the quenching activation energies of the three series diverge with increasing x from the initial ~0.28 eV. The cation-size-mismatch effect may arise from changes in the numbers of coordinated nitrogen and oxygen anions around Eu^{2+}. In other words, the local structure around Eu^{2+} differs from the average structure around M^{2+} cations. It forms local anion clustering which could effectively reduce lattice strains by equalizing the mean cation-cation distances [122].

Since the doped Eu^{2+} ions are larger than the host Ca^{2+}, the lattice strain could be relieved by forming more coordinated oxygen anions around Eu^{2+} on AlO^+ substitution, which at the same time leaves more coordinated nitrogen for Ca^{2+}. The local structure of oxygen clustering results in a large blue shift in PL spectra. On the contrary, Eu^{2+} is smaller than the host Ba^{2+}, and thus, it tends to be coordinated by nitrogen, leaving more oxygen coordinated to Ba^{2+}. Therefore, in M = Ba sample Eu^{2+} could be thought as embedded in a pure nitrogen coordinated environment, which accounts for the differences in thermal luminescence quenching behaviors.

Generally, thermal luminescence quenching measurements for a specific phosphor are conducted by recording PL spectra of the same sample under varying temperatures. It happens for $Sr_{2-x}Si_5N_8:Eu_x$ ($x = 0.10$) that the heating process could cause irreversible material degradation, which brings negative effects on thermal luminescence quenching resistance. In Figure 4.46, Yeh et al. measured the temperature-dependent PL spectra of $M_2Si_5N_8$ (M = Sr, Ba) in air from 25 to 300°C and way back [126]. The PL intensities of Figure 4.46(b) and c are reversible on temperature loop, but irreversible for $Sr_{2-x}Si_5N_8:Eu_x$ ($x = 0.10$), with only half the intensity left on cooling to room temperature. To elucidate the origin of thermal degradation, in situ XRD is performed for the three samples under temperature loop, and the results prove that the intensity loss is closely related to the structural collapse. Transmission electron microscopic (TEM) images of $Sr_{2-x}Si_5N_8:Eu_x$ ($x = 0.10$) show the presence of amorphous surface of the heat-treated sample. In the amorphous area, more Eu^{3+} ions are oxidized from Eu^{2+} during the heating process, as evidenced by ESCA (electron spectroscopy for chemical analysis) measurements. Therefore, $Sr_{2-x}Si_5N_8:Eu_x$ sample with large Eu content ($x = 0.10$) suffers more from the Eu^{3+}–Eu^{2+} conversion than that of $x = 0.02$. On the other hand, thermal degradation disappears in $Ba_{2-x}Si_5N_8:Eu_x$ ($x = 0.10$), possibly caused by the low covalency of Ba–N with regard to Sr–N bond, which retards the connection between Eu^{2+} and O.

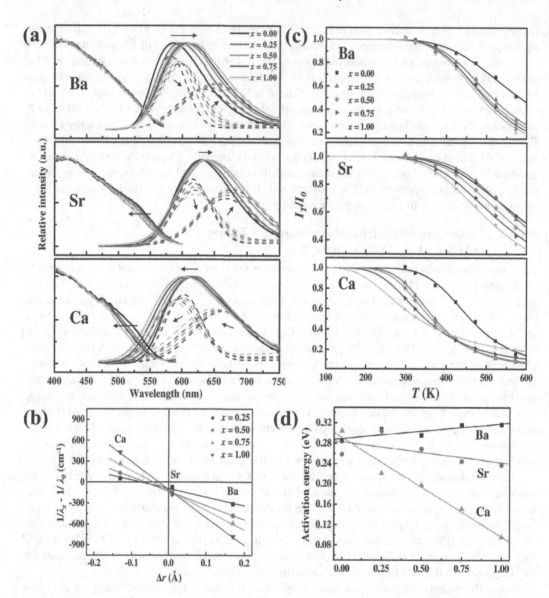

FIGURE 4.45 Photoluminescence and thermal luminescence quenching of $M_{1.95}Eu_{0.05}Si_{5-x}Al_xN_{8-x}O_x$ materials (M = Ca, Sr, Ba). (a) Normalized PLE (lower wavelength contribution) and PL spectra measured at room temperature. The photoluminescence spectra are decomposed into Gaussian contributions from the two M cation sites in the $M_2Si_5N_8$-type structures. The arrows illustrate the energy shift and change in relative intensity of the spectra with increasing x. (b) Energy shift (in wavenumbers) vs $\Delta r = r(M^{2+}) - r(Eu^{2+})$, the difference between ionic radii of the host (M) and dopant (Eu) cations, for several compositions x. The crossover of the constant composition lines near $\Delta r = 0$ demonstrates the size-mismatch effect. (c) PL intensity ratios I_T/I_0 measured at temperatures T = 298–573K. (d) Plot of the activation energies against composition variable x for the M = Ca, Sr and Ba series showing a size-mismatch variation of the slopes. (Chen W.T. et al., *J. Am. Chem. Soc.* 134, 8022–8025, 2012. With permission.)

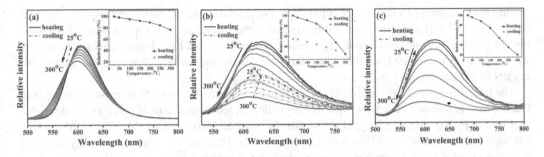

FIGURE 4.46 Temperature dependence of PL spectra of (a) $Sr_{2-x}Si_5N_8:Eu_x$ ($x = 0.02$), (b) $Sr_{2-x}Si_5N_8:Eu_x$ ($x = 0.10$) and (c) $Ba_{2-x}Si_5N_8:Eu_x$ ($x = 0.10$) from 25 to 300°C (solid line) and way back (dash line) under air, with the relative intensity as a function of the temperature plotted in the inset. (Yeh C.W. et al., *J. Am. Chem. Soc.* 134, 14108–14117, 2012. With permission.)

4.3.7 β-SiAlON: Local Structure and Electronic Structure

β-SiAlON:Eu²⁺ is important narrow-emitting phosphor and has been widely applied in pc-wLEDs as advanced wide-gamut LCD backlights [127]. It is important to identify the atomic site of Eu²⁺ dopants in phosphor host to guide the luminescence tuning. It is possible that the luminescent dopants exist in the crystal matrix, the byproduct surface layer or even intergranular boundary. However, the dopant atomic site is unclear by XRD analysis due to its low density. Kimoto et al. managed to improve the detection limit of scanning TEM (STEM) and visualize a single Eu dopant atom in β-SiAlON as shown in Figure 4.47 [128]. The positions of Si atomic columns are schematically overlapped as black circles. Si atomic columns are observed as black and white dots in bright-field (BF) and annular dark-field (ADF) images, respectively. Upper insets in solid-line rectangles show single-scanning BF or ADF images obtained in a conventional manner, resulting in severe quantum noise, unable to detect the dopant location. In the high signal-noise-ratio BF image

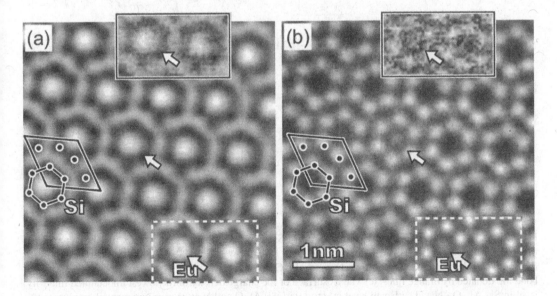

FIGURE 4.47 (a) Bright-field (BF) and (b) Annular dark-field (ADF) STEM images of β-SiAlON:Eu²⁺ observed by improved STEM. Upper insets in solid-line rectangles show single-scanning images obtained in a conventional manner, resulting in severe quantum noise. White arrows show Eu atom positions. (Kimoto K. et al., *Appl. Phys. Lett.* 94, 041908, 2009. With permission.)

(Figure 4.47(a)), no apparent contrast of the dopant is observed, but the high signal-noise-ratio ADF image (Figure 4.47(b)) clearly shows an Eu dopant at the origin, as shown by an arrow. Therefore, Eu^{2+} in β-SiAlON belongs to interstitial dopant, and the estimated bond length between Eu and an anion is roughly estimated from the radius of the atomic channel to be 0.26 nm, comparable to that in EuN (0.251 nm) or EuO (0.257 nm).

The interstitial site of Eu^{2+} in β-SiAlON has also been confirmed by Rietveld refinement method. Placing Eu^{2+} into 2b site of β-Si_3N_4 with six-ring $(Si/Al)N_4$ tetrahedra channel could yield reasonable structure of $Eu_xSi_{6-z}Al_{z-x}O_{z+x}N_{8-z-x}$ [129]. Unlike Eu^{2+}-doped M-α-SiAlON (M = Ca, Li, Mg) which has significant emission band shift with regard to M, the PLE and PL spectra are hardly modified by doping ions of alkaline-earth elements, as depicted in Figure 4.48(a).

Persistent luminescence could be realized in β-SiAlON:Eu^{2+} with the general formula $Eu_{0.015}Si_{5.5}Al_{0.485}O_{0.515}N_{7.485}$ [130]. Under 254 nm irradiation, it has 400 s afterglow. Through DFT calculations, random substitution of Al–O for Si–N in β-Si_3N_4 proves to provide trap levels which are essential to persistent luminescence. The band structures and total and partial densities of states (DOS and PDOS) of the $Si_{5.5}Al_{0.5}O_{0.5}N_{7.5}$ are displayed in Figure 4.48(b). Compared to pure β-Si_3N_4, a small amount of Al–O randomly substituted for Si–N introduces Si–O bonds, resulting in the localized impurity levels being just below the bottom of the conduction band, which can capture electrons as the effective electron traps. The charge density distribution of the lowest unoccupied

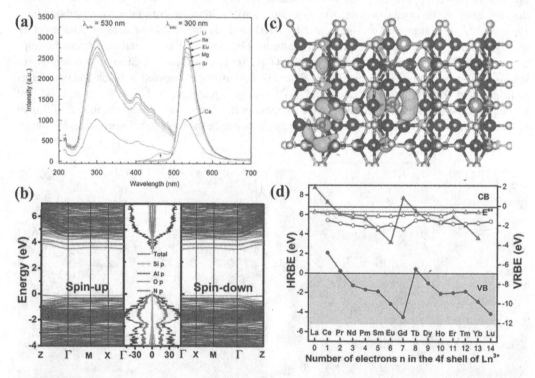

FIGURE 4.48 (a) PLE and PL spectra of $Eu_xM_ySi_{6-z}Al_{z-x-y}O_{z+x+y}N_{8-z-x-y}$ varying with the type of M (M = 2Li, Mg, Ca, Sr, Ba). (Li Y. et al., *J. Solid State Chem. 181* (12). 3200–3210, 2008. With permission.) (b) Computed electronic band dispersion, DOS and PDOS of $Si_{5.5}Al_{0.5}O_{0.5}N_{7.5}$. The impurity levels of the Si–O bond are donated by orange lines. (c) The isosurface of the charge density of the lowest unoccupied impurity level in $Si_{5.5}Al_{0.5}O_{0.5}N_{7.5}$. The Fermi level is set to 0eV. Si, Al, O and N atoms are represented by blue, cyan, red and grey spheres, respectively. (d) The HRBE and VRBE schemes. Red and blue "zigzag" curves connect the energy for electrons in the 4f ground state of divalent and trivalent lanthanides respectively. The other red and blue curves connect the energy for the lowest 5d states of divalent and trivalent lanthanides respectively; (Wang S.X. et al., *J. Mater. Chem. C.* 7, 12544–12551, 2019. With permission.)

impurity level in Figure 4.48(c) can further prove that it is the localized level instead of the conduction band that acts as an electron trap center. The HRBE diagram in Figure 4.48(d) could be constructed using the onset energy of the thermoluminescence excitation spectrum as the energy difference between the 4f ground state and the bottom of the conduction band to pinpoint the 4f energy level location of Eu^{2+}.

4.4 SUMMARY

Luminescence tuning by local structure plays an important role in phosphor research. The local coordination environment around the dopant center may deviate from the average structure, and its luminescence needs to be interpreted with more subtle structure-property relationship. We have summarized several representative phosphors with versatile local structure types, including preferential occupancy, the degree of ordering, split-atom-site and optimal bonding, local structure evolution, remote control of neighboring cation, cation-size-mismatch, chemical substitution/co-substitution. Their effects on luminescence properties are discussed. For the future research work on local structure, it is promising to combine more advanced structure characterization techniques and first-principal calculations to obtain more details about local coordination environment. On the other hand, we also can speculate or analyze the local structure from luminescence spectra because a close relationship exists between local structure and luminescent characteristics. We hope that this review on detailed composition-structure-property relationships of these representative phosphors could help the readers to research on luminescent materials.

REFERENCES

1. Wang SS, Chen WT, Li Y, Wang J, Sheu HS, Liu RS (2013) Neighboring-cation substitution tuning of photoluminescence by remote-controlled activator in phosphor lattice. *J. Am. Chem. Soc.* 135:12504.
2. George NC, Brgoch J, Pell AJ, Cozzan C, Jaffe A, Dantelle G, Llobet A, Pintacuda G, Seshadri R, Chmelka BF (2017) Correlating local compositions and structures with the macroscopic optical properties of Ce^{3+}-doped $CaSc_2O_4$, an efficient green-emitting phosphor. *Chem. Mater.* 29:3538.
3. Denault KA, Brgoch J, Gaultois MW, Mikhailovsky A, Petry R, Winkler H, DenBaars SP, Seshadri R (2014) Consequences of optimal bond valence on structural rigidity and improved luminescence properties in $Sr_xBa_{2-x}SiO_4$:Eu^{2+} orthosilicate phosphors. *Chem. Mater.* 26:2275.
4. Dunn TM, McClure DS, Pearson RG (1965) *Some aspects of crystal field theory.* Harper & Row, New York, NY.
5. Sugano S, Tanabe Y, Kamimura H (1970) *Multiplets of transition-metal ions in crystals.* Academic Press, New York and London.
6. Song Z, Xia ZG, Liu QL (2018) Insight into the relationship between crystal structure and crystal-field splitting of Ce^{3+} doped garnet compounds. *J. Phys. Chem. C* 122:3567.
7. Goldschmidt VM (1926) Die Gesetze der Krystallochemie. *Naturwissenschaften* 14:477.
8. Song Z, Zhou DD, Liu QL (2019) Tolerance factor and phase stability of the garnet structure. *Acta Crystallogr. Sect. C* 75:1353.
9. Song Z, Liu QL (2020) Tolerance factor and phase stability of the normal spinel structure. *Cryst. Growth Des.* 20:2014.
10. Song Z, Liu QL (2020) Tolerance factor, phase stability and order–disorder of the pyrochlore structure. *Inorg. Chem. Front.* 7:1583.
11. Xia ZG, Meijerink A (2017) Ce^{3+}-Doped garnet phosphors: composition modification, luminescence properties and applications. *Chem. Soc. Rev.* 46:275.
12. Menzer G (1929) Die Kristallstruktur der Granate. *Z. Für Krist.-Cryst. Mater.* 69:300.
13. Muñoz-García AB, Artacho E, Seijo L (2009) Atomistic and electronic structure of antisite defects in yttrium aluminum garnet: density-functional study. *Phys. Rev. B* 80:014105.
14. Muñoz-García AB, Barandiarán Z, Seijo L (2012) Antisite defects in Ce-doped YAG ($Y_3Al_5O_{12}$): first-principles study on structures and 4f–5d transitions. *J. Mater. Chem.* 22:19888.
15. Laguta V, Zorenko Y, Gorbenko V, Iskaliyeva A, Zagorodniy Y, Sidletskiy O, Bilski P, Twardak A, Nikl M (2016) Aluminum and gallium substitution in yttrium and lutetium aluminum–gallium garnets: investigation by single-crystal NMR and TSL methods. *J. Phys. Chem. C* 120:24400.

16. Geller S (1967) Crystal chemistry of the garnets. *Z. Für Krist.-Cryst. Mater.* 125:1.
17. De Pape R, Portier J, Gauthier G, Hagenmuller P (1967) Les grLes grenats fluorés des éléments de transition $Li_3Na_3In_2F_{12}$ (M = Ti, V, Cr, Fe ou Co). *CR Acad Sc Paris Sér. C* 265:1244.
18. Wang X, Zhou G, Zhang H, Li H, Zhang Z, Sun Z (2012) Luminescent properties of yellowish orange $Y_3Al_{5-x}Si_xO_{12-x}N_x$:Ce phosphors and their applications in warm white light-emitting diodes. *J. Alloys Compd.* 519:149.
19. Asami K, Ueda J, Shiraiwa M, Fujii K, Yashima M, Tanabe S (2019) Redshift and thermal quenching of Ce^{3+} emission in $(Gd,Y)_3(Al,Si)_5(O,N)_{12}$ oxynitride garnet phosphors. *Opt. Mater.* 87:117.
20. Shannon RD (1976) Revised effective ionic radii and systematic studies of interatomic distances in halides and chalcogenides. *Acta Crystallogr. A* 32:751.
21. Ottonello G, Bokreta M, Sciuto PF (1996) Parameterization of energy and interactions in garnets: End-member properties. *Am. Mineral.* 81:429.
22. Geiger CA, Armbruster T (1997) $Mn_3Al_2Si_3O_{12}$ spessartine and $Ca_3Al_2Si_3O_{12}$ grossular garnet: Structural dynamic and thermodynamic properties. *Am. Mineral.* 82:740.
23. Meagher E (1975) The crystal structures of pyrope and grossularite at elevated temperatures. *Am. Mineral. J. Earth Planet. Mater.* 60:218.
24. Pkandl W (1966) Verfeinerung der Kristallstruktur des Grossulars mit Neutronen-und Röntgenstrahlbeugung. *Z. Für Krist.-Cryst. Mater.* 123:81.
25. Sawada H (1999) Electron density study of garnets: $Z_3Al_2Si_3O_{12}$(Z = Mg, Fe, Mn, Ca) and $Ca_3Fe_2Si_3O_{12}$. *J. Solid State Chem.* 142:273.
26. Gentile A, Roy R (1960) Isomorphism and crystalline solubility in the garnet family. *Am. Mineral. J. Earth Planet. Mater.* 45:701.
27. Novak GA, Gibbs GV (1971) The crystal chemistry of the silicate garnets. *Am. Mineral. J. Earth Planet. Mater.* 56:791.
28. Marin SJ, O'Keeffe M, Young VG, Von Dreele RB (1991) The crystal structure of $Sr_3Y_2Ge_3O_{12}$. *J. Solid State Chem. Fr.* 91:173.
29. Pasiński D, Sokolnicki J (2017) Luminescence study of Eu^{3+}-doped garnet phosphors: relating structure to emission. *J. Alloys Compd.* 695:1160.
30. Li HL, Kuang XY, Mao AJ, Li Y, Wang SJ (2010) Study of local structures and optical spectra for octahedral Fe^{3+} centers in a series of garnet crystals $A_3B_2C_3O_{12}$ (A = Cd, Ca; B = Al, Ga, Sc, In; C = Ge, Si). *Chem. Phys. Lett.* 484:387.
31. Piccinelli F, Speghini A, Mariotto G, Bovo L, Bettinelli M (2009) Visible luminescence of lanthanide ions in $Ca_3Sc_2Si_3O_{12}$ and $Ca_3Y_2Si_3O_{12}$. *J. Rare Earths* 27:555.
32. Yamane H, Nagasawa T, Shimada M, Endo T (1997) $Ca_3Y_2(SiO_4)_3$. *Acta Crystallogr. Sect. C* 53:1367.
33. George NC, Denault KA, Seshadri R (2013) Phosphors for solid-state white lighting. *Annu. Rev. Mater. Res.* 43:481.
34. Blasse G, Bril A (1967) Investigation of Some Ce^{3+}-+vestigat Phosphors. *J. Chem. Phys.* 47:5139.
35. Dorenbos P (2002) 5d-level energies of Ce^{3+} and the crystalline environment. IV. Aluminates and "simple" oxides. *J. Lumin.* 99:283.
36. Mihóková E, Nikl M, Bacci M, Dušek M, Petříček V (2009) Assignment of 4f–5d absorption bands in Ce-doped $RAlO_3$ (R = La, Gd, Y, Lu) perovskites. *Phys. Rev. B* 79:195130.
37. Song Z, Liu QL (2019) Effect of polyhedron deformation on 5d energy level of Ce^{3+} in lanthanide aluminum perovskites. *Phys. Chem. Chem. Phys.* 21:2372.
38. Rotenberg M, Bivins R, Metropolis N, Wooten Jr JK (1959) *The 3-j and 6-j Symbols*. The MIT Press, Cambridge.
39. Nielson CW, Koster GF (1963) *Spectroscopic coefficients for the pn, dn, and fn configurations*. The MIT Press, Cambridge.
40. Tanner PA, Fu LS, Ning LX, Cheng BM, Brik MG (2007) Soft synthesis and vacuum ultraviolet spectra of YAG:Ce^{3+} nanocrystals: reassignment of Ce^{3+} energy levels. *J. Phys. Condens. Matter* 19:216213.
41. Noll W (1963) The silicate bond from the standpoint of electronic theory. *Angew. Chem. Int. Ed. Engl.* 2:73.
42. Etourneau J, Portier J, Menil F (1992) The role of the inductive effect in solid state chemistry: how the chemist can use it to modify both the structural and the physical properties of the materials. *J. Alloys Compd.* 188:1.
43. Kong YW, Song Z, Wang SX, Xia ZG, Liu QL (2018) The inductive effect in nitridosilicates and oxysilicates and its effects on 5d energy levels of Ce^{3+}. *Inorg. Chem.* 57:2320.
44. Kong YW, Song Z, Wang SX, Xia ZG, Liu QL (2018) Charge transfer, local structure, and the inductive effect in rare-earth-doped inorganic solids. *Inorg. Chem.* 57:12376.

45. Van Pieterson L, Heeroma M, De Heer E, Meijerink A (2000) Charge transfer luminescence of Yb^{3+}. *J. Lumin.* 91:177.
46. Nikl M, Yoshikawa A, Fukuda T (2004) Charge transfer luminescence in Yb^{3+}-containing compounds. *Opt. Mater.* 26:545.
47. Nakazawa E (1979) Charge transfer type luminescence of Yb^{3+} ions in RPO_4 and R_2O_2S (R = Y, La, and Lu). *J. Lumin.* 18:272.
48. Mizoguchi H, Eng HW, Woodward PM (2004) Probing the electronic structures of ternary perovskite and pyrochlore oxides containing Sn^{4+} or Sb^{5+}. *Inorg. Chem.* 43:1667.
49. Blasse G, Corsmit AF (1973) Electronic and vibrational spectra of ordered perovskites. *J. Solid State Chem.* 6:513.
50. Dieke GH, Crosswhite HM, Crosswhite H, others (1968) *Spectra and energy levels of rare earth ions in crystals.* John Wiley & Sons Ltd, Interscience Publishers, New York, London, Sydney, Toronto.
51. Dorenbos P (2013) A review on how lanthanide impurity levels change with chemistry and structure of inorganic compounds. *ECS J. Solid State Sci. Technol.* 2:R3001.
52. Zhang RJ, Song Z, He LZ, Xia ZG, Liu QL (2017) Improvement of red-emitting afterglow properties via tuning electronic structure in perovskite-type $(Ca_{1-x}Na_x)[Ti_{1-x}Nb_x]O_3$: Pr^{3+} compounds. *J. Alloys Compd.* 729:663.
53. Dorenbos P (2013) The electronic level structure of lanthanide impurities in $REPO_4$, $REBO_3$, $REAlO_3$, and $RE2O_3$ (RE = La, Gd, Y, Lu, Sc) compounds. *J. Phys. Condens. Matter* 25:225501.
54. Song Z, Liu XL, He LZ, Liu QL (2016) Correlation between the energy level structure of cerium-doped yttrium aluminum garnet and luminescent behavior at varying temperatures. *Mater. Res. Express* 3:055501.
55. Li JG, Sakka Y (2015) Recent progress in advanced optical materials based on gadolinium aluminate garnet ($Gd_3Al_5O_{12}$). *Sci. Technol. Adv. Mater.* 16:014902.
56. Hoshina T, Kuboniwa S (1971) 4f-5d transition of Tb^{3+} and Ce^{3+} in MPO_4 (M = sc, Y and lu). *J. Phys. Soc. Jpn.* 31:828.
57. Robbins DJ (1979) The effects of crystal field and temperature on the photoluminescence excitation efficiency of Ce^{3+} in YAG. *J. Electrochem. Soc.* 126:1550.
58. Tien TY, Gibbons EF, DeLosh RG, Zacmanidis PJ, Smith DE, Stadler HL (1973) Ce^{3+} activated $Y_3Al_5O_{12}$ and some of its solid solutions. *J. Electrochem. Soc.* 120:278.
59. Robbins D, Cockayne B, Glasper J, Lent B (1979) The temperature dependence of rare-earth activated garnet phosphors: I. Intensity and lifetime measurements on undoped and Ce-doped. *J. Electrochem. Soc.* 126:1213.
60. Liu X, Wang X, Shun W (1987) Luminescence properties of the Ce^{3+} ion in yttrium gallium garnet. *Phys. Status Solidi A* 101:K161.
61. Song Z, Wang ZZ, He LZ, Zhang RJ, Liu XL, Xia ZG, Geng WT, Liu QL (2017) After-glow, luminescent thermal quenching, and energy band structure of Ce-doped yttrium aluminum-gallium garnets. *J. Lumin.* 192:1278.
62. Ahn W, Kim YJ (2017) Substitutional solubility limit for Ce^{3+} ions in $Lu_3Al_5O_{12}$: xCe^{3+} and its effect on photoluminescence. *Ceram. Int.* 43:S412.
63. Dorenbos P (2013) Electronic structure and optical properties of the lanthanide activated $RE_3(Al_{1-x}Ga_x)_5O_{12}$ (RE = Gd, Y, Lu) garnet compounds. *J. Lumin.* 134:310.
64. Wu JL, Gundiah G, Cheetham AK (2007) Structure–property correlations in Ce-doped garnet phosphors for use in solid state lighting. *Chem. Phys. Lett.* 441:250.
65. Dorenbos P (2000) The 5d level positions of the trivalent lanthanides in inorganic compounds. *J. Lumin.* 91:155.
66. Dorenbos P (2002) Crystal field splitting of lanthanide $4f^{n-1}5d$-levels in inorganic compounds. *J. Alloys Compd.* 341:156.
67. Wang SX, Song Z, Kong YW, Xia ZG, Liu QL (2018) Crystal field splitting of $4f^{n-1}5d$-levels of Ce^{3+} and Eu^{2+} in nitride compounds. *J. Lumin.* 194:461.
68. Bachmann V, Ronda C, Meijerink A (2009) Temperature quenching of yellow Ce^{3+} luminescence in YAG:Ce. *Chem. Mater.* 21:2077.
69. Chen L, Chen X, Liu F, et al (2015) Charge deformation and orbital hybridization: intrinsic mechanisms on tunable chromaticity of $Y_3Al_5O_{12}$:Ce^{3+} luminescence by doping Gd^{3+} for warm white LEDs. *Sci. Rep.* 5:11514.
70. Ogieglo JM, Zych A, Ivanovskikh KV, Justel T, Ronda CR, Meijerink A (2012) Luminescence and energy transfer in $Lu_3Al_5O_{12}$ scintillators co-doped with Ce^{3+} and Tb^{3+}. *J. Phys. Chem. A* 116:8464.

71. Ueda J (2015) Analysis of optoelectronic properties and development of new persistent phosphor in Ce^{3+}-doped garnet ceramics. *J. Ceram. Soc. Jpn.* 123:1059.
72. Ogieglo JM, Katelnikovas A, Zych A, Justel T, Meijerink A, Ronda CR (2013) Luminescence and luminescence quenching in $Gd_3(Ga,Al)_5O_{12}$ scintillators doped with Ce^{3+}. *J. Phys. Chem. A* 117:2479.
73. Kaminska A, Duzynska A, Berkowski M, Trushkin S, Suchocki A (2012) Pressure-induced luminescence of cerium-doped gadolinium gallium garnet crystal. *Phys. Rev. B* 85:155111.
74. Zhong J, Zhao W, Zhuang W, Xiao W, Zheng Y, Du F, Wang L (2017) Origin of spectral blue shift of Lu^{3+}-codoped YAG:Ce^{3+} phosphor: first-principles study. *ACS Omega* 2:5935.
75. Görller-Walrand C, Binnemans K (1996) Rationalization of crystal-field parametrization. *Handb. Phys. Chem. Rare Earths* 23:121.
76. Liu G, Jacquier B (2006) *Spectroscopic properties of rare earths in optical materials.* Springer Science & Business Media, Tsinghua University Press and Springer-Verlag Berlin Heidelberg.
77. Momma K, Izumi F (2011) VESTA 3 for three-dimensional visualization of crystal, volumetric and morphology data. *J. Appl. Crystallogr.* 44:1272.
78. Seijo L, Barandiarán Z (2013) 4f and 5d Levels of Ce^{3+} in D2 8-fold oxygen coordination. *Opt. Mater.* 35:1932.
79. Song Z, Liu QL (2019) Effects of neighboring polyhedron competition on the 5d level of Ce^{3+} in lanthanide garnets. *J. Phys. Chem. C* 123:8656.
80. Song Z, Liu QL (2020) Structural indicator to characterize the crystal-field splitting of Ce^{3+} in garnets. *J. Phys. Chem. C* 124:870.
81. Blasse G, Grabmaier BC (1994) *Luminescent materials.* Springer Verlag, Berlin, Heidelberg, New York, London, Paris, Tokyo, Hong Kong, Barcelona, Budapest.
82. Ueda J, Tanabe S, Nakanishi T (2011) Analysis of Ce luminescence quenching in solid solutions between $Y_3Al_5O_{12}$ and $Y_3Ga_5O_{12}$ by temperature dependence of photoconductivity measurement. *J. Appl. Phys.* 110:53102.
83. Dorenbos P (2005) Thermal quenching of Eu^{2+} 5d–4f luminescence in inorganic compounds. *J. Phys. Condens. Matter* 17:8103.
84. Uheda K, Hirosaki N, Yamamoto Y, Naito A, Nakajima T, Yamamoto H (2006) Luminescence properties of a red phosphor, $CaAlSiN_3$: Eu^{2+}, for white light-emitting diodes. *Electrochem. Solid-State Lett.* 9:H22.
85. He LZ, Song Z, Xiang QC, Xia ZG, Liu QL (2016) Relationship between thermal quenching of Eu^{2+} luminescence and cation ordering in $(Ba_{1-x}Sr_x)_2SiO_4$:Eu phosphors. *J. Lumin.* 180:163.
86. Barry TL (1968) Fluorescence of Eu^{2+}-Activated phases in binary alkaline earth orthosilicate systems. *J. Electrochem. Soc.* 115:1181.
87. Catti M, Gazzoni G, Ivaldi G, Zanini G (1983) The β-α' phase transition of Sr_2SiO_4. I. Order–disorder in the structure of the α' form at 383K. *Acta Crystallogr. B* 39:674.
88. Catti M, Gazzoni G (1983) The β-α' phase transition of Sr_2SiO_4. II. X-ray and optical study, and ferroelasticity of the β form. *Acta Crystallogr. B* 39:679.
89. Liu GY, Rao GH, Feng XM, Yang HF, Ouyang ZW, Liu WF, Liang JK (2003) Structural transition and atomic ordering in the non-stoichiometric double perovskite $Sr_3Fe_xMo_{2-x}O_6$. *J. Alloys Compd.* 353:42.
90. Matković B, Popović S, Gržeta B, Halle R (1986) Phases in the system Ba_2SiO_4–Ca_2SiO_4. *J. Am. Ceram. Soc.* 69:132.
91. Park K, Kim J, Kung P, Kim SM (2010) New green phosphor $(Ba_{1.2}Ca_{0.8-x}Eu_x)SiO_4$ for white-light-emitting diode. *Jpn. J. Appl. Phys.* 49:020214.
92. Fukuda K, Ito M, Iwata T (2007) Crystal structure and structural disorder of $(Ba_{0.65}Ca_{0.35})_2SiO_4$. *J. Solid State Chem.* 180:2305.
93. He LZ, Song Z, Jia XH, Xia ZG, Liu QL (2018) Consequence of optimal bonding on disordered structure and improved luminescence properties in T-phase $(Ba,Ca)_2SiO_4$: Eu^{2+} phosphor. *Inorg. Chem.* 57:4146.
94. Ji H, Huang Z, Xia Z, Molokeev MS, Atuchin VV, Fang M, Liu Y (2015) Discovery of new solid solution phosphors via cation substitution-dependent phase transition in $M_3(PO_4)_2$: Eu^{2+} (M = Ca/Sr/Ba) quasi-binary sets. *J. Phys. Chem. C* 119:2038.
95. Xia ZG, Liu H, Li X, Liu C (2013) Identification of the crystallographic sites of Eu^{2+} in $Ca_9NaMg(PO_4)_7$: structure and luminescence properties study. *Dalton Trans.* 42:16588.
96. Xia ZG, Liu G, Wen J, Mei Z, Balasubramanian M, Molokeev MS, Peng L, Gu L, Miller DJ, Liu QL (2016) Tuning of photoluminescence by cation nanosegregation in the $CaMg_x(NaSc)_{1-x}Si_2O_6$ solid solution. *J. Am. Chem. Soc.* 138:1158.

97. Chen MY, Xia ZG, Molokeev MS, Wang T, Liu QL (2017) Tuning of photoluminescence and local structures of substituted cations in $xSr_2Ca(PO_4)_2$-$(1-x)Ca_{10}Li(PO_4)_7$: Eu^{2+} phosphors. *Chem. Mater.* 29:1430.

98. Chen MY, Xia ZG, Molokeev MS, Lin CC, Su CC, Chuang YC, Liu QL (2017) Probing Eu^{2+} luminescence from different crystallographic sites in $Ca_{10}M(PO_4)_7$: Eu^{2+} (M = Li, Na, and K) with β-$Ca_3(PO_4)_2$-type structure. *Chem. Mater.* 29:7563.

99. Shang M, Fan J, Lian H, Zhang Y, Geng D, Lin J (2014) A double substitution of Mg^{2+}–Si^{4+}/Ge^{4+} for $Al_{(1)}^{3+}$–$Al_{(2)}^{3+}$ in Ce^{3+}-doped garnet phosphor for white LEDs. *Inorg. Chem.* 53:7748.

100. Huang KW, Chen W-T, Chu C-I, Hu SF, Sheu HS, Cheng BM, Chen JM, Liu RS (2012) Controlling the activator site to tune europium valence in oxyfluoride phosphors. *Chem. Mater.* 24:2220.

101. Li G, Lin CC, Wei Y, Quan Z, Tian Y, Zhao Y, Chan TS, Lin J (2016) Controllable Eu valence for photoluminescence tuning in apatite-typed phosphors by the cation cosubstitution effect. *Chem Commun* 52:7376.

102. Wang T, Yang JJ, Mo YD, Bian L, Song Z, Liu QL (2013) Synthesis, structure and tunable red emissions of $Ca(Al/Si)_2N_2(N_{1-x}O_x)$: Eu^{2+} prepared by alloy-nitridation method. *J. Lumin.* 137:173.

103. Huang ZK, Sun WY, Yan DS (1985) Phase relations of the Si_3N_4-AlN-CaO system. *J. Mater. Sci. Lett.* 4:255.

104. Yang J, Wang T, Chen D, Chen G, Liu Q (2012) An investigation of Eu^{2+}-doped $CaAlSiN_3$ fabricated by an alloy-nitridation method. *Mater. Sci. Eng. B* 177:1596.

105. Piao X, Horikawa T, Hanzawa H, Machida K (2006) Characterization and luminescence properties of $Sr_2Si_5N_8$: Eu^{2+} phosphor for white light-emitting-diode illumination. *Appl. Phys. Lett.* 88:161908.

106. Piao X, Machida K, Horikawa T, Hanzawa H, Shimomura Y, Kijima N (2007) Preparation of $CaAlSiN_3$: Eu^{2+} phosphors by the self-propagating high-temperature synthesis and their luminescent properties. *Chem. Mater.* 19:4592.

107. Wang T, Zheng P, Liu XL, Chen HF, Yang SS, Liu QL (2014) Decay behavior analysis of two-peak emission in $Ca(Al/Si)_2N_2(N_{1-x}O_x)$: Eu^{2+} Phosphors. *J. Electrochem. Soc.* 161:H25.

108. Wang T, Xiang QC, Xia ZG, Chen J, Liu QL (2016) Evolution of structure and photoluminescence by cation cosubstitution in Eu^{2+}-doped $(Ca_{1-x}Li_x)(Al_{1-x}Si_{1+x})N_3$ solid solutions. *Inorg. Chem.* 55:2929.

109. Morrison CA (1980) Host dependence of the rare-earth ion energy separation $4f^N$–$4f^{N-1}$ nl. *J. Chem. Phys.* 72:1001.

110. Dorenbos P (2000) 5d-level energies of Ce^{3+} and the crystalline environment. I. Fluoride compounds. *Phys. Rev. B – Condens. Matter Mater. Phys.* 62:15640.

111. Li Y, Hirosaki N, Xie R, Takeka T, Mitomo M (2009) Crystal, electronic structures and photoluminescence properties of rare-earth doped $LiSi_2N_3$. *J. Solid State Chem.* 182:301.

112. Wu Q, Li Y, Wang X, Zhao Z, Wang C, Li H, Mao A, Wang Y (2014) Novel optical characteristics of Eu^{2+} doped and Eu^{2+}, Ce^{3+} co-doped $LiSi_2N_3$ phosphors by gas-pressed sintering. *RSC Adv.* 4:39030.

113. Piao X, Horikawa T, Hanzawa H, Machida K (2006) Preparation of $(Sr_{1-x}Ca_x)_2Si_5N_8/Eu^{2+}$ solid solutions and their luminescence properties. *J. Electrochem. Soc.* 153:H232.

114. Piao X, Machida K, Horikawa T, Hanzawa H (2007) Synthesis of nitridosilicate $CaSr_{1-x}Eu_xSi_5N_8$ (x = 0–1) phosphor by calcium cyanamide reduction for white light-emitting diode applications. *J. Electrochem. Soc.* 155:J17.

115. Xiaoming T, Yuanhong L, Yuzhu L, Yunsheng H, Huaqiang H, Zhuang W dong (2009) Preparation and luminescence properties of the red-emitting phosphor $(Sr_{1-x}Ca_x)_2Si_5N_8$: Eu^{2+} with different Sr/Ca ratios. *J. Rare Earths* 27:58.

116. Li YQ, de With G, Hintzen HT (2008) The effect of replacement of Sr by Ca on the structural and luminescence properties of the red-emitting $Sr_2Si_5N_8$:Eu^{2+} LED conversion phosphor. *J. Solid State Chem.* 181:515.

117. Wang T, Zheng P, Liu XL, Chen HF, Bian L, Liu QL (2014) Effects of replacement of AlO+ for SiN+ on the structure and optical properties of $Sr_2Si_5N_8$: Eu^{2+} phosphors. *J. Lumin.* 147:173.

118. Xie RJ, Mitomo M, Uheda K, Xu FF, Akimune Y (2002) Preparation and luminescence spectra of calcium-and rare-earth (R = Eu, Tb, and Pr)-codoped α-SiAlON ceramics. *J. Am. Ceram. Soc.* 85:1229.

119. Van Krevel J, Van Rutten J, Mandal H, Hintzen H, Metselaar R (2002) Luminescence properties of terbium-, cerium-, or europium-doped α-sialon materials. *J. Solid State Chem.* 165:19.

120. Xie RJ, Hirosaki N, Sakuma K, Yamamoto Y, Mitomo M (2004) Eu^{2+}-doped Ca-α-SiAlON: a yellow phosphor for white light-emitting diodes. *Appl. Phys. Lett.* 84:5404.

121. Mueller-Mach R, Mueller G, Krames MR, Höppe HA, Stadler F, Schnick W, Juestel T, Schmidt P (2005) Highly efficient all-nitride phosphor-converted white light emitting diode. *Phys. Status Solidi A* 202:1727.

122. Chen WT, Sheu HS, Liu RS, Attfield JP (2012) Cation-size-mismatch tuning of photoluminescence in oxynitride phosphors. *J. Am. Chem. Soc.* 134:8022.
123. Liu XL, Song Z, Kong YW, Wang SX, Zhang SY, Xia ZG, Liu QL (2019) Effects of full-range Eu concentration on $Sr_{2-2x}Eu_{2x}Si_5N_8$ phosphors: a deep-red emission and luminescent thermal quenching. *J. Alloys Compd.* 770:1069.
124. Fuertes A (2006) Prediction of anion distributions using Pauling's second rule. *Inorg. Chem.* 45:9640.
125. Jung YW, Lee B, Singh SP, Sohn KS (2010) Particle-swarm-optimization-assisted rate equation modeling of the two-peak emission behavior of non-stoichiometric $CaAl_xSi_{(7-3x)/4}N_3:Eu^{2+}$ phosphors. *Opt. Express* 18:17805.
126. Yeh CW, Chen WT, Liu RS, Hu SF, Sheu HS, Chen JM, Hintzen HT (2012) Origin of thermal degradation of $Sr_{2-x}Si_5N_8$: Eu_x phosphors in air for light-emitting diodes. *J. Am. Chem. Soc.* 134:14108.
127. Hirosaki N, Xie RJ, Kimoto K, Sekiguchi T, Yamamoto Y, Suehiro T, Mitomo M (2005) Characterization and properties of green-emitting β-SiAlON: Eu^{2+} powder phosphors for white light-emitting diodes. *Appl. Phys. Lett.* 86:211905.
128. Kimoto K, Xie RJ, Matsui Y, Ishizuka K, Hirosaki N (2009) Direct observation of single dopant atom in light-emitting phosphor of β-SiAlON: Eu^{2+}. *Appl. Phys. Lett.* 94:041908.
129. Li Y, Hirosaki N, Xie R, Takeda T, Mitomo M (2008) Crystal and electronic structures, luminescence properties of Eu^{2+}-doped $Si_{6-z}Al_zO_zN_{8-z}$ and $M_ySi_{6-z}Al_{z-y}O_{z+y}N_{8-z-y}$ (M = 2Li, Mg, Ca, Sr, Ba). *J. Solid State Chem.* 181:3200.
130. Wang SX, Liu XL, Qu BY, Song Z, Wang ZZ, Zhang SY, Wang FX, Geng WT, Liu QL (2019) Green persistent luminescence and the electronic structure of β-Sialon:Eu^{2+}. *J. Mater. Chem. C* 7:12544.

5 Design of Narrowband Emission Phosphors

Mu-Huai Fang and Ru-Shi Liu

CONTENTS

DOI: 10.1201/9781003098669-5

5.1   INTRODUCTION

For the past few years, many traditional lighting devices, including mercury lamps and incandescent bulbs, were substituted with light-emitting diodes (LEDs). Compared with traditional devices, LED possesses the advantages of high efficiency, high brightness, long lifetime, etc. Phosphors, the inorganic materials to tune the luminescent properties of LEDs, are an important component during the development of LEDs. The most famous white LED is fabricated with a blue LED chip and the $Y_3Al_5O_{12}:Ce^{3+}$ (YAG) yellow phosphor. Such a device is easy to fabricate; however, the color-rendering index of such a white LED device is low, which cannot represent the practical color correctly. Nowadays, not only the high brightness of LED devices is required, but the color performance of them is also important. In the past decade, a large number of commercial phosphors, including (Ca,Sr) $AlSiN_4:Eu^{2+}$ and $(Sr,Ba)_2Si_5N_8:Eu^{2+}$, are successfully developed to improve the disadvantages of the aforementioned devices. Nevertheless, the full width at half maximum (*fwhm*) of these phosphors is still too broad to achieve strict narrowband emission for high-quality devices. The part of the emission spectrum longer than 650 nm is less sensitive to the human eye, decreasing the luminous efficacy. Furthermore, the broadband emission phosphors will result in serious energy loss after penetrating the color filter in the backlighting devices. Therefore, well-defined strategies to achieve the narrowband emission phosphors are urgent to be established. This chapter will start with the selection of the suitable activators, for which the Eu^{2+} and Mn^{4+} ions are selected and the reasons will be introduced as well. The development of the narrowband emission phosphors in this chapter is shown in Figure 5.1. The key strategies to approach the narrowband properties from the perspective of structural properties will be explained, including the number of dopant sites, the first and second coordinated shell, electron–lattice interaction, and the volume size of the coordinated environment. Then, the structural and luminescent properties of the nitride, oxynitride, oxide, and fluoride phosphors will be introduced to interpret the key strategies. Finally, the perspectives to improve the performance of the currently developed narrowband emission phosphors will be provided.

5.2   KEY POINTS TO NARROWBAND EMISSION

Given that the narrowband emission is important for LED, the control parameters should be introduced in detail. In this section, we will introduce the key points to narrowband emission from the choice of activators to the first coordinated environment and the properties of the host structure.

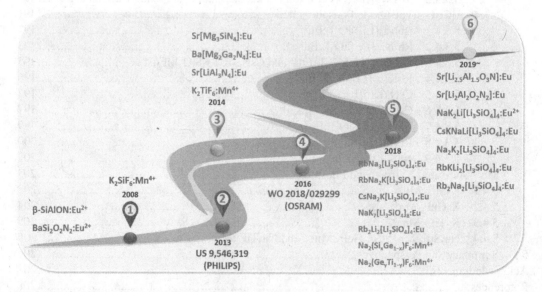

FIGURE 5.1   Development of the narrowband emission phosphors in this chapter.

5.2.1 Choice of the Activators

The phrase "activator" indicates that an ion can emit light after being illuminated by the excitation light. Usually, the transition-metal and rare-earth ions will be selected as the activators. In this section, we will introduce three kinds of transition, $f–f$ transition, $f–d$ transition, and $d–d$ transition and demonstrate the prominent candidates for the narrowband emission.

5.2.1.1 Rare-Earth Ions

The rare-earth ions are provided with the electron configuration of $4f^n5s^25p^6$. For the trivalent lanthanides, from Ce^{3+} to Lu^{3+} ions, there are 1–14 $4f$ electrons, in which their orbitals are in the inner shell and shielded by the $5s^2$ and $5p^6$ orbitals. In this regard, the difference in the $4f$ electron configuration between host lattices is scarce. These properties make the rare-earth ions in obvious contrast with the transition-metal ions, in which the $3d$ orbitals are in the outer shell and will be easily influenced by the host lattices. The rare-earth ions can be further divided into two groups. One is the $4f–4f$ electronic transition group and the other is $5d–4f$ electronic transition group. For the $4f–4f$ electronic transition group, the emission and excitation will show the sharp-line spectra because the ground and the excited states belong to the parallel parabola in the configuration coordinate diagram. In spite of the narrowband emission from the $4f–4f$ electronic transition group, the excitation spectrum also belongs to the sharp-line excitation. That is to say, the excitation wavelength should be very accurate to excite the activator, which is unsuitable for the LED application. Moreover, emission intensity for $4f–4f$ electronic transition group is typically weak due to the forbidden selection rule of $4f–4f$ transition. Therefore, the discussion of the $4f–4f$ electronic transition group will be excluded in this section. In contrast, for the $5d–4f$ electronic transition group, the emission and excitation spectra will be broader than that of the $4f–4f$ electronic transition group due to the allowed-selection rule. Moreover, the emission intensity of the $5d–4f$ electronic transition group will be stronger than that of the $4f–4f$ electronic transition group as well.

To understand the behavior of $5d$ bands, their location can be well predicted by the following formula [1]:

$$E(\text{Ln}, A) = 49,340 \text{ cm}^{-1} - D(A) + \Delta E^{\text{Ln,Ce}},\tag{5.1}$$

where $E(\text{Ln}, A)$ is $f–d$ energy difference with the unit of cm^{-1}, A is the selected compound, $D(A)$ is the crystal field depression, $49,340 \text{ cm}^{-1}$ is the $f–d$ transition energy of Ce^{3+} in the free (gaseous) ion condition. $\Delta E^{\text{Ln,Ce}}$ is defined as the difference in $f–d$ energy of Ln^{3+} with that of the first electric dipole allowed transition in Ce^{3+}, known as $E(\text{Ln}, A) - E(\text{Ce}, A)$. Here, $D(Ce, A)$ is the decreasing of the energy when a Ce^{3+} is incorporated into a selected A compound. The redshift is approximated the same between different lanthanide ions in the selected A compound and one can average $D(\text{Ln}, A)$ to obtain $D(A)$ in the equation. Besides, $\Delta E^{\text{Ln,Ce}}$ is approximately independent of the type of compound. As a result, $\Delta E^{\text{Ln,Ce}}$ is the intrinsic property and determined by averaging the values in different compounds, as shown in Table 5.1.

Within the $5d–4f$ electronic transition group, Eu^{2+} and Ce^{3+} are the most common activators in the phosphor community.

Ce^{3+} possesses only one $4f$ electron, of which the excitation will be from $4f^1$ to $5d^1$ state. The energy of the $5d$ state will be easily affected by the crystal field strength. In the free ion condition, the five d of the lanthanide ions' orbitals possess the same energy. Namely, the energy of d_{xy}, d_{yz}, d_{xz}, d_{x2-y2}, and d_{z2} orbital is the same, which is called degenerate states. However, when the lanthanide ions such as Ce^{3+} ion are put in in the octahedral environment, the electrons around the lanthanide ion will encounter a repulsion force, the d orbitals will separate into a higher energy state of e_g (d_{x2-y2} and d_{z2}) and lower energy state of t_{2g} (d_{xy}, d_{yz}, and d_{xz}), as shown in Figure 5.2. The energy difference between these two states is identified as Δ_o. The ground state of Ce^{3+} consists of two ^{2}F-term states, the $^2F_{7/2}$ and $^2F_{5/2}$ states, separated by typically around 2000 cm^{-1} due to the spin-orbital splitting.

TABLE 5.1

Average $E^{Ln,Ce}$ Values Averaged Over Several Different Compounds

Ln	$\Delta E^{Ln,Ce}$ Allowed fd	N
Pr	$12,240 \pm 750$	64
Nd	$22,700 \pm 650$	18
Pm	$(25,740)$	0
Sm	$26,500 \pm 460$	2
Eu	$35,900 \pm 380$	4
Gd	$45,800$	1
Tb	$13,200 \pm 920$	30
Tb[b]	–	–
Dy	$25,100 \pm 610$	4
Dy[b]	–	–
Ho	$31,800 \pm 1400$	5
Er	$30,000 \pm 1300$	8
Tm	$29,300 \pm 1100$	9
Yb	$38,000 \pm 570$	3
Lu	$49,170$	1

Source: Dorenbos, P. *J. Lumin.* 91, 155, 2000. With permission.

Accordingly, we can obtain two emission peaks from the lowest $5d$ states to the $^2F_{7/2}$ and $^2F_{5/2}$ states. However, this intrinsic property makes Ce^{3+} unsuitable for the narrowband emission activator. The decay time of Ce^{3+} emission is around 10^{-7}–10^{-8} s, which is the shortest in the lanthanide group due to the single f electron configuration.

Another common lanthanide activator is Eu^{2+}. Different from the Ce^{3+} ion that only possesses one $4f$ electron, the Eu^{2+} ion is provided with a $4f^7$ electron configuration. The Eu^{2+} ion can provide either broad- or sharp-line emission spectra due to the crystal field strength. When the crystal

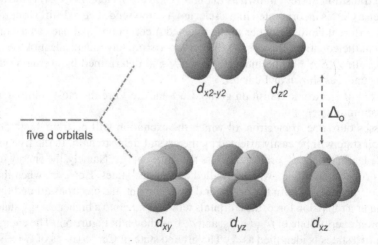

FIGURE 5.2 The splitting of d_{xy}, d_{yz}, d_{xz}, d_{x2-y2}, and d_{z2} orbitals in the octahedrally coordinated environment.

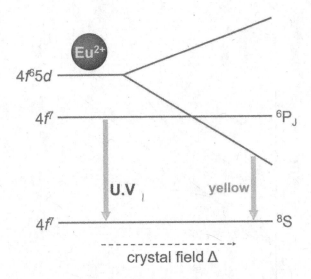

FIGURE 5.3 Schematic of Eu^{2+} energy level with different crystal field strengths Δ.

strength is strong, the electronic transition of the emission will be from the lowest excited $4f^6 5d^1$ state to the $4f^7$ ground state. The stronger the crystal field strength is, the lower the lowest $4f^6 5d^1$ state is, and the longer emission wavelength can be expected. However, if the crystal field strength is too weak, the emission from the lowest excited state will be from the $^6P_{7/2}$ ($4f^7$) to the $^8S_{7/2}$ ($4f^7$) state and the sharp-line emission spectra can be expected, as shown in Figure 5.3 [2]. The lifetime of Eu^{2+} broadband emission is around 10^{-6} s. This value is longer than other allowed-transition lanthanide ions. The ground state of Eu^{2+} is $^8S_{7/2}$ ($4f^7$), which is an octet multiplicity. By contrast, the excited state of $4f^6 5d^1$ might be octet and sextet, where the sextet portion with the spin-forbidden character slows down the transition rate [3].

5.2.1.2 Transition-Metal Ions

The transition-metal ions are provided with the outer electron configuration of $(n-1)d^{1-10}ns^{1-2}$. The emission spectra involved the $d–d$ electronic transition. The most famous narrowband emission activator in the transition-metal group is Mn^{4+}. Mn^{4+} possesses the d^3 electronic configuration. To understand the photoluminescent properties, the Tanabe–Sugano diagram designed by Yukito Tanabe and Satoru Sugano should be introduced [4]. The Tanabe–Sugano diagram utilizes coordination chemistry to realize the electromagnetic spectrum of coordinated compounds, which can be applied in high spin and low spin cases when there are more than $3d$ electrons.

The d^3 Tanabe–Sugano diagram in the octahedral coordinated environment is shown in Figure 5.4. The x-axis is in terms of the ligand-field splitting parameter (D_q) divided by the Racah parameter (B), while the y-axis is in terms of energy (E) divided by the Racah parameter (B). Three Racah parameters, A, B, and C, are used to describe the Coulomb interaction between the electrons. A is the average of the total repulsion between the electrons, which is a constant for a specific d-electron configuration. Alternatively, B and C are related to the individual d-electron repulsion. The transition-metal ions possess nearly the same B/C values around 4.5. When the transition-metal ions are placed in a structure, they will encounter the nephelauxetic effect, reducing the Racah parameter B from the free ion condition (B_0). The degree of reduction can be expressed in terms of the nephelauxetic parameter, β, where the formula is

$$\beta = \frac{B(\text{complex})}{B_0(\text{complex})} \tag{5.2}$$

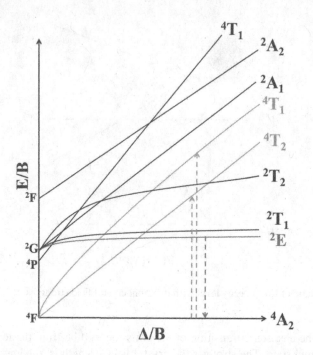

FIGURE 5.4   The d^3 Tanabe–Sugano diagram.

The trend of the degree of the nephelauxetic effect of different ligands can be shown as [5]:

$$F^- < H_2O < NH_3 < en < [NCS]^- < Cl^- < [CN]^- < Br^- < N_3^- < I^- \qquad (5.3)$$

One should notice that the fluoride ions provide the weakest nephelauxetic effect, namely, the minimal decrease of the Racah parameter (B). Given the y-axis is E/B in the Tanabe–Sugano diagram, the minimal decrease of B is a benefit for the researchers to develop the phosphor with the higher emission energy. As a result, the octahedrally coordinated Mn^{4+}-doped fluorides are good candidates for the narrowband emission phosphors with higher emission energy, as shown in Figure 5.5 [6].

The typical excitation and emission spectrum of the Mn^{4+}-doped fluoride phosphors are shown in Figure 5.6(a) and the correlated electronic transitions are shown in Figure 5.4. The excitation

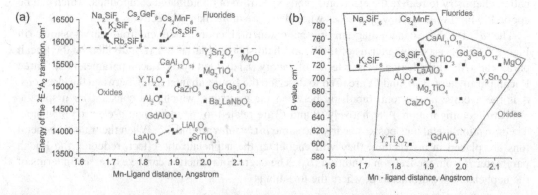

FIGURE 5.5   (a) The relationship between the position of the ^2E_g level. (Brik, M.G. and Srivastava, A.M., *J. Lumin.* 133, 69, 2013. With permission.) (b) Racah parameter with the Mn-ligand bond distance. (Brik, M.G. and Srivastava, A.M., *J. Lumin.* 133, 69, 2013. With permission.)

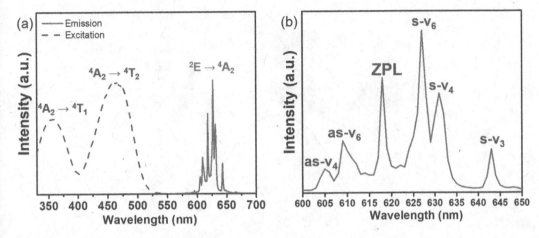

FIGURE 5.6 (a) Typical excitation and emission spectrum of fluoride phosphor and (b) Stokes and anti-Stokes phonon sidebands and zero-phonon line emission.

spectrum belongs to $^4A_2 \rightarrow {}^4T_2$ and $^4A_2 \rightarrow {}^4T_1$ spin-allowed electronic transition. In most cases, the excitation peak maximum of $^4A_2 \rightarrow {}^4T_2$ transition can fit well with the 450 nm LEDs, making the Mn^{4+}-doped fluoride phosphors promising candidates for practical applications. After bumping into the excited state, the electrons will encounter vibrational relaxation to the ^2E state if the Mn^{4+} is in the strong crystal field environment. The sharp-line emission will be observed from the $^2E \rightarrow {}^4A_2$ spin-forbidden electronic transition.

One may notice that the transition violates the Laporte selection rule if the Mn^{4+} is coordinated with six fluoride ions to form the MnF_6 octahedron. To break the selection rule, the symmetric $^4A_2$ state will couple with the asymmetric vibrational mode and then further break the symmetry of the overall wave function. The typical sharp-line emission spectrum is shown in Figure 5.6(b). The peaks longer than 617 nm are attributed to Stokes shift peaks, while the ones shorter than 617 nm are attributed to anti-Stokes shift peaks. These peaks belong to phonon sidebands (PS), which couple with the v_6 (t_{2u} bending), v_4 (t_{1u} bending), and v_3 (t_{1u} stretching) modes. Alternatively, the peak at around 617 nm is the zero-phonon line (ZPL). This emission peak is attributed to the transition that does not encounter vibrational relaxation. The ZPL is also sensitive to the symmetry of the coordinated environment. Typically, the more the asymmetric coordinated environment of Mn^{4+} is, the stronger the ZPL intensity we can observe. It is because the ZPL emission results from static symmetry distortions, which remove the inversion symmetry of the MnF_6^{2-}. That is to say, the distortion leads to an admixture of odd parity components to even parity ^2E and $^4A_2$ wave functions, thus removing the Laporte selection rule that the dipole transitions between states of the same parity are forbidden [7].

5.2.2 Number of the Dopant Sites

After choosing the proper activators, the number of possible crystallographic sites is the first point we should check. The different dopant sites for the activator will provide different nephelauxetic effects, crystal field strength, etc., leading to different emission spectra of the individual site. Given that the overall photoluminescent spectra are the combination of the emission from each dopant site, the emission bandwidth will become broader if there are multi-dopant sites in the phosphor structure, as shown in Figure 5.7. This broadening effect will depend on the concentration of the activators. Take the $Sr_2Si_5N_8$:Eu^{2+} phosphor as an example, there are two dopant sites for the Eu^{2+} ions, which are ten and eight coordinated sites [8]. The Eu^{2+} will prefer to occupy the ten coordinated sites under the low Eu^{2+} concentration condition, while Eu^{2+} will gradually occupy the eight

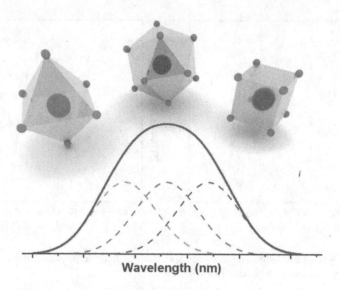

FIGURE 5.7   Scheme of the luminescent spectrum with multi-emission sites.

coordinated sites when the Eu^{2+} concentration increases, leading to the broadening of the emission bandwidth. One should notice that in some special cases, the activators will not only occupy the crystallographic sites originally with the atoms, but they may also dope in the interstitial site. One of the famous examples is the Eu-doped β-$Si_{6-z}Al_zO_zN_{8-z}$ (β-SiAlON) phosphor. The Eu^{2+} ion does not substitute into any crystallographic site. Instead, it occupies the interstitial site along the hexagonal channel directly observed by the scanning transmission electron microscopy, as reported by Kimoto's group [9].

5.2.3   First Coordinated Shell

The composition of the first coordinated shell is extremely important. The first coordinated shell is usually composed of anions. In most cases, the broadband emission spectra will be obtained when there is more than one type of ligand in the first coordinated shell. For the dopant site with only one type of ligand, the coordinated environment is homogeneous, leading to the single emission band, as shown in Figure 5.8(a). By contrast, when there is more than one type of ligand and these ligands are exactly at the same crystallographic site, it will encounter the so-called inhomogeneous broadening effect. Despite the single crystallographic dopant site acquired from the crystallographic information file (CIF), the average information combining all the possibilities is provided. That is to say, the emission spectrum from this single crystallographic dopant site is actually the overlapping of many emission spectra, resulting in the broadband emission, as shown in Figure 5.8(b). Take $SrAl_{2-x}Li_{2+x}O_{2+2x}N_{2-2x}$:$Eu^{2+}$ ($0.12 \leq x \leq 0.66$) as an example; there is only one crystallographic site for Sr^{2+} ion and it is coordinated by eight O/N anions. For a specific Sr^{2+}/Eu^{2+} site, one only knows that the total number of O^{2-} and N^{3-} anions equal eight. However, there are a large number of possibilities and each combination will have its own emission spectrum. The emission from the sum of these possibilities will then lead to a broadband emission.

5.2.4   Second Coordinated Shell

The composition of the second coordinated shell is also important. The second coordinated shell is usually composed of cations. Typically, the broadband emission spectra will be obtained when there is more than one type of cation in the second coordinated shell.

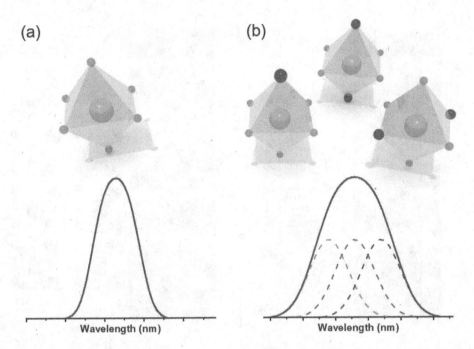

FIGURE 5.8 Scheme of the luminescent spectrum with (a) homogeneous coordinated environment and (b) inhomogeneous coordinated environment in a single crystallographic site.

The first situation is the partial substitution of the cations in the same crystallographic site. When there is a partial substitution of the cations in the structure, the coordinated environment will become diverse. For example, when one Sr^{2+} is substituted by the Eu^{2+}, the second coordinated environment should be Sr–Eu–Sr for the pure Sr-based phosphor. In contrast, for the (Sr,Ba) co-substitution case, the second coordinated environment around Eu^{2+} will be Sr–Eu–Sr, Ba–Eu–Sr, Ba–Eu–Ba, etc. As a result, the emission from the sum of these possibilities will also lead to a broader emission.

The second situation is the ordering properties in the structure. Some phosphors possess a channel structure, which is composed of different tetrahedra. Take $SrLiAl_3N_4:Eu^{2+}$ and $SrMg_2Al_2N_4:Eu^{2+}$ as examples [10, 11]. There are two and one Sr^{2+} sites in $SrLiAl_3N_4:Eu^{2+}$ and $SrMg_2Al_2N_4:Eu^{2+}$, respectively. The channel structure of $SrLiAl_3N_4:Eu^{2+}$ is composed of $[LiN_4]$ and $[AlN_4]$ tetrahedra, while the one of $SrMg_2Al_2N_4:Eu^{2+}$ is composed of $[MgN_4]$ and $[AlN_4]$ tetrahedra. The first coordinated environment of Eu^{2+} is the eight N^{3-} for both structures. Nevertheless, the ordering properties of these tetrahedra are quite different. For the channel structure of $SrLiAl_3N_4:Eu^{2+}$, one $[LiN_4]$ tetrahedra will follow with three $[AlN_4]$ tetrahedra along the channel direction, as shown in Figure 5.9(a). In contrast, $[MgN_4]$ and $[AlN_4]$ tetrahedra are randomly distributed along the channel of $SrMg_2Al_2N_4:Eu^{2+}$, resulting in many possibilities of the practical coordinated environment around Sr^{2+}, as shown in Figure 5.9(b). Even though $SrMg_2Al_2N_4:Eu^{2+}$ possesses less crystallographic dopant site compared to $SrLiAl_3N_4:Eu^{2+}$, the emission of $SrMg_2Al_2N_4:Eu^{2+}$ is actually the sum of these possibilities. As a result, the emission bandwidth of $SrLiAl_3N_4:Eu^{2+}$ (~1180 cm^{-1}) is much narrower than that of $SrMg_2Al_2N_4:Eu^{2+}$ (~1838 cm^{-1}) [10, 11].

The distance between the second coordinated cation is also important. When the activators are too closed, the energy transfer will occur easily, as shown in Figure 5.10(a). During this process, some of the energy will lose or even transfer to the defect sites, following with the redshift in the photoluminescence spectrum compared with the case without the energy transfer. This effect will lead to the overlapping of the luminescent spectra with different energies and the broadband emission can be expected. This effect can be suppressed by low activator concentration, while this may

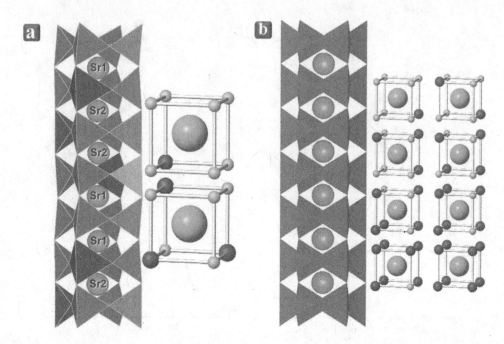

FIGURE 5.9   Channel structure and the microstructure of (a) SrLiAl$_3$N$_4$ (Fang, M.H., Leaño, J.L. and Liu, R.S., *ACS Energy Lett*. 3, 2573, 2018. With permission.) and (b) SrMg$_2$Al$_2$N$_4$. (Fang, M.H., Leaño, J.L. and Liu, R.S., *ACS Energy Lett*. 3, 2573, 2018. With permission.)

also lead to low absorption and external quantum efficiency. Therefore, another approach is to discover the structure with a longer cation distance. Some of the host lattices possess the channel structure, generating the far distance between the channels, as shown in Figure 5.10(b). Phosphors such as the nitride UCr$_4$C$_4$ system and the oxynitride β-SiAlON system are the most famous cases that will be introduced in the following content.

5.2.5   RIGIDITY AND ELECTRON–LATTICE INTERACTION

Structural rigidity is also an important parameter to achieve the narrowband emission spectra. The degree of condensation, which is denoted as κ and defined as the atomic ratio of tetrahedral centers

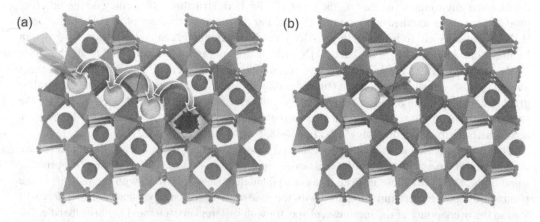

FIGURE 5.10   Phosphor materials with (a) short and (b) far distance between activators.

to vertices within one network, is a common parameter to evaluate rigidity. For $SrLiAl_3N_4:Eu^{2+}$, the degree of concentration $(Li^+ + 3Al^{3+})/4N^{3-}$ equals one, indicating the high rigidity in this structure. Structural rigidity is highly related to phonon energy, which is related to structural vibration. Phonon energy is typically very low and difficult to be observed under room-temperature conditions. As a result, one of the analysis methods is to measure the emission and excitation at a low temperature.

Take the $Y_3Al_5O_{12}:Ce^{3+}$ as an example. Some of the fine emission and excitation features can be observed when the spectra are measured at 7K, as shown in Figure 5.11(a). The interval of these characteristic peaks is around 200 cm^{-1}, which indicates the phonon energy. The intersection peak between the emission and excitation spectra at around 489 nm is the ZPL. Take $SrLiAl_3N_4:Eu^{2+}$ as another example. To avoid the thermal broadening of the emission spectrum, photoluminescence at 6K is shown in Figure 5.11(b). The ZPL energy can be observed at 15,797 cm^{-1}. Moreover, the vibronic structures can be approximately determined to 202 cm^{-1}, indicating the weak electron–lattice interaction in this system [10].

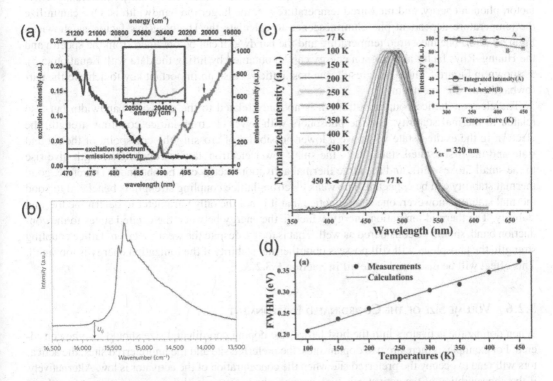

FIGURE 5.11 (a) Emission spectrum of YAG:Ce 0.033% for λ_{exc} = 450 nm taken at 7K; (inset) detail of the emission spectrum measured with higher resolution showing the zero-phonon line (ZPL) under 0.1 MPa. (Bachmann, V., Ronda, C. and Meijerink, A., *Chem. Mater.* 21, 2077, 2009. With permission.) (b) Low-temperature luminescence of next-generation red-emitting $Sr[LiAl_3N_4]:Eu^{2+}$. The emission spectrum for excitation of SLA at λ_{exc} = 470 nm at 6K with the vibronic structure of the emission transitions. The black arrow indicates the ZPL energy U_0 at 15,797 cm^{-1}. (Pust, P., Weiler, V., Hecht, C., Tücks, A., Wochnik, A.S., Henß, A.K., Wiechert, D., Scheu, C., Schmidt, P.J. and Schnick, W., *Nat. Mater.* 13, 891, 2014. With permission.) (c) Normalized emission spectra of $Sr_{7.7}Eu_{0.3}(Si_4O_{12})Cl_8$ at different temperatures from 77K to 450K under excitation at 320 nm; inset shows the temperature dependence of the integrated emission intensity (curve A) and the emission peak height (curve B) of $Sr_{7.7}Eu_{0.3}(Si_4O_{12})Cl_8$. (Liu, C., Qi, Z., Ma, C.G., Dorenbos, P., Hou, D., Zhang, S., Kuang, X., Zhang, J. and Liang, H., *Chem. Mater.* 26, 3709, 2014. With permission.) (d) Experimental (circle dots) and calculated (line) temperature dependence of the *fwhm* of the emission band measured under 320 nm excitation for $Sr_{7.7}Eu_{0.3}(Si_4O_{12})Cl_8$. (Liu, C., Qi, Z., Ma, C.G., Dorenbos, P., Hou, D., Zhang, S., Kuang, X., Zhang, J. and Liang, H., *Chem. Mater.* 26, 3709, 2014. With permission.)

When the electron is excited to the excited state, the phonon interacts with the excited electron and lifts the electron to the higher vibronic states. This effect is the electron–lattice interaction, which will lead to the Stokes shift and the broadening of the emission spectra. To evaluate the strength of the electron–lattice interaction, Huang–Rhys factor (S) is then employed and can be calculated from the following equations [12–14]:

$$E_{Stokes} = (2S - 1)h\upsilon \tag{5.4}$$

$$fwhm = \sqrt{8ln2} \times h\upsilon \times \sqrt{S} \times \sqrt{coth\left(\frac{h\upsilon}{2kT}\right)} \tag{5.5}$$

where E_{Stokes} is the Stoke shift energy; $h\upsilon$ is the phonon energy; $fwhm$ is the full width at half maximum ($fwhm$); T is the measured temperature. E_{Stokes}, as well as $fwhm$, are related to the Huang–Rhys factor and phonon energy, as shown in Equation (5.4). Moreover, the larger the Huang–Rhys factor, phonon energy, and measured temperature are, the larger the bandwidth is. One can utilize the temperature-dependent photoluminescence to calculate the Huang–Rhys factor, as shown in Figure 5.11(c). The measured temperature and the bandwidth can be obtained from the spectra and the Huang–Rhys factor and phonon energy can be obtained by fitting the data with Equation (5.5), as shown in Figure 5.11(d). As a result, the host lattice is also an important key to achieve the narrowband emission spectrum.

The electron–lattice coupling strength is not only related to the emission bandwidth but also related to thermal stability. For the phosphors with lower electron–lattice coupling strength, the electron in the excited state will have a lower possibility of crossing the crosspoint of the ground state and the first excited state. Also, the small lower electron–lattice coupling strength give rise to the small Stokes shift, and the large thermal activation energy can be obtained. Therefore, good thermal stability can be expected. The weak electron–lattice coupling strength is beneficial to good thermal stability; however, one should notice that it is not the only parameter to determine thermal stability. The thermal ionization energy, which is the energy between the excited states to the conduction band, should be considered as well. That is to say, despite the weak electron–lattice coupling strength, the phosphors will still possess poor thermal stability if the ionization energy is too small. This effect will be discussed in detail in **Section 5.3.2.3**.

5.2.6 VOLUME SIZE OF THE COORDINATED ENVIRONMENT

When doping the activators into the host lattice, the doping coordinated size should also be considered. For example, if there are two dopant sites, the preferred site and the unpreferred sites, the activators will tend to occupy the preferred site when the concentration of the activator is low. Alternatively, if the concentration of the activator is high enough, the activator will then occupy the preferred and non-preferred sites at the same time. That is to say, the difference between the two sites is important, especially in the condition of the low activator concentration. If the difference is large enough, the preferably occupied effect will be obvious. This means that it's still possible to obtain the narrowband emission even possessing more than one possible dopant site if the difference between these sites is large enough and the concentration of the activator is low enough. Cases such as the UCr_4C_4 alkali lithosilicate phosphors are good examples, where the coordinated volume in each possible dopant site is quite different. Nevertheless, how to define the "preferred" dopant site? Each ion will have its most suitable chemical bond distance between the given ligands. If the chemical bond distance is either too long or too short, the total energy will increase and lead to an unstable local structure. People are used to comparing the ionic radii of the dopant site with the activators. For example, Eu^{2+} (1.25 Å; CN = 8) (CN denotes coordinated number) may prefer to dope at the Sr^{2+} (1.26 Å; CN = 8) site [15]. Despite the same cation, it will possess different chemical bond distances and coordinated volume in different

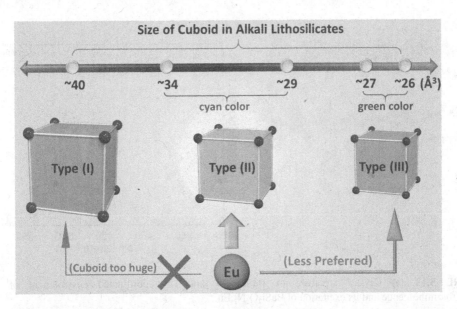

FIGURE 5.12 Scheme of the cuboid size tuning in oxide phosphors. (Fang, M.H., Mariano, C.O.M., Chen, P.Y., Hu, S.F. and Liu, R.S., *Chem. Mater.* 32, 1748, 2020. With permission.)

structures, indicating that only consideration of the ionic radii is not precise, and sometimes it will lead to incorrect analysis results. For example, the coordinated volumes of Rb^+ in $RbNa_2K[Li_3SiO_4]_4$ and $CsNaRbLi[Li_3SiO_4]_4$ are 42.6 and 33.8 $Å^3$, respectively [16, 17]. This difference makes Eu^{2+} prefer to occupy the Rb^+ site in $CsNaRbLi[Li_3SiO_4]_4$ and not prefer to occupy that in $RbNa_2K[Li_3SiO_4]_4$ [18]. The entire concept can be demonstrated using the study of UCr_4C_4 alkali lithosilicate, as shown in Figure 5.12 [18]. The coordinated size can be divided into three sets: (I) $\geq$ 40 $Å^3$, (II) 29–34 $Å^3$, and (III) 26–27 $Å^3$. In this system, the most suitable coordinated volume for Eu^{2+} is around 29–34 $Å^3$. When the Eu^{2+} dope into the smaller size ranging from 26 to 27 $Å^3$, local distortion is expected and Eu^{2+} will be less preferred in this site. Alternatively, for the crystallographic site with a size larger than 40 $Å^3$, a strong shift in the position of eight oxygen ions should be involved in forming a stable Eu–O bond. Nevertheless, the UCr_4C_4 system is highly rigid, making this process difficult. Accordingly, the Eu^{2+} cannot dope into these sites. Structures such as $RbNa_3[Li_3SiO_4]_4$ and $Rb_2Li_2[Li_3SiO_4]_4$ possess more than one crystallographic site; however, only one site is suitable for the Eu^{2+}, leading to the narrowband emission. The detailed discussion will be delivered in **Section 5.3.4**.

In summary, we comprehensively investigate the key parameters to obtain the narrowband emission, from the basic number of dopant sites and selection of the activator to the coordinated environment and structural rigidity. In the following section, we will introduce some of the famous Eu^{2+}-activated (oxy)-nitride phosphors, Eu^{2+}-activated oxide phosphors, and Mn^{4+}-activated fluoride phosphors to understand the key parameters introduced above.

5.3 EU²⁺-ACTIVATED NARROWBAND PHOSPHORS

In this section, we will briefly introduce the Eu^{2+}-activated nitride, oxynitride, and oxide phosphors to demonstrate the aforementioned key strategies for generating the narrowband emission.

5.3.1 TRADITIONAL Eu²⁺-ACTIVATED NARROWBAND PHOSPHOR

5.3.1.1 BaSi₂O₂N₂:Eu²⁺

$BaSi_2O_2N_2:Eu^{2+}$ possesses an orthorhombic structure with the space group of *Pbcn*, as shown in Figure 5.13(a) [19–21]. Si^{4+} is coordinated by three N^{3-} and one O^{2-} ions, forming the $SiON_3$

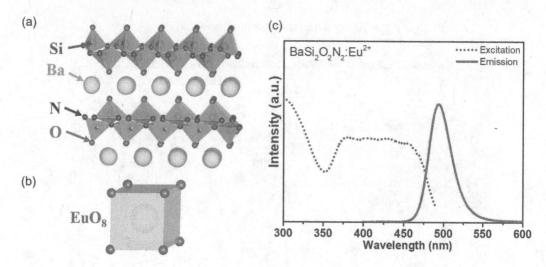

FIGURE 5.13 (a) Crystal structure of $BaSi_2O_2N_2$ and (b) coordinated environment of Eu^{2+}. (c) Photoluminescence and its excitation of $BaSi_2O_2N_2:Eu^{2+}$.

tetrahedral structure. The Eu^{2+} will occupy the Ba^{2+} site and there is only one Ba^{2+} site in the structure, which is beneficial to generate the narrowband emission. Moreover, Eu^{2+} is coordinated with eight O^{2-} ions with a bond distance of 2.73–3.088 Å, forming the cuboid-like coordinated environment, as shown in Figure 5.13(b). The cuboid coordinated environment is a highly symmetrical structure that is beneficial to generate a narrowband emission. Despite two more N^{3-} coordinated ions in the literature, the bond length of Ba–N is 3.343 Å. The degree of crystal field strength (Dq) can be written as Equation (5.6) [22]:

$$D_q = \frac{ze^2r^4}{6R^5} \tag{5.6}$$

where R is the distance between the central ion and the coordinated ligands, z is the valence of the anion, e is the charge of an electron, and r is the radius of the d wave function. Accordingly, the weak influence of N^{3-} could be expected owing to the far Ba–N distance. Under the excitation of 405 nm, $BaSi_2O_2N_2:Eu^{2+}$ exhibits the narrowband emission spectrum with the peak maximum and *fwhm* around 494 and 48 nm (~1985 cm^{-1}), respectively, as shown in Figure 5.13(c). However, the closest Ba–Ba distance is approximately 3.74712 Å, which is relatively short and results in a broader emission band than the other UCr_4C_4 nitrides or β-SiAlON system.

5.3.1.2 β-SiAlON:Eu²⁺

The abbreviation of β-SiAlON:Eu²⁺ is from the chemical formula of β-$Si_{6-z}Al_zO_zN_{8-z}$ ($0 < z \leq 4.2$). This structure is derived from the β-Si_3N_4 by substituting Si–N with Al–O [23]. β-SiAlON:Eu²⁺ belongs to the hexagonal structure with the space group of $P63/m$. Moreover, the structure is composed of the corner-shared (Si,Al)(O,N)₄ tetrahedral, which builds up the channel structure along the $\dot{c}$-axis, as shown in Figure 5.14(a). The Eu²⁺ ion does not occupy the crystallographic site; instead, it occupies the empty sites along the c-axis, which is proved by Kimoto et al. [9] with the aid of scanning transmission electron microscopy. Besides, ninefold coordination around the Eu²⁺ ions (EuN₉) is revealed by Wang et al. with the theoretical calculations, as shown in Figure 5.14(b) [24, 25]. The z values represent the content of the oxygen ions. Based on the z values, the β-SiAlON:Eu²⁺ can be further divided into high-z and low-z β-SiAlON:Eu²⁺ with the z value of 0.18 and 0.03, respectively. Under the excitation of 325 nm, the high-z β-SiAlON can provide the emission wavelength and *fwhm* of 540 and 54 nm (~1870 cm^{-1}), respectively, as shown in Figure 5.14(c) [26].

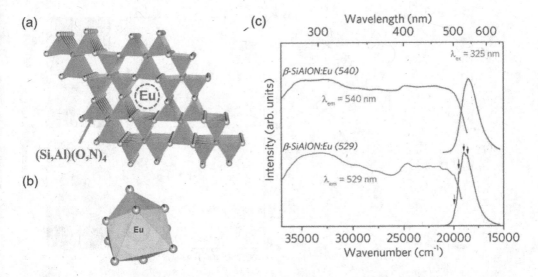

FIGURE 5.14 (a) Crystal structure of β-SiAlON and (b) coordinated environment of Eu^{2+}. (c) Photoluminescence and its excitation of β-SiAlON:Eu^{2+}. (Zhang, X., Fang, M.H., Tsai, Y.T., Lazarowska, A., Mahlik, S., Lesniewski, T., Grinberg, M., Pang, W.K., Pan, F., Liang, C. and Liu, R.S., *Chem. Mater.* 29, 6781, 2017. With permission.)

On the other hand, the low-z β-SiAlON possesses emission wavelength and *fwhm* of 528 and 49 nm ($\sim$1772 cm^{-1}), respectively [26].

Typically, the emission wavelength will be affected by the crystal field effect, nephelauxetic effect, and Stokes shift [27–29]. The nephelauxetic effects for high-z and low-z β-SiAlON:Eu^{2+} are nearly the same due to the predominant Si_3N_4 composition for both structures. The crystal field strength is affected by the bond distance and distortion index between the center ion and the ligands. The distortion index can be written as the following formula [30, 31]:

$$D = \frac{1}{n} \sum_{i=1}^{n} \frac{|l_i - l_{av}|}{l_{av}} \tag{5.7}$$

where l_i is the distance between the center ion and the n ligands; l_{av} is the average bond length; D is the distortion index. The larger the distortion index, the stronger the crystal field strength and the more redshifted luminescent spectrum are. The D increases from 0.06723 (low-z) to 0.06726 (high-z), which is approximately 0.036% and is comparable with the increase in l_{av} (0.087%). As a result, the nephelauxetic effect and crystal field strength are not the main factors to determine their emission wavelength. The host of β-SiAlON mainly consists of Si_3N_4, which shows a high degree of condensation and rigidity. This property is beneficial in narrowband emission. Furthermore, the low-z β-SiAlON possesses a smaller Huang–Rhys factor and electron–lattice coupling strength, which may lead to a narrower emission band, as shown in panels a and b of Figure 5.15. This effect leads to the shorter emission wavelength of the low-z β-SiAlON compared with the high-z one, as shown in Figure 5.14(c).

Debye temperature (θ_D) can be used to determine the rigidity as well with the aid of low-temperature heat capacity measurement, as shown in Figure 5.15(c). The higher the Debye temperature is, the higher rigidity is. The Debye temperature can be calculated using the following formula [30, 32, 33]:

$$C_p \approx \frac{12\, Nk_B\pi^4}{5} \left(\frac{T}{\theta_D}\right)^3 \tag{5.8}$$

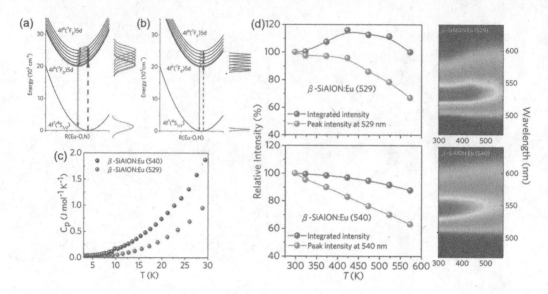

FIGURE 5.15 Configurational coordinate diagram representing the energetic structure of Eu^{2+} system for (a) strong (Zhang, X., Fang, M.H., Tsai, Y.T., Lazarowska, A., Mahlik, S., Lesniewski, T., Grinberg, M., Pang, W.K., Pan, F., Liang, C. and Liu, R.S., *Chem. Mater.* 29, 6781, 2017. With permission.) and (b) weak electron–lattice coupling of the 5d electron with lattice, the respective emission, and absorption line shapes, which are presented on the right side of the figures. (Zhang, X., Fang, M.H., Tsai, Y.T., Lazarowska, A., Mahlik, S., Lesniewski, T., Grinberg, M., Pang, W.K., Pan, F., Liang, C. and Liu, R.S., *Chem. Mater.* 29, 6781, 2017. With permission.) (c) Thermal quenching behavior of photoluminescence for low-z and high-z β-SiAlON. (Zhang, X., Fang, M.H., Tsai, Y.T., Lazarowska, A., Mahlik, S., Lesniewski, T., Grinberg, M., Pang, W.K., Pan, F., Liang, C. and Liu, R.S., *Chem. Mater.* 29, 6781, 2017. With permission.) (d) Low-temperature heat capacities of low-z β-SiAlON and high-z β-SiAlON, which are denoted as β-SiAlON(529) and β-SiAlON(540) in the literature. (Zhang, X., Fang, M.H., Tsai, Y.T., Lazarowska, A., Mahlik, S., Lesniewski, T., Grinberg, M., Pang, W.K., Pan, F., Liang, C. and Liu, R.S., *Chem. Mater.* 29, 6781, 2017. With permission.)

where k_B is the Boltzmann constant; T is the measured temperature; N is the number of atoms in the formula unit multiplied by the Avogadro number. By fitting the obtained heat capacity at the low temperature, the fitted Debye temperatures θ_D of low-z β-SiAlON and high-z β-SiAlON are 901K and 747K, respectively. These results also prove that the low-z β-SiAlON possesses higher rigidity compared with the high-z β-SiAlON. Moreover, due to the lower Stokes shift, the thermal activation energy, which is the energy difference between the lowest excited state and the crossing point between the first excited state and ground state, will be larger. This effect also makes low-z β-SiAlON have better thermal stability.

Considering the first coordinated environment of Eu^{2+}, the inhomogeneous broadening effect may be more serious in the high-z β-SiAlON, resulting in a broader emission peak. On the other hand, considering the second coordinated environment of Eu^{2+}, the distance between the two closest channels is approximately 7.60720 Å, which is quite far and the energy transfer between the Eu^{2+} ions can be suppressed.

5.3.2   UCr_4C$_4$-Type Nitride Phosphors

The UCr$_4$C$_4$-type phosphors were first revealed in the patent of US 9,546,319 with the applicant of Koninklijke Philips N.V. Nitride phosphors, such as SrMg$_3$SiN$_4$:Eu, BaMg$_2$Ga$_2$N$_4$:Eu, and CaLiAl$_3$N$_4$:Eu, are demonstrated in this patent. In this section, we will introduce the three most famous Sr-based nitride UCr$_4$C$_4$-type phosphors, which are SrLiAl$_3$N$_4$:Eu^{2+}, SrMg$_2$Al$_2$N$_4$:Eu^{2+},

TABLE 5.2
Basic Information on Eu-Doped UCr$_4$C$_4$ Nitride Phosphors

Phosphor	Ex (nm)	Em (nm)	fwhm (cm^{-1})	fwhm (nm)	Tetrahedra Alignment	No. of Dopant Sites	Structure	Space Group
SrLiAl$_3$N$_4$:Eu^{2+}	440	650	1180	50	Ordered	2	Triclinic	$P\bar{1}$
SrMg$_2$Al$_2$N$_4$:Eu^{2+}	440	612	1838	68	Disordered	1	Tetragonal	$I4/m$
SrMg$_3$SiN$_4$:Eu^{2+}	450	615	1170	43	Ordered	1	Tetragonal	$I4_1/a$

and SrMg$_3$SiN$_4$:Eu^{2+}, to reveal their relationship between the structural and luminescent properties. The basic properties of SrLiAl$_3$N$_4$:Eu^{2+}, SrMg$_2$Al$_2$N$_4$:Eu^{2+}, and SrMg$_3$SiN$_4$:Eu^{2+} are shown in Table 5.2.

5.3.2.1 SrLiAl$_3$N$_4$:Eu^{2+}

SrLiAl$_3$N$_4$:Eu^{2+} (SLA) possesses the space group of $P\bar{1}$ belonging to a triclinic system, as shown in Figure 5.16(a). Li$^+$ and Al^{3+} ions in this structure form [LiN$_4$]$^{11-}$ and [AlN$_4$]$^{9-}$ tetrahedra, respectively, which built up the channel structure. The degree of condensation, which is denoted as κ and defined as the atomic ratio of tetrahedral centers to vertices within one network, equals one. Besides, these tetrahedra reveal the ordered alignment along the channel direction. Two Sr^{2+} sites are in the host of SLA and both of them are coordinated with eight N^{3-}, which is a highly symmetrical cuboid-like coordination. Eu will dope at the Sr position in the structure. Moreover, the distance between the two Sr^{2+} ions sitting at the two closest channels is around 5.79 Å. These structural properties, including rigid structure, highly symmetrical coordination, ordered tetrahedra alignment, and far Eu–Eu distance, fit the requirement of narrowband emission. SLA can provide the emission wavelength of around 650 nm with the *fwhm* of 50 nm (~1180 cm^{-1}) under the excitation of 440 nm, as shown in Figure 5.16(b) [10]. One should notice that even though there are two emission centers, two of them are quite similar and generate the narrowband emission accordingly. Besides,

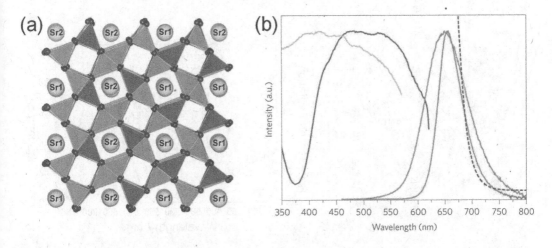

FIGURE 5.16 (a) Crystal structure of SrLiAl$_3$N$_4$. The pink and green polyhedra represent the LiN$_4$ and AlN$_4$ tetrahedrons. The blue ball represents the nitrogen. (b) Excitation (SLA, blue; CaAlSiN$_3$:Eu^{2+}, light gray) and emission spectra for λ_{exc} = 440 nm of SLA (pink) and CaAlSiN$_3$:Eu^{2+} (dark gray). The dotted curve indicates the upper limit of sensitivity of the human eye. (Pust, P., Weiler, V., Hecht, C., Tücks, A., Wochnik, A.S., Henß, A.K., Wiechert, D., Scheu, C., Schmidt, P.J. and Schnick, W., *Nat. Mater.* 13, 891, 2014. With permission.)

the optical bandgap of SLA is 4.7 eV [10]. Typically, the large bandgap is beneficial in the increase of the energy difference between the Eu excited state and the conduction band, which suppresses the thermal ionization process. Due to the high rigidity and large bandgap, SLA maintains around 95% luminescent intensity at 500K compared with that at room temperature. The study also shows that with the optimization of the synthesis condition, the external quantum efficiency can be obviously improved [34]. Further posttreatment such as particle separation or surface modification may further enhance its performance. Therefore, SLA could be a potential candidate for the next-generation narrowband emission phosphor.

5.3.2.2 $SrMg_2Al_2N_4:Eu^{2+}$

$SrMg_2Al_2N_4:Eu^{2+}$ (SMA) possesses the space group of $I4/m$ belonging to a tetragonal system, as shown in Figure 5.17(a). Mg^{2+} and Al^{3+} ions in this structure form $[MgN_4]$ and $[AlN_4]$ tetrahedra, respectively, which built up the channel structure. The degree of condensation equals one. Differently, these tetrahedra reveal the disordered alignment along the channel direction, which is less rigid compared with the structure of the SLA. Only one Sr^{2+} site coordinated with eight N^{3-} is in the host of SMA, which is also a highly symmetrical cuboid-like coordination. Eu^{2+} will dope at the Sr^{2+} position in the structure. Moreover, the distance between the two Sr^{2+} ions sitting at the two closest channels is around 5.96 Å. Structural properties, including rigid structure, highly symmetrical coordination, single emission center, and far Eu–Eu distance, fit the requirement of narrowband emission. Despite the single Sr^{2+} site in the host of SMA, different local coordination environments resulting from the disordered $[MgN_4]$ and $[AlN_4]$ tetrahedra are presented. This effect will lead to varying bond distances (Al^{3+}/Mg^{2+})–N and the broader emission spectrum [11]. SMA can provide the emission wavelength of around 612 nm with the *fwhm* of 68 nm ($\sim$1838 cm^{-1}) under the excitation of 440 nm, as shown in Figure 5.17(b) [11]. One should notice that the bandwidth difference between SLA and SMA calculated in wavenumber (1180 and 1838 cm^{-1}) is much more obvious than in wavelength (50 and 68 nm). It is important to convert the unit from wavelength to wavenumber (or other units in energy scale) to correctly analyze the bandwidth between different phosphors. SMA maintains around 78% luminescent intensity at 200K compared with that at 7K, in which the thermal stability is worse than SLA due to the less rigid structure and smaller bandgap (3.65–3.8 eV) [11].

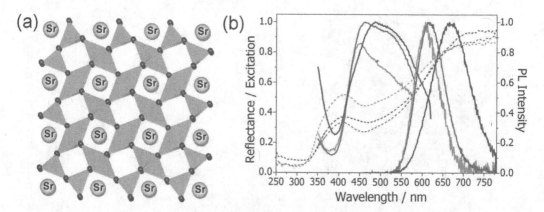

FIGURE 5.17   (a) Crystal structure of $SrMg_2Al_2N_4$. The light green polyhedra represent the $(Li,Al)N_4$ tetrahedra. The blue ball represents the nitrogen. (b) Excitation, reflectance (dashed curves), and emission (λ_{exc} = 440 nm) spectra of $M[Mg_2Al_2N_4]:Eu^{2+}$ (0.1%) (M = Ca, Sr, Ba) bulk samples at room temperature. $Ca[Mg_2Al_2N_4]:Eu^{2+}$ orange, $Sr[Mg_2Al_2N_4]:Eu^{2+}$ green, and $Ba[Mg_2Al_2N_4]:Eu^{2+}$ blue. (Pust, P., Hintze, F., Hecht, C., Weiler, V., Locher, A., Zitnanska, D., Harm, S., Wiechert, D., Schmidt, P.J. and Schnick, W., *Chem. Mater.* 26, 6113, 2014. With permission.)

5.3.2.3 SrMg₃SiN₄:Eu²⁺

SrMg$_3$SiN$_4$:Eu^{2+} (SMS) possesses the space group of $I4_1/a$ belonging to a tetragonal system, as shown in Figure 5.18(a). Mg^{2+} and Si^{4+} ions in this structure form [MgN$_4$] and [SiN$_4$] tetrahedra, respectively, which built up the channel structure. The degree of condensation equals one. The same as the SLA structure, [MgN$_4$] and [SiN$_4$] tetrahedra reveal the ordered alignment along the channel direction. Only one Sr^{2+} site coordinated with eight N^{3-} ions is in the host of SMS, which is also a highly symmetrical cuboid-like coordination. Eu^{2+} will dope at the Sr^{2+} position in the structure. The distance between the two Sr^{2+} ions sitting at the two closest channels is around 5.97 Å. These structural properties, including rigid structure, highly symmetrical coordination, ordered tetrahedra alignment, single emission center, and far Eu–Eu distance, fit the requirement of narrowband emission. SMS can provide the emission wavelength of around 615 nm with the *fwhm* of 43 nm (~1170 cm^{-1}) under the excitation of 450 nm, as shown in Figure 5.18(b) [35]. The photoluminescence of SMS is even narrower than SLA due to the single emission center and ordered structure. However, the optical bandgap of SMS (3.9 eV) is much smaller than that of SLA (4.7 eV), leading to poor thermal stability [35].

5.3.3 UCr₄C₄-TYPE OXYNITRIDE PHOSPHORS

Among the aforementioned UCr$_4$C$_4$-type nitride phosphors, the most promising one is SLA while the emission wavelength is around 650 nm, which is not sensitive to the human eye. Efforts have been made to tune the emission wavelength by the solid-solution method at the cationic sites, such as SrLi(Al$_{1-x}$Ga$_x$)$_3$N$_4$:Eu^{2+}, Sr(LiAl$_3$)$_{1-x}$(SiMg$_3$)$_x$N$_4$:Eu^{2+}, and Sr$_{1-x}$Ca$_x$LiAl$_3$N$_4$:Eu^{2+} [36–38]. However, the emission spectra will suffer from inhomogeneous broadening due to the destruction of the ordered structure. As a result, the better approach is to develop the UCr$_4$C$_4$-type oxynitride phosphors. With the weaker nephelauxetic effect, the position of the excited state of Eu^{2+} will become higher and result in a blueshift in the emission spectra. In this section, we will introduce the Sr[Li$_2$Al$_2$O$_2$N$_2$]:Eu^{2+} and Sr[Li$_{2.5}$Al$_{1.5}$O$_3$N]:Eu^{2+} to reveal the subtle structural properties between these two phosphors and the resulting photoluminescence. The basic information of Sr[Li$_2$Al$_2$O$_2$N$_2$]:Eu^{2+} and Sr[Li$_{2.5}$Al$_{1.5}$O$_3$N]:Eu^{2+} is shown in Table 5.3.

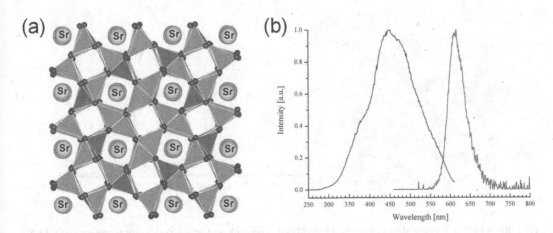

FIGURE 5.18 (a) Crystal structure of SrMg$_3$SiN$_4$. The pink and green polyhedra represent the SiN$_4$ and MgN$_4$ tetrahedra. The blue ball represents the nitrogen. (b) Excitation (blue) and emission (red) spectra of Sr[Mg$_3$SiN$_4$]:Eu^{2+} (2 mol% Eu^{2+}, nominal composition), $\lambda_{exc,max}$ = 450 nm, λ_{em} = 615 nm, *fwhm* ~ 1170 cm^{-1} (~43 nm). (Schmiechen, S., Schneider, H., Wagatha, P., Hecht, C., Schmidt, P.J. and Schnick, W., *Chem. Mater.* 26, 2712, 2014. With permission.)

TABLE 5.3

Basic Information on Eu-Doped UCr$_4$C$_4$ Oxynitride Phosphors

Phosphor	Ex (nm)	Em (nm)	fwhm (cm^{-1})	fwhm (nm)	Tetrahedra Alignment	No. of Dopant Sites	Structure	Space Group	Color
Sr[Li$_2$Al$_2$O$_2$N$_2$]:Eu^{2+}	460	614	1286	48	Ordered	1	Tetragonal	$P4_2/m$	Red
Sr[Li$_{2.5}$Al$_{1.5}$O$_3$N]:Eu^{2+}	460	578	2380	80	Disordered	1	Tetragonal	$I4/m$	Yellow

5.3.3.1   Sr[Li$_2$Al$_2$O$_2$N$_2$]:Eu^{2+}

Sr[Li$_2$Al$_2$O$_2$N$_2$]:Eu^{2+} possesses the space group of $P4_2/m$ belonging to a tetragonal system, as shown in Figure 5.19(a) [39]. Li$^+$ and Al^{3+} ions in this structure form [LiO$_3$N]$^{8-}$ and [AlON$_3$]$^{8-}$ tetrahedra, respectively, which built up the channel structure. The degree of condensation equals one. Besides, these tetrahedra reveal the ordered alignment along the channel direction. Only one Sr^{2+} site is in the host of Sr[Li$_2$Al$_2$O$_2$N$_2$]:Eu^{2+}, which is coordinated with four N^{3-} and four O^{2-} as cuboid-like coordination. Eu^{2+} will dope at the Sr^{2+} position in the structure. Moreover, the distance between the two Sr^{2+} ions sitting at the two closest channels is around 5.85 Å. These structural properties, including rigid structure, highly symmetrical coordination, ordered tetrahedra alignment, single dopant site, and far Eu–Eu distance, fit the requirement of narrowband emission. Sr[Li$_2$Al$_2$O$_2$N$_2$]:Eu^{2+} can provide the emission wavelength of around 614 nm with the *fwhm* of 48 nm (~1286 cm^{-1}) under the excitation of 460 nm, as shown in Figure 5.19(b) [39]. The bandwidth of Sr[Li$_2$Al$_2$O$_2$N$_2$]:Eu^{2+} is comparable with that of SLA. The advantage of Sr[Li$_2$Al$_2$O$_2$N$_2$]:Eu^{2+} compared with SLA is its emission wavelength. The shorter emission wavelength of Sr[Li$_2$Al$_2$O$_2$N$_2$]:Eu^{2+} can provide higher luminous efficiency of radiation (LER) and increases human eye sensitivity. The LER of Sr[Li$_2$Al$_2$O$_2$N$_2$]:Eu^{2+} and SLA are 266 and 77 lm/W$_{opt}$, respectively. Furthermore, Sr[Li$_2$Al$_2$O$_2$N$_2$]:Eu^{2+} can maintain 96% luminescent intensity at 420K compared with that at room temperature. As a result, this phosphor is a promising phosphor with a shorter emission wavelength without reducing the thermal stability or broadening the emission bandwidth.

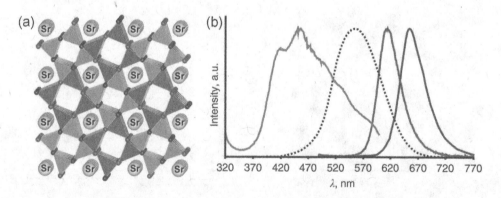

FIGURE 5.19   (a) Crystal structure of Sr[Li$_2$Al$_2$O$_2$N$_2$]. The pink and green polyhedra represent the LiO$_3$N and AlON$_3$ tetrahedra. The red and blue balls represent oxygen and nitrogen. (b) Normalized excitation (gray, for the emission at λ_{max} equals 614 nm) and emission spectrum (red, excited with λ_{exc} equals 460 nm) of Sr[Li$_2$Al$_2$O$_2$N$_2$]:Eu^{2+} in comparison to Sr[LiAl$_3$N$_4$]:Eu^{2+}. (Hoerder, G.J., Seibald, M., Baumann, D., Schröder, T., Peschke, S., Schmid, P.C., Tyborski, T., Pust, P., Stoll, I., Bergler, M., Patzig, C., Reißaus, S., Krause, M., Berthold, L., Höche, T., Johrendt, D. and Huppertz, H., *Nat. Commun.* 10, 1824, 2019. With permission.)

5.3.3.2 Sr[Li$_{2.5}$Al$_{1.5}$O$_3$N]:Eu^{2+}

The chemical formula can be written as SrAl$_{2-x}$Li$_{2+x}$O$_{2+2x}$N$_{2-2x}$, in which the $x \approx 0.5$ (Sr[Li$_{2.5}$Al$_{1.5}$O$_3$N]) sample is utilized to represent the crystal structure. Sr[Li$_{2.5}$Al$_{1.5}$O$_3$N]:Eu^{2+} possesses the space group of *I*4/*m* belonging to a tetragonal system, as shown in Figure 5.20(a) [40]. Li$^+$ and Al^{3+} ions in this structure form (Li,Al)(O,N)$_4$ tetrahedra, which built up the channel structure. The degree of condensation equals one. Only one Sr^{2+} site is in the host of Sr[Li$_{2.5}$Al$_{1.5}$O$_3$N], which is coordinated with eight equidistant O/N-sites as cuboid-like coordination. Eu will dope at the Sr position in the structure. Moreover, the distance between the two Sr^{2+} ions sitting at the two closest channels is around 5.78 Å. Different from the Sr[Li$_2$Al$_2$O$_2$N$_2$]:Eu^{2+} phosphor, statistically mixed occupancy is expected at the Al/Li and N/O sites for Sr[Li$_{2.5}$Al$_{1.5}$O$_3$N]:Eu^{2+}. Despite the rigid structure, highly symmetrical coordination, single dopant site, and far Eu–Eu distance, Sr[Li$_{2.5}$Al$_{1.5}$O$_3$N]:Eu^{2+} will suffer from severe inhomogeneous broadening due to the disordered structure. Sr[Li$_{2.5}$Al$_{1.5}$O$_3$N]:Eu^{2+} can provide an emission wavelength around 578 nm with a *fwhm* of 80 nm (~2380 cm^{-1}) under the excitation of 460 nm, as shown in Figure 5.20(b) [40]. The emission spectra can cover the range from 581 to 672 nm by tuning the x value in SrAl$_{2-x}$Li$_{2+x}$O$_{2+2x}$N$_{2-2x}$:Eu^{2+}. Furthermore, the $x = 0.60$ sample can just maintain 40% luminescent intensity at 200°C compared with that at room temperature.

5.3.4 UCr$_4$C$_4$-Type Oxide Phosphors

The aforementioned patent, US 9,546,319, introduced not only the UCr$_4$C$_4$-type nitride phosphors but also the UCr$_4$C$_4$-type oxide phosphors, such as NaLi$_3$SiO$_4$. This type of oxide phosphor is also called alkali lithosilicate phosphors. The overall chemical formula can be written as Z[Li$_3$SiO$_4$], where Z is selected from the group consisting of monovalent Li$^+$, Na$^+$, K$^+$, Rb$^+$, and Cs$^+$. Based on the patent WO2018/029299, OSRAM Opto Semiconductors reported alkali lithosilicate phosphors at the Phosphor Global Summit conference in 2018. In this section, we will introduce the Eu^{2+}-doped alkali lithosilicate phosphors to reveal the subtle structural properties between these phosphors and the resulting photoluminescence. In **Section 5.2.6**, we have explained that Eu^{2+} will prefer to occupy the sites with the size in 29–34 Å^3, less prefer to occupy that in 26–27 Å^3, and be unable to occupy that in ≥ 40 Å^3. In this section, the mechanism will also be delivered in detail. The basic information of Eu^{2+}-doped alkali lithosilicate phosphors is shown in Table 5.4.

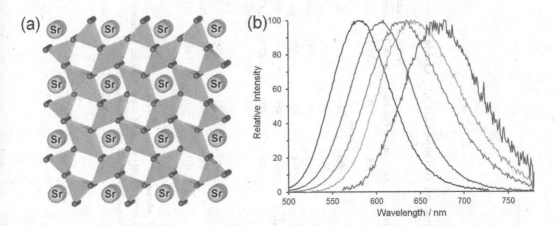

FIGURE 5.20 (a) Crystal structure of Sr[Li$_{2.5}$Al$_{1.5}$O$_3$N]. The light green polyhedra represent the (Al,Li)(N,O)$_4$ tetrahedra. (b) Emission spectra of bulk samples with different nominal compositions of SrAl$_{2-x}$Li$_{2+x}$O$_{2+2x}$N$_{2-2x}$ exemplifying the shift of the emission maximum. Black curve ($\lambda_{max} = 581$ nm, $x = 0.66$), blue curve ($\lambda_{max} = 605$ nm, $x = 0.60$), green curve ($\lambda_{max} = 629$ nm, $x = 0.44$), orange curve ($\lambda_{max} = 642$ nm, $x = 0.32$), and red curve ($\lambda_{max} = 672$ nm, $x = 0.12$). (Hoerder, G.J., Peschke, S., Wurst, K., Seibald, M., Baumann, D., Stoll, I. and Huppertz, H., *Inorg. Chem.* 58, 12146, 2019. With permission.)

TABLE 5.4
Basic Information on Eu-Doped UCr_4C_4 Oxide Phosphors

Eu^{2+}-doped Phosphor	Ex. (nm)	Em. (nm)	fwhm (cm^{-1})	fwhm (nm)	Structure (space group)	Size of the Cuboid Site (Å^3)					Color	Shoulder Emission (nm)
						Cs	Rb	K	Na	Li		
$RbNa_3[Li_3SiO_4]_4$	400	471	~980	22.4	tetragonal ($I4/m$)		40.2	31.5	30.5 29.0		blue	–
$RbNa_2K[Li_3SiO_4]_4$	395	480	~1129	26.0	tetragonal ($I4/m$)		42.6	32.9	26.2		Cyan	Green (530 nm)
$CsNa_2K[Li_3SiO_4]_4$	395	485	~1106	26.0	tetragonal ($I4/m$)	46.9		32.9	26.9		Cyan	Green (530 nm)
$RbKLi_2[Li_3SiO_4]_4$	374	476	~1095	24.8	tetragonal ($I4/m$)		43.2 (Rb/K)			26.4	blue	Green (532 nm)
$CsNaKLi[Li_3SiO_4]_4$	398	526	~1920	53.0	tetragonal ($I4/m$)	45.4		34.2	27.1		Green	Cyan (485 nm)
$CsNaRbLi[Li_3SiO_4]_4$	~390	486	~1017	24	tetragonal ($I4/m$)	44.8	33.8		26.8		Cyan	Green (530 nm)
$Na_2K_2[Li_3SiO_4]_4$	400	486	~877	20.7	tetragonal ($I4/m$)			40.9 31.7 (K/Na)	26.6		Cyan	Green (530 nm)
$Rb_2Li_2[Li_3SiO_4]_4$	460	531	~1492	42.0	Monoclinic ($C2/m$)		40.1			26.0	Green	–
$Rb_2Na_2[Li_3SiO_4]_4$	450	523	~1501	41.0	Monoclinic ($C2/m$)		40.5		26.5		Green	–
$NaK_2Li[Li_3SiO_4]_4$	450	530	~1569	44.0	Monoclinic ($C2/m$)			36.4	26.3 (Na/Li)		Green	–

Source: Fang, M.H., Mariano, C.O.M., Chen, P.Y., Hu, S.F. and Liu, R.S., *Chem. Mater.* 32, 1748, 2020. With permission.

a The bold numbers in the size of the cuboid site indicate the possible Eu^{2+}-substituted sites.

b This site is the position between the two Li^+ ions.

5.3.4.1 RbNa$_3$[Li$_3$SiO$_4$]$_4$:Eu^{2+}

RbNa$_3$[Li$_3$SiO$_4$]$_4$:Eu^{2+} possesses the space group of $I4/m$ belonging to a tetragonal system. Li$^+$ and Si^{4+} ions in this structure form [LiO$_4$]$^{7-}$ and [SiO$_4$]$^{4-}$ tetrahedra, respectively, which built up the ordered channel structure, as shown in Figure 5.21(b). The degree of condensation equals one. Rb$^+$ and Na$^+$ ions are coordinated by eight oxygen ions, forming cuboid-like coordination, as shown in Figure 5.21(c). RbNa$_3$[Li$_3$SiO$_4$]$_4$:Eu^{2+} can provide the emission wavelength of around 471 nm with the *fwhm* of 22.4 nm (~980 cm^{-1}) under the excitation of 400 nm, as shown in Figure 5.21(a). Besides, RbNa$_3$[Li$_3$SiO$_4$]$_4$:Eu^{2+} can maintain around 96% luminescent intensity at 150°C compared with that at room temperature [41]. Structural properties, including rigid structure, highly symmetrical coordination, ordered tetrahedra alignment, and far Eu–Eu distance, fit the requirement of narrowband emission.

The original literature assigns the Eu positions at Na1, Na2, and Rb sites [41]. However, if considering the fact that the low-concentration Eu^{2+} will prefer to occupy the position with proper coordinated size discussed in **Section 5.2.6**, the Eu^{2+} will not occupy all the alkali sites. Eu^{2+} is expected to occupy the Na$^+$ site with the coordinated size of 30.5 and 29.0 Å^3, as shown in Figure 5.21(d) and Table 5.4.[18] Alternatively, the coordinated size of the Rb$^+$ site is 40.2 Å^3, which makes Eu^{2+} unfavorable to dope. As a result, the emission from this phosphor can still be regarded as the single emission center.

5.3.4.2 Rb$_2$Na$_2$[Li$_3$SiO$_4$]$_4$:Eu^{2+}

Rb$_2$Na$_2$[Li$_3$SiO$_4$]$_4$:Eu^{2+} possesses the space group of $C2/m$ belonging to a monoclinic system. Li$^+$ and Si^{4+} ions in this structure form [LiO$_4$]$^{7-}$ and [SiO$_4$]$^{4-}$ tetrahedra, respectively, which built up the ordered channel structure, as shown in Figure 5.22(b). The degree of condensation equals one. Rb$^+$

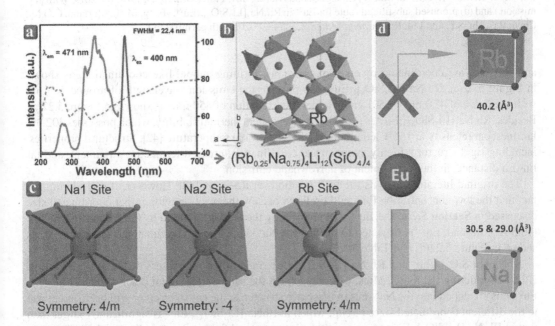

FIGURE 5.21 (a) Photoluminescence and its excitation spectrum, (Liao, H., Zhao, M., Molokeev, M.S., Liu, Q. and Xia, Z., *Angew. Chem. Int. Ed.* 130, 11902, 2018. With permission.) (b) crystal structure, (Liao, H., Zhao, M., Molokeev, M.S., Liu, Q. and Xia, Z., *Angew. Chem. Int. Ed.* 130, 11902, 2018. With permission.) (c) coordinated environment of the Rb and Na ions of RbNa$_3$[Li$_3$SiO$_4$]$_4$:Eu^{2+}, (Liao, H., Zhao, M., Molokeev, M.S., Liu, Q. and Xia, Z., *Angew. Chem. Int. Ed.* 130, 11902, 2018. With permission.) and (d) proposed substituted route for Eu^{2+} in RbNa$_3$[Li$_3$SiO$_4$]$_4$:Eu^{2+}. (Fang, M.H., Mariano, C.O.M., Chen, P.Y., Hu, S.F. and Liu, R.S., *Chem. Mater.* 32, 1748, 2020. With permission.)

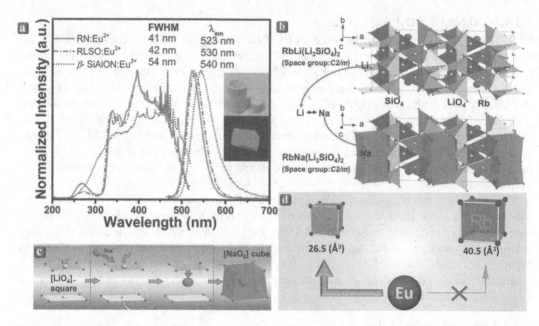

FIGURE 5.22   (a) Photoluminescence and its excitation, (Liao, H., Zhao, M., Zhou, Y., Molokeev, M.S., Liu, Q., Zhang, Q. and Xia, Z., *Adv. Funct. Mater.* 29, 1901988, 2019. With permission.) (b) crystal structure, (Liao, H., Zhao, M., Zhou, Y., Molokeev, M.S., Liu, Q., Zhang, Q. and Xia, Z., *Adv. Funct. Mater.* 29, 1901988, 2019. With permission.) (c) coordinated environment of Na^+ and Li^+ ions of $Rb_2Na_2[Li_3SiO_4]_4$:Eu^{2+} (Liao, H., Zhao, M., Zhou, Y., Molokeev, M.S., Liu, Q., Zhang, Q. and Xia, Z., *Adv. Funct. Mater.* 29, 1901988, 2019. With permission.) and (d) proposed substituted route for Eu^{2+} in $Rb_2Na_2[Li_3SiO_4]_4$:Eu^{2+}. (Fang, M.H., Mariano, C.O.M., Chen, P.Y., Hu, S.F. and Liu, R.S., *Chem. Mater.* 32, 1748, 2020. With permission.)

and Na^+ ions are coordinated by eight oxygen ions, forming cuboid-like coordination, as shown in Figure 5.22(c). $Rb_2Na_2[Li_3SiO_4]_4$:Eu^{2+} can provide the emission wavelength of around 523 nm with the *fwhm* of 41.0 nm (~1501 cm^{-1}) under the excitation of 450 nm, as shown in Figure 5.22(a). Besides, $Rb_2Na_2[Li_3SiO_4]_4$:Eu^{2+} is provided with excellent thermal stability, which possesses 102.3% luminescent intensity at 150°C compared with that at room temperature [42]. Structural properties, including rigid structure, highly symmetrical coordination, ordered tetrahedra alignment, and far Eu–Eu distance, fit the requirement of narrowband emission.

The original literature assigns the Eu^{2+} positions at Rb^+ sites [42]. However, if considering the fact that the low-concentration Eu^{2+} will prefer to occupy the position with proper coordinated size discussed in **Section 5.2.6**, the Eu^{2+} will not occupy the Rb^+ sites due to its large coordinated size (40.5 Å^3). Instead, Eu^{2+} is expected to occupy the Na^+ site with the coordinated size of 26.5 Å^3, as shown in Figure 5.22(d) and Table 5.4.[18]

$RbNa_3[Li_3SiO_4]_4$:Eu^{2+} and $Rb_2Na_2[Li_3SiO_4]_4$:Eu^{2+} are similar but possess different Rb:Na ratios. If just considering the ionic radius, one may expect the same luminescent color. In fact, when the Eu^{2+} ions are doped into the Na^+ sites in these two structures, the emission color will be different due to the different coordinated sizes of Na^+, which in $RbNa_3[Li_3SiO_4]_4$:Eu^{2+} is significantly larger than that in $Rb_2Na_2[Li_3SiO_4]_4$:Eu^{2+}. As a result, the larger crystal field strength in $Rb_2Na_2[Li_3SiO_4]_4$:Eu^{2+} leads to a more redshifted emission spectrum.

Alternatively, if just considering the ionic radius, one may expect that Eu^{2+} is unable to dope in the Rb^+ site due to the large ionic size of Rb^+ (1.61 Å; CN = 8) compared with that of Eu^{2+} (1.25 Å; CN = 8) [15]. However, the coordinated size may change from case to case. In **Section 5.3.4.6**, the $CsNaRbLi[Li_3SiO_4]_4$:Eu^{2+} will be used to demonstrate this perspective, which in this case Eu^{2+} can be doped into the Rb^+ site.

5.3.4.3 RbNa$_2$K[Li$_3$SiO$_4$]$_4$:Eu^{2+} and CsNa$_2$K[Li$_3$SiO$_4$]$_4$:Eu^{2+}

RbNa$_2$K[Li$_3$SiO$_4$]$_4$:Eu^{2+} and CsNa$_2$K[Li$_3$SiO$_4$]$_4$:Eu^{2+} possess the space group of *I4/m* belonging to a tetragonal system. Li$^+$ and Si^{4+} ions in this structure form [LiO$_4$]$^{7-}$ and [SiO$_4$]$^{4-}$ tetrahedra, respectively, which built up the ordered channel structure, as shown in Figure 5.23(c). The degree of condensation equals one. Rb$^+$, K$^+$, and Na$^+$ ions are coordinated by eight oxygen ions, forming cuboid-like coordination, as shown in Figure 5.23(d). RbNa$_2$K[Li$_3$SiO$_4$]$_4$:Eu^{2+} can provide the emission wavelength of around 480 nm with the *fwhm* of 26.0 nm (~1129 cm^{-1}) under the excitation of 395 nm, as shown in Figure 5.23(a) [17]. By contrast, CsNa$_2$K[Li$_3$SiO$_4$]$_4$:Eu^{2+} can provide an emission wavelength around 485 nm with the *fwhm* of 26.0 nm (~1106 cm^{-1}) under the excitation of 395 nm, as shown in Figure 5.23(b). Besides, both RbNa$_2$K[Li$_3$SiO$_4$]$_4$:Eu^{2+} and CsNa$_2$K[Li$_3$SiO$_4$]$_4$:Eu^{2+} possess shoulder emission at around 530 nm. Moreover, RbNa$_2$K[Li$_3$SiO$_4$]$_4$:Eu^{2+} and CsNa$_2$K[Li$_3$SiO$_4$]$_4$:Eu^{2+} are able to maintain around 99% and 92% luminescent intensity at 150°C compared with that at room temperature, respectively. Structural properties, including rigid structure, highly symmetrical coordination, ordered tetrahedra alignment, and far Eu–Eu distance, fit the requirement of narrowband emission.

The original literature assigns the Eu positions at K$^+$, Na$^+$, and Rb$^+$ sites and K$^+$, Na$^+$, and Cs$^+$ sites for RbNa$_2$K[Li$_3$SiO$_4$]$_4$:Eu^{2+} and CsNa$_2$K[Li$_3$SiO$_4$]$_4$:Eu^{2+}, respectively [17]. However, if considering the fact that the low-concentration Eu^{2+} will prefer to occupy the position with proper coordinated size discussed in **Section 5.2.6**, the Eu^{2+} will not occupy all the alkali sites. For both RbNa$_2$K[Li$_3$SiO$_4$]$_4$:Eu^{2+} and CsNa$_2$K[Li$_3$SiO$_4$]$_4$:Eu^{2+}, Eu^{2+} is expected to preferentially occupy the K$^+$ site with the coordinated size of 31.5 and 32.9 Å^3, respectively, as shown in Figure 5.23(e) and Table 5.4.[18] Eu^{2+} will also incorporate into the Na$^+$ sites with the coordinated size of 26.4 and 26.9 Å^3, although this

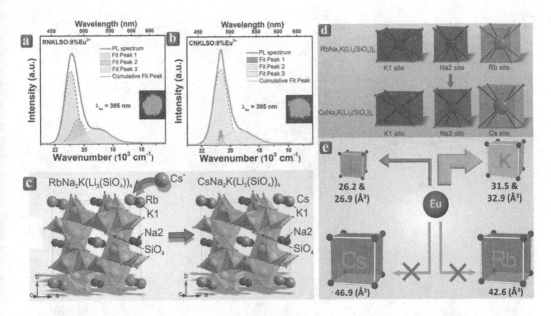

FIGURE 5.23 Photoluminescence spectrum of (a) RbNa$_2$K[Li$_3$SiO$_4$]$_4$:Eu^{2+} (Zhao, M., Zhou, Y., Molokeev, M.S., Zhang, Q., Liu, Q. and Xia, Z., *Adv. Opt. Mater.* 7, 1801631, 2019. With permission.) and (b) CsNa$_2$K[Li$_3$SiO$_4$]$_4$:Eu^{2+}. (Zhao, M., Zhou, Y., Molokeev, M.S., Zhang, Q., Liu, Q. and Xia, Z., *Adv. Opt. Mater.* 7, 1801631, 2019. With permission.) (c) Crystal structure (Zhao, M., Zhou, Y., Molokeev, M.S., Zhang, Q., Liu, Q. and Xia, Z., *Adv. Opt. Mater.* 7, 1801631, 2019. With permission.) and (d) coordinated environment of Rb, Na, and Cs of RbNa$_2$K[Li$_3$SiO$_4$]$_4$:Eu^{2+} and CsNa$_2$K[Li$_3$SiO$_4$]$_4$:Eu^{2+}. (Zhao, M., Zhou, Y., Molokeev, M.S., Zhang, Q., Liu, Q. and Xia, Z., *Adv. Opt. Mater.* 7, 1801631, 2019. With permission.) (e) Proposed substituted route for Eu^{2+} in RbNa$_2$K[Li$_3$SiO$_4$]$_4$:Eu^{2+} and CsNa$_2$K[Li$_3$SiO$_4$]$_4$:Eu^{2+}. (Fang, M.H., Mariano, C.O.M., Chen, P.Y., Hu, S.F. and Liu, R.S., *Chem. Mater.* 32, 1748, 2020. With permission.)

site is less preferred for Eu^{2+} compared with the K^+ sites. On the other hand, the coordinated size of the Rb^+ and Cs^+ sites are 43.2 and 46.9 $Å^3$, respectively, which makes Eu^{2+} unfavorable to dope. These results explain the reason that one can observe the main peak at the blue (cyan) region with a shoulder emission at the green region.

5.3.4.4 $RbKLi_2[Li_3SiO_4]_4:Eu^{2+}$

$RbKLi_2[Li_3SiO_4]_4:Eu^{2+}$ possesses the space group of $I4/m$ belonging to a tetragonal system. Li^+ and Si^{4+} ions in this structure form $[LiO_4]^{7-}$ and $[SiO_4]^{4-}$ tetrahedra, respectively, which built up the ordered channel structure, as shown in Figure 5.24(a). The degree of condensation equals one. Rb^+ and Na^+ ions are coordinated by eight oxygen ions, forming cuboid-like coordination consisting of one channel, as shown in Figure 5.24(b). In another channel, the two Li^+ sites, Li3 and Li4, consist of the channel. Li3 and Li4 are coordinated by four and eight oxygen ions to form a square-planar and cuboid-like coordination, respectively, as shown in Figure 5.24(c). Interestingly, the $RbKLi_2[Li_3SiO_4]_4:Eu^{2+}$ can provide the emission in either green (single crystal) or blue (single crystal and powder) regions. The blue one can provide the main emission wavelength around 476 nm with the *fwhm* of 24.8 nm (~1095 cm^{-1}) [43]. On the other hand, the green one can provide the main emission wavelength around 534 nm with the *fwhm* of 43.5 nm (~1523 cm^{-1}) [43]. Besides, $RbKLi_2[Li_3SiO_4]_4:Eu^{2+}$ is able to maintain around 87% luminescent intensity at 150°C compared with that at room temperature [43]. Structural properties, including rigid structure, highly symmetrical coordination, ordered tetrahedra alignment, and far Eu–Eu distance, fit the requirement of narrowband emission.

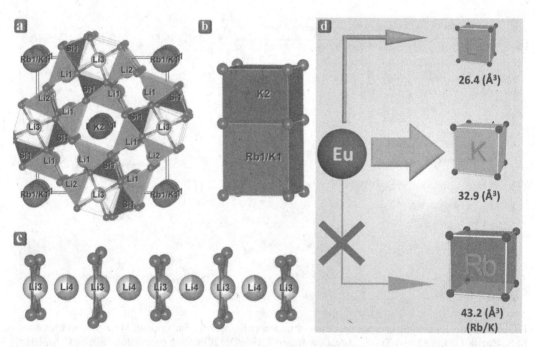

FIGURE 5.24   (a) Crystal structure of $RbKLi_2[Li_3SiO_4]_4:Eu^{2+}$. (Fang, M.H., Mariano, C.O.M., Chen, P.Y., Hu, S.F. and Liu, R.S., *Chem. Mater.* 32, 1748, 2020. With permission.) The coordinated environment of (b) K^+ and Rb^+ (Fang, M.H., Mariano, C.O.M., Chen, P.Y., Hu, S.F. and Liu, R.S., *Chem. Mater.* 32, 1748, 2020. With permission.) and (c) Li^+ in the channel. (Fang, M.H., Mariano, C.O.M., Chen, P.Y., Hu, S.F. and Liu, R.S., *Chem. Mater.* 32, 1748, 2020. With permission.) (d) Proposed substituted route for Eu^{2+} in $RbKLi_2[Li_3SiO_4]_4:Eu^{2+}$. (Fang, M.H., Mariano, C.O.M., Chen, P.Y., Hu, S.F. and Liu, R.S., *Chem. Mater.* 32, 1748, 2020. With permission.)

One should note that in this structure, the occupation values for Li3 and Li4 are less than one and it is unsuitable for Li^+ to simultaneously occupy the Li3 and Li4 sites due to the short distance between them. Besides, the average bond length of Li4–O is around 2.494(1) and 2.649(1) Å, which is longer than the value in the literature [15]. In this research, the authors propose a new model that Eu^{2+} will occupy the Li4 site with the average Eu–O equaling 2.57 Å, which is comparable with that in literature [15]. In addition, after one Eu^{2+} occupies one Li4 site, two Li3 sites near the dopant Eu^{2+} will become unoccupied to prevent the unsuitably short distance between Li3 and Eu^{2+}. Finally, no residual electron density is observed at the Li4 site for the blue one compared with that of the green one, indicating that the blue emission may be from the Eu^{2+} at the Rb^+ and K^+ sites instead of the Li4 site. Considering the theory discussed in **Section 5.2.6**, the Eu^{2+} will prefer to occupy the K^+ site and less preferred to occupy the Li^+ (Li4) site with the coordinated size of 32.9 and 26.4 Å^3, respectively, as shown in Figure 5.24(d). Differently, the Eu^{2+} will not occupy the Rb^+ due to its large coordinated size of 43.2 Å^3. This prediction supports the description and opinion of the original literature that blue emission may be from the Eu^{2+} at the Rb^+ and K^+ sites [43]. Additionally, the emission spectrum in the literature also possesses a weak green emission, which should be the emission from the Eu^{2+} at the Li4 site.

5.3.4.5 CsNaKLi[Li$_3$SiO$_4$]$_4$:Eu^{2+}

CsNaKLi[Li$_3$SiO$_4$]$_4$:Eu^{2+} possesses the space group of $I4/m$ belonging to a tetragonal system. Li^+ and Si^{4+} ions in this structure form $[LiO_4]^{7-}$ and $[SiO_4]^{4-}$ tetrahedra, respectively, which built up the ordered channel structure, as shown in Figure 5.25(c). The degree of condensation equals one. Na^+, K^+, and Cs^+ ions are coordinated by eight oxygen ions, forming cuboid-like coordination. The original research synthesizes a series of a solid solution of CsKNa$_{1.98-y}$Li$_y$[Li$_3$SiO$_4$]$_4$:0.02Eu^{2+} ($0 \leq y \leq 1$) and their photoluminescence properties are shown in Figure 5.25(a) and (b). CsNaK Li[Li$_3$SiO$_4$]$_4$:Eu^{2+} can provide an emission wavelength around 526 nm with the *fwhm* of 53.0 nm (~1920 cm^{-1}) under the excitation of 398 nm [44]. The broader emission bandwidth compared with other UCr$_4$C$_4$ phosphors results from the comparative emission intensity between the green and cyan emissions. Furthermore, when increasing the Li^+ concentration in CsKNa$_{1.98-y}$Li$_y$[Li$_3$SiO$_4$]$_4$:0.02Eu^{2+}, the emission intensity between 562 and 485 nm gradually increases, indicating the change of the Eu^{2+} occupancy between two crystallographic sites. CsNaKLi[Li$_3$Si O$_4$]$_4$:Eu^{2+} can maintain around 90% luminescent intensity at 150°C compared with that at room temperature [44].

The original literature assigns that the Eu will mainly occupy Na^+ and K^+ sites, while only a tiny amount of Eu^{2+} occupy the Cs^+ site. In addition, the Eu^{2+} occupancy ratio in Na^+ sites will increase with increasing Li^+ concentration, increasing emission intensity in the green region, as shown in Figure 5.25(d). Considering the theory discussed in **Section 5.2.6**, the Eu^{2+} will just occupy K^+ and Na^+ sites instead of the Cs^+ site. This result also supports the opinion from the original literature [44]. For CsKNa$_2$[Li$_3$SiO$_4$]$_4$:Eu^{2+}, the Eu^{2+} will prefer to occupy the K^+ site, leading to the cyan emission, as shown in Figure 5.25(e). By contrast, for CsNaKLi[Li$_3$ SiO$_4$]$_4$:Eu^{2+}, the coordinated size of the K^+ site slightly increases to 34.2 Å^3, which is close to the borderline of 29–34 Å^3, changing the Eu^{2+} occupancy ratio between K^+ and Na^+ sites, as shown in Figure 5.25(f). Notably, when synthesizing the CsKNa$_{1.98-y}$Li$_y$[Li$_3$SiO$_4$]$_4$:0.02Eu^{2+}, the precursors were weighted based on the assumption that Eu^{2+} is incorporated into the Na^+ site. This effect also increases the preference for Eu^{2+} at the Na^+ site to prevent the formation of vacancy at the Na^+ site.

5.3.4.6 CsNaRbLi[Li$_3$SiO$_4$]$_4$:Eu^{2+}

CsNaRbLi[Li$_3$SiO$_4$]$_4$:Eu^{2+} possesses a space group of $I4/m$ belonging to a tetragonal system. Li^+ and Si^{4+} ions in this structure form $[LiO_4]^{7-}$ and $[SiO_4]^{4-}$ tetrahedra, respectively, which built up the ordered channel structure, as shown in Figure 5.26(a). The degree of condensation equals one. Cs^+, Na^+, and Rb^+ ions are coordinated by eight oxygen ions, forming cuboid-like coordination, as shown

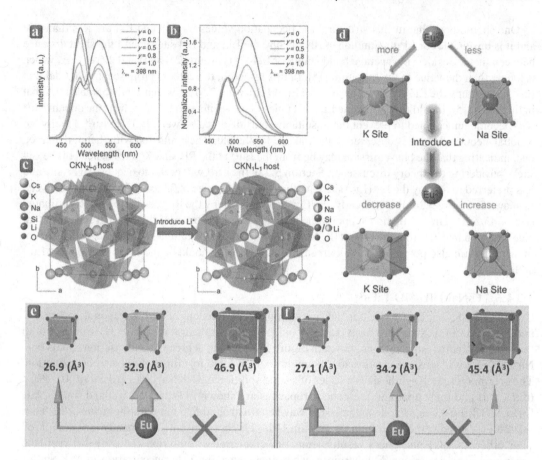

FIGURE 5.25 (a) Photoluminescence spectra (Wang, W., Tao, M., Liu, Y., Wei, Y., Xing, G., Dang, P., Lin, J. and Li, G., *Chem. Mater.* 31, 9200, 2019. With permission.) and (b) normalized photoluminescence spectra of $CsKNa_{1.98-y}Li_y[Li_3SiO_4]_4:0.02Eu^{2+}$ ($0 \leq y \leq 1$) (λ_{ex} = 398 nm). (Wang, W., Tao, M., Liu, Y., Wei, Y., Xing, G., Dang, P., Lin, J. and Li, G., *Chem. Mater.* 31, 9200, 2019. With permission.) (c) Crystal structure of $CsKNa_2[Li_3SiO_4]_4$ and $CsKNaLi[Li_3SiO_4]_4$. (Wang, W., Tao, M., Liu, Y., Wei, Y., Xing, G., Dang, P., Lin, J. and Li, G., *Chem. Mater.* 31, 9200, 2019. With permission.) (d) Coordinated environment of K and Na for $CsKNa_2[Li_3SiO_4]_4$ and $CsKNaLi[Li_3SiO_4]_4$. (Wang, W., Tao, M., Liu, Y., Wei, Y., Xing, G., Dang, P., Lin, J. and Li, G., *Chem. Mater.* 31, 9200, 2019. With permission.) The proposed substituted route for Eu^{2+} in (e) $CsKNa_2[Li_3SiO_4]_4$ (Fang, M.H., Mariano, C.O.M., Chen, P.Y., Hu, S.F. and Liu, R.S., *Chem. Mater.* 32, 1748, 2020. With permission.) and (f) $CsKNaLi[Li_3SiO_4]_4$. (Fang, M.H., Mariano, C.O.M., Chen, P.Y., Hu, S.F. and Liu, R.S., *Chem. Mater.* 32, 1748, 2020. With permission.)

in Figure 5.26(a). $CsNaRbLi[Li_3SiO_4]_4:Eu^{2+}$ can provide an emission wavelength around 486 nm with the *fwhm* of 24 nm (~1017 cm^{-1}) under the excitation of 390 nm [16]. It also possesses green shoulder emission at around 530 nm.

The original patent did not assign the occupation site for Eu^{2+}, while it speculates that the green emission results from the impurity phase of $Rb_2Na_2[Li_3SiO_4]_4:Eu^{2+}$. Considering the theory discussed in **Section 5.2.6**, the Eu^{2+} will just occupy Rb^+ and Na^+ sites with the coordinated size of 33.8 and 26.8 Å^3, respectively, as shown in Figure 5.26(b) and Table 5.4. The Eu^{2+} will prefer to occupy the Rb^+ site compared with the Na site, generating the cyan emission. Meanwhile, the Eu^{2+} at the Na^+ site will generate the green shoulder emission.

Notably, this is why one should not consider the Eu^{2+} dopant site just based on the ionic radius as discussed in **Section 5.3.4.2**. If just considering the ionic radius, one will expect that Eu^{2+} is unable

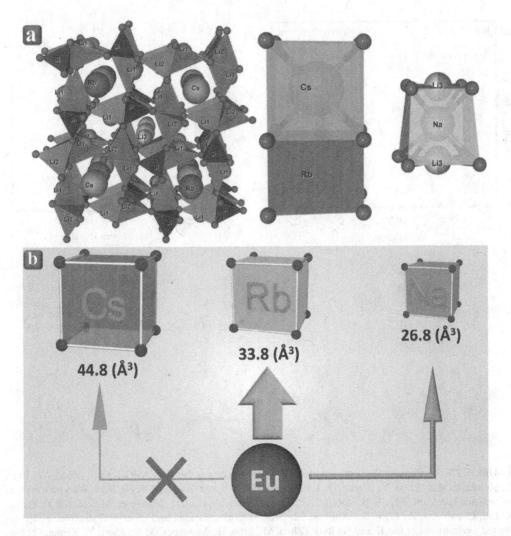

FIGURE 5.26 (a) Crystal structure and coordinated environment of the Cs^+, Rb^+, and Na^+ ions of CsNaRb Li[Li_3SiO_4]$_4$:Eu^{2+}. (Fang, M.H., Mariano, C.O.M., Chen, P.Y., Hu, S.F. and Liu, R.S., *Chem. Mater.* 32, 1748, 2020. With permission.) (b) Proposed substituted route for Eu^{2+} in CsNaRbLi[Li_3SiO_4]$_4$:Eu^{2+}. (Fang, M.H., Mariano, C.O.M., Chen, P.Y., Hu, S.F. and Liu, R.S., *Chem. Mater.* 32, 1748, 2020. With permission.)

to occupy the Rb^+ site; however, if examining the coordinated environment carefully, in some of the cases, the coordinated size of the Rb^+ site is actually suitable for Eu^{2+}.

5.3.4.7 $Na_2K_2[Li_3SiO_4]_4$:Eu^{2+}

$Na_2K_2[Li_3SiO_4]_4$:Eu^{2+} possesses the space group of *I4/m* belonging to a tetragonal system. Li^+ and Si^{4+} ions in this structure form [LiO_4]$^{7-}$ and [SiO_4]$^{4-}$ tetrahedra, respectively, which built up the ordered channel structure. The degree of condensation equals one. Rb^+ and Na^+ ions are coordinated by eight oxygen ions, forming cuboid-like coordination, as shown in Figure 5.27(c). $Na_2K_2[Li_3SiO_4]_4$:Eu^{2+} can provide the emission wavelength of around 486 nm with the *fwhm* of 20.7 nm (~877 cm^{-1}) under the excitation of 400 nm, as shown in Figure 5.27(a) [45]. Besides, $Na_2K_2[Li_3SiO_4]_4$:Eu^{2+} is able to maintain around 93% luminescent intensity at 150°C compared with that at room temperature [45].

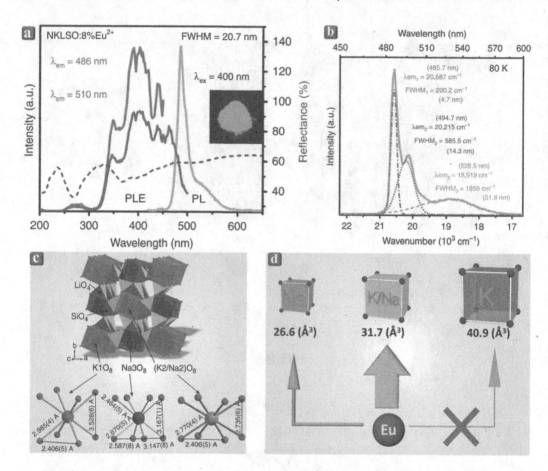

FIGURE 5.27   (a) Photoluminescence and its excitation spectra of Na$_2$K$_2$[Li$_3$SiO$_4$]$_4$:Eu^{2+}, (Zhao, M., Liao, H., Molokeev, M.S., Zhou, Y., Zhang, Q., Liu, Q. and Xia, Z., *Light-Sci. Appl.* 8, 38, 2019. With permission.) (b) emission spectrum of Na$_2$K$_2$[Li$_3$SiO$_4$]$_4$:Eu^{2+} at 80K, (Zhao, M., Liao, H., Molokeev, M.S., Zhou, Y., Zhang, Q., Liu, Q. and Xia, Z., *Light-Sci. Appl.* 8, 38, 2019. With permission.) and (c) crystal structure and coordinated environment of the K and Na ions. (Zhao, M., Liao, H., Molokeev, M.S., Zhou, Y., Zhang, Q., Liu, Q. and Xia, Z., *Light-Sci. Appl.* 8, 38, 2019. With permission.) (d) Proposed substituted route for Eu^{2+} in Na$_2$K$_2$[Li$_3$SiO$_4$]$_4$:Eu^{2+}. (Fang, M.H., Mariano, C.O.M., Chen, P.Y., Hu, S.F. and Liu, R.S., *Chem. Mater.* 32, 1748, 2020. With permission.)

The original literature assigns the Eu positions at three sites, Na3, K1, and Na2/K2 sites, as shown in Figure 5.27(b) and (c). Considering the theory discussed in **Section 5.2.6**, the Eu^{2+} will occupy the K2/Na2 and Na3 sites with the coordinated size that equals 31.7 and 26.6 Å^3, respectively, as shown in Figure 5.27(d). On the other hand, the coordinated size of the K1 site is 40.9 Å^3, which is a little bit too large for the Eu^{2+} ion. Notably, at low temperatures, two independent emissions at the cyan region are observed, which could be attributed to the K2/Na2 mixing the K$^+$ and Na$^+$ ions in the rigid structure. Similar properties will be discussed in detail in **Section 5.3.4.9** for NaK$_2$Li[Li$_3$SiO$_4$]$_4$:Eu^{2+}.

5.3.4.8   Rb$_2$Li$_2$[Li$_3$SiO$_4$]$_4$:Eu^{2+}

Rb$_2$Li$_2$[Li$_3$SiO$_4$]$_4$:Eu^{2+} possesses the space group of *C*2/m belonging to a monoclinic system. Li$^+$ and Si^{4+} ions in this structure form [LiO$_4$]$^{7-}$ and [SiO$_4$]$^{4-}$ tetrahedra, respectively, which built up the ordered channel structure, as shown in Figure 5.28(b). The degree of condensation equals one. Rb$^+$ ion is coordinated by eight oxygen ions, forming cuboid-like coordination, as shown in

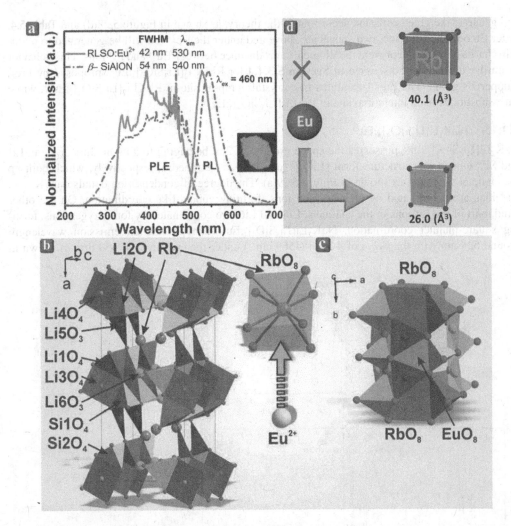

FIGURE 5.28 (a) Photoluminescence, (Zhao, M., Liao, H., Ning, L., Zhang, Q., Liu, Q. and Xia, Z., *Adv. Mater.* 30, 1802489, 2018. With permission.) (b) crystal structure, (Zhao, M., Liao, H., Ning, L., Zhang, Q., Liu, Q. and Xia, Z., *Adv. Mater.* 30, 1802489, 2018. With permission.) and (c) coordinated environment of Rb^+ ions of $Rb_2Li_2[Li_3SiO_4]_4$:Eu^{2+}. (Zhao, M., Liao, H., Ning, L., Zhang, Q., Liu, Q. and Xia, Z., *Adv. Mater.* 30, 1802489, 2018. With permission.) (d) Proposed substituted route for Eu^{2+} in $Rb_2Li_2[Li_3SiO_4]_4$:Eu^{2+}. (Fang, M.H., Mariano, C.O.M., Chen, P.Y., Hu, S.F. and Liu, R.S., *Chem. Mater.* 32, 1748, 2020. With permission.)

Figure 5.28(b). Alternatively, part of the Li^+ ions in the channels, Li3 and Li4, are coordinated by four oxygen ions, forming square-planner coordination. $Rb_2Li_2[Li_3SiO_4]_4$:Eu^{2+} can provide an emission wavelength around 531 nm with the *fwhm* of 22.4 nm (~980 cm^{-1}) under the excitation of 460 nm, as shown in Figure 5.28(a). Besides, $Rb_2Li_2[Li_3SiO_4]_4$:Eu^{2+} is provided with excellent thermal stability, which possesses 103.7% luminescent intensity at 150°C compared with that at room temperature.

The original literature assigns the Eu positions at the Rb site, as shown in Figure 5.28(c) [46]. Considering the theory discussed in **Section 5.2.6**, the Eu^{2+} will not occupy the Rb site with a coordinated size of 40.1 Å³. Moreover, the green emission of $Rb_2Li_2[Li_3SiO_4]_4$:Eu^{2+} also suggests that the Eu^{2+} should occupy a small site with a small size. If examining the crystal structure carefully, there is another possible cuboid-like coordinated site located between the two LiO_4 square-planner structures. Furthermore, the coordinated size of this site is 26.0 Å³, and Eu^{2+} can occupy this site and generate the green emission according to the theory, as shown in Figure 5.28(d) and Table 5.4. After Eu occupies this site, two vacancies above and under the Eu^{2+} ion will be generated to maintain charge balance and prevent the illogical short distance between Eu^{2+} and Li^+ ions. This behavior is similar to the effect discussed in **Section 5.3.4.4** for $RbKLi_2[Li_3SiO_4]_4:Eu^{2+}$ proposed by Prof. Huppertz's group. The Eu^{2+} substitutes in a crystallographic site in $RbKLi_2[Li_3SiO_4]_4:Eu^{2+}$, while Eu^{2+} substitutes in an interstitial site in $Rb_2Li_2[Li_3SiO_4]_4:Eu^{2+}$.

5.3.4.9 $NaK_2Li[Li_3SiO_4]_4:Eu^{2+}$

$NaK_2Li[Li_3SiO_4]_4:Eu^{2+}$ possesses the space group of $C2/m$ belonging to a monoclinic system. Li^+ and Si^{4+} ions in this structure form $[LiO_4]^{7-}$ and $[SiO_4]^{4-}$ tetrahedra, respectively, which built up the channel structure, as shown in Figure 5.29(a). The degree of condensation equals one. K^+ and Na^+ ions are coordinated by eight oxygen ions, forming cuboid-like coordination. On the other hand, part of the Li^+ ions in the channels, Li5 and Li6, are coordinated by four oxygen ions, forming square-planner coordination. $NaK_2Li[Li_3SiO_4]_4:Eu^{2+}$ can provide an emission wavelength around 528 nm with the *fwhm* of 44 nm (1569 cm^{-1}) under the excitation of 450 nm, as shown in

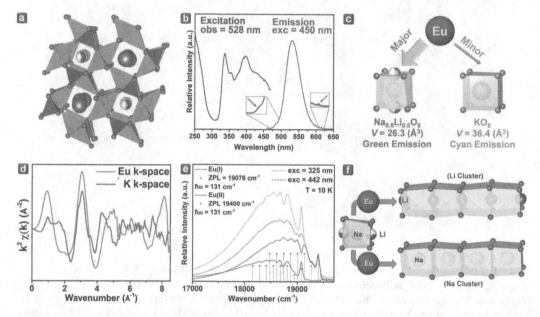

FIGURE 5.29   (a) Crystal structure of $NaK_2Li[Li_3SiO_4]_4:Eu^{2+}$. The purple, yellow, dark-green, and red balls represent K^+, Na^+, Li^+, and O^{2-}, respectively. (Fang, M.H., Mariano, C.O., Chen, K.C., Lin, J.C., Bao, Z., Mahlik, S., Lesniewski, T., Lu, K.M., Lu, Y.R., Wu, Y.J., Sheu, H.S., Lee, J.F., Hu, S.F., Liu, R.S. and Attfield, J.P., *Chem. Mater.* 33, 1893, 2021. With permission.) (b) Emission and excitation spectrum of $Na_{0.87}K_2Li[Li_3SiO_4]_4:0.13Eu$. (Fang, M.H., Mariano, C.O., Chen, K.C., Lin, J.C., Bao, Z., Mahlik, S., Lesniewski, T., Lu, K.M., Lu, Y.R., Wu, Y.J., Sheu, H.S., Lee, J.F., Hu, S.F., Liu, R.S. and Attfield, J.P., *Chem. Mater.* 33, 1893, 2021. With permission.) (c) Proposed Eu substitution mechanism. (Fang, M.H., Mariano, C.O., Chen, K.C., Lin, J.C., Bao, Z., Mahlik, S., Lesniewski, T., Lu, K.M., Lu, Y.R., Wu, Y.J., Sheu, H.S., Lee, J.F., Hu, S.F., Liu, R.S. and Attfield, J.P., *Chem. Mater.* 33, 1893, 2021. With permission.) (d) K *K-edge* and Eu L_3-*edge* k^2-weighted Fourier transform of EXAFS spectra. (Fang, M.H., Mariano, C.O., Chen, K.C., Lin, J.C., Bao, Z., Mahlik, S., Lesniewski, T., Lu, K.M., Lu, Y.R., Wu, Y.J., Sheu, H.S., Lee, J.F., Hu, S.F., Liu, R.S. and Attfield, J.P., *Chem. Mater.* 33, 1893, 2021. With permission.) (e) Emission spectrum measured under 10K for $Na_{0.87}K_2Li[Li_3SiO_4]_4:0.13Eu$. (Fang, M.H., Mariano, C.O., Chen, K.C., Lin, J.C., Bao, Z., Mahlik, S., Lesniewski, T., Lu, K.M., Lu, Y.R., Wu, Y.J., Sheu, H.S., Lee, J.F., Hu, S.F., Liu, R.S. and Attfield, J.P., *Chem. Mater.* 33, 1893, 2021. With permission.) (f) Clustered coordinated environments of Li^+ and Na^+ in the channel. (Fang, M.H., Mariano, C.O., Chen, K.C., Lin, J.C., Bao, Z., Mahlik, S., Lesniewski, T., Lu, K.M., Lu, Y.R., Wu, Y.J., Sheu, H.S., Lee, J.F., Hu, S.F., Liu, R.S. and Attfield, J.P., *Chem. Mater.* 33, 1893, 2021. With permission.)

Figure 5.29(b) [47]. Considering the theory discussed in **Section 5.2.6**, the Eu^{2+} will prefer to occupy the Na^+ site with the coordinated size of 26.3 Å^3, and the green emission can be expected, as shown in Figure 5.29(c). Although the coordinated size of K^+ is out of the range of 29–34 Å^3, the coordinated size of the K^+ site is 36.4 Å^3, which is much smaller than the forbidden sites in other alkali lithosilicate phosphors. In fact, the weak cyan shoulder emission is observed in the photoluminescence spectrum, as shown in Figure 5.29(b). Besides, $NaK_2Li[Li_3SiO_4]_4:Eu^{2+}$ can maintain around 97.1% luminescent intensity at 150°C compared with that at room temperature [47].

To prove the practical occupation site for Eu^{2+}, the extended X-ray absorption fine structure (EXAFS) is measured for Eu and K elements. Due to the low absorption edge of Na, the original literature cannot obtain the EXAFS spectrum for the Na element. The k^2-weighted Fourier transform oscillation patterns of Na and Eu are obviously different, revealing that Eu and Na are not at the same position, as shown in Figure 5.29(d).

When carefully examining the photoluminescence of the green emission peak at low temperature (10K), two similar emission bands can be resolved, as shown in Figure 5.29(e), indicating that there are two emission sites for Eu. Notably, there is only one crystallographic site, Na1, in the $NaK_2Li[Li_3SiO_4]_4:Eu^{2+}$. However, in the Na/Li channel, the occupancy of Na and Li equals 0.5, respectively. Furthermore, the Li–Na distance equals 1.57 Å from the CIF, which is illogically short. As a result, it is speculated that the Li^+ and Na^+ will form clusters in a short range to prevent the illogically short distance between Li^+ and Na^+. Meanwhile, these clusters are disorderly distributed in the long range so that the structural information will show 50% occupancy for both Na and Li sites. When the Eu^{2+} occupy the Na site in the Na-cluster and the interstitial site in the Li-cluster, one can observe two similar emission bands, as shown in Figure 5.29(f).

5.4   MN⁴⁺-ACTIVATED NARROWBAND PHOSPHORS

The Mn^{4+}-doped fluoride phosphors are critical in the history of narrowband emission. In this section, we will introduce some Mn^{4+}-doped fluoride phosphors to reveal their relationship with structural and luminescent properties. The basic properties of these Mn^{4+}-doped fluoride phosphors are shown in Table 5.5.

5.4.1 $K_2SiF_6:Mn^{4+}$

$K_2SiF_6:Mn^{4+}$ possesses the space group of $Fm3m$ belonging to a cubic system. Si^{4+} ion is coordinated with six F^- ions to form the $[SiF_6]^{2-}$ octahedra, as shown in Figure 5.30(a). Mn^{4+} ion will dope at the Si site and there is only one kind of $[MnF_6]^{2-}$ in the $K_2SiF_6:Mn^{4+}$ structure. As a result, the Tanabe–Sugano energy-level diagram with d^3 configuration can be used to explain the photoluminescence

TABLE 5.5
Basic Information of Mn⁴⁺-Doped Fluoride Phosphors

Phosphor	Excitation (nm)	Emission (nm)	ZPL Intensity	Number of Dopant Sites	Structure	Space Group
$K_2SiF_6:Mn^{4+}$	460	632	Weak	1	Cubic	$Fm3m$
$K_2GeF_6:Mn^{4+}$	460	632	Weak	1	Trigonal	$P\bar{3}m1$
$K_2TiF_6:Mn^{4+}$	468	632	Weak	1	Trigonal	$P\bar{3}m1$
$Na_2SiF_6:Mn^{4+}$	460	627	Medium	2	Trigonal	$P321$
$Na_2GeF_6:Mn^{4+}$	460	626	Medium	2	Trigonal	$P321$
$Na_2TiF_6:Mn^{4+}$	460	626	Strong	3	Triclinic	$P1$

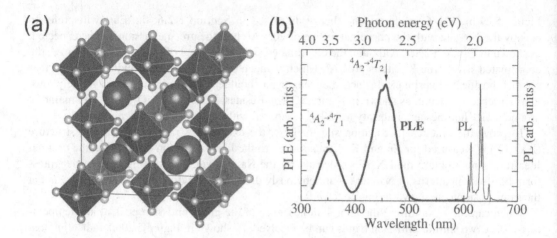

FIGURE 5.30   (a) Crystal structure of K_2SiF_6. The blue octahedra represent the SiF_6^{2-}. The purple ball represents the K^+ ion. (b) PL and PLE spectra for $K_2SIF_6:Mn^{4+}$ phosphor at 300K. The PLE spectrum was measured by monitoring at $\lambda_{em} = 630$ nm, while the PL spectrum was obtained by exciting at $\lambda_{ex} = 450$ nm. (Takahashi, T. and Adachi, S., *J. Electrochem. Soc.* 155, E183, 2008. With permission.)

properties, as shown in Figure 5.4. $K_2SiF_6:Mn^{4+}$ possesses two excitation bands centering around 460 and 350 nm, which can be attributed to the spin-allowed $^4A_2 \rightarrow ^4T_2$ and $^4A_2 \rightarrow ^4T_1$ electronic transition, respectively [48]. According to the Tanabe–Sugano energy-level diagram, the energy of these transitions will be strongly affected by the crystal field. If the crystal field strength increases, the excitation band will blueshift toward a shorter wavelength, and vice versa. Moreover, the strongest excitation band centered at 460 nm can be effectively excited by the LED chip, which means that $K_2SiF_6:Mn^{4+}$ possesses great potential for practical applications. $K_2SiF_6:Mn^{4+}$ can provide the emission wavelength of around 632 nm under the excitation of 460 nm, which can be attributed to the $^2E \rightarrow ^4A_2$ spin-forbidden electronic transition, as shown in Figure 5.30(b). Different from the broadband excitation spectrum, the emission shows sharp-line emission, which provides extremely narrow bandwidth, which is beneficial for the backlighting system. As discussed in **Section 5.2.1.2**, the emission spectrum of $K_2SiF_6:Mn^{4+}$ consists of PS couple with the v_6 (t_{2u} bending), v_4 (t_{1u} bending), and v_3 (t_{1u} stretching) modes. No obvious ZPL emission is observed at around 620 nm because the Mn^{4+} reside in a perfect octahedrally coordinated environment, which will be explained in detail in the following section.

5.4.2 $K_2GeF_6:Mn^{4+}$

$K_2GeF_6:Mn^{4+}$ possesses the space group of $P\bar{3}m1$ belonging to a trigonal system. Ge^{4+} ion is coordinated with six F^- ions to form the $[GeF_6]^{2-}$ octahedra, as shown in Figure 5.31(a). Mn^{4+} ion will dope at the Ge site and there is only one kind of $[MnF_6]^{2-}$ in the $K_2GeF_6:Mn^{4+}$ structure. $K_2GeF_6:Mn^{4+}$ possesses two excitation bands centering around 359 and 465 nm, which can be attributed to the spin-allowed $^4A_2 \rightarrow ^4T_2$ and $^4A_2 \rightarrow ^4T_1$ electronic transition, respectively [49]. The excitation wavelength of $K_2GeF_6:Mn^{4+}$ is slightly longer than that of $K_2SiF_6:Mn^{4+}$, which can be attributed to the bigger ionic radius of Ge^{4+} (0.53 Å; CN = 6) compared with Si (0.4 Å; CN = 6). $K_2GeF_6:Mn^{4+}$ can provide an emission wavelength around 632 nm under the excitation of 460 nm, which can be attributed to the $^2E \rightarrow ^4A_2$ spin-forbidden electronic transition, as shown in Figure 5.31(b). As discussed in **Section 5.2.1.2**, the emission spectrum of $K_2GeF_6:Mn^{4+}$ consists of PS couple with the v_6 (t_{2u} bending), v_4 (t_{1u} bending), and v_3 (t_{1u} stretching) modes. No obvious ZPL emission is observed at around 620 nm.

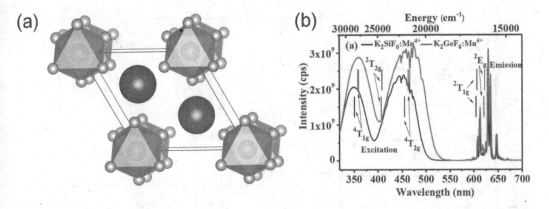

FIGURE 5.31 (a) Crystal structure of K_2GeF_6. The green octahedra represent the GeF_6^{2-}. The purple ball represents the K^+ ion. (b) Experimental excitation and emission spectra of K_2GeF_6:Mn^{4+} and K_2SiF_6:Mn^{4+} phosphors (solid lines) compared with the calculated energy levels of Mn^{4+} (vertical bars). (Wei, L.L., Lin, C.C., Fang, M.H., Brik, M.G., Hu, S.F., Jiao, H. and Liu, R.S., *J. Mater. Chem. C* 3, 1655, 2015. With permission.)

5.4.3 K_2TiF_6:Mn^{4+}

K_2TiF_6:Mn^{4+} possesses the space group of $P\overline{3}m1$ belonging to a trigonal system. Ti^{4+} ion is coordinated with six F^- ions to form the $[TiF_6]^{2-}$ octahedra, as shown in Figure 5.32(a). Mn^{4+} ion will dope at the Ti site and there is only one kind of $[MnF_6]^{2-}$ in the K_2TiF_6:Mn^{4+} structure. K_2TiF_6:Mn^{4+} possesses three excitation bands ranging in 200–500 nm. The excitation band at around 362 and 468 nm can be attributed to the spin-allowed $^4A_2 \rightarrow {}^4T_2$ and $^4A_2 \rightarrow {}^4T_1$ electronic transition, respectively. By contrast, the excitation band at around 255 nm results from the charge transfer between Mn^{4+} and F^- ions [50]. The excitation wavelength of K_2TiF_6:Mn^{4+} is slightly longer than that of K_2SiF_6:Mn^{4+}, which can be attributed to the bigger ionic radius of Ti^{4+} (0.605 Å; CN = 6) compared with Si (0.4 Å; CN = 6). With the longer Mn–F bond length in K_2TiF_6:Mn^{4+}, Mn^{4+} will encounter weaker crystal field strength, and the smaller energy of $^4A_2 \rightarrow {}^4T_2$ and $^4A_2 \rightarrow {}^4T_1$ can be expected. K_2TiF_6:Mn^{4+} can provide the emission wavelength of around 632 nm under the excitation of 468 nm, which can be attributed to the $^2E \rightarrow {}^4A_2$ spin-forbidden electronic transition, as shown in Figure 5.32(b) [51]. As discussed in **Section 5.2.1.2**, the emission spectrum of K_2TiF_6:Mn^{4+} consists

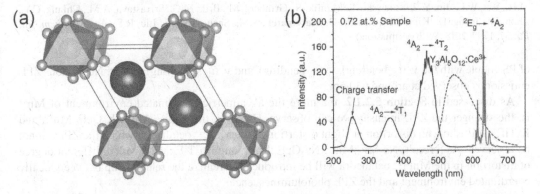

FIGURE 5.32 (a) Crystal structure of K_2TiF_6. The cyan octahedra represent the TiF_6^{2-}. The purple ball represents the K^+ ion. (b) Room-temperature excitation and emission spectra of the K_2TiF_6:Mn^{4+} (0.72 at.%) sample. The dashed line displays the emission spectrum of commercial $Y_3Al_5O_{12}$:Ce^{3+} yellow phosphor. (Zhu, H., Lin, C.C., Luo, W., Shu, S., Liu, Z., Liu, Y., Kong, J., Ma, E., Cao, Y. and Liu, R.S., *Nat. Commun.* 5, 4312, 2014. With permission.)

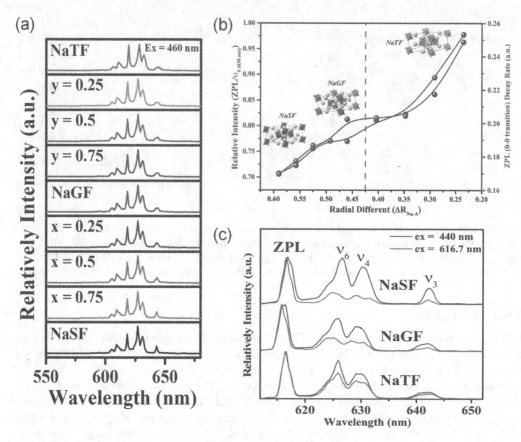

FIGURE 5.33 (a) Emission spectra of $Na_2(Si_xGe_{1-x})F_6:Mn^{4+}$ and $Na_2(Ge_yTi_{1-y})F_6:Mn^{4+}$. (Fang, M.H., Wu, W.L., Jin, Y., Lesniewski, T., Mahlik, S., Grinberg, M., Brik, M.G., Srivastava, A.M., Chiang, C.Y., Zhou, W., Jeong, D., Kim, S.H., Leniec, G., Kaczmarek, S.M., Sheu, H.S. and Liu, R.S., *Angew. Chem. Int. Ed.* 57, 1797, 2018. With permission.) (b) Relation between ZPL (0–0 transition) decay rate and relative intensity of ZPL/v_6. (Fang, M.H., Wu, W.L., Jin, Y., Lesniewski, T., Mahlik, S., Grinberg, M., Brik, M.G., Srivastava, A.M., Chiang, C.Y., Zhou, W., Jeong, D., Kim, S.H., Leniec, G., Kaczmarek, S.M., Sheu, H.S. and Liu, R.S., *Angew. Chem. Int. Ed.* 57, 1797, 2018. With permission.) (c) Luminescence spectra of NaSF, NaGF, and NaTF obtained under excitation 440 nm (blue curves) and 616.7 nm (red curves) at the temperature of 10 K. (Fang, M.H., Wu, W.L., Jin, Y., Lesniewski, T., Mahlik, S., Grinberg, M., Brik, M.G., Srivastava, A.M., Chiang, C.Y., Zhou, W., Jeong, D., Kim, S.H., Leniec, G., Kaczmarek, S.M., Sheu, H.S. and Liu, R.S., *Angew. Chem. Int. Ed.* 57, 1797, 2018. With permission.)

of PS couple with the v_6 (t_{2u} bending), v_4 (t_{1u} bending), and v_3 (t_{1u} stretching) modes. No obvious ZPL emission is observed at around 620 nm.

As discussed in **Section 5.2.1.2**, the more the asymmetric coordinated environment of Mn^{4+} is, the stronger the ZPL intensity we can observe. However, the $K_2SiF_6:Mn^{4+}$, $K_2GeF_6:Mn^{4+}$, and $K_2TiF_6:Mn^{4+}$ show no distortion or slight distortion of the MnF_6^{2-} octahedra, which the ZPL cannot be observed. As a result, $Na_2SiF_6:Mn^{4+}$, $Na_2GeF_6:Mn^{4+}$, and $Na_2TiF_6:Mn^{4+}$ with a different degree of distortion in the MnF_6^{2-} octahedra will be introduced to realize the relationship between locally coordinated environment and the ZPL photoluminescence.

5.4.4 $Na_2SiF_6:Mn^{4+}$, $Na_2GeF_6:Mn^{4+}$, AND $Na_2TiF_6:Mn^{4+}$

$Na_2SiF_6:Mn^{4+}$ (NaSF) and $Na_2GeF_6:Mn^{4+}$ (NaGF) possess the space group of $P\bar{3}m1$ belonging to a trigonal system, while $Na_2TiF_6:Mn^{4+}$ (NaTF) possesses the space group of $P1$ belonging to a triclinic system. Si^{4+}, Ge^{4+}, and Ti^{4+} ions are coordinated with six F^- ions to form the $[SiF_6]^{2-}$, $[GeF_6]^{2-}$, and $[TiF_6]^{2-}$ octahedra. Mn^{4+} ion will dope at the Si^{4+}, Ge^{4+}, and Ti^{4+} sites. Two Si^{4+} sites, two Ge^{4+} sites, and three Ti^{4+} sites are in NaSF, NaGF, and NaTF, respectively. The emission spectra of $Na_2(Si_xGe_{1-x})F_6:Mn^{4+}$ and $Na_2(Ge_yTi_{1-y})F_6:Mn^{4+}$ are shown in Figure 5.33(a) [52]. These phosphors demonstrate similar main emission peaks with a peak maximum at around 626 nm. However, the emission intensity ratio between ZPL and the main peak at 626 nm (v_6) is gradually increased from NaSF to NaTF, as shown in Figure 5.33(b) [52]. Analyzing the locally coordinated environment between NaSF and NaTF, the Si1 site shows nearly perfect $[SiF_6]^{2-}$ octahedra with just slightly bending angles (by 1°–2°) in NaSF. By contrast, the Si2 site in NaSF and the other three Ti sites in NaTF are provided with lower symmetry in the octahedra. The F–M–F (M = Si^{4+}, Ge^{4+}, and Ti^{4+}) information reveals that the deviations in MF_6^{2-} are more significant in NaTF than in NaGF and NaSF. This effect reveals that why the higher ZPL intensity can be observed in NaTF.

Moreover, 10K emission spectra of NaSF, NaGF, and NaTF obtained under excitation at 440 and 616.7 nm (ZPL wavelength) are obtained through time-resolved emission spectra by integrating the emission streak over 10 ms time window, which are shown in Figure 5.33(c) [52]. The excitation at 616.7 nm should more efficiently excite the Mn^{4+} centers with stronger ZPL transition. By contrast, excitation at 440 nm is unselective and will populate each site with equal probability. Notably, one can clearly observe that the obtained emission spectra have a much higher intensity of ZPL compared to that of PS, indicating that there are Mn^{4+} centers with higher and lower ZPL intensity. Moreover, the discrepancy between the intensity of ZPL and PS is gradually decreased from NaSF to NaTF. These results indicate that the Mn^{4+} sites for NaTF can be effectively excited to provide the ZPL emission.

5.5 SUMMARY AND PERSPECTIVES

The demand for high-quality devices, including the lighting and backlighting systems, grows up very quickly. The traditional backlighting devices using broadband emission phosphors will lead to the waste of energy after penetrating the color filter. Furthermore, the emission wavelength longer than 650 nm from the broadband emission phosphor will also result in energy waste. As a result, the narrowband emission phosphors are urgent to be developed. In this chapter, the Eu^{2+}- and Mn^{4+}-doped narrowband phosphors are introduced. The key points to approach the narrowband properties, including the perspectives from the number of dopant sites, the first and second coordinated shell, electron–lattice interaction, and the volume size of the coordinated environment, are explained in detail. Moreover, the structures of the nitride, oxynitride, oxide, and fluoride phosphors are analyzed to reveal the relationship between the structural properties and the aforementioned key points. The resulting photoluminescent properties are also introduced.

To compete with the practical commercial broadband emission phosphors, such as $CaAlSiN_3:Eu^{2+}$, some improvements could be applied to the currently developed narrowband emission phosphors in addition to developing the narrowband emission phosphors. The first one should be improving the quantum efficiency of the narrowband emission phosphors. If the luminescent performance after wasting the energy (penetrating the color filter, emission longer than 650 nm, etc.) of the commercial broadband emission phosphors is still better than the narrowband one, people will not accept the narrowband emission phosphors. The potential methods, including synthesis condition tuning, adding the flux, particle separation, and annealing, can be conducted. The next step is to improve the chemical stability. The poor chemical stability will cause color distortion and a decrease in the luminescent intensity after using a period. The potential methods, including surface passivation and coating, can also be conducted. Then, minimizing the particle size of the narrowband emission phosphors is also encouraged. The smaller LED chips require smaller luminescent materials.

ACKNOWLEDGMENTS

This work was supported by the Ministry of Science and Technology of Taiwan (Contract Nos. MOST 109-2112-M-003-011 and MOST 107-2113-M-002-008-MY3).

REFERENCES

1. Dorenbos P (2000) The 5d level positions of the trivalent lanthanides in inorganic compounds. *J. Lumin.* 91: 155.
2. Blasse G (1978) *Luminescence of Inorganic Solids*, Springer, Boston, pp. 457.
3. Blasse G (1968) Fluorescence of Eu^{2+}-activated alkaline-earth aluminates. *Philips Res. Reps.* 23: 201.
4. Tanabe Y, and Sugano S (1954) On the absorption spectra of complex ions II. *J. Phys. Soc. Jpn.* 9: 766.
5. Housecroft CE, and Sharpe AG (2012) *Inorganic Chemistry (2nd ed.)*, Elsevier, Oxford.
6. Brik MG, and Srivastava AM (2013) On the optical properties of the Mn^{4+} ion in solids. *J. Lumin.* 133: 69.
7. Lesniewski T, Mahlik S, Grinberg M, and Liu RS (2017) Temperature effect on the emission spectra of narrow band Mn^{4+} phosphors for application in LEDs. *Phys. Chem. Chem. Phys.* 19: 32505.
8. Yeh CW, Chen WT, Liu RS, Hu SF, Sheu HS, Chen JM, and Hintzen HT (2012) Origin of thermal degradation of $Sr_{2-x}Si_5N_8:Eu_x$ phosphors in air for light-emitting diodes. *J. Am. Chem. Soc.* 134: 14108.
9. Kimoto K, Xie RJ, Matsui Y, Ishizuka K, and Hirosaki N (2009) Direct observation of single dopant atom in light-emitting phosphor of β-SiAlON:Eu^{2+}. *Appl. Phys. Lett.* 94: 041908.
10. Pust P, Weiler V, Hecht C, Tücks A, Wochnik AS, Henß AK, Wiechert D, Scheu C, Schmidt PJ, and Schnick W (2014) Narrow-band red-emitting $Sr[LiAl_3N_4]:Eu^{2+}$ as a next-generation LED-phosphor material. *Nat. Mater.* 13: 891.
11. Pust P, Hintze F, Hecht C, Weiler V, Locher A, Zitnanska D, Harm S, Wiechert D, Schmidt PJ, Schnick W (2014) Group (III) nitrides $M[Mg_2Al_2N_4](M = Ca, Sr, Ba, Eu)$ and $Ba[Mg_2Ga_2N_4]$ structural relation and nontypical luminescence properties of Eu^{2+} doped samples. *Chem. Mater.* 26: 6113.
12. Liu C, Qi Z, Ma C-G, Dorenbos P, Hou D, Zhang S, Kuang X, Zhang J, and Liang H (2014) High light yield of $Sr_8(Si_4O_{12})Cl_8:Eu^{2+}$ under X-ray excitation and its temperature-dependent luminescence characteristics. *Chem. Mater.* 26: 3709.
13. Bachmann V, Ronda C, and Meijerink A (2009) Temperature quenching of yellow Ce^{3+} luminescence in YAG:Ce. *Chem. Mater.* 21: 2077.
14. Henderson B, and Imbusch GF (2006) *Optical Spectroscopy of Inorganic Solids*, Oxford University Press, Oxford.
15. Shannon RD (1976) Revised effective ionic radii and systematic studies of interatomic distances in halides and chalcogenides. *Acta Crystallogr. Sect. A: Found. Crystallogr.* 32: 751.
16. WO Pat. (2018) WO 2018/029299 A1.
17. Zhao M, Zhou Y, Molokeev MS, Zhang Q, Liu Q, and Xia Z (2019) Discovery of new narrow-band phosphors with the UCr_4C_4-related type structure by alkali cation effect. *Adv. Opt. Mater.* 7: 1801631.
18. Fang MH, Mariano COM, Chen PY, Hu SF, and Liu RS (2020) Cuboid-size-controlled color-tunable Eu-doped alkali-lithosilicate phosphors. *Chem. Mater.* 32: 1748.
19. Li G, Lin CC, Chen WT, Molokeev MS, Atuchin VV, Chiang CY, Zhou W, Wang CW, Li WH, Sheu HS, Chan T S, Ma C, and Liu RS (2014) Photoluminescence tuning via cation substitution in oxonitridosilicate phosphors: DFT calculations, different site occupations, and luminescence mechanisms. *Chem. Mater.* 26: 2991.
20. Kechele JA, Oeckler O, Stadler F, and Schnick W (2009) Structure elucidation of $BaSi_2O_2N_2$—A host lattice for rare-earth doped luminescent materials in phosphor-converted (pc)-LEDs. *Solid State Sci.* 11: 537.
21. Seibald M, Rosenthal T, Oeckler O, and Schnick W (2014) Highly efficient pc-LED phosphors $Sr_{1-x}Ba_xSi_2O_2N_2:Eu^{2+}$-crystal structures and luminescence properties revisited. *Crit. Rev. Solid State Mater. Sci.* 39: 215.
22. Penghui Y, Xue Y, Hongling Y, Jiang T, Dacheng Z, and Jianbei Q (2012) Effects of crystal field on photoluminescence properties of $Ca_2Al_2SiO_7:Eu^{2+}$ phosphors. *J. Rare Earths* 30: 1208.
23. Izhevskiy VA, Genova LA, Bressiani JC, and Aldinger F (2000) Progress in SiAlON ceramics. *J. Eur. Ceram. Soc.* 20: 2275.
24. Wang Z, Ye W, Chu IH, and Ong SP (2016) Elucidating structure–composition–property relationships of the β-SiAlON:Eu^{2+} phosphor. *Chem. Mater.* 28: 8622.

25. Wang Z, Chu IH, Zhou F, and Ong SP (2016) Electronic structure descriptor for the discovery of narrow-band red-emitting phosphors. *Chem. Mater.* 28: 4024.
26. Zhang X, Fang MH, Tsai YT, Lazarowska A, Mahlik S, Lesniewski T, Grinberg M, Pang WK, Pan F, Liang C, and Liu RS (2017) Controlling of structural ordering and rigidity of β-SiAlON:Eu through chemical cosubstitution to approach barrow-band-emission for light-emitting diodes application. *Chem. Mater.* 29: 6781.
27. Tessier F, and Marchand R (2003) Ternary and higher order rare-earth nitride materials: Synthesis and characterization of ionic-covalent oxynitride powders. *J. Solid State Chem.* 171: 143.
28. Jørgensen C (1962) The nephelauxetic series. *Prog. Inorg. Chem.* 4: 73.
29. Van Vleck J (1932) Theory of the bariations in paramagnetic anisotropy among different salts of the iron group. *Phys. Rev.* 41: 208.
30. Denault KA, Brgoch J, Gaultois MW, Mikhailovsky A, Petry R, Winkler H, DenBaars SP, and Seshadri R (2014) Consequences of optimal bond valence on structural rigidity and improved luminescence properties in $Sr_xBa_{2-x}SiO_4$:Eu^{2+} orthosilicate phosphors. *Chem. Mater.* 26: 2275–2282.
31. Sato Y, Kuwahara H, Kato H, Kobayashi M, Masaki T, and Kakihana M (2015) Large redshifts in emission and excitation from Eu^{2+}-activated Sr_2SiO_4 and Ba_2SiO_4 phosphors induced by controlling Eu^{2+} occupancy on the basis on crystal-site engineering. *Opt. Photonics J.* 5: 326.
32. Francisco E, Recio J, Blanco M, Pendás AM, and Costales A (1998) Quantum-mechanical study of thermodynamic and bonding properties of MgF_2. *J. Phys. Chem. A* 102: 1595.
33. Francisco E, Blanco MA, and Sanjurjo G (2001) Atomistic simulation of SrF_2 polymorphs. *Phys. Rev. B* 63: 094107.
34. Fang MH, Tsai YT, Sheu HS, Lee JF, and Liu RS (2018) Pressure-controlled synthesis of high-performance $SrLiAl_3N_4$:Eu^{2+} narrow-band red phosphors. *J. Mater. Chem. C* 6: 10174.
35. Schmiechen S, Schneider H, Wagatha P, Hecht C, Schmidt PJ, and Schnick W (2014) Toward new phosphors for application in illumination-grade white pc-LEDs: The nitridomagnesosilicates $Ca[Mg_3SiN_4]$:Ce^{3+}, $Sr[Mg_3SiN_4]$:Eu^{2+}, and $Eu[Mg_3SiN_4]$. *Chem. Mater.* 26: 2712.
36. Hu WW, Ji WW, Khan SA, Hao LY, Xu X, Yin LJ, and Agathopoulos S (2016) Preparation of $Sr_{1-x}Ca_xLiAl_3N_4$:Eu^{2+} solid solutions and their photoluminescence properties. *J. Am. Ceram. Soc.* 99: 3273.
37. Fang MH, Meng SY, Majewska N, Lesniewski T, Mahlik S, Grinberg M, Sheu H-S, and Liu RS (2019) Chemical control of $SrLi(Al_{1-x}Ga_x)_3N_4$:Eu^{2+} red phosphor at extreme condition for the application in light-emitting diodes. *Chem. Mater.* 31: 4614.
38. Fang MH, Mahlik S, Lazarowska A, Grinberg M, Molokeev MS, Sheu HS, Lee JF, and Liu RS (2019) Structural evolution and neighbor-cation control of photoluminescence in $Sr(LiAl_3)_{1-x}(SiMg_3)_xN_4$:$Eu^{2+}$ phosphor. *Angew. Chem. Int. Ed.* 58: 7767.
39. Hoerder GJ, Seibald M, Baumann D, Schröder T, Peschke S, Schmid PC, Tyborski T, Pust P, Stoll I, Bergler M, Patzig C, Reißaus S, Krause M, Berthold L, Höche T, Johrendt D, and Huppertz H (2019) $Sr[Li_2Al_2O_2N_2]$:Eu^{2+}—A high performance red phosphor to brighten the future. *Nat. Commun.* 10: 1824.
40. Hoerder GJ, Peschke S, Wurst K, Seibald M, Baumann D, Stoll I, and Huppertz H (2019) $SrAl_{2-x}Li_{2+x}O_{2+2x}N_{2-2x}$:$Eu_{2+}$ ($0.12 \le x \le 0.66$)—Tunable luminescence in an oxonitride phosphor. *Inorg. Chem.* 58: 12146.
41. Liao H, Zhao M, Molokeev MS, Liu Q, and Xia Z (2018) Learning from a mineral structure toward an ultra-narrow-band blue-emitting silicate phosphor $RbNa_3(Li_3SiO_4)_4$:Eu^{2+}. *Angew. Chem. Int. Ed.* 130: 11902.
42. Liao H, Zhao M, Zhou Y, Molokeev MS, Liu Q, Zhang Q, and Xia Z (2019) Polyhedron transformation toward stable narrow-band green phosphors for wide-color-gamut liquid crystal display. *Adv. Funct. Mater.* 29: 1901988.
43. Dutzler D, Seibald M, Baumann D, Philipp F, Peschke S, and Huppertz H (2019) $RbKLi_2[Li_3SiO_4]_4$:Eu^{2+} an ultra narrow-band phosphor. *Z. Naturforsch. B* 74: 535.
44. Wang W, Tao M, Liu Y, Wei Y, Xing G, Dang P, Lin J, and Li G (2019) Photoluminescence control of UCr_4C_4-typed phosphors with superior luminous efficiency and high color purity via controlling site-selection of Eu^{2+} activators. *Chem. Mater.* 31: 9200.
45. Zhao M, Liao H, Molokeev MS, Zhou Y, Zhang Q, Liu Q, and Xia Z (2019) Emerging ultra-narrow-band cyan-emitting phosphor for white LEDs with enhanced color rendition. *Light-Sci. Appl.* 8: 38.
46. Zhao M, Liao H, Ning L, Zhang Q, Liu Q, and Xia Z (2018) Next-generation narrow-band green-emitting $RbLi(Li_3SiO_4)_2$:Eu^{2+} phosphor for backlight display application. *Adv. Mater.* 30: 1802489.

47. Fang MH, Mariano CO, Chen KC, Lin JC, Bao Z, Mahlik S, Lesniewski T, Lu KM, Lu YR, Wu YJ, Sheu HS, Lee JF, Hu SF, Liu RS, and Attfield JP (2021) High-performance NaK$_2$Li[Li$_3$SiO$_4$]$_4$:Eu green phosphor for backlighting light-emitting diodes. *Chem. Mater.* 33: 1893.
48. Takahashi T, and Adachi S (2008) Mn^{4+}-activated red photoluminescence in K$_2$SiF$_6$ phosphor. *J. Electrochem. Soc.* 155: E183.
49. Wei LL, Lin CC, Fang MH, Brik MG, Hu SF, Jiao H, and Liu RS (2015) A low-temperature co-precipitation approach to synthesize fluoride phosphors K$_2$MF$_6$:Mn^{4+} (M = Ge, Si) for white LED applications. *J. Mater. Chem. C* 3: 1655.
50. Reisfeld MJ, Matwiyoff NA, and Asprey LB (1971) The electronic spectrum of cesium hexafluoroman-ganese (IV). *J. Mol. Spectrosc.* 39: 8.
51. Zhu H, Lin CC, Luo W, Shu S, Liu Z, Liu Y, Kong J, Ma E, Cao Y, and Liu RS (2014) Highly efficient non-rare-earth red emitting phosphor for warm white light-emitting diodes. *Nat. Commun.* 5: 4312.
52. Fang MH, Wu WL, Jin Y, Lesniewski T, Mahlik S, Grinberg M, Brik MG, Srivastava AM, Chiang CY, Zhou W, Jeong D, Kim SH, Leniec G, Kaczmarek SM, Sheu H-S, and Liu RS (2018) Control of luminescence via tuning of crystal symmetry and local structure in Mn^{4+}-activated narrow band fluoride phosphors. *Angew. Chem. Int. Ed.* 57: 1797.

6 Mn⁴⁺-Doped LED Phosphors

Sisi Liang, Decai Huang, Haomiao Zhu, and Xueyuan Chen

CONTENTS

6.1 INTRODUCTION: BACKGROUND AND DRIVING FORCES

Transition metal ions–doped inorganic compounds, which have been extensively studied for over 150 years, are well-established luminescent and laser materials. For instance, the first working solid-state laser demonstrated by Maiman in 1960 utilized ruby (Al_2O_3: Cr^{3+}) as the gain medium. Mn^{2+}-doped $ZnSiO_4$ was widely used as the green phosphors in fluorescent lamps, cathode ray tubes (CRT) and plasma display panels. With the fast development of solid-state lighting technology, researchers try to exploit new phosphors based on transition metal ions for white light-emitting diodes (LEDs). Among these ions, Mn^{4+} ions are of special interest due to their distinct electronic structures and corresponding optical features, namely, suitable absorption bands in UV-blue region and emission bands in red region. Particularly, Mn^{4+}-activated compounds have attracted

DOI: 10.1201/9781003098669-6

tremendous attention for their great potential as narrow-band red phosphors for backlight units of liquid crystal displays (LCDs) with high color gamut in recent years. In this section, we will focus on the electronic configuration, synthesis, optical property and application of Mn^{4+} ions–activated red phosphors.

6.2   ELECTRONIC STRUCTURE AND OPTICAL TRANSITIONS OF MN⁴⁺ IONS

Transition metal ions have an incompletely filled outside d^n shell ($0 < n < 10$). The Coulomb interaction energy between these electrons is comparable to the crystal field (CF) energy; therefore, the mixing of wave functions derived from different CF configurations must be taken into account when calculating their energy levels. The Hamiltonian operator that describes the electronic center can be expressed as:

$$H = H_O + H' + H_{SO} + H_C \tag{6.1}$$

where H_O represents the orbital Hamiltonian; H' denotes Coulomb interaction between the outer electrons; H_{SO} is the energy of spin-orbit interaction; and H_C is the energy of interaction of the outer electrons with the electrostatic CF. To determine the eigenstates and eigenvalues of H, there are three approaches that depend upon the relative sizes of various terms: (1) weak CF ($H_C \ll H', H_{SO}$); (2) intermediate CF ($H' > H_C > H_{SO}$); (3) strong CF ($H_C > H' > H_{SO}$). The weak field approach is applied for trivalent rare-earth ions since for these ions the 4f electrons are shelled by their outer filled $5s^2 5p^6$ subshells. In this case, the free-ion states are first calculated, and then H_C is taken into account by perturbation theory. For transition metal ions with $H_C > H_{SO}$, we can use either intermediate or strong field schemes. The spin-orbit coupling is neglected first, then the eigenstates of Hamiltonian are calculated by diagonalizing the matrix for $H' + H_C$. Note that the CF term H_C refers to the dominant high-symmetry field (octahedral or tetrahedral). Smaller low-symmetry CF terms and the spin-orbit coupling term H_{SO} are treated afterward by perturbation theory.

The energy levels of $3d^n$ ions were theoretically calculated by Tanabe and Sugano by means of strong CF approach.[1,2] Mn^{4+} has an outer $3d^3$ electron configuration, and Figure 6.1(a) shows the Tanabe-Sugano energy diagram of Mn^{4+} ions in an octahedral CF, which is the most common for transition metal ions. This octahedral field splits the fivefold degenerate d orbital into a twofold degenerate e_g state and a threefold degenerate t_{2g} state. The energy separation between e_g and t_{2g} is labeled as $10D_q$, where D_q is a parameter which characterizes the strength of the octahedral field. The Coulomb interaction between electrons is characterized by three Racah parameters A, B, C, where the parameter A does not enter into the expressions for the separation between energy levels. Thus, we only need to concern B and C. The ratio of C/B is expected to be almost the same as 4.5 for all the transition metal ions. Figure 6.1(a) shows how the energy E varies with the ratio of D_q/B. It shows clearly that the energies of most multiplets are strongly dependent on CF strength except $^2T_1$ and ^2E levels.

Generally, radiative transitions between $3d^n$ states are electric-dipole forbidden because the parity does not change. However, the parity selection rule can be relaxed when the active ions are situated at the site without inversion symmetry by mixing a small amount of opposite-parity wave functions (e.g., $3d^{n-1}4p$ and $3d^{n-1}4f$) into the $3d^n$ wave functions through the interaction with odd-rank CF components. Therefore, the emission intensity of the zero-phonon line was generally higher when the Mn^{4+} ions were situated at crystallographic site with lower local site symmetry. Additionally, the interaction between the electronic transitions with lattice vibrations of odd parity may also render the vibronic transition allowed for electric dipole radiation. The optical characteristics of Mn^{4+} ions with octahedral symmetry can be inferred from their configurational coordinate diagrams. The ^2E, $^2T_1$, $^2T_2$ and $^4A_2$ levels are derived from the t_2^3 electronic orbital, whereas the $^4T_1$ and $^4T_2$ levels are formed from another t_2^2e orbital. As such, a large lateral displacement can be observed between the parabolas of the ground state $^4A_2$ and $^4T_1$ (or $^4T_2$), while a small displacement between

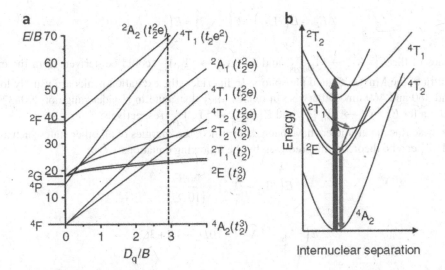

FIGURE 6.1 (a) Tanabe–Sugano energy diagram of a $3d^3$ system in an octahedral crystal field. (b) Configurational coordinate diagram for Mn^{4+} ions in fluorides hosts. (Zhu, H.M., Lin, C.C., Luo W.Q. et al., *Nat. Commun.* 5, 4312, 2014. With permission.)

the parabolas of $^4A_2$ and ^2E (or $^2T_1$, $^2T_2$). Larger displacement implies stronger electron-phonon interaction (namely, larger Huang-Rhys factor) and, thus, larger spectral bandwidth of the transition. Moreover, according to the spin selection rule of $\Delta S = 0$, the transitions between $^4T_1$ (or $^4T_2$) and $^4A_2$ levels are spin-allowed; thus, intense excitation or absorption bands with relatively large bandwidth are expected between these levels. Upon excitation to $^4T_1$ or $^4T_2$ levels, the excited ions usually relax nonradiatively to ^2E followed by the spin-forbidden $^2E \rightarrow {^4A_2}$ transition characterized by sharp emission lines.

6.3 DETERMINATION OF THE RACAH PARAMETERS

The $3d^3$ electronic configuration of Mn^{4+} determines that there are 120 possible distribution modes of electrons in the ten usable states of the 3d shell (including 5 spin states with up and down, respectively). The Coulomb repulsion of electrons causes the splitting of the 120-fold degenerate configuration in the eight LS terms, and the Racah parameters C and B could be used to evaluate the splitting magnitude. The symmetry of CF determines the number of required parameters and the remaining degeneracy of electronic energy levels. For the octahedral CF, the CF parameter of $10Dq$ is used to characterize the CF strength. The Racha parameters B and C can be calculated respectively by the following equations:

$$\frac{Dq}{B} = \frac{15\left(\Delta E_T / (Dq - 8)\right)}{\left(\Delta E_T / Dq\right)^2 - 10\left(\Delta E_T / Dq\right)} \tag{6.2}$$

$$C \approx 0.328\Delta E_E - 2.59B + 0.59B^2 / Dq \tag{6.3}$$

where ΔE_T is the difference in the ZPL energies of the $^4T_{1g}$ and $^4T_{2g}$ states and ΔE_E is the ZPL energy of the ^2E_g state. They are defined as:

$$\Delta E_T = E\left(^4A_{2g} \rightarrow {^4T_{1g}}\right)_{ZPL} - E\left(^4A_{2g} \rightarrow {^4T_{2g}}\right)_{ZPL} = E\left(^4T_{1g}\right)_{ZPL} - E\left(^4T_{2g}\right)_{ZPL} \tag{6.4}$$

$$\Delta E_E = E\left(^2E_{2g}\right) - E\left(^4A_{2g}\right) = E\left(^2E_{2g}\right)_{ZPL} \tag{6.5}$$

The values of the $E\left(^4A_{2g} \rightarrow {}^4T_{1g}\right)_{ZPL}$ and $E\left(^4A_{2g} \rightarrow {}^4T_{2g}\right)_{ZPL}$ could be derived from the excitation spectra of the Mn^{4+}-activated phosphors. In fluorides, the excitation peaks are usually located at around 360 and 460 nm. Otherwise in oxides, they distribute in a wider range of 290–370 and 410–470 nm for $E\left(^4A_{2g} \rightarrow {}^4T_{1g}\right)_{ZPL}$ and $E\left(^4A_{2g} \rightarrow {}^4T_{2g}\right)_{ZPL}$, respectively.

Once these Racah parameters have been determined, the energies of all other states such as $^2T_{1g}$, $^2A_{1g}$ and $^4T_{1g}$ can be theoretically predicted by the following equations:

$$E\left(^2T_{1g} - {}^2E_g\right) = \frac{66B^2}{\left(10D_q\right)} \tag{6.6}$$

$$E\left(^2A_{1g} - {}^4A_{2g}\right) = 10D_q + 4B + 3C \tag{6.7}$$

$$E\left(^4T_{1g}(P) - {}^4A_{2g}\right) = 15D_q + 7.5B - 0.5\left[100D_q^2 - 180D_qB + 225B^2\right]^{1/2}$$

The local crystal-field strength parameter Dq can be determined by the average peak energy of the $^4A_{2g} \rightarrow {}^4T_{2g}$ transition according to the following equation:

$$Dq = \frac{E\left(^4T_{2g} - {}^4A_{2g}\right)}{10} \tag{6.8}$$

6.4 NEPHELAUXETIC EFFECT

The Racah parameters B and C in crystals are considerably smaller than those of free ions because some electrons of the ligands move into the orbitals of the central ion and reduce the cationic valency. Consequently, the d-electron wavefunctions expand toward the ligand to increase the distances between electrons and reduce the interaction between them. This effect is named as nephelauxetic effect. In fact, some $3d$ electrons are known to exist even at the positions of the nuclei of the ligands as evidenced by ESR and NMR measurements. Therefore, the assumption is not strictly hold that the expansion of the $3d$ orbitals is small and could be negligible. That also means the coordination with respect to different ligands results in different values of B and C. The interatomic bond distances and angles between the Mn^{4+} ions and their ligands determine the overlap of the wave functions. Intuitively, it could be concluded that the competition between the covalence and the ionicity of the chemical bonding is the key-factor. Taking the Mn^{4+} with octahedral coordination as an example, the Racah parameters B and C in fluorides will be smaller than those in the oxides since the nephelauxetic effect is weaker in fluorides than in oxides. Correspondingly, the ^2E_g energy of Mn^{4+} in fluorides is usually higher than in oxides; thus, the emission of the Mn^{4+} has a longer wavelength in oxides than that in fluorides.

6.5 SYNTHESIS METHODS OF THE MN⁴⁺ PHOSPHOR

The preparation approaches directly affect the crystallinity, photoluminescent properties, morphology and particle size of phosphors. Therefore, the exploration for more efficient and environmentally friendly synthetic methods is of great significance for improving the performance and also commercial application of phosphors. In terms of the Mn^{4+}-doped fluorides and oxides, their representative synthesis methods are introduced as follows.

6.5.1 Mn⁴⁺-DOPED FLUORIDES

Up to now, the researchers have made great efforts on the synthesis of Mn^{4+}-doped fluoride phosphors. Typically, Mn^{4+}-doped fluoride red phosphors are prepared through the wet chemical routes based on hydrofluoric acid solution such as cation exchange and coprecipitation etc. Compared with the traditional high-temperature solid-state reaction, the wet chemical methods have numerous advantages including lower synthesis temperature (room temperature), easier control of composition, more uniform size distribution and better morphology of phosphor particles.

6.5.1.1 Cation Exchange Method

K_2TiF_6: Mn^{4+} is a typical red phosphor synthesized through cation exchange method. Generally, efficient cation exchange process was only observed in nano-sized materials. However, it was interesting to find that this method can also be used to synthesize micro-sized K_2TiF_6: Mn^{4+} phosphors efficiently at room temperature. The synthetic procedure can be divided into three main steps. First, the K_2MnF_6 crystals were prepared by dissolving the $KMnO_4$ and KF in hydrofluoric acid according to the following chemical equation:

$$2KMnO_4 + 4KF + 10HF + 3H_2O_2 \rightarrow 2K_2MnF_6 + 8H_2O + 3O_2 \uparrow \qquad (6.9)$$

Next, the prepared K_2MnF_6 powders were dissolved in hydrofluoric acid at room temperature to obtain a transparent solution. Second, the blank K_2TiF_6 powders were added into the solution under stirring for about 20 min to form a uniform muddy mixture. Finally, after the cation-exchange reaction was completed, the mixture was dried at 70°C for 3 h in an oven to produce the phosphors. The cation exchange method is also available for preparing other Mn^{4+}-doped phosphors such as K_2SiF_6: Mn^{4+}, but the reaction rate is less efficient as compared with that observed in K_2TiF_6: Mn^{4+} phosphor.

6.5.1.2 Coprecipitation Method

Most of the fluoride phosphors, typically K_2SiF_6: Mn^{4+}, were prepared by the coprecipitation method at room temperature. The two-step coprecipitation method is convenient for massive production. The typical synthetic procedure for K_2SiF_6:·Mn^{4+} is described next. The precursor of K_2MnF_6 crystals were first prepared based on the method presented earlier. Then the K_2MnF_6 powder was dissolved in a solution of SiO_2 in HF by stirring. Subsequently, KF was added slowly to the solution with continuous stirring. The precipitate is formed after the reaction mixture is kept at room temperature for 30 min. When the precipitation was completed, the solution was filtered, and the filtrate was washed with acetone several times and, finally, dried at 100°C to obtain the final products.

6.5.1.3 HF-Free Hydrothermal Route

To develop more environment friendly synthesis method, the researchers tried to prepare Mn^{4+}-doped fluoride phosphors without the using of hydrofluoric acid. For instance, a hydrothermal method for preparing Mn^{4+}-doped K_2SiF_6 using low-toxic H_3PO_4/KHF_2 solution instead of high-toxic HF solution was presented. First, the soluble MnL_2 (L = HPO_4^{2-}) precursor were prepared as the following procedure. According to the stoichiometric ratio, the $KMnO_4$ solution was added dropwise into potassium formate $CH_2O_2 \cdot K$ solution under ultrasonic vibration to produce active $MnO(OH)_2$. The precipitates were collected and dispersed in phosphoric acid followed by ultrasonic vibration. After being diluted with phosphoric acid, the resultant MnL_2 solution was obtained for further synthetic procedure. Next, the raw chemicals of SiO_2, KHF_2 and as-prepared MnL_2 solution are mixed thoroughly and charged into a Teflon vessel. After ultrasonic vibration for 30 min, the Teflon vessel was sealed up in a stainless-steel autoclave and kept at 180°C for 6 h. After reaction, the autoclave was cooled naturally to room temperature. The resulting products were collected by centrifugation and washed by distilled water twice and dried under vacuum at 40°C for 24 h. The reaction could be described as the following equations:

$$2MnO_4^- + 3CHO_2^- + 3H_2O \rightarrow 2MnO(OH)_2 + 3CO_2 + 5OH^- \tag{6.10}$$

$$MnO(OH)_2 + 2H_2L \rightarrow MnL_2 + 3H_2O \tag{6.11}$$

$$MnL_2 + 3[HF_2]^- + H^+ \leftrightarrow [MnF_6]^{2-} + 2H_2L \tag{6.12}$$

$$SiO_2 + 3[HF_2]^- + H^+ \rightarrow [SiF_6]^{2-} + 2H_2O \tag{6.13}$$

$$2K^+ + (1-x)[SiF_6]^{2-} + x[MnF_6]^{2-} \leftrightarrow K_2Si_{1-x}Mn_xF_6 \tag{6.14}$$

In this HF-free preparation process, the H_3PO_4/KHF_2 plays the same key role as HF in the process of stabilizing Mn^{4+} and promoting Mn^{4+} into the K_2SiF_6 lattice. The concentration of Mn^{4+} ions could be controlled by adjusting the amount of KHF_2 in the host K_2SiF_6.

6.5.2 Mn⁴⁺-Doped Oxides

Different from the fluorides, the preparation of Mn^{4+}-doped oxides phosphors are mostly based on high-temperature solid-state reaction method, and the raw materials are usually chemically stable oxides or carbonates. This method is widely used for massively production of most commercial phosphors.

6.5.2.1 The High-Temperature Solid-State Method

The Mn^{4+}-doped oxide phosphors are mostly prepared via the traditional high-temperature solid-state reaction method. MnO_2 is usually chosen as manganese source. The typical procedure is described as follow: all raw materials are weighed based on stoichiometric ratio and ground thoroughly before sintering. Depending on the specific material, the sintering atmosphere could be the air, N_2 and other inert gases. In the calcination process, appropriate amount of flux could be added to reduce the reaction temperature and improve the quality of the synthesized phosphors. The solid-state method is beneficial to the massive production of the phosphors with relatively low cost and high crystallinity. However, further posttreatments such as ball-milling and sieving are usually needed to improve the morphology and particle size distribution of the prepared phosphors for commercial applications.

6.5.2.2 The Sol-Gel Method

The sol-gel method is an important method for preparing phosphors with uniform size distribution under much lower temperature than the solid-state reaction method. The raw materials used in the sol-gel method are first dispersed in the solvent to form a low-viscosity solution; thus, the molecular level uniformity can be obtained in a short time. When the gel is formed, the homogeneity is maintained and the uniformity of the final product is ensured. In particular, the Pechini-type sol-gel method generally uses inorganic salts as precursors, acid as ligand and polyethylene glycol as cross-linking agent. It is a nontoxic, low-cost and easy-to-control method for synthesizing inorganic phosphors. The typical synthetic process is described as follow:

First, the stoichiometric amounts of soluble-salt raw materials were dissolved in deionized water with stirring, and the citric acid was added as a complexing agent with the ratio of 1:2 (metal-to-citric acid), capped and stirred. Then, the pH of the solution was adjusted to 7 by NH_4OH. Finally, a certain amount of polyethylene glycol (PEG; molecular weight = 20,000) was added. The resultant mixtures were stirred for 2 h and then held in a water bath maintained at 75°C until homogeneous gels formed. The obtained gel was dried in an oven at 100°C for 10 h and then is prefired at 500–800°C for 4 h in the muffle furnace in air. The obtained precursors are sintered again at suitable temperature and atmosphere to obtain the final phosphors.

6.6 REPRESENTATIVE MN⁴⁺-DOPED INORGANIC PHOSPHOR

6.6.1 Mn⁴⁺-ACTIVATED FLUORIDE PHOSPHORS

The CF exerted by the host plays a key role on the optical properties of Mn^{4+} ions. Fluorine is the most electronegative element and gives rise to the most ionic ligand. Therefore, the crystal structures of metal fluorides are mainly governed by geometric and electrostatic principles. In most cases, the local symmetry of central metal ions is octahedral. When Mn^{4+} ions are situated at fluoride hosts with octahedral coordination, they exhibit the most intense broadband excitation at ~460 nm with a bandwidth of ~50 nm and sharp emission lines peaking at ~630 nm. As compared to those in oxide hosts, the excitation peaks in Mn^{4+}-doped fluoride phosphors shift to longer wavelength, but the emission peaks shift conversely to shorter wavelength. The red shift of the excitation band is due to the weaker CF strength experienced by Mn^{4+} ions in fluoride host than those in oxide counterparts. Nevertheless, the $^2E_g \rightarrow {}^4A_{2g}$ transition is nearly independent on CF strength. The observed blue shift of the emission band in fluoride hosts can only be ascribed to the so-called nephelauxetic (or electron cloud-expanding) effect, which is generally regarded a measure of the bond covalency between active ions and ligands. As the covalency increases, the interaction between the electrons is decreased due to the expansion of electron cloud. It, thus, leads to the shift of electronic transitions to lower energy. This is consistent with the fact that fluoride complexes are highly ionic crystals and oxide hosts have higher covalency. Obviously, Mn^{4+}-doped fluoride compounds show more desirable optical properties than that of oxide compounds for white LEDs based on blue-diode chips. Particularly, the peak wavelength at ~630 nm is more sensitive to human eyes than that of Mn^{4+}-doped oxides; thus, we anticipate higher luminous efficacy when using Mn^{4+}-doped fluoride compounds as red phosphors in white LEDs.

The biggest challenge of Mn^{4+}-doped fluoride phosphors for commercial applications is their poor water-resistant performance. Most of fluoride phosphors suffer from their poor stability in moisture environment due to the hydrolysis of MnF_6^{2-} group to hydrated manganese dioxide. This drawback causes deterioration of LEDs during long-term operation and hinders commercial application of these materials. Therefore, great efforts have been made to improve the chemical stability of Mn^{4+}-doped fluorides phosphors. The hosts of the Mn^{4+}-doped fluoride phosphors can be roughly divided into three categories, i.e., A–IV–F_6 (A = Ca, Sr, Ba; IV = Si, Ge, Ti, Sn, Zr), I_2–IV–F_6 (I = Li, Na, K, Rb, Cs; IV = Si, Ge, Ti, Sn, Zr) and I_3–III–F_6 (I = Li, Na, K, Rb, Cs, III = Al, Ga).

6.6.1.1 K_2SiF_6:Mn⁴⁺ (I_2–IV–F_6)

The structure of K_2SiF_6 belongs to cubic system and the space group is O_h-$Fm\bar{3}m$. Figure 6.2 shows the excitation and emission spectra of the K_2SiF_6: Mn^{4+}.[3] The optimum excitation wavelength is at 450 nm, and the emission is peaked at 630 nm in red region. It indicates that the strongest absorption of K_2SiF_6: Mn^{4+} match well with the commercial blue LED chips. The photoluminescence quantum yields (PL QYs) of these phosphors were estimated to be higher than ~80%. More importantly, the brightness of the phosphors was observed to increase slightly when the temperature was up to about 200°C, indicating excellent PL thermal stability. It was also found that the PL quenching temperature increased with the decrease in the radius of the cation substituted by Mn^{4+}, which suggests that K_2SiF_6:Mn^{4+} phosphor might have the best PL thermal stability. Currently, K_2SiF_6: Mn^{4+} red phosphors have been commercially used in the backlights of LCDs in mobile phones.

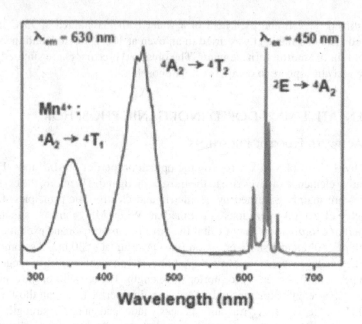

FIGURE 6.2   The excitation and emission spectra of the K_2SiF_6: Mn^{4+} phosphor. (Huang, L., Zhu, Y.W., Zhang X.J. et al., *Chem. Mater.* 28, 1495, 2016. With permission.)

6.6.1.2 K_2TiF_6:Mn^{4+} (I_2–IV–F_6)

K_2TiF_6: Mn^{4+} could be synthesized by a cation-exchange reaction at room temperature.[4] The lattice structure of K_2TiF_6: Mn^{4+} is identified as crystallizing in hexagonal crystal system with $P\bar{3}m1$ space group. Each Ti^{4+} ion is surrounded by six F^- ions that form only one kind of TiF_6^{2-} octahedral structure. Notably, the K^- ion is at the center of 12 nearest neighbor F^- ions forming a nearly regular polyhedron. Under the 468 nm excitation, K_2TiF_6: Mn^{4+} exhibits a highly efficient red light emission at 631 nm. The internal and external QYs can reach as high as 98% and 68%, respectively. The K_2TiF_6: Mn^{4+} also exhibits good thermal stability. At 150°C, it could retain 98% of the initial intensity at room temperature. Similarly, as observed in K_2SiF_6:Mn^{4+} phosphor, the total PL (assigned to the ^2E_g–$^4A_{2g}$ transition) intensity at 300K increases about 40% relative to that at 10K. The increased absorption of excitation light and enhanced radiative transition probability at higher temperatures contributes to the incremental PL intensity with the rise of temperature.

In terms of the poor moisture-resistant performance suffered by K_2TiF_6: Mn^{4+}, researchers have made great efforts to solve this problem. The works mainly focus on the modification of the surface of phosphor particles or constructing the unique structure to improve the hydrophobicity. For instance, Zhu's group designed a convenient reverse cation exchange strategy to construct a core-shell structured K_2TiF_6: Mn^{4+}@K_2TiF_6 phosphor. The outer pure K_2TiF_6 shell can not only act as a shield for preventing moisture in air to hydrolyze inner MnF_6^{2-} group, but also effectively cut off the path of energy migration to surface defects, thus increasing the emission efficiency. Besides, the phosphors could also be coated with organic hydrophobic layer, such as polypropylene glycol (PPG), to improve the moisture-resistant performance. However, it is important to ensure that the coating shell should not react with the silicon, which is usually mixed with phosphors for LEDs packaging.

6.6.1.3 K_3AlF_6: Mn^{4+} (I_3–III–F_6)

The K_3AlF_6 crystal has a cubic structure (δ-phase) with space group $Fm\bar{3}m$, and the doped Mn^{4+} ions were supposed to occupy the six-coordinated Al^{3+} octahedral sites.[5] Meanwhile, the formation of fluorine defects may occur considering charge compensation. Under 463 nm blue-light excitation,

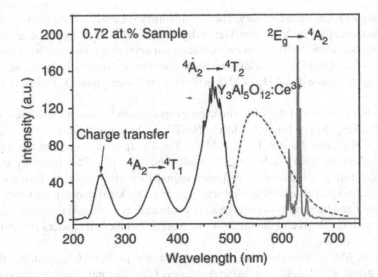

FIGURE 6.3 The solid-lines are the excitation and emission spectra of K_2TiF_6: Mn^{4+}, respectively. The dash line is the emission spectrum of the commercial yellow-emitting $Y_3Al_5O_{12}$:Ce^{3+} phosphor. (Zhu, H.M., Lin, C.C., Luo W.Q. et al., *Nat. Commun.* 5, 4312, 2014. With permission.)

K_3AlF_6: Mn^{4+} shows the typical sharp emission spectra peaked at 630 nm as shown in Figure 6.3. The internal and external PL QYs are 88% and 50.6%, respectively.

6.6.1.4 Na_2GeF_6: Mn^{4+} (I_2–IV–F_6)

The zero-phonon line (ZPL) emission intensity is different in different host lattices. In some fluorides and oxide host, ZPL exhibits extremely weak intensity which is hardly detected in the emission spectra. However, in the host crystals with the form of I_2–IV–F_6, such as Na_2GeF_6, most of

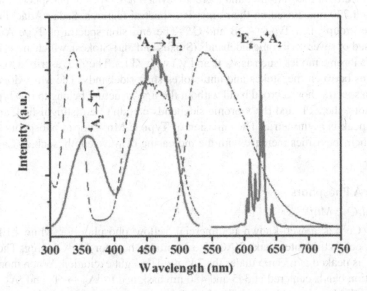

FIGURE 6.4 Excitation and emission spectra of K_3AlF_6: Mn^{4+} (solid line) and commercial phosphor YAG: Ce^{3+} (dashed line). (Song, E.H., Wang, J.Q., and Shi, J.H. et al., *ACS Appl. Mater. Interfaces* 9, 8805, 2017. With permission.)

them show strong ZPL emission intensity. The ZPL intensity is closely related to the local symmetry of the lattice sites occupied by Mn^{4+} ions. Generally, the emission intensity of the zero-phonon line was generally higher when the Mn^{4+} ions were situated at crystallographic site with lower local site symmetry. It was found that the strong ZPL emission could be realized in the hexagonal space group C_{6v}^{4} (P6$_3$mc), trigonal space group D_3^2 (P321), orthorhombic space group D_{2h}^{24} (Fddd), D_{2h}^{16} (Pnma) and C_{2v}^9 (Pna21).

Liu's group have discussed the ZPL emission properties including the PL intensity and lifetimes of the Na_2SiF_6: Mn^{4+}, Na_2GeF_6: Mn^{4+}, Na_2TiF_6: Mn^{4+} and their solid solutions such as $Na_2(Si_xGe_{1-x})F_6$: Mn^{4+} and $Na_2(Ge_yTi_{1-y})F_6$: Mn^{4+}. The PL spectra show that the ZPL intensity continuously increased from Na_2SiF_6 to Na_2GeF_6 and Na_2TiF_6. This intensity enhancement is due to the crystallographic evolution from trigonal to triclinic system. Based on the high-resolution emission spectra and EPR characterizations, the local coordination environment of the MnF_6^{2-} cluster and the distortion degree were evaluated. It was confirmed that the ZPL intensity correlates with the degree of deviation of the MnF_6^{2-} octahedral clusters compared with the ideal octahedral symmetry.

A summary of Mn^{4+}-doped fluorides phosphors is listed in Table 6.1, including the emission and excitation peaks wavelength, internal and external QYs, thermal stability (emission intensity I_{298K}/I_{423K}) as well as the key optoelectronic parameters of fabricated LEDs (CCT, CRI, Luminous efficacy).[4, 6–68]

6.6.2 Representative Oxide Materials

Compared with the fluoride hosts, the potential oxide hosts for Mn^{4+} doping are more structurally diverse, including silicates, germanates, titanate and aluminates. We have mentioned that the Mn^{4+} ions usually occupy the octahedral sites in the lattice structure. Taking the ionic radii into consideration, the Mn^{4+} ions may occupy the Ge^{4+}, Si^{4+}, Ti^{4+}, Sn^{4+}, Al^{3+}, Ga^{3+}, Ta^{5+}, Sb^{5+} sites to form the luminescent centers. Table 6.2 compares the effective ionic radii of the ions that may be substituted by Mn^{4+}.

The emission of Mn^{4+}-doped oxide phosphor covers a wider spectral range than fluorides. Most emission peak locates at 650–715 nm, and there are even a few types of phosphors whose emission peaks are around 720 nm. Based on their emission spectral features, Sadao Adachi et al. divided them into three groups, i.e., Types A, B and C.[69] The emission spectra of Type A phosphor are mainly composed of several vibronic sidebands (Stokes and anti-Stokes), which are originated from the coupling of vibronic modes (such as v_3, v_4 and v_6) with ZPLs. The ZPL ascribed to the $^2E_g \rightarrow {}^4A_2$ transition locates between the Stokes and anti-stokes shift sidebands. The Mn^{4+}-doped phosphors whose emission spectra show a broad band without the fine-structure belong to the Type B. For this kind of phosphors, the ZPL and the vibronic sidebands couldn't be distinguished, and the broadband emission peak is asymmetric. The emission of **Type C** Mn^{4+}-doped phosphors show a series of peaks, and their intensities increase with the increasing of wavelength, such as $Ca_{14}Zn_6Ga_{10}O_{35}$: Mn^{4+} phosphor.

6.6.2.1 Type A Phosphors

6.6.2.1.1 $Y_3Al_5O_{12}$: Mn^{4+}

The $Y_3Al_5O_{12}$: Ce^{3+} is a well-known commercial yellow phosphor for white LEDs. The six-coordinated Al octahedral sites make YAG also a suitable host for Mn^{4+} doping. The emission of $Y_3Al_5O_{12}$: Mn^{4+} is peaked at 673 nm under the 345 nm UV-light excitation. When monitored at 673 nm, two excitation bands centered at 345 and 480 nm assigned to $^4A_2 \rightarrow {}^4T_1$ and $^4A_2 \rightarrow {}^4T_2$ transitions are observed. Moreover, when partial of the Al^{3+} ions are replaced by Ga^{3+} ions, the excitation band of $^4A_2 \rightarrow {}^4T_2$ transition shows a continuous red-shift with increasing Ga^{3+} content. Meanwhile,

TABLE 6.1

Overview of the Mn⁴⁺-Doped Fluorides Phosphors and Corresponding Parameters

Red Phosphors	Materials					Yellow/Green Phosphors	Device				Ref.
	Emission Peak (nm)	Excitation Peak (nm)	Internal PL QY (%)	External PL QY (%)	I_{423K}/I_{298K} (%)		LED Chip (nm)	CCT (K)	CRI	Luminous Efficacy (lm/W)	
$KZnF_3$:Mn^{4+}	633	460			~10						6
$KTeF_5$:Mn^{4+}	629	454			86.2						7
Li_2SiF_6:Mn^{4+}	630	460			95						8
Li_2ZrF_6:Mn^{4+}	631	460			~55	YAG:Ce^{3+}	460	3789	91		9
$LiSrGaF_6$:Mn^{4+}	628	460			~60						10
Na_2SiF_6:Mn^{4+}	627	465			92	YAG:Ce^{3+}	450	4986	86.5	101.5 (120 mA)	11·12
Na_2GeF_6:Mn^{4+}	626	465				YAG:Ce^{3+}	465	3554	92	46.7 (20 mA)	13
Na_2TiF_6:Mn^{4+}	627	465				YAG:Ce^{3+}	465	3841	90.7	34.2 (20 mA)	8
Na_2SnF_6:Mn^{4+}	625	460	46	6	80	YAG:Ce^{3+}	450	3146	90		14
Na_2AlF_6:Mn^{4+}	627	465	69		170	YAG:Ce^{3+}	450	3903	92.7	70 (350 mA)	15
Na_3GaF_6:Mn^{4+}	626	462	69		118.8	YAG:Ce^{3+}	460	2966	81	56.73 (30 mA)	16
$Na_{1.5}Li_{0.5}SiF_6$:Mn^{4+}	620	462	96.1		15	YAG:Ce^{3+}	450	3975	91.06		17
$Na_2(Ge_{0.5}Si_{0.5})F_6$:$Mn^{4+}$	627	465									18
$Na_2(Ge_{0.5}Ti_{0.5})F_6$:$Mn^{4+}$	630	465									
K_2SiF_6:Mn^{4+}	630	460	81.5	56	102	YAG:Ce^{3+}	460	3766	86	79 (60 mA)	19
K_2GeF_6:Mn^{4+}	630	455	93	73	99	YAG:Ce^{3+}	460	3652	94	150 (60 mA)	20
K_2TiF_6:Mn^{4+}	631	455	98	68	98	YAG:Ce^{3+}	455	3510	93	162 (60 mA)	4·21
K_2SnF_6:Mn^{4+}	630	458	59.8	28		YAG:Ce^{3+}					22
K_2ZrF_6:Mn^{4+}	630	460			~60						23
K_3AlF_6:Mn^{4+}	630	465	88	50.6	~200	YAG:Ce^{3+}	450	3665	84	190 (40 mA)	5
K_3GaF_6:Mn^{4+}	626	464	46		70	YAG:Ce^{3+}	460	3691	87.2	92.1 (20 mA)	24
K_3ScF_6:Mn^{4+}	631	466	67		26	YAG:Ce^{3+}	450	3250	86.4	65.6 (20 mA)	25
$K_2(Ge_{0.5}Si_{0.5})F_6$:Mn^{4+}	630	465				β-SiAlON:Eu^{2+}	460			78 (120 mA)	26
$KNaSiF_6$:Mn^{4+}	629	462	90	37	56	YAG:Ce^{3+}	460	3455	90	118	27
$KNaSnF_6$:Mn^{4+}	630	465	84	13	20	YAG:Ce^{3+}	460	3746	90.3	137 (30 mA)	28

(Continued)

TABLE 6.1 (Continued)

Overview of the Mn^{4+}-Doped Fluorides Phosphors and Corresponding Parameters

| | Materials | | | | | | Device | | | | |
Red Phosphors	Emission Peak (nm)	Excitation Peak (nm)	Internal PL QY (%)	External PL QY (%)	I_{423K}/I_{298K} (%)	Yellow/Green Phosphors	LED Chip (nm)	CCT (K)	CRI	Luminous Efficacy (lm/W)	Ref.
K_2LiGaF_6:Mn^{4+}	636	465	20		55	YAG:Ce^{3+}	460	4378	79.6	109.91 (20 mA)	29
K_2NaGaF_6:Mn^{4+}	631	465	82	36	76	YAG:Ce^{3+}	450	3760	92	156 (60 mA)	30
K_2NaAlF_6:Mn^{4+}	631	466	85	22	80	YAG:Ce^{3+}	450	3911	94	89 (60 mA)	31
Rb_2NaAlF_6:Mn^{4+}	631	466	59.2		~53	YAG:Ce^{3+}	450	3870	87		32
Rb_2SiF_6:Mn^{4+}	632	450	73		110						33,34
Rb_2GeF_6:Mn^{4+}	630	460	72.9	57.2	90	YAG:Ce^{3+}	460	3349	90.8	145.21	35
Rb_2TiF_6:Mn^{4+}	630	460	91		120	YAG:Ce^{3+}	460	3730	89	158.8 (20 mA)	36
Rb_2SnF_6:Mn^{4+}	630	458	70.3		88	YAG:Ce^{3+}	460	3936	90	106 (20 mA)	37
Rb_2ZrF_6:Mn^{4+}	629	455	75		~50	YAG:Ce^{3+}	460	3569	85.5	139.1 (20 mA)	38
Rb_2HfF_6:Mn^{4+}	630	458	55.6		~45	YAG:Ce^{3+}	460				39
Cs_2HfF_6:Mn^{4+}	630	462	65.2		~20	YAG:Ce^{3+}	460	3657	92.6	118.5 (20 mA)	39
Cs_2SiF_6:Mn^{4+}	631	462	91	37.9		YAG:Ce^{3+}	460	3667	87.4	130.9 (60 mA)	
Cs_2GeF_6:Mn^{4+}	631	462	98	37.7		YAG:Ce^{3+}	460	3514	81.4	143.9 (60 mA)	40
Cs_2TiF_6:Mn^{4+}	631	462	82	32.2		YAG:Ce^{3+}	460	3212	76.9	129.6 (60 mA)	41
Cs_2ZrF_6:Mn^{4+}	631	462	56.9		~20	YAG:Ce^{3+}	450	3469	82.4	132.2 (20 mA)	42
$CsNaGeF_6$:Mn^{4+}	630	462	95.6		85	YAG:Ce^{3+}	460	3783	92.5	176.3 (120 mA)	43,44
$(NH_4)_2SiF_6$:Mn^{4+}	630	463	64.6		~36						
$(NH_4)_2GeF_6$:Mn^{4+}	630	463									
$(NH_4)_2TiF_6$:Mn^{4+}	630	463	16.4		~10						
$(NH_4)_2SnF_6$:Mn^{4+}	630	463									
$(NH_4)_2NaScF_6$:Mn^{4+}	630	463	21.67		~10	YAG:Ce^{3+}	450	3006	80.6	11.28 (60 mA)	45
$BaSiF_6$:Mn^{4+}	632	468			~120	YAG:Ce^{3+}	460	5903	82	51.73 (20 mA)	46
$BaGeF_6$:Mn^{4+}	635	470			48	YAG:Ce^{3+}	460	3230	87.7	116.1 (20 mA)	47
$BaSnF_6$:Mn^{4+}	631	468	38.9		55	YAG:Ce^{3+}	450	3576	83.6		48
$BaTiF_6$:Mn^{4+}	631	467	44.5		46	YAG:Ce^{3+}	450	4213	83.5	115 (20 mA)	49
$ZnSiF_6$:Mn^{4+}	630	467			7						50

(Continued)

TABLE 6.1 (Continued)
Overview of the Mn⁴⁺-Doped Fluorides Phosphors and Corresponding Parameters

| | Materials | | | | | | Device | | | | |
Red Phosphors	Emission Peak (nm)	Excitation Peak (nm)	Internal PL QY (%)	External PL QY (%)	I_{423K}/I_{298K} (%)	Yellow/Green Phosphors	LED Chip (nm)	CCT (K)	CRI	Luminous Efficacy (lm/W)	Ref.
$ZnGeF_6$:Mn⁴⁺	630	468			8						
$ZnSnF_6$:Mn⁴⁺	630	465			~10						51
$ZnTiF_6$:Mn⁴⁺	631	468	26.4	6.17	5	YAG:Ce³⁺	457	3987	83.1	92.21 (20 mA)	52
K_2NbF_7:Mn⁴⁺	630	465	93.5	26.2	8	β-SiAlON:Eu²⁺	450	3633	81.6	64.52 (120 mA)	53
K_2TaF_7:Mn⁴⁺	630	465	90	24	10	β-SiAlON:Eu²⁺	450	3059	82.2	47.78 (120 mA)	54
$KNaNbF_7$:Mn⁴⁺	627	465									
$KNaTaF_7$:Mn⁴⁺	627	460									
K_3SiF_7:Mn⁴⁺	633	460			88	β-SiAlON:Eu²⁺	450	5686	81		55
K_2ZrF_7:Mn⁴⁺	629	460				YAG:Ce³⁺	460	2970	91.4	74.71 (20 mA)	56
K_3HfF_7:Mn⁴⁺	630	465			10	YAG:Ce³⁺	460	3535	86.2		57
Na_3ZrF_7:Mn⁴⁺	630	467			92						58
Rb_3SiF_7:Mn⁴⁺	630	467									59
$CsRb_2SiF_7$:Mn⁴⁺	631	458									
Cs_3SiF_7:Mn⁴⁺	630	460									
Cs_2RbSiF_7:Mn⁴⁺	631	460									
Ba_2ZrF_8:Mn⁴⁺	623	455									60
K_3HSnF_8:Mn⁴⁺	627	460		92	~36	YAG:Ce³⁺	465	2643	95	322 (40 mA)	61
Na_3HTiF_8:Mn⁴⁺	627	465	~44		~90	β-SiAlON:Eu²⁺	430		83		62
Na_3TaF_8:Mn⁴⁺	627	460					460				63
$Na_2Li_3Ga_2F_{12}$:Mn⁴⁺	629	467	81.2	12.5	~15	YAG:Ce³⁺	450	3986	91		64,65
$Na_2Li_3Al_2F_{12}$:Mn⁴⁺	630	467	28.7	13.5	~20	YAG:Ce³⁺	450	3874	90.6		66
$Na_2Li_3Sc_2F_{12}$:Mn⁴⁺	627	461	66.8	52.2	50	YAG:Ce³⁺	450	3694	92		67
$Na_2Zr_2F_{13}$:Mn⁴⁺	626	466			~50	YAG:Ce³⁺	450	2940	82.91		68
Ba_5AlF_{13}:Mn⁴⁺	627	460			~60						

TABLE 6.2

Overview of Effective Ionic Radii of the Potential Substituted Ions by Mn^{4+}

Ions (CN = 6)	Mn^{4+}	Si^{4+}	Ge^{4+}	Ti^{4+}	Sn^{4+}	Al^{3+}	Ga^{3+}	Ta^{5+}	Sb^{5+}
Effective ionic radius (Å)	0.530	0.540	0.530	0.605	0.690	0.535	0.620	0.640	0.600

the corresponding emission spectra also exhibit a slight red-shift. This phenomenon is owing to the change of the CF induced by cation substitution.

6.6.2.1.2 $LaAlO_3:Mn^{4+}$

Mn^{4+} shows far-red light emission in $LaAlO_3$ crystal. The $LaAlO_3$ crystalizes in rhombohedral system with a space group of R-3cH, and its structure is close to that of cubic perovskite, which involves a rotation of the AlO_6 octahedra with respect to cubic perovskite. The Al sites have a local symmetry of C_{3i}, and the symmetry reduction caused by the AlO_6 octahedra rotation causes a small splitting of the ^2E state. When the Mn^{4+} ions occupy the Al^{3+} sites, the phosphor shows a far-red emission centered at 731 nm under 335 nm NUV excitation (Figure 6.6).[70] Besides, co-doping with other ions such as Cl^-, Na^+, Ca^{2+}, Sr^{2+}, Ba^{2+} and Ge^{4+} were found to be beneficial to improve the emission efficiency of $LaAlO_3:Mn^{4+}$ due to the charge compensation

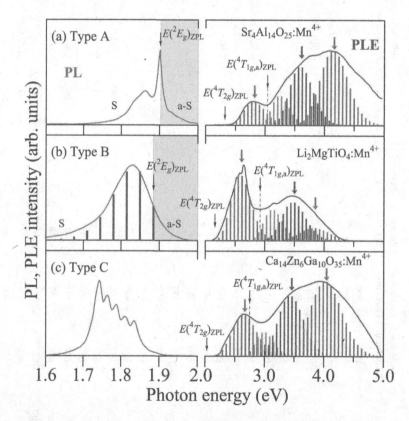

FIGURE 6.5   The PL and PLE spectra for the Mn^{4+}-doped oxide phosphors of (a) Type A, (b) Type B and (c) Type C classified based on their PL spectral features. The spectra are taken for (a) $Sr_4Al_{14}O_{25}$: Mn^{4+} from Peng et al., (b) Li_2MgTiO_4: Mn^{4+} from Chen et al. and (c) $Ca_{14}Zn_6Ga_{10}O_{35}$: Mn^{4+} from Yang et al. (Adachi, S., *J. Lumin.* 202, 263, 2018. With permission.)

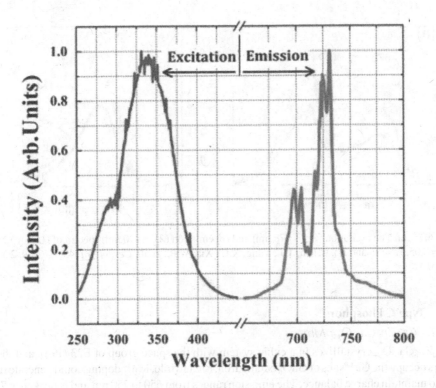

FIGURE 6.6 The excitation (λ_{em} = 731 nm) and emission (λ_{ex} = 335 nm) and spectra of the LaAlO₃:Mn⁴⁺. (Du, J.R., Clercq, O.D. and Korthout, K., *Materials* 10, 1422, 2017. With permission.)

effect brought by the co-doped ions. So far, $LaAlO_3$:Mn⁴⁺ is known as Mn⁴⁺-doped oxide phosphor with the longest emission wavelength.

6.6.2.2 Type B Phosphors

6.6.2.2.1 Li₂MgTiO₄: Mn⁴⁺

Li_2MgTiO_4 is crystallized in a cubic *Fm3m* space group and the lattice contains three kinds of octahedral sites (TiO_6, MgO_6 and LiO_6).[71] The Mn⁴⁺ activators are inferred to substitute the Ti⁴⁺ sites preferentially due to the similar ionic radii and same valence between Ti⁴⁺ and Mn⁴⁺ ions. The PLE spectrum shows two broad bands ranged from 300 to 600 nm when the emission wavelength is monitored at 676 nm. Under the 460 nm blue-light excitation, Li_2MgTiO_4: Mn⁴⁺ shows a broad asymmetric emission band ranged from 625 to 800 nm with the peak at 676 nm. The fine-structure of this rarely Type B emission could not be clearly distinguished. The researchers infer that the relaxation of one set of Ti–O (Mg–O or Li–O) tends to interfere with that of the other set of Ti–O (Mg–O or Li–O). The increased overlap of the corresponding ions leads to the enhancement of chemical bond strength and nephelauxetic effect, which may cause the broadening of the emission of Mn⁴⁺. In addition, the partial substitution of Li⁺ with K⁺ or Na⁺ ions could increase the emission efficiency of the phosphor.

6.6.2.2.2 BaLaMgSbO₆: Mn⁴⁺

Under the 340 nm UV-light excitation, the phosphor exhibits deep-red emission peaked at 700 nm. When the emission wavelength is monitored at 700 nm, the excitation spectrum is in the range of 250–550 nm. Based on the results of Gaussian fitting, the main excitation peak is ascribed to the Mn–O charge transfer band.[72]

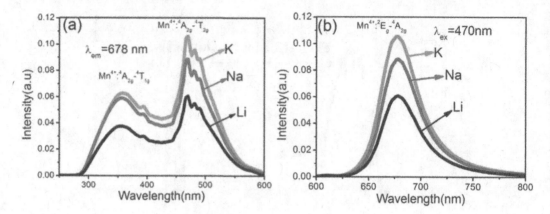

FIGURE 6.7 (a) The excitation (λ_{em} = 678 nm) and (b) emission (λ_{ex} = 470 nm) spectra of $(Li_{1-x}A_x)_2MgTiO_4$: Mn^{4+} (x = 0.05, A = Na and K). (Chen, T.J., Yang, X.L., Xia, W.B., et al., *Ceram. Int.* 43(9), 6949, 2017. With permission.)

6.6.2.3 Type C Phosphors

6.6.2.3.1 $Ca_{14}Zn_6Ga_{10}O_{35}$: Mn^{4+}

The $Ca_{14}Zn_6Ga_{10}O_{35}$ crystallizes in a cubic system with the space group of F23 (196), and the Mn^{4+} activators occupy the Ga^{3+} sites of GaO_6 octahedron. Meanwhile, Mn^{4+} doping could generate interstitial O^{2-} to maintain charge balance. The emission ranges from 650 to 750 nm and is peaked at 711 nm, and the excitation bands are in the range of 250–550 nm (Figure 6.9).[73] Moreover, with partial substitution of Sr^{2+} for Ca^{2+}, the emission intensity exhibits significant enhancement. Based on the XRD and SEM analysis, this emission enhancement was attributed to the crystallinity improvement by the cation substitution.

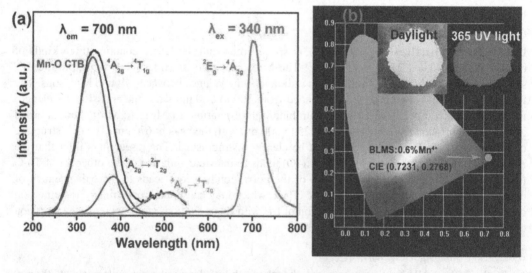

FIGURE 6.8 (a) The excitation and emission spectra of the BaLaMgSbO$_6$: Mn^{4+} monitored at 700 and 340 nm, respectively. The excitation spectrum is fitted by Gaussian function and indicated as the insets. (b) The color coordinate of BaLaMgSbO$_6$: Mn^{4+} on the CIE chromaticity coordinates. The insets are the photographs of the phosphor under daylight and 365 nm UV light illumination, respectively. (Sun, Q., Wang, S.Y., Devakumar, B. et al., *RSC. Adv.* 9, 3303, 2019. With permission.)

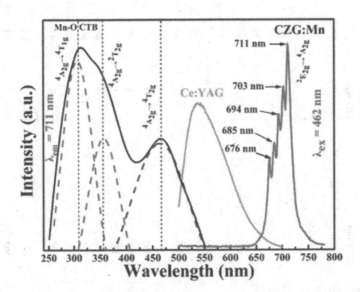

FIGURE 6.9 The excitation (λ_{em} = 711 nm) and emission (λ_{ex} = 462 nm) spectra of the $Ca_{14}Zn_6Ga_{10}O_{35}$: Mn^{4+}, and the excitation are fitted by Gaussian function shown as the dash lines. The emission spectrum (500–700 nm) of YAG: Ce^{3+} phosphor under excited 460 nm is provided as a reference. (Yang, C., Zhang, Z.F., Hu, G.C. et al., *J. Alloys Compd.* 694, 1201, 2017. With permission.)

An overview of Mn^{4+}-doped oxides phosphors is listed in Table 6.3, including the key parameters for characterizing the spectral properties and device performance of the phosphors.[72, 74–155]

6.7 TEMPERATURE-DEPENDENT PL THERMAL QUENCHING PROPERTY

During device operation, the working temperature of phosphors could reach as high as 150°C (or even higher) due to the heat generated by the diode chip and the phosphor itself. Therefore, it is essential to maintain the emission intensity of the phosphor as high as possible with the increasing of temperature. Generally, the emission intensity of most phosphors will decline at higher temperatures, and this phenomenon is called PL thermal quenching. The retained PL intensity at 150°C (I_{423K}) with respect to that at room temperature (I_{298K}) is usually used to evaluate the PL thermal quenching performance of phosphors. The values of I_{423K}/I_{298K} for the Mn^{4+}-doped fluoride and oxide phosphors are also given in Tables 6.1 and 6.3, respectively.

For rare-earth or transitional metal-doped inorganic phosphors, there are several possible PL quenching mechanisms such as multi-phonon relaxation, thermally activated photoionization and thermally activated crossover via the $F^- \rightarrow Mn^{4+}$ charge-transfer (CT) state. However, for Mn^{4+}-doped phosphors, the thermally activated crossover via the $^4T_2$-excited state is the main reason for the PL thermal quenching. This process is illustrated in Figure 6.10.[156] Under photoexcitation (process 1), the electrons of Mn^{4+} are excited from the ground state (curve g) to the excited state (curve e). Normally, most of the excited electrons will return to the ground states along with the releasing of photons (process 2). However, when the temperature is high enough, the relaxed-excited-state may reach the crossing of the two parabolas (point B) and the excited electrons will return to the ground state in a nonradiative manner (process 5). The occurrence probability of this nonradiative transition is strongly dependent on the temperature and the offset between the two parabolas. The thermal quenching process could be described by the Arrhenius equation as:

$$\ln\left(\frac{I_0}{I_T} - 1\right) = -\frac{E_a}{KT} + c \tag{6.15}$$

TABLE 6.3
Overview of the Mn^{4+}-Doped Oxides Phosphors and Corresponding Parameters

Red Phosphors	Material					Yellow/Green Phosphors	Device				Ref.
	Emission Peak (nm)	Excitation Peak (nm)	PL IQY (%)	PL EQY (%)	I_{423K}/I_{298K} (%)		LED Chip (nm)	CCT (K)	CRI	Luminous Efficiency (lm/W)	
La(MgTi)$_{0.5}$O$_3$:Mn^{4+}	708	365[a]*, 487	27.2		53						74
La$_2$MgTiO$_6$:Mn^{4+}	710	330[a] *, 480	58.7								75
Y$_2$MgTiO$_6$:Mn^{4+}	698	350[a]*, 480			~45						76
Gd$_2$ZnTiO$_6$:Mn^{4+}	704	365[a]*, 490	39.7		27.2						77
Sr$_2$TiO$_4$:Mn^{4+}	725	380[a]*, 535	50.5	43.5	10	Non					78
Mg$_2$TiO$_4$:Mn^{4+}, Zn^{2+}	657	366[a]*, 480	38.2		50	YAG:Ce^{3+}	460	4344	79.8	65.3	79
BaMg$_6$Ti$_6$O$_{19}$:Mn^{4+}	660	338[a]*, 475	12			Non					80
Li$_2$MgTiO$_4$:Mn^{4+}	676	330[a]*, 476	34	15	65	YAG:Ce^{3+}	460	5568	71	79.43 (60 mA)	81
LiGaTiO$_4$:Mn^{4+}	668	330[a]*, 465									82
SnMg$_2$La$_2$W$_2$O$_{12}$:Mn^{4+}	708	344[a]*, 470	88		57.5	Non	365			0.03	83
Ca$_3$La$_2$W$_2$O$_{12}$:Mn^{4+}	711	360[a]*, 475	47.9		29	Non	365				84
NaLaMgWO$_6$:Mn^{4+}	700	342[a]*, 465	60		57	Non	365				85
KLaMgWO$_6$:Mn^{4+}	696	348[a]*, 470	43	29	~55	Non	365				86
LiLaMgWO$_6$:Mn^{4+}	713	344[a]*, 469	69		49	Non	365				87
SrLaScO$_4$:Mn^{4+}	690	350[a]*, 513	12.2		15						88
LaAlO$_3$:Mn^{4+}, Bi^{3+}	734	356[a]*, 490	89.3		103	Non	365			0.15	89
CaAl$_2$O$_4$:Mn^{4+}	658	325[a]*, 470	45								90
CaGdAlO$_4$:Mn^{4+}	715	349[a]*, 480	45			Non	365			0.04	91
CaAl$_4$O$_7$:Mn^{4+}	656	335[a]*, 470	44.4		~20						92
Sr$_2$Al$_6$O$_{11}$:Mn^{4+}	652	305[a]*, 450	83.2		30						93
Lu$_3$Al$_5$O$_{12}$:Mn^{4+}	671	327[a]*, 468	42	40.1	95	Non	365	4200	83	81.7	94
Y$_{0.5}$Lu$_{2.5}$Al$_3$Ga$_2$O$_{12}$:Mn^{4+}	673	345[a]*, 478	84.9		~10	YAG:Ce^{3+}	450			92 (350 mA)	95
CaAl$_{12}$O$_{19}$:Mn^{4+}	654	338[a]*, 480	40								96
SrAl$_{12}$O$_{19}$:Mn^{4+}	658	350[a]*, 450									97
LaMgAl$_{11}$O$_{19}$:Mn^{4+}	663	340[a]*, 470	35.8		36						98

(Continued)

TABLE 6.3 (Continued)

Overview of the Mn^{4+}-Doped Oxides Phosphors and Corresponding Parameters

Red Phosphors	Material					Device					Ref.
	Emission Peak (nm)	Excitation Peak (nm)	PL IQY (%)	PL EQY (%)	I_{423K}/I_{298K} (%)	Yellow/Green Phosphors	LED Chip (nm)	CCT (K)	CRI	Luminous Efficiency (lm/W)	
Mg/Sr$_4$Al$_{14}$O$_{25}$:Mn^{4+}	656	350*, 450	60.8		15	Sr$_2$Si$_5$N$_8$:Eu^{2+}	455				99
Sr$_2$MgAl$_{22}$O$_{36}$:Mn^{4+}	659	320*, 480	80			(Sr, Ba)$_2$SiO$_4$:Eu^{2+}	397	5545	62.2		100
SrMgAl$_{10}$O$_{17}$:Mn^{4+}	663	320*, 467	51.4		45	Non	470				101
SrMgAl$_7$Ga$_3$O$_{17}$:Mn^{4+}	662	300*, 460	29.67		20	Non	470				102
BaMgAl$_{10}$O$_{17}$:Mn^{4+}	660	300*, 465			45	YAG:Ce^{3+}	450	3622	86	55.1 (350 mA)	
BaZn$_{1.06}$Al$_{9.94}$O$_{17}$:Mn^{4+}	665	300*, 460	22.7		~30						103
Na$_2$MgAl$_{10}$O$_{17}$:Mn^{4+}	695	300*, 460			~60	Non					104
Na$_{1.57}$Zn$_{0.57}$Al$_{10.43}$O$_{17}$:Mn^{4+}	711	310*, 468		56.2							105
Ca$_{14}$Zn$_6$Al$_{10}$O$_{35}$:Mn^{4+}	714	330*, 460	64.4	56.2	60	YAG:Ce^{3+}	450	4127	83.2	64.83 (20 mA)	106
Ca$_{14}$Zn$_6$Ga$_{10}$O$_{35}$:Mn^{4+}	714	310*, 462	83	53	98	Non	450				107
Ca$_{14-x}$Sr$_x$Zn$_6$Al$_{10}$O$_{35}$:Mn^{4+}	656	310*, 450			5						108
Ca$_2$Mg$_2$Al$_{28}$O$_{46}$:Mn^{4+}		330*, 472			4						
CaMg$_2$Al$_{16}$O$_{27}$:Mn^{4+}	655	338*, 468	35.6								
Ca$_2$YSnMgAlSi$_2$O$_{12}$:Mn^{4+}	683	352*, 521			2						109
Ca$_2$YSnGaGa$_2$SiO$_{12}$:Mn^{4+}	672	351*, 521			1						
Ca$_2$YSn$_2$Ga$_3$O$_{12}$:Mn^{4+}	688	348*, 515			2						
Ca$_3$SnScGaSi$_2$O$_{12}$:Mn^{4+}	683	356*, 526			1						
Mg$_2$Al$_4$Si$_5$O$_{18}$:Mn^{4+}	680	323*, 460			15	Non	370	5844	81.9	25 (20 mA)	110
Sr$_3$Al$_{10}$SiO$_{20}$:Mn^{4+}	663	390*, 490			85						111
SrLaGaO$_4$:Mn^{4+}	716	350*, 520			3.6						112
LaGaO$_3$:Mn^{4+}	698	365*, 515									113
Mg$_6$ZnGeGa$_2$O$_{12}$:Mn^{4+}	660	330, 420*			75						114
Mg$_{14}$Ge$_5$O$_{24}$:Mn^{4+}	659	325*, 440	94		82	YAG:Ce^{3+}	450	3566	87.3	109.42 (60 mA)	115-116
Y$_2$Mg$_3$Ge$_3$O$_{12}$:Mn^{4+}	660	288*, 420	64	51	86						117
Mg$_{3.5}$Ge$_{1.25}$O$_6$:Mn^{4+}	659	300*, 416	49.8	27.5	98	Non	420				118

(Continued)

TABLE 6.3 (Continued)

Overview of the Mn⁴⁺-Doped Oxides Phosphors and Corresponding Parameters

	Material					Device					
Red Phosphors	Emission Peak (nm)	Excitation Peak (nm)	PL IQY (%)	PL EQY (%)	I_{423K}/I_{298K} (%)	Yellow/Green Phosphors	LED Chip (nm)	CCT (K)	CRI	Luminous Efficiency (lm/W)	Ref.
$Mg_3Ga_2GeO_8$:Mn^{4+}	659	330* 419	64.7		72	BAM:Eu^{2+}, Sr_2SiO_4:Eu^{2+}	405	3400	80.3	72.2 (40 mA)	119
$La_3GaGe_5O_{16}$:Mn^{4+}	659	330* 455	68		50						120
$Mg_7Ga_3GeO_{12}$:Mn^{4+}	660	330 420*	28		79	YAG:Ce^{3+}	440	4807			121
$Li_2Ge_4O_9$:Mn^{4+}	671	310* 460	80.57		78	YAG:Ce^{3+}	450	6107	82		122
$BaGe_4O_9$:Mn^{4+}	666	338* 478	28		70						123
$LiGaGe_2O_6$:Mn^{4+}	669	289* 450	15		8						124
$LiAlGe_2O_6$:Mn^{4+}	671	289* 450	32		10						124
$Ba_2TiGe_2O_8$:Mn^{4+}	666	288* 440	35.6		82						125
$Ba_2AlGe_2O_8$:Mn^{4+}	668	288* 440			60						126
$K_2BaGe_8O_{18}$:Mn^{4+}	666	313* 462	32.9								127
Ba_2LaSbO_6:Mn^{4+}	678	360* 465	20.2		20	Non	365	3671	72.1	21 (20 mA)	128
Ba_2GdSbO_6:Mn^{4+}	687	350* 520	27.7		85	$BaMgAl_{10}O_{17}$:Eu^{2+}, $Ba_3La_6(SiO_4)_6$:Eu^{2+}	365	3486	88.3	2.93 (60 mA)	129
$BaSrYSbO_6$:Mn^{4+}	685	355* 528	32.8		85	$BaMgAl_{10}O_{17}$:Eu^{2+}, $Ba_3La_6(SiO_4)_6$:Eu^{2+}	365	3404	86.8	18.6	130
Ca_2LaSbO_6:Mn^{4+}	685	354* 507	52.2		17.1	Non	365	1001	26.6	21.7	131
Ca_2YSbO_6:Mn^{4+}	680	340* 513	62.6		40	$BaMgAl_{10}O_{17}$:Eu^{2+}, $Ba_3La_6(SiO_4)_6$:Eu^{2+}	365	3255	87.5	10.74	132
Ca_2GdSbO_6:Mn^{4+}	676	356* 510	38.9	29.6	22						133
$CaLaMgSbO_6$:Mn^{4+}	708	370 469	88		54	Non	365				134
$BaMgLaSbO_6$:Mn^{4+}	700	340* 500	83		37	Non	365				72
$CaYMgSbO_6$:Mn^{4+}	688	294* 450	51.5		53.2	Non	365	1366	7.6	48 (25 mA)	135
Sr_3LiSbO_6:Mn^{4+}	698	340* 480	52.3		66.2	Non	370				136
Sr_3NaSbO_6:Mn^{4+}	692	320* 470	46		78						137
$Li_3Mg_2SbO_6$:Mn^{4+}	666	325* 465	83	26	55	YAG:Ce^{3+}	454	3254	81	87 (20 mA)	138,139

(Continued)

TABLE 6.3 (Continued)
Overview of the Mn⁴⁺-Doped Oxides Phosphors and Corresponding Parameters

Red Phosphors	Material					Device					Ref.
	Emission Peak (nm)	Excitation Peak (nm)	PL IQY (%)	PL EQY (%)	$I_{423 K}/I_{298 K}$ (%)	LED Chip (nm)	Yellow/Green Phosphors	CCT (K)	CRI	Luminous Efficiency (lm/W)	
Sr_2YNbO_6:Mn⁴⁺	675	316a**			~23						140
Sr_2LaNbO_6:Mn⁴⁺	694	334a**			~10						141
Ca_2AlNbO_6:Mn⁴⁺	712	355a*				365	YAG:Ce³⁺	3727	74.2		142
$BaLaMgNbO_6$:Mn⁴⁺	700	360**	52		13	365	Non			0.02 (60 mA)	143
$SrLaMgNbO_6$:Mn⁴⁺	660	321*			~25	450	YAG:Ce³⁺	4538	83.1	82.8 (30 mA)	144
$LiLa_2NbO_6$:Mn⁴⁺	712	320*			~40						145
$Li_3Mg_2NbO_6$:Mn⁴⁺	670	310*	28.6								146
Ba_2GdNbO_6:Mn⁴⁺	676	350*			~30						147
Ba_2YNbO_6:Mn⁴⁺	695	350*	29.2		~35	380	Non	4131	80.7	35 (20 mA)	148
$Li_2ZnSn_2O_6$:Mn⁴⁺	658	335a*	33.2								149
$Li_2Mg_3SnO_6$:Mn⁴⁺	670	330*	36.3								150
$Li_3Mg_2TaO_6$:Mn⁴⁺	661	349*	23								151
$NaMgLaTeO_6$:Mn⁴⁺	703	350*	57.4		~78						152
$KMgLaTeO_6$:Mn⁴⁺	696	365a*	68.9		90						153
Li_2MgZrO_4:Mn⁴⁺	670	335a*	32.3								154
$Ca_3Al_4ZnO_{10}$:Mn⁴⁺	713	340a*	20.35								155

a Represents the stronger excitation wavelength.

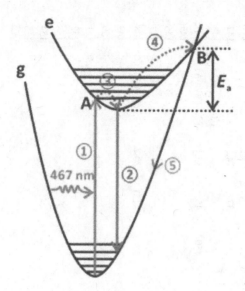

FIGURE 6.10 The configuration coordinate diagram indicated the thermal quenching process of the K_2GeF_6: Mn^{4+}. (Lian, H.Z., Huang, Q.M., Chen, Y.Q. et al., *Inorg. Chem.* 56, 11900, 2017. With permission.)

where I_0 represents the PL intensity when temperature is 0K, I_T is the PL intensity at temperature T, E_a is the activation energy of the thermal quenching or nonradiative transition barrier, c is a constant for a particular host and k is the Boltzmann constant (8.629×10^{-5} eV K^{-1}). The value of the E_a could be used to quantify the probability of thermal quenching occurrence. The larger the activation energy value, the better the PL thermal quenching performance of the phosphor.

For rare earth–doped inorganic phosphors, the PL intensity usually decreases with the rise of temperature due to the larger multiple-phonons relaxation probability at higher temperatures. Interestingly, many Mn^{4+}-doped fluoride phosphors exhibit abnormal temperature-dependent emission behaviors, i.e., the PL intensity increases with the rising of temperature in a certain temperature range, including K_2SiF_6: Mn^{4+}, K_2TiF_6: Mn^{4+}, Na_2GeF_6: Mn^{4+}, K_2GeF_6: Mn^{4+}, Rb_2GeF_6: Mn^{4+} and Cs_2GeF_6: Mn^{4+}. Two facts are mainly responsible for the observed increase of PL intensity. First, the transition probability of the vibronic absorption of $^4A_2 \rightarrow {}^4T_2$ is temperature-dependent and proportional to $\coth(\hbar\omega/2\kappa T)$, where $\hbar\omega$ is the energy of the coupled vibronic mode and κ is the Boltzmann constant. This means the absorption of the excitation light will increase with the increasing of temperature. Second, the radiative transition probability of $^2E \rightarrow {}^4A_2$ will also increase at higher temperature. These two combined effects will induce the increase of PL intensity at a certain temperature range in which the nonradiative transition probability is nonsignificant. Nevertheless, when the temperature is further increased, the nonradiative transition probability will increase drastically and resulting in the decrease of PL intensity. The abovementioned fluoride phosphors usually have very high PL QYs at room temperature, namely, the nonradiative transition probability is negligible. Thus, in a certain temperature range, the PL intensity is observed to increase with the increasing of temperature. For instance, the PL intensity at 160°C could reach 120% of its initial intensity at 20°C (Figure 6.11).[5]

As mentioned earlier, the offset between the two parabolas has a strong influence on the PL thermal stability of Mn^{4+}-doped phosphors. The larger the offset, the easier the nonradiative transition occurs. Furthermore, the offset is closely related to the rigidity of the host crystals. It has been found that the softer surroundings around the luminescent center cause larger offset, thus lower quenching temperature. Herein, for Mn^{4+}-doped oxides with a wide variety of complex structures, enhancing the rigidity of the crystal structure is an effective strategy to improve PL thermal stability. For example, when La^{3+} was partially replaced by Ca^{2+} in $LaAlO_3$ crystal, the lattice structure shrank,

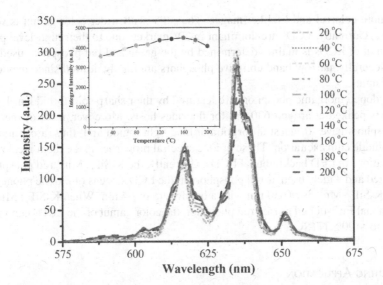

FIGURE 6.11 The temperature-dependent emission spectra of the $BaSiF_6$: Mn^{4+} in the range of 20–200°C, the integrate emission intensity at 160°C could reach 120% of its initial intensity at 20°C. (Song, E.H., Wang, J.Q., Shi, J.H. et al., *ACS Appl. Mater. Interfaces* 9(10), 8805, 2017. With permission.)

and the bonding networks were strengthened, thus leading to an enhancement of the lattice rigidity. The lattice rigidity could be evaluated by the calculation of Debye temperature, and the equation is demonstrated as follow:

$$\Theta_{D,i} = \sqrt{\frac{3h^2 T N_A}{A_i k_B U_{iso,i}}} \tag{6.16}$$

where A_i is the atomic weight of the atom, h is Planck constant, k_B is Boltzmann constant, $U_{iso,i}$ is the atomic average displacement parameter, and it can be obtained by Rietveld refinement. According to the equation, the value of $\Theta_{D,i}$ is inversely proportional to the value of U_{iso}. The values of U_{iso} for the Ca^{2+}-free and 3% Ca^{2+}-doped samples are 0.046 and 0.040, respectively, proving the increase of structure rigidity with the Ca^{2+} substitution. At 150°C, the PL intensity of the Ca^{2+}-co-doped sample increases 50% as compared with the $LaAlO_3$:Mn^{4+} sample.

Another way to improve the PL thermal stability is to form a trap state in the host, which can reserve the excited energy at room temperature and release it slowly with increasing temperature. For example, the substitution of Ge^{4+} by Ti^{4+} ions in $Mg_{14}Ge_5O_{24}$: Mn^{4+} induces lattice distortion. The defects generated by the distortion act as the energy traps to capture a part of the activated electrons. With the increasing of temperature, the energy traps slowly release the reserved electrons, which can compensate the lost photons due to stronger nonradiative transition at higher temperatures.

6.8 THE APPLICATION OF MN⁴⁺-DOPED PHOSPHORS

6.8.1 DISPLAYS APPLICATION

Color gamut is defined as the range of colors that a particular device can produce or record. In display applications, the color gamut is an important parameter to characterize the performance of display devices, namely, the wider a TV's coverage of a color gamut, the more colors a TV will be able to reproduce. This is one of the new developments in TV technology and should help a TV to generate more colorful and lifelike images. Thus, TV with wider color gamut is welcomed by the consumers. Currently, LCDs are dominant in TV markets due to their high-level performance. The color gamut of LCDs is mainly determined by the green and red phosphors used in the LED backlights. Generally, narrow-band emissive phosphors are highly desired since they can generate wider color gamut.

The Mn^{4+}-doped inorganic phosphors are featured by their sharp spin-forbidden $^2E \rightarrow {}^4A_2$ transition, which is peaked at around 630 nm for fluorides hosts. Moreover, most of the Mn^{4+}-doped fluorides phosphors have strongest absorption at blue light region and, thus, are suitable for commercial blue-diode chip excitation. These unique spectral properties make them ideal red phosphor for the applications in LED backlights of LCDs. Currently, the K_2SiF_6: Mn^{4+} red phosphor has been commercialized and widely used as red phosphors in the LCD screens of mobile phones. The emission peak of K_2SiF_6: Mn^{4+} is at 630 nm with a bandwidth of ~4 nm. When K_2SiF_6: Mn^{4+} is used to replace traditional Eu^{2+}-doped nitride red phosphor, the color gamut of the LCDs can increase from ~80% NTSC to ~100% NTSC.

6.8.2   LIGHTING APPLICATION

In terms of lighting applications, the devices must be able to completely cover the visible spectrum to accurately represent the true color of the objects. And the design criterion for the lighting source is to make it as close to the sunlight as possible. However, the spectral components of the lighting source need to be designed and adjusted for specific applications. For example, as for the energy-saving lighting, all invisible radiation outside of the eye sensitivity range can be omitted. For indoor daily lighting, the biological impact of light on the human body should also be considered, and the relative proportion of the spectral components should be scientifically designed. Based on the current market restrictions and requirements on the cost and manufacturing process, the lighting LED devices mostly use phosphor conversion methods to obtain cost-effective lighting sources. The Mn^{4+}-doped red phosphors have important application value in reducing the correlated color temperature of white-LED and compensating the red-light component of specific lighting devices. It is convenient to adjust the emission of LEDs devices combined blue LED chips, yellow phosphors (YAG: Ce^{3+}) and high-efficiency red phosphors.

6.8.3   WLED FOR INDOOR LIGHTING

Currently, solid-state white LEDs have become the dominant lighting devices in people's daily life due to their superior properties such as energy-saving, robust, long-lifetime and environment-friendly features. The YAG: Ce^{3+} yellow phosphor is widely used in commercial white LEDs based on blue-diode chips. However, because of the lack of red-light component in the emission of YAG: Ce^{3+}, it is difficult to fabricate warm white LEDs with high color rendering index (CRI, $R_a > 80$) and low correlated color temperature (CCT < 4500 K), both of which are important for indoor lighting. The discovery of Eu^{2+} or Ce^{3+}-doped (oxy)nitride red phosphors has resolved this problem. However, these red phosphors usually have a very broad emission band and a large part of the spectrum is beyond 650 nm and insensitive to human eyes, thereby decreasing the luminous efficiency of radiation (LER).

The use of Mn^{4+}-doped fluoride phosphor instead of (oxy)nitride phosphors in warm white LEDs can obtain good color rendering index ($R_a > 80$) and enhanced luminous efficacy. For instance, the comparison of the emission spectra of one blue-diode chip and two white LEDs using different phosphor blends is shown in Figure 6.12, and the corresponding optoelectric parameters of the white LEDs are listed in Table 6.4. It is obvious that the addition of the red Mn^{4+}-doped fluoride phosphor can efficiently decreases the CCT and increase the CRI.[46]

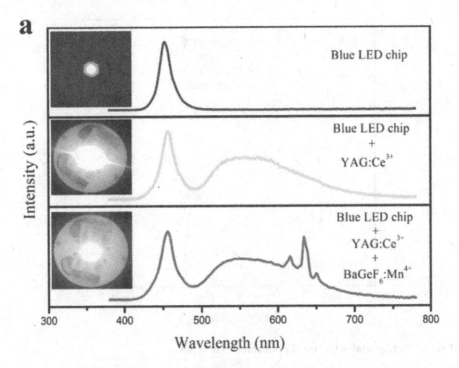

FIGURE 6.12 The electroluminescence spectra of the fabricated LEDs, the corresponding composition are solo blue LED chip; blue LED chip + YAG: Ce^{3+}; blue LED chip + YAG: Ce^{3+} + BaSiF$_6$: Mn^{4+}, respectively. (Zhou, Q., Zhou, Y.Y., Lu, F.Q. et al., *Mater. Chem. Phys.* 170, 32 2015. With permission.)

6.8.4 Plants Cultivation Lighting

To get rid of the natural environment's restrictions on plants cultivation, researchers have developed indoor cultivation technique by simulating and regulating a suitable living environment. Among them, light plays a vital role. It is well known that the photosynthesis process determines the growth of plants, and the phytochromes determine the energy of the photons which can be absorbed by the plants. As shown in Figure 6.13, the absorption spectra of phytochromes in most plants mainly distribute in blue (~450 nm) and deep-red (~660 nm) region. The luminescence properties of Mn^{4+}-doped oxide red phosphors can meet this special requirement. Because the local coordination environment of the oxide matrix is different from fluoride, the Mn^{4+}-doped oxide phosphors have stronger CF, and they usually exhibit deep red-light emission with longer wavelength (650–700 nm). In addition, Mn^{4+}-doped red phosphors can be excited by blue light, which itself is also a light component needed by plants growth. These properties make Mn^{4+}-doped oxide phosphors potential deep-red phosphors for plant-lighting LEDs. Of course, before commercialization, strong absorption of blue light, high PL QYs, good PL thermal quenching performance and excellent chemical stability must be fulfilled first.

TABLE 6.4
Performance of the WLEDs with Different Composition

LED Composition	Color Coordinates	CRI	CCT
Chip + YAG:Ce^{3+}	(0.3426, 0.3647)	74.5	5130
Chip + YAG:Ce^{3+} + BaGeF$_6$:Mn^{4+}	(0.3519, 0.3562)	86.3	4766

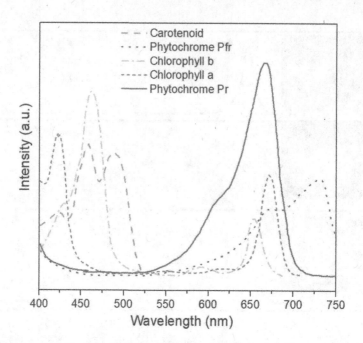

FIGURE 6.13   The absorption spectra of phytochromes in plants.

6.9   CONCLUSION

The rapid advancement of solid-state lighting techniques necessitates the development of more varied phosphors to meet the numerous requirements of various applications such as indoor lighting, automobile lighting, backlighting for LCDs, signs, aircraft lighting, airport lighting, stage lighting, surgical lighting and video flashlights. While rare-earth I–activated inorganic materials play an important role in this area, other non-rare-earth luminescent materials, especially transition metal Mn^{4+}–activated phosphors, have piqued the interest of researchers due to their unique and superior properties such as structural diversity, low cost, ease of spectral modulation or high radiation luminous efficacy. This chapter focuses on the fundamental properties of Mn^{4+} activators, including the Mn^{4+} electronic structure, chemical synthesis methods, optical properties as well as some potential application prospects and efforts in this increasingly growing field.

REFERENCES

1. Tanabe Y., Sugano S. 1954. On the absorption spectra of complex ions .2. *J. Phys. Soc. Jpn.* 9:766–79.
2. Tanabe Y., Sugano S. 1954. On the absorption spectra of complex ions .1. *J. Phys. Soc. Jpn.* 9:753–66.
3. Huang L., Zhu, Y.W., Zhang X.J. et al. 2016. HF-Free Hydrothermal Route for Synthesis of Highly Efficient Narrow-Band Red Emitting Phosphor $K_2Si_{1-x}F_6:xMn^{4+}$ for Warm White Light-Emitting Diodes. *Chem. Mater.* 28: 1495.
4. Zhu H., Lin C.C., Luo W. et al. 2014. Highly efficient non-rare-earth red emitting phosphor for warm white light-emitting diodes. *Nat. Commun.* 5:4312.
5. Song E.H., Wang J., Shi J. et al. 2017. Highly efficient and thermally stable $K_3AlF_6:Mn^{4+}$ as a red phosphor for ultra-high-performance warm white light-emitting diodes. *ACS Appl. Mater. Interfaces* 9:8805–8812.
6. Hu T., Lin H., Lin F. et al. 2018. Narrow- band red- emitting $KZnF_3$: Mn^{4+} fluoroperovskites: insights into electronic/vibronic transition and thermal quenching behavior. *J. Mater. Chem. C* 6:10845–54.
7. Deng T.T., Song E.H., Su J et al. 2018. Stable narrowband red emission in fluorotellurate $KTeF_5:Mn^{4+}$ via Mn^{4+} noncentral-site occupation. *J. Mater. Chem. C* 6:4418–26.

8. Stoll C, Bandemehr J, Kraus F et al. 2019. HF-free synthesis of Li_2SiF_6:Mn^{4+}: a red-emitting phosphor. *Inorg. Chem.* 58:5518–23.

9. Zhang L, Xi L, Pan Y et al. 2018. Synthesis and improved photoluminescence of hexagonal crystals of Li_2ZrF_6:Mn^{4+} for warm WLED application. *Dalton Trans.* 47:16516–23.

10. Zhu M.M., Pan Y.X., Wu M.M. et al. 2018. Synthesis and improved photoluminescence of a novel red phosphor $LiSrGaF_6$: Mn^{4+} for applications in warm WLEDs. *Dalton Trans.* 47:12944–50.

11. Lang T.C., Han T, Peng L.L. et al. 2017. Luminescence properties of Na_2SiF_6:Mn^{4+} red phosphors for high colour-rendering white LED applications synthesized via a simple exothermic reduction reaction. *Mater. Chem. Front.* 1:928–932.

12. Nguyen H.D., Lin C.C., Fang M.H. et al. 2014. Synthesis of Na_2SiF_6:Mn^{4+} red phosphors for white LED applications by co-precipitation. *J. Mater. Chem. C* 2:10268–72.

13. Wang Z, Liu Y, Zhou Y et al.2015. Red-emitting phosphors Na_2XF_6:Mn^{4+} (X = Si, Ge, Ti) with high colour-purity for warm white-light-emitting diodes. *RSC Advances*; 5:58136–58140.

14. Xi L., Pan Y., Chen X. et al. 2017. Optimized photoluminescence of red phosphor Na_2SnF_6:Mn^{4+} as red phosphor in the application in "warm" white LEDs. *J. Am. Ceram. Soc.* 100:2005–2015.

15. Song E.H. Wang J.Q., Ye S. et al. 2016. Room-temperature synthesis and warm-white LED applications of Mn^{4+} ion doped fluoroaluminate red phosphor Na_3AlF_6:Mn^{4+}. *J. Mater. Chem. C* 4:2480–2487.

16. Deng T.T. Song E.H., Sun J. et al. 2017. The design and preparation of the thermally stable, Mn^{4+} ion activated, narrow band, red emitting fluoride Na_3GaF_6:Mn^{4+} for warm WLED applications. *J. Mater. Chem. C* 5:2910–2919.

17. Zhu M, Pan Y, Chen Xa et al. 2018. Formation mechanism and optimized luminescence of Mn^{4+}-doped unequal dual-alkaline hexafluorosilicate $Li_{0.5}Na_{1.5}SiF_6$. *J. Am. Ceram. Soc.* 101:4983–93.

18. Fang M.H. Wu W.L., Jin Y. et al. 2018. Control of luminescence via tuning of crystal symmetry and local structure in Mn^{4+}-activated narrow band fluoride phosphors. *Angew. Chem. Int. Ed.* 57:1797–801.

19. Nguyen H.D., Lin C.C., Liu R.S. 2015. Waterproof alkyl phosphate coated fluoride phosphors for opto-electronic materials. *Angew. Chem. Int. Ed.* 54:10862–66.

20. Zhou W., Fang M.H., Lian S, Liu R.S. 2018. Ultrafast self-crystallization of high-external-quantum-efficient fluoride phosphors for warm white light-emitting diodes. *ACS Appl. Mater. Interfaces* 10:17508–11.

21. Huang D., Zhu H., Deng Z. et al. 2019. Moisture-resistant Mn^{4+}-doped core-shell-structured fluoride red phosphor exhibiting high luminous efficacy for warm white light-emitting diodes. *Angew. Chem. Int. Ed.* 58:3843–3847.

22. Arai Y., Adachi S. 2011. Photoluminescent properties of K_2SnF_6-H_2O:Mn^{4+} hydrate phosphor. *J. Electrochem Soc.* 158:J81–J85.

23. Kasa R., Adachi S. 2012. Mn-activated K_2ZrF_6 and Na_2ZrF_6 phosphors: sharp red and oscillatory blue-green emissions. *J. Appl. Phys.* 112:013506.

24. Deng T.T., Song E.H., Zhou Y.Y. et al. 2017. Stable narrowband red phosphor K_3GaF_6:Mn^{4+} derived from hydrous $K_2GaF_5(H_2O)$ and K_2MnF_6. *J. Mater. Chem. C* 5:9588–9596.

25. Ming H., Liu S., Liu L. et al. 2018. Highly regular, uniform K_3ScF_6:Mn^{4+} phosphors: facile synthesis, microstructures, photoluminescence properties, and application in light-emitting diode devices. *ACS Appl. Mater. Interfaces* 10:19783–95.

26. Jin Y., Liu R., Chen G. et al. 2016. Synthesis and photoluminescence properties of octahedral $K_2(Ge,Si)F_6$:Mn^{4+} red phosphor for white LED. *J. Rare Earths.* 34:1173–1178.

27. Jin Y., Fang M.H., Grinberg M. et al. 2016. Narrow red emission band fluoride phosphor $KNaSiF_6$:Mn^{4+} for warm white light-emitting diodes. *ACS Appl. Mater. Interfaces* 8:11194–203.

28. Yi X.D. Zhu H.M. Gao J., et al. 2017. Room-temperature synthesis and optimized photoluminescence of a novel red phosphor $NaKSnF_6$:Mn^{4+} for application in warm WLEDs. *J. Mater. Chem. C.* 5:9255–9263.

29. Cheng H.M., Song Y., Hong F. et al. 2018. Room-temperature synthesis, controllable morphology and optical characteristics of narrow-band red phosphor K_2LiGaF_6:Mn^{4+}. *Crystengcomm.* 20:2183–2192.

30. Gao J, Zhu H, Li R et al. 2019. Moisture-resistant and highly efficient narrow-band red-emitting fluoride phosphor K_2NaGaF_6:Mn^{4+} for warm white LED application. *J. Mater. Chem. C.* 7:7906–7914.

31. Yi X.D. Zhu H.M. Gao J., et al. 2018. K_2NaAlF_6:Mn red phosphor: room-temperature synthesis and electronic/vibronic structures. *J. Mater. Chem. C* 6:2069–2076.

32. Shi D., Liang Z., Zhang X. et al. 2020. Synthesis, structure and photoluminescence properties of a novel Rb_2NaAlF_6:Mn^{4+} red phosphor for solid-state lighting. *J. Lumin.* 226:117491.

33. Sakurai S., Nakaruma T., Adachi S. 2016. Rb_2SiF_6:Mn^{4+} and Rb_2TiF_6:Mn^{4+} red-emitting phosphors. *ECS J. Solid States Sc.* 5:R206–R210.

34. Fang M.H., Nguyen H.D., Lin C.C. et al. 2015. Preparation of a novel red Rb_2SiF_6:Mn^{4+} phosphor with high thermal stability through a simple one-step approach. *J. Mater. Chem. C* 3:7277–7280.

35. Wu W.L., Fang M.H., Zhou W. et al. 2017. High color rendering index of Rb_2GeF_6:Mn^{4+} for light-emitting diodes. *Chem. Mater.* 29:935–939.

36. Wang L. Y. Zhou Y. Y., Deng T. T. 2018. Efficient and stable narrow band Mn^{4+}-activated fluorotitanate red phosphor Rb_2TiF_6:Mn^{4+} for warm white LED applications. *J. Mater. Chem. C* 6: 8670–8678.

37. Jiang C, Brik M.G., Srivastava A.M. et al. 2019. Significantly conquering moisture-induced luminescence quenching of red line-emitting phosphor Rb_2SnF_6:Mn^{4+} through $H_2C_2O_4$ triggered particle surface reduction for blue converted warm white light-emitting diodes. *J. Mater. Chem. C* 7:247–255.

38. Wang L. Y. Zhou Y. Y., Deng T. T. 2017. Synthesis and warm-white LED applications of an efficient narrow-band red emitting phosphor, Rb_2ZrF_6:Mn^{4+}. *J. Mater. Chem. C* 5: 7253–7261.

39. Yang Z., Wei Q., Rong M. et al. 2017. Novel red-emitting phosphors A_2HfF_6:Mn^{4+} (A = Rb^+, Cs^+) for solid-state lighting. *Dalton Trans.* 46:9451–56.

40. Wang Z., Yang Z., Wang N. et al. 2020. Single-crystal red phosphors: enhanced optical efficiency and improved chemical stability for wLEDs. *Adv. Opt. Mater.* 8:1901512.

41. Zhou Q., Tan H., Zhou Y. et al. 2016. Optical performance of Mn^{4+} in a new hexa-coordinated fluorozirconate complex of Cs_2ZrF_6. *J. Mater. Chem. C* 4:7443–48.

42. Jiang C., Peng M., Srivastava A.M. et al. 2018. Mn^{4+}-doped heterodialkaline fluorogermanate red phosphor with high quantum yield and spectral luminous efficacy for warm-white-light-emitting device application. *Inorg. Chem.* 57:14705–14714.

43. Xi L., Pan Y., Huang S. et al. 2016. Mn^{4+} doped $(NH_4)_2TiF_6$ and $(NH_4)_2SiF_6$ micro-crystal phosphors: synthesis through ion exchange at room temperature and their photoluminescence properties. *RSC Adv.* 6:76251–58.

44. Xu Y.K., Adachi S. 2012. Photoluminescence and Raman scattering spectra in $(NH_4)_2XF_6$-Mn^{4+} (X = Si, Ge, Sn, and Ti) red phosphors. *J. Electro. Chem. Soc.* 159:E11–17.

45. Wang Y., Zhou Y., Song E. 2019. Ammonium salt conversion towards Mn^{4+} doped $(NH_4)_2NaScF_6$ narrow-band red-emitting phosphor. *J. Alloys Compd.* 811:151945.

46. Zhou Q., Zhou Y., Lu F. et al. 2016. Mn^{4+} -activated $BaSiF_6$ red phosphor: Hydrothermal synthesis and dependence of its luminescent properties on reaction conditions. *Mater. Chem. Phys.* 170:32–37.

47. Hong F., Xu H., Pang G. et al. 2020. Moisture resistance, luminescence enhancement, energy transfer and tunable color of novel core-shell structure $BaGeF_6$:Mn^{4+} phosphor. *Chem. Eng. J.* 390:124579.

48. Xi L., Pan Y. 2017. Tailored photoluminescence properties of a red phosphor $BaSnF_6$:Mn^{4+} synthesized from Sn metal at room temperature and its formation mechanism. *Mater. Res. Bull.* 86:57–62.

49. Liu Y., Gao G., Huang L. et al. 2018. Co-precipitation synthesis and photoluminescence properties of $BaTiF_6$:Mn^{4+}: an efficient red phosphor for warm white LEDs. *J. Mater. Chem. C* 6:127–133.

50. Kubus M., Enseling D., Jüstel T. 2013. Synthesis and luminescent properties of red-emitting phosphors: $ZnSiF_6 \cdot 6H_2O$ and $ZnGeF_6 \cdot 6H_2O$ doped with Mn^{4+}. *J. Lumin.* 137:88–92.

51. Hoshino R., Nakamura T., Adachi S. 2016. Structural change induced by thermal annealing of red-light-emitting $ZnSnF_6 \cdot 6H_2O$:Mn^{4+} hexahydrate phosphor. *Jpn. J. Appl. Phys.* 55:052601.

52. Zhong J.S., Chen D.Q., Wang X. et al. 2016. Synthesis and optical performance of a new red-emitting $ZnTiF_6 \cdot 6H_2O$:Mn^{4+} phosphor for warm white-light-emitting diodes. *J. Alloys Compd.* 662:232–239.

53. Lin H., Hu T., Huang Q. et al. 2017. Non-rare-earth K_2XF_7:Mn^{4+} (X = Ta, Nb): A highly-efficient narrow-band red phosphor enabling the application in wide-color-gamut LCD. *Laser Photonics Rev.* 11:1700148.

54. Kumada N., Yanagida S., Takei T. 2019. Hydrothermal synthesis and crystal structure of new red phosphors, $KNaMF_7$:Mn^{4+} (M: Nb, Ta). *Mater. Res. Bull.* 115:170–175.

55. Noh M., Yoon D.H., Kim C.H. et al. 2019. Organic solvent-assisted synthesis of the K^3SiF^7:Mn^{4+} red phosphor with improved morphology and stability. *J. Mater. Chem. C* 7:15014–15020.

56. Tan H., Rong M., Zhou Y. et al. 2016. Luminescence behaviour of Mn(4+) ions in seven coordination environments of K_3ZrF_7. *Dalton Trans.* 45:9654–9660.

57. Qian W., Wei Z.Y, Zhao F Y. et al. 2018. Communication-luminescent properties of Mn^{4+}-activated K_3HfF_7 red phosphor. *ECS J. Solid States Sci.* 7:R39–41.

58. Chen D., Liu Y., Chen J. et al. 2019. $Yb^{3+}/Ln^{3+}/Mn^{4+}$ (Ln = Er, Ho, and Tm) doped Na_3ZrF_7 phosphors: oil–water interface cation exchange synthesis, dual-modal luminescence and anti-counterfeiting. *J. Mater. Chem. C* 7:1321–1329.

59. Kim M., Park W.B., Lee J.W. et al. 2018. Rb_3SiF_7:Mn^{4+} and Rb_2CsSiF_7:Mn^{4+} red-emitting phosphors with a faster decay rate. *Chem. Mater.* 30:6936–6944.

60 Gu M., Tian Y., Cui C. et al. 2018. Two-step synthesis of a novel red-emitting Ba_2ZrF_8:Mn^{4+} phosphor for warm white light-emitting diodes. *Mater. Res. Bull.* 107:242–247.

61 Thomas J., Thomas J. 2018. HK_3SnF_8-Mn^{4+} as a color converter for next generation warm white LEDs. *ECS J. Solid States Sci.* 7:R111–R113.

62 Fang M.H., Yang T.H., Lesniewski T. et al. 2019. Hydrogen-containing $Na_3HTi_{1-x}Mn_xF_8$ narrow-band phosphor for light-emitting diodes. *ACS Energ. Lett.* 4:527–533.

63 Wang Z., Wang N., Yang Z. et al. 2017. Luminescent properties of novel red-emitting phosphor Na_3TaF_8 with non-equivalent doping of Mn^{4+} for LED backlighting. *J. Lumin.* 192:690–694.

64 Zhu M.M., Xi L.Q., Lian H.Z. et al. 2018. Designed synthesis, morphology evolution and enhanced photoluminescence of a highly efficient red dodec-fluoride phosphor, $Li_3Na_3Ga_2F_{12}$:Mn^{4+}, for warm WLEDs. *J. Mater. Chem. C* 6:491–500.

65 Zhu M.M., Xi L.Q., Lian H.Z. et al. 2017. Design, preparation, and optimized luminescence of a dodec-fluoride phosphor $Li_3Na_3Al_2F_{12}$:Mn^{4+} for warm WLEDs application. *J. Mater. Chem. C* 10:1039–51.

66 Li H., Liu Y., Tang S. et al. 2020. Luminescence properties of Mn(4+) with high Eg level energy in the polyfluoride $Na_3Li_3Sc_2F_{12}$. *Dalton Trans.* 49:11613–11617.

67 Xi L., Pan Y., Huang S. et al. 2018. A novel red phosphor of seven-coordinated Mn^{4+} ion-doped trideca fluorodizirconate $Na_5Zr_2F_{13}$ for warm WLED. *Dalton Trans.* 47:5614–5621.

68 Qin L., Cai P., Chen C. et al. 2017. Synthesis, structure and optical performance of red-emitting phosphor Ba_5AlF_{13}:Mn^{4+}. *RSC Adv.* 7:49473–49479.

69 Adachi S. et al. 2018. Photoluminescence properties of Mn^{4+}-activated oxide phosphors for use in white-LED applications: a review. *J. Lumin.* 202:263–281.

70 Du J.R., Clercq, O.D., Korthout, K. et al. 2017. $LaAlO_3$: Mn^{4+} as near-infrared emitting persistent luminescence phosphor for medical imaging: a charge compensation study. *Materials* 10:1422–1430.

71 Chen T.J., Yang, X.L., Xia, W.B. et al. 2017. Deep-red emission of Mn^{4+} and Cr^{3+} in $(Li_{1-x}A_x)_2MgTiO_4$ (A = Na and K) phosphor: potential application as W-LED and compact spectrometer. *Ceram. Int.* 43(9):6949–6954.

72 Sun Q., Wang S.Y., Devakumar B.et al. 2017. Mn^{4+}-activated $BaLaMgSbO_6$ double-perovskite phosphor: a novel high-efficiency far-red-emitting luminescent material for indoor plant growth lighting. *RSC. Adv.* 9, 3303–3310.

73 Yang C., Zhang Z., Hu G. et al. 2017. A novel deep red phosphor $Ca_{14}Zn_6Ga_{10}O_{35}$:Mn^{4+} as color converter for warm W-LEDs: structure and luminescence properties. *J. Alloy Compd.* 694:1201–1208.

74 Zhou Z., Zheng J., Shi R. et al. 2017. Ab initio site occupancy and far-red emission of Mn^{4+} in cubic-phase $La(MgTi)_{1/2}O_3$ for plant cultivation. *ACS Appl. Mater. Interfaces* 9:6177–6185.

75 Takeda Y., Kato H, Kobayashi M. et al. 2015. Photoluminescence properties of Mn^{4+}-activated perovskite-type titanates, La_2MTiO_6:Mn^{4+} (M = Mg and Zn). *Chem. Lett.* 44:1541–1543.

76 Cai P., Qin L., Chen C. et al. 2018. Optical thermometry based on vibration sidebands in Y_2MgTiO_6:Mn^{4+} double perovskite. *Inorg. Chem.* 57:3073–3081.

77 Xiang J., Chen J., Zhang N. et al. 2018. Far red and near infrared double-wavelength emitting phosphor Gd_2ZnTiO_6: Mn^{4+}, Yb^{3+} for plant cultivation LEDs. *Dyes Pigments* 154:257–262.

78 Kato H., Takeda Y., Kobayashi M. et al. 2018. Photoluminescence properties of layered perovskite-type strontium scandium oxyfluoride activated with Mn^{4+}. *Front. Chem.* 6:647.

79 Hu G., Hu X., Chen W. et al. 2017. Luminescence properties and thermal stability of red phosphor Mg_2TiO_4:Mn^{4+} additional Zn^{2+} sensitization for warm W-LEDs. *Mater. Res. Bull.* 95:277–284.

80 Sasaki T., Fukushima J., Hayashi Y., et al. 2014. Synthesis and photoluminescence properties of Mn^{4+}-doped $BaMg_6Ti_6O_{19}$ phosphor. *Chem. Lett.* 43:1061–1063.

81 Jin Y., Hu Y., Wu H. et al. 2016. A deep red phosphor Li_2MgTiO_4:Mn^{4+} exhibiting abnormal emission: potential application as color converter for warm w-LEDs. *Chem. En. J.* 288:596–607.

82 Cao R., Huang J., Ceng X. et al. 2016. $LiGaTiO_4$:Mn^{4+} red phosphor: synthesis, luminescence properties and emission enhancement by Mg^{2+} and Al^{3+} ions. *Ceram. Int.* 42:13296–13300.

83 Xu D., Wu X., Zhang Q. et al. 2018. Fluorescence property of novel near-infrared phosphor Ca_2MgWO_6:Cr^{3+}. *J. Alloys Compd.* 731:156–161.

84 Huang X., Guo H. 2018. Finding a novel highly efficient Mn^{4+}-activated $Ca_3La_2W_2O_{12}$ far-red emitting phosphor with excellent responsiveness to phytochrome P-FR: towards indoor plant cultivation application. *Dyes Pigments* 152:36–42.

85 Huang X., Liang J., Li B. et al. 2018. High-efficiency and thermally stable far-red-emitting $NaLaMgWO_6$:Mn^{4+} phosphors for indoor plant growth light-emitting diodes. *Opt. Lett.* 43:3305–3308.

86 Liang J, Balaji D., Sun L et al. 2019. Mn^{4+}-activated $KLaMgWO_6$: a new high-efficiency far-red phosphor for indoor plant growth LEDs. *Ceram. Int.* 45:4564–4569.

87. Liang J., Balaji D., Wang, S.Y. et al. 2018. Novel Mn^{4+}-activated $LiLaMgWO_6$ far-red emitting phosphors high photoluminescence efficiency, good thermal stability, and potential applications in plant cultivation LEDs. *RSC Adv.* 8:27144–27151.

88. Humayoun U.B., Tiruneh S.N., Yoon D.H. 2018. On the crystal structure and luminescence characteristics of a novel deep red emitting $SrLaScO_4$:Mn^{4+}. *Dyes and Pigments* 152:127–130.

89. Fang S., Lang T., Han T. et al. 2020. Zero-thermal-quenching of Mn^{4+} far-red-emitting in $LaAlO_3$ perovskite phosphor via energy compensation of electrons' traps. *Chem. Eng. J.* 389:124297.

90. Cao R., Zhang F., Cao C. et al. 2014. Synthesis and luminescence properties of $CaAl_2O_4$:Mn^{4+} phosphor. *Opt. Mater.* 38:53–56.

91. Sun Q., Wang S., Li B. et al. 2018. Synthesis and photoluminescence properties of deep red-emitting $CaGdAlO_4$: Mn^{4+} phosphors for plant growth LEDs. *J. Lumin.* 203:371–375.

92. Li P., Peng M., Yin X. et al. 2013. Temperature dependent red luminescence from a distorted Mn^{4+} site in $CaAl_4O_7$:Mn^{4+}. *Opt. Express* 21:18943–18948.

93. Sasaki T., Fukushima J., Hayashi Y. et al. 2018. Synthesis and photoluminescence properties of a novel $Sr_2Al_6O_{11}$:Mn^{4+} red phosphor prepared with a B_2O_3 flux. *J. Lumin.* 194:446–451.

94. Fang S., Lang T., Han T. et al. 2020. A novel efficient single-phase dual-emission phosphor with high resemblance to the photosynthetic spectrum of chlorophyll A and B. *J. Mater. Chem. C* 8:6245–6253.

95. Xu W., Chen D., Yuan S. et al. 2017. Tuning excitation and emission of Mn^{4+} emitting center in $Y_3Al_5O_{12}$ by cation substitution. *Chem. Eng. J.* 317:854–861.

96. Zhu Y., Qiu Z., Ai B. et al. 2018. Significant improved quantum yields of $CaAl_{12}O_{19}$:Mn^{4+} red phosphor by co-doping Bi^{3+} and B^{3+} ions and dual applications for plant cultivations. *J. Lumin.* 201:314–320.

97. Long J., Yuan X., Ma C. et al. 2018. Strongly enhanced luminescence of $Sr_4Al_{14}O_{25}$:Mn^{4+} phosphor by co-doping B^{3+} and Na^+ ions with red emission for plant growth LEDs. *RSC Adv.* 8:1469–1476.

98. Li X., Chen Z., Wang B. et al. 2019. Effects of Impurity Doping on the luminescence performance of Mn^{4+}-doped aluminates with the magnetoplumbite-type structure for plant cultivation. *Materials* 12:1–11.

99. Xiong F.B., Lin L.X., Lin H.F. et al. 2019. Synthesis and photoluminescence of Mn^{4+} in $M_4Al_{14}O_{25}$ (M = Sr or Mg) compounds as red-light phosphors for white LED. *Opt. Laser Technol.* 117:299–303.

100. Cao R., Peng M., Song E. et al. 2012. High efficiency Mn^{4+} doped $Sr_2MgAl_{22}O_{36}$ red emitting phosphor for white LED. *ECS J. Solid States Sc.* 1:R123–126.

101. Gu S., Xia M., Zhou C. et al. 2020. Red shift properties, crystal field theory and nephelauxetic effect on Mn^{4+}-doped $SrMgAl_{10-y}Ga_yO_{17}$ red phosphor for plant growth LED light. *Chem. Eng. J.* 396:125208.

102. Wang B., Lin H., Huang F. et al. 2016. Non-rare-earth $BaMgAl_{10-2x}O_{17}$:xMn^{4+}, xMg^{2+}: A narrow-band red phosphor for use as a high-power warm w-LED. *Chem. Mater.* 28:3515–3524.

103. Peng Q., Cao R., Ye Y. et al. 2017. Photoluminescence properties of broadband deep-red-emitting $Na_2MgAl_{10}O_{17}$:Mn^{4+} phosphor. *J. Alloys Compd.* 725:139–144.

104. Cao R., Ye Y., Peng Q. et al. 2018. Deep-red-emitting $Na_{1.57}Zn_{0.57}Al_{10.43}O_{17}$: Mn^{4+} phosphor: Synthesis and photoluminescence properties. *J. Electro. Mater.* 47:7537–7543.

105. Zhou Z., Wang B., Zhong Y. et al. 2018. PH dependent hydrothermal synthesis of $Ca_{14}Al_{10}Zn_6O_{35}$:0.1 $5Mn^{4+}$ phosphor with enhanced photoluminescence performance and high thermal resistance for indoor plant growth lighting. *Ceram. Int.* 44:19779–19786.

106. Zhou Y., Zhao W., Chen J. et al. 2017. Highly efficient red emission and multiple energy transfer properties of Dy^{3+}/Mn^{4+} co-doped $Ca_{14}Zn_6Ga_{10}O_{35}$ phosphors. *Rsc Adv.* 7:17244–17253.

107. Wu Y., Zhuang Y., Lv Y. et al. 2019. A high-performance non-rare-earth deep-red-emitting Ca14-xSrxZn6Al10O35:Mn4+ phosphor for high-power plant growth LEDs. *J. Alloy Compd.* 781:702–709.

108. Sasaki T., Fukushima J., Hayashi Y. et al. 2017. Synthesis and photoluminescence properties of Mn^{4+}-doped magnetoplumbite-related aluminate X-type $Ca_2Mg_2Al_{28}O_{46}$ and W-type $CaMg_2Al_{16}O_{27}$ red phosphors. *Ceram. Int.* 43:7147–7152.

109. Jansen T., Juestel T., Kirm M. et al. 2018. Thermal quenching of Mn^{4+} luminescence in Sn^{4+}-containing garnet hosts. *Opt. Mater.* 84:600–605.

110. Fu A., Zhou L., Wang S. et al. 2018. Preparation, structural and optical characteristics of a deep red-emitting $Mg_2Al_4Si_5O_{18}$: Mn^{4+} phosphor for warm w-LEDs. *Dyes Pigments* 148:9–15.

111. Sasaki T., Fukushima J., Hayashi Y. et al. 2017. Synthesis and photoluminescence properties of a novel aluminosilicate $Sr_3Al_{10}SiO_{20}$:Mn^{4+} red phosphor. *J. Lumin.* 188:101–106.

112. Jiang C., Zhang X., Wang J. et al. 2019. Synthesis and photoluminescence properties of a novel red phosphor $SrLaGaO_4$:Mn^{4+}. *J. Am. Ceram. Soc.* 102:1269–1276.

113. Srivastava A.M., Camardello S.J., Brik M.G. 2017. Luminescence of Mn^{4+} in the orthorhombic perovskite, $LaGaO_3$. *J. Lumin.* 183:437–441.

114. Ding X., Wang Y. 2017. Structure and photoluminescence properties of rare-earth free narrow-band red-emitting $Mg_6ZnGeGa_2O_{12}$: Mn^{4+} phosphor excited by NUV light. *Opt. Mater.* 64:445–452.

115. Park H.W., Jo H., Anoop G. et al. 2020. Transition metal ion co-doped $MgO–MgF_2$-GeO_2:Mn^{4+} red phosphors for white LEDs with wider color reproduction gamut. *J. Alloys Compd.* 818:152914.

116. Liang S., Li G., Dang P. et al. 2019. Cation substitution induced adjustment on lattice structure and photoluminescence properties of $Mg_{14}Ge_5O_{24}$:Mn^{4+}: optimized emission for w-LED and thermometry applications. *Adv. Opt. Mater.* 7:1900093.

117. Jansen T., Gorobez J., Kirm M. et al. 2018. Narrow band deep red photoluminescence of $Y_2Mg_3Ge_3O_{12}$:Mn^{4+}, Li^+ inverse garnet for high power phosphor converted LEDs. *ECS J. Solid. State. Sci.* 7:R3086–R3092.

118. Deng J., Zhang H., Zhang X. et al. 2018. Ultrastable red-emitting phosphor-in-glass for superior high-power artificial plant growth LEDs. *J. Mater. Chem. C* 6:1738–1745.

119. Ding X., Zhu G., Geng W. et al. 2016. Rare-earth-free high-efficiency narrow-band red-emitting $Mg_3Ga_2GeO_8$:Mn^{4+} phosphor excited by near-UV light for white-light-emitting diodes. *Inorg. Chem.* 55:154–162.

120. Zhang S., Hu Y., Duan H. et al. 2015. Novel $La_3GaGe_5O_{16}$: Mn^{4+} based deep red phosphor: a potential color converter for warm white light. *Rsc Adv.* 5:90499–90507.

121. Wu C., Li J., Xu H. et al. 2015. Preparation, structural and photoluminescence characteristics of novel red emitting $Mg_7Ga_2GeO_{12}$:Mn^{4+} phosphor. *J. Alloy. Compd.* 646:734–740.

122. Cao Y., Fang Y., Zhang G. et al. 2019. High quantum yield red-emission phosphor $Li_2Ge_4O_9$:Mn^{4+} for WLEDs application. *Opt. Mater.* 98:109442.

123. Zhang S., Hu Y. 2016. Photoluminescence spectroscopies and temperature-dependent luminescence of Mn^{4+} in $BaGe_4O_9$ phosphor. *J. Lumin.* 177:394–401.

124. Cao R., Luo W., Xiong Q. et al. 2015. Synthesis and luminescence properties of novel red phosphors $LiRGe_2O_6$:Mn^{4+} (R = Al or Ga). *J. Alloy. Compd.* 648:937–941.

125. Cao R., Ye Y., Peng Q. et al. 2017. Synthesis and luminescence characteristics of novel red-emitting $Ba_2TiGe_2O_8$:Mn^{4+} phosphor. *Dyes Pigments* 146:14–19.

126. Fu S., Tian L. 2019. A novel deep red emission phosphor $BaAl_2Ge_2O_8$:Mn^{4+} for plant growth LEDs. *Optik* 183:635–641.

127. Li K, Zhu D, Van Deun R. 2017. Photoluminescence properties and crystal field analysis of a novel red-emitting phosphor $K_2BaGe_8O_{18}$:Mn^{4+}. *Dyes and Pigments* 142:69–76.

128. Mo F., Lu. Z, Zhou L. 2019. Synthesis and luminescence properties of Mn^{4+}-activated Ba_2LaSbO_6 deep-red phosphor. *J. Lumin.* 205:393–399.

129. Zhong J., Zhou S., Chen D. et al. 2018. Enhanced luminescence of a Ba_2GdSbO_6:Mn^{4+} red phosphor via cation doping for warm white light-emitting diodes. *Dalton Trans.* 47:8248–8256.

130. Zhong J., Chen D., Yuan S. et al. 2018. Tunable optical properties and enhanced thermal quenching of non-rare-earth double-perovskite $(Ba_{1-x}Sr_x)_2YSbO_6$: Mn^{4+} red phosphors based on composition modulation. *Inorg. Chem.* 57:8978–8987.

131. Shi L., Han Y. Zhang Z. et al. 2019. Synthesis and photoluminescence properties of novel $Ca2LaSbO6$:Mn4+ double perovskite phosphor for plant growth LEDs. *Ceram. Int.* 45:4739–4746.

132. Zhong J., Chen D., Chen X. et al. 2018. Efficient rare-earth free red -emitting Ca_2YSbO_6: Mn^{4+}, M(M = Li^+ Na^+ K^+ Mg^{2+}) phosphors for white light -emitting diodest. *Dalton Trans.* 47:6528–6537.

133. Liang J, Devakumar B, Sun L et al. 2019. Novel Mn^{4+} doped Ca_2GdSbO_6 red-emitting phosphor: A potential color converter for light-emitting diodes. *J. Am. Ceram. Soc.* 102:4730–4736.

134. Liang J, Sun L, Devakumar B et al. 2018. Far-red-emitting double-perovskite $CaLaMgSbO_6$:Mn^{4+} phosphors with high photoluminescence efficiency and thermal stability for indoor plant cultivation LEDs. *Rsc Adv.* 8:31666–31672.

135. Shi L, Han YJ, Ji ZX, Zhang Z-W. 2019. Highly efficient and thermally stable $CaYMgSbO_6$:Mn^{4+} double perovskite red phosphor for indoor plant growth. *J Mater Sci-Mater Electron* 30:3107–3113.

136. Shi L, Han Y-j, Zhao Y et al. 2019. Synthesis and photoluminescence properties of novel Sr_3LiSbO_6:Mn^{4+} red phosphor for indoor plant growth. *Opt. Mater.* 89:609–614.

137. Li K., Mara D., Van Deun R. 2019. Synthesis and luminescence properties of a novel dazzling red-emitting phosphor $NaSr_3SbO_6$:Mn^{4+} for UV/n-UV w-LEDs. *Dalton Trans.* 48:3187–3192.

138. Zhong J., Chen X., Chen D. et al. 2019. A novel rare-earth free red-emitting $Li_3Mg_2SbO_6$:Mn^{4+} phosphor-in-glass for warm w-LEDs: synthesis, structure, and luminescence properties. *J. Alloy. Compd.* 773:413–422.

139. Wang S., Sun Q., Devakumar B. et al. 2019. Mn^{4+}-activated $Li_3Mg_2SbO_6$ as an ultrabright fluoride-free red-emitting phosphor for warm white light-emitting diodes. *RSC Adv.* 9:3429–3435.

140. Lu Z., Zhang X., Huang M. et al. 2018. Characterization and properties of red-emitting Sr_2YNbO_6:Mn^{4+} phosphor for white-light-emitting diodes. *J. Mater. Sci. Mater. Elect.* 29:17931–17938.
141. Fu A., Guan A., Yu D. et al. 2017. Synthesis, structure, and luminescence properties of a novel double-perovskite Sr_2LaNbO_6:Mn^{4+} phosphor. *Mater. Res. Bull.* 88:258–265.
142. Luo K., Zhang Y., Xu J. et al. 2019. Double-perovskite Ca_2AlNbO_6: Mn^{4+} red phosphor for white light emitting diodes: synthesis, structure and luminescence properties. *J. Mater. Sci. Mater. Elect.* 30:9903–9909.
143. Sun Q., Wang S., Devakumar B. et al. 2018. Synthesis and photoluminescence properties of novel far-red-emitting $BaLaMgNbO_6$:Mn^{4+} phosphors for plant growth LEDs. *RSC Adv.* 8:28538–28545.
144. Gu X., He Z., Sun X.Y. 2018. The deep red emission of Mn^{4+} doped $SrLaMgNbO_6$ flower-like microsphere phosphors. *Chem. Phys. Lett.* 707:129–132.
145. Qin L., Bi S., Cai P. et al. 2018. Preparation, characterization and luminescent properties of red-emitting phosphor: $LiLa_2NbO_6$ doped with Mn^{4+} ions. *J. Alloy. Compd.* 755:61–66.
146. Cao R., Shi Z., Quan G. et al. 2016. $Li_3Mg_2NbO_6$:Mn^{4+} red phosphor for light-emitting diode: synthesis and luminescence properties. *Opt. Mater.* 57:212–216.
147. Fu A., Zhou C., Chen Q. et al. 2017. Preparation and optical properties of a novel double-perovskite phosphor, Ba_2GdNbO_6:Mn^{4+}, for light-emitting diodes. *Ceram. Int.* 43:6353–6362.
148. Fu A., Pang Q., Yang H., et al. 2017. Ba_2YNbO_6:Mn^{4+}-based red phosphor for warm white light-emitting diodes (WLEDs): Photoluminescent and thermal characteristics. *Opt. Mater.* 70:144–152.
149. Cao R., Liu X., Bai K. et al. 2018. Photoluminescence properties of red-emitting $Li_2ZnSn_2O_6$:Mn^{4+} phosphor for solid-state lighting. *J. Lumin.* 197:169–174.
150. Cao R., Zhang J., Wang W. et al. 2017. Preparation and photoluminescence characteristics of $Li_2Mg_3SnO_6$:Mn^{4+} deep red phosphor. *Mater. Res. Bull.* 87:109–113.
151. Wang S., Sun Q., Liang J. et al. 2020. Preparation and photoluminescence properties of novel Mn^{4+} doped $Li_3Mg_2TaO_6$ red-emitting phosphors. *Inorg. Chem. Commun.* 116:107903.
152. Li K, Lian H, Van Deun R, Brik MG. 2019. A far-red-emitting $NaMgLaTeO_6$:Mn^{4+} phosphor with perovskite structure for indoor plant growth. *Dyes Pigments* 162:214–221.
153. Li K., Lian H., Van Deun R. 2018. A novel deep red-emitting phosphor $KMgLaTeO_6$:Mn^{4+} with high thermal stability and quantum yield for w-LEDs: structure, site occupancy and photoluminescence properties. *Dalton Trans.* 47:2501–2505.
154. Cao R., Shi Z., Quan G. et al. 2017. Preparation and luminescence properties of Li_2MgZrO_4:Mn^{4+} red phosphor for plant growth. *J. Lumin.* 188:577–581.
155. Cao R., Zhang J., Wang W. et al. 2017. Photo-luminescent properties and synthesis of $Ca_3Al_4ZnO_{10}$:Mn^{4+} deep red-emitting phosphor. *Opt. Mater.* 66:293–296.
156. Lian H.Z., Huang Q.M., Chen Y.Q. et al. 2017. Resonance Emission Enhancement (REE) for Narrow Band Red-Emitting A_2GeF_6:Mn^{4+} (A = Na, K, Rb, Cs) Phosphors Synthesized via a Precipitation–Cation Exchange Route. *Inorg. Chem.* 56:11900–11910.

7 Design Principles to Discover Highly Thermally Stable Phosphor

Yoon Hwa Kim, N. S. M. Viswanath, and Won Bin Im

CONTENTS

7.1 INTRODUCTION

The vast development of current technologies is driven by advanced functional materials. Till date, lanthanide ion-activated phosphors or luminescent materials as stable and efficient inorganic emitters have been mainly used in the fields of lighting and display applications [1–7]. However, during the operation of light-emitting diodes (LEDs), it produces a large amount of heat (~150°C); as a result, the luminescence efficacy could be reduced due to the formation of new nonradiative paths. This process is known as thermal quenching (TQ), which seriously effects the white balance of phosphor-converted LEDs. All this pays for a strong motivation for emerging strategies for scheming new, more thermally efficient, and color stable phosphors compared to the ones currently available phosphors. However, success of any such strategy relies on a thorough understanding of the mechanisms of the thermally stable phosphors. For fundamental understanding of TQ mechanisms, two dominant theories have been proposed to explain the TQ behavior in activator substituted phosphor materials. In the 1960s, Blasse et al. proposed that TQ is the result of the non-radiative relaxation of electrons from the excited state to the ground state [8]. This crossover mechanism is represented schematically using the configurational coordinate diagram in Figure 7.1(a), where the energy difference between the relaxed excited state and the crossover point determines the activation barrier for this process. This theory is one of the motivations for the searching of a structurally rigid phosphor host. The fundamental assumption here is that a more rigid host inhibits the soft phonon modes; as a result, the probability of non-radiative relaxation from the excited configuration to the ground-state configuration can be reduced. However, successive experiments have found many violations of this relationship; for example, the $Ca_7Mg(SiO_4)_4:Eu^{2+}$, $CaMgSi_2O_6:Eu^{2+}$, and $Sr_6M_2Al_4O_{15}:Eu^{2+}$ (M = Y,

DOI: 10.1201/9781003098669-7

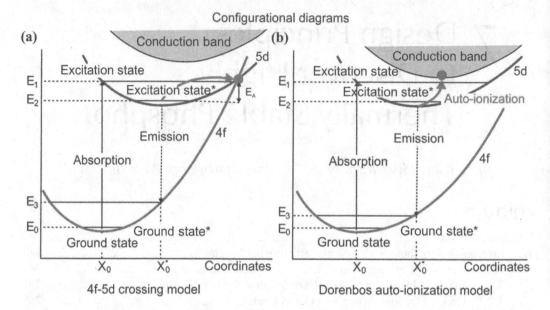

FIGURE 7.1 Comparison between (a) the 4f–5d crossing and (b) the Dorenbos autoionization models to explain the thermal quenching. (S. Poncé, Y. Jia, M. Giantomassi, et al., *J. Phys. Chem. C*, 120, 7, 4040–4047, 2016. With permission.)

Lu, Sc) phosphors suffer from TQ despite their superficial rigid crystal structures, as assessed from their Debye temperatures. The second theory attributed to Dorenbos speculates that TQ is due to the thermal excitation of the excited 5d electron of Ce^{3+}/Eu^{2+} to the conduction band of the host [9]. The activation barrier of this thermal ionization process (Figure 7.1(b)) determines the TQ of a phosphor, and this barrier is in turn associated to the bandgap of the phosphor host. However, the other TQ pathways also have been suggested, for example, the temperature dependence of the conduction band minimum is expected to lower and induce TQ, even though the two contending mechanisms of crossover mechanism and thermal ionization process could dominate in the loss of luminescence as a function of temperature in Ce^{3+}/Eu^{2+}-doped phosphors.

To improve the thermal stability of phosphors, in previous studies, various methods have been employed, such as surface coating [10–12], phosphor-in-glass or silicone [13–15], and ceramic phosphors. However, these methods are substantially not improved the thermal stability up to the expectation. Therefore, it is prerequisite to understand the important parameter that can affect the TQ of phosphors before framing any rules or descriptors. Moreover, it is also needed to develop new methodologies for design thermally stable phosphors. In this chapter, we not only discussed various parameters that can influence the TQ of phosphors but also enclosed various thermally quenching mechanisms that responsible to show high thermal stability at elevated temperatures. Finally, we showcased various parameters that can be useful to develop thermally stable phosphors in solid-state lighting field as demonstrated in Figure 7.2.

7.2   PREREQUISITES TO SHOW HIGH THERMAL STABILITY IN PHOSPHORS

When designing a phosphor, firstly, it is important to select an appropriate activator that suits the desired purpose. Depending on the activator, even the same host has different emission and thermal stability properties. Hence, a careful consideration of the activator is required [16]. Therefore, in this chapter, we explain various parameters that can influence the thermal stability properties of a phosphor material.

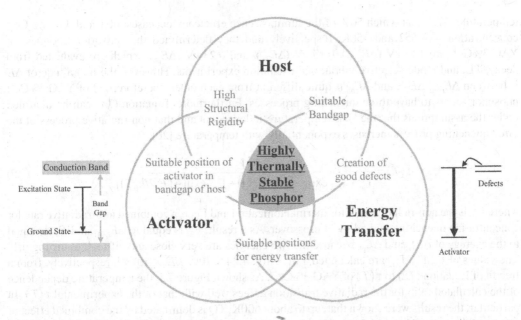

FIGURE 7.2 Schematic diagram for designing of principal elements to discover thermally stable phosphors.

7.2.1 ACTIVATOR CONCENTRATION

The photoluminescence (PL) efficiency is low when the activator concentration is lower than the optimized concentration. On the other hand, the PL efficiency is reduced when the activator concentration is higher than the optimized one due to an increase of a non-radiative transition path. Therefore, the optimization of the activator concentration is a crucial step for designing a phosphor material. However, thermal stability of phosphor is likely to different from the optimization of the activator concentration. For example, in the $Y_3Al_5O_{12}:Ce^{3+}$ (YAG:Ce^{3+}) phosphor, Ce^{3+} ions are substituted at the Y^{3+} site, and a broad yellow emission is obtained under blue excitation by 4f–5d transition of Ce^{3+} ions. The Y^{3+} site of the YAG:Ce^{3+} phosphor combines with eight oxygen ions, forming a distorted dodecahedral structure due to two different Y–O distances. Since the size of CeO_8 dodecahedra is larger than YO_8 dodecahedra, increasing the Ce^{3+} concentration increases the crystal-field splitting from the high distortion of the Ce^{3+} site, resulting in a redshift of the maximum emission wavelength. In essence, when the temperature increases, the maximum emission wavelength is red-shifted with reducing luminescence intensity with increasing the activator concentration by activating phonon modes for non-radiative relaxation mechanisms. Thus, the YAG:Ce^{3+} phosphor with low Ce^{3+} concentration shows high thermal stability.

To understand more detailed TQ phenomenon as a correlation of structural-optical properties, various analysis methods have been employed [17–19]. Karlsson et al. used the combined techniques like PL, thermoluminescence (TL), and mode-selective vibrational excitation, to understand the TQ mechanism in YAG:Ce^{3+}. From the obtained results of the temperature-dependent decay time (τ), curves of YAG:xCe^{3+} ($x = 0.2, 1, 2,$ and 3 mol%) were fitted by a single-barrier quenching model using following equation:

$$\tau(T) = \frac{1}{\Gamma_r + \Gamma_n \exp(-\Delta E / k_B T)} \tag{7.1}$$

where Γ_r is the radiative rate, Γ_n is the attempt rate of non-radiative processes, k_B is the Boltzmann constant, and ΔE is the activation energy for the overall quenching behavior. ΔE is decreased from 0.55 eV for 1 mol% Ce^{3+} concentration to 0.32 eV for 3 mol% Ce^{3+} concentration. The TQ temperatures ($T_{50\%}$), at which 50% of the luminescence efficiency deceases, of 1 and 3 mol% Ce^{3+} concentration was 652 and 556K, respectively and they determined the activation energies for YAG:3%Ce^{3+} are 1.13 eV ($\Delta E_{i,5d1}$), 0.15 eV (ΔE_{TL}), and 0.23 eV (ΔE_{ph}), which are evaluated from decay, TL, and mode-selective vibrational excitation experiments. However, the magnitude of ΔE is between $\Delta E_{i,5d1}$ ΔE_{TL}, and ΔE_{ph} is quite different from each other; therefore, TQ of YAG:3%Ce^{3+} phosphor seems to have three quenching processes. Furthermore, Equation (7.1) can be modified under the assumptions that the Γ_r is temperature-independent and that non-radiative process of the three quenching process increases exponentially with temperature [20]:

$$\tau(T) = \frac{1}{\Gamma_r + \Gamma_{n1}\exp\left(-\left(E_{n1}/k_BT\right)\right) + \Gamma_{n2}\exp\left(-\left(E_{n2}/k_BT\right)\right)} \tag{7.2}$$

where Γ_{n1} is the non-radiative rate for thermal ionization and Γ_{n2} is a combined non-radiative rate for concentration quenching and 5d → 4f crossover. As a result, E_{n1} is equal to $\Delta E_{i,5d1}$ and E_{n2} is equal to the average of ΔE_{TL} and ΔE_{ph} because the two energies are very close and difficult to distinguish. The values of Γ_r, Γ_{n1}, Γ_{n2} are calculated 1.61×10^7, 4.24×10^{15}, 7.72×10^8 s^{-1}, respectively, from a free fit of Equation (7.2) to $\tau(T)$ of YAG:3%Ce^{3+}. As shown Figure 7.3, the temperature dependence of the calculated ratio for the radiative transition agrees well with that of the experimental $\tau(T)$. In particular, the results were shown that, up to about 600K, TQ is dominated by the combined effect of concentration quenching and 5d → 4f crossover quenching, whereas, at higher temperatures, it has lead by thermal ionization. A similar result is also observed in YAG:1%Ce^{3+} (Figure 7.3(b)) using the same method. From these results, we can understand that lowering the Ce^{3+} concentration of YAG phosphor enhance the thermal stability at below 600K by reducing the probability of 5d → 4f crossover quenching as shown in Figure 7.3.

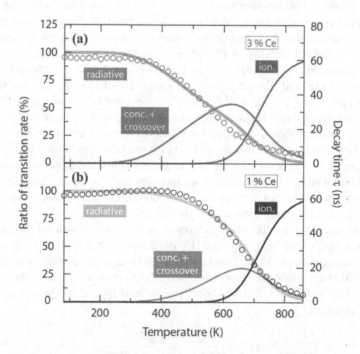

FIGURE 7.3   Left axis: Ratios of the radiative and non-radiative (thermal ionization and combined concentration and 5d → 4f crossover quenching) transition rates of YAG:x%Ce^{3+} with respect to the total transition rate, plotted by solid curves. Right axis: Luminescence decay time of YAG:x%Ce^{3+} ($x = 1$ and 3), plotted by circles. (a) $x = 3$ and (b) $x = 1$. (Y. C. Lin, M. Bettinelli, S. K. Sharma, *J. Mater. Chem. C*, 8, 14015–14027, 2020. With permission.)

In addition, a different TQ process was also observed as a function of activator concentration in other phosphor materials. For example, in $Ca_{1-2x}Li_xAlSiN_3:xCe^{3+}$ phosphors scrutinized with different Ce^{3+} concentrations ($x = 0$–0.05), the TQ temperature ($T_{50\%}$), at which 50% degradation in intensity compared to the intensity at room temperature, is above 300°C at $x \leq 0.02$. As the Ce^{3+} concentration increased, the $T_{50\%}$ gradually decreased like 217 and 202°C for $x = 0.04$ and 0.05, respectively. In addition, the maximum emission wavelength is red-shifted with increasing temperature for $x < 0.04$ (lower concentration) and is blue-shifted for $x > 0.04$ (higher concentration). The TQ at lower Ce^{3+} concentration is mainly altered from the structure relaxation such as the lattice expansion, and expansion of bond lengths around the activator ions can cause the emission band shift to longer wavelength probably due to an increased stokes shift. On the other hand, the energy transfers between Ce^{3+} ions, which is strongly dependent on the bond distance of Ce^{3+}–Ce^{3+} ions. In addition, when the excite state of the activator ion and the bottom of the conduction band of the phosphor are sufficiently close to each other, as a result, the photoionization process is more dominant to reduce the PL efficiency as a function of temperature. Moreover, in the $Ca_{1-2x}Li_xAlSiN_3:_xCe^{3+}$ phosphors, the emission band of Ce^{3+} can be fitted into two peaking at about 556 nm and 618 nm, originated from the transitions of the 5d excited states to its two ground-state configurations of $^2F_{7/2}$ and $^2F_{5/2}$ of Ce^{3+}, respectively. As a result, the blue-shift of emission wavelength with increasing of temperature. Consequently, the emission band in longer wavelength side is thermally quenched faster than that of shorter emission wavelength side. As such, it is known that a high concentration of an activator causes TQ due to an increase in the interaction between ions of the activator. Seshadri's group studied the change of local environment in YAG depending on Ce^{3+} concentration [18]. From the study, not only the non-radiative transition increases, but also greater structural and compositional inhomogeneity occurs as increasing amount of activator in the phosphor because of the size difference between the activator and substituted ions. In addition, they have investigated one reason for the low thermal stability at high Ce^{3+} concentration owing to the decrease of Debye temperature (Θ_D) with increasing Ce^{3+} concentration. The low concentration of activator is reduced the structural inhomogeneity and the energy transfer between activators will be helped to increase thermal stability of the phosphor. However, TQ can also occur at much lower concentration. For example, the TQ of $BaAl_{12}O_{19}:Eu^{2+}$ phosphor was measured at various Eu^{2+} concentrations (1, 5, 10, 20%) [21]. All samples had shown reversible thermal stability due to reduction in crystal rigidity of $BaAl_{12}O_{19}$ host as a function of activator concentration. Hence, the low concentration of the activator does not give any guarantee to improve the TQ properties. In addition, TQ of a given phosphor also depends on the site selectivity of the activator ion. For example, Bi^{3+}-doped $Sr_2P_2O_7$ phosphors had two Sr^{2+} sites coordinated to nine oxygen ions with different average bonding lengths, which is 2.7214 and 2.6794 Å for $Sr^{2+}(1)$ and $Sr^{2+}(2)$, respectively [22]. $Sr_2P_2O_7:Bi^{3+}$ phosphors were broad red-emitting phosphor composed of two Gaussian peaks at 660 and 698 nm. When the excitation was monitored at each emission wavelength, it was observed that the shape of excitation spectrum was similar, but the maximum excitation wavelength was shifted. This behavior suggests that Bi^{3+} ions enter each of the two Sr sites. As the concentrations of Bi^{3+} ions were increased, the emission peak intensity around 660 nm was increased, which means that Bi^{3+} ions were substituted to $Sr^{2+}(1)$ site owing to similar size of the cations. With increasing the Bi^{3+} concentration, Bi^{3+} ions filled the $Sr^{2+}(2)$ site [$Bi^{3+}(2)$] and improved the energy transfer from $Bi^{3+}(1)$ to $Bi^{3+}(2)$, which can improve thermal stability at elevated temperatures.

7.2.2 EFFECT OF PHASE PURITY AND MIXED-VALANCE STATES IN PHOSPHORS

The formation of secondary phase owing to unreacted species during the synthesis and unreduced activator ions remains in the phosphor that provides an additional non-radiative path; as a result, the PL efficiency was reduced a function of temperature. Therefore, reduction of secondary phases and mixed oxidation states of activator ions is paramount to improve thermal stability of phosphors, as it is noticed that Mn^{4+}-doped fluoride-based phosphors exhibit promising applications. However, synthesizing the fluoride-based phosphors remains challenging because these phosphors are produced by wet chemical synthesis involving hydrogen fluoride (HF). A drawback of these methods is proneness to form the impurity phases that may affect thermal stability of the phosphor, as it was noticed K_2SiF_6 host thermally stable at LED operation conditions (<200°C). However, at above 400°C, the K_2SiF_6 endures a surface hydrolysis and bulk decompositions as observed from in situ X-ray diffraction analysis. In both processes, K_2SiF_6 partly converts into K_3SiF_7 with the release of HF. To dope Mn^{4+} ions into the K_2SiF_6 requires a use of wet chemical synthesis methods with a risk of formation of impurity phases such as $K_2MnF_5·H_2O$ and $KMnF_4·H_2O$. Therefore, it is pivotal to minimize the impurity phases to obtain a desired property. It was found that complete evaporation of the solvent led to impure phosphor whereas decantation, filtration, and washing led to impurity-free fluoride phosphors [23]. Thermal stability of K_2SiF_6:Mn^{4+} phosphor was examined as a function of synthesis conditions and undesirable impurity phase and this study shows that phase pure K_2SiF_6:Mn^{4+} retains 90% of its initial PL intensity at 250°C, whereas impure K_2SiF_6:Mn^{4+} sample retains only 70% of its PL intensity. Therefore, phase purity matters in order to develop thermally stable phosphors.

In addition, $β$-SiAlON:Eu^{2+} phosphor is paid attention for wide color gamut white LED backlights because of narrow green emission. However, $β$-SiAlON:Eu^{2+} has a limitation to improve luminescence properties due to the substitution of an extremely small amount of Eu^{2+} ions at the interstitial site formed by AlN_4 and SiN_4 tetrahedra. Moreover, due to the smaller ion size of Eu^{3+} than Eu^{2+} ions, it is not surprise for coexistence of Eu^{2+} and Eu^{3+} ions in $β$-SiAlON host. Thus, in order to improve the luminescence properties, it is necessary to reduce Eu^{3+} to Eu^{2+} ions in the phosphor. For the reduction of Eu^{3+} to Eu^{2+} ions, the synthesized phosphor was subjected to additional heat treatment at a temperature lower than the synthesis temperature in 10%H_2/90%N_2 reducing atmosphere [24]. This additional heat treatment process improves the luminescence properties 2.3 times higher than initial PL intensity (Figure 7.4(a)) and the TQ property was improved by 11% at 300°C due to the reduction of Eu^{3+} into Eu^{2+} ions (Figure 7.4(b)).

Furthermore, controlling the compositions in the host of phosphor can also help the reduction of the activator by charge compensation. For instance, $NaAlSiO_4$:0.01Eu^{2+} phosphor exhibits a broad emission spectrum due to substitution of activator ions into two Na sites, and their maximum emission wavelengths under excitation at 365 nm are ~450 nm (from Na1 site) and 535 nm (from Na2 site), respectively. The compositions of $NaAlSiO_4$ were altered by the addition of Si and Li and reduction of the concentrations of Na and Al; this can be denoted as $Na_{(0.99-x)}\square_xAl_{(1-x)}Si_{(1+x)}O_4$:$yLi$,

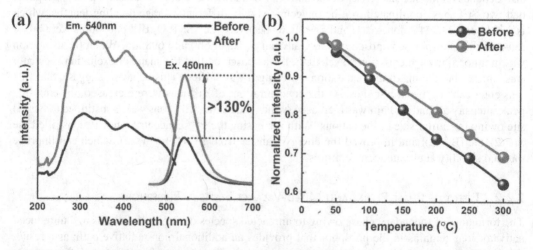

FIGURE 7.4   (a) PLE and PL spectra and (b) temperature-dependent PL intensity of as-synthesized and thermally treated $β$-sialon:Eu^{2+} phosphors. (S. Li, L. Wang, D. Tang, et al., *Chem. Mater.*, 30, 494–505, 2018. With permission.)

$0.01Eu^{2+}$ ($x = y$, $0 \leq y \leq 0.15$, where $\square$ represents Na vacancy, NASO:yLi^+,Eu^{2+}) [25]. The cell volume increased nonlinearly with increasing y ratio. This behavior indicates Li^+ ions could be partly located (or inserted) at sites other than Na^+ sites because the ion radius of Li^+ ions ($LiO_8 = 0.92$ Å) is smaller than Na^+ ions ($NaO_8 = 1.18$ Å). To confirm the substitution of Li^+ ions, ^7Li, ^{27}Al, and Si solid-state nuclear magnetic resonance was measured, and density functional theory (DFT) calculations were also performed. These results suggested that Li^+ ions preferred to be substituted at the Al site (Li_{Al}) and the Na vacancy site (Li_{VNa1}) rather than Na site and Si_{Al} substitutions that help to compensate the reduced charge caused by the Na vacancies with increasing y ratio. These substitutions affected the luminescence properties by increasing the emission from the Na1 site owing to the rise the ratio of Eu^{2+}/Eu^{3+}. Besides, with increasing y ratio, the thermal stability of NASO:yLi^+,Eu^{2+}($y = 0.15$) phosphors was increased about four times compared with $y = 0$. This is ascribed to thermal release of more electrons from Li-related Na vacancy site (Li_{VNa1}) to recombine with Eu^{3+}.

7.2.3 COACTIVATOR IONS ACTING AS A DEFECT CENTER

Generally, dual activators in the phosphor host are not suggested due to reduction of its emission efficiency as a function of temperature, owing to creation of new non-radiative paths by coactivator ions. However, several cases have been reported that thermal stability of a phosphor can be enhanced by forming a defect states in the structure [26–28]. For example, Sr_3SiO_5:Eu^{2+},Tm^{3+} phosphor had shown zero-thermal-quenching property by supporting of creation of deep traps by creating Tm^{3+} ions. Upon excitation at 468 nm, Sr_3SiO_5:Eu^{2+},Tm^{3+} phosphor had shown broad emission spectra located at 578 nm, as depict Figure 7.5(a). The Tm^{3+} concentration was optimized at 0.02 mol with 0.5 mol of Eu^{2+}. On contrary to the Sr_3SiO_5:Eu^{2+} phosphor, Sr_3SiO_5:Eu^{2+},Tm^{3+} phosphor did not show decrease in emission intensity (Figure 7.5(b)) as a function of temperature. This improved TQ property attributed to the formation of defect sites in the phosphor by doping Tm^{3+} ions analyzed from TL spectroscopy as shown in Figure 7.5(c). TL curve of Sr_3SiO_5:Eu^{2+},Tm^{3+} phosphor fitted into four Gaussian profiles (348, 390, 448, and 551K), indicating the phosphor had four defect sites that were like Sr_2SiO_5:Eu^{2+} phosphor as evidenced from the electron paramagnetic resonance. Long persistent luminescence (LPL) was applied to clarify release processes of the carriers at room temperature. Generally, the phosphor materials have the trap centers, which shows an LPL phenomenon. To verify this, the LPL spectra of Sr_3SiO_5:Eu^{2+},Tm^{3+} phosphor were measured that is similar to PL spectra. From this point, it is emphasized that the charge carriers could be released from the traps, which is combined with $4f^65d^1$ state of Eu^{2+} contributing to the LPL. However, the LPL charging

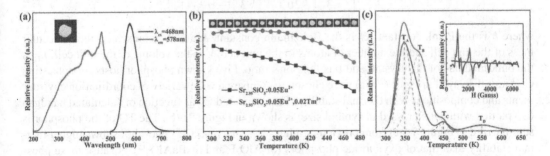

FIGURE 7.5 (a) PLE and PL spectra of $Sr_{2.93}SiO_5$:$0.05Eu^{2+}$,$0.02Tm^{3+}$; the inset is a photograph of $Sr_{2.93}SiO_5$:$0.05Eu^{2+}$,$0.02Tm^{3+}$ under UV radiation. (b) Temperature-dependent luminous intensity changes of $Sr_{2.93}SiO_5$:$0.05Eu^{2+}$, $0.02Tm^{3+}$ and $Sr_{2.95}SiO_5$:$0.05Eu^{2+}$ under 478 nm excitation; the inset is a series of photographs of the $Sr_{2.93}SiO_5$:$0.05Eu^{2+}$,$0.02Tm^{3+}$ phosphor at the corresponding temperatures. (c) TL curves and EPR spectra (shown in the inset) of the Sr_3SiO_5 (black lines), $Sr_{2.95}SiO_5$:$0.05Eu^{2+}$ (blue lines), and $Sr_{2.93}SiO_5$:$0.05Eu^{2+}$,$0.02Tm^{3+}$ (red lines) phosphors. (X. Fan, W. Chen, S. Xin, et al., *J. Mater. Chem. C*, 6, 2978–2982, 2018. With permission.)

spectrum is consisted of a broadband that is ranging from 250 to 400 nm. Therefore, shallow traps only can capture the electrons excited to a higher $4f^6 5d^1$ state. From these results, it was assumed the coactivator ions serving in two different ways. In the first step, under 460 nm excitation, the excited electron is captured through the conduction band with the assistance of thermal disturbance, and the second step is the combination, leading to radiative process, of the captured carrier with $4f^6 5d^1$ state of Eu^{2+} by supplying excessive energy that can help to improve the thermal stability of the phosphor.

7.3 DESCRIPTORS FOR IMPROVING THE THERMAL STABILITY OF A PHOSPHOR BASED ON A SELECTION OF A HOST MATERIAL

Generally, the thermal stability of a phosphor is closely related to the selection of a host material. In this section, we discuss the factors that can influence the thermal stability of phosphor.

7.3.1 STRUCTURE-PROPERTY RELATIONSHIPS IN THERMALLY STABLE PHOSPHORS

One feasible method for identifying materials with a high photoluminescence quantum yield (PLQY) and high thermal stability is Debye temperature (Θ_D), which represents crystal rigidity of and atomic connectivity of host material. These characteristics show high thermal stability of phosphor by constraining soft phonon modes that lead high thermal stability at elevated temperature. For instance, aluminum garnet phosphor ($X_3Al_2Al_3O_{12}$, X = trivalent rare-earth ions like Y, Lu, Gd, Tb, or La) represented by YAG:Ce^{3+} accommodates two sub-lattices that one formed of AlO_4 tetrahedra and AlO_6 octahedra and a second of Y^{3+} ions. These networks composed of interpenetrating three-dimensional connections. Due to these features, the garnet phosphor shows not only the excellent thermal stability but also high luminescence efficiency.

As mentioned, Θ_D was proposed as a method of determining the rigidity and degree of connectivity of the phosphor host through a prototypical example of diamond [29]. The larger value of Θ_D suggests that a structure is rigid as well as liable to be highly connected. *Ab initio* DFT calculations are operated to find out elastic constants of crystal [30]:

$$\Theta_D = \frac{\hbar}{k_B} \left[6\pi^2 V^{\frac{1}{2}} N \right]^{1/3} \sqrt{\frac{B_\Pi}{M}} f(\upsilon) \tag{7.3}$$

$$f(\upsilon) = \left\{ \left[2 \left(\frac{2}{3} \cdot \frac{1+\upsilon}{1-2\upsilon} \right)^{3/2} + \left(\frac{1}{3} \cdot \frac{1+\upsilon}{1-\upsilon} \right)^{3/2} \right]^{-1} \right\}^{1/3} \tag{7.4}$$

where $\hbar$ is the Plank constant, k_B is the Boltzmann constant, M and N represented the molecular mass of the unit cell and the number of atoms in the unit cell, V is the volume of the unit cell, B_Π is the bulk modulus of the crystal, and υ is Poisson's ratio. Five known phosphor hosts for subsisting Ce^{3+} ions displayed in Figure 7.6(a) were chosen to compare a wide variety of coordination environments and compositions with thermal stability. The calculated Θ_D as function of calculated bandgap (E_g) plotted with Φ (expressed in symbol size) is shown in Figure 7.6(b). The TQ of the phosphor is summarized in Figure 7.6(c) from previous studies [31]. The YAG:Ce^{3+} phosphor shows higher thermal stability than that of oxyfluoride phosphors (Sr_3AlO_4F and Sr_2BaAlO_4F), because these phosphors lower Θ_D than that of YAG:Ce^{3+}. Among silicate phosphors (Sr_3SiO_5 and Ba_2SiO_4), the Sr_3SiO_5 phosphor shows higher thermal stability than Ba_2SiO_4 due to higher Θ_D of Sr_3SiO_5. Moreover, the nitride phosphor formed anions with nitrogen ions is another typical phosphor with high thermal stability [32–34]. Because the nitrogen in host composition has less electronegativity (3.04) than oxygen (3.44), which can be connected metal with covalent or ionic-covalent bonding, which possesses smaller Stokes shift and larger crystal-field splitting than oxide phosphors. Besides, nitride phosphor has high structural stiffness and high condensed edge- and/or vertex-sharing tetrahedra

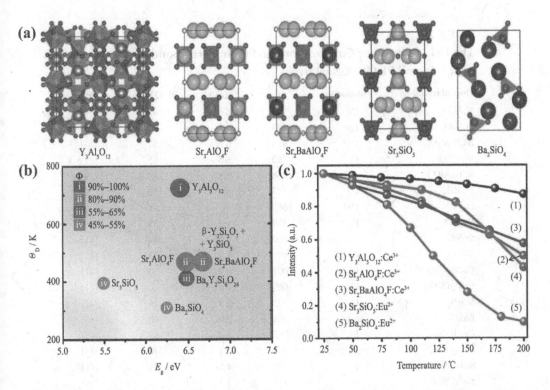

FIGURE 7.6 (a) Crystal structures of five different inorganic hosts: $Y_3Al_5O_{12}$, Sr_3AlO_4F, Sr_2BaAlO_4F, Sr_3SiO_5, and Ba_2SiO_4; (b) a schematic representation showing the capability of using Debye temperature and bandgap parameters for identifying eight high-efficiency phosphors; and (c) temperature-dependent normalized emission spectra of $Y_3Al_5O_{12}$:Ce^{3+}, Sr_3AlO_4F:Ce^{3+}, Sr_2BaAlO_4F:Ce^{3+}, Sr_3SiO_5:Eu^{2+}, and Ba_2SiO_4:Eu^{2+}. (J. Qiao, J. Zhao, Q. Liu, et al., *J. Rare Earths*, 37, 6, 565–572, 2019. With permission.)

with the metal-nitrogen bonding (MN_4, M = Si, Al, Li, Mg, Ge, Ga, etc.). Table 7.1 summarizes the calculated Θ_D and $T_{50\%}$ for each phosphor compositions. From this result, a material that has high Θ_D is the most reliable proxy for screening highly thermally stable phosphors.

From this study, one can understand the rigid structure of host material of a phosphor that probably shows high PL efficiency and high thermal stability of the phosphor up to some extent. However, other impacts such as the E_g and location of energy levels of the activator ions can influence the thermal properties of the phosphor. As mentioned earlier, in TQ mechanisms, the distance (energy difference) between the conduction band of the phosphor host and the energy level of the activator that directly affects the emission is related to the temperature at which TQ occurs. For instance, $Ba_3Si_6O_9N_4$:Eu^{2+} phosphor has bluish-green emission (~480 nm) and $Ba_3Si_6O_{12}N_2$:Eu^{2+} phosphor has green emission (~530 nm) under blue excitation. The calculated Θ_D of $Ba_3Si_6O_9N_4$ is 489K higher than that of $Ba_3Si_6O_{12}N_2$ (433K) [60]. However, $Ba_3Si_6O_{12}N_2$:Eu^{2+} phosphor has weak luminescence at room temperature which indicates that this phosphor shows a strong TQ. This phenomenon can be interpreted in conjunction with centroid shift, crystal-field splitting, and E_g. When Eu^{2+} ions were substituted in Ba^{2+} site, $Ba_3Si_6O_{12}N_2$:Eu^{2+} phosphor may have larger centroid shift as well as larger crystal-field splitting of Eu^{2+} 5d levels from calculated the Ba–N coordinate bond, which can be inferred from shorter coordinate bond length (about 3.0 Å) than in $Ba_3Si_6O_9N_4$ (about 3.2 Å). Thus, the short emission wavelength of $Ba_3Si_6O_9N_4$:Eu^{2+} can be predicted by assuming the positions of the Eu^{2+} 4f level are similar for the two phosphors. Moreover, the calculated E_g of $Ba_3Si_6O_{12}N_2$ from computation calculation is 6.46 eV, which is larger than that of $Ba_3Si_6O_9N_4$ (6.79 eV) [61]. Succinct all these results together, in conclusion, the strong TQ in $Ba_3Si_6O_9N_4$:Eu^{2+}

TABLE 7.1

The List of Phosphor Compositions and their Corresponding Values of Θ_D, E_g, and PL QY, Respectively

Host Material	Activator Ion	Θ_D	E_g (eV)	PL QY (%)	Ref.
$Y_3Al_5O_{12}$	Ce^{3+}	726	6.4	90	[29]
$SrAl_2O_4$	Eu^{2+}	475	4.5	65	[35]
$Ba_9Y_2Si_6O_{24}$	Ce^{3+}	408	6.5	57	[36]
Sr_3AlO_4F	Ce^{3+}	465	6.5	85	[37]
Sr_2BaAlO_4F	Ce^{3+}	466	6.8	83	[38]
Sr_3SiO_5	Ce^{3+}	395	4.4	55	[29]
$La_3Si_6N_{11}$	Ce^{3+}	660	4	100	[39]
Ba_2SiO_4	Eu^{2+}	307	4.2	88	[40]
Sr_2SiO_4	Eu^{2+}	360	4.5	44	[41]
Y_2SiO_5	Ce^{3+}	491	4.7	33	[41]
$Ca_7Mg(SiO_4)_4$	Eu^{2+}	601	6.9	30	[41]
$CaMg(SiO_3)_3$	Eu^{2+}	665	7.0	5	[41]
$SrSiN_2$	Eu^{2+}	375	2.9	25	[42]
$BaSiN_2$	Eu^{2+}	360	2.9	40	[42]
$CaAlSiN_3$	Eu^{2+}	787	3.4	95	[43]
$SrSc_2O_4$	Eu^{2+}	604	3.4	90	[44]
$Sr_2MgSi_2O_7$	Eu^{2+}	476	4.5	95	[45]
$Sr_2Al_2SiO_7$	Eu^{2+}	514	4.2	28	[46]
$BaLu_2Si_3O_{10}$	Eu^{2+}	406	4.8	35	[47]
$BaSc_2Si_3O_{10}$	Eu^{2+}	529	4.7	48	[47]
$BaZrSi_3O_9$	Eu^{2+}	493	4.7	28	[48]
$K_3YSi_2O_7$	Eu^{2+}	515	3.7	56	[49]
Sr_2LiAlO_4	Eu^{2+}	467	4.2	28	[50]
$Ba_3Y_2B_6O_{15}$	Ce^{3+}	379	4.5	34	[51]
$NaSrBO_3$	Ce^{3+}	732	5.3	75	[52]
$NaCaBO_3$	Ce^{3+}	655	4.2	68	[51]
$NaBaB_9O_{15}$	Eu^{2+}	729	5.5	98	[53]
$KSrPO_4$	Eu^{2+}	292	5.1	95	[54]
$KBaPO_4$	Eu^{2+}	392	4.9	90	[55]
$Ca_6BaP_4O_{17}$	Eu^{2+}	507	4.3	35	[56]
$Na_3Sc_2(PO_4)_3$	Eu^{2+}	485	4.6	80	[57]
$RbLi(SiO_4)_2$	Eu^{2+}	556	4.8	98	[58]
$SrLiAl_3N_4$	Eu^{2+}	560	5.0	95	[59]

may lead to the photoionization mechanism because of small gap between the Eu^{2+} 5d-excited states and the conduction band bottom of $Ba_3Si_6O_9N_4$ as illustrated in Figure 7.7.

Furthermore, the thermal stability of the aluminum garnet phosphors is also investigated by substitution of different trivalent cation ions [62, 63]. The luminescence intensity of YAG:Ce^{3+}, $Tb_3Al_5O_{12}$:Ce^{3+}, and $Gd_3Al_5O_{12}$:Ce^{3+} phosphors were studied with linking the distance change between trivalent cation (Y^{3+}, Tb^{3+}, and Gd^{3+}) and oxygen in dodecahedral site depends on temperature. The average distance between the trivalent cations and oxygen is longer in the order of Y^{3+}, Tb^{3+}, and Gd^{3+}. And the crystal-field splitting of 5d state of Ce^{3+} ions was increasing with increasing the average distance of dodecahedral site. Additionally, the thermal stability of those phosphors is higher in the order of Y^{3+}, Tb^{3+}, and Gd^{3+}. These results implied that the increase of average size in

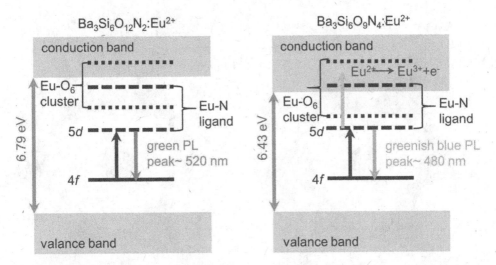

FIGURE 7.7 Schematic illustration of electronic structure of $Ba_3Si_6O_{12}N_2:Eu^{2+}$ and $Ba_3Si_6O_9N_4:Eu^{2+}$. Autoionization process ($Eu^{2+} \rightarrow Eu^{3+} + e^-$) is denoted for $Ba_3Si_6O_9N_4:Eu^{2+}$ (right). (M. Mikami, *Opt. Mater.*, 35, 11, 1958–1961, 2013. With permission.)

dodecahedral site the local environment around activator ion upon substitution of trivalent cations is allayed; as a result, the electron-lattice interaction has been enhanced owing to that the phosphor material shows a reduction in PL efficiency as a function of temperature. Moreover, to understand the relation with TQ and structural changes in the garnet structure, the series of samples were synthesized by doping of Ga^{3+} ions into YAG:Ce^{3+} [63]. As the more Ga^{3+} ions are substituted at the Al^{3+} ion sites, the stronger the TQ occurs because of reduction of E_g of ~6.5 to ~5.5 eV by Ga^{3+} ion substitution; as a result, the probability of thermal ionization TQ mechanism was more dominated.

7.3.2 Crystal-Site Engineering

From Section 7.2.1, one can recognize that the luminescence properties of Ce^{3+}- and Eu^{2+}-doped host materials are very sensitive to the local environment of the activator ions. The crystal rigidity directly relates to the PL efficiency and thermal stability of the phosphor. Hence, TQ of phosphors could be improved by modifying the local environment around the activator using crystal-site engineering. Therefore, material scientists paid a more attention to improve thermal stability of previously known structures by the crystal-site engineering in terms of (1) tuning of a bandgap, (2) enhancement of structure rigidity, and (3) an increase of thermal barrier distance. For instance, $CaAlSiN_{3-4/3x}C_x:Eu^{2+}$ phosphor was developed with improved thermal stability owing to expansion of $(Ca/Eu)–(N/C)_5$ polyhedron [64]. In the structure, Ca^{2+} ions form a polyhedron with five nitrogen ions, and tetrahedra of SiN_4/AlN_4 are connected to each other around the CaN_5 polyhedron and forms a ring shape as displayed in Figure 7.8(a) and (b). The expansion of $(Ca/Eu)–(N/C)_5$ polyhedron led to wider E_g and large crystal-field splitting of Eu^{2+} ions. Therefore, $CaAlSiN_{3-4/3x}C_x:Eu^{2+}$ phosphor ($x = 0.24$) showed improved thermal stability by inhibiting both the thermal crossover and photoionization mechanism. Furthermore, Liu's group have investigated the relationship between structure-property relationship in $CaAlSiN_3$ phosphors by substitutions of $[La^{3+}/Al^{3+}]$ for $[Ca^{2+}/Si^{4+}]$ and $[Li^+/Si^{4+}]$ for $[Ca^{2+}/Al^{3+}]$ pairs at Ca^{2+} site in $CaAlSiN_3$ [65]. When $[La^{3+}/Al^{3+}]$ was substituted at Ca^{2+} site, maximum emission wavelength was blue-shifted due to an increase of covalency from the larger valence of La^{3+} ion than Ca^{2+} ions and shorten average cation-nitrogen distance. Formerly, the Eu^{2+} ions were surrounded with nitride anions, which neighborhood La^{3+} and Si^{4+}/Al^{3+} equivalent coordination. On contrary, when $[Li^+/Si^{4+}]$ was substituted, the emission spectra were decomposed into two Gaussian peaks from different Eu^{2+} environments. The Eu^{2+} ions were located around Li^+

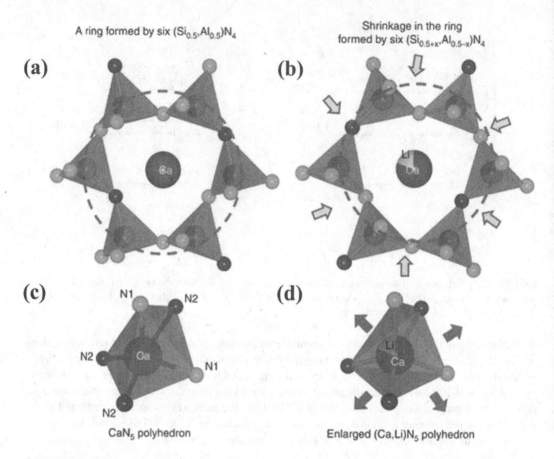

FIGURE 7.8 Schematics of the second coordination spheres of (a) $CaAlSiN_3$ and (b) $CaAlSiN_3$–$LiSi_2N_3$; the polyhedron of (c) CaN_5 and (d) $(Ca,Li)N_5$. (L. Wang, R. J. Xie, Y. Li, et al., *Light. Sci. Appl.*, 5, e16155, 2016. With permission.)

and Si^{4+}-rich coordination, which indicates the enhancement of structure rigidity, when $[Li^+/Si^{4+}]$ was substituted. From this structural change, $[Li^+/Si^{4+}]$-substituted phosphor shows higher thermally stability than $[La^{3+}/Al^{3+}]$-substituted phosphor. Besides, the improvement of thermal stability with increasing structure rigidity is also be observed in $BaY_2Si_3O_{10}$:Eu^{2+} phosphor with the valence mismatch between Mn^{2+}/Mg^{2+} and Y^{3+} site and $Ba_2Zr_2Si_3O_{12}$:Eu^{2+} with Ca^{2+}/Sr^{2+} cation substitution for Ba^{2+} site [66].

Furthermore, $CaAlSiN_3$:Eu^{2+} solid-solution phosphor developed by substitution of $[Li,Si]^{5+}$ for $[Ca,Al]^{5+}$, resulting with thermally stability [67]. The $LiSi_2N_3$ structure is an isostructural with $CaAlSiN_3$ structure, orthorhombic $Cmc2_1$, that can form a solid-solution in the form of $Ca_{1-x}Li_xAl_{1-x}Si_{1+x}N_3$ ($x = 0$–0.22). With increasing x ratio in $Ca_{1-x}Li_xAl_{1-x}Si_{1+x}N_3$, the average distance of Ca–N increased as the substitution of larger Ca^{2+} ions by smaller Li^+ ions. Thus, $(Ca,Li,Eu)N_5$ polyhedron could be expanded. On the other hand, as the x ratio increases, the tetrahedral of SiN_4/AlN_4 around $(Ca,Li,Eu)N_5$ decreases further by increasing the proportion of Si^{4+} ions (0.26 Å, CN = 4) which is smaller than Al^{3+} ions (0.39 Å, CN = 4) (Figure 7.7(c) and (d)). The E_g of $Ca_{1-x}Li_xAl_{1-x}Si_{1+x}N_3$ solid-solution was increasing from ~4.91 eV for the $x = 0$ to 5.08 eV for $x = 0.22$ because of $LiSi_2N_3$ with a larger E_g than $CaAlSiN_3$. Thermal stability is gradually enhanced with increasing LiS_2N_3 content. This enhancement of thermal stability originating from the wider band-gap that can be further separated the distance between the excitation state of Eu^{2+} and the bottom of the conduction band, minimizing photoionization mechanism.

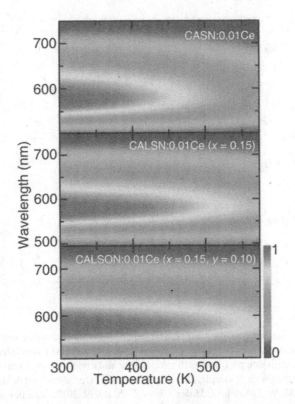

FIGURE 7.9 Temperature-dependent spectra of $CaAlSiN_3{:}0.01Ce^{3+}$ (CASN:0.01Ce), $Ca_{1-x}Li_xAl_{1-x}Si_{1+x}N_3{:}0.01Ce^{3+}$ (CALSN:0.01Ce^{3+}, $x = 0.15$), and $Ca_{0.85-y}Li_{0.15}Al_{0.85-y}Si_{1.15+y}N_{3-y}O_y{:}0.01Ce^{3+}$ (CALSON:0.01Ce^{3+}, $x = 0.15$, $y = 0.10$). (S. You, S. Li, L. Wang, et al., *Chem. Eng. J.*, 404, 126575, 2021. With permission.)

In addition, a ternary solid-solution of $CaAlSiN_3$–$xLiSi_2N_3$–ySi_2N_2O was developed and investigated to understand the structure-property relationship of a phosphor material. The TQ of $CaAlSiN_3{:}0.01Ce^{3+}$ (CASN:0.01Ce), $Ca_{1-x}Li_xAl_{1-x}Si_{1+x}N_3{:}0.01Ce^{3+}$ (CALSN:0.01Ce^{3+}, $x = 0.15$), and $Ca_{0.85-y}Li_{0.15}Al_{0.85-y}Si_{1.15+y}N_{3-y}O_y{:}0.01Ce^{3+}$ (CALSON:0.01Ce^{3+}, $x = 0.15$, $y = 0.10$) is displayed in Figure 7.9. CALSON:0.01Ce^{3+} showed higher thermal stability than CALSN:0.01Ce^{3+}, which is due to weak covalency that occurs when nitrogen ions are partially substituted with oxygen ions along with the increasing of E_g.

To further validate the abovementioned concept, the β-$Si_{6-z}Al_zO_zN_{8-z}{:}Eu^{2+}$ ($z = 0.050$, 0.0075, and 0.125) was combined to elucidate the role of Al/O content (z) in luminescence and thermal stability properties [68]. The Eu state according to the z content change was confirmed by X-ray absorption near-edge structure of the Eu L$_3$ edge, and it was found that the observed luminescence properties were not the difference in Eu valence. As increase z content, the emission spectra were red-shifted with broadening the emission spectra which is due to distortion of [Eu(N/O)$_9$] polyhedra. Figure 7.10 displays temperature-dependent emission spectra that exhibit an enhanced thermal stability with increasing Al/O content. This is because of, upon increasing of Al/O content in the host material that causes to widen the E_g; as a result, the photoionization process was suppressed and displays high thermal stability.

In addition of available theories to improve the thermal stability of phosphor in the literature, Liu's group proposed a novel strategy to improve the thermal stability of by causing cation disorder in a host lattice. For instance, $Sr_{2-x}Si_5N_8{:}Eu^{2+}$ ($x = Ca^{2+}$ and Ba^{2+} ions) phosphor shows a cation disorder effect upon replacing of Sr^{2+} ions by Ca^{2+} ions and Ba^{2+} ions, which have the similar ionic size of Sr^{2+} ions [69]. To quantify a disorder effect, the cation size variance $\left[\sigma^2 = \langle r^2 \rangle - \langle r \rangle^2\right]$ was

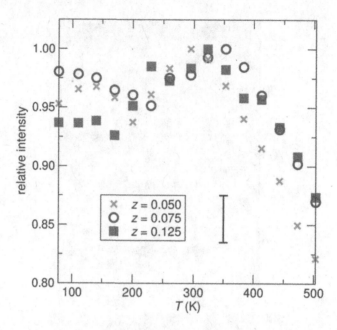

FIGURE 7.10 Relative intensities of the temperature-dependent emission demonstrate higher relative emission intensity at higher temperatures for $z = 0.075$ and $z = 0.125$. Relative emission intensity was computed by taking the integral of the emission profile at each temperature and dividing by the maximum value for a given sample, which was different for each sample. The shared standard error (displayed next to the legend) is ±0.02. (C. Cozzan, G. Laurita, M. W. Gaultois. *J. Mater. Chem. C,* 5, 10039, 2017. With permission.)

varied as a function x. Moreover, the value of σ^2 (change in lattice strain) also increases in proportion to x. The local distortions of the coordination environments around Eu^{2+} ions with increasing of x content could be attributed to enhancement of thermal stability. This is due to increasing lattice disorder create a large number of deep traps, which can compensate thermal losses by supplying excessive energy to the luminescent centers. Taking an advantage from this concept, $Lu_3Al_5O_{12}$–$Lu_2CaMg_2Si_3O_{12}$ solid-solution phosphor was developed with substitution of Ba^{2+} ions in dodecahedra and octahedral sites in garnet structure [70]. With increasing of Ba^{2+} ions concentration, the calculated σ^2 values increased. In particular, the calculated σ^2 was changed larger at the substitution in the octahedral site than at the substitution in the dodecahedral site because ionic size of Ba^{2+} ion is larger than the octahedral site. Therefore, a large number of Ba^{2+} ions are preferred to occupy dodecahedral sites than octahedral sites. In previous studies, co-doping of Ba^{2+} ions in the host structure has helped to increase cation disorder that assisted to reduce Eu^{3+} ions into Eu^{2+} ions, which can work a luminescence killer, in the phosphor host. However, the TQ of Ba^{2+}-substituted sample was improved due to the reduction of probability of thermal ionization process by E_g tuning of a host material. Finally, the optimized sample has shown a high thermal stability (95%@150°C). Later, Xia's group developed the mineral-inspired prototype evolution for discovery of new phosphors. Eu^{2+}-doped $RbNa_3(Li_3(SiO_4))_4$, $RbNa_2K(Li_3(SiO_4))_4$, and $CsNa_2K(Li_3(SiO_4))_4$ silicate phosphors [58, 71, 72] retain 90% of original PL efficiency as a function of temperature. This is due to of their condensed crystal structure and high Θ_D. Later, Ce^{3+}-doped $CsNa_2K(LiSiO_4)_4$ (CNKLSO) cuboid phosphor was developed that shows remarkably, no drop in the emission intensity even at 200°C, making the first Ce^{3+}-based zero-thermal-quenching phosphor [73]. The origin of zero-TQ behavior was mainly attributed to the structural features such as zero volume change, symmetry transformation (low symmetry to high symmetry) of Na and preserving the symmetry of K at elevated temperature can suppress the additional non-radiative emission paths, owing to that CNLKSO:Ce^{3+} preserves its emission properties during the heating process.

7.3.3 POLYMORPHIC NATURE OF A HOST MATERIAL

A phosphor material which is capable of modifying its phase during the heating is advantageous for retaining its original emission properties at elevated temperature. However, finding such kind of material is unusual. Although Im's group reported a new blue-emitting phosphor (Figure 7.11(a)) which shows virtually no TQ up to 200°C. Moreover, the $Na_3Sc_2(PO_4)_3$ (NSPO) phosphor with lower Eu^{2+} concentration (0.01 and 0.03 mol) exhibited anti-TQ (Figure 7.11(b)), which could be due to changing the local environment as a function of temperature. To understand local environment of this phosphor material, temperature-dependent X-ray diffraction, temperature-dependent solid-state nuclear magnetic resonance, and temperature-dependent Raman spectroscopy techniques were

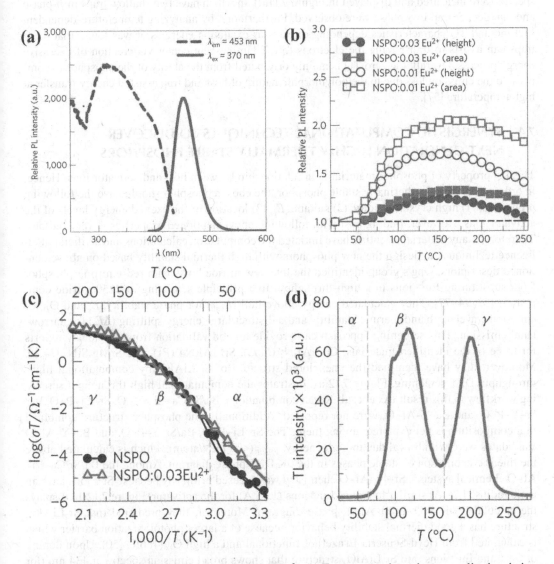

FIGURE 7.11 (a) PLE and PL spectra of NSPO:0.03Eu²⁺. (b) Temperature-dependent normalized emission spectra of NSPO:0.01Eu²⁺ and NSPO:0.03Eu²⁺ (in terms of emission area and height) under 370 nm excitation upon heating from 25 to 200°C. (c) Temperature-dependent impedance spectra of the NSPO host and NSPO:0.03Eu²⁺ upon heating from 25 to 230°C, and subsequent cooling to 25°C. (d) Thermoluminescence curves of NSPO host in the temperature range 25–300°C. (Y. H. Kim, P. Arunkumar, B. Y. Kim, et al., *Nat. Mater.*, 16, 543–550, 2017. With permission.)

employed. From these characterizations, it was found that NSPO structure has phase transition from α-phase (monoclinic) to β-phase (hexagonal), and to γ-phase (hexagonal) under 200°C. Besides, from impedance analysis it was found that the presence of Eu^{2+} ions in NSPO structure shows a high ionic conductivity as shown in Figure 7.11(c). In both cases, the conductivity increases with temperature in the following order: α-phase (low defect crystals), weakly disordered ion-conducting β-phase, and highly disordered superionic γ-phase, which is owing to an increase in Na^+ disorder in the lattice. In addition, NSPO:0.07Eu^{2+} (higher Eu^{2+} concentration) phosphor shows a zero-TQ property (the stabilized emission intensity) because of its highly disordered-disordered transition of Na^+ ions which forms relatively low concentration traps yielding a stable blue emission at elevated temperatures. For further confirmation for the existence of trap states in the E_g of the material, TL spectra were measured and displayed in Figure 7.11(d), in which have two shallow traps in β-phase and one deeper trap in γ-phase were observed. Furthermore, by analyzing temperature-dependent lifetime and TQ characteristics depending on various excitation energies, it was found that these traps can assists to transferring the electrons to the luminescent center via creation of excessive energy process. Overall, the zero TQ is mainly originated from the ability of the phosphor to compensate the emission loss due to the polymorphic nature of host and trap assisted energy transfer at high temperature [57].

7.4 SYNERGISTIC COMPUTATIONAL TECHNIQUES TO DISCOVER NEXT-GENERATION HIGHLY THERMALLY STABLE PHOSPHORS

The TQ property of phosphor material is in relationship between host and activator ions. Hence, in order to develop the thermally stable phosphor, the chosen phosphor should have the following properties: (1) high crystal rigidity, (2) suitable E_g, (3) location of the excited energy levels of the activator ions. However, finding host with following abovementioned properties is very tedious. Therefore, many material scientists have initiated the computation calculations and artificial intelligence techniques to design the new phosphors with high thermal stability based on abovementioned descriptors. Ong's group identified the five new nitride hosts with red-emitting phosphor from substituting Eu^{2+} ions in a high-throughput first-principle screening of 2259 nitride compounds [53, 74–77]. They calculated energy above hull (E_{hull}) for phase stability, E_g and Θ_D for emission wavelength and thermal stability, and a distinct large energy splitting (ΔE_s) for narrowband emission. This screening approach can provide a good validation from previously reports for three of the identified materials, $CaLiAl_3N_4$ ($I4_1/a$), $SrLiAl_3N4$ ($P\bar{1}$), and $SrMg_3SiN_4$ ($I4_1/a$). Moreover, they have identified the unexplored structure of Sr_2LiAlO_4 by computational high-throughput DFT screening. Figure 7.12(a) illustrates the computational high-throughput screening workflow in this result. As a result, seven combinations, Ba/Sr/Ca–Li–Al–O, Sr–Li–P–O, Ba/Sr–Y–P–O, and Ba–Y–Al–O, were not reported. Additionally, the phosphor structure combined in a composition is not explored among them. The Sr–Li–P–O, Ba/Sr–Y–P–O, and Ba–Y–Al–O candidates were withdrawal due to high energy, E_{hull} (<35 meV/atom), which is calculated above the linear combination of stable phases in the 0K DFT phase diagram. Among the Ba/Sr/Ca–Li–Al–O chemical systems, Sr–Li–Al–O chemistry was focused primarily because Sr^{2+} ions have an ionic radius (1.26 Å) similar to that of Eu^{2+} ions (1.25 Å) for the activator. Figure 7.12(b) displays the calculated 0K $SrO–Li_2O–Al_2O_3$ phase diagram. Moreover, they predicted that Sr_2LiAlO_4 structure has a good thermal stability behavior because of a larger photoionization barrier which is calculated from Heyd-Scuseria-Ernzerhof functional and a high Θ_D (466K) [50]. Upon doping of Ce^{3+} and Eu^{2+} ions into Sr_2LiAlO_4 structure that shows broad emission spectra at 434 nm (for Ce^{3+} ions) and 512 nm emission (for Eu^{2+} ions), respectively. Besides, Sr_2LiAlO_4:Eu^{2+}/Ce^{3+} phosphor exhibits high thermal stability and retains their 88% of its original intensity at 150°C. From these results, one can observe that the unexplored phosphor with thermal stability compositions successfully screened using computational techniques.

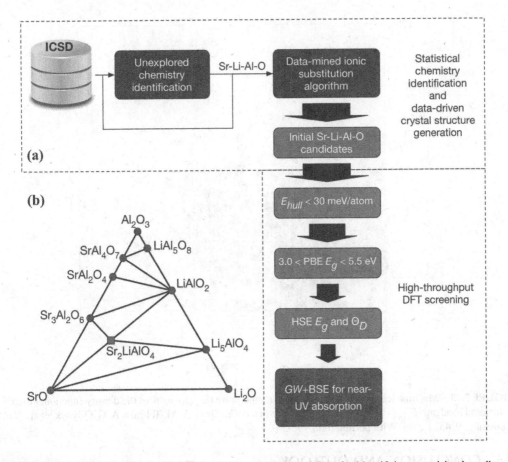

FIGURE 7.12 (a) Computational high-throughput screening workflow for identifying promising broadband emitters. E_g is the host bandgap calculated with both Perdew-Burke-Ernzerhof (PBE) functional and Heyd-Scuseria-Ernzerhof (HSE) functional. Θ_D is calculated Debye temperature. GW + BSE denotes the Bethe-Salpeter equation (BSE) calculation performed on top of G_0W_0. (b) Calculated 0K SrO–Li$_2$O–Al$_2$O$_3$ phase diagram. Blue circles, known stable phases in the Materials Project database; red square, new stable quaternary phase, Sr$_2$LiAlO$_4$. (Z. Wang, J. Ha, Y. H. Kim, et al., *Joule*, 2, 914–926, 2018. With permission.)

However, the high-throughput DFT screening method has limitation because DFT cannot handle the atomic disordered complex inorganic solids and the unit cells are currently allowed less than a few hundred atoms. Hence, acceleration of machine learning offers a new alternative that can significantly drive material development beyond DFT capabilities. Brgoch and coworkers developed a sorting diagram composed of the plotted results the 2071 compounds as shown in Figure 7.13. The darker regions of this plot represent a higher density of potential phosphor hosts. Furthermore, breaking down of this sorting diagram into compositional information reveals about the several trends related to the potential phosphor hosts. Furthermore, breaking down of this sorting diagram into compositional information reveals the several trends related to the potential phosphor hosts. Among the other host materials, borate host material has a wide E_g and high Θ_D. Therefore, Eu^{2+} doped borate phosphor of NaBaB$_9$O$_{15}$ was synthesized using solid-state reaction. Moreover, NaBaB$_9$O$_{12}$:Eu^{2+} phosphor shows a bright blue emission with high PL efficiency and high thermal stability properties under ultraviolet excitation. Hence, these experimental results suggest that machine learning is an important tool for finding the next generation of inorganic phosphors.

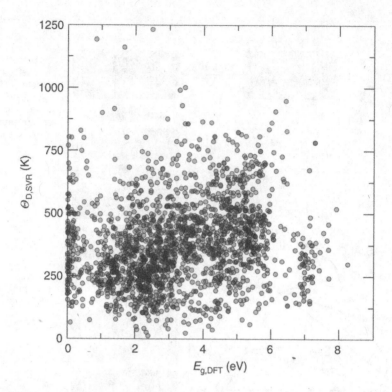

FIGURE 7.13 Machine learning predicted Debye temperature ($\Theta_{D,SVR}$) against the density functional theory calculated bandgap ($E_{g,DFT}$) for 2071 compounds predicted. (Y. Zhuo, A. M. Tehrani, A. O. Oliynyk, et al., *Nat. Commun.*, 9, 4377 2018. With permission.)

7.5 CONCLUSION AND OUTLOOK

In this chapter, we have discussed the various parameters that can influence the TQ at elevated temperatures. Also, various methodologies to develop thermally stable phosphors are thoroughly explained. From the proposed theories, it is evident that mechanisms of TQ of luminescence are a highly thought-provoking and ambiguous subject, because different types of radiative and nonradiative relaxation paths are involved. Furthermore, we envisage a great chance for new basic science to develop various types of local vibrational modes on the PL properties. Significantly, identifying the nature of these vibrational modes would provide insights into which local coordination environments are prone to show TQ of luminescence. Therefore, such knowledge might have useful for developing new efficient phosphors. Although we need to thank the chemists, physicists, and material sciences because of their continuous attempts to understand the TQ mechanisms of phosphor materials. Moreover, there is no feasible method for improving or designing new thermally stable phosphor because of the complex chemical nature of bonding in inorganic solid-state materials that make us challenging to improve the thermal stability of phosphors. Therefore, it is only possible through the combined use of new theoretical and experimental approaches, as well as the investigation of completely new materials that can provide a clear mechanistic understanding of TQ mechanisms. In the present research, many of the aspects related to TQ needs to be improved, including but not inadequate to:

1. Designing a new phosphor material with a high structural rigidity and suitable bang gap of host materials are one of the powerful descriptors to discover a thermally stable phosphor. Introduction of cation mismatchement is also one of the effective tools to improve the thermal stability of phosphor materials up to some extent.

2. Energy transfer from defect centers to luminescent centers is an effective approach to sustain the original luminescence properties, even it can show the zero TQ properties. However, the nature of these good defects remains unclear and it should be elucidated. For suppose, if one can understand the physical methodologies for creation of such defects, a large number of phosphors will comprehend a zero TQ by creating suitable traps in host lattice.

3. Finding a phosphor hosts using mining or machine learning techniques by constructing the sort diagram is also effective approach to develop highly thermally stable phosphors.

Ample research progress has been carried out over the past years to improve or design new thermally stable phosphors for high-power white light-emitting diodes (WLEDs) and LD applications. However, based on the current scientific research and progress of LEDs, it requires a further effort to improve or increasing the current library of highly thermally stable phosphors. We hope further research on thermally stable phosphors should be continued intensively to rapidly develop solid-state lighting technology. We are hopeful that this will motivate further attempts to pay novel methodological methods for evolving a better understanding of the TQ behavior of luminescent materials.

REFERENCES

1. Chen L, Lin CC, Yeh CW, and Liu RS (2010) Light converting inorganic phosphors for white light-emitting diodes. *Materials* 3: 2172.
2. Denault KA, Cantore M, Nakamura S, DenBaars SP, and Seshadri R (2013) Efficient and stable laser-driven white lighting. *AIP Adv.* 3: 072107.
3. Tian Y (2014) Development of phosphors with high thermal stability and efficiency for phosphor-converted LEDs. *J. Solid State Light.* 1(1): 1–15.
4. McKittrick J, and Shea-Rohwer LE (2014) Review: Down conversion materials for solid-state lighting. *J. Am. Ceram. Soc.* 97(5): 1327–1352.
5. Tsao JY, Crawford MH, Coltrin ME, Fischer AJ, Koleske DD, Subramania GS, Wang GT, Wierer JJ, and Karlicek Jr RF (2014) Toward smart and ultra-efficient solid-state lighting. *Adv. Opt. Mater.* 2: 809.
6. Piquette A, Bergbauer W, Galler B, and Mishra K (2015) On choosing phosphors for near-UV and blue LEDs for white light. *ECS J. Solid State Sci. Technol.* 5: R3146.
7. Xia Z, and Liu Q (2016) Progress in discovery and structural design of color conversion phosphors for LEDs. *Prog. Mater. Sci.* 84: 59.
8. Blasse G, and Bril A (1970) Characteristic luminescence. *Philips Tech. Rev.* 31: 303.
9. Poncé S, Jia Y, Giantomassi M, Mikami M, and Gonze X (2016) Understanding thermal quenching of photoluminescence in oxynitride phosphors from first principles. *J. Phys. Chem. C* 120: 4040.
10. Zhang Y, and Li Y (2004) Synthesis and characterization of monodisperse doped ZnS nanospheres with enhanced thermal stability. *J. Phys. Chem. B* 108: 17805.
11. Yu M, Lin J, and Fang J (2005) Silica spheres coated with $YVO_4:Eu^{3+}$ layers via sol–gel process: a simple method to obtain spherical core–shell phosphors. *Chem. Mater.* 17: 1783.
12. Li Z, and Zhang Y (2006) Monodisperse silica-coated polyvinylpyrrolidone/$NaYF_4$ nanocrystals with multicolor upconversion fluorescence emission. *Angew. Chem.* 118: 7896.
13. Allen SC, and Steckl AJ (2008) A nearly ideal phosphor-converted white light-emitting diode. *Appl. Phys. Lett.* 92: 128.
14. Lee YK, Lee JS, Heo J, Im WB, and Chung WJ (2012) Phosphor in glasses with Pb-free silicate glass powders as robust color-converting materials for white LED applications. *Opt. Lett.* 37: 3276.
15. Zhang R, Lin H, Yu Y, Chen D, Xu J, and Wang Y (2014) A new-generation color converter for high-power white LED: transparent Ce^{3+}: YAG phosphor-in-glass. *Laser Photonics Rev.* 8: 158.
16. Dorenbos P (2012) A review on how lanthanide impurity levels change with chemistry and structure of inorganic compounds. *ECS J. Solid State Sci. Technol.* 2: R3001.
17. Chiang CC, Tsai MS, and Hon MH (2008) Luminescent properties of cerium-activated garnet series phosphor: structure and temperature effects. *J. Electrochem. Soc.* 155: B517.
18. George NC, Pell AJ, Dantelle G, Page K, Llobet A, Balasubramanian M, Pintacuda G, Chmelka BF, and Seshadri R (2013) Local environments of dilute activator ions in the solid-state lighting phosphor Y_{3-x} $Ce_xAl_5O_{12}$. *Chem. Mater.* 25: 3979.

19. Lin YC, Bettinelli M, and Karlsson M (2019) Unraveling the mechanisms of thermal quenching of luminescence in Ce^{3+}-doped garnet phosphors. *Chem. Mater.* 31(11): 3851–3862.

20. Di Bartolo B. 2013. *Advances in nonradiative processes in solids*. Vol. 249: Springer Science & Business Media, Berlin/Heidelberg, Germany.

21. Wei Y, Cao L, Lv L, Li G, Hao J, Gao J, Su C, Lin CC, Jang HS, and Dang P (2018) Highly efficient blue emission and superior thermal stability of $BaAl_{12}O_{19}$:Eu^{2+} phosphors based on highly symmetric crystal structure. *Chem. Mater.* 30: 2389.

22. Li L, Peng M, Viana B, Wang J, Lei B, Liu Y, Zhang Q, and Qiu J (2015) Unusual concentration induced antithermal quenching of the Bi^{2+} emission from $Sr_2P_2O_7$:Bi^{2+}. *Inorg. Chem.* 54: 6028.

23. Verstraete R, Sijbom HF, Joos JJ, Korthout K, Poelman D, Detavernier C, and Smet PF (2018) Red Mn^{4+}-doped fluoride phosphors: why purity matters. *ACS Appl. Mater. Interfaces* 10: 18845.

24. Li S, Wang L, Tang D, Cho Y, Liu X, Zhou X, Lu L, Zhang L, Takeda T, and Hirosaki N (2018) Achieving high quantum efficiency narrow-band β-sialon: Eu^{2+} phosphors for high-brightness LCD backlights by reducing the Eu^{3+} luminescence killer. *Chem. Mater.* 30: 494.

25. Zhao M, Xia Z, Huang X, Ning L, Gautier R, Molokeev MS, Zhou Y, Chuang Y-C, Zhang Q, and Liu Q (2019) Li substituent tuning of LED phosphors with enhanced efficiency, tunable photoluminescence, and improved thermal stability. *Sci. Adv.* 5: eaav0363.

26. Fan X, Chen W, Xin S, Liu Z, Zhou M, Yu X, Zhou D, Xu X, and Qiu J (2018) Achieving long-term zero-thermal-quenching with the assistance of carriers from deep traps. *J. Mater. Chem. C* 6: 2978.

27. Peng L, Qinping Q, Cao S, Chen W, and Han T (2020) Enhancement of luminescence and thermal stability in $Sr_{2.92-x}Tm_xSiO_5$:Eu^{2+} for white LEDs. *Phys. Scr.* 96: 025704.

28. Wei Y, Yang H, Gao Z, Liu Y, Xing G, Dang P, Kheraif AAA, Li G, Lin J, and Liu RS (2020) Strategies for designing antithermal-quenching red phosphors. *Adv. Sci.* 7: 1903060.

29. Brgoch J, DenBaars SP, and Seshadri R (2013) Proxies from ab initio calculations for screening efficient Ce^{3+} phosphor hosts. *J. Phys. Chem. C* 117: 17955.

30. Francisco E, Recio J, Blanco M, Pendás AM, and Costales A (1998) Quantum-mechanical study of thermodynamic and bonding properties of MgF_2. *J. Phys. Chem. A* 102: 1595.

31. Qiao J, Zhao J, Liu Q, and Xia Z (2019) Recent advances in solid-state LED phosphors with thermally stable luminescence. *J. Rare Earths* 37: 565.

32. Wang L, Xie R-J, Suehiro T, Takeda T, and Hirosaki N (2018) Down-conversion nitride materials for solid state lighting: recent advances and perspectives. *Chem. Rev.* 118: 1951.

33. Xie R-J, and Hirosaki N (2007) Silicon-based oxynitride and nitride phosphors for white LEDs—a review. *Sci. Technol. Adv. Mater.* 8: 588.

34. Hoerder GJ, Seibald M, Baumann D, Schröder T, Peschke S, Schmid PC, Tyborski T, Pust P, Stoll I, and Bergler M (2019) $Sr[Li_2Al_2O2N_2]$:Eu^{2+}—a high performance red phosphor to brighten the future. *Nat. Commun.* 10: 1.

35. Denault KA, Brgoch J, Kloß SD, Gaultois MW, Siewenie J, Page K, and Seshadri R (2015) Average and local structure, Debye temperature, and structural rigidity in some oxide compounds related to phosphor hosts. *ACS Appl. Mater. Interfaces* 7: 7264.

36. Brgoch J, Borg CK, Denault KA, Mikhailovsky A, DenBaars SP, and Seshadri R (2013) An efficient, thermally stable cerium-based silicate phosphor for solid state white lighting. *Inorg. Chem.* 52: 8010.

37. Im WB, George N, Kurzman J, Brinkley S, Mikhailovsky A, Hu J, Chmelka BF, DenBaars SP, and Seshadri R (2011) Efficient and color-tunable oxyfluoride solid solution phosphors for solid-state white lighting. *Adv. Mater.* 23: 2300.

38. Im WB, Brinkley S, Hu J, Mikhailovsky A, DenBaars SP, and Seshadri R (2010) $Sr_{2.975-x}Ba_xCe_{0.025}AlO_4F$: a highly efficient green-emitting oxyfluoride phosphor for solid state white lighting. *Chem. Mater.* 22: 2842.

39. George NC, Birkel A, Brgoch J, Hong BC, Mikhailovsky AA, Page K, Llobet A, and Seshadri R (2013) Average and local structural origins of the optical properties of the nitride phosphor $La_{3-x}Ce_xSi_6N_{11}$ ($0 < x \leq 3$). *Inorg. Chem.* 52: 13730.

40. Han J, Hannah M, Piquette A, Talbot J, Mishra K, and McKittrick J (2012) Nano-and submicron sized europium activated silicate phosphors prepared by a modified co-precipitation method. *ECS J. Solid State Sci. Technol.* 1: R98.

41. Ha J, Wang Z, Novitskaya E, Hirata GA, Graeve OA, Ong SP, and McKittrick J (2016) An integrated first principles and experimental investigation of the relationship between structural rigidity and quantum efficiency in phosphors for solid state lighting. *J. Lumin.* 179: 297.

42. Duan CJ, Wang XJ, Otten WM, Delsing AC, Zhao JT, and Hintzen H (2008) Preparation, electronic structure, and photoluminescence properties of Eu^{2+}-and Ce^{3+}/Li^+-activated alkaline earth silicon nitride $MSiN_2$ (M = Sr, Ba). *Chem. Mater.* 20: 1597.

43. Uheda K, Hirosaki N, Yamamoto Y, Naito A, Nakajima T, and Yamamoto H (2006) Luminescence properties of a red phosphor, $CaAlSiN_3:Eu^{2+}$, for white light-emitting diodes. *Electrochem. Solid-State Lett.* 9: H22.

44. Müller M, Volhard M-F, and Jüstel T (2016) Photoluminescence and afterglow of deep red emitting $SrSc_2O_4: Eu^{2+}$. *RSC Adv.* 6(10): 8483–8488.

45. Zhang X, Tang X, Zhang J, Wang H, Shi J, and Gong M (2010) Luminescent properties of $Sr_2MgSi_2O_7:Eu^{2+}$ as blue phosphor for NUV light-emitting diodes. *Powder Technol.* 204: 263.

46. Wei Z, Xinxu M, Zhang M, Yi L, and Zhiguo X (2015) Effect of different RE dopants on phosphorescence properties of $Sr_2Al_2SiO_7:Eu^{2+}$ phosphors. *J. Rare Earths* 33: 700.

47. Brgoch J, Hasz K, Denault KA, Borg CK, Mikhailovsky AA, and Seshadri R (2015) Data-driven discovery of energy materials: efficient $BaM_2Si_3O_{10}:Eu^{2+}$ (M = Sc, Lu) phosphors for application in solid state white lighting. *Faraday Discuss.* 176: 333.

48. Komukai T, Takatsuka Y, Kato H, and Kakihana M (2015) Luminescence properties of $BaZrSi3O_9$:Eu synthesized by an aqueous solution method. *J. Lumin.* 158: 328.

49. Qiao J, Amachraa M, Molokeev M, Chuang YC, Ong SP, Zhang Q, and Xia Z (2019) Engineering of $K_3YSi_2O_7$ to tune photoluminescence with selected activators and site occupancy. *Chem. Mater.* 31: 7770.

50. Wang Z, Ha J, Kim YH, Im WB, McKittrick J, and Ong SP (2018) Mining unexplored chemistries for phosphors for high-color-quality white-light-emitting diodes. *Joule* 2: 914.

51. Duke AC, Hariyani S, and Brgoch J (2018) $Ba_3Y_2B_6O_{15}:Ce^{3+}$—a high symmetry, narrow-emitting blue phosphor for wide-gamut white lighting. *Chem. Mater.* 30: 2668.

52. Zhang X, Song J, Zhou C, Zhou L, and Gong M (2014) High efficiency and broadband blue-emitting $NaCaBO_3:Ce^{3+}$ phosphor for NUV light-emitting diodes. *J. Lumin.* 149: 69.

53. Zhuo Y, Tehrani AM, Oliynyk AO, Duke AC, and Brgoch J (2018) Identifying an efficient, thermally robust inorganic phosphor host via machine learning. *Nat. Commun.* 9: 1.

54. Lin CC, Xiao ZR, Guo GY, Chan TS, and Liu RS (2010) Versatile phosphate phosphors $ABPO_4$ in white light-emitting diodes: collocated characteristic analysis and theoretical calculations. *J. Am. Chem. Soc.* 132: 3020.

55. Zhang S, Nakai Y, Tsuboi T, Huang Y, and Seo HJ (2011) The thermal stabilities of luminescence and microstructures of Eu^{2+}-doped $KBaPO_4$ and $NaSrPO_4$ with β-K_2SO_4 type structure. *Inorg. Chem.* 50: 2897.

56. Komuro N, Mikami M, Shimomura Y, Bithell EG, and Cheetham AK (2015) Synthesis, structure and optical properties of cerium-doped calcium barium phosphate—a novel blue-green phosphor for solid-state lighting. *J. Mater. Chem. C* 3: 204.

57. Kim YH, Arunkumar P, Kim BY, Unithrattil S, Kim E, Moon S-H, Hyun JY, Kim KH, Lee D, and Lee J-S (2017) A zero-thermal-quenching phosphor. *Nat. Mater.* 16: 543.

58. Zhao M, Liao H, Ning L, Zhang Q, Liu Q, and Xia Z (2018) Next-generation narrow-band green-emitting $RbLi(Li_3SiO_4)_2:Eu^{2+}$ phosphor for backlight display application. *Adv. Mater.* 30: 1802489.

59. Pust P, Weiler V, Hecht C, Tücks A, Wochnik AS, Henß A-K, Wiechert D, Scheu C, Schmidt PJ, and Schnick W (2014) Narrow-band red-emitting $Sr[LiAl_3N_4]:Eu^{2+}$ as a next-generation LED-phosphor material. *Nat. Mater.* 13: 891.

60. Mikami M (2013) Response function calculations of $Ba_3Si_6O_{12}N_2$ and $Ba_3Si_6O_9N_4$ for the understanding of the optical properties of the Eu-doped phosphors. *Opt. Mater.* 35: 1958.

61. Poncé S, Jia Y, Giantomassi M, Mikami M, and Gonze X (2016) Understanding thermal quenching of photoluminescence in oxynitride phosphors from first principles. *J. Phys. Chem. C* 120: 4040.

62. Ivanovskikh K, Ogiegło J, Zych A, Ronda C, and Meijerink A (2012) Luminescence temperature quenching for Ce^{3+} and Pr^{3+} 5d-4f emission in YAG and LuAG. *ECS J. Solid State Sci. Technol.* 2: R3148.

63. Ueda J, Tanabe S, and Nakanishi T (2011) Analysis of Ce^{3+} luminescence quenching in solid solutions between $Y_3Al_5O_{12}$ and $Y_3Ga_5O_{12}$ by temperature dependence of photoconductivity measurement. *J. Appl. Phys.* 110: 053102.

64. Tsai YT, Chiang CY, Zhou W, Lee JF, Sheu HS, and Liu RS (2015) Structural ordering and charge variation induced by cation substitution in $(Sr,Ca)AlSiN_3$:Eu phosphor. *J. Am. Chem. Soc.* 137: 8936.

65. Wang SS, Chen WT, Li Y, Wang J, Sheu HS, and Liu RS (2013) Neighboring-cation substitution tuning of photoluminescence by remote-controlled activator in phosphor lattice. *J. Am. Chem. Soc.* 135: 12504.

66. Zhang Y, Li X, Li K, Lian H, Shang M, and Lin J (2015) Interplay between local environments and photoluminescence of Eu^{2+} in $Ba_2Zr_2Si_3O_{12}$: blue shift emission, optimal bond valence and luminescence mechanisms. *J. Mater. Chem. C* 3: 3294.

67. Wang L, Xie RJ, Li Y, Wang X, Ma CG, Luo D, Takeda T, Tsai YT, Liu RS, and Hirosaki N (2016) $Ca_{1-x}Li_xAl_{1-x}Si_{1+x}N_3$ Eu^{2+} solid solutions as broadband, color-tunable and thermally robust red phosphors for superior color rendition white light-emitting diodes. *Light Sci. Appl.* 5: e16155.

68. Cozzan C, Laurita G, Gaultois MW, Cohen M, Mikhailovsky AA, Balasubramanian M, and Seshadri R (2017) Understanding the links between composition, polyhedral distortion, and luminescence properties in green-emitting β-$Si_{6-z}Al_zO_zN_{8-z}$:Eu^{2+} phosphors. *J. Mater. Chem. C* 5: 10039.

69. Lin CC, Tsai YT, Johnston HE, Fang MH, Yu F, Zhou W, Whitfield P, Li Y, Wang J, and Liu RS (2017) Enhanced photoluminescence emission and thermal stability from introduced cation disorder in phosphors. *J. Am. Chem. Soc.* 139: 11766.

70. Kim YH, Kim HJ, Ong SP, Wang Z, and Im WB (2020) Cation-size mismatch as a design principle for enhancing the efficiency of garnet phosphors. *Chem. Mater.* 32: 3097.

71. Liao H, Zhao M, Molokeev MS, Liu Q, and Xia Z (2018) Learning from a mineral structure toward an ultra-narrow-band blue-emitting silicate phosphor $RbNa_3(Li_3SiO_4)_4$:Eu^{2+}. *Angew. Chem.* 130: 11902.

72. Zhao M, Cao K, Liu M, Zhang J, Chen R, Zhang Q, and Xia Z (2020) Dual-shelled RbLi (Li_3SiO_4) 2:Eu^{2+}@Al_2O_3@ODTMS phosphor as a stable green emitter for high-power LED Backlights. *Angew. Chem.* 132(31): 13038–13043.

73. Viswanath N, Grandhi GK, Huu HT, Choi H, Kim HJ, Kim SM, Kim HY, Park CJ, and Im WB (2020) Zero-thermal-quenching and improved chemical stability of a UCr_4C_4-type phosphor via crystal site engineering. *Chem. Eng. J.* 420: 127664.

74. Wang Z, Chu IH, Zhou F, and Ong SP (2016) Electronic structure descriptor for the discovery of narrow-band red-emitting phosphors. *Chem. Mater.* 28: 4024.

75. Wang Z, Ye W, Chu IH, and Ong SP (2016) Elucidating structure–composition–property relationships of the β-SiAlON:Eu^{2+} phosphor. *Chem. Mater.* 28: 8622.

76. Amachraa M, Wang Z, Chen C, Hariyani S, Tang H, Brgoch J, and Ong SP (2020) Predicting thermal quenching in inorganic phosphors. *Chem. Mater.* 32: 6256.

77. Hariyani S, Duke AC, Krauskopf T, Zeier WG, and Brgoch J (2020) The effect of rare-earth substitution on the Debye temperature of inorganic phosphors. *Appl. Phys. Lett.* 116: 051901.

8 Combinatorial Chemistry Approach and the Taguchi Method for Phosphors

Lei Chen and Chen Gao

CONTENTS

8.1 COMBINATORIAL CHEMISTRY APPROACH

8.1.1 BRIEF HISTORY AND DEVELOPMENT OF COMBINATORIAL CHEMISTRY

Combinatorial chemistry, which is a branch of modern chemistry, has emerged as a new method to rapidly synthesize vast numbers of samples in a parallel way and then high-throughput characterize them to screen for desirable properties. Initially, combinatorial chemistry was developed by biological chemists to accelerate drug discovery in pharmaceutical research by creating and making use of diversity [1]. As is known, the discovery of a new medicine is time-consuming and very expensive.

DOI: 10.1201/9781003098669-8

Typically, the research and development (R&D) period of a medicine is about 10–15 years, and the total cost toward the R&D of each drug is about US$897 million to US$1.9 billion on an average, owing to the high costs of R&D and human clinical tests [2]. To reduce the R&D cost and duration, chemists developed variant methods to synthesize multiple components to accelerate the discovery of new drugs by increasing the possibility of new clinical candidates [3].

The origin of combinatorial chemistry could date back to 1963, when a biochemist from Rockefeller University named Robert Bruce Merrifield developed the portioning-mixing (also named split-mix) solid-phase synthesis method for multicomponent peptides [4]. Currently, the technique of solid-phase synthesis, invented by Merrifield, has become a standard method for the synthesis of peptides and nucleotides. Consequently, Robert Bruce Merrifield has been awarded with the Nobel Prize in Chemistry in 1984 for his invention of solid-phase peptide synthesis [5].

The concept of combinatorial chemistry was conceived and developed in the mid-1980s due to the co-contribution of many scientists. A series of methods, such as Frank's segmental solid supports [6], Geysen's multi-pin technique [7], Houghten's tea-bag technology [8], and Smith's phage-display peptide method (one-phage one-peptide) [9], were developed for multicomponent synthesis. Later in 1991, Lam [10] reported the one-bead-one-compound strategy for the synthetic peptide library, and Houghten [11] pioneered the solution-phase synthesis of combinatorial peptide libraries. Subsequently, in 1992, Bunin and Ellman reported a small-molecule combinatorial library for the first time [12]. In the 1990s, we witnessed a surge in combinatorial chemistry that penetrated the academic and industrial laboratories [13].

The synthesis of combinatorial libraries focused on creating a diversity of chemical molecules, which have been used in peptides, non-peptide oligomers, peptidomimetics, small-molecules, m-RNA display, and DNA-encoded libraries [1–4, 10–14]. However, the combinatorial chemistry approach of each kind of material has its own unique high-throughput screening and encoding strategy. Besides preparing factual combinatorial libraries for experiments, the development of computational chemistry has enabled the design and screening of virtual chemical libraries that are available, which are more efficient and cost-effective than their conventional counterparts [14]. Therefore, combinatorial chemistry approach revolutionized the research paradigms on drug discovery and had a profound impact on other disciplines of scientific researches.

Compared to organic or small-molecular compounds, the synthesis and characterization of inorganic functional materials in terms of thermal, optical, electronic, optoelectronic, and magnetic properties is far more complex. The multiple-sample concept, i.e., synthesizing, analyzing, testing, and evaluating large parts of multicomponent systems in single steps, was also put forward by Joseph J. Hanak, a professor at RCA Laboratories, Princeton, when he carried out superconductor research in 1970. In one of his published papers [15], he wrote that *"The present approach to the search for new materials suffers from a chronic ailment, that of handling one sample at a time in the processes of synthesis, chemical analysis and testing of properties. It is an expensive and time-consuming approach, which prevents highly trained personnel from taking full advantage of its talents and keeps the tempo of discovery of new materials at a low level"* [15]. However, his research was restricted to the instruments and technologies at that time, and hence, his proposal did not attract enough attention.

In 1995, Dr. Xiao-Dong Xiang and Prof. Peter G. Schultz at Lawrence Berkeley extended the use of combinatorial chemistry approach from biological and organic molecules to materials science [16]. They demonstrated the availability of the methodology with parallel synthesis of spatially addressable libraries of solid-state superconductors through thin-film deposition assisted with binary physical masks [16]. They reported the application of combinatorial chemistry approach to the discovery of solid-state materials for the first time [16]. Soon, this approach was employed to investigate magnetoresistance materials [17], luminescent materials [18, 19], catalysts [20] and electrocatalysts [21], and ferroelectric/dielectric materials [22]. The application of combinatorial chemistry approach in materials formed a new discipline named as combinatorial materials science. Currently, the combinatorial chemistry approaches are widely used to study structural materials, electronic-, magnetic-, optical-, and energy-related functional materials, and phase mapping. Herein, we focused the research of combinatorial chemistry approach on phosphors.

8.1.2 PRINCIPLES OF COMBINATORIAL CHEMISTRY APPROACH

For convenience, materials were usually synthesized and characterized on an individual basis. In practice, a useful material applied in the industry was discovered via countless trials and errors and optimized with tens of thousands of tests. To some extent, the synthesis of materials, in a practical way, is quite similar to cooking, for which the tools, raw materials, and recipes must be meticulously selected and optimized by tasting at various stages [23]. Such a conventional approach is named as trial-and-error or one-by-one method. The synthesis of materials based on the one-by-one method could be described as:

$$A + B \rightarrow AB \tag{8.1}$$

in which the reaction of A and B forms AB.

As for the strategy of parallel synthesis adopted in combinatorial chemistry, multiple components A_i (i = 1 − m) react with B_j (j = 1 − n) to form A_iB_j, which could be described as:

$$
\begin{bmatrix}
A_1 \\
\vdots \\
\vdots \\
A_i \\
\vdots \\
\vdots \\
A_m
\end{bmatrix}
\times
\begin{bmatrix}
B_1 \\
\vdots \\
\vdots \\
B_j \\
\vdots \\
\vdots \\
B_n
\end{bmatrix}^T
\rightarrow
\begin{bmatrix}
A_1B_1 & \cdots\cdots & A_1B_n \\
\vdots & \vdots & \vdots \\
\vdots & \vdots & \vdots \\
A_iB_1 & A_iB_j & A_iB_n \\
\vdots & \vdots & \vdots \\
\vdots & \vdots & \vdots \\
A_mB_1 & \cdots\cdots & A_mB_n
\end{bmatrix}
\tag{8.2}
$$

$$i = 1 \cdots m \qquad j = 1 \cdots n \qquad\qquad m \times n$$

Since there are m kinds of A and n kinds of B, accordingly, m × n kinds of AB compounds are obtained by the cross reaction of A_i (i = 1 − m) with B_j (j = 1 − n). The m × n kinds of A_iB_j compounds were integrated on a substrate to form arrays of dense samples, named as a "combinatorial library" (also named as an "integrated materials chip"). Such a rapid synthesis of a large amount of samples in a parallel way is named as parallel synthesis.

Nearly every chemist has been trained to prepare single compounds in a phase as pure as possible. However, multicomponent mixtures are produced and used for optimization in combinatorial chemistry. In fact, the development of combinatorial chemistry has been able to create diverse combinatorial libraries with a wide range of chemical components, including inorganic functional materials, structural materials, catalysts, and organic materials. Therefore, we need an efficient strategy to identify the component that may be present in the synthetic mixtures with the most desirable objective properties.

In combinatorial chemistry, the technique that is used to rapidly characterize samples is called "high-throughput characterization". Through high-throughput characterization, some components which have met specific performance requirements are screened out. These kinds of materials are

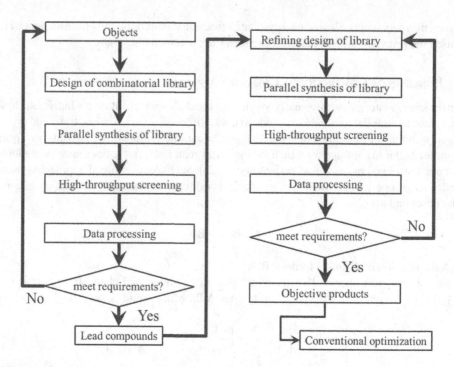

FIGURE 8.1 Schematic diagram of research procedures carried out with the combinatorial approach.

named clue materials or lead compounds. Regarding the optical, electronic, magnetic, ferroelectric, dielectric, superconductor, and magnetoresistance materials, each kind of combinatorial chemistry approach has its own unique high-throughput screening technique and encoding strategy. In contrast to the properties of electronic, magnetic, and catalytic materials among others, the luminescence property is facile to be characterized due to the intuition of visual observation.

The research procedures carried out with the combinatorial chemistry approach could be described using Figure 8.1, i.e., the target selection or the determination of objectives, the design of combinatorial library, the parallel synthesis of combinatorial library, the high-throughput characterization of combinatorial library, and data processing. If the experimental results did not meet the requirements, the combinatorial library could be modified or redesigned based on the previously obtained results. If the requirements are met, we could perform a further optimization on lead compounds by designing a combinatorial library with a higher density of samples through amplifying the concentration interval or the component precision of lead compounds. Finally, the lead compounds screened with the combinatorial chemistry approach could be verified and optimized using conventional methods. The combinatorial chemistry model is remarkably analogous to that of the biological immune system. When an organism is attacked by a virus, the immune system immediately produces a large number of antibodies. Among these antibodies, the immune system will determine the ones that can destroy the viruses most efficiently and multiply them immediately. The lead compounds can be considered to be the antibodies that could destroy the viruses most efficiently in immune systems. The combinatorial chemistry approaches that adopt parallel synthesis and high-throughput screening techniques have superseded the traditional model of materials research that employ "one-at-a-time" synthesis and characterization successively.

8.1.3 DESIGN OF COMBINATORIAL LIBRARY

There are many objectives for phosphor research, such as exploring a new phosphor for a specific application, developing a new recipe of phosphors, including optimization of host components and the concentration of activators and sensitizers, optimizing synthesis processes including the reaction temperature and reaction time and the amount of fluxes added to improve luminous efficiency, identifying a new compound, discovering a new structure and identifying a new symmetry of a compound according to emission spectra of activators, and revealing the luminescence mechanism. Aimed at fulfilling different objectives, there are different optimal strategies to conduct phosphor research using combinatorial chemistry approach. The combinatorial library is presented in a two-dimensional (2D) way, and thus, we need to present the array of phosphor samples in 2D combinatorial libraries.

The unitary composition is considered as one point, the binary composition is represented in a line, and the quaternary composition spreads in a plane. If the composition in the combinatorial library does not exceed ternary, it is convenient to express them in the 2D combinatorial library. However, there are some compositions such as the quaternary and quintuple among others that need to be optimized. Before performing parallel synthesis, we need to map the composition of the combinatorial library in 2D. Herein, two examples about the design of quaternary combinatorial library are given: one regarding the optimization of the Eu^{3+} activator and Bi^{3+} sensitizer in the $(Y,Gd)VO_4$ host which forms the Y–Gd–Bi–Eu quaternary system, and the other is regarding the optimization of the Eu^{3+} photoluminescence in the Y_2O_2–B_2O_3–P_2O_5–SiO_2 quaternary host.

The optimization of the concentrations of activators and sensitizers is prevalent in the study of luminescent materials. The sensitized luminescence of Eu^{3+} by Bi^{3+} has been investigated by many scholars in variant hosts. If the concentrations of Eu^{3+} and Bi^{3+} were merely optimized, the composition of the combinatorial library could be designed as depicted in Figure 8.2(a). Here, the Eu^{3+} is arranged along the vertical Y-axis with its concentration increasing from 0 via 0.005, 0.015, 0.030, and 0.045 to 0.060, and the Bi^{3+} is arranged along the horizontal X-axis with its concentration increasing from 0 via 0.005, 0.015, 0.025, and 0.040 to 0.050. However, the luminescence of Bi^{3+} and Eu^{3+} depends on the relative content of the Y/Gd ratios in the host of $(Y,Gd)VO_4$. Eu^{3+} emits red while Bi^{3+} emits green in the $(Y,Gd)VO_4$ host. The role of Gd^{3+} as an intermediate in energy transfer from the Bi^{3+} sensitizer to the Eu^{3+}/Tb^{3+} activators has been observed previously, and the energy transfer paths of $VO_4^{3-} \rightarrow Eu^{3+}$, $Gd^{3+} \rightarrow Eu^{3+}$, and $Gd^{3+} \rightarrow VO_4^{3-} \rightarrow Eu^{3+}$ were also demonstrated [24, 25]. In the case where a special yellow emission is required that consists of the simultaneous emission of Bi^{3+} and Eu^{3+} through a combination with the blue emission of the LED chips to produce white light, various hosts can be considered such as pure $GdVO_4$, YVO_4, doping Y into $GdVO_4$ (denoted as $(Gd,Y)VO_4$), or doping Gd into YVO_4 (denoted as $(Y,Gd)VO_4$). Determining the best host among these is essential. For this purpose, the ratios of Y/Gd in the $(Y,Gd)VO_4$ host have to be optimized. A further optimization could be classified into two groups: (1) by doping Y into $GdVO_4$ for which the combinatorial library was designed as shown in Figure 8.2(b); (2) by doping Gd into YVO_4 for which the combinatorial library was designed as depicted in Figure 8.2(c). Figure 8.2(a)–(c) present the three subsets of the composition maps and photoluminescence of the combinatorial libraries under the excitation of 365 nm. Besides composition, the optimization of phosphors may involve excitation wavelengths. Figure 8.2(d)–(f) present the three subsets of the composition maps and photoluminescence of the combinatorial libraries irradiated with 254 nm. The dynamic and authentic variation of the emission color and intensity of the samples upon a change of the relative concentrations of Bi^{3+}/Eu^{3+} and Y^{3+}/Gd^{3+} and excitation wavelengths are clearly displayed in the images of the luminescence of the combinatorial libraries in Figure 8.2 [26].

A facile way to map quaternary compositions which construct a tetrahedron is to unfold it, as illustrated in Figure 8.3. The tetrahedron M1M2M3M4 was cut off along the edges of M1M4, M4M3, M3M2, and M2M1, respectively, and subsequently, the forces of the tetrahedron were transferred into two connected rhomboids, as shown Figure 8.3(a). With a deformation of 45 degrees of the rhomboids, two squares were obtained. Besides, the repeated components, namely, the

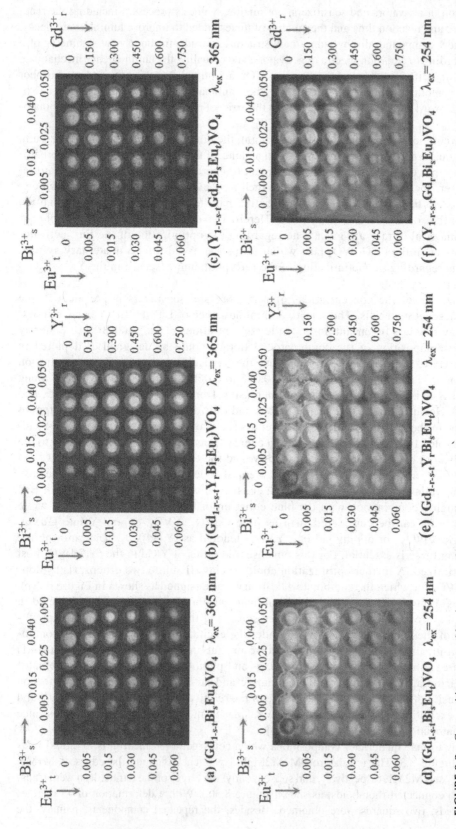

FIGURE 8.2 Composition map and luminescence photographs of combinatorial libraries under 365 and 254 nm excitation, respectively. (a) $GdVO_4$:Bi^{3+},Eu^{3+} excited with 365 nm, (b) by doping Y into YVO_4:Bi^{3+},Eu^{3+} excited with 365 nm, (c) by doping Gd into YVO_4:Bi^{3+},Eu^{3+} excited with 365 nm, (d) $GdVO_4$:Bi^{3+},Eu^{3+} excited with 254 nm, (e) by doping Y into YVO_4:Bi^{3+},Eu^{3+} excited with 254 nm, (f) by doping Gd into YVO_4:Bi^{3+},Eu^{3+} excited with 254 nm. (Chen, L., Chen, K. J., Hu, S. F., and Liu, R. S. 2011. Combinatorial chemistry approach to searching phosphors for white light-emitting diodes in (Gd-Y-Bi-Eu)VO_4 quaternary system. *J. Mater. Chem.* 21: 3677–3685. With permission.)

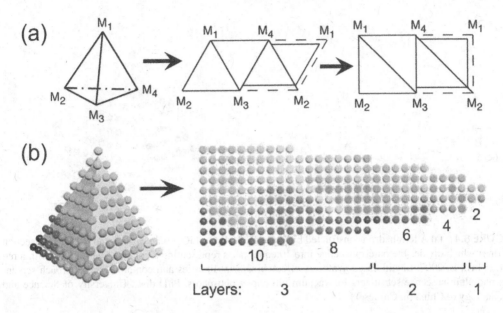

FIGURE 8.3 Expressing a quaternary system in two-dimensional combinatorial library. (a) Unfolding a tetrahedron and transferring it into two squares, (b) composition map for the arrangement of a tetrahedron from the outer to inner layers in the 2D plane.

overlapping binary edges, i.e., M4M1, M1M2, and M2M3, as represented by the dashed lines, were removed. This way, layer-by-layer, from the inside to the outside of the tetrahedron, whose outermost edge consists of 10 points, the tetrahedron was transferred into squares with the length consisting of 2, 4, 6, 8, and 10 points, respectively, as shown in Figure 8.3(b), corresponding to layers 1, 2, and 3. These squares are easily presented in the 2D combinatorial library. A detailed example regarding the design of the Y_2O_2–B_2O_3–P_2O_5–SiO_2 quaternary library, activated by Eu^{3+}, is described as later.

Previous researches reveal that the borate, phosphate, and silicate host could absorb vacuum ultraviolet efficiently [27]. To screen an efficient component that can emit effectively under the excitation of vacuum ultraviolet, a combinatorial library comprising the Y_2O_2–B_2O_3–P_2O_5–SiO_2 quaternary compositions, activated by the activator of Eu^{3+}, was designed [28]. With Y_2O_2, B_2O_3, P_2O_5, and SiO_2 fixed at four corners, a tetrahedron was constructed, as shown in Figure 8.4(a). Each of the binary edges, i.e., Y_2O_3–B_2O_3, Y_2O_3–P_2O_5, Y_2O_3–SiO_2, B_2O_3–P_2O_5, B_2O_3–SiO_2, and SiO_2–P_2O_5, in the triangle pyramid was divided by drawing parallel lines along an adjacent line. Accordingly, there are 11 points in the outermost edge of the triangle. An intersection of any two lines represents one component. From the vertex to the base of the triangle, the number of samples increases one after the other from layer 1 to 11. In this manner, the components in the triangular pyramid could be depicted with the dots shown in Figure 8.4(b). Regarding the base of the pyramid which comprises the B_2O_3–SiO_2–P_2O_3 ternary, there is no Y_2O_3 present. Since there is no site of Y^{3+} for Eu^{3+} to occupy, all the samples in this base are omitted. Instead of B_2O_3, SiO_2, and P_2O_3, the components that correspond to the three vertices of the base triangle are modified to $B_2O_3 + 0.1\ Y_2O_3$, $SiO_2 + 0.1\ Y_2O_3$, and $P_2O_3 + 0.1\ Y_2O_3$, respectively. By processing in this manner, there were 10 points in the outermost binary edge of the pyramid. Regarding the quaternary components presented by the red dots in Figure 8.4(b), there are 220 samples in total, i.e., one unitary (i.e., the Y_2O_3 on top vertices), 27 binary, 136 ternary, and 56 quaternary compositions. Figure 8.5 presents the composition map and photoluminescence of the combinatorial library. Besides variant compositions and excitation wavelengths, the same designed combinatorial library could be synthesized at different

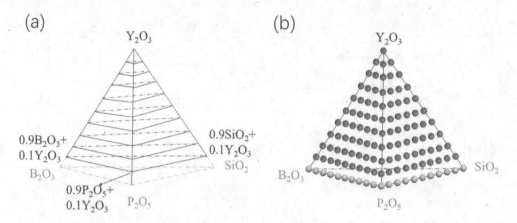

FIGURE 8.4　(a) A tetrahedron constructed by Y_2O_3–B_2O_3–P_2O_5–SiO_2 compositions and (b) the intersection of lines which divides the outermost edge into 10 equal parts representing a composition marked with a red dot. (Chen, L. 2007. Liquid-based synthesis of combinatorial libraries and combinatorial approach screening photoluminescence sensitizers for vacuum ultraviolet phosphors. PhD diss., University of Science and Technology of China (in Chinese).)

temperatures. Figure 8.5 also exhibits different effects of composition, excitation wavelength, and synthesis temperature on luminescence. A total of 220 samples were processed per batch, and 660 samples were presented in Figure 8.5 for three batches. It is hard to accomplish with a convenient approach in a short duration.

8.1.4　PARALLEL SYNTHESIS OF COMBINATORIAL LIBRARIES OF PHOSPHORS

There exist two primary methods for the parallel synthesis of combinatorial libraries: one is thin-film deposition by combining with physic masks, and the other is the liquid-based approach. In addition, we developed a third method to synthesize array samples of phosphors from powder sources, by combining with the Taguchi method to reduce the number of samples. As for the processes of combinatorial synthesis, there are two key steps involved in the parallel synthesis: one is the delivery of raw materials and the other is the controllable reaction of raw materials.

8.1.4.1　Solid-State Thin-Film Synthesis—Preparation of Combinatorial Libraries with Thin-Film Deposition

8.1.4.1.1　Instrument Used for Thin-Film Deposition

In semiconductor industry and laboratories, there are many kinds of instruments used to deposit thin films, such as MOCVD (metal-organic chemical vapor deposition), PLD (pulse laser deposition), MBE (molecular beam expitaxy), electron beam evaporation, magnetron sputtering, and thermal evaporation. By employing the thin-film deposition technique to prepare combinatorial libraries, each component of multiple components was sequentially deposited on a substrate. This technique has rigid requirements on the purity and uniformity of multiple-layer thin films since they are not in-situ heat-treated to form high-density discrete material arrays. As for the fabrication of multiple-component combinatorial libraries, the contamination originating from multiple targets by repeated sputtering or thermal evaporation in a high vacuum cave has to be considered. Thus, commercially available instruments that are generally used to prepare thin films are not directly used to prepare combinatorial libraries. Instead, an appropriate modification to the instruments is needed for the fabrication. For example, by employing a magnetic deflector electron beam in the vacuum evaporation system, it will be helpful to reduce the pollution caused by the evaporation of

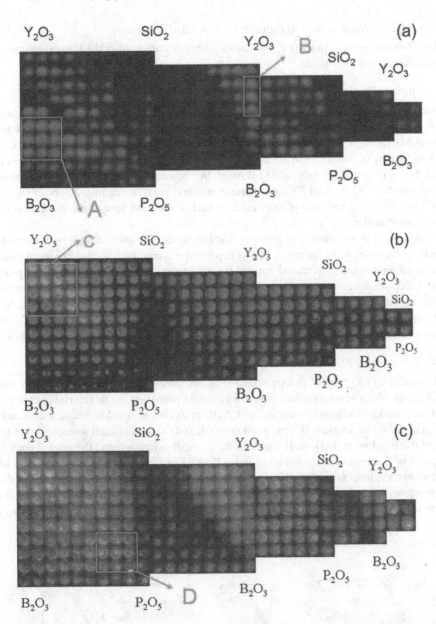

FIGURE 8.5 Composition map and photoluminescence of the quaternary Y_2O_3–B_2O_3–P_2O_5–SiO_2 combinatorial library synthesized under variant temperatures and excited with 147 and 254 nm, respectively, in which the concentration of Eu^{3+} in each sample was set at 5% of Y^{3+}. (a) Fired at 1150°C for 3 h, excited with 147 nm; (b) fired at 1250°C, excited with 254 nm; (c) fired at 1350°C, excited with 254 nm. (Chen, L. 2007. Liquid-based synthesis of combinatorial libraries and combinatorial approach screening photoluminescence sensitizers for vacuum ultraviolet phosphors. PhD diss., University of Science and Technology of China (in Chinese).)

variant target materials; by enlarging the spot area of the electron beam, the uniformity of thin films in the system of the electron beam evaporation can be improved. In the PLD system, the energy of the laser beam and its uniformity are the key factors that affect the uniformity of thin films, and the uniformity of the beam energy density can be improved by adjusting the optical paths. Above all, the preparation of high-quality samples is the most significant factor affecting the selection of the instrument for combinatorial synthesis.

8.1.4.1.2 Physical Masks Used to Prepare Combinatorial Libraries

The physical masks used to prepare combinatorial libraries are mainly the binary, quaternary, and in-situ movable continuous masks.

8.1.4.1.3 Binary Physical Masks

Figure 8.6(a) shows the schematic diagram of a group of binary masks. Regarding the mask M1, the mask is divided into two parts: one half is hollow-carved and the other is used for masking. By rotating M1 through 90° toward the right, the mask M2 is obtained. As for the mask M3, it is divided into four parts. Two of them are hollow-carved and the other two parts are used for masking. The mask M4 is obtained by rotating M3 through 90° toward the right. Similarly, the mask could be divided into 6, 8, 10, ..., and 2N (N is natural number) parts as required to obtain high-density samples. The hollow-carved area of each mask is half of the total area of the mask, and the other half is used for masking.

The usage of binary mask to prepare combinatorial libraries could be illustrated with Figure 8.6(b). By employing the mask M1 to deposit the component A, the sample A and a non-deposited area are obtained. With M1 rotated by 90° toward the right side, i.e., by employing the mask M2 to deposit the component B, the samples A, AB, B, and a nondeposited area are obtained. By employing the mask M3 to deposit the component C, the samples of ABC, AC, AB, A, BC, C, B, and a nondeposited area are obtained. Similarly, more components could be deposited stepwise. After the occurrence of n steps by depositing n layers, a total of 2^n samples could be obtained.

8.1.4.1.4 Quaternary Physical Masks

Figure 8.7(a) displays a group of strictly self-similar and precisely positioned quaternary masks, A, B, C, D, E...., in which several ones consist of a group. In preparing combinatorial libraries, a group of quaternary masks were jointly used. In mask A, there exists one window whose area is one-fourth of the area of the mask. In mask B, four windows exit, and the area of each window is one-fourth of the area of the window in mask A. In mask C, there are 16 windows, and the area of each window is one-fourth of the area of the window in mask B. And so forth, there are 4^{r-1} windows in the rth mask. However, the total area of all the windows in each mask is one-fourth of the area of the mask. All the windows have the same configuration, and they are strictly self-similar. As for the mask A, it is divided into four parts along its symmetry center by the horizontal X-axis and the vertical

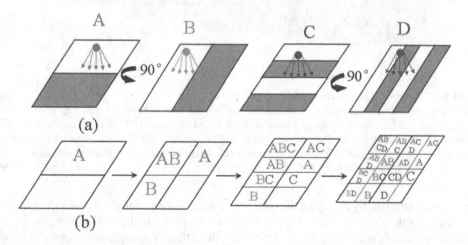

FIGURE 8.6 Schematic diagram of binary masks (a) and their usage for preparing combinatorial libraries (b).

Y-axis, and subsequently, one part is selected and used as the window. In the mask B, each window is one-fourth of the mask A, and the mask is also divided into four parts along its symmetry center by the horizontal X-axis and the vertical Y-axis, and one part is selected as the window. Similarly, for mask C, each window can be considered as one-fourth of the mask B. The mask C is divided into four parts along its symmetry center by the horizontal X-axis and the vertical Y-axis, and one part is selected as the window. Operating with a similar method, the higher order r^{th} masks are obtained. In the rth mask, there are 4^{r-1} windows.

For preparing combinatorial libraries, the mask A is used at first. As shown in Figure 8.7(b), when the physical mask A is used, the component A_1 is deposited at the position of A_1, and then, the components A_2, A_3, and A_4 are deposited sequentially, with the mask rotated by 90° in the clockwise direction. After depositing the first layer, the mask B is employed to deposit B_1, B_2, B_3, and B_4, sequentially. On the basis of the first layer of A, the 16 samples of A_1B_1, A_1B_2, A_1B_3, A_1B_4, A_2B_1,

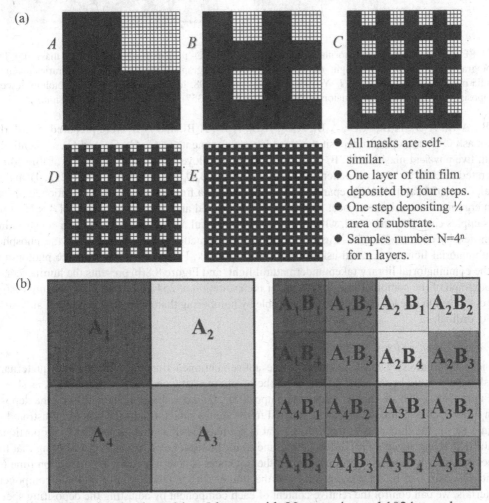

● All masks are self-similar.
● One layer of thin film deposited by four steps.
● One step depositing ¼ area of substrate.
● Samples number $N=4^n$ for n layers.

Likewise, by depositing 5 layers with 20 steps, in total 1024 samples, i.e., $A_mB_nC_hD_jE_k$ (m, n, h, j, k = 1, 2, 3, 4) for 4^5, are fabricated.

FIGURE 8.7 Schematic diagram of quaternary masks (a) and their application in preparing combinatorial libraries (b).

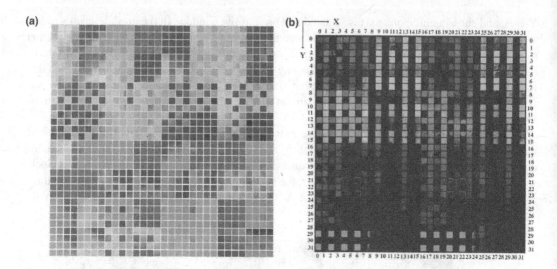

FIGURE 8.8 Photograph of phosphors combinatorial libraries prepared with quaternary mask. (a) The photograph taken under natural light, (b) the luminescent photograph of the combinatorial libraries taken by irradiating with 254 nm. (Wang, J., Yoo, Y., Gao, C., et al. 1998. Identification of a blue photoluminescent composite material from a combinatorial library. *Science* 279 (5357): 1712–1714. With permission.)

A_2B_2, A_2B_3, A_2B_4, A_3B_1, A_3B_2, A_3B_3, A_3B_4, A_4B_1, A_4B_2, A_4B_3, and A_4B_4 were prepared. Similarly, the mask C is employed to deposit the third layer components of C_1, C_2, C_3, and C_4, sequentially. With five physical masks of A, B, C, D, and E being employed to deposit five layers of thin films, operated in 20 steps, the components covered will be $A_mB_nD_hC_jE_k$ (m, n, h, j, k = 1, 2, 3, 4), and in total, $4^5 = 1024$ samples are prepared. As long as there is a finer mask, the abovementioned operation continues toward finer space. If N layers are deposited and operated in a total of $4 \times N$ steps, 4^N samples can be synthesized, which exhibits the powerful ability of this method in constructing high-density discrete samples. Figure 8.8 illustrates a practical example regarding the phosphor combinatorial library prepared using the quaternary mask. Figure 8.8(a) presents the photograph of the combinatorial library taken under natural light, and Figure 8.8(b) presents the luminescence photograph of the combinatorial library under the excitation of 254 nm. As shown in Figure 8.8(b), the composition of each component is accessible by numbering them along the horizontal and vertical coordinates.

8.1.4.1.5 In-Situ Movable Physical Masks

Besides the stationary masks, similar to the abovementioned illustrated binary and quaternary masks, there could be movable masks. With the in-situ masking and continuous shifting, as shown in Figure 8.9(a), the continuous spread of composition could be obtained by controlling the deposition time. Such a technique is widely adopted in the semiconductor industry [29]. As illustrated in Figure 8.9(b), the thickness of the film (L) that is not masked by the movable mask is proportional to the depositing speed (V) and depositing time (t) of the film, i.e., $L = V \cdot t$ [29]. Assuming that the speed of the film deposition is constant (V), the thickness is determined by the deposition time (t), which is controlled by the moving speed of the shutter. In the fabrication of multiple-component materials, we can control the relative content of each component by adjusting the depositing speed (V) and depositing time (t) of each component. Figure 8.9(b) illustrates the growth of gradient films of A, B, and C. As shown in Figure 8.9(a), the three targets of A, B, and C are distributed at the corners of a regular triangle. The film A is deposited initially. With the shift of the shutter along the X-axis from the right to the left, the film of A is prepared with continuously increasing thickness. Next, the film of B is deposited with continuously increasing thickness by rotating the target clockwise by 120°. Finally, the film C is deposited in a similar fashion [29].

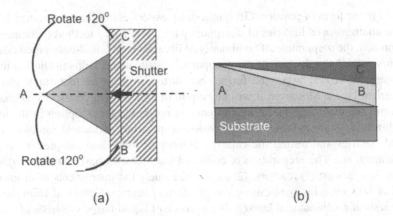

FIGURE 8.9 Schematic diagram of continuously movable masks (a) and their application in depositing multiple layers with the spread of gradient composition (b). (Xiang, X., Wang, G., Zhang, X., Xiang, Y., and Wang, H. 2015. Individualized pixel synthesis and characterization of combinatorial materials chips. *Engineering* 1: 225–233. With permission.)

Regarding the preparation of combinatorial libraries with thin film deposition, the spatially accessible high-density samples were formed with in-situ sequential deposition of multiple layers through a combination with the movement of the masks. With the masking effect of the masks and their movement, the distribution of a specific component on the substrate is controlled. The approach of thin film deposition by combining with in-situ movable physical masks is particularly efficient in mapping continuous phase diagrams and constructing gradient materials of multiple-component compounds but is rarely adopted in luminescent materials. The physical masks used to prepare the combinatorial libraries of luminescent phosphors are mainly the binary mask and the quaternary mask.

8.1.4.1.6 Processing of Thin-Film Combinatorial Libraries

The thin-film samples prepared with sequential deposition techniques have sandwich structures. When employing the sequential deposition technique to prepare combinatorial libraries, the quality of multiple-composition materials depends intensively on the controlled solid reaction between the contacted layers. Among the deposited multiple layers, metastable phases are easily formed between the two contacted layers. Once the metastable phases are formed, they will prevent the formation of a homogeneous compound with a single phase. To avoid forming hetero-phases, it is better to anneal the thin films of multiple layers at a low or medium temperature for a long duration before the high-temperature reaction. A long-duration annealing will promote the diffusion of atoms and help multilayer films to mix homogeneously. Theoretically speaking, a long duration of annealing at a medium and low temperature promisingly enables the films to mix at a scale of micron thickness but at the expense of time. The more the number of layers and the thicker the films, the longer are the time required for the homogeneous diffusion of atoms. To save time, the thickness of a layer should be reduced appropriately, which is helpful to promote the even diffusion of atoms. After low- or medium-temperature diffusion, the deposited films were fired at a high temperature to form the array samples of the combinatorial library. The distribution of the atoms in the mixture of the variant components can be tested by using the Rutherford backscattering technique, and the thin-film quality of the samples can be examined using X-ray diffraction analysis.

8.1.4.2 Liquid-State Synthesis—Preparation of Combinatorial Libraries with Liquid-Based Methods

The previous section elaborates the preparation of combinatorial libraries with thin-film techniques. However, there are some compounds, such as catalysts, molecular sieves, and high-molecular polymers, which are not suitable to be synthesized using thin-film techniques. The phosphors are most popularly used in the form of powders. To synthesize powders of luminescent materials, it is better to prepare the combinatorial libraries of phosphors with liquid-based methods. Compared with the thin-film approach, the preparation of combinatorial libraries with liquid-based methods could mix variant precursors at the molecular or the nanoparticle scale, which helps to shorten the diffusion of atoms among the variant layers and further benefit to eliminate the metastable phases formed among the variant layers. Moreover, it will be helpful to overcome the difficulty in characterizing the properties of the materials due to lack of enough resolution for sampling in the high-density samples of thin films, such as collecting the emission spectrum of discrete samples in phosphors combinatorial libraries and testing the catalytic activity of individual samples in catalysts combinatorial libraries, etc. The preparation of combinatorial libraries based on liquid-state methods involves three key factors: (1) reactors, (2) the tools, namely the instruments of droplets delivery, (3) solutions or inks used for synthesis of combinatorial libraries. Classified from the precursors used for synthesis of combinatorial library, the approach of liquid-based synthesis of combinatorial libraries could be classified into three categories: (1) synthesized from the solutions of soluble compounds, (2) synthesized from the suspensions of insoluble compounds, (3) synthesized both from the solutions and suspensions.

8.1.4.2.1 Reactors

The reactors used to prepare combinatorial libraries with liquid-based methods should be in the form of arrays. If the phosphors were synthesized at a low temperature without a further need of high-temperature treatment, the array reactors could be made from plastic, nylon, polytetrafluoroethylene, or glass materials. If the heat treatment is done at a relatively low temperature, the array reactors could be made from quartz, clay ceramic, etc. In general, quartz can bear temperatures that are no higher than 1200°C. With a reaction temperature that is higher than 1200°C, the array reactors could be made from Al_2O_3 or BN ceramics. Plastic, nylon, polytetrafluoroethylene, or quartz substrates have the disadvantage of low-temperature tolerance. However, they have some merits like easy processing and being waterproof. On the surface of plastic, nylon, or polytetrafluoroethylene substrates, the array wells could be fabricated by drilling, grinding, or pressure die forming. The Al_2O_3 and BN substrates with dwelled wells are the most widely used reactors in the synthesis of phosphors combinatorial libraries at a high temperature. The Al_2O_3 and BN substrates possess merits of high strength, high hardness, high temperature tolerance, and excellent chemical inertness, which can resist acid and alkali corrosions. However, it is hard to drill and grind array wells on the surface of Al_2O_3 or BN substrates. The array wells on the Al_2O_3 and BN substrates usually were formed using pressure die on ceramic embryo, and subsequently, the ceramic embryo were sintered at a high temperature to form ceramic. However, reactors made from Al_2O_3 and BN ceramics are often accompanied with numerous pores and cracks. The leakage caused by the pores and cracks have to be handled before employing ceramic reactors to synthesize combinatorial libraries. Once the Al_2O_3 or BN substrates with wells have been sintered at high temperature, it is forbidden to polish, drill, or grind the surface of the reactors in usage, since such processing will cause the pores and cracks to be exposed to air and result in intensified leakage.

Figure 8.10 illustrates an example related to the synthesis of organic luminescent material libraries using the fluorophlogopite substrate, which is a kind of processable glass consisting of mainly the $KMg_3(AlSi_3O_{10})F_2$ microcrystal [30]. The 6×6 array wells on the fluorophlogopite substrate were used as the reactors [30]. In Figure 8.10(a), the sensitized photoluminescence of $Eu(DBM)_3Phen$ (DBM = dibenzoylmethane; phen = 102 phenanthroline) by rare-earth complexes $(Re(DBM)_3Phen$ $(Re^{3+} = Tb^{3+}, La^{3+}, Gd^{3+}, Dy^{3+}, Y^{3+}, Ce^{3+})$ in poly(methyl methacrylate)(PMMA, Mw 350,000)) was examined [30]. The rare-earth complexes including $Eu(DBM)_3Phen$ and PMMA were dissolved in the cyclopentanone solvent to control the viscosity and evaporation, whose concentrations are at 5 g/L, respectively. The solutions were deposited into the reactors using a microliter pipette, as shown in Figure 8.11. The as-deposited samples in the combinatorial libraries were evaporated at room temperature in air ambient to complete the reaction. The photoluminescence photograph of

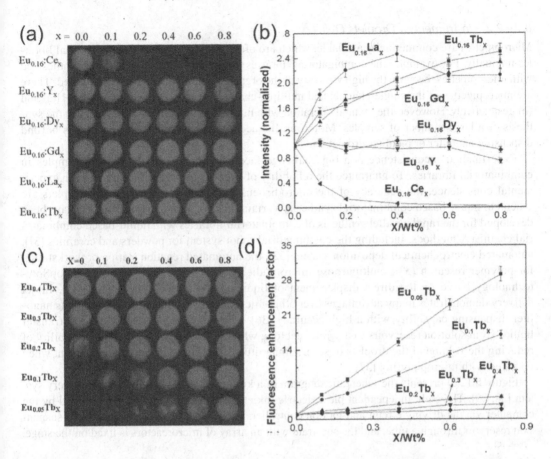

FIGURE 8.10 The organic luminescent combinatorial libraries synthesized using fluorophlogopite reactors. (a) The photograph of Eu(DBM)₃Phen photoluminescence sensitized by Re(DBM)₃Phen (Re³⁺ = Tb³⁺, La³⁺, Gd³⁺, Dy³⁺, Y³⁺, Ce³⁺) and (b) its relative intensity as a function of rare-earth content; (c) the photograph of Eu(DBM)₃Phen photoluminescence sensitized by Tb(DBM)₃Phen and (d) fluorescence enhancement factors as a function of Tb content. (Jiu, H., Ding, J., Sun, Y., Bao, J., Gao, C. and Zhang, Q. 2006. Fluorescence enhancement of europium complex co-doped with terbium complex in a poly(methyl methacrylate) matrix. *J. Non-Cryst. Solids* 352: 197–202. With permission.)

each sample in the combinatorial libraries was recorded using a digital camera, and the emission spectrum of each sample under the UV excitation of 365 nm was collected using a multiple-parallel luminescence measurement system (as shown in Figure 8.19). The relative intensity presented in Figure 8.10(b) shows that Tb^{3+} is an efficient sensitizer among the candidates of Tb^{3+}, La^{3+}, Gd^{3+}, Dy^{3+}, Y^{3+}, and Ce^{3+}, for which the luminescence maximizes at the Tb^{3+} concentration of 0.8%, with the Eu^{3+} concentration at 0.16%. Then, the combinatorial library was designed renewably, and the relative concentration ratios of Eu^{3+} and Tb^{3+} were further optimized by refining the Eu^{3+} concentration. As the photoluminescence photograph shown in Figure 8.10(c), the luminance of the samples as a function of the Tb concentration is described along the horizontal axis, and the luminance as a function of the Eu concentration is described along the vertical axis. With an increase of the Eu^{3+} concentration, the relative luminescence of Eu(DBM)₃Phen intensifies while the efficiency of energy transfers from Tb(DBM)₃Phen to Eu(DBM)₃Phen decreases. With the concentration of Eu(DBM)₃Phen maintained at 5 wt% of PMMA, as shown in Figure 8.10(d), the highest fluorescence enhancement factor F, a factor of luminescence enhancement used to describe energy transfer efficiency, reaches up to 27 at the concentration of Tb(DBM)₃Phen to PMMA at 0.8 wt%.

8.1.4.2.2 Instrument of Droplets Delivery

Micropipettes are commercially available, which are often used to synthesize combinatorial libraries manually. For example, the combinatorial libraries presented in Figure 8.10 were synthesized with micropipettes, because the high viscosity of the organic compounds cannot be ink-jetted. There are micropipettes of the single-channel and multiple-channels type, as shown in Figure 8.11(a) and (b), respectively. However, the manual operation with micropipettes is time-consuming for the synthesis of a large number of samples. Moreover, the accuracy of the microliter pipettes is beyond dispensing nanoliter or picoliter droplets.

Experimental inconsistence is a big challenge faced for the synthesis of micro-samples in combinatorial libraries. To guarantee the reliability of experimental results and maintain experimental consistence, the accuracy of the microdispenser and the delivery of microdroplets are crucial to liquid-based combinatorial synthesis. Variant types of automatic instruments have been developed for the rapid parallel synthesis of combinatorial libraries with liquid-based combinatorial chemistry methods, including the combinatorial robot system for powders and ceramics [31], automated electrochemical deposition system [32], the automated reaction station or workstation for polymer research [33], multipurpose microfluidic perfusion system [34], and the TopSpot-technology based on the direct displacement of liquids using an elastomer-stamp [35]. Inkjet delivery demonstrates many advantages over other microdispensing apparatus, owing to its nano-liter dispensing capability with a high accuracy [36]. Micropipetting is widely used to transfer liquids from solution reservoirs to discrete reactors, while inkjet printing offers the possibility of reducing the volume of the droplets to several magnitudes smaller than the volume present when operated with micropipetting [37].

Figure 8.12(a) presents the schematic diagram of a kind of drop-on-demand inkjet delivery system [38, 39]. The eight independent piezoelectric inkjet heads and x–y stage are controlled by the computer via the driving circuit and motion controller. Each inkjet head is connected to a suspension reservoir through a tube, and the substrate with an array of microreactors is fixed on the stage.

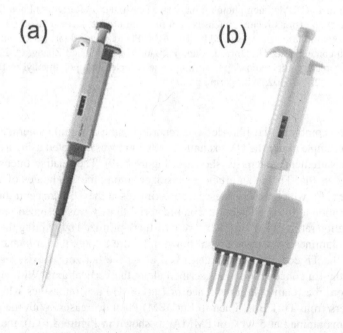

(a)   (b)

FIGURE 8.11 Single-channel (a) and multiple-channels (b) micropipettes for manually operated liquid-based synthesis of combinatorial library.

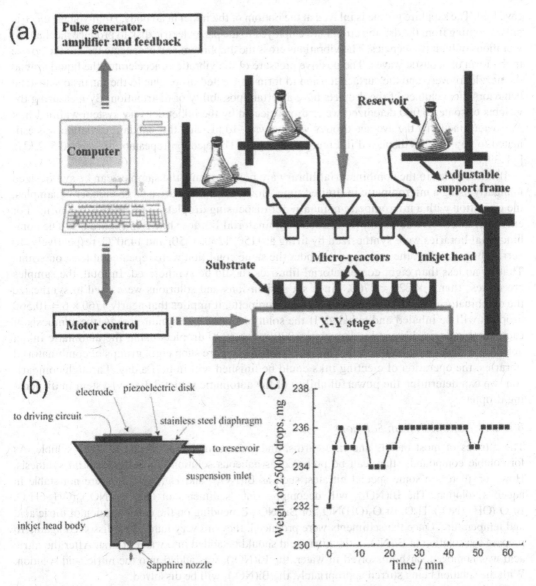

FIGURE 8.12 Schematic diagram of a drop-on-demand inkjet delivery system for automatic liquid-based combinatorial synthesis (a), schematic diagram of the inkjet head (b), and the stability with time of the drop-on-demand inkjet delivery system (c). (Chen, L., Bao, J., Gao, C., Huang, S., Liu, C., and Liu, W. 2004. Combinatorial synthesis of insoluble oxide library from ultrafine/nanoparticles suspension using a drop-on-demand inkjet delivery system. *J. Comb. Chem.* 6(5): 699–702; Chan, T.-S., Kang, C. C., Liu, R. S., et al. 2007. Combinatorial study of the optimization of Y_2O_3:Bi,Eu red phosphors. *J. Comb. Chem.* 9: 343–346. With permission.)

The software automatically coordinates the ejection of the inkjet heads and the position of the stage, according to the concentration of the solutions, the interval of the reactors, substrate size, and composition map, to deliver an appropriate amount of solution into the microreactors. The inkjet head is one of the core components of the drop-on-demand inkjet delivery system, which determines the dispensing ability of the microdroplets volume and their accuracy and consistence. As shown in Figure 8.12(b), the inkjet head mainly consists of the inkjet head body, a piezoelectric disk, a stainless-steel diaphragm, and a sapphire nozzle. The inkjet head body is made of stainless with cave [38]. The sapphire nozzle is inlayed at the bottom of the inkjet head body. High-voltage electric pulses coming from the driving circuit are employed to the piezoelectric disk to produce mechanical vibrations of high frequencies. The vibrations cross the diaphragm and propagate toward the nozzle in the form of acoustic waves. The positive pressure of the vibrations accelerates the liquid around the nozzle to overcome the surface tension to form an ejected drop. Due to the intrinsic statistical behaviors, the volumes of the droplets have a certain possibility of distribution. By measuring the weights of some 25,000 deionized water drops ejected by the inkjet delivery system within 1 h, it was determined that the average droplet volume was ~10 nL, and the standard deviation was estimated to be ~10%, as displayed in Figure 8.12(c) [38]. The ejection repeatability was ~0.5–2 kHz [38, 39].

If the samples in the combinatorial library are not too many, the samples can be synthesized manually using a micropipette in limited time. However, if there are a large number of samples, the operation with a micropipette manually for dispensing droplets is too time-consuming. For example, there were 220 samples in each combinatorial library shown in Figure 8.5. The combinatorial libraries were synthesized by firing at 1150, 1250, 1350, and 1450°C, respectively. To verify the reliability, the experiments under the same condition were repeated at least one time. That is, no less than eight combinatorial libraries should be synthesized. In total, the samples are no less than $8 \times 220 = 1760$. Since six suspensions and solutions were used to synthesize the combinatorial libraries, operated with micropipette, it implies that nearly $1760 \times 6 = 10,560$ droplets will be inhaled and exhaled. If the solutions or suspension are ready, the synthesis of each combinatorial library by dispensing $220 \times 6 = 1320$ droplets using the automatic inkjet delivery system could be finished within 30 min. To prepare such eight groups of combinatorial libraries, the operation of ejecting inks could be finished within half a day. Through comparison, we can determine the powerful ability of the automatic inkjet delivery system in dispensing droplets.

8.1.4.2.3   Synthesis from Solutions

The nitrates of most metals and rare earths, and some compounds like H_3BO_3 are soluble. As for soluble compounds, they can be prepared as aqueous solutions for combinatorial synthesis. However, there are some special nitrates, such as $Bi(NO_3)_3$ and $Hg(NO_3)_2$ that are not stable in aqueous solution. The $Bi(NO_3)_3$ will decompose into sediments of $(Bi_6O_6)_2(NO_3)_{11}(OH) \cdot 6H_2O$, $Bi_6O_4(OH)_4(NO_3)_6 \cdot H_2O$, $Bi_2O_2(OH)NO_3$, or $BiONO_3$, depending on the concentration of nitric acid and temperature. Once the sediments were produced, they are very hard to be dissolved again. To prepare the solution of $Bi(NO_3)_3$, the nitric acid should be added into water at first. After the nitric acid was homogeneously dissolved in water, the $Bi(NO_3)_3$ was added into the nitric acid solution. With the solution being stirred appropriately, the $Bi(NO_3)_3$ will be dissolved.

Bi^{3+} is an efficient sensitizer, which has been used to sensitize the photoluminescence of many activators like Eu^{3+} and Gd^{3+}. The following gives an example about the synthesis of the $Y_2O_3:Bi^{3+}$, Eu^{3+} library with a liquid-based combinatorial chemistry approach.

The $(Y_{2-x-y}Eu_xBi_y)$ combinatorial library was synthesized from the nitrate solutions of Y^{3+}, Bi^{3+} and Eu^{3+} by employing $Y(NO_3)_3 \cdot 6(H_2O)$, Eu_2O_3 and $Bi(NO_3)_3 \cdot 5(H_2O)$ as precursors, whose concentrations are 0.1, 0.01, and 0.01 M, respectively [39]. The $Y(NO_3)_3 \cdot 6(H_2O)$ was dissolved in deionized water to form the Y^{3+} solution, whereas Bi^{3+} and Eu^{3+} solutions were prepared by dissolving Eu_2O_3 and $Bi(NO_3)_3 \cdot 5(H_2O)$ in nitric acid solutions with a pH = 1.0. Then, the solutions Y^{3+}, Bi^{3+}, and Eu^{3+} solutions were ejected into the wells on an Al_2O_3 substrate, by using the drop-on-demand inkjet delivery system shown in Figure 8.12. All ink-jetting operations are automatically performed with the assistance of a programmed computer. The substrate size is 110×110 mm, and the diameter of the well which is used as the reactor is 8 mm. The substrate which contained solutions was first dried at 100°C for 1 h in an oven with a cover. Then, it was transferred to an electric furnace and slowly heated to 500°C at the rate of 3°C/min. Finally, the dried samples were pulverized by using a hemisphere-head glass bar and then fired at 1300°C for 3 h in an ambient air atmosphere.

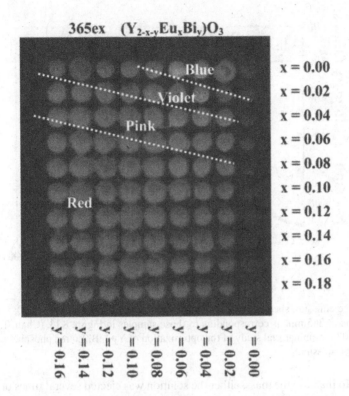

FIGURE 8.13 Composition map of the $(Y_{2-x-y}Eu_xBi_y)$ combinatorial library and its photoluminescence photograph under the excitation of 365 nm. (Chan, T. S., Kang, C. C., Liu, R. S. et al. 2007. Combinatorial study of the optimization of Y_2O_3:Bi,Eu red phosphors. *J. Comb. Chem.* 9: 343–346. With permission.)

Figure 8.13 presents the composition map of $(Y_{2-x-y}Eu_xBi_y)$ combinatorial library and its photoluminescence photograph under the excitation of 365 nm, from which we can observe that the luminescence color changes from blue, violet, and pink to red. Moreover, the relative brightness of the photoluminescence of the samples changes obviously with the variation of Bi^{3+} and Eu^{3+} concentrations, indicating that there is an energy transfer from Bi^{3+} to Eu^{3+}. This conclusion is also verified by the comparison of the emission spectra of the samples of $(Y_{2-x-y}Eu_xBi_y)$ (x = 0, y = 0.04; x = 0.16, y = 0; x = 0.16, y = 0.08). As shown in Figure 8.14, the broad band located in 450–575 nm region with a peak at about 500 nm was clearly observed in $(Y_{2-x-y}Eu_xBi_y)$ (x = 0, y = 0.04), which must be emitted by the Bi^{3+} ion, since only the activator Bi^{3+} is involved. Besides the intense $^5D_0-^7F_2$ emission line of Eu^{3+} at 612 nm, such a broad band was not observed in $(Y_{2-x-y}Eu_xBi_y)$ (x = 0.16, y = 0), since no Bi^{3+} is contained. However, the intense emission line of Eu^{3+} at 612 nm was significantly enhanced in $(Y_{2-x-y}Eu_xBi_y)$ (x = 0.16, y = 0.08) in contrast to that in (x = 0.16, y = 0). These comparisons provide key evidences on approving the energy transfer from Bi^{3+} to Eu^{3+}.

8.1.4.2.4 Synthesis from Inks or Suspensions

On employing solutions to the synthesis of combinatorial libraries, there are some shortcomings that exist. First, the solubility of variant precursors is different. During the drying process, the solved ions will precipitate along the inner wall of the reactor from the top surface of the solution to the bottom of the reactor. The smaller the solubility of the compound, the foremost it precipitates preferentially, which will result in composition segregation. Second, a large number of compounds in nature are insoluble. Third, there some compounds whose solubility is very low. After drying, the solid substance left in microreactors is very less, which will bring with enlarged errors

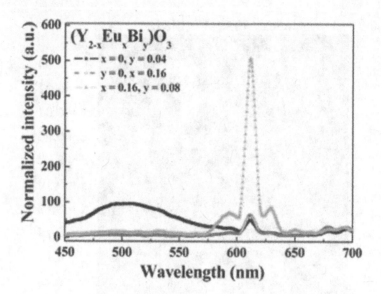

FIGURE 8.14   The emission spectra of $(Y_{2-x-y}Eu_xBi_y)$ (x = 0, y = 0.04; x = 0.16, y = 0; x = 0.16, y = 0.08) under the excitation of 365 nm, as corresponding to those samples in Figure 8.13. (Chan, T. S., Kang, C. C., Liu, R. S. et al. 2007. Combinatorial study of the optimization of Y_2O_3:Bi,Eu red phosphors. *J. Comb. Chem.* 9: 343–346. With permission.)

in experiments. To increase the mass, either the solution was ejected several times or a large volume of the solution was ejected.

Nevertheless, a lot of insoluble compounds in the ceramic industry have been transferred into suspensions and used for injection molding. To extend the application of liquid-based combinatorial chemistry approach, we developed the synthesis of combinatorial libraries synthesis from suspensions of insoluble compounds. As the composition map and photoluminescence photograph shown in Figure 8.15, the combinatorial library was synthesized by ejecting the suspensions of Y_2O_3, Eu_2O_3, and Tb_4O_7, respectively, using the drop-on-demand inkjet delivery system shown in Figure 8.12. To satisfy the requirements of the inkjet delivery instrument for the liquid to be ejected reliably, the suspensions must be stable (no sedimentation and agglomeration) enough during the ink-jetting

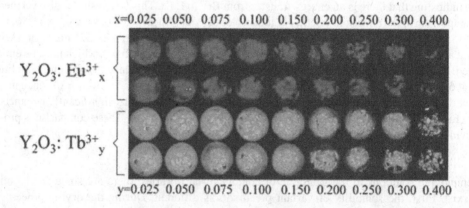

FIGURE 8.15   Composition map of the combinatorial library and its photoluminescence photograph under 254 nm excitation, synthesized by ink-jetting Y_2O_3, Eu_2O_3, and Tb_4O_7 suspensions. (Chen, L., Bao, J., Gao, C., Huang, S., Liu, C., and Liu, W. 2004. Combinatorial synthesis of insoluble oxide library from ultrafine/nano particles suspension using a drop-on-demand inkjet delivery system. *J. Comb. Chem.* 6(5): 699–702. With permission.)

process. Moreover, the suspension must have a high surface tension to let the droplets to be easily formed, with a low viscosity and fewer additives to promote the postprocessing. There are two main methods to prepare insoluble suspensions: one is by dispersing ultrafine/nanoparticles with a small amount of dispersant, stabilizer, and binder in solvents, while the other is by agglomeration from a chemical reaction in sol-gel. However, the suspensions made from both the methods could not fully satisfy the critical requirements of the inkjet delivery instrument, mainly because the organic additives will decrease the surface tension of the suspensions and sol-gels. Either by operating with micropipettes manually or by ejecting with inkjet delivery automatically, the surface tension of the solutions, suspensions, and sol-gels causes the droplets to not be formed properly. Regarding one side, the solutions, suspensions, or sol-gels will spread along the needles of the micropipettes or the head of the inkjet delivery system; on the other side, even though the droplets have deposited into the microwells on the ceramic substrate they will "come out" of the wells, due to the spread caused by low surface tension. In ceramic industry, the slurry of insoluble compounds generally was prepared by long-time ball mill. We applied the same route to prepare the suspensions for liquid-based combinatorial synthesis. The details for preparation of the suspensions applied to synthesis the combinatorial library in Figure 8.15 are described as given next. A 4-g rare-earth oxide powder (Y_2O_3, Eu_2O_3, and Tb_4O_7 with specification 99.99% and particle size distribution ~0.5–6 μm), an equal weight of deionized water, and agate balls (diameter 6 and 10 mm) with sample-to-ball weight ratio of 1:10 were loaded into an agate pot, then oxide powders were ground for 60–80 h at 250 rpm (or 40 h at 400–500 rpm) in a planetary high-energy ball mill. During ball milling, if the slurry dried, another 1–2 mL of deionized water was added into the agate pot to continue grinding. Finally, the deionized water to bring the total amount of water to 30 mL was supplied into the pot, and an additional 2 h of ball milling was carried out to form the 11.76 wt% suspensions. After long-time ball mill, the micron particles have been turned into nanoparticles. The suspensions could withstand long enough time to finish the ink-jetting of suspensions for synthesis of combinatorial library. Our results show that the suspensions of a family of the 17 rare-earth oxides could be prepared by using this way. Among them, the suspension of La_2O_3 and CeO_2 are most stable, which can stand exceeding 24 h without precipitating and agglomerating.

Subsequently, the Y_2O_3, Eu_2O_3, and Tb_4O_7 suspensions were ejected into the microwells on the Al_2O_3 substrate. Before being ejected, these suspensions were infused with argon gas for 2–3 h to expel the air dissolved within them, which tends to agglomerate bubbles and further block the ink-jet nozzle. After the drying process, the library was fired at 1450°C for 3 h to form red $Y_2O_3:Eu^{3+}_x$ and green $Y_2O_3:Tb^{3+}_x$ phosphors. From the composition map shown in Figure 8.15, each group of $Y_2O_3:Eu^{3+}_x$ and $Y_2O_3:Tb^{3+}_x$ was synthesized in two rows, and nearly the same photoluminescence of $Y_2O_3:Eu^{3+}_x$ and $Y_2O_3:Tb^{3+}_x$ was observed in both the rows, which confirmed that the experimental consistence was acceptable. The emission spectra of $Y_2O_3:Eu^{3+}_x$, corresponding with the first row of samples in the combinatorial library shown in Figure 8.15, are shown in Figure 8.16. Here, the $Y_2O_3:Eu^{3+}_x$ sample for x = 0.05 exhibits the strongest luminescence. This conclusion is consistent with that in the case of phosphor synthesized via the conventional method [24], which further validates the applicability of the approach.

8.1.4.3 Parallel Synthesis from Powders Sources

During the early stage of development of combinatorial materials science, scientists expended significant efforts in developing new methodologies to improve the sample density (i.e., the number of samples per unit size) of combinatorial libraries by synthesizing a large number of micro-content samples. The record thereof has been reported by Earl Danielson, who synthesized 25,000 samples on a substrate three inches in diameter [19]. The sample density reached a value of up to 600 variant compositions per square centimeter. The combinatorial libraries were deposited in thin films via electron beam (8 kV) evaporation in vacuum (base pressure, ~5 × 10⁻⁶ Torr) on unheated 3-inch-diameter silicon wafers by jointly using stationary and movable physical masks [19]. Although a large number of diversities were created and a large number of samples was synthesized, a useful

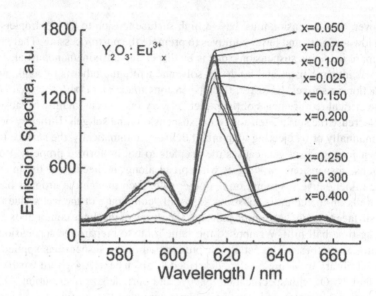

FIGURE 8.16   Emission spectra of $(Y_{1-x}Eu_x)_2O_3$ corresponding to different Eu^{3+} contents in the first row of samples in the combinatorial library shown in Figure 8.15. (Chen, L., Bao, J., Gao, C., Huang, S., Liu, C., and Liu, W. 2004. Combinatorial synthesis of insoluble oxide library from ultrafine/nano particles suspension using a drop-on-demand inkjet delivery system. *J. Comb. Chem.* 6(5): 699–702. With permission.)

material was rarely screened from them. Therefore, it makes more sense to develop new materials rather than to focus on the research methodology.

In combinatorial libraries, the micro-mass samples facilitate saving of expenditure on raw materials; however, these smaller amounts induce bigger errors and inconsistencies of the experiment. Particularly for phosphors, the more is the mass of phosphor synthesized, the more luminous it is. On the contrary, the surface defects induced by the micro-content samples do not improve luminescence. Moreover, it must be stated that the combinatorial approach does not imply that merely micro-contents of each sample should be synthesized. In recent years, the development of materials genomics has resulted in new requirements for multiple-sample synthesis techniques arising. Specifically, simultaneous synthesis of multiple samples, each with a large mass, should be possible.

We developed a Taguchi-method-assisted creative and intelligent combinatorial approach (as elaborated in the next section). When designed with the Taguchi method, 4, 8, 9, 16, 16, 25, 27, and 18 samples should be synthesized corresponding to the standard orthogonal tables of $L4(2^3)$, $L8(2^7)$, $L9(3^4)$, $L16(2^{15})$, $L16(4^5)$, $L25(5^6)$, $L27(3^{13})$, and the mixed-levels orthogonal table of $L18(2^1 \times 3^7)$, instead of a large number of samples being synthesized. On this basis, we developed a method to synthesize these samples via a parallel strategy and performed a Taguchi-method-assisted analysis on them.

We designed a group of nylon rollers, where each roller contained eight holes, as shown in Figure 8.17(a). As shown in Figure 8.17(b), every hole could be filled with an Al_2O_3 or BN crucible, whose dimensions could be adjusted according to the requirement. The cylinder crucible shown in Figure 8.17(c) has the size with outer diameter 21 mm, inner diameter 17 mm, and height 25 mm. The cover of the crucible is specially designed with a raised step used to seal, so lest powders leakage during ball milling. In the experiments, the powder raw materials with the desired stoichiometric ratios were first weighted and then filled into the crucibles, as shown in Figure 8.17(d). Some Si_3N_4 balls were also loaded into the crucibles and used as the grinding medium. Then, the crucibles with covers were placed into the holes of the roller, as shown in Figure 8.17(b). Each roller was filled with 8 samples—two rollers were filled with 16 samples, three were filled with 24 samples, and, in turn, 8N samples were loaded into the N rollers. As the Taguchi-method-assisted design using the

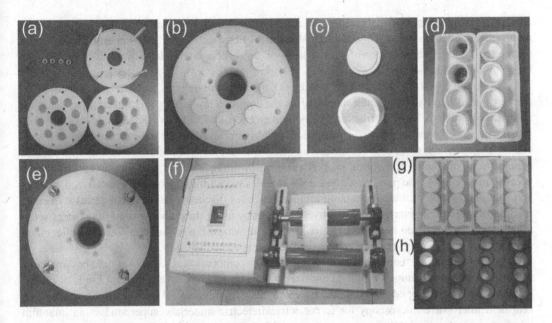

FIGURE 8.17 Instrument employed for parallel synthesis of multiple samples from powder sources. (a) Rollers with holes used to contain crucibles and a base; (b) crucibles with covers in the holes of a roller; (c) the inner shape of a cover paired with a crucible; (d) cylindrical crucibles filled with balls or powders; (e) top view of two tightened rollers on a base with 4 screws; (f) ball mixing on a jar mill; (g) a group of 16 samples designed as the L16($_4^5$) orthogonal table for firing in furnace; (h) the luminescence picture of the 16 samples synthesized by this way.

L16($_4^5$) orthogonal table for 5 factors with 4 levels each, 16 samples should be synthesized, for which two rollers suffice. As shown in Figure 8.17(e), the two rollers were stacked together and tightened using four screws, forming a cylinder. Next, the cylinder was rolled on the jar mill for a certain amount of time (typically 2–3 h), as shown in Figure 8.17(f), through which the raw materials were homogenously mixed. With the cylinder opened up, the crucibles were removed from the holes, the Si_3N_4 balls were taken out from the crucibles, and then, the 16 crucibles placed on 4 rectangular corundum boats were transferred into the furnace for the high-temperature reaction, as shown in Figure 8.17(g). Figure 8.17(h) displays the luminescence picture of a group of 16 samples synthesized by this way, designed with Taguchi-method using the L16($_4^5$) orthogonal table for 5 factors each with 4 levels.

The volume of the crucibles displayed in Figure 8.17 is approximately 5.2 mL; using the 5.2 mL crucible, phosphor with a mass of 1–2 g could be synthesized each time. When compared with the masses of the phosphor displayed in Figures 8.2, 8.5, 8.13, and 8.15 (which were of the order of milligrams in every reactor), the significantly increased mass could clearly reduce experimental errors and improve experiment consistence.

8.1.5   HIGH-THROUGHPUT CHARACTERIZATION OF COMBINATORIAL LIBRARIES OF PHOSPHORS

8.1.5.1   General Methods

The aim of high-throughput characterization is to realize rapid screening from a group of material combinations with specific properties for specific applications. There is no uniform pattern for the high-throughput characterization of combinatorial libraries, because the detection techniques vary with the measured properties, which differ according to different disciplines. Except for a few microscopic methods that can be directly used to measure the physical properties of samples in combinatorial libraries, conventional measurement techniques cannot easily meet the measurement requirements of microarray samples in combinatorial libraries. As for each specific property, the testing tools and techniques thereof vary. Therefore, it is crucial to develop novel testing techniques for combinatorial materials.

As far as the objects being tested are concerned, the high-throughput characterization techniques can be divided into two categories: one involves the common tools and general techniques used for all materials, such as X-ray diffraction, X-ray fluorescence, and SEM and TEM analyses integrated into the devices for testing the array samples in combinatorial libraries; the other involves the special technologies and tools used for measuring the physical properties of a certain material, such as the electrical, magnetic, and optical properties. High-throughput testing can follow two sequences: parallel and serial. As the parallel sequence, all samples were simultaneously characterized, such as by using photography for characterizing luminescence (as discussed later); in the series sequence, the samples were characterized consecutively, such as by using the combinatorial fluorescent scanning system (as discussed later). If measured in a serial way, thereby, the time taken to collect data should be short enough to maintain a practical representational velocity.

The high-throughput techniques used to characterize combinatorial libraries should be capable of detecting micro-area samples rapidly, nondestructively, and quantitatively. Several kinds of instruments have been developed for characterizing samples in combinatorial libraries, such as scanning near-field microwave microscopy for ferroelectric/dielectric materials, superconducting quantum interference device (SQUID) for magnetic materials, nano-indentation for hardness assessment, and electron probe microanalysis and electron backscattering diffraction for composition and structure analysis. To meet the requirements for variant properties, more characterization techniques are developed. Herewith, we introduce two methods for high-throughput characterization that have been successfully applied to the luminescent materials library.

8.1.5.2   Photography

Photoluminescence properties and those properties that can be derived from emission spectra can be detected easily, in contrast to other properties such as superconductivity, magnetics, electronics, thermal conductivity, and catalytic properties. Photography is associated with the characteristics of parallel, rapid, intuitive, and high spatial resolution (~10 μm) and has already been digitized. Thus, the facile method of photography has become one of the most common techniques used for high-throughput characterization of combinatorial libraries. From the luminescence depictions in Figures 8.2, 8.5, 8.8, 8.10, 8.13, and 8.15, the photographs could be captured under the excitation of variant wavelengths and used to examine the selective-excitation properties of phosphors. Moreover, the photograph of invisible luminescence in the infrared wavelength range could be captured using an infrared camera. In addition, the temperature of the phosphors could be detected via thermal imaging and used to examine the thermal stability of photoluminescence in combination with the emission spectra [40]. In addition, the infrared thermal imaging technique was used to rapidly evaluate the catalytic activities of each member of large encoded catalyst libraries [41].

Figure 8.18 presents a schematic of an imaging system used for high-throughput characterization of the vacuum ultraviolet–excited photoluminescence of combinatorial libraries in a vacuum. The system is equipped with an RF-powdered Xe spectrum lamp (model no. XeLM-L, Resonance Ltd), which simulates the discharge radiation in the plasma display panels (PDPs) and Hg-free lamps, used as an excitation light source with line 147 nm emission. The 147 nm vacuum ultraviolet illuminates the combinatorial library in the vacuum chamber, and the emission color and relative brightness of each sample in the combinatorial library are recorded using a digital CCD camera through the optical window of the system. The chamber is evacuated to eliminate air absorption via VUV radiation. Figure 8.18 shows a simple schematic, but the system is much more complex in practice. To work effectively, the system must be equipped with a vacuum pump and vacuum monitor. The vacuum monitor is controlled via an outer connected control instrument, and the vacuum light source is controlled via a stabilized voltage supply. More importantly, the luminance of each sample

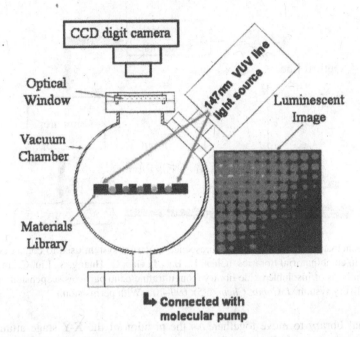

FIGURE 8.18 Schematic of imaging system used to record photoluminescence of combinatorial libraries excited with vacuum-ultraviolet in a vacuum. (Chen, L., Fu, Y., Zhang, G., Bao, J. and Gao, C. 2008. Optimization of Pr^{3+}, Tb^{3+}, and Sm^{3+} co-doped $(Y_{0.65}Gd_{0.35})BO_3:Eu^{3+}_{0.05}$ VUV phosphors through combinatorial chemistry approach. *J. Comb. Chem.* 10: 401–404. With permission.)

in the combinatorial library must be high enough to ensure that a clear photograph of the photoluminescence can be captured, which depends on the intensity of the 147 nm excitation light, sensitivity of the digital camera, and optical distance both from the excitation light source to the samples and from the luminescent samples to the camera. The system must be carefully designed to account for this.

The relative luminescence intensity of each sample in the combinatorial library could be evaluated from the relative brightness of the photograph depicting the photoluminescence with the aid of a computer, via conversion of a color image into gray scale image, as Equation (8.12) shown later. However, the errors due to differences in optical distance of every sample should be calibrated for. In our study, the relative intensity was calibrated using only one kind of commercially available phosphor. Before the measurement, a photograph of the combinatorial library filled with commercial phosphor in each reactor was captured using the same instrument at the same position.

8.1.5.3 Combinatorial Fluorescence Spectrum System

The luminescence photography employed for characterization of combinatorial libraries has the merits of enabling rapid evaluation of emission color and brightness intuitively from the photographs. However, the photoluminescence parameters in terms of color coordinates, color rendering index, and color temperature are all calculated from the emission spectrum. To compensate for the deficiency of the photography approach in terms of collecting the spectra, we developed a combinatorial fluorescence spectrum system [38]. As shown in Figure 8.19, this system mainly consisted of an Hg lamp with 254- and 365 nm emissions available, a portable optical fiber spectrometer (Ocean Optics, Inc., Model: OCEAN-HDX-XR; Detector: Back-thinned CCD image sensor; Entrance slit: 10 μm; Optical Resolution: 1.10 nm; Wave Length Range: 200–1100 nm), and an electric X-Y stage. The combinatorial library was fixed on the moving table of the electric X-Y stage, which drives

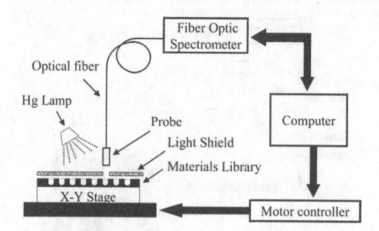

FIGURE 8.19 Schematic of combinatorial fluorescence spectrum system used to collect emission spectra of each phosphor in combinatorial libraries. (Chen, L., Bao, J., Gao, C., Huang, S., Liu, C., and Liu, W. 2004. Combinatorial synthesis of insoluble oxide library from ultrafine/nano particles suspension using a drop-on-demand inkjet delivery system. *J. Comb. Chem.* 6(5): 699–702. With permission.)

the combinatorial library to move together. As the position of the X-Y stage attained balanced, one emission spectrum was collected. The emission from the phosphor in the microreactors of the combinatorial libraries was recorded when the fiber-optic probe focused on the bottom of the microreactor. The movement of the X-Y stage and the spectrum collection by the optical fiber spectrometer were automatically controlled using a computer. To prevent interference from surrounding the samples, a light shield was used. Each time, only the emission from one sample was exposed to a probe detector. Thus, the emission spectra from the array samples in the combinatorial libraries were collected consecutively in the serial model; the emission spectra in Figures 8.14 and 8.16 were obtained using this system. Previously, the system was equipped with USB2000 and USB4000 optical fiber spectrometer models in our work; however, an advanced model is currently available. Moreover, the development of LEDs leads to light sources with more excitation wavelengths replacing the 254- and 365 nm Hg lamps.

8.1.6 COMBINATORIAL CHEMISTRY INFORMATICS ON PHOSPHORS

Much research effort has been expended on exploring the composition-structure-property relationship, but obtaining an accurate prediction of the luminescence, including excitation and emission wavelength and quantum efficiency of phosphors, is still very difficult. Studied via a combinatorial chemistry approach, a large number of samples were synthesized and characterized. All experiments are useful, and the experimental data thereof can serve as a valuable reference to others. To reasonably process and fully exploit these valuable data, the combinatorial materials informatics, as a branch of combinatorial chemistry, has been developed. As for the task of combinatorial materials informatics, not only do large amounts of data obtained in combinatorial experiments need to be reasonably processed and stored, but also these data need to be fully exploited to realize rational designs and develop new materials. Some useful databases, such as the "COD" (Crystallography Open Database); "The Materials Project", which provides open web-based access to computed information on known and predicted materials as well as powerful analysis tools to inspire and design novel materials; "Citrine Informatics", which is an AI-powered materials data platform created on application of data science, materials science, and machine learning to the materials and chemicals space, have been constructed. The task of combinatorial chemistry informatics of phosphors is arduous and involves building a database on photoluminescence properties in terms of emission spectra, excitation spectra, absorption/reflection spectra, external and internal quantum efficiencies, photoluminescence stability, hosts, activators, site occupation, site coordination, site symmetry, Stokes shift, etc.

8.1.7 Applications of Combinatorial Chemistry Approach in Phosphors

Typically, there are two avenues to develop new materials: one is by searching for new compounds with new structures; the other is by optimizing conventional materials using new recipes, new compositions, or new synthesis processing to achieve novel applications with improved properties. Herein, we provided an example on the application of combinatorial chemistry for developing novel phosphor via modification of a traditional one.

$(Y_{0.65}Gd_{0.35})_{0.95}Eu_{0.05}BO_3$ [usually abbreviated as $(Y,Gd)BO_3$:Eu] is the commercially available red phosphor for PDPs. The excitation spectrum of $(Y,Gd)BO_3$:Eu comprises two bands: one in the 125–175 nm range with a peak at 166 nm, and the other in the 175–275 nm range with a peak at 220 nm. In PDPs, the photoluminescence of phosphors is excited mainly by the 147 nm line emission, and minor continuous emission peaked at 172 nm, radiated upon the discharge of Xe gas. We believe that with the 147 nm VUV light incident on the phosphor, the ionized process will occur in the $(Y,Gd)BO_3$:Eu phosphor along with charger carrier formation and release. One key factor that determines photoluminescence efficiency is energy transfer from the host to activator. This transfer is strongly affected by the defects, including the types of carriers and the depth of traps, in the host. From this point of view, traces of co-dopants that have different affinities should have the potential to change trap configuration and consequently improve luminescence efficiency. However, a systematic investigation on the effect of variant co-dopants on $(Y,Gd)BO_3$:Eu photoluminescence is time-consuming. Herein, the combinatorial chemistry approach was employed to examine the photoluminescence of $(Y,Gd)BO_3$:Eu co-doped by Pr^{3+}, Tb^{3+}, and Sm^{3+}, respectively. Figure 8.20 presents the composition map of the combinatorial libraries [42].

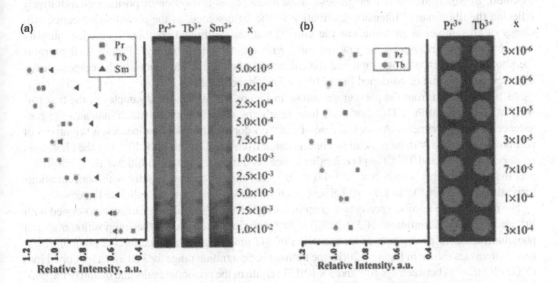

FIGURE 8.20 Composition maps and photoluminescence photographs of $Pr^{3+}/Tb^{3+}/Sm^{3+}$ co-doped (Y,Gd) BO_3:Eu^{3+} combinatorial libraries and relative photoluminescence intensity of samples as a function of Pr^{3+}, Tb^{3+}, or Sm^{3+} concentrations under the excitation of 147 nm: red square, Pr^{3+} co-doped $(Y_{0.65}Gd_{0.35})_{0.95}BO_3$: $Eu^{3+}_{0.05}$; green circle, Tb^{3+} co-doped $(Y_{0.65}Gd_{0.35})_{0.95}BO_3$:$Eu^{3+}_{0.05}$; blue triangle, Sm^{3+} co-doped $(Y_{0.65}Gd_{0.35})_{0.95}$ BO_3:$Eu^{3+}_{0.05}$. (a) The primary and (b) secondary combinatorial libraries. (Chen, L., Fu, Y., Zhang, G., Bao, J. and Gao, C. 2008. Optimization of Pr^{3+}, Tb^{3+}, and Sm^{3+} co-doped $(Y_{0.65}Gd_{0.35})BO_3$:$Eu^{3+}_{0.05}$ VUV phosphors through combinatorial chemistry approach. *J. Comb. Chem.* 10: 401–404. With permission.)

The Pr^{3+}, Tb^{3+}, and Sm^{3+} co-doped (Y,Gd)BO$_3$:Eu combinatorial library was synthesized via a solution-based method involving ejection of the Y_2O_3, Gd_2O_3, and Eu_2O_3 suspensions, Pr^{3+}, Tb^{3+}, and Sm^{3+} nitrate solutions, and aqueous solution of H_3BO_3, using the automatic drop-on-demand inkjet delivery system shown in Figure 8.12. The Y_2O_3, Gd_2O_3, and Eu_2O_3 suspensions were prepared using the high-energy ball mill in deionized water, as described in Section 8.1.4.2; the Pr^{3+}, Tb^{3+}, and Sm^{3+} nitrate solutions were prepared by dissolving Pr_6O_{11}, Tb_4O_7, and Sm_2O_3 in nitric acid with pH = 1.0. Owing to traces of Pr^{3+}, Tb^{3+}, and Sm^{3+} being co-doped, their nitrate solutions were prepared in very low concentrations. However, the solubility of H_3BO_3 is very low, namely, ~5 g in 100 g water at 20°C. To balance the droplet volume, the suspensions of Y_2O_3, Gd_2O_3, and Eu_2O_3 were also prepared at low concentrations, which help to improve the stability of the suspensions. However, to increase the total mass of the phosphors in the microreactors, each suspension or solution was doubly ejected as the sequence of B–Y–Gd–Eu–Pr–Tb–Sm–B–Y–Gd–Eu–Pr–Tb. The amount of H_3BO_3 was in stoichiometric 20 mol% excess to compensate for high-temperature evaporation during the reaction. Before the ink-jetting, the suspensions and solutions were degassed by injecting argon gas for 2 h, and the surfaces of the microreactors were pretreated via coating with a thin layer of liquid paraffin to prevent fluid leakage. The as-prepared libraries were first dried in ambient air to evaporate the solution, heated at 176°C (i.e., nearly the melting point of H_3BO_3) for 30 min to allow for the diffusion of H_3BO_3, annealed at 900°C for 120 min, and finally sintered at 1150°C for 180 min.

The photoluminescence of the combinatorial libraries was characterized via photography, as the system shown in Figure 8.18. The relative photoluminescence intensity of each sample was evaluated with the assistance of a computer according to their brightness in the luminescence photograph. The errors in luminance due to uneven illumination of the point-like source, owing to variant optical distance in space, was eliminated by calibrating with a "standard library" (the microreactors on the substrate were manually filled with the same amount of the commercially available PDP red phosphor (Y,Gd)BO$_3$:Eu^{3+}). At first, a photoluminescence photograph of the "standard library" was captured, in which the relative brightness of each sample as a function of position quantitatively reflected the illumination intensity distribution of the light source, as the photoluminescence efficiency of all samples at each site was the same. Then, the calibrated photoluminescence intensity was obtained by division of the raw intensity with the "standard library". Figure 8.20 presents the photoluminescence photograph and the calibrated intensity of phosphors in the combinatorial libraries as a function of co-doped Pr^{3+}, Tb^{3+}, or Sm^{3+} concentrations [42].

As discriminated from the photograph shown in Figure 8.20(a), the three samples in the first row without co-doping of Pr^{3+}, Tb^{3+}, and Sm^{3+} have nearly the same brightness, which indicates the consistence of the experiments. As for the case of Sm^{3+} co-doping, the photoluminescence brightness of phosphors decreased with increase in its concentration from 5.0×10^{-5} to 1.0×10^{-2}. For the phosphors co-doped with Pr^{3+} and Tb^{3+}, the photoluminescence brightness increased within 5×10^{-5}–2.5×10^{-4}, was maintained stably within 5×10^{-4}–1.0×10^{-3}, and then decreased rapidly with concentrations increasing from 2.5×10^{-3} to 1.0×10^{-2}. These analyses are verified using the calibrated intensities.

For more precise results, a secondary combinatorial library of (Y,Gd)BO$_3$:Eu^{3+} was co-doped with Pr^{3+} and Tb^{3+} in concentrations of 3×10^{-6}–3×10^{-4} with finer intervals. As the composition map and photoluminescence photograph under excitation of 147 nm along with the calibrated photoluminescence intensities shown in Figure 8.20(b), the optimal concentration range for Pr^{3+} and Tb^{3+} doped into (Y,Gd)BO$_3$:Eu^{3+} is between 7×10^{-6} and 3×10^{-4}. To confirm the previous results and quantify the photoluminescence enhancement, bulk samples of $(Y_{0.65}Gd_{0.35})_{0.95}Eu_{0.05}BO_3$ and $(Y_{0.65}Gd_{0.35})_{0.95}Eu_{0.05}BO_3$ doped with 1×10^{-4} Pr^{3+} or Tb^{3+} were synthesized via solid-state reactions under the same conditions. First, the undoped/co-doped $[(Y_{0.65}Gd_{0.35})_{0.95}Eu_{0.05}]_2O_3$ oxides were prepared by firing their coprecipitated oxalates at 1000°C for 120 min. Then, these oxides were wet ball-milled with a stoichiometric 10 mol% excess of H_3BO_3 in additive ethanol. Thereafter, the ball-milled mixtures were sintered at 600°C for 120 min and ground in an agate mortar. These mixtures were then annealed at 900°C for 120 min and fired at 1200°C for 180 min. Finally, the fired products were wet ball-milled in deionized water and washed with 80–100°C deionized water several times to remove excess B_2O_3 and then dried at 120°C. The emission and excitation spectra of the bulk samples were measured using the National Synchrotron Radiation Laboratory (NSRL, China) VUV spectroscopy endstation on the U24 beamline. The emission spectra show that the photoluminescence intensity of $(Y,Gd)BO_3:Eu^{3+}$ was enhanced by approximately 15% by co-doping with 1×10^{-4} Pr^{3+} or Tb^{3+}. However, no characteristic excitation bands of Pr^{3+} and Tb^{3+} were observed upon monitoring the emission of Eu^{3+} from the excitation spectra of Pr^{3+} or Tb^{3+} co-doped $(Y,Gd)BO_3:Eu^{3+}$ phosphors. Therefore, Pr^{3+} and Tb^{3+} are deemed not to be the typical sensitizers for the activator of Eu^{3+} in the $(Y,Gd)BO_3$ host. A similar example is the improved red luminescence of $Y_2O_2S:Eu^{3+}$ by traces of Pr^{3+} and Tb^{3+} under cathode ray excitation [43]. Besides the improved photoluminescence, this work presents a significant phenomenon to reveal the mechanism of VUV-excited $(Y,Gd)BO_3:Eu^{3+}$ photoluminescence, which must be related with defects and the generation and migration of carriers.

8.2 TAGUCHI-METHOD-ASSISTED COMBINATORIAL CHEMISTRY APPROACH—AN INTELLIGENT APPROACH TO RESEARCH ON PHOSPHORS

8.2.1 Application of Intelligent Methods in Combinatorial Chemistry

Recent advances in combinatorial chemistry have enabled rapid synthesis and high-throughput screening of large numbers of diverse compounds to find new functional materials, including phosphors. Nevertheless, even the largest combinatorial libraries (the record being 25,000 samples in a library) that can be synthesized and screened cover only a very small proportion of all possible molecules. In the periodic table, there are more than 65 elements that could be used to synthesize phosphors. Suppose these phosphors were to be synthesized from 3, 4, 5, or even more elements, the potential number of compounds that can be synthesized from readily available components/monomers is extremely large (at least $>10^{65}$). To reduce the cost and time expenditure in the synthesis and characterization of innumerous compounds, some intelligent approaches, including the genetic algorithm, Monte Carlo technique, simulated annealing guided evaluation, and artificial neural network algorithms, have been adopted to compensate for the weaknesses of the combinatorial chemistry approach and promote the discovery of new materials. Undeniably, the application of intelligent algorithms in combinatorial chemistry has significantly accelerated the discovery and optimization of lead compounds, identification of biological targets, design of proteins and nucleic acids, and evaluation of data in pharmaceutical industries.

With respect to inorganic functional materials, genetic algorithms [44], heuristics-assisted combinatorial screening [45], neural network algorithms, and machine learning based on big data have also been utilized in combinatorial chemistry to search for new phosphors. For example, combinatorial chemistry assisted with genetic algorithms, based on a global optimization strategy in which the evolutionary process is imitated with elitism, selection, crossover, and random mutation operations, has been employed to optimize new phosphors for light-emitting diodes and PDPs. However, convincing progress has not been made in terms of evaluating the contribution of a component to newly constructed materials, identifying useful compositions of a particular set of components, and estimating their expected performance. Otherwise, the numbers of samples synthesized can be greatly reduced by excluding useless elements and concentrations. Correspondingly, the cost and time of synthesizing and charactering numerous compounds can be reduced substantially. Prof. Ru-Shi Liu and Lei Chen creatively applied the Taguchi method to combinatorial chemistry, and achieved improvements in the intelligence of the combinatorial chemistry approach via the Taguchi method for evaluating the effects of components on an objective property, identifying the factors that have the strongest effects, reducing the experimental dimensions of multifactors by excluding factors that have harmful or minor effects, and estimating the expected performance under the optimal conditions demonstrated in our research [46].

8.2.2 Brief Introduction to the Taguchi Method

The Taguchi method, as proposed by Dr. Genichi Taguchi, has become a standardized approach to utilize the design of experiment (DOE) technique to enhance the quality of products and processes [46]. The DOE technique is an experimental strategy in which the effects of multiple factors are studied simultaneously by running tests with those factors at various levels. Taguchi's emphasis on loss to society, techniques for investigating variation in experiments, and over-all strategy of the system, parameter, and tolerance design have been influential in improving the quality of manufacturing worldwide. Taguchi's work includes three principal contributions to statistics, namely a specific loss function, the philosophy of off-line quality control, and the innovation in experimental design [46]. In the Taguchi approach, experiments are designed using special orthogonal arrays that allow the whole parameter space to be studied by perform-ing a minimum number of experiments.

Orthogonal arrays are a set of tables containing information on how to determine the least number of experiments required and their conditions. The fixed orthogonal tables, called standard orthogonal arrays, are available to design simple experiments in which factors can vary among a fixed number of levels. Experimental designs with mixed levels of multifactors require knowledge on statistics to modification of the standard arrays. In orthogonal array tables, columns present factors that can be accommodated, while rows present trial conditions. The standard notation for orthogonal arrays is $Ln(_X{}^Y)$, where n represents the number of experiments (rows in the orthogonal table), Y represents the number of factors (columns in the orthogonal table), and X represents the levels of a factor. The commonly used orthogonal tables include $L4(_2{}^3)$, $L8(_2{}^7)$, $L9(3^4)$, $L16(_2{}^{15})$, $L16(_4{}^5)$, $L25(_5{}^6)$, $L27(_3{}^{13})$, and the mixed-levels $L18(_2{}^1 \times _3{}^7)$. Tables 8.1 and 8.2 show two examples on the design of orthogonal tables of $L9(3^4)$ and $L16(4^5)$.

8.2.3 Mathematical Principles of the Taguchi Method

The factors in the column of the orthogonal tables are orthogonal, for which the orthogonality ensures that each level of every two factors is irrelevant in statistics. The orthogonality is manifested in two ways:

1. Uniform dispersion: In each column of an orthogonal table, different numbers appear the same number of times. For example, in the $L9(3^4)$ orthogonal table, the numbers 1, 2, and 3 all appear three times in each column.

TABLE 8.1

The L9(3^4) Orthogonal Table

Experiments No	Factor A	Factor B	Factor C	Factor D
	\multicolumn{4}{c}{Column}			
1	1	1	1	1
2	1	2	2	2
3	1	3	3	3
4	2	1	2	3
5	2	2	3	1
6	2	3	1	2
7	3	1	3	2
8	3	2	1	3
9	3	3	2	1

TABLE 8.2
The L16(4^5) Orthogonal Table

Experiments No	Column				
	Factor A	Factor B	Factor C	Factor D	Factor E
1	1	1	1	1	1
2	1	2	2	2	2
3	1	3	3	3	3
4	1	4	4	4	4
5	2	1	2	3	4
6	2	2	1	4	3
7	2	3	4	1	2
8	2	4	3	2	1
9	3	1	3	4	2
10	3	2	4	3	1
11	3	3	1	2	4
12	3	4	2	1	3
13	4	1	4	2	3
14	4	2	3	1	4
15	4	3	2	4	1
16	4	4	1	3	2

2. Neat comparability: For any two columns of an orthogonal table, two numbers in the same row are treated as an ordered pair, and each pair appears an equivalent number of times. For example, there are nine ordered pairs in the L9(3^4) orthogonal table, i.e., (1,1), (1,2), (1,3), (2,1), (2,2), (2,3), (3,1), (3,2), and (3,3), each of them occurs once.

Herein, we provide an example on the orthogonal design of multiple-factor experiments, illustrated via an experiment on the three factors, each with three levels. Let us suppose there are three factors A, B, and C, and each factor has the levels 1, 2, and 3. If the examination were to be conducted with the full-levels-of-full-factors experiments, 27 points should be examined. Nevertheless, if the examination were to be carried out via the orthogonal method, only nine points need to be studied. The distribution of the examined nine points in three-dimension space is shown in Figure 8.21. In Figure 8.21, the cube is divided into three planes along any of the X, Y, or Z axes. In each plane, there are three points, and every factor appears the same time (once). The nine points in Figure 8.21 show the uniform dispersion and neat comparability characters of orthogonality. The multiple-factor experiments of the three factors, each with three levels, expressed as in L9(3^3), should adopt the L9(3^4) orthogonal table. We should bear in mind that the L9(3^4) orthogonal table could accommodate at most four factors, each with three levels, as in the L9(3^4) orthogonal table displayed in Table 8.1. By using the L9(3^4) orthogonal table to perform the L9(3^3) orthogonal experiment, one column blank is permitted. However, the blank column could be used to set the interaction of A × B, A × C, or B × C.

Experiments conducted to determine the best levels based on orthogonal arrays are balanced with respect to all control factors and are yet minimal in number, as has been proved mathematically. This in turn implies that the resources (such as materials and time) required for the experiments are minimal. For example, consider the optimization of a new material with five elements, with one of four concentrations of each element; the synthesis and characterization of $4^5 = 1024$ samples should be carried out one by one in the case of full factorial experiments. In practice, 4^5 experiments are too many, especially when conventional methods of synthesis and characterization are employed. When taken with the Taguchi method, however, experiments on 16 samples

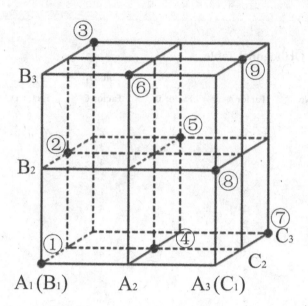

FIGURE 8.21 Schematic of distribution of the nine points in the L9(3^4) orthogonal table.

are sufficient to efficiently optimize the five factors, each with four-levels of concentration. The 16 samples are easily integrated in a combinatorial library through parallel synthesis and can be characterized with the high-throughput screening of the combinatorial chemistry approach. Moreover, the achievements triggered by the Taguchi method have rendered automatic design and data analysis of Taguchi experiments possible with the aid of computers. Software, including Qualitek-4, STATISTICA 8.0, and Assistant for Orthogonal Experimental Design II (in Chinese), have been developed for performing such Taguchi experiments.

Studied via the Taguchi method, the data are analyzed with three models: standard average of results, standard deviation of results, and signal-to-noise analysis. The experimental quality characteristics (QCs) are classified into three forms: larger-the-better (such as agricultural yield), smaller-the-better (such as defects), and nominal-the-best (such as mechanical fittings). As for the analysis of the standard deviation of values, we always choose the QCs of smaller-the-better. This method is widely used to analyze tolerances and fits in machining parts, for which a smaller deviation indicates a higher quality of products.

The analysis of average values could help to identify the main effects and quantitatively evaluate the influence of each level on objective property. Table 8.3 gives an example on experimental designs and results on synthesis of the K_2SiF_6:Mn^{4+} phosphors using the Taguchi method and a results analysis with the standard average model of the Taguchi method. In Table 8.3, the T_1, T_2, and T_3 are the sum of the results of all factors at the same levels of 1, 2, and 3, respectively. For example,

$$T_{1A} = y_1 + y_2 + y_3 = 203 + 136 + 199 = 538 \tag{8.3}$$

$$T_{2C} = y_2 + y_4 + y_9 = 136 + 388 + 1940 = 2464 \tag{8.4}$$

The average values of $\bar{T}_1$, $\bar{T}_2$, and $\bar{T}_3$ are obtained with T_1, T_2, and T_3 divided by 3. Then, the range R for factors A, B, and C are concluded via the maximum of the average minus the minimum of the average. The higher value of the range R indicates that the factor has a higher influence on the objective property. In Table 8.3, $R_A = 1412$, $R_B = 566$, and $R_C = 495$, which suggests the factor A among the three factors of A (Mn^{4+} concentration), B (HF content with respect to solid compounds), and C (ions-exchange time) has the biggest influence on luminescence intensity of K_2SiF_6:Mn.

TABLE 8.3
The Orthogonal Table for Experiment Conditions of K_2SiF_6:Mn Phosphors Synthesized with Ion-Exchange Approach, Experimental Results, and Results Analysis on Average Values

	Column			
Experiments No	Factor A (Mn⁴⁺ Concentration)	Factor B (HF Content with Respect to Solid Compounds)	Factor C (Ions-Exchange Time)	Results (y)
1	1 (0.03)	1 (1:2)	1 (15 min)	203
2	1 (0.03)	2 (1:3)	2 (30 min)	136
3	1 (0.03)	3 (1:4)	3 (45 min)	199
4	2 (0.06)	1 (1:2)	2 (30 min)	388
5	2 (0.06)	2 (1:3)	3 (45 min)	743
6	2 (0.06)	3 (1:4)	1 (15 min)	912
7	3 (0.09)	1 (1:2)	3 (45 min)	761
8	3 (0.09)	2 (1:3)	1 (15 min)	2073
9	3 (0.09)	3 (1:4)	2 (30 min)	1940
T_1	538	1352	3188	Sum of every
T_2	2043	2464	2464	factor at each
T_3	4774	3051	1703	level
$\overline{T}_1$	179	451	1063	Mean every factor
$\overline{T}_2$	681	984	821	at each level
$\overline{T}_3$	1591	1017	568	
R	1412	566	495	$\overline{T}_{max} - \overline{T}_{min}$

The Taguchi method divides all optimization problems into two categories: static or dynamic. Dynamic problems have a signal factor, while static problems do not. In dynamic problems, the optimization is performed using two S/N ratios: slope and linearity. In static problems, the optimization is achieved using three forms of S/N ratio: larger-the-better (such as agricultural yield), smaller-the-better (such as defects), and nominal-the-best (such as mechanical fittings).

Analyzed with the standard average of results, the larger difference between the maximum and the minimum average values of the factor suggests a bigger influence on the objective property. However, the bigger variance does not indicate that the result has a higher accuracy and higher precision, as the schematic shown on the left in Figure 8.22. To overcome these shortcomings, an analysis of variance (ANOVA, a form of statistical analysis) was performed to determine the significance of control factors on the objective property. The results of the ANOVA are displayed for the values of:

- DOF (f): degree of freedoms, which is one fewer than the number of levels of each factor.
- Sum of Sqrs (S) in total: sum of squared deviations from the mean. For n values of y_i, and mean of $\overline{y}$,

$$S = \sum_{i=1}^{n}(y_i - \overline{y})^2 \qquad (8.5)$$

- Sum of Sqrs (S_A) of factor A: sum of squared deviations of from the mean. The mean result $\overline{T}_i$ (as corresponding to Table 8.3) for n_i times experiments under the level i, where a is the level number of factor A.

$$S_A = \sum_{i=1}^{a} n_i(\overline{T}_i - \overline{y}) \qquad (8.6)$$

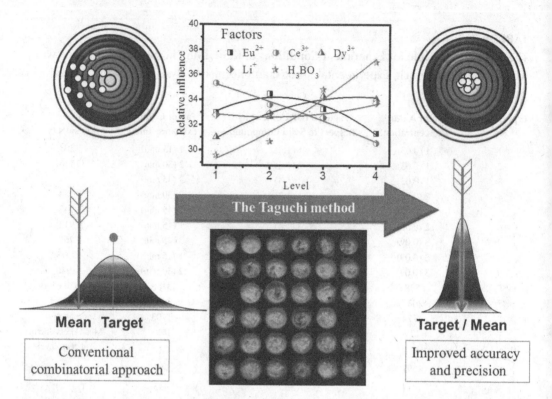

FIGURE 8.22   Schematic of intelligence of combinatorial chemistry approach improved by Taguchi method.

Sum of Sqrs (S) of other factors could be expressed in a likewise way.

- Sqrs of error (S_{error}), difference between the sum of squared deviations and the squared deviations of all factors.

$$S_{error} = S - S_A - S_B - S_C - \ldots \tag{8.7}$$

- Pure Sum (S′): the mean of squares minus the mean of squares error

$$S' = S - S_{error} \tag{8.8}$$

- Variance (V): mean of squares per DOF

$$V = \frac{S}{f} \tag{8.9}$$

- F-ratio (F): F is the ratio between the mean of squares and the mean of squares error.

$$F = \frac{S}{S_{error}} \tag{8.10}$$

The right-most column in the ANOVA table presents the relative effects of the factors on the results. The sum of all percentage effects always adds up to 100%. The percentages are determined by taking the ratio of the column variation (S) to the total variation. A small percentage effect means that the factor tolerance can be relaxed, while the tolerances for factors with higher percentage effects have to be tightened or monitored carefully.

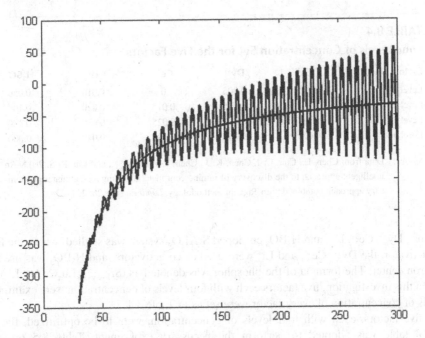

FIGURE 8.23   Illustration of signal-to-noise in curves.

In the experiments, we need to identify main factors which have key effects on objective properties. Nevertheless, the main factor usually was evaluated by the variation range of average values of experimental results (as seen Table 8.3). If the factor is a main factor, its alteration must produce significant variation on objective properties. But, in turn, a big variation may be in terms of noise, rather than in terms of useful information, as illustrated in Figure 8.23. In Figure 8.23, the bigger magnitude of the vibration suggests higher noise whereas the useful information is the average value of the vibration curve. Before optimization, the experimental results usually exhibit high variation, but after optimization, the results usually are stable with small variation (as seen Figure 8.22). Based on previous analysis, Taguchi's signal-to-noise (the magnitude of the mean of a process compared to its variation) ratios (S/N), which are log functions of desired output, serve as objective functions for optimization and help in data analysis and prediction of optimum results. The function proposed by Taguchi for calculating the S/N ratio is:

$$\frac{S}{N} = -10\log_{10}\left[\frac{1}{n}\sum_{i=1}^{n}\frac{1}{y_i^2}\right] \qquad (8.11)$$

where y_i denotes the n observations of the responding variable. As stated earlier, the Taguchi method divides all optimization problems into two categories-static or dynamic. Dynamic problems have a signal factor, while static problems do not. In static problems, the optimization is achieved using three forms of S/N ratio: larger-the-better (such as agricultural yield), smaller-the-better (such as defects), and nominal-the-best (such as mechanical fittings).

8.2.4   IMPROVED INTELLIGENCE OF COMBINATORIAL CHEMISTRY APPROACH BY ASSISTING WITH THE TAGUCHI METHOD

Herein, we provide an example on the optimization of long-lasting phosphor by combining the combinatorial chemistry approach with the Taguchi method, used to demonstrate the improved intelligence of combinatorial chemistry approach by assistance from the Taguchi method.

TABLE 8.4

Four Levels of Concentration Set for the Five Factors

Factors Levels	Eu^{2+}	Dy^{3+}	Ce^{3+}	Li^+	H_3BO_3
Level 1	0.01	0.005	0	0.010	0.050
Level 2	0.02	0.015	0.015	0.020	0.100
Level 3	0.03	0.030	0.025	0.030	0.200
Level 4	0.05	0.050	0.040	0.040	0.300

Source: Data from Chen, L., Chu, C.-I, Chen, K. J., Chen, P. Y. Hu, S. F., and Liu, R. S. 2011. An intelligent approach to the discovery of luminescent materials using a combinatorial chemistry approach combined with Taguchi methodology. *Luminescence* 26: 229–238.

The Eu^{2+}, Dy^{3+}, Ce^{3+}, Li^+, and H_3BO_3 co-doped $SrAl_2O_4$ system was studied, where the Eu^{2+} was used as activator, the Dy^{3+}, Ce^{3+}, and Li^+ were used as co-activators, and H_3BO_3 was used as flux and electron center. The formula of the phosphor was denoted as $(Sr_{1-x-y-z-s}Eu_xCe_yDy_zLi_s)Al_2O_4 + tH_3BO_3$. In this investigation, five factors each with four levels of concentration were examined. The four levels of concentration of every factor were set as in Table 8.4.

Since five factors, each with four levels of concentration, were to be optimized, the L16(4^5) orthogonal table was adopted to perform the necessary experiment. Table 8.5 presents the

TABLE 8.5

Experimental Design Using the L16(4^5) Orthogonal Table and Experimental Results

Experimental Design					Experimental Results (Relative Luminance, a.u.)			
Orthogonal Arrays Type L-16					Persisting for 1 s after Stopping Excitation			
Columns	1	2	3	4	5	Statistical		
Factors	Eu	Ce	Dy	Li	B	1	2	3	Average
Experiment 1	0.010	0	0.005	0.010	0.050	30.25	30.59	29.77	30.20
Experiment 2	0.010	0.015	0.015	0.020	0.100	35.01	34.46	34.64	34.70
Experiment 3	0.010	0.025	0.030	0.030	0.200	57.60	57.82	58.44	57.95
Experiment 4	0.010	0.040	0.050	0.040	0.300	65.48	66.70	64.11	65.43
Experiment 5	0.020	0	0.015	0.030	0.300	100	100	100	100
Experiment 6	0.020	0.015	0.005	0.040	0.200	58.60	62.14	60.59	60.45
Experiment 7	0.020	0.025	0.050	0.010	0.100	41.40	42.45	42.92	42.26
Experiment 8	0.020	0.040	0.030	0.020	0.050	31.10	30.20	31.09	30.80
Experiment 9	0.030	0	0.030	0.040	0.100	58.52	57.79	57.55	57.95
Experiment 10	0.030	0.015	0.050	0.030	0.050	36.12	35.16	36.55	35.94
Experiment 11	0.030	0.025	0.005	0.020	0.300	55.08	52.97	55.32	54.46
Experiment 12	0.030	0.040	0.015	0.010	0.200	40.37	39.83	39.83	40.01
Experiment 13	0.050	0	0.050	0.020	0.200	67.66	67.24	64.11	66.34
Experiment 14	0.050	0.015	0.030	0.010	0.300	72.33	69.74	69.45	70.51
Experiment 15	0.050	0.025	0.015	0.040	0.050	25.04	24.45	24.46	24.65
Experiment 16	0.050	0.040	0.005	0.030	0.100	16.33	16.04	15.82	16.06

Source: Data from Chen, L., Chu, C. I, Chen, K. J., Chen, P. Y. Hu, S. F., and Liu, R.-S. 2011. An intelligent approach to the discovery of luminescent materials using a combinatorial chemistry approach combined with Taguchi methodology. *Luminescence* 26: 229–238.

experimental design, in which the 16 trial conditions represent the compositions of the 16 samples. A combinatorial library that integrates the 16 samples was synthesized by ink-jetting Sr^{2+}, Al^{3+}, Eu^{2+}, Ce^{3+}, Dy^{3+}, and Li^+ nitrite solutions and boric acid solution into microreactors using a home-made ink-jet delivery system as shown in Figure 8.12. After drying at 60–70°C, the microreactors that contained the precursors were pre-fired at 600°C for 3 h, and then fired at 1250°C for 4 h, to yield the final combinatorial library. The emission spectrum of each sample in the combinatorial library was obtained by using a scanning fluorescence characterization system, as shown in Figure 8.19. The long-lasting luminescence was characterized by taking a photograph of the phosphor luminescence that persisted for certain time after the excitation light source was turned off. The relative brightness of luminescence from samples in the photograph was determined by a computer. The combinatorial library was synthesized twice (as the luminescence photograph shown in Figure 8.22) to verify experimental consistence and every combinatorial library was characterized three times to improve the statistical accuracy of results.

Figure 8.24 presents the luminescence photograph of the combinatorial library, which integrated the 16 samples, designed using the aforementioned Taguchi method. The 17th sample, $(Sr_{0.98}Eu_{0.02})Al_2O_4$, in the combinatorial library had no $Ce^{3+}/Dy^{3+}/Li^+/H_3BO_3$ co-doping and was used as a reference for comparison with other samples. Strong green luminescence was observed under the excitation of 365 nm, and a significant variation among the luminance of samples was presented. Figure 8.25 presents emission spectra of every sample in the combinatorial library, as compared with the reference sample of $(Sr_{0.98}Eu_{0.02})Al_2O_4$. These broad emission bands are all attributed to the 5d–4f transition of Eu^{2+}, and no other emission from Ce^{3+} or Dy^{3+} was observed.

To investigate the persistent luminescence, samples in the combinatorial library were firstly excited at 365 nm for 20 min, and then the light source was turned off. Figure 8.26(a)–(c) present the photographs taken after 1, 2, and 5 s of luminescence, respectively. The comparison of Figure 8.24 with Figure 8.26 shows the significant difference between the luminescence under 365 nm excitation and without excitation, especially for the luminescence from the 17th sample. Strong green emission from $(Sr_{0.98}Eu_{0.02})Al_2O_4$ is observed under 365 nm excitation, but no emission is observed after the excitation has been stopped. This difference reveals that Ce^{3+}, Dy^{3+}, Li^+, or H_3BO_3 play an important role on the long-lasting luminescence of Eu^{2+} in $SrAl_2O_4$ host.

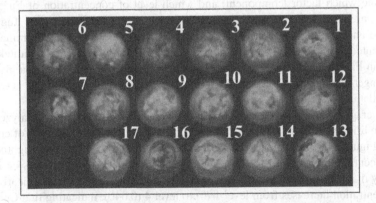

FIGURE 8.24 Photograph of luminescence of $SrAl_2O_4$:$Eu^{2+}/Dy^{3+}/Ce^{3+}/Li^+/H_3BO_3$ combinatorial library under 365 nm excitation; number of each sample corresponds to number of corresponding experiment in Table 8.5. (Chen, L., Chu, C. I, Chen, K. J., Chen, P. Y. Hu, S. F., and Liu, R. S. 2011. An intelligent approach to the discovery of luminescent materials using a combinatorial chemistry approach combined with Taguchi methodology. *Luminescence* 26: 229–238. With permission.)

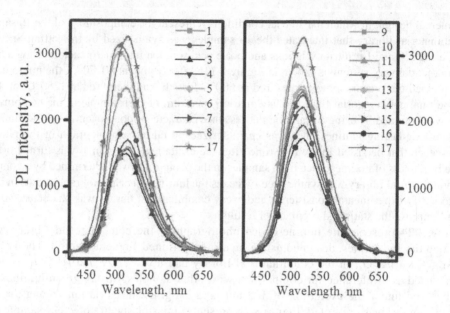

FIGURE 8.25 Emission spectra of $SrAl_2O_4$:Eu^{2+}/Dy^{3+}/Ce^{3+}/Li^+/H_3BO_3 samples in combinatorial library under 365 nm excitation. Samples number corresponds to the experiment number in Table 8.5. (Chen, L., Chu, C.-I, Chen, K. J., Chen, P. Y. Hu, S. F., and Liu, R. S. 2011. An intelligent approach to the discovery of luminescent materials using a combinatorial chemistry approach combined with Taguchi methodology. *Luminescence* 26: 229–238. With permission.)

According to computer graphics, the relative brightness in the photograph corresponds to the value of gray, and the value of gray can be obtained by calculating with Equation (8.12).

$$\text{Gray} = \frac{(9788 \times \text{Red} + 19,235 \times \text{Green} + 3735 \times \text{Blue})}{32,678} \tag{8.12}$$

Thus, the relative brightness of luminescence that persisted for 1 s was obtained by statistical average of the brightness of all pixels, as presented in area of every sample luminescence in Figure 8.26(a), is presented in Table 8.5. The calculation is conducted automatically using a computer.

To determine which factor (component) and which level of concentration of Eu^{2+}, Ce^{3+}, Dy^{3+}, Li^+ and H_3BO_3 are most desirable for achieving the objectives of this project, the Taguchi method is employed to analyze the data concerning the relative brightness of 1 s of luminescence from samples, presented in Table 8.5. The results were statistically analyzed using Qualitek-4 software. As displayed in Figure 8.26, higher brightness of luminescence that persists for a given time corresponds to longer-lasting luminescence. Therefore, the "bigger-the-better" of QC is selected in the processing of data.

First, the average effects of every control factors on objective property were analyzed by averaging the relative brightness of all samples that contains each one of the four levels of concentration. As shown in Figure 8.27, Li^+ has a minor effect on the variation of relative brightness with the change of concentrations, while Eu^{2+}, Ce^{3+}, Dy^{3+}, and H_3BO_3 have strong impacts on the luminescence intensity that persists over a particular period. The luminescence almost linearly declines as the Ce^{3+} concentration increases from level 1 (0) to level 4 (0.040), indicating that Ce^{3+} is harmful to obtain long-lasting luminescence. Hence, Ce^{3+} should be excluded from the raw materials in selecting elements to construct a new material. Luminescence initially increases and then decreases as the concentrations of Eu^{2+} and Dy^{3+} increase, reaching its maximum concentrations of 0.02 and 0.03, respectively, indicating that Eu^{2+} and Dy^{3+} contribute to long-lasting luminescence and do so most at concentrations of 0.02 and 0.03. H_3BO_3 clearly promotes long-lasting luminescence.

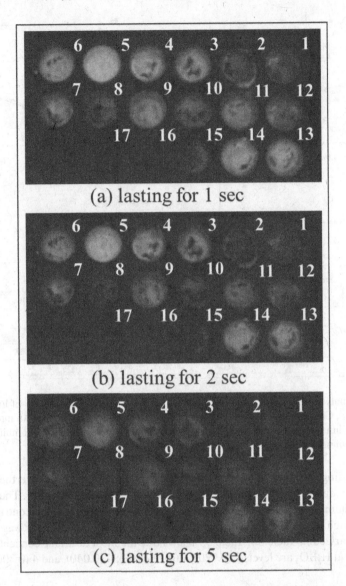

FIGURE 8.26 Luminescence of combinatorial library over 1 s (a), 2 s (b), and 5 s (c) after excitation at 365 nm for 20 min followed by the shutting off of the light source. (Chen, L., Chu, C. I., Chen, K. J., Chen, P. Y. Hu, S. F., and Liu, R.-S. 2011. An intelligent approach to the discovery of luminescent materials using a combinatorial chemistry approach combined with Taguchi methodology. *Luminescence* 26: 229–238. With permission.)

Second, ANOVA (a form of statistical analysis) was performed to determine the significance of control factors on the luminescence intensity over 1 s. Table 8.6 presents the results obtained by calculating with Qualitek-4 software. The right-most column in the ANOVA table presents the relative effects of the factors on objective property. The sum of all percentage effects always adds up to 100%. Table 8.6 reveals that the relative effects of Eu^{2+}, Ce^{3+}, Dy^{3+}, Li^+, and H_3BO_3 on the results are 8.22%, 19.48%, 11.40%, 1.26%, and 59.64%, respectively.

The F-Ratio test was conducted to evaluate the effect of each factor on signal-to-noise ratio. As shown in the final row of Table 8.6, the error in the DOF is zero, which represents the experiential error/the DOF of all other unconsidered factors. Therefore, the F-Ratio cannot be obtained by

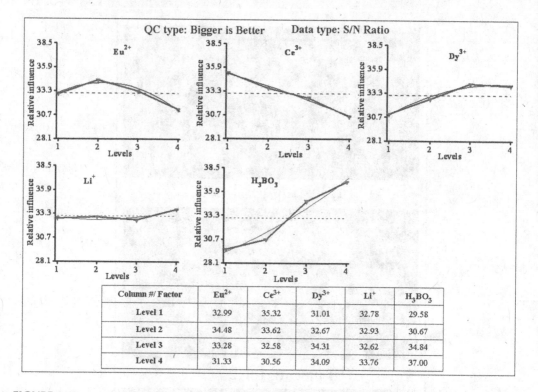

FIGURE 8.27 Analysis of effects of control parameters on relative average intensity of luminescence lasting for 1 s (Chen, L., Chu, C. I, Chen, K. J., Chen, P. Y. Hu, S. F., and Liu, R. S. 2011. An intelligent approach to the discovery of luminescent materials using a combinatorial chemistry approach combined with Taguchi methodology. *Luminescence* 26: 229–238. With permission.)

calculating according to Equation (8.10) because the denominator is zero. Nevertheless, the operation of parameters optimizing was further conducted with Qualitek-4 software. Thus, the optimum conditions for obtaining best long-lasting luminescence were optimized, the contribution of every factor was evaluated, and the expected performance under optimal conditions was estimated. The results are summarized in Table 8.7. Table 8.7 shows that the most desirable concentrations of Eu^{2+}, Ce^{3+}, Dy^{3+}, Li^+, and H_3BO_3 are levels 2 (0.020), 1 (0), 3 (0.03), 4 (0.040), and 4 (0.300), respectively,

TABLE 8.6
Analysis of Variance (ANOVA)

Column No./Factor	DOF (f)	Sum of Sqrs (S)	Variance (V)	F-Ratio (F)	Pure Sum (S')	Percent P (%)
1/Eu^{2+}	3	20.12	6.71	–	20.12	8.22
2/Ce^{3+}	3	47.64	15.88	–	47.64	19.48
3/Dy^{3+}	3	27.88	9.29	–	27.88	11.40
4/Li^+	3	3.07	1.02	–	3.07	1.26
5/H_3BO_3	3	145.89	48.63	–	145.89	59.64
Other/Error	0					
Total	15	244.60				100

Source: Data from Chen, L., Chu, C. I, Chen, K. J., Chen, P. Y. Hu, S. F., and Liu, R. S. 2011. An intelligent approach to the discovery of luminescent materials using a combinatorial chemistry approach combined with Taguchi methodology. *Luminescence* 26: 229–238.

TABLE 8.7
Optimal Conditions and Expected Performance

Column No./Factor	Level Description	Level	Contribution
1/Eu^{2+}	0.02	2	1.45
2/Ce^{3+}	0	1	2.30
3/Dy^{3+}	0.03	3	1.29
4/Li^+	0.04	4	0.74
5/H_3BO_3	0.30	4	3.98

Total contribution from all factors 9.75
Current grand average of performance 33.02
Expected results at optimum condition 42.77

Source: Data from Chen, L., Chu, C. I, Chen, K. J., Chen, P. Y. Hu, S. F., and Liu, R. S. 2011. An intelligent approach to the discovery of luminescent materials using a combinatorial chemistry approach combined with Taguchi methodology. *Luminescence* 26: 229–238.

and that the contributions of Eu^{2+}, Ce^{3+}, Dy^{3+}, Li^+, and H_3BO_3 to the objective property are 1.45, 2.30, 1.29, 0.74, and 3.98, respectively. The total contribution from all factors is 9.75. Current grand average of performance is 33.02, while the expected performance under optimum conditions is estimated to 42.77. The comparison of objective property between before and after optimizing demonstrates that the distribution of variations is reduced greatly. Thus, the accuracy and precision of the targets that search for new materials with the combinatorial chemistry approach is improved greatly by combining with the Taguchi method.

The error of DOF is zero, which does not suggest that the percentage effect of the error term is none. In practice, an experimental error may still exist. In this case, pooling factors is strongly urged until the error term of DOF is approximately half the total DOF of the experiment, so as to reduce the probability of "calling something important while it is not" no matter the size of the experiment and nature of results. Here, the operation of pooling factor was implemented. In view of the lowest sum of squares and the smallest effect of Li^+, Li^+ was pooled. Table 8.8 presents the results of ANOVA obtained after pooling Li^+ and reveals that H_3BO_3 most strongly affects S/N ratio, followed by Ce^{3+} and Dy^{3+}, and finally Eu^{2+} in that order. The optimal concentrations of the Eu^{2+}, Ce^{3+}, Dy^{3+}, and H_3BO_3 factors are levels 2 (0.020), 1 (0), 3 (0.03), and 4 (0.300), and the contributions at these

TABLE 8.8
Analysis of Variance (ANOVA) after Pooling the Factor of Li^+

Column No./Factor	DOF (f)	Sum of Sqrs (S)	Variance (V)	F-Ratio (F)	Pure Sum (S′)	Percent P (%)
1/Eu^{2+}	3	20.12	6.71	6.55	17.05	6.97
2/Ce^{3+}	3	47.65	15.88	15.51	44.57	18.22
3/Dy^{3+}	3	27.88	9.29	9.07	24.80	10.14
4/Li^+	(3)	(3.07)		POOLED	(CL=*NC*)	
5/H_3BO_3	3	145.90	48.63	47.48	142.82	58.39
Other/Error	3	3.07	1.02			6.28
Total	15	244.61				100

Source: Data from Chen, L., Chu, C. I, Chen, K. J., Chen, P. Y. Hu, S. F., and Liu, R. S. 2011. An intelligent approach to the discovery of luminescent materials using a combinatorial chemistry approach combined with Taguchi methodology. *Luminescence* 26: 229–238.

TABLE 8.9

Optimal Conditions and Expected Performance after Pooling the Factor of Li⁺

Column No./Factor	Level Description	Level	Contribution
1/Eu^{2+}	0.02	2	1.45
2/Ce^{3+}	0	1	2.30
3/Dy^{3+}	0.03	3	1.29
5/H_3BO_3	0.30	4	3.98

Total contribution from all factors 9.02
Current grand average of performance 33.02
Expected results at optimum condition 42.04

Source: Data from Chen, L., Chu, C. I., Chen, K. J., Chen, P. Y. Hu, S. F., and Liu, R. S. 2011. An intelligent approach to the discovery of luminescent materials using a combinatorial chemistry approach combined with Taguchi methodology. *Luminescence* 26: 229–238.

optimal levels are 1.45, 2.30, 1.29, and 3.98, respectively, as shown in Table 8.9. The total contribution of all factors is 9.02. The current grand average performance of the relative brightness of the 16 samples luminescence that persisted for 1 s is 33.02, and the expected performance at optimum conditions is estimated to be 42.04.

In choosing various components from which to construct a new material, the components that can contribute substantially to objective property rather than those that are harmful, or have no effect, or only have a minor effect on objective property, should be considered. Therewith, Ce^{3+} and Li^+ should be excluded from the raw materials. In Figure 8.27, the brightness decreases with an increase of Ce^{3+} concentration suggests Ce^{3+} is a quenching center, which detrimentally affects persistent luminescence. The optimal concentration of Ce^{3+} in Tables 8.7 and 8.9 is zero, further indicating that Ce^{3+} should not be used as a co-dopant.

H_3BO_3 not only promotes the crystallization of $SrAl_2O_4$ for its role of flux during sintered at high temperature, but also forms electron center by the oxygen vacancy associated with BO_3^{3-}, i.e., $[BO_3–V_O]^{3-}$, and substitutional defect complexes with Dy^{3+}, i.e., $[Dy–BO_4–V_{Sr}]^{2-}$ in $SrAl_2O_4$ host. Induced with near ultraviolet light, the photoinduced electron center is formed as $[BO_3–V_O(e')]^{4-}$. The hole is released from $[Dy–BO_4–V_{Sr}(h')]^{2-}$ under thermal excitation at room temperature. Energy that is released continuously upon the recombination of electrons with holes is expected to excite Eu^{2+} to produce long-lasting luminescence. Therefore, H_3BO_3 is an essential component for the synthesis of the long-lasting phosphor of $SrAl_2O_4$:Eu^{2+}, Dy^{3+}.

The optimum concentrations, 0.020 and 0.030, for Eu^{2+} and Dy^{3+} co-doped in $SrAl_2O_4$, that yield the most desired long-lasting luminescence is highly consistent with other studies, but the optimal concentration, 0.300, of H_3BO_3 is slightly exceeds that obtained by Wang et al. [47] because the amount of evaporated H_3BO_3 is correlated with the total weight of sample, the sintering temperature, and the holding time. Here, every sample in combinatorial library is about several milligram, which is far less than that normally several grams synthesized with the conventional solid-state reaction method. Therefore, the concentration of H_3BO_3 obtained by optimizing with the combinatorial chemistry approach here is, reasonably, a little higher than that synthesized by the solid-state reaction method elsewhere.

8.3 CONCLUSION

We safely conclude that the intelligence of combinatorial chemistry approach in evaluating the effects of components on an objective property, identifying the factors that have the strongest effect, reducing the experimental dimensions of multifactors by excluding those factors that have harmful or minor effects, and estimating the expected performance under the optimal conditions is improved by the Taguchi method. The marriage of the combinatorial chemistry approach with the Taguchi method is promising to reduce the cost and time of the synthesis and characterization of innumerous compounds by reducing the dimensions of multifactor in combinatorial library design and, thus, reducing the number of constructed samples. Accordingly, this scheme is promising to promote the efficient discovery of new materials. Thereby, the availability of the smart-performance of the combinatorial chemistry approach improved by the Taguchi method is demonstrated perfectly herein.

ACKNOWLEDGMENTS

The authors acknowledge the National Natural Science Foundation of China (21875058, U1332133, and 51002043), the National High-Tech R&D Program (863program) (2013AA03A114), and the Projects of Science and Technology of Anhui Province (202004a05020179, 1301022062, 1301022067, 12010202004) and Guangdong Province (2017B090901070) for long-term financial support.

REFERENCES

1. Lebl M (1999) Parallel Personal Comments on "Classical" Papers in Combinatorial Chemistry. *J. Comb. Chem.* 1: 3.
2. Rasheed A, and Farhat R (2013) Combinatorial Chemistry: A Review. *Int. J. Pharm. Sci. Res.* 4: 2502.
3. Liu R, Li X, and Lam K (2017) Combinatorial Chemistry in Drug Discovery. *Curr. Opin. Chem. Biol.* 38: 117.
4. Shastri S, and Narang H (2017) Combinatorial Chemistry – Modern Synthesis Approach. *PharmaTutor* 5: 37.
5. Merrifield B (1984) Solid Phase Synthesis. Nobel Lecture (December 8). www.nobelprize.org/prizes/chemistry/1984/merrifield/lecture/.
6. Frank R, Heikens W, Heisterberg-Moutsis G, and Blöcker H (1983) A New General Approach for The Simultaneous Chemical Synthesis of Large Numbers of Oligonucleotides: Segmental Solid Supports. *Nucleic Acids Res.* 11: 4365.
7. Geysen HM, Meloen RH, and Barteling SJ (1984) Use of Peptide Synthesis to Probe Viral Antigens for Epitopes to a Resolution of a Single Amino Acid. *Proc. Natl. Acad. Sci. U.S.A.* 81: 3998.
8. Houghten RA (1985) General-Method for The Rapid Solid-Phase Synthesis of Large Numbers of Peptides—Specificity of Antigen–Antibody Interaction at The Level of Individual Amino Acids. *Proc. Natl. Acad. Sci. U.S.A.* 82: 5131.
9. Smith GP (1985) Filamentous Fusion Phage: Novel Expression Vectors That Display Cloned Antigens on the Virion Surface. *Science* 228: 1315.
10. Lam KS, Salmon SE, Hersh EM, Hruby VJ, Kazmierski WM, and Knapp RJ (1991) A New Type of Synthetic Peptide Library for Identifying Ligand-Binding Activity. *Nature* 354: 82.
11. Houghten RA, Pinilla C, Blondelle SE, Appel JR, Dooley CT, and Cuervo JH (1991) Generation and Use of Synthetic Peptide Combinatorial Libraries for Basic Research and Drug Discovery. *Nature* 354: 84.
12. Bunin BA, and Ellman JA (1992) A General and Expedient Method for The Solid-Phase Synthesis of 1,4-Benzodiazepine Derivatives. *J. Am. Chem. Soc.* 114: 10997.
13. Kennedy JP, Williams L, Bridges TM, Daniels RN, Weaver D, and Lindsley CW (2008) Application of Combinatorial Chemistry Science on Modern Drug Discovery. *J. Comb. Chem.* 10: 345.
14. Cavasotto CN, Aucar MG, and Adler Natalia S (2019) Computational Chemistry in Drug Lead Discovery and Design. *Int. J. Quantum Chem.* 119: e25678.
15. Hanak JJ (1970) The "Multiple-Sample Concept" in Materials Research: Synthesis, Compositional Analysis and Testing of Entire Multicomponent Systems. *J. Mater. Sci.* 5: 964.
16. Xiang XD, Sun X, Briceno G, Lou Y, Wang KA, Chang H, Wallace-Freedman WG, Chen SW, and Schultz PG (1995) A Combinatorial Chemistry Approach to Materials Discovery. *Science* 268: 1738.
17. Briceño G, Chang H, Sun X, Schultz PG, and Xiang XD (1995) A Class of Cobalt Oxide Magnetoresistance Materials Discovered with Combinatorial Synthesis. *Science* 270: 273.
18. Wang J, Yoo Y, Gao C, Takeuchi I, Sun X, Chang H, Xiang XD, and Schultz PG (1998) Identification of a Blue Photoluminescent Composite Material from a Combinatorial Library. *Science* 279: 1712.

19. Danielson E, Golden JH, McFarland EW, Reaves CM, Weinberg WH, and Wu XD (1997) A Combinatorial Chemistry Approach to the Discovery and Optimization of Luminescent Materials. *Nature* 389: 944.

20. Boussie TR, Coutard C, Turner H, Murphy V, and Powers TS (1998) Solid-Phase Synthesis and Encoding Strategies for Olefin Polymerization Catalyst Libraries. *Angew. Chem. Int. Ed.* 37: 3272.

21. Reddington E, Sapienza A, Gurau B, Viswanathan R, Sarangapani S, Smotkin ES, and Mallouk TE (1998) Combinatorial Electrochemistry: A Highly Parallel, Optical Screening Method for Discovery of Better Electrocatalysts. *Science* 280: 1735.

22. Chang H, Gao C, Takeuchi I, Yoo Y, Wang J, Schultz PG, Xiang XD, Sharma RP, Downes M, and Venkatesan T (1998) Combinatorial Synthesis and High Throughput Evaluation of Ferroelectric/Dielectric Thin-Film Libraries for Microwave Applications. *Appl. Phys. Lett.* 72: 2185.

23. Koinuma H, and Takeuchi I (2004) Combinatorial Solid-State Chemistry of Inorganic Materials. *Nat. Mater.* 3: 429.

24. Blasse G, and Grabmaier BC (1994) *Luminescent Materials.* Springer-Verlag: Berlin, Heidelberg: p. 92.

25. Chen L, Jiang Y, Chen S, Zhang G, Wang C, and Li G (2008) The Intermediate Role of Gd^{3+} in Energy Transfer to Give Light under VUV Excitation. *J. Lumin.* 128: 2048.

26. Chen L, Chen KJ, Hu SF, and Liu RS (2011) Combinatorial Chemistry Approach to Searching Phosphors for White Light-Emitting Diodes in $(Gd-Y-Bi-Eu)VO_4$ Quaternary System. *J. Mater. Chem.* 21: 3677.

27. Kim CH, Kwon IE, Park CH, Hwang YJ, Bae HS, Yu BY, Pyun CH, and Hong GY(2000) Phosphors for Plasma Display Panels. *J. Alloys Cmpds.* 311: 33.

28. Chen L (2007) Liquid-Based Synthesis of Combinatorial Libraries and Combinatorial Approach Screening Photoluminescence Sensitizers for Vacuum Ultraviolet Phosphors. PhD Diss. University of Science and Technology of China (in Chinese).

29. Xiang X, Wang G, Zhang X, Xiang Y, and Wang H (2015) Individualized Pixel Synthesis and Characterization of Combinatorial Materials Chips. *Engineering* 1: 225.

30. Jiu H, Ding J, Sun Y, Bao J, Gao C, and Zhang Q (2006) Fluorescence Enhancement of Europium Complex Co-Doped with Terbium Complex in a Poly(Methyl Methacrylate) Matrix. *J. Non-Cryst. Solids* 352: 197.

31. Wang J, and Evans JRG (2005) London University Search Instrument: A Combinatorial Robot for High-Throughput Methods in Ceramic Science. *J. Comb. Chem.* 7: 665.

32. Jaramillo TF, Baeck SH, Kleiman-Shwarsctein A, Choi KS, Stucky GD, and McFarland EW (2005) Automated Electrochemical Synthesis and Photoelectrochemical Characterization of $Zn_{1-x}Co_xO$ Thin Films for Solar Hydrogen Production. *J. Comb. Chem.* 7: 264.

33. Schmatloch S, Meier MA, and Schubert US (2003) Instrumentation for Combinatorial and High-Throughput Polymer Research: A Short Overview. *Macromol. Rapid Commun.* 24: 33.

34. Cooksey GA, Sip CG, and Folch A (2009) A Multi-Purpose Microfluidic Perfusion System with Combinatorial Choice of Inputs Mixtures Gradient Patterns and Flow Rates. *Lab Chip* 9: 417.

35. Steinert CP, Goutier I, Gutmann O, Sandmaier H, Daub M, de Heij B, and Zengerle R (2004) A Highly Parallel Picoliter Dispenser with an Integrated, Novel Capillary Channel Structure. *Sens. Actuators, A* 103: 88.

36. Sun XD, Wang KA, Yoo Y, Wallace-Freedman WG, Gao C, Xiang XD, and Schultz PG (1997) Solution-Phase Synthesis of Luminescent Materials Libraries. *Adv. Mater.* 9: 1046.

37. Maier WF, Stöwe K, and Sieg S (2007) Combinatorial and High-Throughput Materials Science. *Angew. Chem. Int. Ed.* 46: 6016.

38. Chen L, Bao J, Gao C, Huang S, Liu C, and Liu W (2004) Combinatorial Synthesis of Insoluble Oxide Library from Ultrafine/Nano Particles Suspension Using a Drop-on-Demand Inkjet Delivery System. *J. Comb. Chem.* 6: 699.

39. Chan TS, Kang CC, Liu RS, Chen L, Liu XN, Ding JJ, Bao J, and Gao C (2007) Combinatorial Study of the Optimization of Y_2O_3:Bi,Eu Red Phosphors. *J. Comb. Chem.* 9: 343.

40. Chen L, Fei M, Zhang Z, Jiang Y, Chen S, Dong Y, Sun Z, Zhao Z, Fu Y, He J, and Li C (2016) Understanding the Local and Electronic Structures towards Enhanced Thermal Stable Luminescence of $CaAlSiN_3$:Eu^{2+}. *Chem. Mater.* 28: 5505.

41. Taylor SJ, and Morken JP (1998) Thermographic Selection of Effective Catalysts from an Encoded Polymer-Bound Library. *Science* 280: 267.

42. Chen L, Fu Y, Zhang G, Bao J, and Gao C (2008) Optimization of Pr^{3+} Tb^{3+} and Sm^{3+} Co-Doped $(Y_{0.65}Gd_{0.35})BO_3$:$Eu^{3+}_{0.05}$ VUV Phosphors through Combinatorial Chemistry Approach. *J. Comb. Chem.* 10: 401.

43. Yamamoto H, and Kano T (1979) Enhancement of Cathodoluminescence Efficiency of Rare Earth Activated Y_2O_2S by Tb^{3+} or Pr^{3+}. *J. Electrochem. Soc.* 126: 305.

44. Jung YS, Kulshreshtha C, Kim JS, Shin N, and Sohn KS (2007) Genetic Algorithm-Assisted Combinatorial Search for New Blue Phosphors in a $(Ca,Sr,Ba,Mg,Eu)_xB_yP_zO_8$ System. *Chem. Mater.* 19: 5309.
45. Park WB, Shin N, Hong KP, Pyo M, and Sohn KS (2012) A New Paradigm for Materials Discovery: Heuristics-Assisted Combinatorial Chemistry Involving Parameterization of Material Novelty. *Adv. Funct. Mater.* 22: 2258.
46. Chen L, Chu CI, Chen KJ, Chen PY, Hu SF, and Liu RS (2011) An Intelligent Approach to the Discovery of Luminescent Materials Using a Combinatorial Chemistry Approach Combined with Taguchi Methodology. *Luminescence* 26: 229.
47. Wang Y, Song X, and Zhang S (2006) $SrAl_2O_4$:Eu^{2+}, Dy^{3+}: Synthesis by Flux Method and Long Afterglow Properties. *Chin. J. Inorg. Chem.* 22: 41.

9 Energy Transfer between Lanthanide to Lanthanide in Phosphors for Bioapplications

Shuai Zha, Celina Matuszewska, and Ka-Leung Wong

CONTENTS

9.1 INTRODUCTION

The term *energy transfer* (ET) stands for numerous types of processes, including both sensitization and quenching of luminescence, that occurs between the donor and acceptor species in a wide range of materials, such as inorganic/organic phosphors, inorganic/metal nanoparticles, quantum dots (QD), metal–organic complexes, biomolecules, proteins etc. The aim of this chapter is to introduce a general classification and description of the types of ETs, which are possible between lanthanide ions.

Upon the type of mechanism and characteristics of the energy being transferred, three types of transfers are distinguished: hole transfer, electron transfer and ET. The first and second type of transfer occurs during an interaction of a charged (positively or negatively) molecule with another molecule, resulting in the second molecule becoming charged as well. On the other hand, ET is a photophysical process based on interactions between two entities: an excited donor and acceptor in its electronic ground state (or metastable excited state in upconversion [UC] ET) leading to the

DOI: 10.1201/9781003098669-9

situation when the acceptor becomes excited at the expense of donor returning to its electronic ground state.

Lanthanide-based nanophosphors have attracted increasing attention in biological applications due to their characteristic emission peaks, tunable lifetime/emission wavelength, non-photobleaching, minimized autofluorescence and desirable biocompatibility. The optical traits of nanophosphors could be adjusted through ET among multiple energy levels in lanthanide ions. To date, the derivatives of emerging research on lanthanide-doped nanophosphors in bioapplications could be categorized into cancer treatment, diagnostic imaging, biosensing and other applications such as *in vivo* information storage and encoding/decoding.

In this chapter, we will elaborate and discuss the concept of various categories of ET in lanthanide ions, including radiative and non-radiative ET, resonant ET, nonresonant ET, Förster–Dexter theory, unified theory of radiative and non-radiative ET, host–Ln^{3+} and Ln^{3+}–Ln^{3+} ETs, UC and downconversion (DC) between Ln^{3+} ions. Afterward, we will summarize and emphasize a wide range of biomedical applications of lanthanide-doping nanophosphors, such as bioimaging, therapeutics, biosensors and others. The bigger picture is that by evaluating and gaining deeper insights into the material science and biological activities of lanthanide-based nanomaterials, we will be able to extend promising biomedical applications to clinical translation potential.

9.2 ENERGY TRANSFER

9.2.1 RADIATIVE AND NON-RADIATIVE ENERGY TRANSFER

Generally, ETs are divided into two categories – radiative and non-radiative ones, which are considered as optical processes since both of them are mediated by photons. The former requires the release of the excessive energy of the donor in a radiative manner, namely by emitting photon, and subsequent absorption of the resulting photon by the acceptor, provided that there is spectral overlap between the donor's emission spectrum and absorption spectrum of the acceptor. Unlike non-radiative ET, radiative ET is independent of the distance between donor and acceptor species and their luminescence lifetimes; however, it relies on the donor's quantum yield as well as the molar absorption coefficient and concentration of the acceptor. This type of ET manifests itself in the decrease of the donor's emission intensity in the regions of spectral overlap observed as the characteristic dips in the resulting emission spectrum. The process of radiative ET can be described accordingly:

$$D^* \rightarrow D + h\nu \ h\nu + A \rightarrow A^*$$

Reversely, non-radiative ET is governed by the luminescence lifetime of the donor and the distance between donor and acceptor entities, as it can occur only within the time when the donor exists in its excited state and, moreover, the creation of the collision complex between species is required. Thus, non-radiative ET can be presented as:

$$A + D^* \rightarrow [AD^*] \rightarrow [A^*D] \rightarrow A^* + D$$

9.2.2 RESONANT ENERGY TRANSFER

The presence of ET between a pair of donor and acceptor can be identified by a change in characteristics, such as luminescence decay, donor emission intensity and also, in some cases, by observation of donor absorption bands in the acceptor excitation spectrum. Upon the characteristics of the donor–acceptor pair, ET mechanism can occur *via* resonant, semi-resonant or double-resonant process. Undeniably, the resonant ETs are the most commonly studied and utilized among all ETs. However, in this case, using the term *resonance* might be misleading, since it does not mean that the differences between stated energy levels in both donor and acceptor species are equal. The actual requirement for the ET to happen is a spectral overlap between the emission of the donor entity and the absorption of the acceptor entity, which also involves the participation of vibrational modes [1]. In theory, when resonant ET is considered, backwards ET from A to D should occur simultaneously; however, in reality, it is energetically unfavorable. This is due to the fact that after absorption of photon by the donor, some of the acquired energy is instantaneously lost, mainly through the process of vibrational dissipation, where the energy is converted to heat. At that point, if the excited donor is in proximity with the acceptor bearing similar or slightly lower energy levels, that acceptor can acquire energy through resonance ET before any other type of deexcitation takes place in the donor.

9.2.3 NONRESONANT ENERGY TRANSFER

On the contrary, ETs without fulfilling the resonance condition are still possible and were previously proven both theoretically and experimentally. In the case of a small energy mismatch between the donor and acceptor energy levels, such ET is regarded as semi-resonance type, where the difference in energy is compensated by the simultaneous absorption or emission of one or more phonons [2, 3]. A phonon-assisted ET was experimentally proven between Ln^{3+} with energy mismatch as big as 10^3 cm^{-1}, e.g. between Yb^{3+} ($^2F_{5/2}$) and Tm^{3+} ($^3H_5$) energy levels with ~1600 cm^{-1} energy difference [4]. The rate of such ET increases with the rise of temperature as a result of more intensified phonon population [3]. Furthermore, ET is also possible within three non-resonating entities. If one of the species is characterized by $2E$ energy, while two others have excited states at E energy, the following ET between them can be expressed as [5]:

$$D* + A + A \rightarrow D + A* + A*$$

This type of ET is a second-order process known as double-resonance ET and can be referred to as cooperative DC or cooperative UC (CUC), where the entity with excited state at $2E$ serves as donor or acceptor, respectively. In order for this ET to occur, close proximity of donor and acceptor entities is required, thus the two species with equal energy must be present within the host material at a high concentration. These types of processes are connected with exchange interactions as well as higher order interactions (dipole–quadrupole, quadrupole–quadrupole), which can be explained using modern quantum electrodynamics. More detailed explanation of these processes is provided further in the chapter.

9.2.4 FÖRSTER–DEXTER THEORY

The process of non-radiative transfer of energy from an excited donor to an acceptor occurs through dipolar coupling between the involved species, as shown in Figure 9.1 [6]. The first observation of this phenomena was reported by Cario and Frank in 1922 [7]. Later on, after discoveries of J. Perrin and his son F. Perrin, Förster presented the first theory concerning mechanism of non-radiative resonance ET [7, 8]. He reported that ET governed by dipole–dipole coupling strongly relies on two factors, the distance between entities as well as spectral overlap of their emission and absorption spectrum. Based on the knowledge that there is an inverse proportionality between static dipole–dipole interactions and the cube of species separation, the rate of such ET scales with the square of this coupling. Förster formulated the famous distance-dependence law, known also as the Förster equation [9]:

$$w_F(R) = \frac{9\kappa^2 c^2}{8\pi\tau_D n^4 R^6} \int \frac{F_D(\omega)\sigma_A(\omega)}{\omega^4} d\omega \tag{9.1}$$

where κ is orientation factor, n is the refractive index of the host, τ_D is the radiative lifetime of the donor in the absence of acceptor, R is D–A separation, $F_D(\omega)$ stands for normalized emission spectra

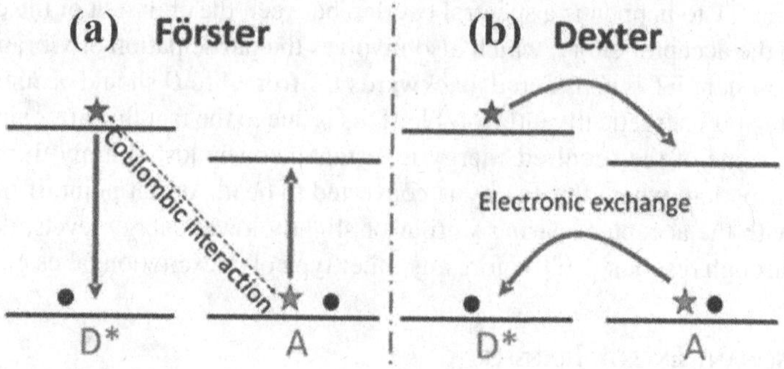

FIGURE 9.1   Schematic representation of (a) Förster resonance energy transfer and (b) Dexter energy transfer [6].

of the donor and $\sigma_A(\omega)$ denotes linear absorption cross section of the acceptor represented in radians per unit time. Thus, if donor and acceptor can be approximated using point dipoles, ET rate between them scales with R^{-6}. By proposing a value of critical distance R_0, known also as a Förster distance, Eq. (9.1) is presented as [10]:

$$w_F(R) = \frac{1}{\tau_D}\left(\frac{R_0}{R}\right)^6 \tag{9.2}$$

With critical distance defined as:

$$R_0 = \frac{9\kappa^2 c^2}{8\pi n^4}\int\frac{F_D(\omega)\sigma_A(\omega)}{\omega^4}d\omega \tag{9.3}$$

According to the equation concerning rate of ET, the transfer efficiency Φ_T is expressed as:

$$\Phi_T = \frac{1}{1+(R/R_0)^6} = 1 - \frac{\tau_{fl}}{\tau_D} = 1 - \frac{I_{fl}}{I_D} \tag{9.4}$$

where τ_{fl} and τ_D is the fluorescence lifetime of the donor with and without the acceptor, while I_{fl} and I_D is the intensity of donor's fluorescence with and without acceptor, respectively. For the separation $R = R_0$ the transfer efficiency Φ_T equals 50%. The following equations illustrate the propensity of ET with respect to the distance between the donor and acceptor. If the excited donor is situated within the R_0 distance from a suitable acceptor, then the ET will dominate over the deexcitation of donor by means of radiative decay, usually fluorescence [7, 9].

It is worth noting that although both donor decay and acceptor excitation electronic transitions are usually considered as electric dipole – allowed, there are several possible exceptions. Since the theory of ET proposed by Förster is based on the electric dipole approximation, significant problem arises in the case when electronic transition happens to be electronic dipole-forbidden but, at the same time, allowed by the magnetic dipole or multipoles of higher order [11, 12]. In such situation, ET is governed by interactions weaker than Coulombic interactions, considered in Förster theory. The rate of the ET based on different types of interactions can be calculated using the following formula [12]:

$$K_{DA}(R) = \frac{\alpha_{dd}}{R^6} + \frac{\alpha_{dq}}{R^8} + \frac{\alpha_{qq}}{R^{10}} + \cdots = \sum_{i=6,8,10}\frac{\alpha_i}{R^i} \tag{9.5}$$

where dd, dq and qq stand for dipole–dipole, dipole–quadrupole and quadrupole–quadrupole interactions, respectively. Eq. (9.5) can be used to acquire the dominating mechanism governing ET process. As the magnitude of α_i generally follows the sequence of $\alpha_{dd} > \alpha_{dq} > \alpha_{qq}$, in most cases only d–d interactions are considered. However, since both magnetic dipole interactions and electric or magnetic multipole interactions are highly dependent on distance (ET scales with R^{-8} and R^{-10} for d–q and q–q interactions), their value becomes significant for very small donor–acceptor separations. Furthermore, multipole interactions should be likewise considered when dipole–dipole interactions are not allowed, as it is observed for f–f transitions in lanthanide ions. In such cases, when there is an overlap between the donor and acceptor wavefunctions, exchange interactions between them dominate the ET and thus, an alternative mechanism should be considered.

The theory of ET firstly introduced by Förster in the 1940s was a decade later extended by Dexter to include interactions of higher order multipoles. ET mechanism proposed by Dexter was explained using electron exchange based on quantum mechanics rules [11]. The rate of such ET can be defined as:

$$w_{dex}(R) = \frac{2\pi}{\hbar} K^2 \exp\left(\frac{-2R}{L}\right) \int F_D(\omega)\sigma_A(\omega)d\omega \tag{9.6}$$

where K is a constant describing the overlap between wavefunctions of donor and acceptor, while L is the Bohr radius of the system. Dexter type of transfer depends exponentially on the distance L, which is a direct reflexion of radial distribution of overlapping wavefunctions. Within the Dexter theory, singlet–triplet type of ET is also allowed, unlike in Förster mechanism, and can be described accordingly [7]:

$$^3D^* + {}^1A \rightarrow {}^3D + {}^1A^*$$

However, as Dexter ET requires spatial overlap of D–A wavefunctions, it is only applicable at very short distances, namely less than 10 Å, and therefore, its practical applications are much reduced. A thorough comparison, highlighting the differences and similarities between Förster and Dexter ET, was presented thoroughly in the review of Tanner et al. [1].

9.2.5 UNIFIED THEORY OF RADIATIVE AND NON-RADIATIVE ENERGY TRANSFER

As far as values of D–A separations are concerned, Förster ET is generally considered within the 10–100 Å range. According to this limit, it was originally assumed that ET undergoes radiative mechanism for a D–A distance beyond 100 Å, resulting from the well-established belief treating Förster ET as a solely non-radiative transfer. Notwithstanding the fact that ET would be correctly described in the radiative manner for very long D–A separations, these two types of ET – nonradiative Foster ET and radiative ET – cannot be treated as two distinct mechanisms, separated by the established value of 100 Å. In fact, they are just two extremes of a broader mechanism, which apart from Förster type of ET and "radiative" ET embraces at the same time intermediate range between these extremes, in which neither of previously presented mechanisms is completely valid [7]. The rate of unified ET mechanism for both near-field and far-field conditions can be expressed as [13]:

$$w = w_F + w_I + w_{rad} \tag{9.7}$$

where w_F is considered as Förster ET rate described by Eq. (9.1), and thus applicable only for nearfield conditions ($R \ll \lambda$), w_{rad} is proportional to R^{-2} and dominates in a far field ($R \gg \lambda$), while w_I operates within the range where neither of these conditions is fulfilled and scales with R^{-4}. The second and third term of Eq. (9.7) can be described accordingly [13]:

$$w_I = \frac{9c^2}{8\pi\tau_D n^2 R^4}(\kappa^2 - 2\kappa\kappa')\int F_D(\omega)\sigma_A(\omega)d\omega \tag{9.8}$$

$$w_{rad} = \frac{9\kappa'^2}{8\pi\tau_D R^2} \int F_D(\omega)\sigma_A(\omega)\frac{d\omega}{\omega^2} \tag{9.9}$$

The unified theory of ET, which is based on modern quantum electrodynamics, was firstly presented by Avery et al. [14] in the 1960s and subsequently developed by Andrews et al. [13]. To conclude, with the increase of D–A separation, participation of the radiative character in overall ET increases and eventually photons dominate the ET mechanism. Unlike the Förster–Dexter theory, the unified theory of ET embraces radiative and non-radiative ET under one coherent theory, which is applicable for all distance ranges.

9.2.6 Host–Ln³⁺ and Ln³⁺–Ln³⁺ Energy Transfers

Processes of ET in practical systems are very diverse and include interactions between different entities, such as metal ion centers (mostly lanthanide ions – Ln^{3+} ions), defects, QD, plasmonic nanostructures and molecules. In this section, interactions between lanthanide ions and the resulting UC, DC (or quantum cutting [QC]) and downshifting (DS) processes will be discussed.

In the case of some phosphors, photoluminescence emission from a dopant ion is the result of host sensitization occurring under sufficiently high energy excitation light (usually UV) through bandgap transition. The host lattice efficiently captures excitation photons, creating excitons, which are more probable to pass their energy to ions incorporated in the host matrix rather than recombine and emit excessive energy by themselves. However, generally ET processes in most luminescent materials occur without sensitization by the host, which solely acts as an ET platform. This type of ET requires a donor ion, more commonly known as a sensitizer, and should be characterized with large absorption cross section as well as strong and broadband emission. Frequently used Ln^{3+} ions that act as effective sensitizers are those with allowed d–f transitions, such as Ce^{3+} and Eu^{2+} [15, 16]. On the other hand, in the NIR region, Yb^{3+} is the most commonly exploited sensitizer (e.g. in Er^{3+}, Tm^{3+} and Ho^{3+} co-doped materials) because of its high absorption cross section even for forbidden f–f transition $^2F_{7/2} \rightarrow {}^2F_{5/2}$, simple system of energy levels and relatively long luminescence lifetime of its only excited state ($^2F_{7/2}$). This is due to the fact that Ln^{3+} ions embedded in the crystal host lattice are affected by its crystal field resulting in the relaxation of selection rules thus enhancing the probability of f–f transitions [17]. Therefore, Ln^{3+} ions incorporated in the host lattice can act as active luminescent centers, responsible for the absorption of excitation energy (sensitization), ET processes as well as resulting emission (Figure 9.2) [18].

Due to their unique optical properties, such as well-separated and narrow emission lines, long luminescence lifetimes (usually in microsecond to millisecond range), variety of possible transitions covering the UV, visible and NIR regions (except for Yb^{3+} and Ce^{3+}) and large possible effective shift between excitation and emission wavelengths, Ln^{3+} ions are perfect candidates in the studies of ET processes [19,20]. Up to now, ETs were investigated within disparate pairs of triple-positive Ln^{3+}, such as Yb^{3+}–Er^{3+} [21, 22], Yb^{3+}–Pr^{3+} [23, 24], Yb^{3+}–Ho^{3+} [25], Ce^{3+}–Tb^{3+} [26], Tb^{3+}–Eu^{3+} [26, 27], Dy^{3+}–Tb^{3+} [28] and Sm^{3+}–Eu^{3+} [28, 29].

9.2.7 Upconversion and Downconversion Between Ln³⁺ Ions

Photon UC in Ln^{3+}-doped materials is a nonlinear process in which at least two low-energy photons are absorbed by long-lived intermediate energy levels of Ln^{3+} ion resulting in the emission of high-energy photons. Although the idea of UC in ion-doped materials dates back to 1959, when the suggestion of sequential absorption of NIR photons was proposed by Bloembergen [30], it was not until 1966 that the remarkable phenomenon of UC was fully recognized by independent discoveries of Auzel [31], Ovsyankin and Feofilov [32]. Auzel, for the first time, suggested that ET is possible between two ions, both in their excited states, unlike previously assumed that transfers only take place between excited

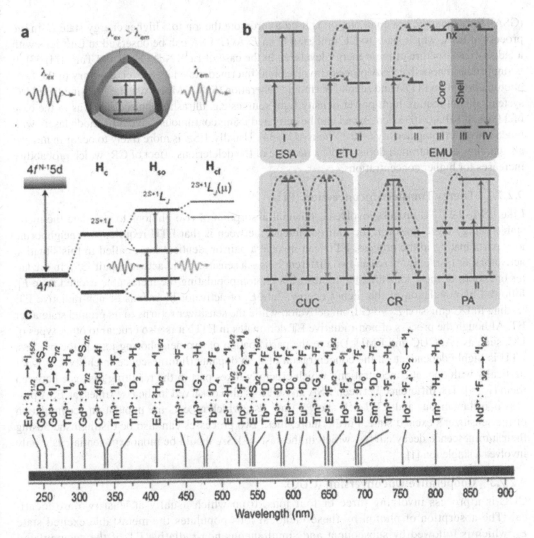

FIGURE 9.2 (a) Schematic illustration of an upconverting nanoparticle with core–shell structure, with its characteristic emission $\lambda_{ex} > \lambda_{em}$ (top graph) energy level splitting in a 4f shell on Ln^{3+} ion (bottom graph); (b) mechanisms of the most typical Ln^{3+}-based UC processes: excited-state absorption (ESA), energy transfer upconversion (ETU), energy migration-mediated upconversion (EMU), cooperative upconversion (CUC), cross-relaxation (CR) and photon avalanche (PA). Continuous and dotted arrows represent radiative and nonradiative transitions, respectively. (c) Characteristic Ln^{3+} transitions in a UV, Vis and NIR range presented in a wavelength scale [18]. (Zheng KZ, Loh KY, Wang Y, Chen QS, Fan JY, Jung TY, Nam SH, Suh YD, Liu X. *Nano Today* 29, 100797, 2018. With permission.)

state of first ion to the second ion in the ground state [31]. On the other hand, Ovsyankin and Feofilov proved the migration mechanism of excitation energy leading to emission of photon with doubled or tripled energy compared to the excitation energy [32]. In general, we can distinguish the following types of Ln^{3+}-based UC processes: excited-state absorption (ESA), ET UC (ETU), CUC, cross-relaxation (CR), energy migration-mediated UC (EMU) and photon avalanche (PA).

9.2.7.1 Excited-State Absorption (ESA)

ESA is the simplest single-center UC mechanism involving sequential absorption of two photons. Firstly, if the excitation energy matches the transition energy between the ground state and the metastable excited state E_1, then E_1 is populated through the process of ground-state absorption

(GSA). Subsequently, a second photon is used to promote the ion to a higher energy state E_2 in the process of ESA, which leads to UC emission from E_2 to G. ESA can be observed in Ln^{3+} ions with a ladder-like structure of their energy levels, as in the case of Er^{3+}, Nd^{3+}, Ho^{3+} and Tm^{3+} [31, 33]. It is important to note that ESA process involves real intermediate energy level, contrary to the two-photon absorption (TPA) and second harmonic generation (SHG). Moreover, the Ln^{3+}-mediated UC systems do not require high-power density light sources e.g. ultrashort pulsed laser, as in the case of TPA and SHG ($>10^6$ W/cm^2), and can be activated using continuous-wave NIR diode lasers with much lower pumping density (~10^{-1} W/cm^2) [18, 34]. Usually, ESA is more likely to occur in materials with low concentration doping (<1%), because of the deleterious effect of CR, which probability increases for higher concentrations.

9.2.7.2 Energy Transfer Upconversion (ETU)

Like ESA, the ETU process involves sequential absorption of two photons to populate the metastable energy level; however, the main difference between is that ETU requires two neighboring ions participating in the process. ETU can involve a pair of identical ions, called in this situation activators, or take place between two different ions – a sensitizer and activator pair [35]. In the latter one, both of these ions are able to absorb a photon populating the metastable excited state E_1, followed by population of the higher energy state E_2 of activator by means of non-radiative ET, leading to UC emission $E_2 \rightarrow G$ from activator, while the sensitizer returns to its ground state after ET. Although the process of non-radiative ET dominates in ETU, it is also crucial to other types of UC, such as PA, CUC and EMU. Due to the participation of two neighboring ions in the process, ETU is highly dependent on the concentration of both dopants (luminescence intensity vary quadratically with dopant concentration), which crucially takes part in the resulting distance between them [1, 34]. The difference between ESA and ETU is also observed in the rise times, as ESA process is instantaneous and ETU will show a time rise, which reflects the population accumulation of the sensitizer's excited state. Furthermore, these two processes can also be distinguished using their luminescence decay curves, which in the case of ESA should be monoexponential, as it only involves a single ion [1].

9.2.7.3 Cooperative Upconversion (CUC)

CUC is a process involving three or four ions, from which usually at least two are identical. The absorption of photon by these identical ions populates the metastable excited state E_1, which is followed by subsequent and simultaneous non-radiative ET of the accumulated energy to the E_2 excited state of the second ion. Radiative relaxation to the ground state results in UC emission. Two types of CUC can be generally distinguished, cooperative luminescence and cooperative sensitization [31]. In the former, the sensitizer and activator are identical ions, e.g. Yb^{3+} ions pair, while the latter occurs for two different Ln^{3+} ions, such as Yb^{3+}–Pr^{3+}, Yb^{3+}–Tb^{3+} or Yb^{3+}–Eu^{3+}. As transitions in the process of CUC involve quasi-virtual pair levels, described using higher order of perturbation by quantum mechanics; therefore, the efficiency of this process is generally several orders of magnitude lower than observed for ESA and ETU [33].

9.2.7.4 Energy Migration-Mediated Upconversion (EMU)

On the other hand, EMU is a process characteristic of nanoparticles with a core–shell structure. It involves energy migrators, which are able to pass the excitation energy through the multilayer structure of the nanoparticle. The process of EMU can be illustrated in the following way. Firstly, the sensitizer (I) absorbs NIR photon and successively populates the excited energy levels E_1 and E_2 of the accumulator ion by means of non-radiative ET. Consequently, the excited accumulator passes excessive energy to the neighboring migrator ion (III), followed by random process of energy hopping from one migrator ion to another across the multilayered structure, until it reaches the activator ion (IV) in the nanoparticle shell. Finally, the activator obtains the migrating energy that induces UC emission. The process of EMU involves as many as four different types of ions, from which the sensitizer and accumulator are situated in the nanoparticles core, the activator in its shell, and the migrator ions (most commonly Gd^{3+} ions) are localized in both core and shell of the nanoparticle [35]. Such strategy reduces the possibility of undesirable CR between distinct ions in the matrix, leading to significant increase in UC efficiency without the need of activators with ladder-like energy levels [17].

9.2.7.5 Cross-Relaxation (CR)

CR is a type of ET possible between the same or different Ln^{3+} ions, provided that they have matching energy levels. It is usually considered as the main factor responsible for concentration quenching of Ln^{3+} luminescence; however, this effect might also be used to tune the emission profile or even enhance the UC efficiency, as a part of PA process.

9.2.7.6 Photon Avalanche (PA)

The PA phenomenon was discovered for the first time over 40 years ago by Chivian et al. in $LnCl_3$ crystals doped with Pr^{3+} ions after the observation of an unexpected increase in UC efficiency under excitation beyond the critical intensity of laser pumping [36]. PA is a positive feedback system initiated by weak GSA followed by sequential ESA and CR events between Ln^{3+} ions, resulting in UC emission. There are several prerequisites for the PA to occur, namely an excitation intensity above-stated threshold value, a much higher absorption cross section of ESA than GSA and matching of Ln^{3+} energy levels for effective CR can take place. The exponential increase in the E_2 excited-state population is achieved accordingly. Initially, nonresonant GSA allows some primary population of E_1 energy state, while subsequent strong resonant ESA populates higher excited energy state E_2. Then, a CR event occurs between two neighboring ions, resulting in both ions occupying the intermediated energy level E_1, in which their population allows further ESA processes followed by CR ETs. The population of E_2 level increases after each feedback loop, which finally leads to strong UC emission by a PA process [37, 38].

9.2.7.7 Efficiency of Upconversion Processes

Generally, efficiency of UC processes is governed by three parameters, (i) the host material, (ii) the concentration of Ln^{3+} ions (or more generally dopants) and (iii) surface quenching and other related quenching mechanisms [33]. Furthermore, it significantly varies with the type of UC process occurring in the material. PA exhibits the highest efficiency among all UC processes; however, due to several drawbacks, namely high excitation threshold, dependence on the pump power and slow response time (seconds to minutes), the range of its potential applications is limited [31]. Moreover, up-to-date reports on PA cover mainly bulk materials and aggregates, with only a few reports in nanoparticles [37, 38]. Nevertheless, ETU offers the second-highest efficiency, without the abovementioned disadvantages, as it is instantaneous and independent of the pump power, it is therefore the most widely used type of UC. Compared with ETU, both ESA and CUC have lower efficiency, approximately two or three (or more) orders of magnitude, respectively.

9.2.7.8 Downconversion and Downshifting

It is crucial to note that generally, along with UC, there are several different processes competing over excitation energy, mainly DC (QC) and DS. DC, contrary to UC, refers to a process in which out of one higher energy photon, two photons with lower energy are produced [39]. DS, on the other hand, is a process in which one emitted photon is generated with a lower energy compared to the excitation photon. Owing to the fact that Ln^{3+} ions have numerous energy levels, some of them exhibit transitions with lower energy than the excitation energy used for generating UC, e.g. Er^{3+} ions with characteristic transition $^4I_{13/2} \rightarrow {}^4I_{15/2}$ at 1.54 μm, utilized in optoelectronics [40]. Since the most common strategy of improving performance of UC materials involves decreasing the forbidden character of Ln^{3+} transitions by modification of crystal field strength of matrix, DC luminescence is a significant challenge on this way, as it simultaneously enhances UC luminescence. Up to now, there is no approach that leads to an effective rise in UC intensity and a decline in DC intensity at the same time. However, DC luminescence is not always considered as a detrimental process and nowadays is mainly applied in solar cells [41].

9.3 BIOIMAGING

Bioimaging is of great significance in biomedical fields because it can achieve tumor therapy visualization and better monitor the functionalized nanophosphors in the living system. Noninvasive visualization is the most desired type of bioimaging during the biological processes in real time. Bioimaging aims to visualize the processes of live with as little interference as possible. Furthermore, it is conveniently utilized to obtain the information on the three-dimensional (3D) structure of the observed specimen from the outside, i.e. without physical interference or damage. Bioimaging spans the observation of subcellular structures and entire cells over tissues up to entire cellular organisms. In general, light, fluorescence, electrons, ultrasound, X-ray, magnetic resonance and positrons are utilized as sources for imaging.

Nowadays, there are numerous well-developed imaging techniques such as computed tomography (CT), magnetic resonance imaging (MRI), positron emission tomography (PET), single-photon emission computed tomography (SPECT), fluorescent imaging (FI), photoacoustic imaging (PAI), and ultrasound imaging (USI). Moreover, there is a summary on merits and shortcomings of recent imaging methods listed in Table 9.1.

Generally, there are three tissue transparency windows in the NIR region: NIR-I, 650–950 nm; NIR-II, 1000–1350 nm; NIR-III, 1500–1800 nm [51]. Because the light absorption of tissue is at a relatively minimum level in these ranges, it assures deeper tissue penetration depth. In the last few years, researchers reported many fabricated nanophosphors, for instance, UC nanoparticles (UCNP), the emission wavelengths of which cover the first [52] and second windows of NIR [53], and both NIR-I and NIR-II wavelengths region as well [54]. Higher resolution and deeper tissue penetration depth is possible with the NIR-II window due to its lower scattering coefficient when compared to the NIR-I window [55].

Due to the rich energy level structures of lanthanide ions that are doped in nanophosphors, they can serve as imaging contrast agents for bioimaging applications. The well-developed

TABLE 9.1

Summary of Bioimaging Techniques

Imaging Mode	Spatial Resolution	Merits	Drawbacks	Examples
CT	25~200 μm	3D data	Low sensitivity	AuNPs[42]; Ln[3+]NP[43]
MRI	25~100 μm	Soft tissue imaging	High toxicity from leaky metal	Gd/Fe based NPs[44]
PET	Less than 1 mm	Organ functional information	Short half-life of radioactive tracers	[124]I-labeled C dots[45]
SPECT	0.5~2 mm	Long half-life of radioactive tracers	Long scan time	^{99m}Tc-labeled iron oxide NPs[46]
FI	2~3 mm	High sensitivity	No 3D data	Quantum dots; C dots; Ln[3+]NP[47]
PAI	5 μm~1 mm	Low scattering of sound	Interfered by acoustic absorbing substance	AuNPs[48]
UI	10~100 μm	Handy	Limited view	Nanobubbles[49]; Silica NPs[50]

imaging modalities of lanthanide-based nanophosphors include luminescent imaging, X-ray/CT and MRI. Taking advantage of the luminescent nanophosphors developed recently, nanophosphors have been applied for fluorescence cell imaging. Compared to traditional organic fluorophore, inorganic nanophosphors doping with lanthanide ions (e.g. UCNPs) have displayed intriguing fluorescent property for bioimaging such as deep penetration depth, nontoxic nature, good biocompatibility, tunable lifetimes/emission wavelength, high photostability and desirable quantum yields. For instance, Hao et al. designed the poly(acrylic acid) (PAA)-modified nanorods β-NaLuF$_4$:40%Gd/20%Yb/2%Er (Ce^{3+} doping) with sufficient brightness DS upon excitation 980 nm wavelength (Yb $^2F_{7/2} \to {}^2F_{5/2}$) to beyond 1500 nm (Er $^4I_{13/2} \to {}^4I_{15/2}$); therefore, it can achieve *in vivo* fluorescence imaging in the NIR-IIb second biological windows on the bioapplication of small tumor/metastatic tiny tumor detection, tumor vessel visualization and brain vessel imaging, as shown in Figure 9.3(a). The 1525 nm DS emission is improved by suppressing the UC pathway in Lu hosts *via* Ce^{3+} doping. Hence, the nanorod was fabricated with the excellent characteristics of high imaging sensitivity, desirable spatial resolution, good biocompatibility and high quantum yield [56].

In another case, Zhang et al. developed the lanthanide-doped β-NaGdF$_4$@NaGdF$_4$:Yb/Er (Tm, Ho)@NaYF$_4$:Yb@NaNdF$_4$:Yb NIR-II agent with good photostability and excellent biocompatibility for deep tissue imaging by multiplexed imaging. The penetration depth of the NIR-II agent can reach up to 8 mm in biological tissues with the engineered luminescence lifetimes and desirable signal-to-noise ratio. Under the excitation at 808 nm (Nd $^4I_{9/2} \to {}^4F_{5/2}$), 1525 nm (Er $^4I_{13/2} \to {}^4I_{15/2}$) or 1475 nm (Tm $^3H_4 \to {}^3F_4$) [57], *in vivo* luminescence lifetime imaging was achieved and this noninvasive imaging method with distinct lifetime channels can be translational for cancer and disease diagnosis.

More recently, Zhang et al. reported a glutathione-modified nanoprobe β-NaGdF$_4$:5%Nd@NaGdF$_4$ with cross-linking strategy to gain a rapid and responsive imaging on inflammation area of reactive oxygen species (ROS) in NIR-II window (Figure 9.3(b)). The nanoprobes were able to be excreted out of the mice bodies because of their ultrasmall size. Through NIR-II emission guiding at 1064 nm (Nd $^4F_{3/2} \to {}^4I_{11/2}$) under 808 nm excitation (Nd $^4I_{9/2} \to {}^4F_{5/2}$), the high efficiency of DC with ROS-responding *in vivo* crosslinking method provides the long-lasting bioimaging. (Zhao M, Wang R, Li B, Fan Y, Wu Y, Zhu X, Zhang F *Angew. Chem. Int. Ed.* 58: 2050, 2019. With permission)

In addition, nanophosphors can also be used as X-ray/CT contrast agents. Lanthanide-doping nanomaterials could be used as X-ray/CT for biomedical imaging modality in consideration of the high atomic number i.e. Gd (Z = 64) element. X-ray/CT is able to provide 3D tomography of the anatomic structure based on the differential X-ray absorptions between the lesions and tissues [59]. Previously, Xing et al. reported rare earth-doped nanoprobes (Er, Yb dopants) in applications of biomedical imaging, which exhibited shortwave infrared light under X-ray excitation and provided clear biological structure owing to minimized autofluorescence and photobleaching. After coating with the shell and PEG ligand, the nanoprobes were able to achieve dual-modality molecular imaging with desired biocompatibility [60].

Apart from the luminescence imaging and X-ray-CT, MRI has been broadly applied in clinically diagnostic imaging because of the high spatial resolution, excellent contrast difference of 3D soft tissues and noninvasive characteristics [61]. Integration of functional molecules with nanophosphors paves a new road for multifunctionalization of nanocomposites, which extends the applications in diagnostic imaging fields. For instance, Gao and co-workers investigated the luminescent property of PEGylated NaGdF$_4$:Yb,Er nanocrystal. The major visible emission peaks were recorded upon NIR I 980 nm laser excitation and achieved by energy transferring between Yb^{3+} and Er^{3+}, which are attributed to radiative relaxations from $^2H_{11/2}$, $^4S_{3/2}$ and $^4F_{9/2}$ states to the $^4I_{15/2}$ state of Er^{3+} ion, respectively. Because of the paramagnetic properties of Gd ions, MRI and luminescence imaging of NaGdF$_4$:Yb,Er with PEG ligand are both achieved *in vivo* tumor imaging, which demonstrated lanthanide-based functionalized UCNP

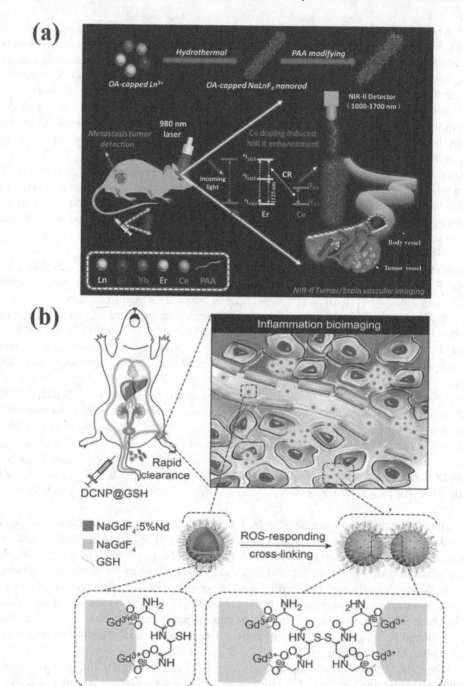

FIGURE 9.3 (a) Schematic illustration of the enhanced NIR-IIb emission of PAA-Ln-NRs *via* Ce³⁺ doping for noninvasive tumor metastasis/vascular visualization and brain vessel imaging [56]; (b) Schematic illustration of bioimaging for acute local epidermal inflammation in mice utilizing ultrasmall DCNP@GSH nanoprobes. DCNP@GSH cross-link at inflamed area in response to ROS, meanwhile they would be excreted from the body rapidly [58]. (With permission)

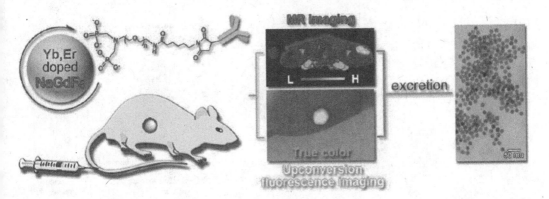

FIGURE 9.4   Magnetic/upconversion fluorescent $NaGdF_4$:Yb,Er nanoparticle-based dual-modal molecular probes for imaging tiny tumors *in vivo*. (Liu C, Gao Z, Zeng J, Hou Y, Fang F, Li Y, Qiao R, Shen L, Lei H, Yang W, Gao M *ACS Nano* 7, 7227, 2013. With permission.)

with exceptional biocompatibility which has promising prospect in multimodal imaging, as shown in Figure 9.4 [62].

9.4   THERAPEUTIC APPLICATIONS

Nowadays, chemotherapy, radiotherapy, photodynamic therapy (PDT) and photothermal therapy (PTT) are the most common cancer treatments associated with lanthanide-doped nanomaterials. Additionally, they can work separately and collectively, and combining two or three of them is a trend for cancer therapy. Chemotherapy means that antitumor drugs are directed to tumor sites, normally encompassed by nanoparticles, and are released when they are close to the tumor. Radiotherapy uses a kind of agent that is responsive to ionizing radiation and then kills the tumor. PDT and PTT have similar principles, both use a type of agent which generates toxic ROS, such as singlet oxygen ($^1O_2$) and heat respectively under light excitation to kill tumors. These categories of chemical agents are called photosensitizers (PS).

Lanthanide-doping UCNP have a wide range of applications such as multimodal bioimaging, biodetection assays and visualized tumor treatment. Previously, PDT with UCNPs is reported in many research papers [63–65], also some reports demonstrated the potential of a combination of multimodal imaging with PTT [66], imaging-guided PDT and PTT [67], chemotherapy, radiotherapy combined with PDT dual-modal imaging [68].

Recently, Lin et al. reported a UC luminescent system, which is desirable example of UCNP for biomedical application, PDT and PTT. PDT and PTT agent moieties are coated on the surface of UCNP for *in vivo* optical imaging and cancer inhibition [69]. Nowadays, the bioconjugate with different functional compounds or the UCNPs for underlying the life span is very mature for multimodal *in vivo* imaging. Hence, Lin et al. reported the $Na_{0.52}YbF_{3.52}$: Er UC nanoprobes for MR/CT imaging and its core–shell structure can perform PDT as well [70]. Zhao et al. introduced a core–shell structure UCNP that can achieve trimodal imaging, i.e. UCL/MR/CT imaging, and it is able to kill tumor after loading two types of anticancer drugs [71]. They showed responsive signal and emission as well and guided the drugs to the cancer cells.

More recently, Chen and co-workers developed a type of lanthanide nanoparticles that can trigger PDT treatment on drug-resistant *Acinetobacter baumannii* (Figure 9.5(a)). Upon NIR light excitation, the PS loaded in UCNPs ($LiYF_4$:Yb/Er) were released, and generated ROS to damage the cell membrane of the bacteria. Strong antibacterial capability is proven with a dose of 50 $\mu g/mL$ under 1 W/cm^2 980 nm light. In addition, UC emission can be observed due to ET from Yb^{3+} to Er^{3+} and excellent therapeutic performances on deep tissue bacterial infections are achieved in murine

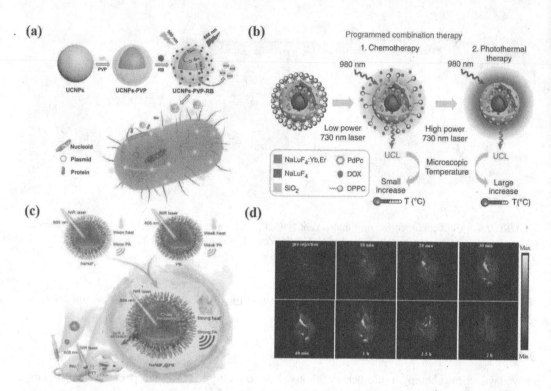

FIGURE 9.5 (a) Schematic diagram of synthesis and antibacterial action of UCNPs-PVP-RB nanosystem. (Liu W, Zhang Y, You W, Su J, Yu S, Dai T, Huang Y, Chen X, Song X, Chen Z. *Nanoscale* 12, 13948, 2020. With permission.) (b) Schematic illustration of temperature-responsive UCNPs nanocomposite for precise programming combination cancer therapy. (Zhu X, Li J, Qiu X, Liu Y, Feng, W, Li F, *Nat. Commun.* 9, 2176, 2018. With permission.) (c) Fabrication of NdNdF4@PB core/shell structure to generate new cross-relaxation pathways to enhance photoacoustic imaging signal and photothermal therapy efficacy [74]. (d) *In vivo* PA imaging by using the nanocomposite. (Yu Z, Hu W, Zhao H, Miao X, Guan Y, Cai W, Zeng Z, Fan Q, Tan TTY *Angew. Chem. Int. Ed.* 58, 8536, 2019. With permission.)

model. The lanthanide-based nanoparticles open up new avenues for combating severe bacterial infection in deep tissues [72].

In another example, Li's group designed a temperature-sensitive UC nanoprobe for chemotherapy-PTT combinational therapy (Figure 9.5(b)). PTT performance can be tuned to realize temperature-triggered combination therapy in a sequence of chemotherapy, subsequently PTT through monitoring the microscopic temperature of the nanocomposite with UC luminescence (ET from Yb^{3+} to Er^{3+}) spectroscopy. The desired therapeutic performance *in vivo* is observed and the dosage of chemodrug is retained at a relatively low level. This work provides a noninvasive new strategy of programmed combination therapy, which can achieve the maximum therapeutic effect of each treatment mode. More importantly, the dose of chemodrug can be lowered, hence the side effect could be minimized and this strategy has great potential for development on personalized treatment [73].

Yu et al. reported that CR among sensitizers is able to improve PTT conversion efficiency and therapeutic efficacy (Figure 9.5(c)). After coating PTT agents Prussian blue (PB) on $NaNdF_4$ nanoparticles, CR pathways can be generated between the multiple energy levels of Nd^{3+} ($^4F_{3/2} \rightarrow ^4I_{15/2}$ and $^4I_{9/2} \rightarrow ^4I_{15/2}$ states in Nd^{3+}) and continuous energy band of PB. The nanocomposite with good biocompatibility has exhibited promising potential as efficient PTT agent evidenced by the results of *in vitro* and *in vivo* PAI and PTT performance. The shorter distance of energy relaxation in new CR pathways and higher non-irradiative energy loss resulted in the enhancement of PTT therapeutic performance [74].

Recently, Zha et al. developed $NaGdF_4:Yb^{3+}, Er^{3+}@NaGdF_4$ core–shell UCNP coating with Epstein–Barr nuclear antigen 1 (EBNA1)-specific peptide, as shown in Figure 9.6(a). The nano-probes exhibited characteristic photophysical property due to the doping of lanthanide ions, the UC emission peaked at 520, 540 and 654 nm from Yb^{3+} to $Er^{3+} \rightarrow {}^2H_{11/2}$, ${}^4S_{3/2} \rightarrow {}^4I_{15/2}$ and ${}^4F_{9/2} \rightarrow {}^4I_{15/2}$ ET transitions. The viability of Epstein–Barr virus (EBV)-infected cells was signif-icantly prohibited, whereas the nanoprobes do not harm non-EBV-related cells or normal cells. Notably, UC emission with a twofold enhancement was observed only upon addition of EBNA1. The results of the dimerization assays and *in vivo* tumor inhibition suggested that the nanoprobes can hinder the process of EBNA1 dimerization; therefore, an excellent inhibitory performance can be achieved in EBV-positive cell xenograft tumor. Additionally, the data on *in vitro* confocal imag-ing also evidenced that the nanoprobes were able to get inside the nucleus of EBV-infected cells and bind with EBNA1. Based on these, it can be proved that this newly developed dual-functional, lanthanide-based, luminescent EBNA1-specific bioprobes, with the ability to visualize EBNA1 and inhibit its dimerization, would become the ultimate weapon toward eradicating EBV-associated tumors with a responsive UC enhancement for *in vitro* imaging [75].

Later on, the same group reported a safe UC nanoplatform with dual-targeting peptides to upgrade the nanoprobes in terms of cellular uptake, biostability and inhibitory performance, and make nano-probes more universal toward EBV-associated cancer cells. The strategy on integration of another

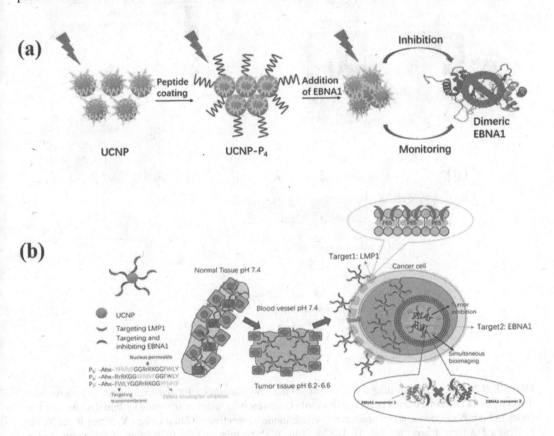

FIGURE 9.6   (a) Schematic illustration of the dual-function for imaging and inhibition of UCNP-P$_4$ in EBNA1 dimerization. (Zha S, Fung YH, Chau HF, Ma P, Lin J, Wang J, Chan LS, Zhu G, Lung HK, Wong KL *Nanoscale* 10, 15632, 2018. With permission.) (b) Schematic illustration of the path of entry of the nanoprobe UCNP-P$_n$, into an EBV-infected cancer cell from normal tissues through sequential and selective targeting. (Zha S, Chau HF, Chau WY, Chan LS, Lin J, Lo KW, Cho WC, Yip YL, Tsao SW, Farrell PJ, Feng L, Di JM, Law GL, Lung HK, Wong KL *Adv. Sci.* 8, 2002919, 2021. With permission.)

critical transmembrane oncoprotein latent membrane protein (LMP1) and EBNA1 were proposed to achieve nuclear localization of UC nanoplatform, as shown in Figure 9.6(b). The target EBV-infected cells can be uptake more new-generation nanoprobes since there is a pH linker designed between UCNP and dual-targeting peptide which is cleavable in tumor microenvironment weak acidic condition. Inhibition on downstream effect of nanoplatforms toward LMP1 was observed by western blotting as well. In addition, subcellular location and cellular uptake rate of nanoprobes were investigated through confocal imaging at different time intervals, which indicated that the nanoplatforms possess rapid cellular uptake rate and the final location is nucleus where EBNA1 located. More importantly, excellent inhibitory performance of the nanoprobes can be observed in EBV-infected *in vitro* and *in vivo*, especially in LMP1-positive cell lines. Hence, a new generation of pH-responsive lanthanide-based UC nanoplatforms with dual-targeting protein-specific peptide and

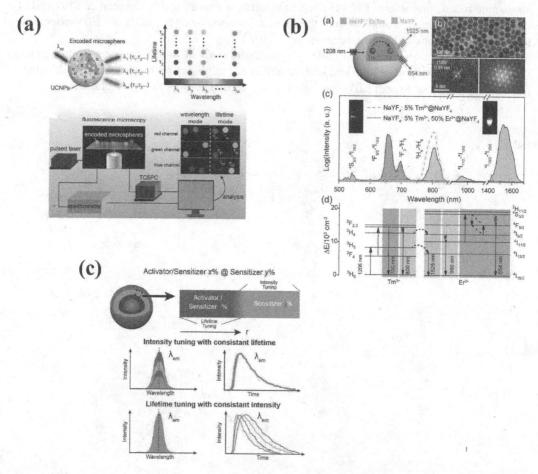

FIGURE 9.7 (a) Schematic diagram of upper: Schematic illustration for the high-capacity upconversion wavelength and lifetime binary encoding; bottom: Time-resolved imaging scanning system decodes the barcodes in both fluorescence wavelength and lifetime mode, respectively. (Zhou L, Fan Y, Wang R, Li X, Fan L, Zhang F *Angew. Chem. Int. Ed.* 57, 12824, 2018. With permission.) (b) Illustration of schematic design, TEM, HRTEM images, the corresponding Fourier transform diffraction pattern, upconversion and downshifted luminescence of the as-synthesized nanocrystals and simplified energy transfer pathway from Tm^{3+} to Er^{3+} respectively. (Zhang H, Fan Y, Pei P, Sun C, Lu L, Zhang F *Angew. Chem. Int. Ed.* 58, 10153, 2019. With permission.) (c) Schematic illustration of sensitizer gradient doping for independent intensity and lifetime tuning strategy. (Liu, X., Chen, Z.H., Zhang, H., Fan, Y. and Zhang, F. *Angew. Chem. Int. Ed*, 60, 7041, 2021. With permission.)

responsive emission signal for suppression and monitoring of EBV-infected cancers was developed. This study has opened up a new road for the bioapplication of lanthanide-based UC nanoprobes and is vitally significant for imaging agents and imaging-guided treatment design and development on the eradication of EBV-related cancers [76].

9.5 BIOSENSORS AND OTHERS

Beyond the various applications of nanophosphors in bioimaging and therapeutics, they are also able to be employed as smart biosensing system for detection of metal ions, biomolecules or bioeffects with ultrasensitive detection limit. Recently, Chen et al. reported the ultrasmall luminescent nanoprobes for highly sensitive detection of prostate cancer-specific antigen, with a lower detection limit of 0.52 pg/mL in clinical serum samples when compared with the commercial kit [77]. The enhanced UC emission is attributed to the efficient energy back transfer process from Er^{3+} to Yb^{3+} which has huge potential in early diagnosis of prostate cancers.

In another example, Bao et al. developed a cell-permeable ytterbium bioprobe to detect Hg^{2+} in aqueous solution and *in vitro* by reversible signal with visible and NIR emission. The low limit of detection 150 nM of Hg^{2+} is achieved by FRET and the lanthanide antenna effect. Therefore, these desired properties make it practical for biosensing applications [78].

In addition, Zhang and co-workers reported a new concept on dual-channel UC emission and lifetime encoding for biodetection, through ET in multiple layers from $Nd^{3+} \rightarrow Yb^{3+}$, which can be utilized for precise detection of human papillomavirus (HPV) subtypes from patient sample. Subsequently, Pr^{3+} is co-doped and introduced in the system and the ET between Er^{3+} and Pr^{3+} ($Er^{3+} \rightarrow Pr^{3+}$, $^4I_{13/2} \rightarrow {}^3F_3$, $^4I_{11/2} \rightarrow {}^1G_4$) is achieved for better monochromaticity [79]. Later on, the same group developed the lanthanide-doped nanocrystal for information storage and decoding *via* time-gated fluorescence imaging, which can achieve *in vivo* implantable device encoding and decoding through efficient ET pathway from Tm^{3+} to Er^{3+}, as shown in Figure 9.7(b) [80]. In 2020, tuning the UC luminescence and lifetime by doping lanthanide ions *via* gradient doping strategy from Yb^{3+} to Tm^{3+} was reported, multiplexed encoding capacity is expanded and this model has promising potential on being a multi-dimensional biological probe in the near future [81].

9.6 CONCLUSION

Nanophosphors can be extensively applied in a wide range of biological and biomedical fields due to their outstanding and tunable electronic, optical and magnetic properties. After conducting appropriate surface modification approaches on nanomaterials, nanophosphors with desirable properties (e.g. lanthanide characteristic emission) are able to be manufactured and used for specific biomedical applications through efficient ET between lanthanide ions. The excellent capabilities of nanophosphors have been verified in biomedical applications, including bioimaging agents, therapeutics, biosensors and information storage or decoding. More continuous efforts should be devoted to developing novel nanophosphors for future biomedical applications.

In addition, thanks to the desirable merits of UCNP, such as deep penetration depth, nontoxic nature, tunable lifetimes/emission wavelength and taking advantage of the properties of lanthanide ions from UCNP, luminescence emission is supplied and solid support to antitumor drugs is provided to achieve the aims of visualization and inhibition respectively. In fact, lanthanide-based nanomaterials have a bright future for cancer treatment and bioimaging. There are numerous strengths for nanomaterials for cancer drug delivery, they include increase in solubility and stability and pharmacokinetics of the drugs in blood, prevention on nonspecific interactions during circulation, enhancement of drug accumulation in tumor tissue through passive and active targeting, minimization on side effects, reduction of long-term immunotoxicity.

REFERENCES

1. Tanner PA, Zhou L, Duan C, Wong KL (2018) Misconceptions in electronic energy transfer: Bridging the gap between chemistry and physics. *Chem. Soc. Rev.* 47: 5234.
2. Liu X, Qiu J (2015) Recent advances in energy transfer in bulk and nanoscale luminescent materials: From spectroscopy to applications. *Chem. Soc. Rev.* 44: 8714.
3. Miyakawa T, Dexter DL (1970) Phonon sidebands, multiphonon relaxation of excited states, and phonon-assisted energy transfer between ions in solids. *Phys. Rev. B* 1: 2961.
4. Pandozzi F, Vetrone F, Boyer JC, Naccache R, Capobianco JA, Speghini A, Bettinelli MA (2005) Spectroscopic analysis of blue and ultraviolet upconverted emissions from $Gd_3Ga_5O_{12}$:Tm^{3+}, Yb^{3+} nanocrystals. *J. Phys. Chem. B* 109: 17400.
5. Andrews DL, Jenkins RD (2001) Quantum electrodynamical theory of three-center energy transfer for upconversion and downconversion in rare earth doped materials. *J. Chem. Phys.* 114: 1089.
6. Martins TD, Ribeiro ACC, Souza GA, Cordeiro DS, Silva RM, Colmati F, Lima RB, Aguiar LF, Carvalho LL, Reis RGCS, Santos WDC (2018) New materials to solve energy issues through photochemical and photophysical processes: The kinetics involved. *Adv. Chem. Kinet.* doi:10.5772/intechopen.70467.
7. Andrews DL, Bradshaw DS, Dinshaw R, Scholes GD (2015) Resonance energy transfer. *Photonics Sci. Found. Technol. Appl.* 4: 101.
8. Clegg RM (2006) The history of FRET: From conception through the labors of birth. *Rev. Fluoresc.* 3: 1.
9. Andrews DL, Bradshaw DS (2009) Resonance energy transfer. *Digital Encycl. Appl. Phys.* 533. doi:10.1002/3527600434.eap685.
10. Lakowicz JR (2006) *Principles of Fluorescence Spectroscopy.* Springer, Boston, MA.
11. Dexter DL (1953) A theory of sensitized luminescence in solids. *J. Chem. Phys.* 21: 836.
12. Nakazawa K, Shionoya S (1967) Energy transfer between trivalent rare-earth ions in inorganic solids. *J. Chem. Phys.* 47: 3267.
13. Andrews DL (1989) A unified theory of radiative and radiationless molecular energy transfer. *Chem. Phys.* 135: 195.
14. Avery JS (1966) Resonance energy transfer and spontaneous photon emission. *Proc. Phys. Soc.* 88: 1.
15. Yang WJ, Luo L, Chen TM, Wang NS (2005) Luminescence and energy transfer of Eu- and Mn-coactivated $CaAl_2Si_2O_8$ as a potential phosphor for white-light UVLED. *Chem. Mater.* 17: 3883.
16. Chang CK, Chen TM (2007) $Sr_3B_2O_6$: Ce^{3+}, Eu^{2+}: A potential single-phased white-emitting borate phosphor for ultraviolet light-emitting diodes. *Appl. Phys. Lett.* 91: 1.
17. Qin X, Xu J, Wu Y, Liu X (2019) Energy-transfer editing in lanthanide-activated upconversion nanocrystals: A toolbox for emerging applications. *ACS Cent. Sci.* 5: 29.
18. Zheng KZ, Loh KY, Wang Y, Chen QS, Fan JY, Jung TY, Nam SH, Suh YD, Liu X (2019) Recent advances in upconversion nanocrystals: Expanding the kaleidoscopic toolbox for emerging applications. *Nano Today* 29:100797.
19. Bünzli JCG (2010) Lanthanide luminescence for biomedical analyses and imaging. *Chem. Rev.* 110: 2729.
20. Eliseeva SV, Bünzli JCG (2010) Lanthanide luminescence for functional materials and bio-sciences. *Chem. Soc. Rev.* 39: 189.
21. Da Vila LD, Gomes L, Tarelho LVG, Ribeiro SJL, Messadeq Y (2003) Mechanism of the Yb-Er energy transfer in fluorozirconate glass. *J. Appl. Phys.* 93: 3873.
22. Wang M, Hou W, Mi CC, Wang WX, Xu ZR, Teng HH, Mao CB, Xu SK (2009) Immunoassay of goat antihuman immunoglobulin G antibody based on luminescence resonance energy transfer between near-infrared responsive $NaYF_4$:Yb, Er upconversion fluorescent nanoparticles and gold nanoparticles. *Anal. Chem.* 81: 8783.
23. Wen H, Tanner PA (2011) Energy transfer and luminescence studies of Pr^{3+}, Yb^{3+} co-doped lead borate glass. *Opt. Mater. (Amst).* 33: 1602.
24. Naccache R, Vetrone F, Speghini A, Bettinelli M, Capobianco JA (2008) Cross-relaxation and upconversion processes in Pr^{3+} singly doped and Pr^{3+}/Yb^{3+} codoped nanocrystalline $Gd_3Ga_5O_{12}$: The sensitizer/activator relationship. *J. Phys. Chem. C* 112: 7750.
25. Peng B, Izumitani T (1995) Optical properties, fluorescence mechanisms and energy transfer in Tm^{3+}, Ho^{3+} and Tm^{3+}-Ho^{3+} doped near-infrared laser glasses, sensitized by Yb^{3+}. *Opt. Mater. (Amst).* 4: 797.
26. Zhang X, Zhou L, Pang Q, Shi J, Gong M (2014) Tunable luminescence and Ce^{3+} +nab^{3+} +nab^{3+} energy transfer of broadband-excited and narrow line red emitting Y_2SiO_5:Ce^{3+}, Tb^{3+}, Eu^{3+} phosphor. *J. Phys. Chem. C* 118: 7591.

27. Luo Y, Liu Z, Wong HT, Zhou L, Wong KL, Shiu KK, Tanner PA (2019) Energy transfer between Tb^{3+} and Eu^{3+} in $LaPO_4$: Pulsed versus switched-off continuous wave excitation. *Adv. Sci.* 6: 1.

28. Lin H, Pun EYB, Wang X, Liu X (2005) Intense visible fluorescence and energy transfer in Dy^{3+}, Tb^{3+}, Sm^{3+} and Eu^{3+} doped rare-earth borate glasses. *J. Alloys Compd.* 390: 197.

29. Min X, Huang Z, Fang M, Liu YG, Tang C, Wu X (2014) Energy transfer from Sm^{3+} to Eu^{3+} in red-emitting phosphor $LaMgAl_{11}O_{19}$:Sm^{3+}, Eu^{3+} for solar cells and near-ultraviolet white light-emitting diodes. *Inorg. Chem.* 53: 6060.

30. Bloembergen N (1959) Solid state infrared quantum counters. *Phys. Rev. Lett.* 2: 84.

31. Auzel F (2004) Upconversion and anti-stokes processes with f and d ions in solids. *Chem. Rev.* 104: 139.

32. Ovsyankin VV, Feofilov PP (1966) Two-quantum cooperative frequency conversion of weak light fluxes. *JETP Lett.* 14: 377.

33. Li D, Ågren H, Chen G (2018) Near infrared harvesting dye-sensitized solar cells enabled by rare-earth upconversion materials. *Dalton Trans.* 47: 8526.

34. Wang F, Liu X (2009) Recent advances in the chemistry of lanthanide-doped upconversion nanocrystals. *Chem. Soc. Rev.* 38: 976.

35. Nadort A, Zhao J, Goldys EM (2016) Lanthanide upconversion luminescence at the nanoscale: Fundamentals and optical properties. *Nanoscale* 8: 13099.

36. Chivian JS, Case WE, Eden DD (1979) The photon avalanche: A new phenomenon in Pr^{3+}-based infrared quantum counters. *Appl. Phys. Lett.* 35: 124.

37. Lee C, Xu E, Liu Y, Teitelboim A (2021) Giant nonlinear optical responses from photon-avalanching nanoparticles. *Nature* 589: 230.

38. Li M, Hao ZH, Peng XN, Li JB, Yu XF, Wang QQ (2010) Controllable energy transfer in fluorescence upconversion of NdF_3 and $NaNdF_4$ nanocrystals. *Opt. Express* 18: 3364.

39. Wegh RT, Donker H, Oskam KD, Meijerink A (1999) Visible quantum cutting in $LiGdF_4$:Eu^{3+} through downconversion. *Science* 283: 663.

40. Valenta J, Repko A, Greben M, Nižňanský D (2018) Absolute up- and down-conversion luminescence efficiency in hexagonal Na (Lu/Y/Gd)F_4:Yb, Er/Tm/Ho with optimized chemical composition. *AIP Adv.* 8: 0.

41. Trupke T, Green MA, Würfel P (2002) Improving solar cell efficiencies by down-conversion of high-energy photons. *J. Appl. Phys.* 92: 1668.

42. Zhuang Y, Katayama Y, Ueda J, Tanabe S (2014) A brief review on red to near-infrared persistent luminescence in transition-metal-activated phosphors. *Adv. Opt. Mater.* 36: 1907.

43. Song L, Lin XH, Song XR, Chen S, Chen XF, Li J, Yang HH (2017) Repeatable deep-tissue activation of persistent luminescent nanoparticles by soft X-ray for high sensitivity long-term in vivo bioimaging. *Nanoscale* 9: 2718.

44. Maldiney T, Lecointre A, Viana B, Bessière A, Bessodes M, Gourier D, Richard C, Scherman D (2011) Controlling electron trap depth to enhance optical properties of persistent luminescence nanoparticles for in vivo imaging. *J. Am. Chem. Soc.* 133: 11810.

45. Abdukayum A, Chen JT, Zhao Q, Yan XP (2013) Functional near infrared-emitting Cr^{3+}/Pr^{3+} Co-doped zinc gallogermanate persistent luminescent nanoparticles with superlong afterglow for in vivo targeted bioimaging. *J. Am. Chem. Soc.* 135: 14125.

46. Maldiney T, Richard C, Seguin J, Wattier N, Bessodes M, Scherman D (2011) Effect of core diameter, surface coating, and PEG chain length on the biodistribution of persistent luminescence nanoparticles in mice. *ACS Nano* 5: 854.

47. Li ZJ, Zhang YW, Wu X, Wu XQ, Maudgal R, Zhang HW, Han G (2015) Persistent luminescence: In vivo repeatedly charging near-infrared-emitting mesoporous SiO_2/$ZnGa_2O_4$:Cr^{3+} persistent luminescence nanocomposites. *Adv. Sci.* 2: 1500001.

48. Penas C, Pazos E, Mascareñas JL, Vázquez ME (2013) A folding-based approach for the luminescent detection of a short RNA hairpin. *J. Am. Chem. Soc.* 135: 3812.

49. Bessière A, Jacquart S, Priolkar K, Lecointre A, Viana B, Gourier D (2011) $ZnGa_2O_4$:Cr^{3+}: A new red long-lasting phosphor with high brightness. *Opt. Express* 19: 10131.

50. Li ZJ, Zhang YW, Wu X, Huang L, Li DS, Fan W, Han G (2015) Direct aqueous-phase synthesis of sub-10 nm "luminous pearls" with enhanced in vivo renewable near-infrared persistent luminescence. *J. Am. Chem. Soc.* 137: 5304.

51. Wang J, Ma Q, Wang Y, Shen H, Yuan Q (2017) Recent progress in biomedical applications of persistent luminescence nanoparticles. *Nanoscale* 9: 6204.

52. Liu F, Yan W, Chuang YJ, Zhen Z, Xie J, Pan Z (2013) Photostimulated near-infrared persistent luminescence as a new optical read-out from Cr^{3+}-doped $LiGa_5O_8$. *Sci. Rep.* 3: 1554.

53. Qin X, Li Y, Zhang R, Ren J, Gecevicius M, Wu Y, Sharafudeen K, Dong GP, Zhou SF, Ma ZJ, Qiu JR (2016) Hybrid coordination-network-engineering for bridging cascaded channels to activate long persistent phosphorescence in the second biological window. *Sci. Rep.* 6: 20275.

54. Yu N, Liu F, Li X, Pan Z (2009) Near infrared long-persistent phosphorescence in $SrAl_2O_4$:Eu^{2+}, Dy^{3+}, Er^{3+} phosphors based on persistent energy transfer. *Appl. Phys. Lett.* 95: 231110.

55. Xu J, Murata D, Ueda J, Tanabe S (2016) Near-infrared long persistent luminescence of Er^{3+} in garnet for the third bio-imaging window. *J. Mater. Chem. C* 4: 11096.

56. Li YB, Zeng SJ, Hao JH (2019) Non-invasive optical guided tumor metastasis/vessel imaging by using lanthanide nanoprobe with enhanced down-shifting emission beyond 1500 nm. *ACS Nano* 13: 248.

57. Fan Y, Wang P, Lu Y, Wang R, Zhou L, Zheng X, Li X, Piper JA, Zhang F (2018) Lifetime-engineered NIR-II nanoparticles unlock multiplexed in vivo imaging. *Nat Nanotechnol.* 13: 941.

58. Zhao M, Wang R, Li B, Fan Y, Wu Y, Zhu X, Zhang F (2019) Precise in vivo inflammation imaging using in situ responsive cross-linking of glutathione modified ultra-small NIR-II lanthanide nanoparticles. *Angew. Chem. Int. Ed.* 58: 2050.

59. Kalender WA (2006) X-ray computed tomography. *Phys. Med. Biol.* 51: R29.

60. Naczynski DJ, Sun C, Türkcan S, Jenkins C, Koh AL, Ikeda D, Pratx G, Xing L (2015) X-ray-induced shortwave infrared biomedical imaging using rare-earth nanoprobes. *Nano Lett.* 15: 96.

61. Terreno E, Castelli DD, Viale A, Aime S (2010) Challenges for molecular magnetic resonance imaging. *Chem. Rev.* 110: 3019.

62. Liu C, Gao Z, Zeng J, Hou Y, Fang F, Li Y, Qiao R, Shen L, Lei H, Yang W, Gao M (2013) Magnetic/upconversion fluorescent $NaGdF_4$: Yb, Er nanoparticle-based dual-modal molecular probes for imaging tiny tumors in vivo. *ACS Nano* 7: 7227.

63. Zhang P, Steelant W, Kumar M, Scholfield M (2007) Versatile photosensitizers for photodynamic therapy at infrared excitation. *J. Am. Chem. Soc.* 129: 4526.

64. Ungun B, Prud'homme RK, Budijono SJ, Shan J, Lim SF, Ju Y, Austin R (2009) Nanofabricated upconversion nanoparticles for photodynamic therapy. *Opt. Express* 17: 80.

65. Qian HS, Guo HC, Ho PCL, Mahendran R, Zhang Y (2009) Mesoporous-silica-coated up-conversion fluorescent nanoparticles for photodynamic therapy. *Small* 5: 2285.

66. Cheng L, Yang K, Li Y, Chen J, Wang C, Shao M, Lee ST, Liu Z (2011) Facile preparation of multifunctional upconversion nanoprobes for multimodal imaging and dual-targeted photothermal therapy. *Angew. Chem. Int. Ed.* 32: 7523.

67. Chen Q, Wang C, Cheng L, He W, Cheng Z, Liu Z (2014) Protein modified upconversion nanoparticles for imaging-guided combined photothermal and photodynamic therapy. *Biomaterials* 35: 2915.

68. Fan WP, Shen B, Bu W, Chen F, He Q, Zhao K, Zhang S, Zhou L, Peng W, Xiao Q, Ni D, Liu J, Shi J (2014) A smart upconversion-based mesoporous silica nanotheranostic system for synergetic chemo-/radio-/photodynamic therapy and simultaneous MR/UCL imaging. *Biomaterials* 35: 8992.

69. Liu B, Li CX, Xing BG, Yang PP, Lin J (2016) Multifunctional UCNPs@PDA-ICG nanocomposites for upconversion imaging and combined photothermal/photodynamic therapy with enhanced antitumor efficacy. *J. Mater. Chem. B* 4: 4884.

70. Huang Y, Xiao Q, Hu H, Zhang K, Feng Y, Li F, Wang J, Ding X, Jiang J, Li Y, Shi L, Lin H (2016) 915 nm light-triggered photodynamic therapy and MR/CT dual-modal imaging of tumor based on the nonstoichiometric $Na_{0.52}YbF_{3.52}$: Er upconversion nanoprobes. *Small* 12: 4200.

71. Tian G, Yin W, Jin J, Zhang X, Xing G, Li S, Gu Z, Zhao Y (2014) Engineered design of theranostic upconversion nanoparticles for tri-modal upconversion luminescence/magnetic resonance/X-ray computed tomography imaging and targeted delivery of combined anticancer drugs. *J. Mater. Chem. B* 2: 1379.

72. Liu W, Zhang Y, You W, Su J, Yu S, Dai T, Huang Y, Chen X, Song X, Chen Z (2020) Near-infrared-excited upconversion photodynamic therapy of extensively drug-resistant Acinetobacter baumannii based on lanthanide nanoparticles. *Nanoscale* 12: 13948.

73. Zhu X, Li J, Qiu X, Liu Y, Feng, W, Li F (2018) Upconversion nanocomposite for programming combination cancer therapy by precise control of microscopic temperature. *Nat. Commun.* 9: 2176.

74. Yu Z, Hu W, Zhao H, Miao X, Guan Y, Cai W, Zeng Z, Fan Q, Tan TTY (2019) Generating new cross relaxation pathways by coating Prussian blue on $NaNdF_4$ for enhanced photothermal agents. *Angew. Chem. Int. Ed.* 58: 8536.

75. Zha S, Fung YH, Chau HF, Ma P, Lin J, Wang J, Chan LS, Zhu G, Lung HK, Wong KL (2018) Responsive upconversion nanoprobe for monitoring and inhibition of EBV-associated cancers via targeting EBNA1. *Nanoscale* 10:15632.

76. Zha S, Chau HF, Chau WY, Chan LS, Lin J, Lo KW, Cho WC, Yip YL, Tsao SW, Farrell PJ, Feng L, Di JM, Law GL, Lung HK, Wong KL (2021) Dual-targeting peptide-guided approach for precision delivery and cancer monitoring by using a safe upconversion nanoplatform. *Adv. Sci.* 8: 2002919.
77. Xu J, Zhou S, Tu D, Zheng W, Huang P, Li R, Chen Z, Huang M, Chen X (2016) Sub-5 nm lanthanide-doped lutetium oxyfluoride nanoprobes for ultrasensitive detection of prostate specific antigen. *Chem. Sci.* 7: 2572.
78. Bao G, Zha S, Liu Z, Fung YH, Chan CF, Li H, Chu PH, Jin D, Tanner PA, Wong KL (2018) Reversible and sensitive Hg^{2+} detection by a cell-permeable ytterbium complex. *Inorg. Chem.* 57: 120.
79. Zhou L, Fan Y, Wang R, Li X, Fan L, Zhang F (2018) High-capacity upconversion wavelength and lifetime binary encoding for multiplexed biodetection. *Angew. Chem. Int. Ed.* 57:12824.
80. Zhang H, Fan Y, Pei P, Sun C, Lu L, Zhang F (2019) Tm^{3+} sensitized 1208 nm excitation and 1525 nm emission NIR-II fluorescent nanocrystals for in vivo information storage and decoding. *Angew. Chem. Int. Ed.* 58: 10153.
81. Liu X, Chen ZH, Zhang H, Fan Y, Zhang F (2020) Independent luminescent lifetime and intensity tuning of upconversion nanoparticles by gradient doping for multiplexed encoding. *Angew. Chem. Int. Ed.* 60: 7041–7045.

10 Design Growth of Nanophosphors and Their Applications

Ming-Hsien Chan, Wen-Tse Huang, Michael Hsiao, and Ru-Shi Liu

CONTENTS

10.1 INTRODUCTION

Multifunctional nanophosphors have received much attention because of their ability to have multiple applications such as lighting [1], imaging [2], and treatment [3]. This term is used to describe the ability of nanophosphors to serve as a platform for different fields such as illumination [4], promoting plant growth [5], and biological application at the same time [6]. In recent years, scientists have developed some different templates and scaffolds, such as multifunctional mesoporous silica nanoparticles (MSNs) as a carrier for dual-effect imaging and treatment. Compared with traditional nanocarriers with imaging functions, silica nanoparticles have a variety of potential advantages that can be served as a template of nanophosphors, such as high specific surface area and pore volume, adjustable pore structure, particle size, good physical and chemical stability, and excellent biocompatibility; besides, through different multifunctional silica nanoparticles synthesized by the design can improve the sensitivity of light-emitting diode (LED) generation, photo modulating detection and enhance the efficiency of targeted therapy [7–9].

Nanotechnology has developed rapidly in recent years. The most representative of these is the development and application of fluorescent nanophosphors. Because nanophosphors are so small, they can be regarded one macroscopically. Generally, nanocrystals smaller than 10 nm have obvious quantum effects, and their properties are very different from those of the same large-sized crystals [10–12].

DOI: 10.1201/9781003098669-10

Taking semiconductor nanophosphors as an example, when the crystal size is reduced to a few nanometers, the material properties (such as band structure and the composition of the crystal surface) will change with size; furthermore, the size can affect the emission of the nanophosphors. [13–15]. Also, because the size of nanophosphors can be adjusted under a size controlling template materials, the carrying space (such as have hole structures or a space that can hold nanocrystals) in the template materials are restricted the growth of nanophosphors and provide a space for crystallization [14]. In the space, the probability of carrier recombination is greatly increased, making the carrier recombination efficiency of some materials close to 100%. After forming nanophosphors through space limitations, most of the input energy can be converted into light energy, and the energy loss is extremely low [16]. These two properties – material properties vary with size and composition. Therefore, semiconductor nanophosphors have many new physical properties and applications, and many documents have provided many new research directions [17]. At present, most of them are based on optoelectronic materials or components, such as nanophosphors electrical excitation light components [16], nanophosphors biosensors [17], nanophosphors solar cells [18], white LEDs [19], and drug delivery system [20], among which the most attention is the application of LEDs and biological applications (Figure 10.1). The former has gradually entered the application stage, and the latter is currently the rapid development of scientific research in recent years.

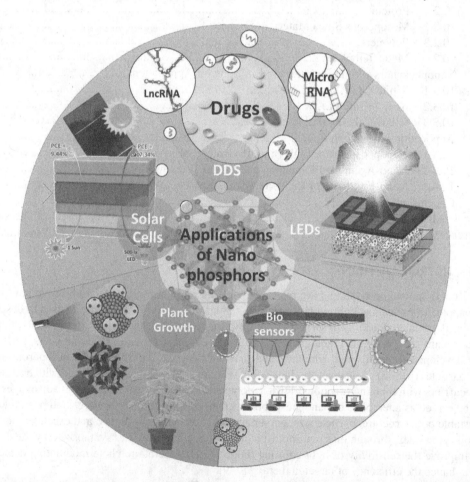

FIGURE 10.1 Nanophosphors have broad applications. Most of them are based on optoelectronic materials or biological components, such as electrical excitation light components, biosensors, solar cells, LEDs, and drug delivery system, among which the most attention is the application of illumination and biological applications.

This chapter focuses on the recent developments in the field of nanophosphors. The synthesis method of a nanophosphor can be classified into chemical synthesis methods such as a solution method and physical synthesis methods such as mechanical grinding. In the chemical synthesis method, the nanophosphor is obtained at low temperature and has low crystallinity and high-temperature annealing, which leads to grain growth. In the physical method, the luminescence efficiency decreases by the formation of the surface defects with the mechanical contact.

10.2 SYNTHESIS OF NANOPHOSPHORS WITH LIMITED SPACE

Nanostructured templates usually refer to materials with porosity, and their pore sizes mostly fall in the nanoscale range (1–100 nm) [21]. The template is used to prepare nanostructures, because the template's pore size, pore distribution, porosity and pore aspect ratio, and pore surface characteristics (hydrophilic or hydrophobic characteristics) can be adjusted to achieve production with special performance [22]. Because of the uniformity of materials, the use of nanotemplates to prepare nanostructures has become one of the most popular methods for synthesizing nanofluorescent powders in recent years (Figure 10.2). Herein, this section will give a detailed introduction to the classification of templates, preparation methods, and the technology of using templates to synthesize nanostructures, so that readers can have a preliminary understanding of template nanotechnology.

10.2.1 PROTEIN

When the protein is folded, some pores are formed inside, and nanoparticles or structures can be made with this characteristic [23–25]. Take a protein that exists on the surface of bacteria (S-layer) as an example. S-layer protein has many biological functions, such as forming protective coats, providing

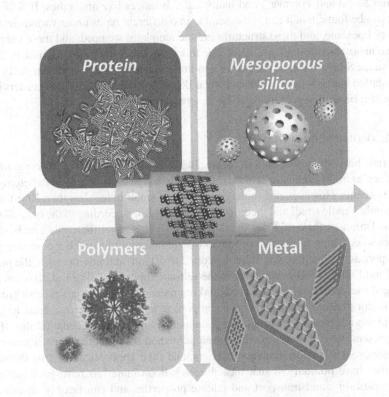

FIGURE 10.2   The classification of templates and scaffolds. The nanotechnology of using templates to synthesize nanostructures and limited the growth space of nanophosphors.

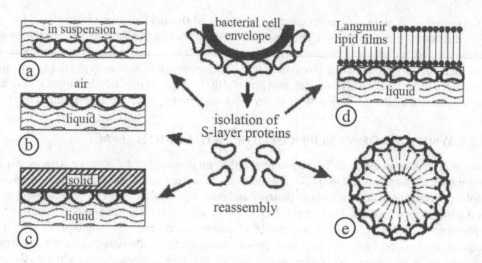

FIGURE 10.3 Self-assembly of S-layer protein. This technique is also called targeted protein immobilization (a) suspension, (b) separation air-liquid layer, (c) solid layer addition, (d) lipid film degeneration and (e) Self-assembly. Through the boronic acid group, the metal coordination center and the aptamer-based immobilization are covalently immobilized [30]. (Mark, S. S.; Bergkvist, M.; Yang, X.; Angert, E. R.; Batt, C. A., *Biomacromolecules* **2006**, 7, (6), 1884–1897. With permission)

bacteria attachment and identification, and maintaining cell shape [26–28]. Because S-layer protein has the same characteristics, it can be combined with many substances, including some materials like "Silicon Wafer Metal and Polymer", and biological substances like antibodies. If S-layer protein is separated, it can be found that it can self-assemble into different types under various interfaces, such as lipid film or liposome, and these structures are the templates we need, and these templates can be used to create nanostructures (Figure 10.3). The following takes the preparation of Pt clusters as an example. First, the S-layer sheet on the suspension-ureae cell is separated, and then the Pt in K_2PtC_4 is reduced and settled on the S-layer by the molecular deposition method to form Pt clusters [29]. Similar reactions can also be used to prepare Ag, CdS nanoparticles.

10.2.2 Mesoporous Silica Material

Porous materials have already existed in nature. In 2002, M. E. Davis *et al.* discovered that zeolite had many irregular microporous structures [31]. Zeolite has been widely used in heterogeneous catalysts and adsorbents. However, due to the limitations of small holes (less than 2 nm) and slow diffusion, it can only handle small and medium-sized molecules. According to the definition of IUPAC (International Union of Pure and Application Chemistry), porous materials can be divided into three categories according to the size of the pore size: macroporous, with a pore size greater than 50 nm; mesoporous, with a pore size ranging from 2 to 50 nm; and microporous, the pore diameter is less than 2 nm [32]. In recent years, due to the self-assembly capabilities of different surfactants, neatly arranged mesopores can be prepared, making mesoporous materials receive great attention, and their properties have a stable and consistent pore structure, high surface area, adjustable pore size, clear and modifiable surface properties, nontoxicity, and good biocompatibility (Figure 10.4) [33–35]. At present, many documents have been published using mesoporous silicon materials or modified mesoporous silicon materials as carriers, and their applications in drug delivery systems [36–38]. Studies have pointed out that mesoporous silicon materials have good structural properties, biocompatibility, and transport and release properties and can flexibly absorb and release drug molecules and be used in biological tests [39–41]. Also, mesoporous materials can control the release rate of drugs in different solutions due to their special surface characteristics and structural

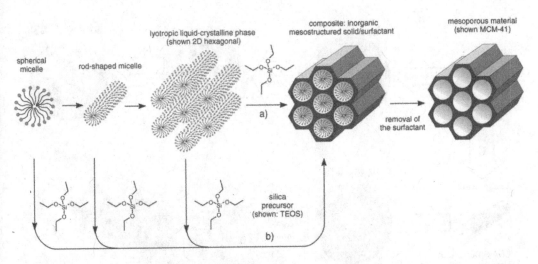

FIGURE 10.4 Synthesis path of mesoporous silicon material formed by interface active agent as a template. (a) Interface the self-assembly process of surfactants. (b) Silica shell formation [45]. (Hoffmann, F.; Cornelius, M.; Morell, J.; Froba, M., *Angew. Chem. Int. Ed.* **2006**, 45, (20), 3216–3251. With permission)

regularity. Before loading the drug molecules, the mesoporous material is first calcined or refluxed to remove the template scaffold in the hole channel, so that the drug to be loaded can be absorbed into the hole channel [42–44].

In recent years, in addition to drug delivery functions, mesoporous materials can be modified by functional groups on their surface to give the materials different functional properties, which can be applied to various types of biomedicine and biotechnology [40]. Above, such as cell-type identification, disease diagnosis, intracellular visualization, and drug and gene delivery, there have been researching results that have confirmed that mesoporous materials can enter cells through cell endocytosis [46, 47]. These grafted molecules will use various environmental factors, such as light-cutting effect, redox reaction, enzyme presence, or different pH value, to produce different chemical reactions, and then control the opening and closing of pores to achieve controlled release of drugs [46–48]. The effect in Figure 10.5 is the difference between traditional and novel drug delivery systems. For instance, MSNs limit the growth space of crystals, add metal ions with long afterglow characteristics, and sinter the two at high temperatures to synthesize persistent nanoparticles [49–51].

10.2.3 Polymers

The preparation methods of polymer templates can be divided into two categories. One is to use the difference between crystalline and non-crystalline reactions to obtain porosity; the other is to form copolymers from two or more different monomers [53–55]. There are differences in polarity, chemical reactivity, thermal stability, and photoreactivity, and a hole is formed by removing one of the components. The primary key to preparing a polymer template is to master its dispersion, to produce a porous template with uniform pore size and uniform distribution. Besides, other control variables include polymer molecular weight distribution, polymer thermal stability parameters (Tg, Tm, Tc), and mechanical strength [56–58]. The common production method is to use polycarbonate, polyester, and other polymer films with a thickness of about 6–20 μm to cause damage marks on the surface of the film in various ways and then use chemical etching to make these marks into holes. The characteristic of the template prepared in this way is that the pores are randomly arranged in a cylindrical shape. Due to the limitation of the preparation method, the pores and the membrane surface are mostly oblique. Therefore, there is a phenomenon that the pore channels cross in the thick membrane, and the pore density is about 10^9 pieces/cm^2 [59–61]. Figure 10.6

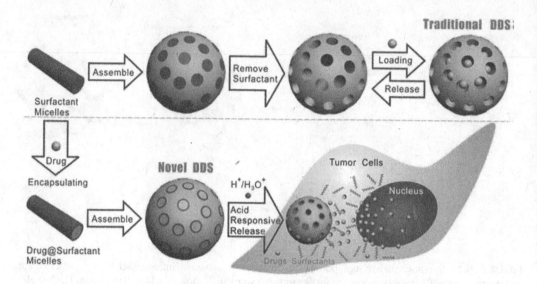

FIGURE 10.5 Comparison of traditional and novel nanodrug delivery systems with drug loading and release. It can be seen from the figure that the new nanodrug delivery system can achieve drug control and release through pH value [52]. (He, Q. J.; Shi, J. L., *J. Mater. Chem.* **2011**, 21, (16), 5845–5855. With permission)

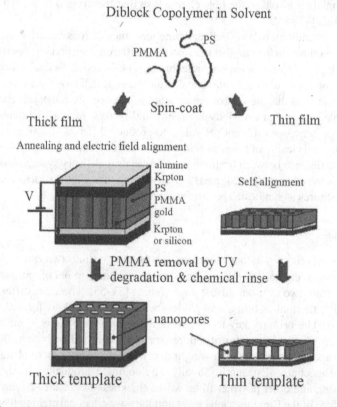

FIGURE 10.6 Schematic diagram of polymer template formation. Using molecular self-assembly methods to prepare organic-inorganic nano-hybrid systems and organic-inorganic nanocomposites is a very effective strategy. The key lies in the use of special secondary forces between organic and inorganic molecules such as polarity, hydrogen bonding, or miscombination, use organic materials as template molecules to achieve orderly hybridization [63]. (Berman, D.; Shevchenko, E., *J. Mater. Chem. C* **2020**, 8, (31), 10604–10627. With permission)

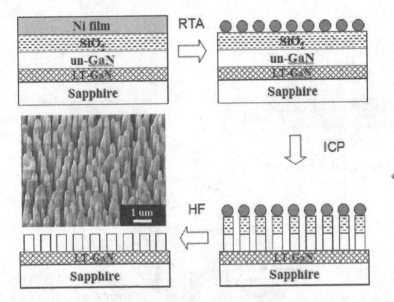

FIGURE 10.7 Schematic of the process used to prepare the metal nanorod template. Schematic diagram of fabricating nanowire arrays. Synthesizing fabricating nanowire arrays first anodized template to compose high-pressure injection of molten metal to the vacuum treated template via acid etching to produce the metal composite film [64]. (Masuda, H.; Fukuda, K., *Science* **1995,** 268, (5216), 1466–1468. With permission)

shows the polymer template with straight-through holes developed by Farani *et al.* [62] The author uses two kinds of polystyrene (PS) and polymethylmethacrylate (PMMA) with different photoreactivity to form a copolymer, and by controlling the interface effect, an electric field is applied to form a cylindrical hole perpendicular to the substrate, and the thickness of the template can be controlled from 20 nm to 10 μm.

10.2.4 METAL TEMPLATE

Masuda and Fukuda developed a two-stage replication method to prepare metal templates (Figure 10.7) [64]. First, a layer of this metal film (such as Pt and Au), and then inject organic monomers under vacuum to fill the pores, and then use UV light or heating to polymerize the monomers into polymers. At this time, as long as the original inorganic template is removed by chemical methods, the negative type can be obtained. The polymer template is catalyzed by the metal film at the bottom by electroless plating, the metal will gradually fill the holes of the negative template, and finally, the organic negative template is washed away with an organic solvent to obtain the metal template [65–67].

10.3 NANOPHOSPHORS WITH DIFFERENT LIGHT EMISSION FOR APPLICATIONS

Optical detection has become one of the most important techniques for qualitative and quantitative judgment in the field of medical diagnosis because of its noninvasive and high precision. Principles of light detection technology in this field include absorbance, fluorescence, chemiluminescence, interferometry, Raman scattering, and surface plasmon resonance. Based on the abovementioned principles, optical diagnostic and analytical instruments are based on different spectral bands. In this section, we briefly describe the application of infrared light in medical diagnostics and therapeutics: such as blood monitoring and tumor light therapy.

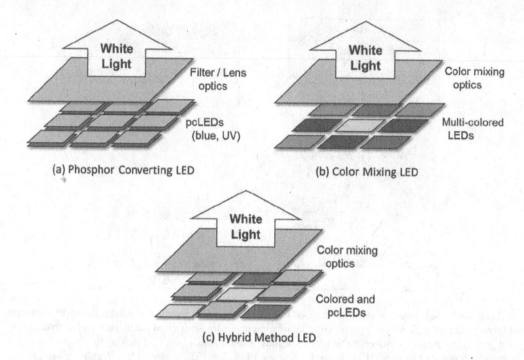

FIGURE 10.8 General Types of White Light–Emitting Diode (LED) Devices Source: DOE, 2012b. Solid-State Lighting Research and Development: Multi-Year Program Plan. (a) Phosphor Converting LED, (b) color mixing LED, (c) hybrid method LED [77]. (Chang, S. J.; Wang, S. M.; Chang, P. C.; Kuo, C. H.; Young, S. J.; Chen, T. P., *Ieee Photonic. Tech. L.* **2010**, 22, (9), 625–627. With permission)

10.3.1 WHITE LIGHT

The white light spectrum is continuous. It requires at least two colors (wavelengths) or more to be mixed to form white light visible to the naked eye. For example, the three primary colors "red light + green light + blue light" are mixed to become white light, or complementary colors are used, or "Blue + Yellow" can also form white light [68–70]. According to the principle of white light formation, white light LEDs are generally divided into two types. The first type is three primary color white light LEDs (RGB white LEDs), which are composed of red, green, and blue semiconductor chips, also known as multi-chip white LED or called triple wavelength white LED or three-band white LED. The other is the complementary color white light LED because it only uses a single chip, so it is also called single-chip white LED (Figure 10.8) [71–73].

According to the theory of quantum confinement effect, the emission wavelength of nanophosphor powder will shift to a short wavelength with the size, from blue shift to visible light, and its energy gap and emission color can be controlled by the size. The current preparation technology of nanophosphor powder can be adjusted by precise size control to adjust its emission wavelength from blue to red (460–650 nm), and its peak half-height width is between 24 and 35 nm. With shape control, we can make nanophosphors produce white light and then make white commercial light micro-LEDs [74–76].

10.3.2 NEAR-INFRARED LIGHT

The near-infrared light window (or biological tissue optical window and bio-light treatment window) refers to the wavelength range in which the penetration depth of light in the biological tissue reaches a maximum, generally in the near-infrared wavelength range [78]. According to the literature, the near-infrared region has two transparent windows of organisms, located at 650–950 nm (first near-infrared light window) and 1000–1350 nm (second near-infrared light window), which can be obtained by the organization [79]. Good energy absorption and low scattering provide maximum tissue penetration and reduce auto-fluorescence. In the visible-near-infrared band, also known as the infrared I region, scattering is the most important form of interaction between light and tissue, causing light to diffuse rapidly during propagation [80]. Since scattering increases the distance traveled by photons within the tissue, the probability that photons will be absorbed by the tissue also increases. The scattering effect varies little with wavelength. Therefore, the range of the biological tissue optical window is mainly limited by the absorption effect in the tissue [81]. The lower limit (the short-wavelength end) is determined by the blood absorption, and the upper limit (the long-wavelength end) is the water. The absorption is determined [82]. For applications such as optical imaging and photothermal therapy, selecting a suitable light source located in the optical window of the infrared region I is of great significance for improving imaging (treatment) efficiency, increasing penetration depth, and reducing photoinduced tissue damage. Infrared light I is used in a wide range of applications. The most commonly used is blood monitoring, which analyzes blood samples by different particles/molecules in the blood to reflect and absorb different wavelengths of light. The instrument in the infrared zone I monitor the target component in the blood. Infrared light with a wavelength of 1000–1350 nm is called infrared light II because it has the potential as a biological window [83]. Compared with the first near-infrared light window (NIR-I, 750–900 nm), the blood and tissue have a deeper absorption and scattering of the second near-infrared light, so that it has a deeper penetration of the living tissue [84]. Transmissive ability is showing a higher noise ratio when used for live imaging. Therefore, the second near-infrared light window fluorescent material has greater advantages in biological detection and *in vivo* imaging. However, water molecules have a great absorption effect in the infrared region above 1000 nm, and the infrared region II is more inclined to the application between organs and tissues. The absorption of the near-infrared light region in each organ can be obtained in Figure 10.9, indicating that the infrared light II region is also highly bioavailable [85].

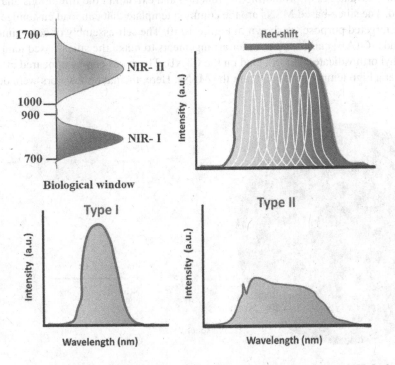

FIGURE 10.9 Classification of the infrared windows. Schematic representation NIR-I phosphors with type 1, and type-II emissions. The red-shifting of the type I emission and combining in a single LED package lead to a broad emission highlighted by the pink region.

10.3.3 GLOW-IN-THE-DARK

When an object is stimulated by light, heat, electricity, or chemical action, it emits light without heat [86]. This phenomenon is usually called "luminescence" (sometimes called cold light because it does not radiate any heat). When an object is exciting, the electrons in the object reach a high energy level (excited state). When it returns to its original state (the ground state), the excess energy is radiated in the form of light [87]. Among all kinds of luminescence, what is excited by light is called "photoluminescence", and according to the mechanism of light emission, it can be divided into "fluorescence" and "phosphorescence". Fluorescence is reflected light, and phosphorescence refers to continuous light emission after leaving the illuminating light source [88].

Nanoscale light-storing self-luminous materials came out in 2010 and are the most representative materials are $SrAl_2O_4$:$Eu^{2+}Dy^{3+}$ [89–91]. Compared with the previous generations of self-luminous materials, this material has significant advantages such as nontoxic and nonradioactive. Its outstanding characteristics are: it absorbs and stores all kinds of visible light during the day, such as daylight, fluorescent light, light, ultraviolet light, and other stray light. After 10–20 minutes, it can self-illuminate, and it can continue to emit light for more than 12 hours in the dark [92]. Its luminous intensity is continuous. The time is 30–50 times that of traditional luminescent materials, and the stability and weather resistance are excellent, which completely changes the shortcomings of traditional self-luminescent materials in the application.

10.4 APPLICATIONS OF NANOPHOSPHORS

Nanotechnology of revolutionary applications with manipulating the molecular structure and intrinsic properties of materials provides gigantic possibilities for contemporary science and industry. The branch of nanotechnology involves electronics, energy, biomedicine, environment, food, and textile, which possess advantages of promoting the function and extending the limitations and capabilities in each field. The silica-based MSNs are the common template that can load customized matters to achieve the targeted purpose, as shown in Figure 10.10. The self-assembly of cetyltrimethylammonium bromide (CTAB) can form the order arrangements to make the silica-based template precursor, tetraethyl orthosilicate (TEOS), clad on the CTAB. The pores can be appeared after removing the CTAB at a high temperature to obtain the MSNs. Here, the nanophosphors were demonstrated

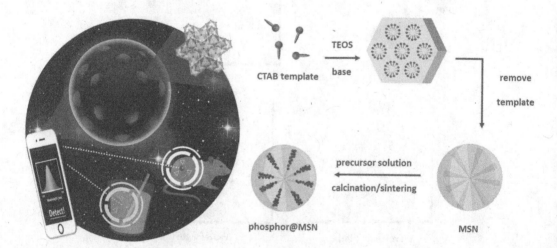

FIGURE 10.10 Scheme of template method to synthesize nanophosphor. (Huang, W. T.; Cheng, C. L.; Bao, Z.; Yang, C. W.; Lu, K. M.; Kang, C. Y.; Lin, C. M.; Liu, R. S., *Angew. Chem. Int. Edit.* **2019**, 58, (7), 2069–2072. With permission.)

with $ZnGa_2O_4$ as the model. The precursor solution was prepared with suitable concentration and mixed evenly with MSNs. After high-temperature treatments (calcination or sintering), the form of nanosized $ZnGa_2O_4$ can be observed in the pores of MSNs. Different applications of MSN-typed nanophosphors will be mentioned in detail in the following paragraphs, including lighting, theranostics, and agriculture.

10.4.1 Nanophosphor for Lightening Package in Mini- or Micro-LED

Lighting is an indispensable part of life. Especially, near-infrared LEDs have highly attracted considerable attention due to their wide applications, such as solar energy utilization, bioimaging, photo-biomodulation, cancer diagnostic, and therapy. With the micro-miniaturization of technology, electronic devices are constantly advancing with the times. In recent years, the chip size of LEDs has made comprehensive breakthroughs in standards, such as mini-LEDs (100–200 μm) and micro-LEDs (<100 μm), which have made tremendous contributions to display technology. The better color rendering and finer high dynamic range imaging can be approached through local dimming. With flexible substrate, it can also achieve the form of a high-curved backlight on the micro-device. The high quantum efficiency phosphors ($CaAlSiN_3:Eu^{2+}$ and $Y_3Al_5O_{12}:Ce^{3+}$) provided in the commercial market all present micro-sized particle distribution. However, it is unsuitable for mini- or micro-LEDs. In contrast, nanophosphors are considered to be good candidates for materials using micro-LEDs [93].

In Figure 10.11, the nanophosphor is applied on the mini- or micro-LEDs in an electronic device. To apply on mini- or micro-LED chips successfully, the NIR emissive co-doping $ZnGa_2O_4:Cr^{3+}$, Sn^{4+} (ZGOCS) was attempted to shrink the size to nanoscale level by template method. The size and morphology can be easily controlled to improve and crystallization and dispersion. In Figure 10.11(a), the MSNs still maintained the spherical morphology from scanning electron microscopy (SEM) before/after the cladding of ZGOCS. It also showed the robust physical properties of MSNs under the high-temperature treatment around 1000°C. Crystal formation of ZGOCS can be distinguished with dark spots from the images of transmission electron microscopy (TEM) in Figure 10.11(b). The size distribution of ZGOCS@MSN was showed 70 nm.

Detailed information about the status of pores can be obtained through nitrogen adsorption and desorption. The process measured under 77K, as shown in Figure 10.12(a), and it is known that the adsorption and desorption isotherm of MSNs is type IV. When the partial pressure is 0.9–1, because nitrogen condensed in the material pores, the phenomenon of inconsistent adsorption and desorption isotherms occurred. Calculated by Barrett-Joyner-Halenda (BJH) method, the pore size of MSNs was about 2.3 nm. After the nitrogen desorption curve is calculated by the Brunauer-Emmett-Teller (BET) method, it can be seen that the specific surface area of MSNs is 258.4 m^2/g and the pore volume is 0.265 m^3/g. It is proved that the synthesized MSN is of the mesoporous-type and has the function of loading ZGOCS. The adsorption and desorption isotherm of ZGOCS@MSN decreased significantly when the partial pressure is 0–0.8. In Figure 10.12(b), the pore volume can't be detected, confirming the pores of MSNs were indeed loading ZGOCS.

The excitation band of ZGOCS spanned ultraviolet light and visible light at 280, 406, and 550 nm, respectively (Figure 10.13(a)). The corresponding transition states are $^4A_2 \rightarrow {}^4T_1$ (4P), $^4A_2 \rightarrow {}^4T_1$ (4F), and $^4A_2 \rightarrow {}^4T_2$ (4F). The emission spectrum of relative intensity can be obtained by using 406 nm as the excitation source. Its wavelength is in the near-infrared region of 600–850 nm, and the corresponding transition state is $^2E \rightarrow {}^4A_2$. With Cr^{3+} doped into the octahedral position of Ga^{3+}, the R line can be observed at 686 nm. The phonon sideband (PSB) can be found on the left and right sides of the R line. The anti-Stokes shift (AS-PSB) was blue-shifted at 678 nm, and the Stokes shift (S-PSB) was red-shifted at 706 and 712 nm. At the same time, the anti-site of the adjacent Ga^{3+} octahedron produced a peak at 692 nm due to defects, which was the N2 line. The wavelength of commercial blue chips is generally located at 460 nm. To present the best performance, the LED chips can be changed to 405 nm or enhance the luminous intensity. In this research, the co-doping ratio

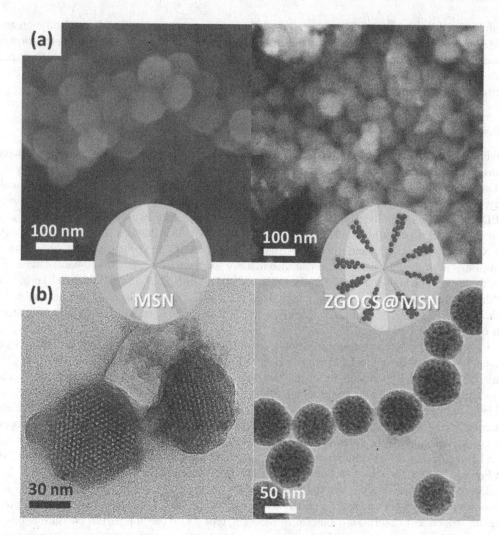

FIGURE 10.11 Morphological characterization of MSN and ZGOCS@MSN with (a) SEM and (b) TEM images. (Huang, W. T.; Cheng, C. L.; Bao, Z.; Yang, C. W.; Lu, K. M.; Kang, C. Y.; Lin, C. M.; Liu, R. S., *Angew. Chem. Int. Ed.* **2019**, 58, (7), 2069–2072. With permission.)

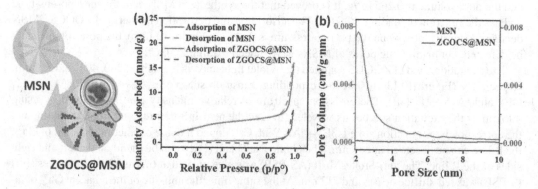

FIGURE 10.12 BET analysis of MSN and ZGOCS@MSN with (a) nitrogen adsorption/desorption isotherm and (b) pore volume. (Huang, W. T.; Cheng, C. L.; Bao, Z.; Yang, C. W.; Lu, K. M.; Kang, C. Y.; Lin, C. M.; Liu, R. S., *Angew. Chem. Int. Edit.* **2019**, 58, (7), 2069–2072. With permission.)

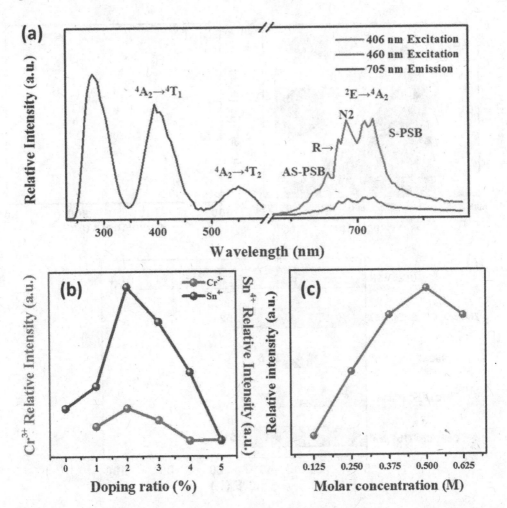

FIGURE 10.13 The luminous properties of ZGOCS@MSN with (a) emission, excitation, (b) co-doping ratio, and (c) molar concentration. (Huang, W. T.; Cheng, C. L.; Bao, Z.; Yang, C. W.; Lu, K. M.; Kang, C. Y.; Lin, C. M.; Liu, R. S., *Angew. Chem. Int. Ed.* **2019**, 58, (7), 2069–2072. With permission.)

of Cr^{3+}/Sn^{4+} and concentration of precursor solution were optimized to have the best light-emitting performance, as shown in Figure 10.13(b) and (c).

The optimized ZGOCS@MSN was packaged on a 9 × 5 mil mini-LED chip. The performance of the device emitted the near-infrared light with 625–950 nm, as shown in Figure 10.14(a). With a 45-mA input current, the output voltage can be obtained by 3.3 mW (light conversion efficiency of 37%). In general, the smaller the particle size, the higher the concentration of surface defects because smaller particles have a larger surface-to-volume ratio. This type of defect will function as killer sites, reducing the luminescent intensity and quantum efficiency of luminescent materials. Therefore, gallium oxide (Ga_2O_3) is a potential host structure to approach higher quantum efficiency nanophosphors. Consequently, the IQE of $Ga_2O_3:Cr^{3+}$ in MSN (GOC@MSN) was measured and compared with those of other nanosized phosphors, as shown in Figure 10.14(b). GOC@MSN was performed with a high IQE of 91.4%, which is considerably higher than those of $ZnGa_2O_4:Cr^{3+}$, Sn^{4+}@MSN, $NaCeF_4$: Er^{3+}/Yb^{3+}, $CaF_2:Ce^{3+}/Mn^{2+}$, and $CaF_2:Ce^{3+}/Tb^{3+}/Na^+$. The IQE value of GOC@MSN was even comparable with that of bulk $Ga_2O_3:Cr^{3+}$ phosphor (92.4%).

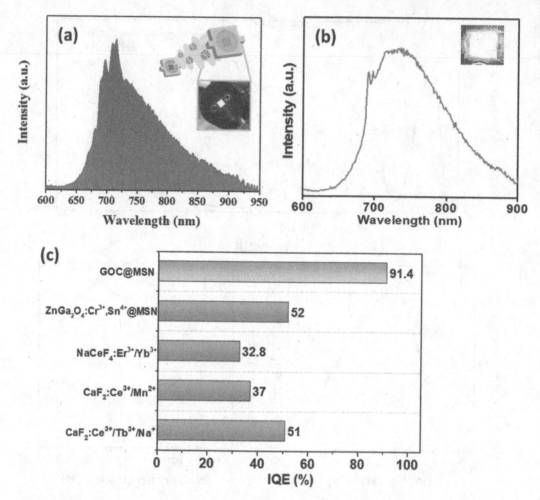

FIGURE 10.14 The device performance of nanophosphor with (a) PL package spectrum of ZGOCS@MSN, (b) GOC@MSN, and (c) different nanophosphor comparison. (Huang, W. T.; Cheng, C. L.; Bao, Z.; Yang, C. W.; Lu, K. M.; Kang, C. Y.; Lin, C. M.; Liu, R. S., *Angew. Chem. Int. Ed.* **2019**, 58, (7), 2069–2072. With permission.)

10.4.2 NANO-AGENT FOR BIOIMAGING AND SIGNAL TRACKING

The leading cause of death in the world is malignant tumors. In nowadays, various cancer detection technologies are developed with their advantages and shortcomings. Two tricky problems in cancer treatments are the inability to do specific targeting or to trace targeting drugs for effective therapy and the poor accumulation of antitumor drugs due to the blood-brain barrier. ZGOCS@ MSN, a kind of NIR persistent luminescence nanoparticles (PLNs), can be used as tracking agents. The unique luminous properties of PLNs involve the absorption and retention of photons for several hours, followed by the emission of long-term luminescence for at least an hour. The following paragraphs will introduce how PLNs can solve the abovementioned tricky problems with different strategies, as shown in Figure 10.15 [94].

The treatment of brain diseases has aroused considerable interest in research fields all over the world. Glioblastoma is the most malignant and aggressive brain cancer. Although many chemotherapeutic drugs can be used for glioma in the market, the blood-brain barrier prevents drugs from entering the tumor treatment area and the accumulation of drugs at the tumor site is poor. Therefore, it is effective and safe to open the blood-brain barrier and the target of the drug The tropism is the

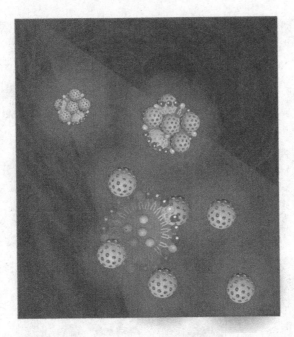

FIGURE 10.15   ZGOCS@MSN applied on cancer treatments with different strategies.

main problem of drug delivery in the treatment of brain tumors. In this research, a new type of nano-composite material combining temozolomide (TMZ), nanobubbles (NBs), PLNs (ZGOCS@MSN), an aptamer AS1411 (AAp) was developed for bio-tracking imaging and treatment of glioblastoma and proved Its excellent therapeutic effect, as shown in Figure 10.16.

In Figure 10.17, the morphology changes in the synthetic process were observed by SEM and TEM. However, the surface of TMZ-NB wasn't conductive, which need to sputter platinum on the

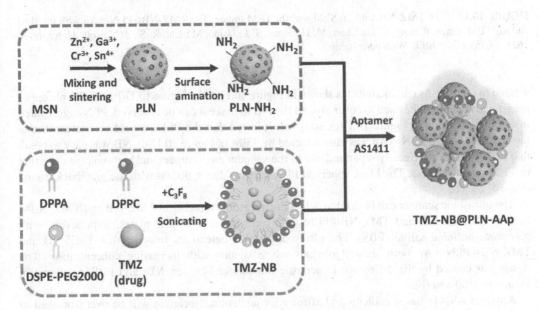

FIGURE 10.16   Scheme of TMZ-NB@PLN-AAp synthesis. (Cheng, C. L.; Chan, M. H.; Feng, S. J.; Hsiao, M.; Liu, R. S., *ACS Appl. Mater. Inter.* **2021,** 13, (5), 6099–6108. With permission.)

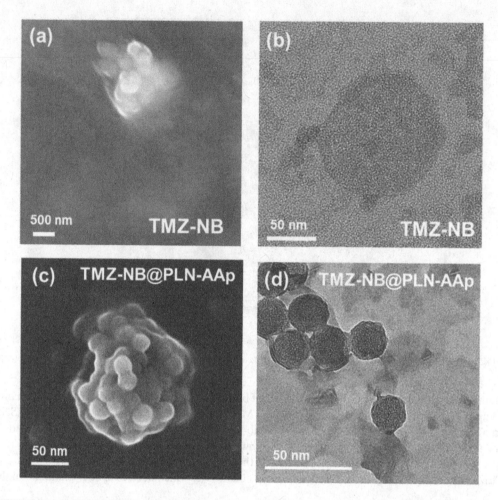

FIGURE 10.17 The TMZ-NB with (a) SEM and (b) TEM image. The TMZ-NB@PLN-AAp with (c) SEM and (d) TEM image. (Cheng, C. L.; Chan, M. H.; Feng, S. J.; Hsiao, M.; Liu, R. S., *ACS Appl. Mater. Inter.* **2021,** 13, (5), 6099–6108. With permission.)

surface to increase its conductivity, as shown in Figure 10.17(a). In Figure 10.17(b), the 2% phospho-tungstic acid needs to be used to cover around the NB to make it can be observed. PLNs aggregated in a spherical shape in Figure 10.17(c), so it was inferred that the PLNs should be coated outside the NB. TMZ-NB@PLN-AAp was also detected by TEM in Figure 10.17(d). NB was easy to break that PLN and NB were easily separated due to the vacuum environment and high-voltage detection in TEM measurements. The black spherical PLNs can be distinguished with the gray background shape of the NB.

The ultrasonic scanner can be used to detect the ultrasonic imaging of TMZ-NB and TMZ-NB@PLN-AAp. TMZ-NB and TMZ-NB@PLN-AAp were put into 96-grid plates, respectively, with phosphate-buffered saline (PBS). The ultrasound probe detects its imaging that TMZ-NB and TMZ-NB@PLN-AAp both showed obvious white signals with increasing concentration. The signal was caused by the difference in acoustic impedance between NBs and PBS, as shown in Figure 10.18(a) and (b).

Aptamer AS1411 has specificity and affinity for nucleolin. Nucleolin will be overexpressed in glioblastoma cells. The cell line, SVGp12, A172, D54MG, U87, and U251 were selected to analyze by the Western blotting method. The β-actin antibody was used in the control group for protein

(a) TMZ-NB

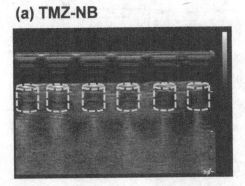

(b) TMZ-NB@PLN-AAp

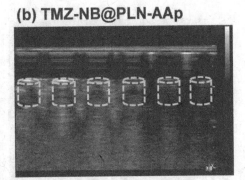

FIGURE 10.18 The ultrasound images of (a) TMZ-NB and (b) TMZ-NB@PLN-AAp. (Cheng, C. L.; Chan, M. H.; Feng, S. J.; Hsiao, M.; Liu, R. S., *ACS Appl. Mater. Inter.* **2021**, 13, (5), 6099–6108. With permission.)

expression. Among them, SVGp12 cells were human fetal glial cell lines, as a normal cell group, while A172, D54MG, U87, and U251 cells were all human glioblastoma cell lines. Western blot analysis showed that U87 and D54MG cells had higher nucleolin expression. Especially, U87 has relatively higher nucleolin expression, as shown in Figure 10.19.

To evaluate the targeting ability of the aptamer AS1411 and the imaging function of the NIR PLNs in cells, SVGp12, D54MG, and U87 cells were cultured with NB@PLN and NB@PLN-AAp and cell images were taken. To avoid cell apoptosis, this culturing process did not add temozolomide drug. The cell images showed that SVGp12 cells in the normal cell group did not show NIR signals, indicating that NB@PLN and NB@PLN-AAp were almost not endocytosed by SVGp12 cells. D54MG and U87 cultured with NB@PLN and NB@PLN-AAp that cell images displayed a clear NIR signal in the cell area. Compared with D54MG and U87 cells cultured with NB@PLN, the NIR signals of D54MG and U87 cells cultured with NB@PLN-AAp was clearer, indicating that the aptamer AS1411 can assist the ability of NB@PLN to be endocytosed by cells and enhance the NIR signal of PLNs in the cell, as shown in Figure 10.20. The uptake of NB@PLN and NB@PLN-AAp in cells was quantified by flow cytometry, also shown in Figure 10.20. The uptake of NB@PLN and NB@PLN-AAp in cells was quantified by flow cytometry, also shown in Figure 10.20. The P3 area represented the signal of NIR in the cell. The distribution of SVGp12 cells in the normal group showed that NB@PLN and NB@PLN-AAp almost entered the cells without endocytosis, while the distribution of D54MG and U87 cells cultured with NB@PLN showed that the cells contained NIR. The distribution of D54MG and U87 cells cultured by NB@PLN-AAp showing obvious NIR in the

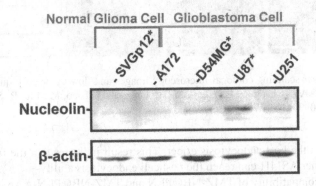

FIGURE 10.19 Western blot analysis of SVGp12, A172, D54MG, U87, and U251 cell lines. (Cheng, C. L.; Chan, M. H.; Feng, S. J.; Hsiao, M.; Liu, R. S., *ACS Appl. Mater. Inter.* **2021**, 13, (5), 6099-6108. With permission.)

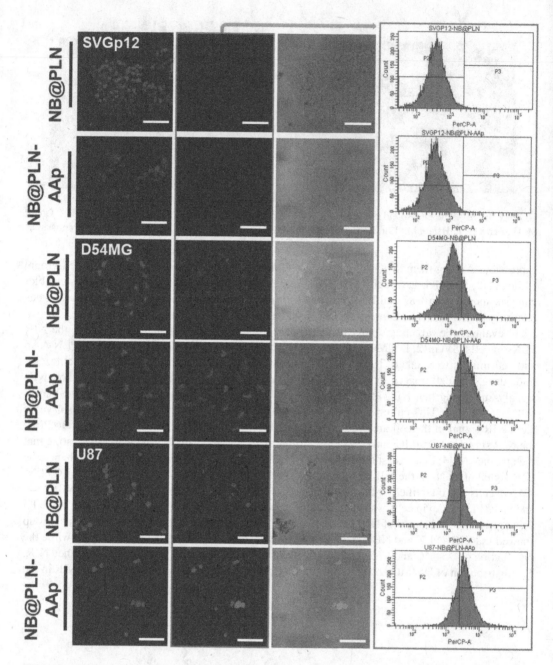

FIGURE 10.20 Laser scanning confocal microscopy images and flow cytometric quantification of NB@ PLN and NB@PLN-AAp. (Cheng, C. L.; Chan, M. H.; Feng, S. J.; Hsiao, M.; Liu, R. S., *ACS Appl. Mater. Inter.* **2021**, 13, (5), 6099–6108. With permission.)

cells, which showed the best endocytosis effect. This result indicates that the functionalization of NB@PLN by aptamer AS1411 can reach the best cell endocytosis ability.

To test the cytocompatibility of TMZ-NB@PLN and TMZ-NB@PLN-AAp on normal SVGp12 cells and the cytotoxic effect on D54MG and U87 cells in the experimental group. TMZ-NB@PLN and TMZ-NB@PLN-AAp were taken concentrations of 3, 9, 27, 83, 250 mg mL^{-1} were cultured for three groups of cells for 24 hours to detect their cell viability. Even if TMZ-NB@PLN and

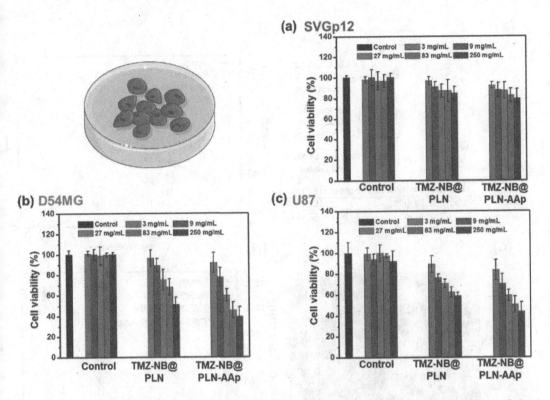

FIGURE 10.21 (a) Cytocompatibility of SVGp12, (b) cytotoxicity of D54MG and (c) U87. (Cheng, C. L.; Chan, M. H.; Feng, S. J.; Hsiao, M.; Liu, R. S., *ACS Appl. Mater. Inter.* **2021**, 13, (5), 6099–6108. With permission.)

TMZ-NB@PLN-AAp were cultured at a high concentration of 250 mg mL^{-1}, the effect on the cell survival rate of SVGp12 cells in the normal group wasn't significant. Both TMZ-NB@PLN and TMZ-NB@PLN-AAp have excellent cell compatibility with more than 80% cell survival rate, as shown in Figure 10.21(a). Regarding the cytotoxicity of glioblastoma D54MG and U87 cells, the cell survival rate of D54MG and U87 cells both decreased with the increase of the concentration of TMZ-NB@PLN and TMZ-NB@PLN-AAp, as shown in Figure 10.21(b), and due to the targeting effect of the aptamer AS1411, TMZ-NB@PLN-AAp at a concentration of 83 mg mL^{-1} makes the cell survival rate of D54MG and U87 cells both approximately 50%, even less than 250 mg mL^{-1}. At the concentration, D54MG and U87 cells only have about 40% cell survival rate, as shown in Figure 10.21c.

U87 cells were subcutaneously implanted into the legs of NOD-SCID mice, and TMZ-NB@PLN and TMZ-NB@PLN-AAp were intravenously injected into the mice on the 21st, 25th, and 29th days after tumor implantation. With each injection of 100 μL of material into each mouse, the material concentration was 10 mg mL^{-1}, and the tumor site was irradiated by ultrasound to guide the rupture of NB to release TMZ. The tumor and organs of 12 mice were collected 60 days after tumor implantation and performed the physiological analysis of mice. The experimental procedure was shown in Figure 10.22(a). Within 60 days of the experiment, the control group (indicating no treatment measures), mice treated with TMZ-NB@PLN and TMZ-NB@PLN-AAp, there was no significant difference in weight changes, as shown in Figure 10.22(b). However, after treatment with TMZ-NB@PLN or TMZ-NB@PLN-AAp, the tumor volume was greatly reduced. The tumor image was shown in Figure 10.22(c). The material passively accumulated in the tumor by the high permeability and long retention effect. Compared with the control group, the treatment effectively reduced the tumor volume by about 60% at 5 weeks after treatment, especially when

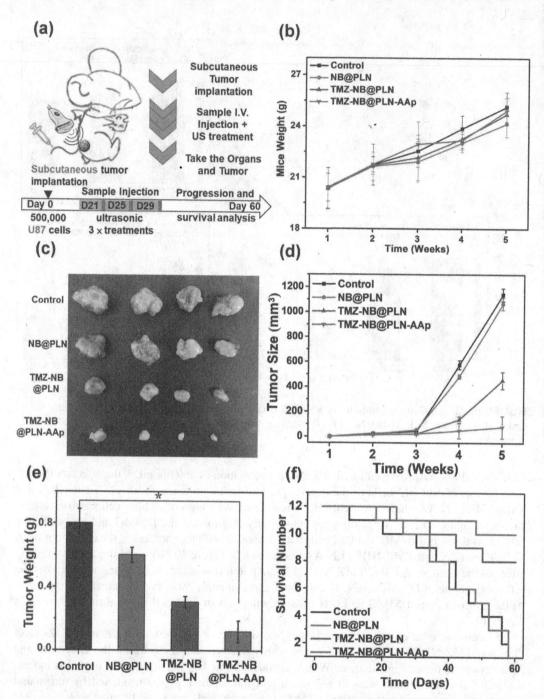

FIGURE 10.22 Therapeutic effect in mice. (a) Flow chart of *in vivo* treatment experiment. (b) Bodyweight, (c) U87 tumor image, (d) tumor size, (e) tumor weight, and (f) survival curve of mice. (Cheng, C. L.; Chan, M. H.; Feng, S. J.; Hsiao, M.; Liu, R. S., *ACS Appl. Mater. Inter.* **2021,** 13, (5), 6099–6108. With permission.)

treated with TMZ-NB@PLN-AAp. Because the targeting of the aptamer AS1411 can promote the active accumulation of PLNs in the tumor, the tumor volume after treatment was reduced by about 90% compared with the control group, as shown in Figure 10.22(d). After weighing the tumor, the tumor weight after TMZ-NB@PLN treatment was reduced by about 60% compared with the control group, and the tumor weight after TMZ-NB@PLN-AAp treatment was reduced by about 85% compared with the control group, as shown in Figure 10.22(e). Also, the mortality rate of untreated mice was 100%, while the mortality rate of mice treated with TMZ-NB@PLN was 34%, and the mortality rate of mice treated with TMZ-NB@PLN-AAp was 17%. As shown in Figure 10.22(f), the results showed that TMZ-NB@PLN-AAp has a good *in vivo* therapeutic ability.

In situ tumor implantation of D54MG cells into the brains of NOD-SCID mice, TMZ-NB@PLN and TMZ-NB@PLN-AAp were injected intravenously into the mice on the 7, 11, and 15 days after tumor implantation. *In vivo*, the device for tumor implantation in mice is shown in Figure 10.23(b). Each time 100 μL of material is injected into the brain of each mouse, and the material concentration is 10 mg mL^{-1}, and one day after the material injection, that is, on the 8, 12, and 16 days after tumor implantation, the brain was irradiated with ultrasound to guide the NB to rupture, causing a cavitation effect to open the blood-brain barrier and release TMZ and PLN into the brain. The brain was collected 40 days after tumor implantation. To confirm the therapeutic effect of TMZ-NB@PLN-AAp and perform image tracking, the experimental process is shown in Figure 10.23(a). The TMZ-NB@PLN-AAp was monitored by IVIS by intravenous injection into mice, and its performance in the mouse brain after circulating in the mouse body, the PLN can be continuously radiated by near-infrared light within 8 hours, and in the vein, two hours after the injection, the intensity of PLN was the highest. This result indicates that the accumulation of TMZ-NB@PLN-AAp in the brain can reach the mouse brain through the circulation process in the body, as shown in

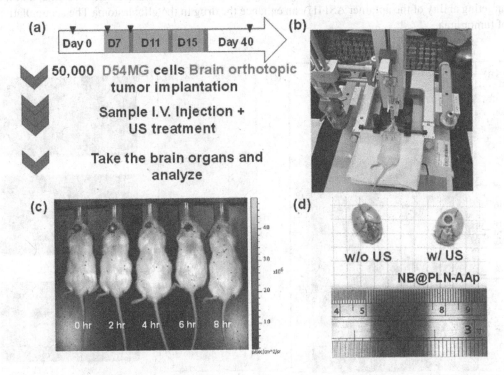

FIGURE 10.23 Brain orthotopic therapeutic effect in mice. (a) Flow chart of *in vivo* orthotopic treatment experiment. (b) Stereotactic machine. (c) Long-term tracking NIR signals. (d) Tumor localization with ultrasonic treatment. (Cheng, C. L.; Chan, M. H.; Feng, S. J.; Hsiao, M.; Liu, R. S., *ACS Appl. Mater. Inter.* **2021**, 13, (5), 6099–6108. With permission.)

Figure 10.23(c). The cavitation effect caused by ultrasound-guided NB rupture is regarded a key factor that can open the blood-brain barrier. The result can be checked in the brain of mice, as shown in the yellow circle in Figure 10.23(d).

To evaluate the growth inhibitory effect of glioblastoma, IVIS was used to track the brain changes of mice for three weeks. In Figure 10.24(a), three groups of materials are used to treat tumors, namely, the control group (without any treatment), TMZ-NB@PLN, and TMZ-NB@PLN-AAp to treat glioblastoma, after three weeks of treatment, glioblastomas in TMZ-NB@PLN and TMZ-NB@PLN-AAp shrank. Besides, the brain was stained with Ki67 protein (a marker of cell proliferation) and terminal deoxynucleotidyl transferase deoxyuridine triphosphate nicked end labeling (TUNEL) to detect changes in glioblastoma. Ki67 protein is related to cell proliferation and can be detected in mitotic and intermittent cells (G1, S, G2, and M phase), while Ki67 protein cannot be detected in mitotic cells (G0 phase). As shown in Figure 10.24(b), according to Ki67 immunohistology, the control group showed obvious cell division and glioblastoma cell metabolism. The TMZ-NB@PLN group showed less Ki67 protein than the control group. The red arrow indicates the glial edge of treatment for blastoma. Compared with the TMZ-NB@PLN group, TMZ-NB@PLN-AAp shows the ability to actively target glioblastoma, so Ki67 staining is not shown in the results. TUNEL staining also proved similar results. TUNEL is a general method for detecting DNA fragments generated by the apoptotic signal cascade. The nicks in the DNA can be recognized by terminal deoxynucleotidyl transferase (TdT), which will catalyze the addition of dUTP to the incision, dUTP can then be marked by the remaining markers. In Figure 10.24(c), the TMZ-NB@PLN group shows that the TUNEL signal is only located at the edge of the glioblastoma. Also, after breaking through the blood-brain barrier, the TMZ-NB@PLN-AAp group shows excellent penetration. These results indicate that TMZ-NB@PLN can break through the blood-brain barrier and deliver the drug and near-infrared light long afterglow nanoparticles to glioblastoma, and the targeting ability of the aptamer AS1411 can enhance the drug in the glioblastoma The accumulation of tumor area.

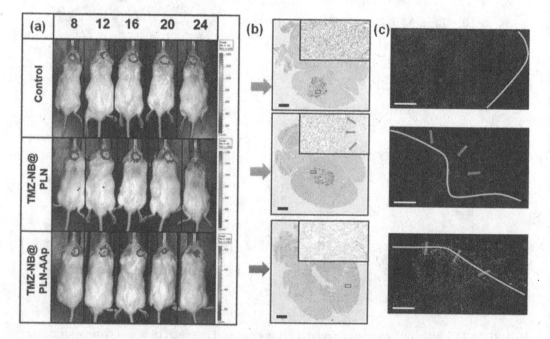

FIGURE 10.24 Tumor tissue tracking and staining. (a) IVIS signal shows the size of tumors. (b) Ki67 staining presents the region of tumors. (c) TUNEL assay of brain tissue. (Cheng, C. L.; Chan, M. H.; Feng, S. J.; Hsiao, M.; Liu, R. S., *ACS Appl. Mater. Inter.* **2021**, 13, (5), 6099–6108. With permission.)

The nanocomposite material TMZ-NB@PLN-AAp developed in this research for the treatment of glioblastoma uses nanobubbles as nanocarriers to coat the multidrug drug, and the nanobubbles can be induced by ultrasound to produce voids. The acupoint effect temporarily opens the tight connection of the blood-brain barrier, thereby delivering the Dimeng multidrug to the brain. Combined with the continuous near-infrared light marking of the near-infrared light long afterglow nanoparticles, it can improve the sound-to-noise ratio of biological images and track the accumulation of nanocomposites in the brain and the distribution of organs. With the targeting ability of aptamer AS1411, it can actively accumulate nanocomposites in glioblastoma, prolong the life of mice and make it difficult for tumors to grow.

10.4.3 Nano-Fertilizer for Light-Harvesting and Photosystem Enhancement

Facing the issues of global climate change and limited resources, nanotechnology has brought a glimmer of light to the instability of agricultural production. Nowadays, in addition to using specific radiation wavelengths to increase the overlap of chlorophyll absorption, thereby enhancing the efficiency of photosynthesis, nanoparticles with different functions also promote nutrient transport and seed germination. However, silica nanoparticles are an effective and versatile nanofertilizer, but they have not been fully utilized in the agricultural industry. MSNs have a positive effect on plant growth, as shown in Figure 10.25. The previous literature mentioned that their pores are loaded with phosphors and still have high biocompatibility [95].

The wide excitation range and specific emission wavelength of nanophosphor have unique advantages in agricultural applications. With the addition of chromium and tin ions, the emission wavelength can reach the NIR region, which effectively collects photons from the environment and converts them to their own growth needs. In Figure 10.25(a), the excitation and emission spectrum are showed. The commercial T4 tube was regarded the common-used light source in the market. In the plant experiment, this T4 tube demonstrated the light conversion efficiency of nanophosphors (ZGOCS@MSN) under different excitation wavelengths (Figure 10.25(c)). The corresponding spectrum can be observed in Figure 10.25(b). The reason why NIR nanophosphor can promote plant

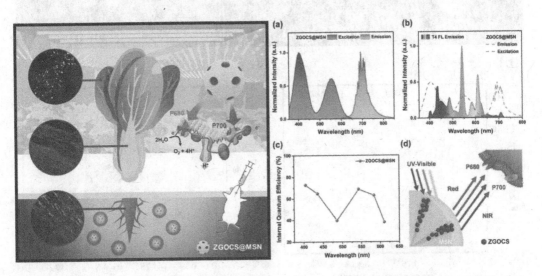

FIGURE 10.25 The photosynthetic enhancement of luminescent nanophosphor in agriculture applications and luminescent properties of ZGOCS@MSN. (a) Excitation and emission spectrum of ZGOCS@MSN. (b) The emission spectrum of the T4 tube. (c) The light conversion efficiency of ZGOCS@MSN under different wavelengths. (d) The mechanism of plant growth under NIR irradiation. (Huang, W. T.; Su, T. Y.; Chan, M. H.; Tsai, J. Y.; Do, Y. Y.; Huang, P. L.; Hsiao, M.; Liu, R. S., *Angew. Chem. Int. Ed.* **2021**, 60, (13), 6955–6959. With permission.)

growth is highly related to the Emerson effect, as shown in Figure 10.25(d). When the photosynthetic centers (P680 and P700) are activated by the NIR, the electron transfer rate will be accelerated for growth.

From the photograph of the plants, the growth situation can be distinguished by the color of the leaves of the plants, and it is presented in the data of fresh weight and chlorophyll content. At the same time, the chlorophyll fluorescence of leaves is measured in real time through portable equipment to observe the current electron transmission rate. Under the appropriate concentration of nanophosphor, the best photosynthesis efficiency can be achieved, effectively reducing the nitrate content in plants (Figure 10.26).

The harvested plants were prepared as an aqueous solution and fed to mice by oral gavage to assess their biological toxicity. IVIS images showed that most of the nanophosphor accumulated in the stomach with an excellent NIR signal (Figure 10.27). In summary, this research synthesizes a new type of high-light conversion nanofertilizer, establishes and studies it with a systematic animal and plant experimental model, and provides future nontoxic and eco-friendly strategies to promote plant growth and solve the food crisis.

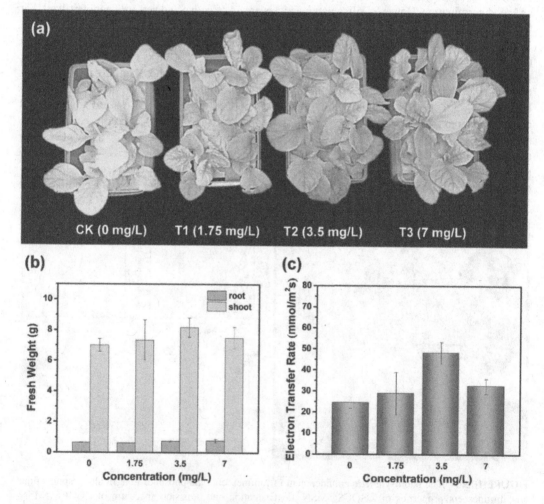

FIGURE 10.26 Plant experiment under different concentrations of ZGOCS@MSN treatments. (a) Photograph of plant with (b) chlorophyll content and (c) electron transfer rate. (Huang, W. T.; Su, T. Y.; Chan, M. H.; Tsai, J. Y.; Do, Y. Y.; Huang, P. L.; Hsiao, M.; Liu, R. S., *Angew. Chem. Int. Ed.* **2021,** 60, (13), 6955–6959. With permission.)

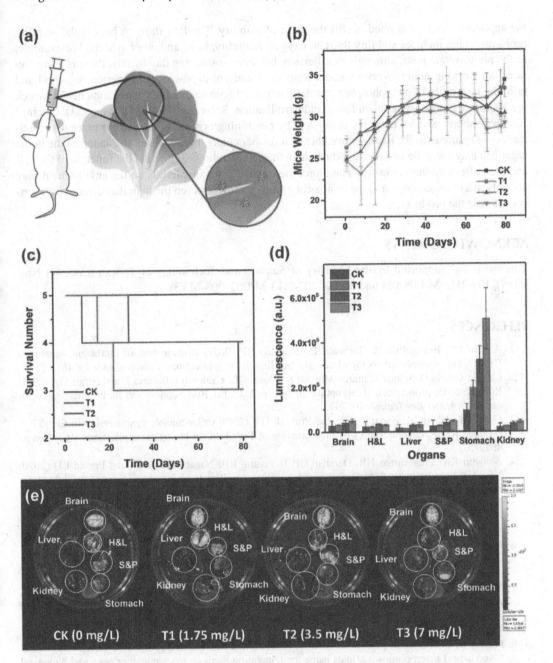

FIGURE 10.27 Animal test of nanophosphor-treated plants. (a) Scheme of experiments with (b) mice weight, (c) survival curve, and (d) IVIS imaging of organs distribution. (Huang, W. T.; Su, T. Y.; Chan, M. H.; Tsai, J. Y.; Do, Y. Y.; Huang, P. L.; Hsiao, M.; Liu, R. S., *Angew. Chem. Int. Ed.* **2021,** 60, (13), 6955–6959. With permission.)

10.5 CONCLUSION

In this chapter, we have summarized the basic properties of nanophosphors with an emphasis on the emerging applications that take advantage of these multiple applications. Despite the prominent performance of nanophosphors such as tunable emission and excellent stability, they still face severe challenges: first, there is no application without stability. The main bottleneck of whether the application can be applied is still the issue of stability. The first thing to bear is the stability problem, which includes stability to light, oxygen, humidity, heat, and other aspects. For example, in the photovoltaic field, although its efficiency has been improving day by day, there is a big gap in actual working hours. Working under long-term sunlight exposure and experiencing wind and rain, the instability of nanophosphor powder itself is difficult to meet the application needs, which has become the primary problem facing industrialization. Secondly, poison? Or is it good medicine? Nanophosphors can also enter the ecosystem by combining organic substances or other toxic substances in nature to make them more stable. If it further enters the organism, the nanoparticle in the organism may enter the cell through the hole in the cell membrane, affecting the function of the cell and then affecting the organism. Therefore, the impact of nanomaterials on the environment may not only affect the ecology and the biological groups within but even the manufacturers or consumers who use the products.

ACKNOWLEDGMENTS

This work was supported by the Ministry of Science and Technology of Taiwan (Contract Nos. MOST 109-2112-M-003-011 and MOST 107-2113-M-002-008-MY3).

REFERENCES

1. Azeem PA, Evangeline B, Haranath D, and Rao RP (2021) Investigation of lanthanum-sensitized $CaZrO_3$ blue nanophosphors for white light-emitting diode applications. *Luminescence* 36: 481.
2. Chavez-Garcia D, Sengar P, Juarez-Moreno K, Flores DL, Calderon I, Barrera J, and Hirata GA (2021) Luminescence properties and cell uptake analysis of Y_2O_3:Eu, Bi nanophosphors for bio-imaging applications. *J Mater Res Technol* 10: 797.
3. Gupta M, Adnan M, Nagarajan R, and Prakash GV (2019) Color-tunable upconversion in Er^{3+}/Yb^{3+}-codoped $KLaF_4$ nanophosphors by incorporation of Tm^{3+} ions for biological applications. *ACS Omega* 4: 2275.
4. Ashwini KR, Premkumar HB, Darshan GP, Basavaraj RB, Nagabhushana H, and Prasad BD (2020) Near UV-light excitable $SrAl_2O_4$:Eu^{3+} nanophosphors for display device applications. *J Sci-Adv Mater Dev* 5: 111.
5. Munirathnam K, Rajavaram R, Nagajyothi PC, Thiyagaraj S, and Srinivas M (2020) Synthesis and optimization of Dy-doped $SrZr_4(PO_4)_6$ nanophosphors for plant growth light-emitting diodes. *Solid State Sci* 109: 106455.
6. Hong AR, Kim Y, Lee TS, Kim S, Lee K, Kim G, and Jang HS (2018) Intense red-emitting upconversion nanophosphors (800 nm-driven) with a core/double-shell structure for dual-modal upconversion luminescence and magnetic resonance in vivo imaging applications. *ACS Appl Mater Inter* 10: 12331.
7. Mehmood Y, Khan IU, Shahzad Y, Khan RU, Iqbal MS, Khan HA, Khalid I, Yousaf AM, Khalid SH, Asghar S, Asif M, Hussain T, and Shah SU (2020) In-vitro and in-vivo evaluation of velpatasvir-loaded mesoporous silica scaffolds. A prospective carrier for drug bioavailability enhancement. *Pharmaceutics* 12: 307.
8. Pandey B, Chatterjee S, Parekh N, Yadav P, Nisal A, and Sen Gupta S (2018) Silk-mesoporous silica-based hybrid macroporous scaffolds using ice-templating method: mechanical, release, and biological studies. *ACS Appl Bio Mater* 1: 2082.
9. Boccardi E, Liverani L, and Boccaccini AR (2019) Bioactive behavior of mesoporous silica particle (MCM-41) coated bioactive glass-based scaffolds. *Int J Appl Ceram Technol* 16: 1753.
10. Zako T, Hyodo H, Tsuji K, Tokuzen K, Kishimoto H, Ito M, Kaneko K, Maeda M, and Soga K (2010) Development of near infrared-fluorescent nanophosphors and applications for cancer diagnosis and therapy. *J Nanomater* 2010: 491471.
11. Li JT, Zhao Q, Shi F, Liu CH, and Tang YL (2016) NIR-mediated nanohybrids of upconversion nanophosphors and fluorescent conjugated polymers for high-efficiency antibacterial performance based on fluorescence resonance energy transfer. *Adv Healthc Mater* 5: 2967.
12. Song XQ, Guo QY, Cai ZL, Qiu JR, and Dong GP (2019) Synthesis of multi-color fluorescent carbon quantum dots and solid state $CQDs@SiO_2$ nanophosphors for light-emitting devices. *Ceram Int* 45: 17387.

13. Derbenyova NV, Konakov AA, and Burdov VA (2021) Recombination, multiplication, and transfer of electron-hole pairs in silicon nanocrystals: effects of quantum confinement, doping, and surface chemistry. *J Lumin* 233: 117904.

14. Gongalsky MB, Tsurikova UA, Gonchar KA, Gvindgiliiia GZ, and Osminkina LA (2021) Quantum-confinement effect in silicon nanocrystals during their dissolution in model biological fluids. *Semiconductors* 55: 61.

15. Alet F, Hanada M, Jevicki A, and Peng C (2021) Entanglement and confinement in coupled quantum systems. *J High Energy Phys* v2: 34.

16. Wei J, Han BN, Liu XH, Ge DY, Zhang W, and Yang YM (2016) Single-component white light emission from Eu^{3+} doped $BaIn_6Y_2O_{13}$ nanophosphors. *Nanosci Nanotech Let* 8: 459.

17. Zhai Y, Liu DL, Jiang YD, Chen X, Shao L, Li J, Sheng K, Zhang XR, and Song HW (2019) Near-infrared-light-triggered photoelectrochemical biosensor for detection of alpha-fetoprotein based on upconversion nanophosphors. *Sensor Actuat B-Chem* 286: 468.

18. Jia JB, Dong J, Lin JM, Lan Z, Fan LQ, and Wu JH (2019) Improved photovoltaic performance of perovskite solar cells by utilizing down-conversion $NaYF_4$:Eu^{3+} nanophosphors. *J Mater Chem C* 7: 937.

19. Shanbhag VV, Prashantha SC, Shekhar TRS, Nagabhushana H, Naik R, Girish KM, Ashwini S, Rangappa D, and Prasanna DS (2021) Enhanced photoluminescence of SiO_2 coated $CaTiO_3$:Dy^{3+},Li^+ nanophosphors for white light emitting diodes. *Ceram Int* 47: 10346.

20. Liu B, Chen YY, Li CX, He F, Hou ZY, Huang SS, Zhu HM, Chen XY, and Lin J (2015) Poly(acrylic acid) Modification of Nd^{3+}-sensitized upconversion nanophosphors for highly efficient UCL imaging and pH-responsive drug delivery. *Adv Funct Mater* 25: 4717.

21. Fang MH, Li TY, Huang WT, Cheng CL, Bao Z, Majewska N, Mahlik S, Yang CW, Lu KM, Leniec G, Kaczmarek SM, Sheu HS, and Liu RS (2021) Surface-protected high-efficiency nanophosphors via space-limited ship-in-a-bottle synthesis for broadband near-infrared mini-light-emitting diodes. *ACS Energy Lett* 6: 659.

22. Suresh C, Darshan GP, Sharma SC, Venkataravanappa M, Premkumar HB, Shanthi S, Venkatachalaiah KN, and Nagabhushana H (2020) Imaging sweat pore structures in latent fingerprints and unclonable anti-counterfeiting patterns by sensitizers blended LaOF: Pr^{3+} nanophosphors. *Opt Mater* 100: 109625.

23. Ardini M, Huang JA, Caprettini V, De Angelis F, Fata F, Silvestri I, Cimini A, Giansanti F, Angelucci F, and Ippoliti R (2020) A ring-shaped protein clusters gold nanoparticles acting as molecular scaffold for plasmonic surfaces. *Bba-Gen Subjects* 1864: 129617.

24. Benavides BS, Valandro S, and Kurtz DM (2020) Preparation of platinum nanoparticles using iron(ii) as reductant and photosensitized H_2 generation on an iron storage protein scaffold. *RSC Adv* 10: 5551.

25. Kwon KC, Jo EJ, Kwon YW, Lee B, Ryu JH, Lee EJ, Kim K, and Lee J (2017) Superparamagnetic gold nanoparticles synthesized on protein particle scaffolds for cancer theragnosis. *Adv Mater* 29: 1701146.

26. Stel B, Cometto F, Rad B, De Yoreo JJ, and Lingenfelder M (2018) Dynamically resolved self-assembly of S-layer proteins on solid surfaces. *Chem Commun* 54: 10264.

27. Wang XY, Wang DB, Zhang ZP, Bi LJ, Zhang JB, Ding W, and Zhang XE (2015) A S-layer protein of *Bacillus anthracis* as a building block for functional protein arrays by in vitro self-assembly. *Small* 11: 5826.

28. Teixeira LM, Strickland A, Mark SS, Bergkvist M, Sierra-Sastre Y, and Batt CA (2010) Entropically driven self-assembly of *Lysinibacillus sphaericus* S-layer proteins analyzed under various environmental conditions. *Macromol Biosci* 10: 147.

29. Bolla PA, Sanz A, Huggias S, Ruggera JF, Serradell MA, and Casella ML (2020) Regular arrangement of Pt nanoparticles on S-layer proteins isolated from *Lactobacillus kefiri*: synthesis and catalytic application. *Mol Catal* 481: 110262.

30. Mark SS, Bergkvist M, Yang X, Angert ER, and Batt CA (2006) Self-assembly of dendrimer-encapsulated nanoparticle arrays using 2-D microbial S-layer protein biotemplates. *Biomacromolecules* 7: 1884.

31. Davis ME (2002) Ordered porous materials for emerging applications. *Nature* 417: 813.

32. Havard DC, and Wilson R (1976) Pore measurements on Sci-Iupac-Npl Meso-porous silica surface-area standard. *J Colloid Interf Sci* 57: 276.

33. Deng YD, Zhang XD, Yang XS, Huang ZL, Wei X, Yang XF, and Liao WZ (2021) Subacute toxicity of mesoporous silica nanoparticles to the intestinal tract and the underlying mechanism. *J Hazard Mater* 409: 124502.

34. von Baeckmann C, Kahlig H, Linden M, and Kleitz F (2021) On the importance of the linking chemistry for the PEGylation of mesoporous silica nanoparticles. *J Colloid Interf Sci* 589: 453.

35. Didwal PN, Singhbabu YN, Verma R, Sung BJ, Lee GH, Lee JS, Chang DR, and Park CJ (2021) An advanced solid polymer electrolyte composed of poly(propylene carbonate) and mesoporous silica nanoparticles for use in all-solid-state lithium-ion batteries. *Energy Storage Mater* 37: 476.

36. Akram Z, Daood U, Aati S, Ngo H, and Fawzy AS (2021) Formulation of pH-sensitive chlorhexidine-loaded/mesoporous silica nanoparticles modified experimental dentin adhesive. *Mat Sci Eng C-Mater* 122: 111894.

37. Pinna A, Baghbaderani MT, Hernandez VV, Naruphontjirakul P, Li SW, McFarlane T, Hachim D, Stevens MM, Porter AE, and Jones JR (2021) Nanoceria provides antioxidant and osteogenic properties to mesoporous silica nanoparticles for osteoporosis treatment. *Acta Biomater* 122: 365.

38. Cai GR, Zhu FY, Liu W, Zhu TF, Duan L, Tan H, Pan J, Xiong JY, and Wang DP (2021) Preparation and characterization of a three dimensional porous poly (lactic-co-glycolic acid)/mesoporous silica nanoparticles scaffold by low-temperature deposition manufacturing. *Mater Today Commun* 26: 101824.

39. Mi ZM, Zhang DW, Wang DM, and Liu ZX (2021) Mixed matrix membranes embedded with Janus mesoporous silica nanoparticles for highly efficient water treatment. *Express Polym Lett* 15: 289.

40. Ma JZ, He Y, Liu JW, Chen DW, and Hu HY (2021) Cloaking mesoporous silica nanoparticles with phenylboronic acid-conjugated human serum albumin-co-polydopamine films for targeted drug delivery. *J Drug Deliv Sci Tec* 62: 102392.

41. Wu YS, Wang J, Zhu YC, Yu XW, Shang ZK, Ding Y, and Hu AG (2021) Controlled synthesis of conjugated polymers in dendritic mesoporous silica nanoparticles dagger. *Chem Commun* 57: 4146.

42. Dau TAN, Le VMH, Pham TKH, Le VH, Cho SK, Nguyen TNU, Ta TKH, and Tran TTV (2021) Surface functionalization of doxorubicin loaded MCM-41 mesoporous silica nanoparticles by 3-amino-propyltriethoxysilane for selective anticancer 9 effect on A549 and A549/DOX cells. *J Electron Mater* 50: 2932.

43. Hosseinpour S, Cao YX, Liu JY, Xu C, and Walsh LJ (2021) Efficient transfection and long-term stability of rno-miRNA-26a-5p for osteogenic differentiation by large pore sized mesoporous silica nanoparticles. *J Mater Chem B* 9: 2275.

44. Liu YK, Zhou L, Tan J, Xu WQ, Huang GL, and Ding J (2021) Ent-11 alpha-hydroxy-15-oxo-kaur-16-en-19-oic acid loaded onto fluorescent mesoporous silica nanoparticles for the location and therapy of nasopharyngeal carcinoma. *Analyst* 146: 1596.

45. Hoffmann F, Cornelius M, Morell J, and Froba M (2006) Silica-based mesoporous organic-inorganic hybrid materials. *Angew Chem Int Ed* 45: 3216.

46. Li XN, Hu S, Lin Z, Yi J, Liu X, Tang XP, Wu QH, and Zhang GL (2020) Dual-responsive mesoporous silica nanoparticles coated with carbon dots and polymers for drug encapsulation and delivery. *Nanomedicine-UK* 15: 2447.

47. Chen WD, Chen Q, Chen YN, Zhang WJ, Huang QQ, Hu MR, Peng DY, Peng C, and Wang L (2020) Acidity and glutathione dual-responsive polydopamine-coated organic-inorganic hybrid hollow mesoporous silica nanoparticles for controlled drug delivery. *Chemmedchem* 15: 1940.

48. Carvalho GC, Sabio RM, Ribeiro TD, Monteiro AS, Pereira DV, Ribeiro SJL, and Chorilli M (2020) Highlights in mesoporous silica nanoparticles as a multifunctional controlled drug delivery nanoplatform for infectious diseases treatment. *Pharm Res-Dordr* 37: 191.

49. Almomen A, El-Toni AM, Badran M, Alhowyan A, Abul Kalam M, Alshamsan A, and Alkholief M (2020) The design of anionic surfactant-based amino-functionalized mesoporous silica nanoparticles and their application in transdermal drug delivery. *Pharmaceutics* 12: 1035.

50. Juere E, Caillard R, and Kleitz F (2020) Pore confinement and surface charge effects in protein-mesoporous silica nanoparticles formulation for oral drug delivery. *Micropor Mesopor Mater* 306: 110482.

51. Murugan B, Sagadevan S, Lett JA, Fatimah I, Fatema KN, Oh WC, Mohammad F, and Johan MR (2020) Role of mesoporous silica nanoparticles for the drug delivery applications. *Mater Res Express* 7: 102002.

52. He QJ, and Shi JL (2011) Mesoporous silica nanoparticle based nano drug delivery systems: synthesis, controlled drug release and delivery, pharmacokinetics and biocompatibility. *J Mater Chem* 21: 5845.

53. Pan DY, Zheng XL, Chen M, Zhang QF, Li ZQ, Duan ZY, Gong QY, Gu ZW, Zhang H, and Luo K (2021) Dendron-polymer hybrid mediated anticancer drug delivery for suppression of mammary cancer. *J Mater Sci Technol* 63: 115.

54. Shi Z, Wang YZ, Sun Y, Wu XY, Xiao TY, Dong SB, and Lan TY (2021) Facile one-pot synthesis of magnetic targeted polymers for drug delivery and study on thermal decomposition kinetics. *Chemistryselect* 6: 1191.

55. Castillo-Henriquez L, Castro-Alpizar J, Lopretti-Correa M, and Vega-Baudrit J (2021) Exploration of bioengineered scaffolds composed of thermo-responsive polymers for drug delivery in wound healing. *Int J Mol Sci* 22: 1408.

56. Boncu TE, and Ozdemir N (2021) Electrospinning of ampicillin trihydrate loaded electrospun PLA nanofibers I: effect of polymer concentration and PCL addition on its morphology, drug delivery and mechanical properties. *Int J Polym Mater Polyem* 70: 1876057.

57. Aranda-Lara L, Garcia BEO, Isaac-Olive K, Ferro-Flores G, Melendez-Alafort L, and Morales-Avila E (2021) Drug delivery systems-based dendrimers and polymer micelles for nuclear diagnosis and therapy. *Macromol Biosci* 21: e2000362.

58. He SN, Zhang LP, Bai SK, Yang H, Cui Z, Zhang XF, and Li YP (2021) Advances of molecularly imprinted polymers (MIP) and the application in drug delivery. *Eur Polym J* 143: 110179.

59. Jana P, Shyam M, Singh S, Jayaprakash V, and Dev A (2021) Biodegradable polymers in drug delivery and oral vaccination. *Eur Polym J* 142: 110155.

60. Wen WQ, Guo C, and Guo JW (2021) Acid-responsive adamantane-cored amphiphilic block polymers as platforms for drug delivery. *Nanomaterials-Basel* 11: 188.

61. Mahzabin A, and Das B (2021) A review of lipid-polymer hybrid nanoparticles as a new generation drug delivery system. *Int J Pharm Sci Res* 12: 65.

62. Farani RD, Teles WM, Pinheiro CB, Guedes KJ, Krambrock K, Yoshida MI, de Oliveira LFC, and Machado FC (2008) Spectroscopic and structural analyses of the copper(II) 2-D coordination polymer [Cu_2(BPP)(4)(NCS)(4)]}(n) (BPP=1,3-bis(4-pyridyl)propane) comprising interpenetrated layers of (4,4) topology. *Inorg Chim Acta* 361: 2045.

63. Berman D, and Shevchenko E (2020) Design of functional composite and all-inorganic nanostructured materials via infiltration of polymer templates with inorganic precursors. *J Mater Chem C* 8: 10604.

64. Masuda H, and Fukuda K (1995) Ordered metal nanohole arrays made by a 2-step replication of honeycomb structures of anodic alumina. *Science* 268: 1466.

65. Kong LY, Zhao YS, Dasgupta B, Hippalgaonkar K, Li XL, Chim WK, and Chiam SY (2017) Minimizing isolate catalyst motion in metal-assisted chemical etching for deep trenching of silicon nanohole array. *ACS Appl Mater Inter* 9: 20981.

66. Kumar R, and Mujumdar S (2016) Intensity correlations in metal films with periodic-on-average random nanohole arrays. *Opt Commun* 380: 174.

67. Lu BR, Xu C, Liao JF, Liu JP, and Chen YF (2016) High-resolution plasmonic structural colors from nanohole arrays with bottom metal disks. *Opt Lett* 41: 1400.

68. Devi S, Taxak VB, and Khatkar SP (2021) Structural and optical characterizations of cool white light emitting $Ba_2Zn_2La_4O_{10}$: Dy^{3+} nanophosphor for advanced optoelectronic applications. *Chem Phys Lett* 765: 138289.

69. Hooda A, Khatkar A, Chahar S, Singh S, Dhankhar P, Khatkar SP, and Taxak VB (2020) Photometric features and typical white light emanation via combustion derived trivalent dysprosium doped ternary aluminate oxide based nanophosphor for WLEDs. *Ceram Int* 46: 4204.

70. Rashmi, and Dwivedi Y (2019) Optical interactions and white light emission in Eu:Y_2O_3/YAG:Ce nanophosphor. *Appl Phys A—Mater* 125: 198217570.

71. Geng DL, Lozano G, and Miguez H (2019) Highly efficient transparent nanophosphor films for tunable white-light-emitting layered coatings. *ACS Appl Mater Inter* 11: 4219.

72. Dwivedi RY (2019) White light color tuning ability of hybrid Dibenzoylmethane/YAG: chock for Ce nanophosphor. *Spectrochim Acta A* 206: 141.

73. Samanta B, Dey AK, Bhaumik P, Manna S, Halder A, Jana D, Chattopadhyay KK, and Ghorai UK (2019) Controllable white light generation from novel BaWO4: Yb^{3+}/Ho^{3+}/Tm^{3+} nanophosphor by modulating sensitizer ion concentration. *J Mater Sci-Mater El* 30: 1068.

74. van der Gon DD, Timmerman D, Matsude Y, Ichikawa S, Ashida M, Schall P, and Fujiwara Y (2020) Size dependence of quantum efficiency of red emission from GaN:Eu structures for application in micro-LEDs. *Opt Lett* 45: 3973.

75. Carreira JFC, Griffiths AD, Xie E, Guilhabert BJE, Herrnsdorf J, Henderson RK, Gu E, Strain MJ, and Dawson MD (2020) Direct integration of micro-LEDs and a SPAD detector on a silicon CMOS chip for data communications and time-of-flight ranging. *Opt Express* 28: 6909.

76. Wierer J, and Tansu N (2019) III-nitride micro-LEDs for efficient emissive displays. *Laser Photonics Rev* 13: 1900141.

77. Chang SJ, Wang SM, Chang PC, Kuo CH, Young SJ, and Chen TP (2010) GaN metal-semiconductor-metal photodetectors prepared on nanorod template. *IEEE Photonic Tech L* 22: 625.

78. Tsai MF, Chang SHG, Cheng FY, Shanmugam V, Cheng YS, Su CH, and Yeh CS (2013) Au nanorod design as light-absorber in the first and second biological near-infrared windows for in vivo photothermal therapy. *ACS Nano* 7: 5330.

79. Ryu MH, Kang IH, Nelson MD, Jensen TM, Lyuksyutova AI, Siltberg-Liberles J, Raizen DM, and Gomelsky M (2014) Engineering adenylate cyclases regulated by near-infrared window light. *Proc Natl Acad Sci USA* 111: 10167.

80. Wu MC, Shi Y, Li RY, and Wang P (2018) Spectrally selective smart window with high near-infrared light shielding and controllable visible light transmittance. *ACS Appl Mater Inter* 10: 39819.

81. Wang FF, Wan H, Ma ZR, Zhong YT, Sun QC, Tian Y, Qu LQ, Du HT, Zhang MX, Li LL, Ma HL, Luo J, Liang YY, Li WJ, Hong GS, Liu LQ, and Dai HJ (2019) Light-sheet microscopy in the near-infrared II window. *Nat Methods* 16: 545.

82. Golovynskyi S, Golovynska I, Stepanova LI, Datsenko OI, Liu LW, Qu JL, and Ohulchanskyy TY (2018) Optical windows for head tissues in near-infrared and short-wave infrared regions: Approaching transcranial light applications. *J Biophotonics* 11.

83. Vasilopoulou M, Kim HP, Kim BS, Papadakis M, Gavim AEX, Macedo AG, da Silva WJ, Schneider FK, Teridi MAM, Coutsolelos AG, and Yusoff AB (2020) Efficient colloidal quantum dot light-emitting diodes operating in the second near-infrared biological window. *Nat Photonics* 14: 50.

84. Ren JJ, Zhang L, Zhang JY, Zhang W, Cao Y, Xu ZG, Cui HJ, Kang YJ, and Xue P (2020) Light-activated oxygen self-supplied starving therapy in near-infrared (NIR) window and adjuvant hyperthermia-induced tumor ablation with an augmented sensitivity. *Biomaterials* 234: 119771.

85. Cao YB, Ouyang BS, Yang XW, Jiang Q, Yu L, Shen S, Ding JD, and Yang WL (2020) Fixed-point "blasting" triggered by second near-infrared window light for augmented interventional photothermal therapy. *Biomater Sci-UK* 8: 2955.

86. Johnsen G, Candeloro M, Berge J, and Moline M (2014) Glowing in the dark: discriminating patterns of bioluminescence from different taxa during the Arctic polar night. *Polar Biol* 37: 707.

87. Christiani DC (2014) Radiation risk from lung cancer screening glowing in the dark? *Chest* 145: 439.

88. Ckurshumova W, Caragea AE, Goldstein RS, and Berleth T (2011) Glow in the dark: fluorescent proteins as cell and tissue-specific markers in plants. *Mol Plant* 4: 794.

89. Aliabadi HM, Zargoosh K, Afshari M, Dinari M, and Maleki MH (2021) Synthesis of a luminescent g-C_3N_4-WO_3-Bi_2WO_6/$SrAl_2O_4$:Eu^{2+},Dy^{3+} nanocomposite as a double z-scheme sunlight activable photocatalyst. *New J Chem* 45: 4843.

90. Cai T, Guo ST, Li YZ, Peng D, Zhao XF, Wang WZ, and Liu YZ (2020) Ultra-sensitive mechanoluminescent ceramic sensor based on air-plasma-sprayed $SrAl_2O_4$:Eu^{2+}, Dy^{3+} coating. *Sensor Actuat A-Phys* 315: 112246.

91. Duan XX, Yi LX, and Huang SH (2020) Structural and optical properties of size and morphology controllable nanoscale $SrAl_2O_4$: Eu^{2+}, Dy^{3+}. *Integr Ferroelectr* 210: 73.

92. Renaud EJ (2012) Glowing in the dark: should the time of day determine radiographic imaging in the evaluation of abdominal pain in children? Reply. *J Pediatr Surg* 47: 438.

93. Huang WT, Cheng CL, Bao Z, Yang CW, Lu KM, Kang CY, Lin CM, and Liu RS (2019) Broadband Cr^{3+}, Sn^{4+}-doped oxide nanophosphors for infrared mini light-emitting diodes. *Angew Chem Int Ed* 58: 2069.

94. Cheng CL, Chan MH, Feng SJ, Hsiao M, and Liu RS (2021) Long-term near-infrared signal tracking of the therapeutic changes of glioblastoma cells in brain tissue with ultrasound-guided persistent luminescent nanocomposites. *ACS Appl Mater Inter* 13: 6099.

95. Huang WT, Su TY, Chan MH, Tsai JY, Do YY, Huang PL, Hsiao M, and Liu RS (2021) Near-infrared nanophosphor embedded in mesoporous silica nanoparticle with high light-harvesting efficiency for dual photosystem enhancement. *Angew Chem Int Ed* 60: 6955.

11 Near-Infrared Phosphors with Persistent Luminescence over 1000 nm for Optical Imaging

Jian Xu, Michele Back, and Setsuhisa Tanabe

CONTENTS

11.1 INTRODUCTION

Optical imaging, also commonly referred as fluorescence imaging, is one of the fastest-growing fields for biomedical research and clinical practice [1–3], which is also expected to be an alternative for other well-established imaging techniques utilizing magnetic, radiation and/or fluorescence-based approaches (*e.g.*, magnetic resonance imaging [MRI], computed tomography [CT], single-photon emission CT [SPECT], positron emission tomography [PET] and ultrasound imaging [USI] [4–6]). This fluorescence-based noninvasive imaging technique simultaneously features a series of salient merits such as rapid feedback, high sensitivity, multiple signal acquisition capability as well as absence of ionizing radiation, enabling the acquisition of real-time data at high-speed for dynamic visualization of physiological and metabolic processes within specific bio-tissues [7, 8]. However, in spite of recent advances in fluorescence microscopy instruments and optical contrast agents, the main limitation of optical imaging is the restricted tissue penetration depth. In fact, the acquisition of optical signals is strongly dependent on the interaction between incident photons (light) and biological tissues (matter), such as reflection, refraction, absorption, scattering, and autofluorescence (intrinsic fluorescence by the bio-entity itself). In the past, the wavelengths used in classical optical imaging for both excitation and emission were mainly located in the visible-light range (380–750 nm), where significant absorption and scattering effects induced by biological tissues as well as strong autofluorescence originating from certain biomolecules (*e.g.*, flavins,

DOI: 10.1201/9781003098669-11

lipofuscin, reticulin, and exogenous foods) result in relatively low tissue penetration (<3 mm) [9, 10]. Such low penetration depth further limits the real performance of optical imaging in clinical trials, in which high-quality visualization of deeply embedded anatomical structures is essential. Fortunately, this drawback has been largely overcome by the fast development of biocompatible optical probes operating in the near-infrared (NIR) (780–3000 nm) spectral region, where high-contrast imaging with detailed spatial and temporal resolution is available at deeper penetration depth owing to a remarkable reduction in light absorption and scattering.

Recently, the discovery of biological-tissue transparency windows and advances in NIR optical probes suggest exciting prospects for *in vivo* optical imaging [2, 3]. Wavelengths ranging from 650 to 950 nm is regarded as the first biological transparency window (denoted as NIR-I), and this range exhibits significantly increased tissue penetration compared with visible light. Till now, various optical probes with emission bands matching well with the NIR-I window, such as organic dyes [11, 12], fluorescent proteins [13, 14], single-walled carbon nanotubes (SWCNTs) [15–21], semiconductor quantum dots (QDs) [22–27], have been successfully developed and widely exploited to investigate biological systems. However, most of them usually suffer from either high photo-bleaching rate (photochemical destruction of fluorophores under long-term light exposure), strong autofluorescence (*e.g.*, body fluids and proteins are well known to give emission under ultraviolet [UV] light exposure), short luminescence lifetimes (typically with the order of nano/submicron seconds) or poor biocompatibility (most heavy-metal-based QDs are readily toxic). Besides, lanthanide/transition metal ions-activated inorganic nanoparticles (NPs) [28–32] utilizing either spontaneous downshifting (*e.g.*, Nd^{3+}, Er^{3+}, Cr^{3+}, Mn^{4+}) [33–36] or nonlinear multiphoton up-conversion processes (*e.g.*, Er^{3+}–Yb^{3+}, Tm^{3+}–Yb^{3+}, Ho^{3+}–Yb^{3+}) [37–48] exhibit superior luminescent properties than other mentioned bio-probes due to their higher photochemical stability (rigid lattice structure), tunable emission bands ($4f$–$4f$ intra-configuration narrow-band transitions or d–d spin-allowed broadband transitions), longer luminescence lifetimes (order of μs or ms for parity/spin-forbidden $4f$–$4f$/d–d transitions), lower photo-bleaching potentials and lower cytotoxicity (~1000 times higher LD_{50} value compared to QDs [49, 50]). However, in most cases, real-time illumination by high-order coherent excitation, for instance, laser sources (*e.g.*, 980 or 808 nm are commonly used for *in vivo* imaging) still limits the practical use of these NPs since lasers are costly and may create autofluorescence [51, 52] and non-negligible heating effects even under a low output power [53] during imaging operation. Therefore, novel NIR bio-probes/agents with "autofluorescence-free" and "non-heating effects" features are considered to be a new avenue to overcome these complicated problems in current optical imaging fields.

Persistent phosphors, a kind of specific luminescent materials that can exhibit "self-sustained" persistent luminescence (PersL) for minutes, hours, or even days after ceasing excitation in the dark is considered to be one of the best answers for this problem [54–65]. Recently, the emission wavelength of persistent phosphors has been successfully extended from visible to NIR, and demonstrated the superior power to obtain high quality *in vivo* bioimaging due to the avoidance of real-time external excitation after injecting into living bodies [66–71]. The exclusion of external illumination can totally remove the autofluorescence as background noise, avoid the heating effects by conventional laser excitation, and thus improve the signal-to-noise ratio (SNR) remarkably [72–90]. Furthermore, simply considering the Rayleigh scattering (varying as λ^{-4}, where λ is the wavelength) decreases with the increase of wavelength, longer wavelength is expected to achieve a much lower optical scattering coefficient. Once the emission band of NIR-persistent phosphors can be successfully extended over 1000 nm (sometimes referred to as over-1000 nm [OTN]-NIR [2, 43–45] or short-wave infrared [~900–1700 nm], SWIR) into the so-called second (NIR-II, ~1000–1350 nm) and third (NIR-III, ~1500–1800 nm) NIR windows, deeper tissue penetration depth and improved imaging contrast can be expected compared with that from the shorter wavelength NIR-I window [2, 3, 16–21, 43–45].

In this chapter, we will give comprehensive insights into the state-of-the-art of NIR phosphors with PersL over 1000 nm, starting from the introduction of basic principles of light-matter interactions (Section 11.1.1), followed by the description of NIR autofluorescence phenomena of bio-tissues and related filtering approaches (wavelength-based *vs.* lifetime-based) (Section 11.1.2). Then, the mechanism of PersL, including charge carrier trapping-detrapping phenomena, persistent energy transfer (ET) process, and photostimulation-induced trap redistribution approaches, will be explained, to give a general picture of how PersL works and how we can play with the information of energy-level locations of conduction/valence bands (CB/VB), activators (energy donor/acceptors) and traps (electron/hole trapping centers) (Section 11.1.3). The reported over 1000 nm NIR-persistent phosphors, basically divided into lanthanide (Section 11.2.1) and transition metal ions (Section 11.2.2)-activated ones, will be summarized with highlighted examples to give vivid explanation about the design concept of these phosphors. New post-charging concepts by NIR or visible-light lasers trying to compensate the fading effect of PersL along with decay time will also be introduced. Finally, we draw a prospective of key challenges and feasible improvements in the future following the current trends of this research field (Section 11.3).

11.1.1 Basic Principles of Light-Matter Interactions

Depending on the properties of the incident light (*e.g.*, wavelength, pulse duration, and power density), interactions between light and matter (here referred to bio-tissues) can result in the modification of the incident light (tissue properties remain unaltered) or in the modification of the tissue (*e.g.*, laser ablation or thermal therapy). Considering the purpose of this section, we will only focus on the first case, in which describing the basic principles of the phenomena that occur when light encounters a tissue: absorption, scattering, reflection, and refraction [3, 51].

Absorption accounts for the radiation that is dissipated without re-emission of light. For a single particle, it can be estimated by measuring the amount of power loss due to its interaction with an incident light. The ratio of incident light energy $|\langle S^{(inc)} \rangle|$ (W) and absorbed power $\overline{P_{abs}}$ (W·cm²) defines the absorption cross-section σ_a:

$$\sigma_a = \frac{\overline{P_{abs}}}{|\langle S^{(inc)} \rangle|} \qquad (11.1)$$

The absorption cross-section of a particle represents the effective cross-section compared to its geometrical cross-section in area units. This value represents only the absorption for a single particle, while in the case of optical imaging, it is much more useful to calculate the equivalent for a statistical ensemble of N particles per unit volume V, represented by the absorption coefficient μ_a:

$$\mu_a = \rho \sigma_a \qquad (11.2)$$

where

$$\rho = \frac{N}{V} \qquad (11.3)$$

Absorption is the main limiting factor in terms of how deep light can penetrate bio-tissues, since it removes energy of the incident light when propagating within the medium. Absorption also attenuates the incident light, converting it into either vibration of the medium heating through non-radiative processes, or emission of light with different wavelengths through radiative processes. The re-emission of the incident radiation is known as scattering, and practical cases of light propagation theory with *in vivo* applications, only light reemitted with the same frequency is considered. In this case, the phenomenon is specifically termed elastic scattering. In addition to the arrangement of the particles, the amount of scattering of a medium depends also on the size, shape, and spatial distribution of these scatterers. According to the Mie theory [91], the probability for incident light of being scattered into a certain direction depends on the relative size of the particle with respect to

the incident wavelength. When particles in the medium are smaller, light is scattered as an outgo-
ing spherical wave. This solution to the Mie theory is known as Rayleigh scattering. When there
is a slight mismatch of refractive index, such as between cells and the surrounding extracellular
medium, as the size of the particle increases the scattered radiation deviates from that of an outgo-
ing spherical wave to the forward scattering direction.

The scattering cross-section σ_s quantifies the scattering efficiency of a particle. This can be
obtained by calculating the ratio between the scattered power $\overline{P_{sc}}$ (W·cm^2) and the incident light
energy. However, given the directional dependence of scattering, it can also be estimated as the
integral of the scattering amplitude $\left| f(\hat{s}, \hat{s}_0) \right|$ over all angles. The scattering amplitude accounts for
the contribution of the scattered wave to a certain direction $\hat{s}$, given an incident direction $\hat{s}_0$. Hence,
the scattering cross-section gives information about the probability of light being scattered, no matter the direction, and is defined as:

$$\sigma_s = \frac{\overline{P_{sc}}}{\left| \langle S^{(inc)} \rangle \right|} = \int\limits_{(4\pi)} \left| f(\hat{s}, \hat{s}_0) \right|^2 d\Omega \tag{11.4}$$

The scattering coefficient μ_s can be estimated similarly to the absorption coefficient for a known
density of particles in the medium.

$$\mu_s = \rho \sigma_s \tag{11.5}$$

In biological tissues, cells are considered to be the main scatterers since they are quite transparent and their average size is usually greater than the wavelengths used for imaging. The angular
distribution of the scattered light becomes highly anisotropic mainly due to the small mismatch
of refractive index between different bio-tissues. Therefore, the description of scattering in tissues
must account for this effect, and the anisotropy factor g represents the average cosine of the angle
between the incident light and the scattered one. This unitless parameter takes values from −1 to 1,
being "−1" fully backward scattering and "1" only forward. A particle with perfect isotropic scattering would have a value of "0". In terms of scattering and neglecting absorption, this coefficient
informs about how much transparent is the medium.

The anisotropy factor is included in the model through the reduced scattering coefficient μ_s', with
the expression:

$$\mu_s' = (1 - g) \mu_s \tag{11.6}$$

In general, biological tissues have very high scattering anisotropy, taking the g value around
0.8–0.9 [92]. The dependence of the angular distribution of scattering on the relative size between
the wavelength and the particles makes the scattering coefficient value to follow a negative power
law with respect to the wavelength, thus for tissues it decreases as wavelength increases with a
monotonic dependence.

Reflection (the returning of the light to the incidence medium) and refraction (direction change
of the light that is transmitted into the second medium) are strongly related phenomena that occur
when light hits the interphase of two media with different refractive indexes. When the irregulari-
ties of interphase are smaller than the wavelength of the incident light, specular reflection takes
place, meaning that the angle that forms the incident light with the normal to the separation surface
is the same as the angle of the reflected light with the normal [93]. However, this situation is quite
rare when dealing with the complexities of biological tissues, and in general, diffuse reflection is
observed, meaning that the reflected light does not possess a preferential direction but in a rather
random case. The main interaction between light and tissues is experienced by the refracted light,
which will be attenuated due to the existence of absorption and scattering phenomena.

There are two strategies to describe the light transmission in the tissue: analytical and transport theory. The analytical approach consists in solving Maxwell's equations, which is usually unpractical due to the mathematical challenge of obtaining exact analytical solutions when introducing the complexities of biological tissue. However, the electromagnetic theory is useful to introduce the processes occurring when light is traveling across a tissue, by analyzing the energy conservation principle, which is summarized by the Poynting theorem for electromagnetic waves:

$$\frac{\partial W}{\partial t} + \frac{d\overline{P}_{abs}}{dV} + \vec{\nabla} \cdot \vec{S} = 0 \tag{11.7}$$

where W represents the energy density and S represents the Poynting vector, which is related to the flux of energy to or from the medium. From this conservation law, the radiative transfer equation can be deduced, which in differential form has the following expression [94]:

$$\frac{1}{c}\frac{\partial}{\partial t}I_v + \hat{\Omega} \cdot \nabla I_v + \left(k_{v,s} + k_{v,a}\right)I_v = j_v + \frac{1}{4\pi}k_{v,s}\int_{\Omega} I_v d\Omega \tag{11.8}$$

where c is the speed of light (in vacuum), j_v is the emission coefficient, $k_{v,s}$ is the scattering opacity, $k_{v,a}$ is the absorption opacity, I_v is the spectral radiance at a frequency v, and Ω is the solid angle that the radiation is crossing. The last term represents radiation scattered from other directions into a surface. Hence, the meaning of the radiative transfer equation is that as a beam of radiation travels, its energy is lost due to absorption, emission, and redistributed by scattering. Solving this equation is the objective of the transport theory, which is based on introducing certain assumptions on the behavior of absorbed and scattered radiation.

The most important hindrance for *in vivo* optical imaging in the visible range when moving away from microscopy is not tissue autofluorescence but absorption and scattering effects, affecting both the excitation light and the emitted fluorescence inside the tissues (see Figure 11.1(a)–(c) for a schematic representation of all three effects). To introduce these phenomena, the Beer-Lambert (B-L) law can be used for didactic purposes. In its common form, B-L law assumes that the attenuation of the detected light is completely explained by the absorption effects. Hence, if a beam of light passes through a tissue, the log term of the ratio between the transmitted light intensity I and the initial light intensity I_0 is supposed to be equal to the opposite of the product of the tissue's absorption coefficient μ_a with the total path length L of light (which in a first approximation coincides with the thickness of the tissue). This product, in turn, is defined as the absorbance A of the tissue. However, since the absorption coefficient μ_a itself is the sum of the product between the molar extinction coefficients ε_i of the chromophore in the tissue with their respective concentrations $[C_i]$, this relation can be summarized by the following equation:

$$A = -\log\frac{I}{I_0} = \mu_a L = L\sum_i \varepsilon_i \left[C_i\right] \tag{11.9}$$

Nevertheless, Eq. (11.9) is an approximation that is only valid when the tissue and the distribution of absorbing chromophores inside it are assumed to be homogeneous and scattering is considered to be negligible. The latter condition is hardly true in bio-tissues. After all, the presence of scattering is determined by changes in the refractive index due to media variation and inhomogeneities in tissues. To consider scattering effects, one needs to take into account a modified version of Eq. (11.9) known as the modified B-L law. As opposed to the simplified rendering, it (i) includes a function f that accounts for the scattering effects and (ii) multiplies L by a parameter called differential pathlength factor (DPF). The DPF accounts for increases in the optical path due to scattering,

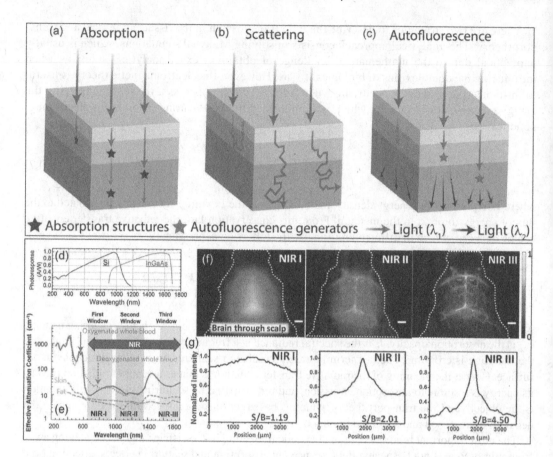

FIGURE 11.1 Schematic diagram of different interactions between incident light and biological tissues: (a) absorption, (b) scattering, and (c) autofluorescence (reproduced with permission from Ref. [51], copyright 2020, American Institute of Physics); (d) response curves of the Si (S-025-H, Electro-Optical System Inc.,) and InGaAs (IGA-030-H, Electro-Optical System Inc.,) photodiodes at *RT*; (e) relevance of the wavelength for three biological windows in some biological tissues and fluids; (f) fluorescence images of the cerebrovascular of mice without craniotomy in three biological windows with the corresponding signal-to-background ratio (SBR) analysis shown in part (g) (scale bars: 2 mm) (reproduced with permission from Ref. [57], copyright 2019, Elsevier).

the source-detector configuration, and the geometry of the medium. The result of the modified B-L law is mathematically represented by:

$$\log \frac{I}{I_0} = -f\left(\mu_s, V\right) - \mu_a \times L \times \mathrm{DPF}\left(\mu_s, \mu_a, V\right) \tag{11.10}$$

where V is the abovementioned particles per unit volume, accounting for the volumetric region defined by the tissue. As one can readily observe from Eq. (11.10), both the excitation and tissue fluorescence will be attenuated by the presence of the absorption and scattering effects, which presents an easy approach to recognize the dependence of the detected light on internally occurring effects.

Based on these fundamental considerations on light-tissue interactions, specific regions for *in vivo* optical imaging where attenuation and scattering are minimized were identified within the NIR region of the electromagnetic spectrum. In 2001, R. Weissleder proposed the term NIR biological windows (BWs) for them [95], and later in 2009, A.M. Smith *et al.* gave a more detailed classification of BWs together with the spectral sensitivity curves of various commercial photodetectors [96].

Traditionally, NIR-I, II, and III windows are defined as the spectral regions of (~650–950 nm), (~1000–1350 nm), and (~1500–1800 nm), respectively. However, it is worth noting that there are still no official definitions for the exact wavelength regions of these three BWs, and in some cases, NIR-II and NIR-III are combined to be NIR-II, while the wavelength region of 1300–1400 nm is defined as NIR-IIa and that of 1500–1700 nm is defined as NIR-IIb [9, 16–21]. Actually, the gaps between NIR-I and NIR-II, as well as NIR-II and NIR-III, are mainly dominated by the endogenous absorption of photons induced by water and hemoglobins (including both oxygenated and deoxygenated hemoglobins) (see Figure 11.1(e)) [43]. Liquid water has a number of vibrational overtone and related absorption bands shown as local maxima at around 970, 1450, and beyond 1800 nm [9], thus sets the inherent barriers for different NIR BWs. Therefore, the high percentage of water in almost any living organism precludes deep tissue imaging in these BWs. On the other hand, hemoglobin molecules, being the most abundant chromophores in mammals, absorb intensely in the visible spectral region with multiple absorption bands extending to ~650 nm. As a consequence, the NIR-I window starts from the low absorption edge of hemoglobin molecules and ends to the first main absorption band of water around ~950 nm, while the NIR-II window starts from ~1000 nm until the onset of the second strong absorption band of water around ~1350 nm, and the NIR-III window is then defined as the wavelength region between 1500 and 1800 nm. It is also easy to associate these BWs with the "optical transmittance windows" for fiber-optic telecommunication [97] or even inter-satellite quantum communication [98] with similarly "desirable" wavelength regions [99], in which the water absorption plays an "undesirable" role for the high-fidelity data transportation. Also note that together with the development of any new imaging techniques, more advanced detecting apparatus with higher sensitivity and broader response regions are always pursued for the improvement of imaging qualities. The sensitivity limitation of the commonly used crystalline Si (c-Si)-based photodiodes or charge-coupled device (CCD) cameras in the NIR region is ~1100 nm (bandgap around 1.12 eV) at room temperature (RT), and it will be shifted to shorter wavelength when cooling systems (e.g., Peltier cooling or liquid N_2 cooling) are applied for highly sensitive operation in bio-experiments, which is generally invalid to monitor NIR light over 1000 nm. Because beyond 1000 nm, the Si semiconducting material becomes transparent to photons as the corresponding photon energy drops below its bandgap energy. Instead, photodetectors based on germanium (Ge), indium antimonide (InSb), or mercury cadmium telluride (HgCdTe) are more sensitive but still suffer from low quantum efficiency over 1000 nm [100]. Fortunately, thanks to fast development and availability of the indium gallium arsenide (InGaAs) detector, monitoring NIR light located in the NIR-II/III windows becomes feasible taking into account its high response curve ranging from ~1000 to 1700 nm with relatively high quantum efficiency (see Figure 11.1(d)). Here, we would like to emphasize that special attention should be paid to cooling type InGaAs detectors, especially in the case of liquid N_2 cooling, as considerable bandgap broadening of InGaAs semiconductors at low temperatures will give rise to a large blueshift of response edge to ~1600 nm [101], which sometimes losses the real spectral information of emission bands within the wavelength region between 1600 and 1700 nm, e.g., broadly split bands of the $Er^{3+}:^4I_{13/2} \rightarrow {}^4I_{15/2}$ transition under high crystal-field strength [102–104].

One of the most important advantages using NIR light over 1000 nm for optical imaging is to acquire the detailed structure of the brain at cellular resolution in neuroscience [105–107]. The brain is a unique tissue, made up of a massive network of neural cells and well distinguished from other tissue types in living animals and humans [108]. It contains mostly water and has twice as much lipid while less than half the protein of muscle tissue, thus, complexities of all these components make brain imaging a big challenge. Although the widely used MRI technique offers infinite observation depth, it can only image brain tissue at the resolution level of millimeter scale. Optical imaging is still the only applicable way to precisely study neural tissues with resolution at the micrometer or sub-micrometer scale [43]. To improve imaging contrasts of blood vessels within the brain and in vivo extracellular space, optical probes with intense NIR light over 1000 nm, especially in the NIR-III window (also named "golden window" for brain imaging [107]), provide the optimal transmittance for deep-brain imaging, which in turn motivate the rapid growth of both NIR-emitting materials and new-type InGaAs detectors [16–21, 33–36]. A typical example goes to the recent report by H. Dai *et al.* using SWCNTs as a bio-probe for high-resolution brain imaging in mice as shown in Figure 11.1(f) [21]. They evaluated the wavelength-dependent noninvasive fluorescence imaging of mouse brain vessels at depths of up to 3 mm through the intact scalp and skull of the mice, by intravenously injecting a mixture of biocompatible high-pressure carbon monoxide conversion (HiPCO) and semiconducting SWCNTs. The *in vivo* brain imaging in the longer wavelength NIR-III window shows the obvious benefit of minimized photon scattering and deep tissue penetration. Besides, the signal-to-background ratio (SBR) obtained by imaging in the NIR-III window was found to be much higher than that in the NIR-I and NIR-II windows (4.50 in NIR-III *vs.* 2.01 in NIR-II and 1.19 in NIR-I), suggesting that *in vivo* imaging beyond 1500 nm indeed benefits from both increased SNR and improved spatial resolution (see Figure 11.1(g)). Furthermore, simultaneous single-vessel-resolved blood-flow speed mapping for multiple hindlimb arterial vessels can also be achieved by video-rate fluorescence imaging in the NIR-III region using SWCNTs, which strongly demonstrated the powerful performance of optical probes working in the NIR-II/III window for *in vivo* (brain) imaging.

11.1.2 NEAR-INFRARED AUTOFLUORESCENCE AND FILTERING APPROACHES

Bio-tissues by default possess endogenous fluorescent components, each with different absorption/emission properties and lifetimes. In optical imaging modalities, the excitation light can also activate these molecules, creating an endogenous background as autofluorescence in the images that decreases the SNR, the sensitivity, and the specificity of applied optical probes. This is especially critical for the cases where the optical probe is located deeply in the tissues and nearly no excitation illumination can reach it. In this case, simply increasing the exposure time of the sensor can also amplify the background noise, thus NIR light, especially over 1000 nm, presents two advantages regarding tissue autofluorescence: (i) much lower autofluorescence intensity than that in the visible and NIR-I region. (ii) Increased penetration depth and reduced light scattering allow the excitation of target probes with improved emission intensity and local accuracy. However, it is found that, depending on the intended application, autofluorescence even in the NIR is not completely negligible, contradictory to what has been assumed based on the tissue autofluorescence observed in the visible [109–111]. Up to that point, very few mentions of tissue autofluorescence in the NIR from endogenous fluorophores have been reported in the literature, compared with the knowledge of visible-light autofluorescence: *e.g.*, flavins and pyridine nucleotides exhibit 500 nm autofluorescence upon UV excitation [3], and series of cellular proteins emit from UV to blue light under UV excitation because of the presence of aromatic amino acids (*e.g.*, tryptophan, tyrosine, and phenylalanine) and lipopigments. Generally, similar with the cases in visible light, it has not changed a lot and so far, only a few possible sources of tissue autofluorescence in the NIR have been identified unequivocally or at a molecular level. This, in turn, could lead to erroneous conclusions concerning the detailed information (*e.g.*, health status or composition) of the studied tissues even within the NIR-II/III BWs. Therefore, although one could say that, relative to the whole electromagnetic spectrum, autofluorescence, scattering, and absorption are all minimized in the NIR, they still play important but undesirable roles in this particular spectral region.

A typical example goes to the feeding diet (Figure 11.2(a)), a primary source of observable autofluorescence in animal NIR imaging, even though the special feedstuff for laboratory mice or rats is often labeled as "autofluorescence-free". Figure 11.2(b)–(d) shows NIR autofluorescence images of different types of laboratory animal feedstuff [51]. All five imaged food pellets present a declining NIR fluorescence signal that reaches into the NIR-II window up to 1200 nm under 808 nm excitation with a power density of 50 mW/cm^2 (15-s integration time, 5 nm spectral resolution). The origin of the optical signal is most likely due to chlorophyll molecules stemming from the plant

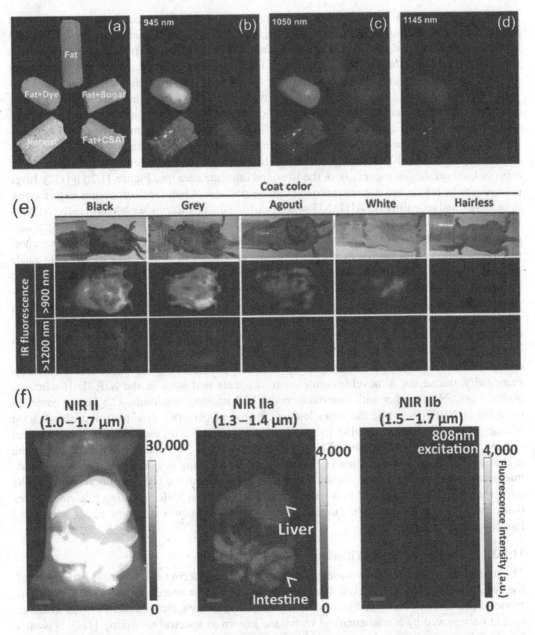

FIGURE 11.2 Autofluorescence investigation of different types of laboratory mouse feedstuff: (a) the optical photo of different food pellets (fat, fat+dye, normal food, fat+CAST, and fat+sugar); (b)–(d) the NIR autofluorescence image under 808 nm laser excitation (power density: 50 mW/cm², integration time: 15 s, spectral resolution: 5 nm) visualized at 945, 1050, and 1145 nm, respectively (reproduced with permission from Ref. [51], copyright 2020, American Institute of Physics); (e) autofluorescence of different mice. Top: photographs of the mice strains. Center and bottom: infrared fluorescence images under 808 nm laser excitation (power density: 0.2 W/cm²) without any optical agents (reproduced with permission from Ref. [53], copyright 2016, Wiley); (f) autofluorescence images from a mouse (abdomen skin opened to reveal inner organs) in the NIR-II, NIR-IIa, and NIR-IIb regions under 808 nm laser excitation (reproduced with permission from Ref. [112], copyright 2015, Springer).

ingredients (*e.g.*, alfalfa) in the food pellets, and the feedstuff's fluorescence has been observed through the gastrointestinal tract and in the feces of the animals. Besides diet, hair color and skin pigmentation of animals also play significant roles in NIR tissue autofluorescence. This was very effectively illustrated by B. del Rosal *et al.* in their study of various mice strains (see Figure 11.2(e)) [53]. The autofluorescence of skin tissue can basically be attributed to the main skin pigment: melanin. Melanin is a copolymer based on indole subunits that are obtained in pigment cells through biochemical polymerization reactions of the amino acid tyrosine. Due to the large size of the polymer, its aromaticity and lipophilicity, the separation from peptides and fatty tissues is quite difficult. In addition to diet and skin pigment, as demonstrated by H. Dai *et al.*, high autofluorescence over 1000 nm can also be well detected from the inner organs of mice under 808 nm laser excitation, especially in the liver and intestine area (see Figure 11.2(f)) [112]. Liver is known to be highly autofluorescent owing to the richness of natural fluorophores (*e.g.*, flavins, lipofuscins, and reticulin fibers) [113]. However, shifting the monitoring wavelength to the NIR-IIb region (1500–1700 nm) is able to significantly decrease the autofluorescence signal, which suggests ultralow autofluorescence of various organs and tissues in this region. As a consequence, when performing NIR imaging even over 1000 nm, effectively discriminating the optical signals emitted from bio-tissue itself and/or from the injected optical probes becomes one of the most important issues during the real-time imaging operation.

In many situations, the most common result of background autofluorescence is an inconvenient increment in signal that is hard to circumvent. It meddles with the detection of luminescence emitted by the bio-probe of interest and complicates the detection of weak optical signals. Hence, two technical approaches (*i.e.*, wavelength-based and lifetime-based) are introduced to overcome these hindrances and lead *in vivo* optical imaging into a practical technology. Principally, the advent of novel imaging/contrast agents that work in the NIR-II/III windows makes it possible, together with improvements in NIR imaging apparatus. The former comes in the form of NIR NPs, while the latter leads to a wave of cameras, detectors, and illumination sources (diodes and lasers), owing to semiconducting diodes with emissions in these spectral regions becoming more available. The combination of these two factors basically covers the excitation, emission, and detection in the NIR regions, allowing to ignore the issues with autofluorescence and opacity in the visible. Thus, the focus shifts to the NIR autofluorescence as the remaining stumbling block and how to drastically improve the SNR regarding this endogenous tissue background. To do so, the currently investigated techniques can most easily be divided into two filtering approaches.

11.1.2.1   Wavelength-Based Filtering

The first filtering approach is a wavelength-based one, considering one is looking for quantitative results while fluorescence from injected agents and autofluorescence from endogenous fluorophores are usually overlapped with each other. In the most generic approach, autofluorescence should be removed by a mathematical technique known as spectral unmixing [114]. It is most successfully applied when the spectral signatures of the targeted signal(s) as well as the undesired light are known, while the mathematical task is made more computationally challenging for most *in vivo* situations when unwanted background light is not so easily determined. Instead, practical experimental steps can be taken into account, and the most straightforward is the use of laser light and optical filters (often placed in a filter wheel) as illustrated in Figure 11.3(a) and (b), by selecting either suitable excitation wavelength of continuous-wave (CW) laser to only activate the optical probes, or selecting suitable band-pass filters to only collect emission signals from injected probes. The nature of selected filters (*e.g.*, cut-on wavelength) is determined by the intended target as well as the overlap of its excitation/emission bands with potential autofluorescence spectra.

Even though background autofluorescence is targeted for reduction in most experiments, the artifacts induced by the excitation light itself can also be a problem, for instance, when determining the

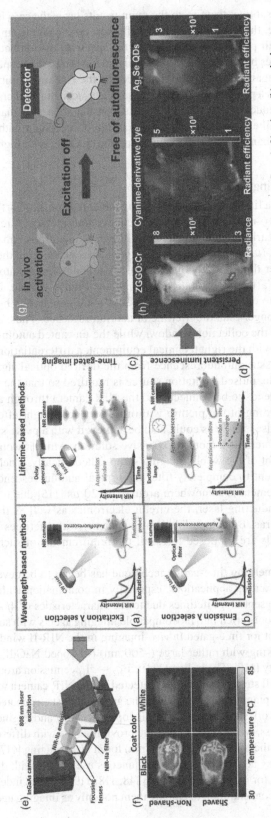

FIGURE 11.3   Schematic representation of the existing approaches for the removal of tissue autofluorescence: (a) excitation wavelength filtering; (b) emission wavelength filtering; (c) time-gated temporal imaging; (d) persistent luminescence imaging (reproduced with permission from Ref. [52], copyright 2018, Wiley); (e) a schematic of NIR-II imaging setup consisting of InGaAs camera, 808 nm laser, focus lenses, and band-pass filters (reproduced with permission from Ref. [19], copyright 2014, Nature Publishing Group); (f) *in vivo* thermographic imaging: thermal images after a 1-min exposure to 808 nm laser excitation (power density: 0.2 W/cm²) corresponding to a black C57 mouse (left) and an albino CD1 mouse (right) before (top row) and after (bottom row) hair removal (reproduced with permission from Ref. [53], copyright 2016, Wiley); (g) illustration of two steps in persistent luminescence imaging; (h) *in vivo* mice imaging with ZGGO:Cr ($x = 0.2$) nanoparticles, cyanine derivative dye, and Ag₂Se quantum dots. The injected dosage is kept at 1 μg (reproduced with permission from Ref. [89], copyright 2017, American Chemical Society).

autofluorescence spectra to later subtract them from the main results (as is the case in the spectral unmixing technique). In principle, the light sources used in fluorescence imaging emit in a narrow range of wavelengths close to their peak of emission, while the spectral tails (on both ends) could not only activate other fluorophores but also be added to the signal measured in the spectrum. Thus, similar to the precautions taken against background fluorescence, the blocking ability of thin-film interference filters reveals them as a primary tool to prevent such artifacts. Since filters with limited blocking capability ultimately reduce image fidelity, careful consideration is demanded in their selection. Basically, the applied filters should spectrally select the desired light with high optical transmittance and prevent out-of-band light by providing superior blocking with high optical density (OD) levels.

11.1.2.2 Lifetime-Based Filtering

The second filtering approach is a lifetime-based one, taking advantage of the short fluorescence lifetimes of organic fluorophores (~ns), which includes the endogenous molecules responsible for the autofluorescence described earlier. In contrast, some inorganic materials can present lifetimes orders of magnitude longer (μs to ms), which means that fluorescence from these materials can still be observed "long" after the excitation has stopped, while any tissue autofluorescence has almost disappeared directly after the excitation ceasing (see Figure 11.3(c)). It means, the triggered time delay is tailored to ensure complete removal of autofluorescence by postponing camera acquisition with a time longer than the lifetime of autofluorescence. Thus, only the long lifetime signal is able to fall into the collection window, while the unwanted autofluorescence is ruled out completely. Therefore, with the right technical equipment, a differentiation in lifetimes will also allow a separation of tissue autofluorescence from the desired optical signal of a lifetime contrast agent. First, a stable, pulsed excitation source is required so that the fluorescence can decay in the off-periods. It needs to be connected to the NIR camera through an electronic circuit that can trigger the recording of the picture within a short delay right after the laser's pulse but before the following pulse [52]. This could also be achieved with a pulse-synchronized chopper wheel in front of the camera as a more inexpensive solution, while it would introduce more insecurities in the alignment of the whole imaging setup. Theoretically, a delay of a hundred nanoseconds will be sufficient to completely filter the typical autofluorescence out. From a practical viewpoint, the requirement is somewhere around 10–20 μs [115] due to the delayed response of commercial semiconducting laser, the circuit electronics as well as the characteristics of employed InGaAs cameras. Biocompatible contrast agents with lifetimes greater than the abovementioned microseconds are supposed to fulfill the requirements of such time-gated imaging.

Over the last few years, this emerging time-gating technique has begun to be developed for *in vivo* imaging, resulting in some exciting applications, especially in combination with lanthanide-activated NPs, which generally present long lifetimes due to the characteristics of *4f–4f* forbidden transitions and are readily enough converted into biocompatible imaging agents with sufficient light brightness. A first proof-of-concept for time-gated *in vivo* imaging in the NIR-II window was presented by B. del Rosal *et al.*, working with rather large (~500 nm) Nd-doped NaGdF$_4$ particles to achieve sufficient emission intensity [116]. The utilized Nd^{3+}:$^4F_{3/2} \rightarrow {}^4I_{11/2}$ emission around 1060 nm after 1 μs of acquisition delay is still strong enough to be detected by the NIR camera while autofluorescence from mice diet and skin pigmentation after this time gate is almost eliminated because of its fast decay (ns) feature. A later work showed that a careful dopant strategy and engineering of the NPs further improved the brightness, while at the same time providing NPs with different lifetimes in controlled manner, and allowed the use of real NPs (~10 nm) for *in vivo* imaging [117]. Currently, this field is rather expanding, taking advantage of the simultaneous use of NPs with different lifetimes at the same NIR wavelength for multiplexed imaging [48, 118, 119], and the independence of the fluorophore's lifetime from its concentration to employ them not only as imaging agents but also as *in vivo* nano-sensors [120].

The abovementioned three methods (excitation/emission signal filtering, time-gated signal collection) are all based on an imaging system that typically consists of laser sources (either CW or pulse mode) and InGaAs cameras as illustrated in Figure 11.3(e) [19], which definitely requires a real-time laser excitation for signal acquisition. However, the non-negligible heating effects even under a low output power (e.g., 808 nm laser with 0.2 W/cm^2 output power) should be taken into account especially when we look at the severe temperature increase for the mouse with darker colored hair (shown in Figure 11.3(f)) due to the strong heat absorption of pigments [53]. To solve this problem, the other lifetime-based filtering approach using PersL NPs (PLNPs) was firstly proposed by Le Masne de Chermont et al., in 2007 [72], taking advantages of superlong lifetime of persistent phosphors (see Figure 11.3(g)) [89]. The principle usually occurs in two steps: (i) the material is illuminated, particularly with UV light, which promotes electrons either from the VB of the host or from the ground state of activators to traps (defects with associated long-lifetime energy levels), where the electrons can stay for a long time; (ii) the illumination source is removed and the trapped electrons are gradually released from the trapping centers, thanks to thermally stimulated energy or an external NIR excitation source, so delayed light emission can be monitored for a certain period of time [54–65]. Depending on the material and performance of applied detectors, the duration (i.e., the time when luminescent signal is still detectable) can last for several hours or even days. The interest in these materials for bioimaging stems from the fact that it is possible to develop such long persistent materials with emission wavelength in the BWs. For this application, the as-prepared NPs are charged outside the body with UV light (in most cases) and injected afterward, and then persistent emission allows the long-term tracking of NPs inside the body. As no excitation light is used during the acquisition of images, no autofluorescence signals are detected and high-contrast images can be obtained. A typical in vivo imaging comparison of PLNPs (Cr^{3+}-doped $Zn_{1+x}Ga_{2-2x}Ge_xO_4$, $0 \leq x \leq 0.5$, ZGGO:Cr) with conventional cyanine-derivative dye and Ag_2Se NPs is shown in Figure 11.3(h). The emission signal from the injected ZGGO:Cr NPs is clearly visualized without any autofluorescence interference, while for the cyanine-derivative dye (middle panel) and Ag_2Se (right panel) injected mice, strong autofluorescence is observed, and emission signals from these two injected agents can hardly be distinguished. Obviously, these images show the potential ability of PLNPs in eliminating autofluorescence interference.

11.1.3 Persistent Luminescence over 1000 nm

"Self-sustained" light emission lasting for a relatively long time from seconds to even days after switching off certain excitation sources, typically UV light, electron beams or high energy radiation (X-, α-, β-, or γ-rays, etc.), is generally called "afterglow" [121]. However, it worth to note that afterglow is a well-known phenomenon in the scintillation field [122–126], mainly due to shallow charge carrier traps existing in the forbidden band, leading to delayed luminescence with μs or even ms lifetime order. It is orders of magnitude longer than the fast radiative transitions of, for instance, lanthanide activators (Ce^{3+} or Pr^{3+} are widely adopted in scintillators thanks to their $5d \rightarrow 4f$ transitions) [127] or host excitons (typically in the case of BGO, $Bi_4Ge_3O_{12}$) with ns lifetime order [128]. In this case, afterglow works as an "undesirable" component accounting for the so-called ghost imaging during the high energy detection, such as in PET or X-ray CT [126, 129]. Therefore, when using afterglow, it should be paid more attention if the time duration is short, up to a limit when it is still a nuisance but not yet long enough to be exploited for something useful [58]. Contrarily, in this chapter, when we talk about "persistent luminescence", it should be a more preferable name in the case of prolonging the time duration as long as possible, because they are highly "desirable" for certain applications as we favor. One strong evidence about this definition can be traced back to the voting result in the 1st International Workshop on Persistent and Photostimulable Phosphors (IWPPP, at that time, was named "Phosphoros 2011 conference" [121]) held in Ghent (Belgium, 2011). A poll on specific aspects of "afterglow" was conducted during the workshop, and it definitely included the choice of a suitable name for such kind of "self-sustained" luminescent phenomenon. About 75% of the participants favored the use of a single name, with clearly the most support for "persistent luminescence" (68%), followed by "long-lasting phosphorescence" (20%). Thus, inorganic materials showing "persistent luminescence" are referred to as "persistent phosphors". The term "afterglow" then becomes much less used in this research field, although it can be a general name describing the delayed luminescence phenomenon in all substances or just "the decay of persistent luminescence [64]". Note that "persistent luminescence" is usually abbreviated as "PersL" [57] or "PeL" [130, 131] in most literatures, to distinguish from photoluminescence (PL).

Another terminological misconception about PersL is the difference between PersL and phosphorescence since they are usually mixed with each other in many literatures. Especially for phosphorescence, it was often equally applied in both organic and inorganic substances, which are governed by completely different mechanistic origins. In principle, for organic molecules, phosphorescence is referred to as the radiative triplet-to-singlet transition, *i.e.*, from the first triplet excited state (T_1) to the singlet ground state (S_0) [132, 133]. Considering that such "forbidden" radiative transition is ascribed to the spin-flip feature between the states with two different electronic spin multiplicities in quantum mechanics, phosphorescence usually exhibits longer emission lifetime in a range of 10^{-6} s to even seconds, different from the spin-allowed transition of short-lived fluorescence with lifetime in a range of 10^{-8}–10^{-9} s [134]. To distinguish phosphorescence from PersL, three distinct features of phosphorescence should be addressed: (i) lifetime increasing with decreasing temperature, (ii) lifetime in the time window from microseconds up to seconds, (iii) red-shifted emission band compared with the relevant fluorescence band upon excitation [57]. Since the detailed discussion of phosphorescence is out of the scope of this chapter, some specific reviews [65, 133–135] and related chapters in this book concerning about the history, mechanism, and state-of-the-art of phosphorescent organic materials will be much better choices.

Interestingly, the rich and longstanding history of PersL can be traced back as early as 1000 years ago, when it was considered as a mysterious phenomenon, and firstly described in a Chinese miscellaneous note called *"Xiang-shan Ye Lu"* published in the Song dynasty (960–1279 A.D.) [136, 137]. In 1602, an Italian shoemaker, V. Casciarolo observed bright PersL from a mineral barite ($BaS:Cu^+$) in darkness, later to be known as the famous Bologna stone [138, 139]. Since then, numerous persistent phosphors have been reported, while it was only in 1993 that this research field began to attract much wider interest due to the discovery of the new-generation green persistent phosphor, $SrAl_2O_4:Eu^{2+}$–Dy^{3+} (SAO:Eu–Dy) by Nemoto & Co., Ltd. [140]. This aluminate phosphor gives nearly perfect properties: (i) extremely bright and long duration PersL in the dark (over 30 h before the emission intensity dropping upon 0.32 mcd/m², which is roughly 100 times the sensitivity of the dark-adapted human eyes [140]); (ii) green emission band peaking at ~520 nm that matches well with the human's photopic vision (maximum response at 555 nm with 683 lm/W); (iii) broad excitation band suitable for the charging process *via* conventional fluorescent lamps and large absorption cross-section of the parity-allowed $Eu^{2+}:4f^7 \rightarrow 4f^65d^1$ transition; (iv) highly chemical and physical stability without any radioactive elements. As a consequence, $SrAl_2O_4:Eu^{2+}$–Dy^{3+} together with the famous brand name, LumiNova® [141] rapidly replaced its predecessor, $ZnS:Cu^+$–Co^{2+} in the business market with similar emission wavelength but ten-fold higher brightness and longer duration. Three years after intensive press releases and reports by domestic news and academic journals in Japanese, the R&D group of Nemoto published the milestone research article in August, 1996, entitled "*A New Long Phosphorescent Phosphor with High Brightness, $SrAl_2O_4:Eu^{2+},Dy^{3+}$*" [140] and definitely sent a huge shockwave to the world's academia concerning this relatively unpopular field at that time. Since then, the wheel of persistent phosphor history has begun to speed up around the world.

Twenty-five years after this discovery, many lanthanide-ion-doped aluminates and silicates such as $CaAl_2O_4:Eu^{2+}$–Nd^{3+} ($\lambda_{em} = 440$ nm) [142], $Sr_4Al_{14}O_{25}:Eu^{2+}$–$Dy^{3+}$ ($\lambda_{em} = 490$ nm) [143, 144], and $Sr_2MgSi_2O_7:Eu^{2+}$–Dy^{3+} ($\lambda_{em} = 470$ nm) [145] with emission colors mainly in blue and green regions have achieved a big commercial success for civil applications, especially in emergency signage, safety indication, luminous paints, and watch dials [146]. On the contrary, only very limited reddish orange- and red-persistent phosphors such as $Y_2O_2S:Eu^{3+}$, Mg^{2+}, Ti^{4+} [147, 148], and $(Ca_{1-x}Sr_x)S:Eu^{2+}$ [149–151] have gained the access to markets. The oxysulfides or sulfides rather than conventional oxides are chosen only because of the lack of really powerful candidates emitting in red, especially taking into account the insensitivity of human's visual perception in the red spectral region [152]. Together with the intensive R&D of improving known visible light persistent phosphors or searching for really new (red) ones, much more attention has been paid to deep-red and NIR ones, especially in the form of NPs as we mentioned earlier, due to their potential application as contrast agents for new-generation *in vivo* optical imaging.

In this section, we will explain the mechanism of PersL, including charge carrier trapping-detrapping phenomena, persistent ET process, and photostimulation-induced trap redistribution approaches, to give a general picture of how PersL works and how we can play with the information of energy-level locations of CB/VB, activators and traps. Before starting, some standard physical quantities to evaluate the real performance of PersL properties (*i.e.*, emission intensity and persistent decay) are introduced at the beginning.

It is well known that visible light from 380 to 780 nm can be observed and distinguished by human eyes while the sensitivity or response of human eyes (known as luminous efficacy, in unit of lm/W) for different colors is quite different, and it is mainly dominated by two different visual perceptions [57, 152] as shown in Figure 11.4. One is the photopic vision and the other is scotopic

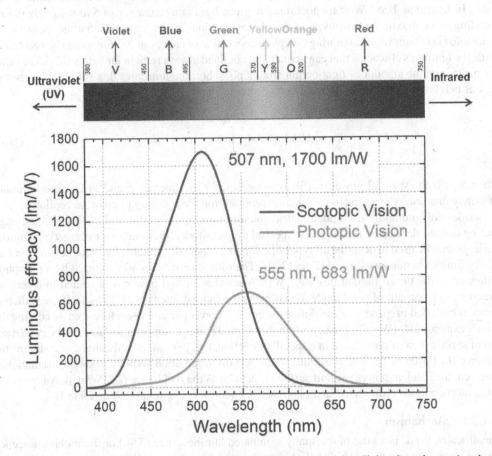

FIGURE 11.4 Luminous efficacy (lm/W) of human eyes in photopic vision (light-adapted; green) and scotopic vision (dark-adapted; blue), the peak intensities of photopic and scotopic visions are 683 lm/W at 555 nm and 1700 lm/W at 507 nm, respectively (reproduced with permission from Ref. [57], copyright 2019, Elsevier).

vision, which actually are mediated by two different cells in the retina: cone cells controlling the photopic vision and rod cells controlling the scotopic vision. The sensitivity of cone cells to optical radiation is relatively low in comparison with that of rod cells. The cone cells can be fully activated when there is a field of view with luminance (in unit of cd/m²) higher than ~1.0 cd/m², while the rod cells can be fully activated in oppositely dark environments, when luminance is less than ~0.01 cd/m². In both cases, human eyes are always sensitive to visible light especially in green or blue regions, while insensitive to the wavelength shorter than 400 nm in violet, UV and longer than 650 nm in deep-red and NIR regions. Note that, under intermediate conditions, where the luminance range in the field of view is between 0.01 and 1.0 cd/m², it is called mesopic vision that both cone and rod cells are partially active. The absolute maximum value of the luminous efficacy at the peak wavelength of 555 nm is 683 lm/W for the photopic vision and 1700 lm/W at 507 nm for the scotopic vision. The mesopic vision is thought to have intermediate spectral sensitivity characteristics, while no accurate definition has been established for it since it varies gradually as the luminance changes. Therefore, photometric quantities such as luminous flux (Φ_p [lm]), luminous exitance (M_p [lm/m²]), luminous intensity (I_p [lm/sr] or [cd]), luminance (L_p [cd/m²] or [nit]), illuminance (E_p [lm/m²] or [lx]), and luminous efficacy (lm/W) are derived after integrating the energy of the optical radiation at different wavelengths over the standard spectral efficiency function for the photopic vision. For example, luminous intensity (cd) is defined as luminous flux (lm) that is emitted in a specified direction per solid angle, and luminance is defined as luminous flux (lm) that is emitted from a surface and falls into a specified direction per unit solid angle per unit area (cd/m²). The lumen is defined as 1/683 W of monochromatic green light at a frequency of 540×10^{12} Hz (corresponding to the maximum sensitivity of the photopic vision), which indicates that the theoretically attainable maximum value, assuming complete conversion of energy at 555 nm would be 683 lm/W, while the luminous efficacies that can actually be obtained always remain far below this ideal value. For measuring the amount of light output from a phosphor, the luminous flux (Φ_p) is the power of light as perceived by human eyes and can be expressed next:

$$\Phi_p = 683 \text{ lm/W} \times \int_{380 \text{ nm}}^{780 \text{ nm}} V(\lambda)S(\lambda)d\lambda \tag{11.11}$$

where $S(\lambda)$ is the spectral power distribution measured experimentally, and $V(\lambda)$ is the luminous efficiency function or eye sensitivity function representing the intensity of color perception.

Besides the commonly used photometric quantities and units in the field of illumination engineering that are defined by means of the spectral eye sensitivity, radiometric or energetic quantities can in general be used to describe all types of radiation, especially suitable for those beyond the sensitivity limit of human eyes, *i.e.*, UV and NIR. The total integrated power, emitted by a phosphor, is defined as the (total) radiant flux (Φ_E [W]). When the emitted power is restricted to a specific direction, *i.e.*, per unit of solid angle, one obtains the radiant intensity (I_E [W/sr]). If the radiation source is extended in space, one can define the emitted power per unit of surface area, becoming the radiant exitance (M_E [W/m²]). Finally, when the emitted power of an extended source, expressed per unit of surface area, is considered in a specific direction, *i.e.*, per unit of solid angle, one obtains the radiance (L_E [W/(sr m²)]). The above quantities pertain to radiation sources, and when radiation is observed, the incident power per unit surface is defined as the irradiance (E_E [W/m²]). An overview of the most common radiometric and photometric quantities is summarized in Table 11.1.

11.1.3.1 Mechanism

Scientifically, PersL is a kind of thermally stimulated luminescence (TSL) or thermoluminescence (TL) at given temperatures [153, 154]. Although some emerging applications using high-temperature PersL (>450 K) have been recently introduced for optical information storage [155, 156] or approaching the so-called zero-thermal-quenching phosphors [157–162], most of the persistent

TABLE 11.1

Overview of Radiometric and Photometric Quantities

Radiometric	Definition	Unit	Photometric	Definition	Unit
Radiant flux	Φ_E	W	Luminous flux	Φ_p	lm
Radiant exitance	$M_E = \dfrac{d\Phi_E}{dS}$	W/m²	Luminous exitance	$M_p = \dfrac{d\Phi_p}{dS}$	lm/m²
Radiant intensity	$I_E = \dfrac{d\Phi_E}{d\Omega}$	W/sr	Luminous intensity	$I_p = \dfrac{d\Phi_p}{d\Omega}$	lm/sr (cd)
Radiance	$L_E = \dfrac{d\Phi_E}{dS\cos\phi}$	W/sr/m²	Luminance	$L_p = \dfrac{d\Phi_p}{dS\cos\phi}$	cd/m² (nit)
Irradiance	$E_E = \dfrac{d\Phi_E}{dS}$	W/m²	Illuminance	$E_p = \dfrac{d\Phi_p}{dS}$	lm/m² (lx)

ϕ Represents the angle between the surface normal and the line of sight.

phosphors are working around *RT* (~20–25°C or 293–298K). Its basic principles are well understood by now while the detailed mechanisms are still the subject of debates [163, 164]. In general, two kinds of activation centers are involved in persistent phosphors, *i.e.*, emission centers and trap centers. Emission centers can be lanthanide ions (*e.g.*, Ce^{3+}, Eu^{2+}, Nd^{3+}, Er^{3+}) with $5d \rightarrow 4f$ or $4f \rightarrow 4f$ transitions, transition metal ions (*e.g.*, Cr^{3+}, Mn^{2+}, Mn^{4+}, Ni^{2+}) with $d \rightarrow d$ transitions or main group/post-transition metal ions (*e.g.*, Pb^{2+}, Bi^{3+}) with $p \rightarrow s$ transitions, *etc*. Trap centers can be lattice/intrinsic defects (*e.g.*, oxygen vacancies, *F*-centers, anti-site defects), impurities (*e.g.*, Cu^+, Co^{2+}, Ti^{3+}), or intentionally introduced aliovalent or isovalent co-dopants (*e.g.*, Dy^{3+} in $SrAl_2O_4$:Eu^{2+} [140], Nd^{3+} in $CaAl_2O_4$:Eu^{2+} [142], Cr^{3+} in $Y_3Al_2Ga_3O_{12}$:Ce^{3+} [165, 166]), *etc*. In some cases, the emitting centers can also play the role of trap centers, such as Cr^{3+} or Bi^{3+}, thanks to their multivalence features. Both of the emission/trap centers are located inside the bandgap (forbidden band), and the trap centers are mainly few electron volts (eV), or more precisely less than 1 eV (depending on different estimation methods of trap depth [54, 153, 154]), just below the bottom of CB in the case of electron traps or just above the top of VB in the case of hole traps. The physical mechanism can be briefly divided into four main processes:

i. When the persistent phosphor is excited by external excitation, the charge carriers (electrons and/or holes) are liberated upon illumination at specific wavelengths.

ii. Instead of being radiatively relaxed, the excited electrons or generated holes can be non-radiatively captured by the electron or hole traps through CB or VB, respectively, or by the quantum tunneling process through the forbidden band, which is called trapping process. The traps usually do not emit electromagnetic radiation, but they store the excitation energy for a long time, which is the reason why this phenomenon can also be referred to as an "optical battery" [61, 167].

iii. After ceasing the excitation, the captured charge carriers can be released mainly by thermal stimulation energy or intentionally applied NIR light, which is called detrapping process.

iv. Finally, the released charge carriers move back to the emission center yielding the delayed luminescence due to the electron-hole recombination, which is called recombination process.

Therefore, the characteristics of the emission center mainly determine the emitting wavelength of interest, while those of the trap center such as trap depth, trap density, or concentration generally determine the intensity and duration of PersL at certain temperatures. Previously, several possible models, *e.g.*, electron/hole trapping model, quantum tunneling model, or intrinsic defect (oxygen vacancy) model [58, 60], were proposed to give convincible explanations for the mechanism, while the improvement of known persistent phosphors and/or the discovery of new ones are still a matter of the "Edisonian approach": trial-and-error. Recently, thanks to electronic level schemes of the host CB/VB together with the energy-level locations of ground and excited states of activators proposed by P. Dorenbos [168–176], one becomes possible to explain, even predict PersL and charge carrier trapping-detrapping phenomena of persistent phosphors.

Concerning the efforts on underlying the mechanism of PersL, the discovery of $SrAl_2O_4$:Eu^{2+}–Dy^{3+} by Matsuzawa *et al.* also marked the beginning of a renewed search for it. Until then, relatively little research had been done on this subject. The proposed mechanism at that time was described by the hole trapping-detrapping model as follows: when the Eu^{2+} ion is excited by an incident photon, a hole escapes to VB, thereby leaving behind an Eu^+ ion, while the liberated hole is then captured by the co-doping Dy^{3+} ion creating a Dy^{4+} ion [140, 177]. However, it seems to be quite improbable considering the required huge amount of energy to form Eu^+ under ambient conditions. Instead, in the same year (1996), S. Tanabe proposed a new model [178], in which Eu^{3+} was created after leaving an electron to CB while the Dy^{3+} ion captured the escaped electron through CB to be Dy^{2+}. Nowadays, it is well known as the electron trapping-detrapping model and has been widely used as one of the most acceptable mechanisms to explain PersL. This hypothesis was strongly supported by the existence of Eu^{3+} species under excitation through the XANES results [179], and the observation of the current rise by photocurrent excitation (PCE) measurements also confirmed the photoionization process of excited electrons from the Eu^{2+}:$5d$ state into CB at *RT* [180]. Although the reduction of Dy valence state from 3+ to 2+ serves as an open question for a long time, very recently, a dedicated resonant inelastic X-ray scattering (RIXS) study with the help of synchrotron radiation demonstrated by P.F. Smet *et al.* gave a strong evidence of valence state change from Dy^{3+} to Dy^{2+} in a similar strontium aluminate host ($Sr_4Al_{14}O_{25}$) during charging process [181], which answered this long-standing question of the electron trapping-detrapping model. Below, two popular charge carrier trapping models of PersL, *i.e.*, electron trapping-detrapping and hole trapping-detrapping, are introduced.

The schematic illustrations of the electron trapping-detrapping mechanism are shown in Figure 11.5 in the case of (a) high energy band-to-band excitation by mercury UV (254 nm) lamp, electron beams, or radiation sources as well as (b) low-energy UV or visible-light excitation directly pumping the emission center. Note that we assume the simplest isovalent co-doping situation for the discussion, possible defect-related electron/hole traps created by aliovalent co-doping can be inspired by this simple model once their energy-level locations are precisely determined. In the case of part (a), when the persistent phosphor is charged by band-to-band excitation, the electron is excited into CB leaving the generated hole in VB (process 1). Then, both the electron and hole can freely move through CB and VB, respectively, followed by being trapped by the electron trapping center and the ground state of the emission center (process 2). After ceasing the excitation, the captured electron can be subsequently released by thermal activation energy, and its detrapping probability is directly correlated to the trap depth (ΔE) which represents the energy gap between the bottom of CB and the trap state. If the thermal activation energy at *RT* is high enough to "bridge" the energy gap (ΔE), the captured electron can be gradually released from the trap center back to the excited state of the emission center through CB (process 3). The recombination of the "returned" electron at the excited state with the "waiting" hole at the ground state finally gives rise to PersL (process 4). This mechanism is quite similar with the afterglow process in scintillation while the main difference is the value of trap depth (ΔE). In the case of scintillation, ΔE is usually quite small so that the main detrapping temperature is far below *RT* [125, 126]. Therefore, the captured electron in the shallow trap can be totally released within a very short time at *RT* leading to the slow emission component with μs or ms order, as an "undesirable" afterglow. However, in the case of PersL, the intentionally introduced co-dopant usually has much higher trapping capacities, *e.g.*, Dy^{3+} in $SrAl_2O_4$:Eu^{2+} [140] or Nd^{3+} in $CaAl_2O_4$:Eu^{2+} [142], and the main detrapping temperature is either

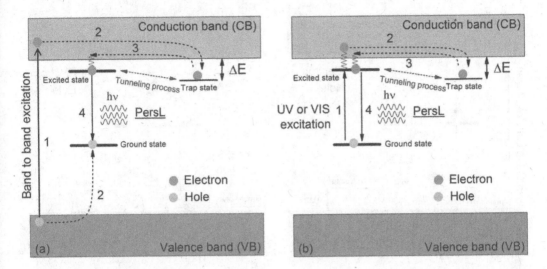

FIGURE 11.5 The schematic illustrations of the electron trapping-detrapping model in the case of (a) high energy band-to-band excitation (b) low-energy UV or visible-light excitation only pumping the emission ions (reproduced with permission from Ref. [57], copyright 2019, Elsevier).

close to *RT* or slightly higher than *RT* so that the captured electron can be slowly detrapped by the relevant thermal assisted energy, giving rise to "desirable" PersL at *RT*.

The trapping-detrapping mechanism under low-energy excitation such as UV or visible light shown in Figure 11.5(b) is only different from that of band-to-band excitation for the process 1. Rather than the hole being captured by the ground state of the emission center through VB, it is directly localized at the ground state while the excited electron is delocalized into the excited state under excitation. When the energy gap between the excited state and the bottom of CB is small enough or even they are overlapped with each other (often the case for the 5*d* excited state of Ce^{3+} and Eu^{2+} in LED phosphors [182–187]), the excited electron can "jump" into CB with thermal assisted energy (thermal-ionization process [188]), then captured by the electron trap (process 1). The following processes from 2 to 4 are nearly the same with that described in Figure 11.5(a) if we ignore the different dynamics and thermal equilibriums of the trapping-detrapping process under different excitation sources. It is worth noting that an athermal (not totally [189]) tunneling process [190] through the forbidden band between the excited state and the trap state possibly occurs due to the overlapping of their vibrational parabolas. In some material systems, this tunneling process was considered as the main trapping-detrapping mechanism accounting for PersL rather than the classical route through CB and/or VB [191]. Also note that giving a strong evidence for the tunneling process is not easy since it is usually hidden behind the main process through CB and/or VB and much less pronounced, especially considering the non-negligible thermal effects when performing persistent decay measurements at *RT*. Therefore, to give a convincing evidence of the tunneling process (*i.e.*, inverse decay intensity [I^{-1}] that is linear proportional to the evolution of decay time [192]), it is recommended to (i) measure the decay curves at a very low temperature (*e.g.*, liquid He) able to avoid the disturbance of thermal assisted energy; (ii) select low-energy excitation sources pumping low excited levels (*e.g.*, the $^4T_1$ (^4F) level of Cr^{3+} by 460 nm blue light excitation [193]) to avoid the possible thermal-ionization process of the excited electron "jumping" from the upper excited state into CB.

Besides electron trapping-detrapping model, in which electrons being released from the electron trapping center can recombine with a luminescent center through CB, holes can also be released from a hole trapping center to recombine with a luminescent center through VB. It is the so-called hole trapping-detrapping model. Although this model probably fails to explain the PersL mechanism

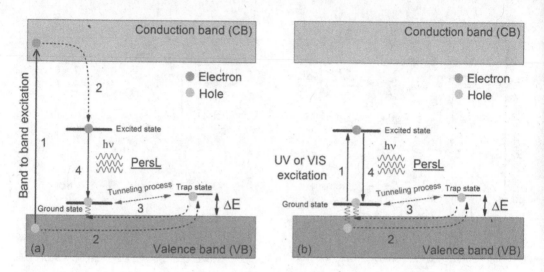

FIGURE 11.6 The schematic illustrations of the hole trapping-detrapping model in the case of (a) high energy band-to-band excitation (b) low-energy UV or visible-light excitation only pumping the emission ions (reproduced with permission from Ref. [57], copyright 2019, Elsevier).

of the Eu^{2+}–Dy^{3+}/Eu^+–Dy^{4+} redox couple in $SrAl_2O_4$:Eu^{2+}–Dy^{3+} [179, 181], it can be applicable for other persistent phosphors once energy-level locations of both emission and hole trapping centers are carefully evaluated. According to the schematic illustrations of the hole trapping-detrapping mechanism shown in Figure 11.6, one can easily recognize that the hole picture is exactly a mirror image of the electron picture shown in Figure 11.5. Under either (a) high energy band-to-band excitation or (b) low-energy excitation directly pumping the emission center, the electron and hole are generated (process 1). Similar to the moving path of electrons in CB, VB can be regarded as the electronic state free of holes so that the excited hole can "freely" move through VB, then being trapped by the hole trapping center while the generated electron is captured by the excited state of the emission center (process 2). After ceasing the excitation, the captured hole can be subsequently released through VB if the thermal activation energy at *RT* is high enough to "bridge" the energy gap (ΔE) between the top of VB and the trap state (process 3). Finally, the recombination of the "returned" hole at the ground state with the "waiting" electron at the excited state gives rise to PersL (process 4).

Here, one may recognize that rather than "schematic illustration" exhibited previously, if the absolute energy of VB, CB, excited/ground states of activators and traps can be fairly determined in one energy-level diagram, it will be much easier to explain charge carrier trapping phenomena of persistent phosphors in a more precise way, or even design novel ones *via* manipulating host's band structure. In 2005, concerning about almost 10 years debate of the mechanism behind $SrAl_2O_4$:Eu^{2+}–Dy^{3+}, P. Dorenbos constructed the famous energy-level scheme of the $SrAl_2O_4$ host with two zigzag curves representing the ground states of divalent and trivalent lanthanide ions (see Figure 11.7(a) [194], the detailed construction flow of host referred binding energy [HRBE]/ vacuum referred binding energy [VRBE] diagrams refer to [168–176, 195–201] and related chapter in this book). From the energy-level locations of lanthanides in this semiempirical diagram (more precisely, that is the HRBE diagram because the energy of the top of VB was set to be 0 eV), it is suggested that Eu^{2+} can act as a stable hole trapping center, while some trivalent ions such as Nd^{3+}, Dy^{3+}, Ho^{3+}, Er^{3+} are expected to be potential electron trapping centers, since the ground states of their divalent species are located below the bottom of CB with suitable trap depths (~1 eV). Based on this diagram, the improved electron trapping-detrapping mechanism involving the ionization process of the Eu^{2+}:$5d$ excited electron to CB and subsequent trapping by

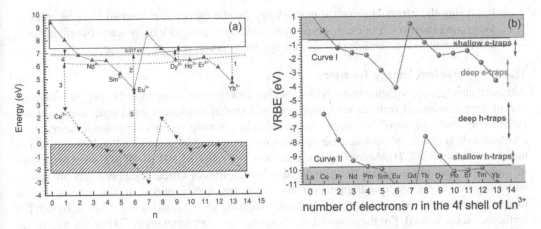

FIGURE 11.7   (a) Energy-level scheme of the SrAl$_2$O$_4$ host (reproduced with permission from Ref. [194], copyright 2005, Electrochemical Society); (b) vacuum referred binding energy (VRBE) diagram of YPO$_4$ with various Ln traps, two zigzag curves represent the ground states of divalent (Curve I) and trivalent (Curve II) lanthanide ions (reproduced with permission from Ref. [202], copyright 2018, Royal Society of Chemistry).

Dy^{3+} (the Eu^{2+}–Dy^{3+}/Eu^{3+}–Dy^{2+} redox couple) was proposed. Judging from this diagram, it worth noting that, besides the efficient sensitization of Dy^{3+} as co-dopant in SrAl$_2$O$_4$:Eu^{2+}–Nd^{3+} may also work as a sensitizer considering the similar trap depth between the ground state of Nd^{2+} and the bottom of CB with that of Dy^{2+}, which is in accordance with the experimental results reported in Matsuzawa's paper [140]. Although this model is probably difficult to give the direct evidence of valence state change of co-dopant lanthanide ions, especially the trap ions, the energy-level scheme that can be constructed by specific spectroscopy data is definitely a big breakthrough to explain the luminescence mechanism of (persistent) phosphors. With this diagram, it becomes possible to predict at least the following:

i. The nature of the charging process by electron and/or hole transfer: Taking the VRBE diagram of a wide bandgap host, YPO$_4$ as an example (see Figure 11.7(b)) [202], when we consider the electron trapping mechanism, shallow electron traps together with deep hole traps as recombination centers will be a preferred choice, such as the popular combinations of Eu^{2+}–Dy^{3+}, Eu^{2+}–Nd^{3+}, Eu^{2+}–Ho^{3+}, or Eu^{2+}–Er^{3+} [54, 55, 57, 60]. On the contrary, shallow hole traps together with deep electron traps as recombination centers should be the prior combination of the hole trapping mechanism, in which the hole trapped at the hole trapping center is easier to move to VB when compared with moving a photo-induced electron into CB upon illumination.

ii. Identification of the candidate emission/trap centers for PersL and the role of co-dopants. For example, Eu^{2+} and Ce^{3+} usually act as emission centers since they prefer to be stable hole traps with large energy gap between the ground state and the top of VB, while Nd^{3+}, Dy^{3+}, Ho^{3+}, Er^{3+} (Pm^{3+} is not preferred because of its strong radioactive nature) are often the case as efficient electron trapping centers considering their suitable trap depths (specifically in many oxide phosphors) for the electron detrapping process at *RT* [54, 55, 57, 60]. On the contrary, Eu^{3+} and Yb^{3+} ions are much less concerned as co-doping electron trapping centers because the ground states of their divalent species are far below the bottom of CB, which needs much higher thermal energy to "bridge" the energy gap (ΔE) at higher temperature than that at *RT*. They can be sufficiently utilized as sensitizers only when the CB edge of the host is considerably lowered [203–205], or just serve as trapping centers with deep trap depth for dosimetry or storage phosphors.

iii. Predicting the essential delocalization energy for the carrier (*i.e.*, excited electron from the ground state to CB or excited hole from the excited state to VB), which is beneficial to quickly identify the possible charging energy for a certain persistent phosphor.

11.1.3.2   Persistent Energy Transfer

Although fluorescence imaging using NIR light over 1000 nm has been proved to permit visualization of deep anatomical features with unprecedented spatial resolution, and PersL imaging using PLNPs with "self-sustained" NIR light is expected to be even superior to conventional fluorescence imaging, only very limited persistent phosphors with intense and long duration PersL over 1000 nm have been reported [57, 71]. According to the summary of reported NIR-II/III persistent phosphors in Table 11.2, one can recognize that compared to big families of visible-light persistent phosphors with a large variety of emitting/trapping centers, the choices in NIR-persistent phosphors are rather limited, and they can be generally classified into two groups: (i) transition metal-ion activated; (ii) lanthanide-ion activated. For the group (i), the "shining star" Cr^{3+} ion working extremely well in the NIR-I window [57, 60, 71, 206–209] becomes much less popular in the NIR-II/III windows, since only in very rare cases (*e.g.*, $La_3Ga_5GeO_{14}$ [210, 211]), the main peak of its PersL band can be successfully shifted over 1000 nm. According to the Tanabe-Sugano (d^3) diagram [212, 213], realizing broadband emission of Cr^{3+} over 1000 nm due to the spin-allowed $^4T_2$ (^4F) $\rightarrow$ $^4A_2$ (^4F) transition usually requires a quite small value of D_q/B. In such a case, the crystal-field parameter (D_q) should be relatively small and/or the Racah parameter (B) should be relatively large. Besides Cr^{3+}, other transition metal ions, especially Ni^{2+} and Co^{2+}, become better choices thanks to the broadband $^3T_2$ (^3F) $\rightarrow$ $^3A_2$ (^3F) and $^4T_2$ (^4F) $\rightarrow$ $^4A_2$ (^4F) transitions in the NIR (over 1000 nm) region, respectively [214, 215]. Specific examples of persistent phosphors with Ni^{2+} PersL at ~1290 nm will be highlighted in Section 11.2.2.

For the group (ii) that is the lanthanide-ion-activated system, particular attentions have been paid to those typical lanthanide ions with promising NIR emitting features that already been well exploited for solid-state lasers (*e.g.*, Yb^{3+}, Nd^{3+}, Ho^{3+}, or Er^{3+}), optical amplifiers in fiber telecommunication (*e.g.*, Pr^{3+}, Tm^{3+}, or Er^{3+}), quantum cutting (Yb^{3+}), and up-conversion fluorescence (*e.g.*, Yb^{3+}, Tm^{3+}, or Er^{3+}), *etc.* Owing to the characteristics of $4f$–$4f$ intra-configuration transitions, the emitting wavelengths of these NIR activators are almost kept constant in spite of varied crystal fields in different host compounds [28–31]. Although the spontaneous transition probabilities and branching ratios change with different $4f$–$4f$ transitions from various "ladder-like" excited levels of lanthanide ions, they can be well predictable from the famous Judd-Ofelt theory [216–218], which in turn makes the potentially useful optical transitions of these NIR emitters for the NIR-II/III BWs quite clear. For example, the Yb^{3+}:$^2F_{5/2}$ $\rightarrow$ $^2F_{7/2}$ transition (~1000 nm), the Nd^{3+}:$^4F_{3/2}$ $\rightarrow$ $^4I_{11/2}$ transition (~1064 nm), the Ho^{3+}:$^5I_6$ $\rightarrow$ $^5I_8$ transition (~1200 nm), the Pr^{3+}:$^1G_4$ $\rightarrow$ $^3H_5$ transition (~1310 nm), $^1D_2$ $\rightarrow$ $^3F_{3,4}$ transition (~1040 nm), and the Tm^{3+}:$^1G_4$ $\rightarrow$ $^3H_4$ transition (~1200 nm) are considered to be main choices for the NIR-II window, while the Er^{3+}:$^4I_{13/2}$ $\rightarrow$ $^4I_{15/2}$ transition (~1550 nm) is matching well with the NIR-III window as well as the high sensitivity region of commercial InGaAs detectors. Despite these advantages of lanthanide ions, in fact, the photoionization processes of NIR emitting ions are quite difficult to be realized under ambient conditions, especially for the formation of their tetravalent (4+) states upon photoexcitation, which impedes their feasibilities as effective recombination centers for NIR PersL.

To solve this problem, persistent ET process is able to perform in a smart way to either transfer the energy from other PersL centers in the visible-light region (*e.g.*, Ce^{3+}, Eu^{2+}, Cr^{3+} with allowed transitions and broad emission bands) or from the lattice-related defect emission in the high-energy UV region [54, 57]. Since PersL intensity and duration of NIR emitters are directly correlated with those of energy donor in visible light, once a material exhibits high-performance PersL in visible, long-term PersL in NIR is also expected if ET efficiency is considerably high between donor and acceptor. For example, in 2014, J. Ueda and S. Tanabe *et al.* reported a series of novel garnet persistent phosphors with compositions of $Y_3Al_{5-x}Ga_xO_{12}$ (YAGG):Ce^{3+}–Cr^{3+} ($x = 0$–5), which

TABLE 11.2
Near-Infrared (NIR, over 1000 nm) Persistent Phosphors

Host Material	Activator	Trap	Charging Band (nm)	Emission Peak (nm)	Duration by Detector	Publishing year	Ref.
$SrAl_2O_4$	$Eu^{2+} \rightarrow Er^{3+}$	Dy^{3+}/Er^{3+}	<500 [180]	1530	>10 min	2009	[101]
	$Eu^{2+} \rightarrow Nd^{3+}$	Dy^{3+}/Nd^{3+}	<500 [180]	1064	>15 min	2011	[224]
$La_3Ga_5GeO_{14}$	Cr^{3+}	Dy^{3+}	<400	~1000	>8 h	2010	[211]
Sr_2SnO_4	Nd^{3+}	Defect	<450	1079	>1 h	2014	[225]
$Sr_3Sn_2O_7$	Nd^{3+}	Defect	–	1079	>1 h	2017	[226]
$Y_3Al_2Ga_3O_{12}$	$Ce^{3+} \rightarrow Nd^{3+}$	Cr^{3+}	<500 [165]	1064	>10 h	2015	[219]
	$Ce^{3+} \rightarrow Er^{3+}$	Cr^{3+}	<500 [165]	1532	>10 h	2016	[103]
	$Cr^{3+} \rightarrow Er^{3+}$	Cr^{3+} + defect	<450	1532	>10 h	2017	[102]
	Ni^{2+}	Defect	–	~1400	>30 min	2020	[227]
$GdY_2Al_3Ga_2O_{12}$	$Cr^{3+} \rightarrow Nd^{3+}$	Defect	–	1060	>15 min	2020	[228]
$LaAlO_3$	$Cr^{3+} \rightarrow Er^{3+}$	Sm^{3+}	<350 [223]	1553	>10 h	2017	[104]
	$Cr^{3+} \rightarrow Ho^{3+}$	Sm^{3+}	<350 [223]	1200	>2 h	2018	[222]
$LaGaO_3$	$Cr^{3+} \rightarrow Ho^{3+}$	Defect	<300	1200	>1 h	2018	[222]
$Zn_{1-x}Ca_xGa_2O_4$	$Cr^{3+} \rightarrow Nd^{3+}$	Defect	–	1064	>5 min	2016	[229]
$Zn_2Ga_{3-x-y}Ge_{0.75}O_8$	$Cr^{3+} \rightarrow Nd^{3+}$	Defect	–	1067	>10 min	2020	[230]
$Ca_3Ga_2Ge_3O_{12}$	$Cr^{3+} \rightarrow Nd^{3+}$	Defect	–	1062	–	2016	[231]
$(Y/Lu/La/Gd)PO_4$	$Bi^{3+} \rightarrow Nd^{3+}$	$Tb^{3+}/Pr^{3+}/Bi^{3+}$	–	1062	>1 h	2019	[232]
$Zn_3Ga_2Ge_2O_{10}$	Ni^{2+}	Defect	<450	1290	>12 h	2016	[214]
$Zn_{1+y}Sn_yGa_{2-x-2y}O_4$	Ni^{2+}	Defect	<410	1270–1430	>10 min	2017	[233]
$Zn_2Ga_2Sn_{0.5}O_6$	$Yb^{3+} \rightarrow Ni^{2+}$	Defect	–	1349	>1 h	2020	[234]
$La_3Ga_5GeO_{14}$	Ni^{2+}	Defect	–	1430	>300 s	2016	[214]
$LiGa_5O_8$	Ni^{2+}	Defect	–	1220	>300 s	2016	[214]
$ZnGa_2O_4$	Ni^{2+}	Defect	–	1275	>5 min	2018	[215]
	Co^{2+}	Defect	–	~1000	>40 min	2018	[215]
	$Cr^{3+} \rightarrow Yb^{3+}$	Defect	<650 [73]	~1000	>10 h	2019	[235]
$Gd_2O_2CO_3$	Yb^{3+}	Defect	–	~1000	>144 h	2014	[236]
$LiGa_5O_8$	Yb^{3+}	Defect	–	1010	>100 s	2016	[237]
$Zn_3Ga_2GeO_8$	Yb^{3+}	Defect	–	1007	>100 s	2016	[237]
$La_3Ga_5GeO_{14}$	Yb^{3+}	Defect	–	1009	>100 s	2016	[237]
$SrZrO_3$	Yb^{3+}	Defect	<300	~1000	>20 min	2018	[238]
$Ca_2Ga_2GeO_7$	$Pr^{3+} \rightarrow Yb^{3+}$	Defect	–	~1000	>100 h	2016	[239]
$CaTiO_3$	$Bi^{3+} \rightarrow Yb^{3+}$	Defect	<395	~1000	>80 h	2017	[240]
$Ca_3Ga_2Ge_3O_{12}$	$Pr^{3+} \rightarrow Yb^{3+}$	Defect	–	~1000	>100 h	2017	[241]
$MgGeO_3$	Yb^{3+}	Defect	<330	~1000	>100 h	2016	[237]
	Pr^{3+}	Defect	<360	1085	>120 h	2017	[242]
$CdSiO_3$	Pr^{3+}	Defect	–	1100	>120 h	2017	[242]
Ca_2SnO_4	Defect $\rightarrow Er^{3+}$	Defect	<400	1533	>10 h	2017	[243]
	Defect $\rightarrow Yb^{3+}$	Defect	<400	1000	>10 h	2017	[243]
	Defect $\rightarrow Pr^{3+}$	Defect	<400	1115	>10 h	2017	[243]
	Defect $\rightarrow Nd^{3+}$	Defect	<450	1080	~1 h	2017	[243]
	Defect $\rightarrow Ho^{3+}$	Defect	<400	1180	~30 min	2017	[243]
	Defect $\rightarrow Tm^{3+}$	Defect	<400	1195	~20 min	2017	[243]

"$\rightarrow$" represents persistent energy transfer process, "charging band" represents persistent luminescence excitation (PersLE) spectrum, and "duration by detector" represents persistent luminescent decay monitored by certain detector instead of conventional fluorescence decay (the detailed information of detector refers to the corresponding reference).

exhibit very bright green PersL (>10 h) due to the Ce^{3+}:$5d \rightarrow 4f$ transition that able to be efficiently charged by blue light [165, 166]. Inspired by the efficient ET processes of Ce^{3+} to Nd^{3+}/Er^{3+} that have been widely used in solar/flash lamp-pumped solid-state laser [103, 219] (see Figure 11.8(a)), J. Xu *et al.* developed YAGG:Nd^{3+}–Ce^{3+}–Cr^{3+} and YAGG:Er^{3+}–Ce^{3+}–Cr^{3+} persistent phosphors, in which intense Nd^{3+} and Er^{3+} PersL can be achieved *via* the persistent ET process from Ce^{3+} in the same host (see Figure 11.8(b) and (c)). The main PersL bands from the Nd^{3+}:$^4F_{3/2} \rightarrow {}^4I_{11/2}$ transition (~1064 nm) and the Er^{3+}:$^4I_{13/2} \rightarrow {}^4I_{15/2}$ transition (~1532 nm) match well with the spectral regions of the NIR-II and NIR-III BWs, respectively. Especially for the Er^{3+}:$^4I_{13/2} \rightarrow {}^4I_{15/2}$ transition, similar to that in the famous $Y_3Al_5O_{12}$ (YAG) host [220], broadly split bands ranging from 1450 to 1670 nm in the YAGG host are also observed. The luminescence intensity of the Er^{3+}:$^4I_{11/2} \rightarrow {}^4I_{15/2}$ transition peaking at 966 and 1013 nm is almost quenched, which can be attributed to the ET from Er^{3+}:$^4I_{11/2} \rightarrow {}^4I_{13/2}$ to Ce^{3+}:$^2F_{7/2} \leftarrow {}^2F_{5/2}$, the so-called cross-relaxation process in the Ce^{3+}–Er^{3+} pair [221]. This process can efficiently facilitate the population of the $^4I_{13/2}$ level and simultaneously improves the emission intensity of the Er^{3+}:$^4I_{13/2} \rightarrow {}^4I_{15/2}$ transition dramatically [103]. Besides Nd^{3+} and Er^{3+} in garnets, intense Ho^{3+} PersL at ~1200 nm in perovskites can also be realized utilizing the similar persistent ET concept [222], indicating this idea serves a rather effective way to acquire PersL over 1000 nm. Specific examples of persistent phosphors with Nd^{3+}, Ho^{3+}, and Er^{3+} long PersL using this concept will be highlighted in Section 11.2.1.2.

11.1.3.3   Photostimulation Induced Trap Redistribution

Although NIR optical imaging using PLNPs is intended as a promising technique for biomedical research and clinical practice, it still suffers from several limitations, including the signal detecting and long-term feasibility. First, as the release of trapped charge carriers (electrons and/or holes) is thermally activated, long time accumulations of the signal (from seconds to minutes) are usually required during recording, thus materials with intensive PersL on one side and low noise detector on the other side should be guaranteed. The second point concerns the excitation mechanisms of PLNPs: being excited *ex vivo* by UV light prior to systemic administration, which prevents long-term imaging in living bodies due to the fading effect of PersL together with monitoring time. Depending on the characteristics of NPs, slow accumulation of stealth nanocarriers within malignant stroma by the enhanced permeability and retention effect usually requires from 2 to 24 h [244]. This is far too long in relation to the light duration or decay of injected PLNPs, which hardly exceeds 1 h *in vivo*. To overcome this major restriction, developments of new materials and of new modalities have been undertaken, in which one important point is to charge PLNPs at the lower energy region inside BWs instead of UV light, and if possible, long time after the injection. The new modality has been proposed in different ways as shown in Figure 11.9 [90], summarizing various approaches that can be employed for long-term *in vivo* imaging. At the beginning, a suspension of PLNPs in a biological buffer is pre-activated *ex vivo*, followed by being injected into the animal and placed under a photon-counting system to detect PersL signals (step 1). Then, PLNPs can be activated or reactivated *in situ* (after the injection) by visible light through animal tissues (step 2), which allows the most simple and convenient recovery of PersL signals in optical imaging whenever required. Although the charging efficiency is quite lower compared with conventional UV precharging (step 1), it is indeed sufficient to be detected and to localize the probe *in vivo*, especially in some Cr^{3+}-doped spinel materials (*e.g.*, $ZnGa_2O_4$:Cr^{3+}) [73]. Furthermore, for some PLNPs, when deeper traps are observed in TL glow curves, PersL can be stimulated by NIR light, as the traps can be depopulated by low-energy light stimulation through the so-called photostimulated PersL (PSPL) (step 3) [73, 90]. The photostimulation technique has been used over the years for UV/X-ray dosimetry, as well as for dating geological and archeological materials [245]. One can adjust the depth of related traps responsible for PersL and thus carefully control the release of trapped carriers. In that case, the release of trapped charge carriers and the delayed emission could be started at the convenience of the user using a red/NIR LED, for instance, which will be inside the BWs and generate negligible heating effects. Besides, the photostimulation or excitation by red/NIR light can

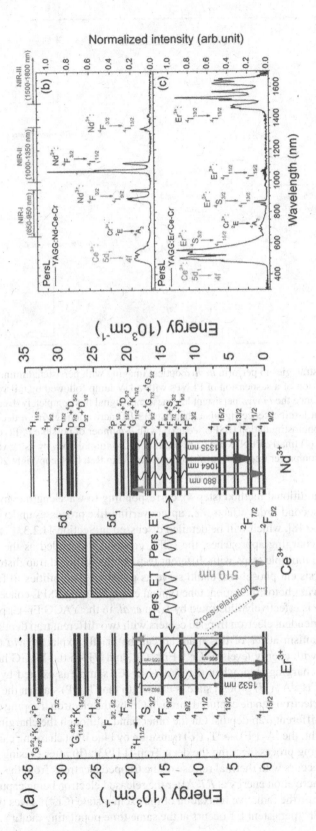

FIGURE 11.8 (a) Energy-level diagrams of Er³⁺, Ce³⁺, and Nd³⁺ in the Y₃Al₂Ga₃O₁₂ (YAGG) host as well as the persistent luminescence spectra of (b) YAGG:Nd³⁺–Ce³⁺–Cr³⁺ and (c) YAGG:Er³⁺–Ce³⁺–Cr³⁺ after ceasing 460 nm LED excitation (reproduced with permission from Ref. [57], copyright 2019, Elsevier).

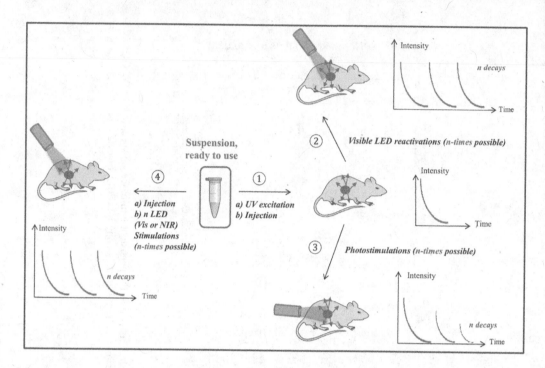

FIGURE 11.9   Possible strategies to perform *in vivo* optical imaging with persistent luminescence nanoparticles (PLNPs): (1) Excitation of a suspension of PLNPs with a UV lamp followed by tail vein injection and then by optical imaging. Once the *in vivo* persistent luminescence signal has completely disappeared, PLNPs can give further persistent luminescence signals either by (2) LED activation (giving *n* decay signals of the same intensity) or by (3) photostimulation (PSPL see text) (with *n*-number of possible PSPL corresponding to detrapping of deep traps); (4) finally, PLNPs can be re-excited, in the animal body by using visible or infrared photons with high excitation power (reproduced with permission from Ref. [90], copyright 2017, Elsevier).

also be carried out by additional method (step 4), corresponding to wavelength conversion, or possible charging through second-order effects, *i.e.*, up-converting-like processes under high excitation power (*e.g.*, laser) [246–248], which will be detailed discussed in Section 11.2.3.

Among these three charging approaches, the PSPL concept is regarded as the most common and available way since multiple traps with different trap depths or broad trap distribution, which are often existed in persistent phosphors, could serve as potential possibilities to fully utilize the trapped photon energy *via* photostimulation. One typical example using PSPL concept for prolonging PersL over 1000 nm is effectively illustrated by J. Xu *et al.* in the YAGG:Er–Cr persistent phosphor, in which two independent electron trapped centers with two different trap depths are observed [102]. The PersL mechanism at *RT* of this material can be briefly explained *via* the abovementioned VRBE diagram with energy levels of Cr^{3+}, Er^{3+}, CB, and VB in the YAGG host (see Figure 11.10(a)–(c)): (a) in the charging step, when the YAGG:Er–Cr sample is excited by UV light, the electron located in the $Cr^{3+}:^4A_2$ (^4F) ground state is excited to the $^4T_1$ (^4P) state in the vicinity of the CB. Hence, the excited electron can be captured through CB by the electron trapping centers (Trap-I and Trap-II) with two different trap depths. On the other hand, although the charging efficiency is much lower than UV light, the $^4A_2$ (^4F) → $^4T_1$ (^4F) transition by blue light charging can also give rise to PersL through tunneling process feeding the deep Trap-II [193]; (b) after ceasing the excitation, the detrapping process occurs with thermal release of the trapped electrons from the shallow Trap-I assisted by the thermal activation energy at *RT*. After the released electron being captured by Cr^{3+} in the recombination process, the radiative relaxation of the excited state $(Cr^{3+})^*$ gives the typical Cr^{3+} PersL, and the Cr^{3+} → Er^{3+} persistent ET occurs at the same time populating the $Er^{3+}:^4I_{9/2}$ level. The rapid multi-phonon relaxation (MPR) from $^4I_{9/2}$ down to the $^4I_{11/2}$ and $^4I_{13/2}$ excited levels takes place followed by the PersL of Er^{3+}:$^4I_{11/2} \rightarrow {}^4I_{15/2}$ (966 nm) and $^4I_{13/2} \rightarrow {}^4I_{15/2}$ (1532 nm) in the NIR region; (c) after a long fading time at *RT*, nearly all of the trapped electrons in Trap-I will be thermally released (trap cleaning) accompanied with the intensity decrease of both Cr^{3+} and Er^{3+} PersL, while those captured by Trap-II are almost unchanged since the thermal activation energy at *RT* is not sufficient to empty Trap-II. Instead, a proper light source such as 660 nm LED light (located in the NIR-I window) could photostimulate the deeper trap (Trap-II) as demonstrated in Figure 11.10(d). Then, Trap-II can be redistributed toward Trap-I through CB (Figure 11.10(e)) during *in situ* photostimulation. As a consequence, the refilled or recharged Trap-I after photostimulation can again contribute to the Cr^{3+}/Er^{3+} PersL as shown in Figure 11.10(f), which is the so-called photostimulation induced trap redistribution concept. This method is rather useful to recover both Cr^{3+} and Er^{3+} PersL by fully making use of the trapped electrons in the deep trap. Detailed experimental results can refer to Section 11.2.1.2.

11.2   NIR-II/III PERSISTENT PHOSPHORS

11.2.1   Lanthanide Ions Activated

Owing to the characteristics of $4f$–$4f$ intra-configuration transitions, the emission wavelengths of lanthanide ions are almost kept constant in different host compounds in spite of varied crystal fields [28–31]. Therefore, the candidates such as Yb^{3+}, Pr^{3+}, Nd^{3+}, Ho^{3+}, Er^{3+} with NIR emitting features are considered to be highly desirable for acquiring PersL over 1000 nm.

11.2.1.1   Single-Ion Doping

Yb^{3+} owns only two energy levels in the $4f$ shell, *i.e.*, the $^2F_{7/2}$ ground state and the $^2F_{5/2}$ excited state, producing a typical emission at ~1000 nm with a fine-structure shape, due to the Stark splitting of two manifolds. Generally, Yb^{3+} ion is extensively used for other photonic applications, *e.g.*, optical sensitizer to enlarge the absorption cross-section of other lanthanide ions (*e.g.*, Er^{3+} or Tm^{3+}) for up-conversion luminescence (UCL) [37–48] or spectral converter in Yb^{3+}–Ln^{3+} co-doped quantum cutting materials for high-efficiency silicon photovoltaics [249, 250]. Since the emission band of Yb^{3+} is partially overlapped with the first vibrational overtone of water (~970 nm), it is generally much less considered for *in vivo* optical imaging. However, according to the previous report [251], the spectral region between 900 and 1000 nm was still recognized to be suitable for optical imaging with markedly enhanced SNR despite the strong water absorption, which suggests the potential possibility using Yb^{3+} for optical imaging. Recently, by utilizing $MgGeO_3$ as the host, Z. Pan *et al.* have successfully developed the new function of Yb^{3+} ion, in which acts as an efficient emitter in the SWIR region and exhibits the super-long PersL at ~1000 nm for more than 100 h (see Figure 11.11(a) and (b)) [237]. The Yb^{3+}-activated $MgGeO_3$ persistent phosphor also exhibits the PSPL capability, indicating that the UV light pre-charged sample can be reactivated by low-energy NIR light to prolong duration of PersL. Through analyzing TL glow curves by the initial rise method [153, 154], they revealed that there are two types of traps in the $MgGeO_3$ host: the shallow trap (TRAP-1) with continuous trap distribution and the photochromic-center-related deep trap (TRAP-2), thus a schematic model was proposed to explain the mechanism of PersL and related PSPL approach in $MgGeO_3$:Yb^{3+} (see Figure 11.11(c)). Moreover, Y. Wang *et al.* also observed superlong Yb^{3+} PersL in perovskite $CaTiO_3$ host that able to be efficiently charged by not only UV lamp (254 or 365 nm) but also sunlight [240]. Due to the introduction of more Bi^{3+}-related defects and the possible PersL quantum cutting process, Yb^{3+} PersL can be efficiently sensitized by Bi^{3+} co-dopants and exhibits a recordable duration for more than 80 h.

Besides Yb^{3+}, Pr^{3+} also acts as a feasible PersL emitter in the SWIR region, *e.g.*, $^1D_2 \rightarrow {}^3H_6$ (~900 nm), $^3F_{3,4}$ (~1080 nm), or $^1G_4 \rightarrow {}^3H_5$ (~1310 nm) transitions falling into the NIR-I or NIR-II/III spectral regions, although most attention regarding Pr^{3+} PersL had been paid to its dominant red

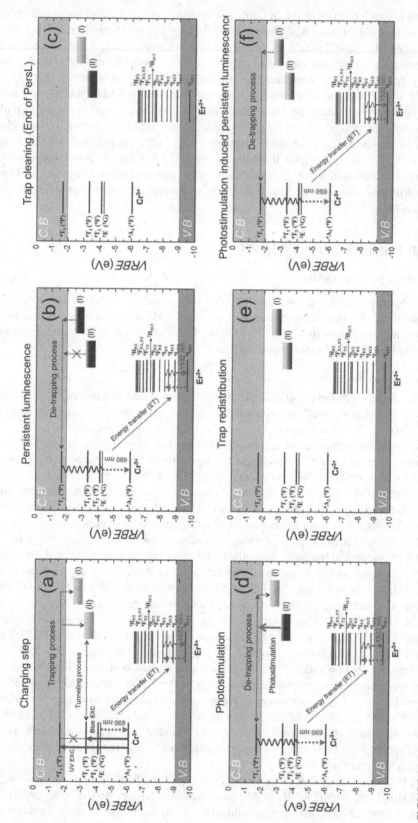

FIGURE 11.10 Mechanism of the persistent luminescence and photostimulation-induced persistent luminescence using the VRBE diagram including selected energy levels of Cr^{3+} and Er^{3+} in the $Y_3Al_2Ga_3O_{12}$ (YAGG) host: (a) charging step by UV or blue light excitation, (b) persistent luminescence at room temperature (RT), (c) shallow trap cleaning after detrapping process at RT, (d) photostimulation, (e) trap redistribution after photostimulation, (f) photostimulation-induced persistent luminescence at RT (reproduced with permission from Ref. [102], copyright 2018, American Chemical Society).

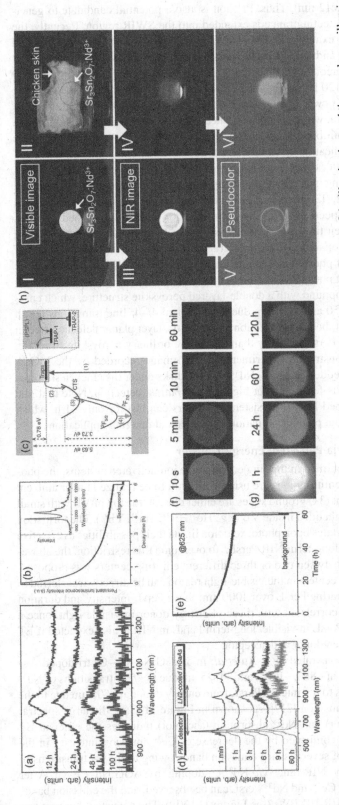

FIGURE 11.11 (a) Persistent luminescence emission spectra of MgGeO₃:Yb³⁺ phosphor at 12–100 h after the stoppage of the illumination, and the sample was illuminated by a 254 nm UV lamp for 15 min; (b) persistent luminescence decay curve monitored at 1019 nm after illumination by a 254 nm UV lamp for 15 min. The inset shows the persistent luminescence emission spectrum recorded at 30 min after stopping the illumination; (c) a schematic representation of SWIR PersL and SWIR PSPL mechanisms in MgGeO₃:Yb³⁺ (reproduced with permission from Ref. [237], copyright 2016, Nature Publishing Group); (d) normalized persistent luminescence emission spectra recorded at different decay times from 1 min to 60 h. A PMT detector (for <850 nm) and an InGaAs detector (for >850 nm) were used to acquire the spectra; (e) persistent luminescence decay curve obtained by monitoring at 625 nm emission. The recording lasted for 60 h; (f) digital images showing the red PersL recorded using a digital camera at different decay times from 10 s to 60 min. The imaging parameters are 10 s decay, ISO 400/15 s; for 5 and 10 min decays, ISO 400/30 s; and for 60 min decay, ISO 1600/30 s; (g) persistent luminescence images taken using a night-vision monocular at different decay times from 1 to 120 h. The imaging parameter is ISO 400/10 s. For all the measurements, the samples were illuminated by using a 254 nm UV lamp for 10 min (reproduced with permission from Ref. [242], copyright 2017, Royal Society of Chemistry); (h) NIR PersL penetration by using Sr₃Sn₂O₇:Nd³⁺ ceramic disk within a chicken skin (The thickness of a chicken skin was 2 mm); (I) visible image of NIR PersL from Sr₃Sn₂O₇:Nd³⁺, which recorded by using complementary metal-oxide-semiconductor (CMOS) camera; (II) visible image of NIR PersL from Sr₃Sn₂O₇:Nd³⁺ in the skin of a chicken, where broken red circles indicated a position of the ceramic disk; (III) the NIR PersL image from Sr₃Sn₂O₇:Nd³⁺; (IV) the NIR PersL image from Sr₃Sn₂O₇:Nd³⁺ in the skin of a chicken. Both the NIR images were recorded by using InGaAs camera; (V) pseudocolor images of part (III); (VI) pseudocolor images of part (IV) (reproduced with permission from Ref. [226], copyright 2017, Ceramic Society of Japan).

emission ($^1D_2 \rightarrow {}^3H_4$ transition at ~612 nm). Thus, Pr^{3+} ion is also a potential candidate to generate PersL over 1000 nm, once the detecting range is extended into the SWIR region. Recently, the usage of Pr^{3+} is rather expanding, for example, Z. Pan *et al.* have reported two Pr^{3+}-doped multiband persistent phosphors, *i.e.*, $MgGeO_3$ and $CdSiO_3$, which exhibit not only the well-pronounced red PersL at ~612 nm, but also simultaneously intense and long-term SWIR PersL at ~900 and 1080 nm [242], able to be detected over 120 h (see Figure 11.11(d)–(g)). They also demonstrated the *ex vivo* experiment that the penetration power of these three wavelengths in chicken breast follows the order of 1085 nm > 900 nm > 625 nm, which is in accordance with the abovementioned knowledge of light-tissue interactions. These multiband persistent phosphors may find potential applications as optical probes or barcodes for identification or multi-channel optical imaging.

Nd^{3+} ion, widely used as the emitter of solid-state laser, is also regarded as one of the best candidates for the NIR-II optical imaging, since the electromagnetic transitions from the $^4F_{3/2}$ excited state to the ^4I_J states (J = 9/2, 11/2, 13/2) [252, 253] could exhibit line emission bands located in ~900, ~1060, and ~1340 nm, respectively. However, similar with other singly doped systems, Nd^{3+} also suffers the parity-forbidden transition feature and relatively low absorption coefficient, thus is often employed as the co-dopant (*e.g.*, electron trapping center) for enhancing PersL of other lanthanide-activated persistent phosphors, such as $CaAl_2O_4:Eu^{2+}-Nd^{3+}$ [142]. To explore the potential of Nd^{3+}-activated persistent phosphors as a new class of NIR-II PersL probes, C. Xu *et al.* demonstrated the $Sr_3Sn_2O_7:Nd^{3+}$ compound with a double-layered perovskite structure, which emitted intense NIR-II PersL between 850 and 1450 nm due to its typical $4f$–$4f$ line transitions [226]. The unique structure with oxygen octahedral tilt/rotation and the intra-layer planar defect existed in $Sr_3Sn_2O_7:Nd^{3+}$ contributed to higher PersL intensity than that of the ordinary perovskite structure, $SrSnO_3$. As a simple proof of demonstration experiments, in the image recorded by the InGaA camera, the NIR-II PersL of the embedded $Sr_3Sn_2O_7:Nd^{3+}$ ceramic disks under the chicken skin was clearly detectable, which suggested its feasibility for high image contrast detection. Note that, the reported PersL duration of Nd^{3+}-doped NIR-II persistent phosphors can last for only ~1 h, which still remains a problem considering the practical situation in biomedical imaging applications.

11.2.1.2 Multi-Ions Co-Doping via Persistent Energy Transfer

Although NIR-II/III PersL can be obtained in the single lanthanide ion-activated systems, the photoionization processes of these NIR emitting ions are usually difficult to be realized under ambient conditions since most of their trivalent (3+) ground states are either close to the top of VB with small energy gaps (*e.g.*, Nd^{3+}) or directly embedded inside VB (*e.g.*, Ho^{3+}, Er^{3+}, Tm^{3+}, Yb^{3+}). The formation difficulty of relevant tetravalent (4+) states upon photoexcitation limits their possibilities as effective recombination centers for producing long-term NIR PersL. To overcome this restriction, the abovementioned strategy of persistent ET between two or three different emitting centers was proposed, in which the energy from other PersL centers in the visible-light region can be directly transferred to the target NIR emitting centers generating PersL over 1000 nm. Since PersL intensity and duration of NIR energy acceptor are directly correlated with those of energy donor in visible light, once a material exhibits high-performance PersL in visible, long-term PersL in NIR is also expected if ET efficiency is considerably high between donor and acceptor.

This strategy was effectively demonstrated by J. Xu *et al.* in YAGG:Nd–Ce–Cr tri-doped system, in which multiwavelength PersL at ~880, ~1064, and ~1335 nm due to $4f$–$4f$ transitions of Nd^{3+} can be well detected over 10 h, thanks to the intense PersL from energy donor at ~505 nm due to the $Ce^{3+}:5d_1 \rightarrow 4f$ parity/spin-allowed transition [219]. As shown in Figure 11.12(a), the large spectral overlap between Ce^{3+} emission (donor) and Nd^{3+} absorption (acceptor) indicates the efficient ET process from Ce^{3+} to Nd^{3+}, which is confirmed by the decrease of Ce^{3+} emission intensity in the visible-light range and the presence of several sharp emission bands owing to the f–f transitions of $Nd^{3+}:^4F_{3/2} \rightarrow {}^4I_{9/2}$, $^4I_{11/2}$, and $^4I_{13/2}$ in the NIR range *via* Nd^{3+} co-doping in YAGG:Nd–Ce–Cr. After ceasing the blue-light excitation, both Ce^{3+} and Nd^{3+} PersL can be observed, and the emission bands of Nd^{3+} indeed match well with the NIR-I/II BWs (see Figure 11.12(b)). The persistent luminescent

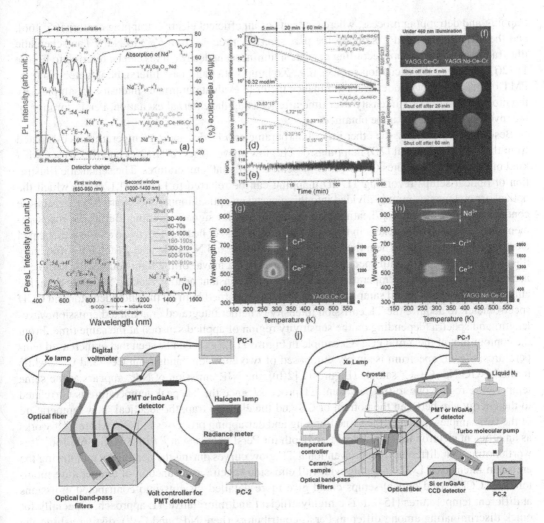

FIGURE 11.12 (a) PL spectra (λ_{ex} = 442 nm) of the YAGG:Ce–Cr and YAGG:Nd–Ce–Cr ceramics as well as the diffuse reflectance of the YAGG:Nd ceramic; (b) PersL spectra of the YAGG:Ce–Cr and YAGG:Nd–Ce–Cr ceramics (integrating time: 10 s) after ceasing blue light excitation; persistent decay curves of the YAGG:Nd–Ce–Cr ceramic: (c) luminance monitoring Ce^{3+} (YAGG:Ce–Cr and $SrAl_2O_4$:Eu^{2+}–Dy^{3+} ceramics as references); (d) radiance monitoring Nd^{3+} ($ZnGa_2O_4$:Cr^{3+} ceramic as a reference); (e) Nd^{3+}/Ce^{3+} radiance ratio (%) against the monitoring time of the decay curve; (f) photographs of the YAGG:Ce–Cr and YAGG:Nd–Ce–Cr ceramics under and after blue LED lamp (460 nm, 3 W output) illumination; wavelength-temperature (λ–T) contour plots of the (g) YAGG:Ce–Cr and (h) YAGG:Nd–Ce–Cr ceramics (reproduced with permission from Ref. [219], copyright 2015, American Institute of Physics); (i) schematic illustration of optical setups recording persistent luminescent decay curves; (j) schematic illustration of optical setups recording thermo-luminescence 2D mapping.

decay curves simultaneously monitoring Ce^{3+} emission (475–650 nm) and Nd^{3+} NIR emission (>800 nm) shown in Figure 11.12(c) and (d), respectively, give clear evidence of long-term PersL (>10 h) both in visible and NIR regions (also demonstrated from the photographs of Nd^{3+} doped and non-Nd^{3+}-doped ceramic pellets under and after blue LED illumination in Figure 11.12(f)). Considering the decay profiles of Ce^{3+} and Nd^{3+} follow a similar fading trend, the persistent radiance ratio (Nd^{3+}/Ce^{3+}) is plotted against the monitoring time of the whole decay curve (see Figure 11.12(e)). The result clearly suggests that the intensity ratio remains almost constant (around 112–114%) with time, which supports that PersL from both ions originates from the expected electron trapping and detrapping process, where Cr^{3+} acts as an efficient electron trapping center [165, 166], and the NIR PersL of Nd^{3+} is due to the persistent ET process from Ce^{3+} to Nd^{3+}. The schematic illustration of optical setups recording persistent luminescent decay curves can be referred to Figure 11.12(i), including the excitation sources (*i.e.*, Xe lamp with band-pass filters), photodetectors (*e.g.*, PMT or InGaAs photodiodes), and luminance meter to evaluate luminous intensity of samples. By monitoring the emission intensity over time after ceasing the external excitation, PersL duration of the investigated sample can be obtained.

Besides decay profiles, the other important parameter, *i.e.*, detrapping temperature, is estimated from the two-dimensional (2D) mappings of TL glow curves (Figure 11.12(g) and (h)), to see what kind of emission contributes to the TL glow peak at different temperatures. The schematic illustration of optical setups recording TL 2D mapping can be referred to Figure 11.12(j), in which the optical arrangement is mainly divided into three parts: an optical input for excitation, a temperature controller for sample cooling/heating, and optical output for signal recording. Before TL measurement, the sample should be kept in the dark for enough time or heating to release all the pre-trapped electrons/holes in nature. Then, the sample is cooled by liquid N_2 or He in cryostat and sufficiently illuminated by the light with specific wavelength. After removal of the light source, the sample is heated by a programmed temperature controller with a constant heating rate. At the same time, the emission of the investigated sample is simultaneously recorded both by the photodetector and CCD spectrometer, to acquire the TL emission intensity (whether integrated or at a fixed emission wavelength) and spectra (depending on the sensitivity region of applied sensors) at the same time. From the contour plot of the YAGG:Ce–Cr sample in Figure 11.12(g), it can be seen that at increased temperatures, the TL spectrum is simply composed of two emission bands from Ce^{3+} and Cr^{3+}, while in the YAGG:Nd–Ce–Cr sample (Figure 11.12(h)), the NIR emission of Nd^{3+} appears at the same temperature region due to the persistent ET process. Since the TL peak temperature is correlated to the energy gap between the bottom of CB and the electron trap, the identical glow temperature of the two samples indicates the same trapping and detrapping processes in both, where Cr^{3+} works as an efficient electron trap with ideal trap depth for PersL working at *RT* in YAGG [165, 166]. It is worth noting that different from conventional TL glow curves during the heating by monitoring the emission intensity, TL 2D mapping with full emission spectra as a function of temperature monitored by CCD-based optical setups could give more detailed insights into contributed emissions at different temperatures [154]. It is a highly efficient and informative TL approach, especially for quick discriminating among different PersL contributors (here Nd^{3+} and Ce^{3+}), distinguishing the real emission centers in the target composition from those in impurity phases, revealing the temperature dependence of PersL from activators at different crystal sites, or identifying black-body radiation from the tested sample and/or measurement setup [214], *etc.*

The success of long-term Nd^{3+} PersL by utilizing the efficient persistent ET from Ce^{3+} in garnets is further extended by J. Xu *et al.*, to the Cr^{3+}–Ho^{3+} pair in perovskites, in which Cr^{3+} with emission around 734 nm acts as energy donor and Ho^{3+} with emission around 1200 nm acts as energy acceptor [222]. Similar with garnets involving three cation sites able to be accommodated by lanthanide and/or transition metal ions, perovskite with a general formula of ABX_3 is also considered as a very flexible luminescent matrix with two types of cation sites able to accommodate different luminescent activators, *e.g.*, lanthanide ions at the A sites and/or transition metal ions at the B sites, leading to various perovskite phosphors with attractive luminescent properties. In fact, 1200 nm emission from Ho^{3+} is regarded to have weaker attenuation effect in the aqueous environment compared with Nd^{3+} and Er^{3+} because of less spectral overlap with vibrational overtone of water, which is beneficial to obtain high-resolution optical imaging. However, because of the small energy gap ($\sim$3500 cm^{-1}) between the first ($^5I_7$) and second ($^5I_6$) excited levels of Ho^{3+}, the excited electrons in the $^5I_6$ level easily suffer from non-radiative relaxation to the next $^5I_7$ level by MPR processes. As a consequence, compared with the $^5I_6 \rightarrow {}^5I_8$ transition at $\sim$1.2 μm, the emission intensity of the $^5I_7 \rightarrow {}^5I_8$ transition at $\sim$2.0 μm is generally much higher. In Xu's work, by carefully evaluating the host phonon energy of two perovskite compounds, *i.e.*, $LaAlO_3$ and $LaGaO_3$, Ho^{3+}–Cr^{3+}–Sm^{3+} tri-doped

LaAlO$_3$ (LAO:Ho–Cr–Sm) and Ho^{3+}–Cr^{3+} co-doped LaGaO$_3$ (LGO:Ho–Cr) are presented, in which Ho^{3+} PersL at ~1200 nm can be observed *via* the persistent ET process from Cr^{3+} as shown in Figure 11.13(b) and (c). Furthermore, taking into account the wavelength regions of NIR-I/II BWs, and the response curves of commercial Si and InGaAs detectors shown in Figure 11.13(a), the whole Cr^{3+} PersL matches well with the NIR-I window and the high sensitivity region of commercial Si detectors while the Ho^{3+} PersL matches well with the NIR-II window and the high sensitivity region of commercial InGaAs detectors.

To clarify the contributed emission bands to the TL glow curves at different temperatures, the TL 2D contour plots of the LAO:Ho–Cr–Sm and LGO:Ho–Cr samples are employed again as shown in Figure 11.13(d)–(g), in particular, covering a wider spectral range from 600 to 1350 nm. From the contour plots of the LAO:Ho–Cr–Sm sample recorded by the Si CCD in Figure 11.13(d) at increased temperatures, the TL spectrum is composed of only one Cr^{3+} emission band at around 734 nm, and no sharp *f*–*f* emission bands from Sm^{3+} (*i.e.*, $^4G_{5/2} \rightarrow {}^6H_{7/2, 9/2}$ transitions located at around 600–650 nm) are observed indicating that Sm^{3+} ions act only as electron traps for Cr^{3+} PersL [223]. When monitored by the InGaAs detector in the wavelength region from 1000 to 1350 nm, besides the intense black-body radiation from the sample and partially from the measurement setup in the NIR region [214] starting from ~500K, the NIR *f*–*f* emission band (1130–1220 nm) from Ho^{3+} appears at the same temperature range as Cr^{3+} luminescence shown in Figure 11.13(e). It indicates that Sm^{3+} works as an efficient electron trap with suitable trap depth for both Cr^{3+} and Ho^{3+} PersL, in which Cr^{3+} acts as energy donor while Ho^{3+} acts as energy acceptor through the persistent ET process. Similar with LAO:Ho–Cr–Sm, in the TL 2D plots of the LGO:Ho–Cr sample (see Figure 11.13(f) and (g)), both Cr^{3+} and Ho^{3+} emission bands are observed at the same temperature range suggesting the similar mean trap depth in both.

The PersL and persistent ET mechanism of two perovskite phosphors can be briefly explained by the constructed VRBE diagram involving Cr^{3+}, Ho^{3+}, Sm^{2+}, CB, and VB energy levels in both LAO and LGO hosts given in Figure 11.13(h). When the two perovskites are charged by UV light, the electron located at the Cr^{3+}:$^4A_2$ ground state is excited to the $^4T_1$ (^4P) state either close on the bottom of CB in LAO or within CB in LGO. Then, the excited electron can be captured by the electron trapping center through CB (Sm^{3+} ions in LAO or intrinsic defects in LGO) with thermal activation energy at *RT* to bridge the small energy gap between the $^4T_1$ (^4P) state and the bottom of CB in the case of LAO. Simultaneously, Cr^{3+} is photo-oxidized into Cr^{4+} or (Cr^{3+} + h$^+$) and Sm^{3+} is formed to be Sm^{2+} or (Sm^{3+} + e$^-$) after capturing one electron (process 1). After ceasing UV excitation, the detrapping process occurs with thermal release of the trapped electron from the Sm^{2+} (Sm^{3+} + e$^-$) or the intrinsic electron trap, and finally the excited state (Cr^{3+})* appears after capturing the released electron in the recombination process (process 2). The radiative relaxation gives the Cr^{3+} sharp luminescence in the deep-red region, and the persistent ET process occurs to Ho^{3+} by populating the $^5I_5$ level (process 3), which is followed by rapid MPR processes down to the $^5I_6$ and $^5I_7$ excited levels result in the emission bands of Ho^{3+}:$^5I_6 \rightarrow {}^5I_8$ (1200 nm) and $^5I_7 \rightarrow {}^5I_8$ (1960 nm) in the NIR region.

Besides Nd^{3+} and Ho^{3+}, taking into account the potential application of brain imaging within the "golden window", the Er^{3+}:$^4I_{13/2} \rightarrow {}^4I_{15/2}$ transition located in ~1500 nm is considered to be the best choice among various optical transitions of lanthanide ions. Z. Pan *et al.* [101] firstly observed the PersL from Er^{3+} through the persistent ET from Eu^{2+} in the well-known SAO:Eu–Dy persistent phosphor. However, compared with super long PersL (>10 h) from Eu^{2+} in green region, the PersL intensity and duration of Er^{3+} emission peaked at 1530 nm was much weaker and shorter (less than 10 min) after ceasing UV excitation. Recently, revisiting the same concept and utilizing efficient persistent ET process from both Ce^{3+} and/or Cr^{3+} to Er^{3+}, J. Xu *et al.* obtained Er^{3+} PersL in YAGG garnets and LAO perovskites, which exhibit long duration (>10 h) after ceasing external excitation (blue light for Ce^{3+} and UV light for Cr^{3+}) as shown in Figure 11.14 [102, 103]. The persistent luminescent decay curve monitoring the Ce^{3+} emission of the YAGG:Er–Ce–Cr sample after ceasing the blue light illumination is shown in Figure 11.14(a), in which the decay curves of the standard YAGG:Ce–Cr ceramic phosphor and a compacted ceramic pellet made of the commercial

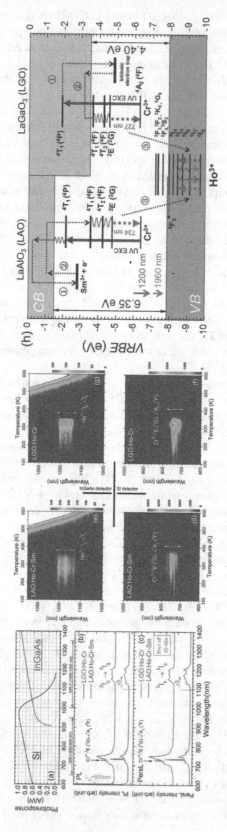

FIGURE 11.13 (a) Response curves of the Si and InGaAs photodiodes at room temperature (*RT*); (b) PL spectra (λ_{ex} = 590 nm) and (c) PersL spectra of the LAO:Ho—Cr—Sm and LGO:Ho—Cr ceramic samples; contour mappings of the thermoluminescence intensity as a function of emission wavelength (λ) and temperature (*T*) of the parts (d), (e) LAO:Ho—Cr—Sm and (f), (g) LGO:Ho—Cr samples; (h) the vacuum referred binding energy (VRBE) diagram with selected energy levels of Cr^{3+}, Ho^{3+}, and Sm^{2+} (ground state) in the LAO and LGO hosts (1: trapping process; 2: detrapping process; 3: persistent energy transfer) (reproduced with permission from Ref. [222], copyright 2018, Royal Society of Chemistry).

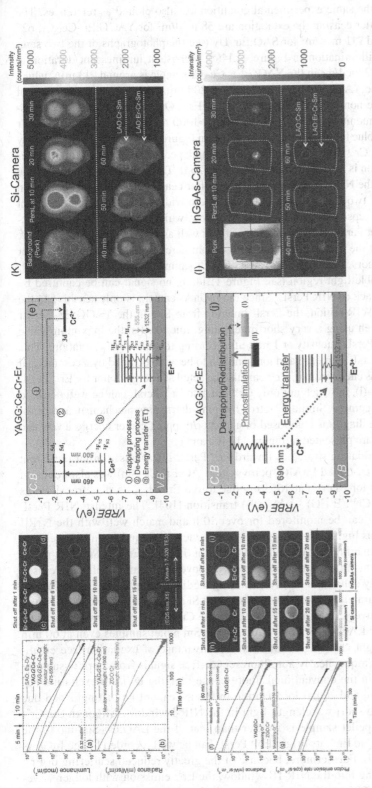

FIGURE 11.14 Persistent luminescent decay curves of the YAGG:Er–Ce–Cr ceramic sample: (a) luminance monitoring Ce[3+] emission (YAGG:Ce–Cr and SAO:Eu–Dy ceramic samples as references); (b) photon energy monitoring Er[3+] emission (ZGO:Cr ceramic sample as a reference); photo images of the YAGG:Ce–Cr and YAGG:Er–Ce–Cr ceramic samples after blue LED (455 nm, 1 W output) illumination for 5 min: (c) taken by a digital camera (EOX kiss X5) with exposure time: 1 s, ISO value: 1600, aperture value (F value) 5.0; (d) taken by a SWIR camera (Xeva-1.7-320 TE3) with integrating time: 0.04 s; (e) the VRBE diagram, including selected energy levels of Ce[3+], Er[3+], and Cr[2+] in the YAGG host (reproduced with permission from Ref. [103], copyright 2016, Royal Society of Chemistry); persistent luminescence decay curves (f) in unit of radiance (mW/(sr m[2]) and (g) in unit of photon emission rate (cps/Sr/m[2]) of the YAGG:Er–Cr ceramic sample monitoring Cr[3+] and Er[3+] emission (YAGG:Cr and ZnGa₂O₄:Cr[3+] [ZGO:Cr] ceramic samples as references); photographs of the YAGG:Cr and YAGG:Er–Cr ceramic samples after 254 nm (6 W output) illumination for 5 min (h) taken by a Si CCD camera (integrating time: 1.0 s, the tail in the 5 min image is due to the intensity saturation of the camera) and (i) taken by an InGaAs camera (integrating time: 0.04 s); (j) the VRBE diagram, including selected energy levels of Cr[3+], Er[3+], and Cr[2+] in the YAGG host (reprinted with permission from Ref. [102], copyright 2018, American Chemical Society); *ex vivo* imaging through raw pork tissues (thickness of 1 cm for the Si camera and 0.5 cm for the InGaAs camera with integrating time: 5.0 s) monitoring (k) Cr[3+] and (l) Er[3+] PersL in the LaAlO₃ (LAO):Er–Cr–Sm pellet sample after ceasing 254 nm (6 W output) charging for 10 min (LAO:Cr–Sm pellet sample as a reference) (reproduced with permission from Ref. [104], copyright 2017, Royal Society of Chemistry).

SAO:Eu–Dy phosphor under the same experimental condition are also plotted as references. The luminance values at 10 min after ceasing the excitation are 58 mcd/m^2 for YAGG:Er–Ce–Cr, 627 mcd/m^2 for YAGG:Ce–Cr, and 211 mcd/m^2 for SAO:Eu–Dy (see the photographs of the two samples after ceasing blue LED illumination in Figure 11.14(c)). Persistent luminescent duration to reach a luminance value of 0.32 mcd/m^2 in the YAGG:Er–Ce–Cr sample is around 213 min, which is much shorter than that of the YAGG:Ce–Cr sample (about 808 min), due to quenching effect of the green Ce^{3+} emission by the non-radiative persistent ET to Er^{3+}. On the other hand, the persistent luminescent decay curve monitoring Er^{3+} NIR emission (>1000 nm) of the YAGG:Er–Ce–Cr sample after ceasing the same blue light illumination is shown in Figure 11.14(b), in which the decay curve of the standard ZnGa$_2$O$_4$:Cr^{3+} (ZGO:Cr, emitting wavelength peaked at 695 nm) sample under the same experimental condition is also plotted as a reference [254]. The NIR photon energy of the YAGG:Er–Ce–Cr sample for the NIR-III window at 10 min after ceasing the blue light excitation (8.33 × 10^{17} cps/Sr/m^2) is over two times higher than that of the widely used deep-red persistent phosphor, ZGO:Cr (3.30 × 10^{17} cps/Sr/m^2) for the NIR-I window, which indicates that this persistent phosphor exhibits excellent PersL in NIR region by Er^{3+} as well as visible light region by Ce^{3+}. In Figure 11.14(d), they exhibit the first PersL imaging by a commercial InGaAs camera [103] for the two garnet ceramic phosphors. Although the YAGG:Ce–Cr sample shows bright green PersL by a digital camera in the visible light region (see Figure 11.14(c)), no signal can be captured by the InGaAs camera due to its lack of NIR PersL. Since the InGaAs camera is only sensitive to the luminescence located in the SWIR region, the PersL imaging from Er^{3+} in the YAGG:Er–Ce–Cr sample was nicely recorded even using a very short integrating time (0.04 s, the maximum value to avoid the saturation of Er^{3+} PersL intensity at 1 min after ceasing the blue light excitation). This result clearly proves that Er^{3+} PersL is intense and long enough to be well recorded by a commercial InGaAs camera. Besides Ce^{3+} as energy donor, Cr^{3+} can also be acted as energy donor for Er^{3+} PersL as presented in Figure 11.14(f)–(i), in which not only Cr^{3+} but also Er^{3+} PersL can be well detected over 10 h and imaged by a NIR camera. Similar electron trapping-detrapping mechanism and pesistent ET processes using VRBE diagrams composed of activator/trapping center energy levels and band structures in YAGG host are presented in Figure 11.14(e) and (j).

Besides these dopant manipulating strategies toward Er^{3+} PersL in garnets, similar persistent ET concept in Er^{3+}, Cr^{3+}, Sm^{3+} tri-doped LaAlO$_3$ perovskite (LAO:Er–Cr–Sm) was also reported by J. Xu et al., which exhibits long PersL at 1553 nm due to the Er^{3+}:^{4}I$_{13/2}$ → ^{4}I$_{15/2}$ transition as well as at 734 nm due to the Cr^{3+}:^{2}E (^{2}G) → ^{4}A$_2$ (^{4}F) transition [104]. The intense NIR PersL bands from both Cr^{3+} and Er^{3+} can be monitored for over 10 h and match well with the NIR-I and NIR-III windows as well as high response regions of commercial Si and InGaAs detectors. Additionally, *ex vivo* optical imaging of Cr^{3+} and Er^{3+} PersL in the LAO:Er–Cr–Sm bulk material was also carried out, in which raw pork tissues placed above the ceramic pellet were used as a practical model of optically dense and highly scattering media. The optical signals of PersL through pork tissues were separately recorded by commercial Si and InGaAs cameras shown in Figure 11.14(k) and (l), respectively. In Figure 11.14(k), intense Cr^{3+} PersL is clearly captured by the Si CCD camera both in the LAO:Cr–Sm and LAO:Er–Cr–Sm pellet samples even at 60 min after ceasing UV (254 nm) light charging. Thanks to the advantage of excitation-free condition given by this imaging technique, nearly autofluorescence-free signals from pork tissues are observed compared with that in the conventional fluorescence imaging by laser excitation [2]. However, the detailed structure and signal localization in tissues cannot be distinguished due to the strong scattering of such short-wavelength light in the NIR-I window. Although the Cr^{3+} PersL of the LAO:Er–Cr–Sm pellet sample is weaker than that of the LAO:Cr–Sm due to the persistent ET process from Cr^{3+} to Er^{3+}, its bright Er^{3+} PersL can be well recorded by the InGaAs camera shown in Figure 11.14(l). Furthermore, because of the greatly reduced scattering of Er^{3+} NIR light above 1500 nm falling into the NIR-III window, the Er^{3+} emission with longer wavelength can achieve higher spatial resolution and better optical contrast compared with the Cr^{3+} emission in the NIR-I window.

As one can notice in Figure 11.14(j), two independent electron traps (*i.e.*, the shallow trap [Trap-I] and the photochromic-center-related deep trap [Trap-II]) are presented in YAGG:Cr–Er, which indicates its potential PSPL capability by low-energy photostimulation light. Thus, to clarify the feasibility of the mentioned "trap redistribution" principle in Section 11.1.3.3, TL glow curves monitoring Cr^{3+} emission in the YAGG:Er–Cr sample performed under different conditions are given in Figure 11.15(a). The TL glow curve of the same sample recorded after UV charging at 100K (black-dotted line) is also plotted as a reference. Compared with the TL glow curve (curve (i)) after UV charging at 100K, the glow curve obtained after UV charging at 300K (curve (ii)) gives a different shape in which the low-temperature part, below 300K, belonging to Trap-I almost disappeared because of the immediate electron detrapping during the charging process. When extending the waiting time to 2 h at 300K (curve (iii)), the Trap-I related glow curve disappeared as predicted in the PSPL mechanism (Figure 11.10(c)) which proves that most of the trapped electrons in Trap-I are cleaned up by thermal activation energy at *RT* while those are remained in Trap-II. On the other hand, when a 660 nm LED lamp (~0.14 W/cm²) is used as a photostimulation light source at 100K (curve (iv)), it can be clearly seen that, the TL glow curve at lower temperature belonging to Trap-I rises again while that at higher temperature belonging to Trap-II decreases, which confirms the redistribution of trapped electrons from Trap-II to Trap-I after photostimulation. The shift on the TL glow peaks can be explained by different charging temperatures and equilibrium between trap levels [255]. It is also worth noting that since trap depths in the energy-level diagram are estimated from the TL results, the phonon-assisted thermal energy required to reach upper levels should be smaller than the required optical energy [256]. To obtain intense photostimulation induced PersL through "trap redistribution" in YAGG, higher energy photons at 660 nm, for instance, are therefore desirable and much more efficient to liberate the trapped electrons from Trap-II to CB than lower energy ones such as 808- and 977 nm lasers, or the 850 nm LED lamp demonstrated as a NIR photostimulation light source in curve (v).

The rechargeable ability of the YAGG:Er–Cr sample by the 660 nm red LED is further investigated by the photostimulation induced PersL decay curves (red line) separately monitoring the Cr^{3+} and Er^{3+} emission shown in Figure 11.15(b) and (c), in which the corresponding normal decay (black line) without photostimulation is also plotted as a reference. After ceasing UV illumination for 5 min and following 20-min natural decay, the YAGG:Er–Cr sample is then triggered by the red LED for 1 min with 20-min interval for four cycles. Although it is very difficult to totally avoid the intensity saturation during photostimulation ("on" mode) from the normal Cr^{3+} or Er^{3+} PL by absorbing the excitation light from LED, the enhancement of both Cr^{3+} and Er^{3+} PersL can be clearly observed after ceasing the LED photostimulation light ("off" mode) due to the contribution of repopulated shallow Trap-I through trap redistribution from Trap-II. On the contrary, when the same sample is photostimulated by the 850 nm NIR LED as presented in Figure 11.15(d) and (e), the enhancement of Cr^{3+}/Er^{3+} PersL intensity is rather smaller than that by the 660 nm red LED (emphasized by the blue zone), which further proves that relatively high energy photons are desirable to give rise to efficient trap redistribution by overcoming the large energy barrier between Trap-II and CB.

Optical imagings through pork tissues with 1-cm thickness by the *in situ* 660 nm photostimulation under the similar experimental condition as presented in Figure 11.15(b) and (c) are also given in Figure 11.15(f) and (g) monitored by commercial Si and InGaAs cameras, respectively. The ceramic sample is firstly charged by 254 nm light for 10 min before covering by the raw-pork tissue, then after natural decay for 20 min (first cycle), it is *in situ* photostimulated by 660 nm LED for 1 min followed by another 20-min natural decay (second cycle), *etc.* Because of the enhanced PersL by photostimulation, both Cr^{3+} and Er^{3+} emission can be monitored over 104 min and the SNR of Cr^{3+} PersL at 104 min after ceasing the initial UV excitation can reach ~31 thanks to this unique autofluorescence-free imaging technology. Moreover, compared with the Cr^{3+} emission, the Er^{3+} emission can achieve higher spatial resolution and more accurate signal localizations owing to the reduced light scattering at longer wavelengths, as expected from the light-matter interaction theory. This proof-of-concept using the "photostimulation-induced trap redistribution" approach gives a

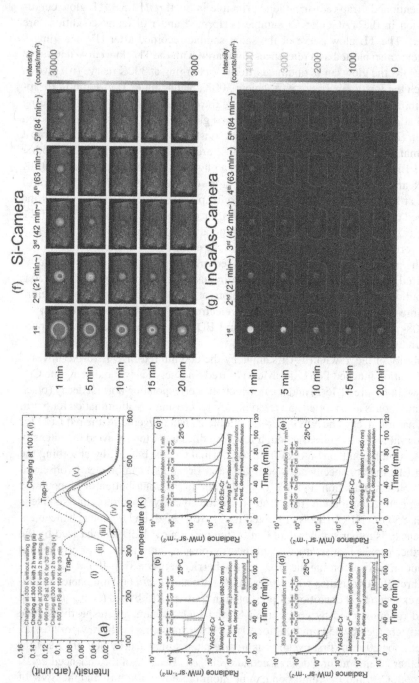

FIGURE 11.15 (a) Thermoluminescence (TL) glow curves monitoring Cr^{3+} emission of the YAGG:Er-Cr sample after various charging by different wavelengths at different temperatures: (i) after UV charging at 100K (black-dotted line); (ii) after UV charging at 300K without waiting (red line) down to 100K; (iii) after UV charging at 300K with 2-h waiting (blue line) down to 100K and photostimulation by (iv) 660 nm LED (power density: ~0.14 W/cm², FWHM: 20 nm) for 30 min at 100K (green line) (v) 850 nm LED (power density: ~0.14 W/cm², FWHM: 48 nm) for 30 min at 100K (magenta line); persistent luminescence decay curves of the YAGG:Er-Cr ceramic sample parts (b), (d) monitoring Cr^{3+} emission (c), (e) monitoring Er^{3+} emission (black line: normal decay; red line: with photostimulation by 660 nm/850 nm LED for 1 min every 20 min); optical imaging through raw-pork tissues with 1-cm thickness of the YAGG:Er-Cr ceramic sample after 254 nm (6 W output) illumination for 10 min (f) Cr^{3+} emission monitored by the Si CCD camera (integrating time: 1 s; the tail in the 1-min image is due to the intensity saturation of the camera); (g) Er^{3+} emission monitored by the InGaAs camera (integrating time: 5 s). The ceramic sample is *in situ* photostimulated by 660 nm LED for 1 min every 20 min after the initial 20-min natural decay (reprinted with permission from Ref. [102], copyright 2018, American Chemical Society).

vivid example of PersL recovery *via* post-charging of low-energy photostimulation light, which is particularly important to acquire long-term PersL for *in vivo* optical imaging.

11.2.2 Transition Metal Ions Activated

For the transition metal ions-activated system, the widely used Cr^{3+} ion in the NIR-I BW becomes less popular in the NIR-II/III BWs, since its main emission band can be hardly shifted over 1000 nm. Besides Cr^{3+}, thanks to the broadband emission in the NIR-II/III region, other transition metal ions, *e.g.*, Ni^{2+} and Co^{2+}, have become better choices than Cr^{3+} for acquiring NIR PersL over 1000 nm in recent years.

In 2016, Z. Pan *et al.* firstly reported a series of NIR-II/III persistent phosphors by doping Ni^{2+} ions into the gallate-based hosts, including $Zn_3Ga_2Ge_2O_{10}$ (ZGGO), $LiGa_5O_8$, and $La_3Ga_5GeO_{14}$ [214]. All these NIR-II/III persistent phosphors exhibited a single-band PersL in the range from 1000 to 1600 nm, which can be attributed to the Ni^{2+}:$^3T_2$ (^3F) $\rightarrow$ $^3A_2$ (^3F) transition. Long-term duration (>12 h) was observed in the ZGGO:Ni^{2+} persistent phosphor as shown in Figure 11.16(a), and was also confirmed by B. Viana *et al.*, in the similar ZGO and ZGGO hosts (see Figure 11.16(b)) [215]. In addition, inspired by Pan's work, J. Qiu *et al.* demonstrated another example of Ni^{2+}-activated NIR-II/III persistent phosphors by utilizing the ZGO as an effective host again. Through the strategy of element substitution (Sn^{4+} and Zn^{2+} for Ga^{3+}) to control the local crystal-field surrounding activators, tunable emission bands of Ni^{2+} peaking from 1270 to 1430 nm in the NIR-II window had been successfully achieved [233]. Similar crystal-field-driven strategy for tuning Ni^{2+} PersL band was also adopted by Y. Hu *et al.* in the YAGG garnet host. The emission band of SWIR PersL of Ni^{2+} can be tuned from 1357 to 1424 nm [227], and PersL behaviors can also be varied in different band structures *via* the simple adjustment of Ga^{3+}/Al^{3+} ratio, referred to the so-called crystal-field engineering [193] and bandgap engineering [166, 191, 203, 205], respectively.

Benefiting from the typical background-free feature, persistent phosphors have shown great potential in the field of anti-counterfeiting technology. However, the current anti-counterfeiting technology based on visible-light faces the inherent disadvantages of low security and inconvenience. Very recently, Z. Li *et al.* demonstrated the nano-sized $Zn_2Ga_2Sn_{0.5}O_6$:Ni^{2+}, Yb^{3+} (ZGSO:Ni–Yb) with intense PersL in the NIR-II region to address this challenge, in which Ni^{2+} ions act as the emitter of NIR-II PersL at around 1350 nm, and Yb^{3+} ions act as co-doping sensitizers to improve the emission intensity [234]. To further emphasize the advantages of NIR-II PersL, "U" patterns composed of as-prepared ZGSO:Yb–Ni phosphors and commercial phosphors with visible PersL were evaluated. As shown in Figure 11.16(c), the visible-light PersL of commercial phosphors can be observed by eyes for at least 15 s after excitation was switched off. In contrast, the pattern made from ZGSO:Yb–Ni phosphors is invisible for naked eyes, while it can be clearly detected by the SWIR detector, which ensures the security of anti-counterfeiting patterns. Additionally, in the proof-of-concept experiment, by combining the ZGSO:Ni–Yb NIR-II PersL and visible PersL from commercial phosphors, multichannel light anti-fake effect can be achieved under different excitation sources or different time gating, which showed higher concealment, flexibility, and SNR compared with current applied visible-light phosphors (see Figure 11.16(d)–(g)). The designed numbers are used as anti-counterfeiting marker, by choosing different excitation sources (Figure 11.16(e)), multichannel light anti-fake effect (Figure 11.16(f)) can be achieved. For example, under 808 nm laser excitation, the marker exhibits both NIR-II PersL and fluorescence, and a sign of "888" was observed. Upon excitation with 254 nm UV light instead of 808 nm laser for 5 min, only PersL was maintained, and the sign of "123" was clearly observed, which presented the potential of NIR-II PersL in anti-counterfeiting applications. Besides, thanks to the broadband $^4T_2$ (^4F) $\rightarrow$ $^4A_2$ (^4F) transition in the NIR-II/III region, Co^{2+} is also a potential activator to obtain PersL over 1000 nm. In the famous ZGO spinel host, the PersL properties of $ZnGa_2O_4$:Co^{2+} were reported by B. Viana *et al.* for the first time [215]. The Co^{2+} PersL in bulk form of $ZnGa_2O_4$ can last for longer than 30 min, whereas the NIR PersL signal of relevant NPs was too weak to be detected by the SWIR detector.

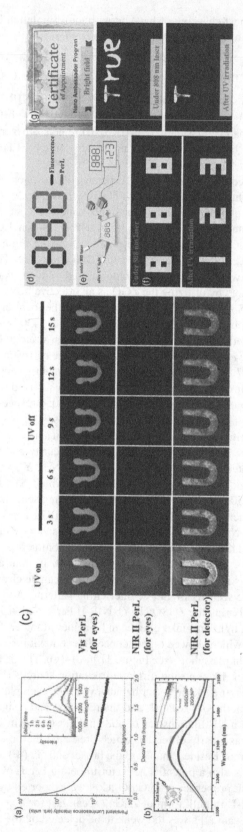

FIGURE 11.16 (a) PersL decay curve monitored at 1290 nm after irradiation by 300 nm light for 20 min, the inset shows the PersL emission spectra recorded at 1, 2, 6, and 12 h after the stoppage of the irradiation (reprinted with permission from Ref. [214], copyright 2016, Wiley); (b) optical features and persistent luminescence of $ZnGa_2O_4:Ni$ and $Zn(GeGa)_2O_4:Ni$: decay profiles and scheme of excitation (reprinted with permission from Ref. [215], copyright 2018, SPIE); (c) "U" patterns composed of commercial visible persistent phosphors and ZGSO:Yb–Ni phosphors with the NIR-II persistent luminescence; (d) and (e) schematic anti-counterfeiting marker fabricated from phosphors with photoluminescence and persistent luminescence; (f) different images of markers recorded with photoluminescence under irradiation of 808 nm laser (3.82 mW/cm²) and persistent luminescence after 254 nm UV light radiation for 5 min, respectively; (g) anti-counterfeiting mark of a certificate (reprinted with permission from Ref. [234], copyright 2020, American Chemical Society).

Though the reported works concerning transition metal ions-activated persistent phosphors showing long-term NIR PersL over 1000 nm are still limited, the flexibility of emission band tuning *via* manipulating local crystal fields definitely provides more possibilities from *d*–*d* transitions of transition metal ions that able to acquire much broader emission band covering wider range of BWs compared with sharp 4*f*–4*f* transitions of lanthanide ions.

11.2.3   New Trap Filling Concepts

In most cases, the occurrence of PersL requires high excitation energy, usually UV light, to facilitate the delocalization process of charge carriers through electronic bands (*e.g.*, excited electron to CB or excited hole to VB) toward trap levels or directly toward trap levels through the quantum tunneling process, which is higher than the energy of emission light in visible or NIR regions. Yet, the requirement of high-energy UV excitation restricts the practical use of PersL, especially for *in vivo* optical imaging, UV light is not only absorbed and scattered by water and biological tissues limiting the penetration depth, it also brings the risk of (i) phototoxicity on biological tissue and (ii) tissue autofluorescence. Therefore, it needs the injected PersL nano-probes to be re-excitable both *in situ* and *in vivo* by high-tissue-penetration and low-energy light such as red or even NIR one. To make persistent phosphors re-excitable or rechargeable by low-energy light sources, above-mentioned PSPL or optical-stimulation-induced PersLs are commonly used [90, 102], in which enhanced NIR PersL can be repeatedly obtained in a UV pre-charged phosphor *via* low-energy photostimulation light within short illumination time. The principle of this method is to fully utilize the stored energy, particularly in the deep trap that is difficult to be totally released at *RT*. However, this method is also restricted by the limited amount of trapped energy during UV pre-charging, and it is more likely to using stored energy rather than receiving extra energy from external excitation. Recently, Z. Pan *et al.* [246–248] have proposed three new charging concepts shown in Figure 11.17, to realize the "real" visible/NIR light charging for persistent phosphors as presented in step 4 of Figure 11.9.

The first concept was named up-converted PersL (UCPL) (see Figure 11.17(a)), inspired by the well-known multiphoton UCL [37]. The concept itself is rather simple: the low-energy incident photons can promote the up-conversion ion system (*e.g.*, Er^{3+}–Yb^{3+}) from the ground state to the high-energy delocalized state *via* an up-conversion excitation channel, followed by filling of traps. When the stored excitation energy is gradually released after removal of excitation sources, NIR PersL with higher energy than the excitation light can be generated. A typical example goes to the famous Cr^{3+}-doped $Zn_3Ga_2GeO_8$ (ZGGO) compound added with the Er^{3+}–Yb^{3+} up-conversion ion pair. When the material is excited by NIR light, *e.g.*, 980 nm laser, the green up-conversion light from Er^{3+} can provide extra energy to fill the high-energy trap levels, which gives rise to intense Cr^{3+} PersL at ~700 nm. The combination of UCL and LPL to be UCPL [246] can totally overcome the drawbacks in the biological field from both the UCL (autofluorescence and non-negligible heating effects under real-time NIR laser excitation as well as optical scattering and absorption of generated up-converted short-wavelength light) and the LPL (short-time window less than few hours after initial UV charging) processes, which makes it possible to realize long-term PersL able to receive external NIR excitation for optical imaging.

Furthermore, based on UCPL, they proposed the second charging concept by NIR light utilizing the phonon energy (*i.e.*, lattice vibration energy caused by temperature) of material hosts. Inspired by the excitation feature of the phonon-assisted anti-Stokes PL, phonon energy was used as a part of excitation agent to lower the needed photon excitation energy for generating PersL [247]. As shown in Figure 11.17(b), thermal excitation boosts the luminescence system to an intermediate state (*I*), from where the system is further excited to the delocalized state (*D*) upon excitation with low-energy photons. According to the scheme, the excitation photon energy can be tuned, depending on the temperature of the material so that the excitation wavelength may be longer than the emission wavelength of PersL if the phonon energy is sufficiently high. It is actually a kind of UCPL

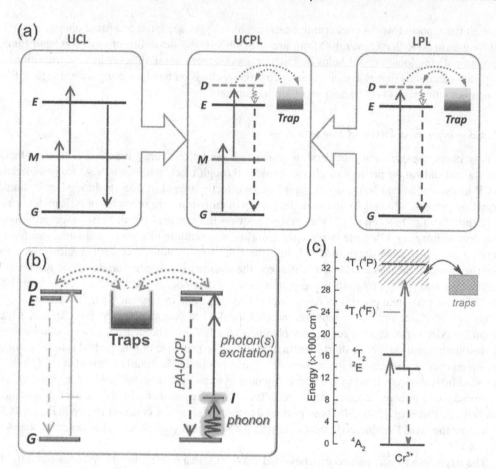

FIGURE 11.17 (a) Schematic diagrams of up-conversion luminescence (UCL), up-converted persistent luminescence (UCPL), and long-persistent luminescence (LPL) with hypothetical energy-level schemes; (b) schematic diagrams of general persistent luminescence (left panel) and phonon-assisted (PA)-UCPL (right panel) with hypothetical energy-level schemes; (c) schematic diagram of the two-photon up-conversion charging (UCC) process in a Cr^{3+}-activated persistent luminescence system under the excitation of a visible-light laser (reproduced with permission from Ref. [57], copyright 2019, Elsevier).

process, thus they named it as phonon-assisted UCPL (*i.e.*, PA-UCPL). The feasibility of this new charging concept was demonstrated again in the same ZGGO:Cr^{3+} compound, while at this time, the Er^{3+}–Yb^{3+} up-conversion pair is no longer needed. As a consequence, Cr^{3+} PersL at ~700 nm in $Zn_3Ga_2GeO_8$ can be achieved through the PA-UCPL approach by 800- or 980 nm laser at ~70°C. Although the simultaneous detrapping/depopulation process of trapped electrons by thermal-stimulation at high temperatures and photostimulation by NIR laser should be carefully evaluated, the power-dependent trap filling behavior by either the one-photon excitation process or the two-photon excitation process gives new possibilities to controllably manipulate the charging process of PersL, especially by low-energy NIR light.

The UCPL and PA-UCPL concepts were further extended to use high-intensity visible-light laser (*e.g.*, 450 or 532 nm) as an efficient charging source, which is named two-photon up-conversion charging (UCC) process (see Figure 11.17(c)) [248]. Cr^{3+}-doped compound ($LiGa_5O_8$:Cr^{3+}) was selected as an example to demonstrate this concept by utilizing long lifetime of Cr^{3+} excited state through excited-state absorption (ESA), instead of up-conversion ion pairs in UCPL or high operation temperatures in PA-UCPL. In the UCC concept, the first absorption of a visible-light photon excites the system to the $^4T_2$ state (*i.e.*, the intermediate state) of Cr^{3+} ions, followed by a fast non-radiative relaxation to the ^2E excited state, which can be regarded as a metastable state because of its long lifetime (usually >1 ms). The high pumping intensity of laser enables the population of the metastable state to be high enough to produce substantial ESA, by absorbing another visible-light photon to further pump the system to the high-energy $^4T_1$ (^4P) state. Subsequently, the electrons will delocalize with a significant probability from the $^4T_1$ (^4P) state to feed the high-energy electron traps. This UCC concept was confirmed not only in transition metal ions (*e.g.*, Cr^{3+} or Mn^{2+}) that usually possess spin-forbidden transitions with long lifetime, but also in rare-earth ions (*e.g.*, Pr^{3+}) with similar metastable excited states. Although strong light scattering and absorption of visible-light laser should be carefully considered when applied into *in vivo* imaging, the UCC process appears to be a common phenomenon in persistent phosphors containing UCC-enabling activators. Thus, the reported persistent phosphors that were previously well-studied using UV excitation can be revisited using visible-light laser excitation, which may give new insights into their luminescent properties as well as potential new applications.

These three charging concepts proposed by Z. Pan *et al.* are a kind of special UCL processes, using traditional UCL ion pairs, phonon energy of host materials or long lifetime of metastable excited states. In particular, with the knowledge of persistent ET process, one can expect the recovery of SWIR PersL by simply utilizing the transferred energy from the reactivated visible light (*e.g.*, Pr^{3+}) or NIR-I light (*e.g.*, Cr^{3+} or Mn^{2+}) *via* these post-charging approaches. The results of persistent phosphors chargeable by NIR light definitely provide exciting solutions to the intractable problem of optical imaging using PersL nano-probes and already confirmed to be applicable for some practical bioimaging experiments [257, 258]. Although in the current stage, charging efficiency by NIR light cannot be comparable to that by conventional high-energy UV light, the connection or combination of PersL with other luminescence phenomena (*e.g.*, UCL or phonon-assisted PL) gives new inspiration for the future development and understanding of trapping-detrapping mechanism in PersL and suggests rather convenient recovery of PersL, whenever required.

11.3   SUMMARY AND PERSPECTIVE

Considering the fast development of cutting-edge optical imaging operating at NIR-II/III BWs, we summarized the state-of-the-art of NIR phosphors with PersL over 1000 nm, starting from the introduction of basic principles of light-matter interactions, followed by the description of NIR autofluorescence phenomena of bio-tissues and related filtering approaches. The mechanisms of PersL, including charge carrier trapping-detrapping phenomena, persistent ET process, and photo-stimulation-induced trap redistribution approaches, are also introduced, to give a general concept of how PersL works and how we can play with the information of energy-level locations of electronic bands, activators, and traps. Vivid examples of lanthanide/transition metal ions-activated NIR-II/III persistent phosphors are presented together with new post-charging techniques able to recover PersL *in vivo*. Thus, the importance, benefits, and urgency of shifting PersL wavelength from the visible light/NIR-I window to the NIR-II/III windows have been approved toward the new-generation autofluorescence-free and real-time excitation-free optical imaging. Prospectively, although substantial and rapid progresses have been made on this subject, further improvements are definitely demanded, and future works include but not limited to these aspects:

i. Standard physical quantities. Fair comparison of different persistent phosphors prepared by different labs/persons needs universal physical quantities to evaluate the real performance of their PersL properties (*i.e.*, emission intensity and persistent decay), which is extremely important for screening potential candidates, especially in the bulk form before pushing them into nanoscale. It is well known that the measured "performances" of persistent phosphors are directly correlated with the real performance of applied detectors, which is even more critical when monitoring weak NIR PersL imaging from PLNPs by

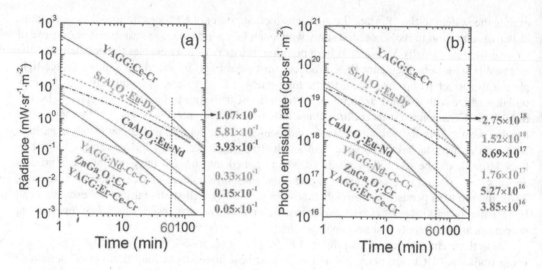

FIGURE 11.18 Persistent luminescent decay curves of various persistent phosphors plotted in unit of (a) radiance (mW/(sr·m²)) and (b) photon emission rate (cps/Sr/m²) (monitoring wavelengths for both Nd³⁺ and Er³⁺ in YAGG are over 1000 nm) (reproduced with permission from Ref. [57], copyright 2019, Elsevier).

various CCD/CMOS cameras with different sensitivities (timescale of detection is directly correlated with the parameters of applied imaging cameras such as response sensitivity, quantum efficiency at certain wavelengths, cooling system for CMOS or CCD arrays, pixel correction, and exposure time, *etc.*). Similar to luminance (mcd/m²) that takes into account the sensitivity of human eyes, radiance (mW/Sr/m²) that does not consider human visual perception can be adopted in the evaluation of deep red to NIR PersL. Unfortunately, very limited reports on NIR-persistent phosphors provide the radiance value as a function of decay time after ceasing excitation sources, which makes such comparison rather difficult. Besides, due to the nature of eye insensitivity, the standard minimum value or "minimum visibility" of radiance becomes not as clear as that of luminance (0.32 mcd/m²), while the decay time still can be roughly defined as the time when radiance has decayed to a certain level, such as 1×10^{-3} mW/Sr/m² (this value corresponds to the luminance of 0.32 mcd/m² for the green emission of the $SrAl_2O_4$:Eu^{2+}–Dy^{3+} phosphor, LumiNova® GLL-300FFS) [59]. Then, the duration upon 1×10^{-3} mW/Sr/m² after removal of excitation corresponds to the persistent decay time, either in a time order of minutes or even hours. In addition to radiance, when we take into account the photon response of different detectors working for different spectral regions, the number of emitted photons from persistent phosphors in a certain time (*i.e.*, photon emission rate, in unit of cps/Sr/m²) is therefore directly proportional to the "received" photons for detectors. In such a case, instead of radiance, photon emission rate using the absolute radiance divided by photon energy (photon energy: $E = hc/\lambda$, where h is the Planck constant, c is the speed of light in vacuum, and λ is the wavelength) becomes a more suitable physical quantity than radiance [57, 103], especially when multiple NIR emitting centers with totally different centroid spectral wavelengths are forced for comparison (*e.g.*, Ce^{3+} *vs.* Er^{3+} [103], Ce^{3+} *vs.* Nd^{3+} [219], or Cr^{3+} *vs.* Er^{3+} [102, 104]). For example, in Figure 11.18, compared to the persistent radiance of Cr^{3+} PersL in $ZnGa_2O_4$ for the NIR-I window, that of Er^{3+} PersL in YAGG for the NIR-III window is around one third at 60 min after ceasing excitation. However, when the persistent luminescent decay is plotted in unit of photon emission rate instead of radiance, the intensity of Er^{3+} PersL is comparable to that of Cr^{3+} PersL due to the lower photon energy of Er^{3+} emission than that of Cr^{3+} emission.

ii. Defect identification and optimization. From Table 11.2, it is easy to conclude that large amount of reported NIR-persistent phosphors is defect dominating while the exact defect species are still unknown. These unknown defects are involved in the compound acting as electron and/or hole trapping centers, or even persistent energy donor to generate NIR (over 1000 nm) PersL. For instance, PersLs of Er^{3+}, Yb^{3+}, Pr^{3+}, Nd^{3+}, Ho^{3+}, and Tm^{3+} ranging from ~1000 to 1533 nm lasting for at least 20 min have been observed in Ca_2SnO_4, in which intrinsic defects play vital roles as electron trapping centers to induce the photoionization process [243]. However, except for some compounds (*e.g.*, Dy^{3+} in $SrAl_2O_4$:Eu^{2+}–Er^{3+} [101], Dy^{3+} in $La_3Ga_5GeO_{14}$:Cr^{3+} [211], Cr^{3+} in $Y_3Al_2Ga_3O_{12}$:Ce^{3+}–Er^{3+} [103], or Sm^{3+} in $LaAlO_3$:Cr^{3+}–Er^{3+} and $LaAlO_3$:Cr^{3+}–Ho^{3+} [222]), the uncertainty of various defects limits the deep understanding of the PersL mechanism, which further precludes the reproducibility of these materials once they are synthesized into nanoscale. Because the so-called intrinsic/lattice defects become even more complicated in the form of NPs, then the identification as well as optimization of those useful but "tricky" trapping defects becomes quite necessary. Alternatively, sensitization of NIR PersL by lanthanide-ion or transition metal-ion co-doping strategies *via* the energy-level diagram of certain host material (previously shown in Section 11.2.1.2) gives a powerful prediction tool to further prolong the duration of NIR PersL, which can be a promising approach just as the same concept previously used for the enhancement of visible-light persistent phosphors.

iii. Evaluation of NIR PersL in nanoscale. When the luminescent material is shifted from bulk form into NPs for the practical bio-applications, weaker emission intensity will be obtained due to the lower crystallization, limited dispersion in biological environment without surface functionalization and more surface defects as luminescence "killer" centers in nanoscale. Such optical attenuation will be even severe in PersL, especially in the case of co-doping systems using persistent ET processes, the distance-dependent interactions between donor and acceptor ions (*e.g.*, dipole-dipole, dipole-quadrupole, or quadrupole-quadrupole interactions [259]) become much more complicated compared with that in the bulk materials. All of these complexities should be carefully considered, and some surface modification or core-shell structures can be definitely effective methods to improve the PersL properties of NIR-persistent phosphors, just as the similar concept used for up-conversion NPs. Moreover, there is still a lack of lanthanide-ion-doped and/or transition metal-ion-doped PLNPs in clinical trials. In fact, despite the tremendous effort invested in nanomedicine, only a few material systems made the transition from the research laboratory to the clinic (trial or approved status). Most of them are organic compounds, often nanostructures based on polymers such as PEG or liposomes, while there are only a few inorganic NPs listed, namely, iron oxide, NBTXR3 (hafnium oxide), gold colloids, and Cornell dots (silica NPs) [43]. It is important to identify drawbacks and challenges that hinder multifunctional nanostructures entering the clinical trial status, then to eventually bring the most promising candidates closer to application. Possibility, one of the reasons is the remaining toxicity concerns that must be dispelled through more rigorous *in vivo* toxicology, chemical stability, physiological integrity, and clearance studies to confirm their safety. Recent results by C. Richard *et al.* [260] and H. Zhang *et al.* [261] suggested that at least for the $ZnGa_2O_4$:Cr^{3+} PLNPs working in the NIR-I window, no significant toxicities were observed in mice for a relatively long-term investigation (*i.e.*, 6 months and 60 days, respectively). Although the lack of obvious toxicity both *in vitro* and *in vivo* based on these reports encourages future development of NIR PLNPs for biomedical research, further toxicology studies are definitely necessary to make sure of their biocompatibilities for living bodies.

iv. Improvement of new-type detectors. Together with the fast development of NIR-emitting optical probes, *in vivo* NIR imaging is also simultaneously stimulated by the advent of new, more advanced imaging cameras with higher sensitivity and broader spectral ranges within or even beyond the current NIR BWs [9]. Besides the InGaAs-based semiconducting sensor commonly used for sensitive photodetection in the NIR-II/III windows, other semiconductor-based cameras such as InSb (bandgap ~0.17 eV) or HgCdTe (bandgap tunable between 0 and 1.5 eV) [262, 263] may offer new possibilities to NIR PersL imaging that are otherwise impossible with current InGaAs cameras, once their sensitivities can be greatly enhanced. For example, the development of sensitive HgCdTe focal okabe array (FPA) cameras with spectral responsivity covering a large NIR range (from 850 nm to 2.5 μm) may allow for *in vivo* imaging with even deeper penetration and depth-resolved 3D reconstruction on the basis of multicolor imaging [9]. Additionally, InSb-based photodetector arrays with up to a 1-kHz full-window frame-rate in the 1.5- to 5.0-μm spectral range may allow the photodetection in the so-called NIR-IV window (~2100–2300 nm, centered at 2200 nm) with even lower optical scattering coefficient than other three BWs [105–107]. Some emerging detectors, for example, those based on two "fever" materials: graphene [264] and organic-inorganic hybrid perovskites [265] have also been reported for the NIR spectral region, which may give more choices in the future. Note that, compared with conventional NIR fluorescence imaging, the performances of available commercial InGaAs cameras, at least at the current stage, are far below the requirements for monitoring weak PersL without real-time excitation. Much more powerful InGaAs cameras are certainly essential for this subject, especially with higher quantum efficiency, we can significantly decrease the exposure time for acquiring optical signals, from the time order of seconds or even minutes (commonly used for PersL imaging) to that of microseconds (commonly used for fluorescence imaging) [2, 3], which will be a highly expected advance for the real-time PersL imaging monitoring physiological processes with fast dynamics in living bodies.

REFERENCES

1. Weissleder R, and Pittet MJ (2008) Imaging in the era of molecular oncology. *Nature* 452: 580.
2. Jaque D, Richard C, Viana B, Soga K, Liu X, and Solé JG (2016) Inorganic nanoparticles for optical bioimaging. *Adv. Opt. Photonics* 8: 1.
3. Benayas A, Hemmer E, Hong G, and Jaque D (2020) *Near infrared-emitting nanoparticles for biomedical applications.* Springer.
4. Smith BR, and Gambhir SS (2017) Nanomaterials for in vivo imaging. *Chem. Rev.* 117: 901.
5. Brenner DJ, and Hall EJ (2007) Computed tomography – an increasing source of radiation exposure. *N. Engl. J. Med.* 357: 2277.
6. Kobayashi H, Ogawa M, Alford R, Choyke PL, and Urano Y (2010) New strategies for fluorescent probe design in medical diagnostic imaging. *Chem. Rev.* 110: 2620.
7. Yao J, Yang M, and Duan Y (2014) Chemistry, biology, and medicine of fluorescent nanomaterials and related systems: new insights into biosensing, bioimaging, genomics, diagnostics, and therapy. *Chem. Rev.* 114: 6130.
8. Martinić I, Eliseeva SV, and Petoud S (2017) Near-infrared emitting probes for biological imaging: organic fluorophores, quantum dots, fluorescent proteins, lanthanide(III) complexes and nanomaterials. *J. Lumin.* 189: 19.
9. Hong G, Antaris A, and Dai H (2017) Near-infrared fluorophores for biomedical imaging. *Nat. Biomed. Eng.* 1: 0010.
10. Li C, Chen G, Zhang Y, Wu F, and Wang Q (2020) Advanced fluorescence imaging technology in the near-infrared-II window for biomedical applications. *J. Am. Chem. Soc.* 142: 14789.
11. Luo S, Zhang E, Su Y, Cheng T, and Shi C (2011) A review of NIR dyes in cancer targeting and imaging. *Biomaterials* 32: 7127.
12. Montalti M, Prodi L, Rampazzo E, and Zaccheroni N (2014) Dye-doped silica nanoparticles as luminescent organized systems for nanomedicine. *Chem. Soc. Rev.* 43: 4243.
13. Shaner NC, Steinbach PA, and Tsien RY (2005) A guide to choosing fluorescent proteins. *Nat. Methods* 2: 905.

14. Giepmans BNG, Adams SR, Ellisman MH, and Tsien RY (2006) The fluorescent toolbox for assessing protein location and function. *Science* 312: 217.

15. Iijima S (1991) Helical microtubules of graphitic carbon. *Nature* 354: 56.

16. Welsher K, Liu Z, Sherlock SP, Robinson JT, Chen Z, Daranciang D, and Dai H (2009) A route to brightly fluorescent carbon nanotubes for near-infrared imaging in mice. *Nat. Nanotechnol.* 4: 773.

17. Welsher K, Sherlock SP, and Dai H (2011) Deep-tissue anatomical imaging of mice using carbon nanotube fluorophores in the second near-infrared window. *Proc. Natl. Acad. Sci. U.S.A.* 108: 8943.

18. Hong G, Lee JC, Robinson JT, Raaz U, Xie LM, Huang NF, Cooke JP, and Dai H (2012) Multifunctional in vivo vascular imaging using near-infrared II fluorescence. *Nat. Med.* 18: 1841.

19. Hong G, Diao S, Chang J, Antaris AL, Chen C, Zhang B, Zhao S, Atochin DN, Huang PL, Andreasson KI, Kuo CJ, and Dai H (2014) Through-skull fluorescence imaging of the brain in a new near-infrared window. *Nat. Photonics* 8: 723.

20. Hong G, Diao S, Antaris AL, and Dai H (2015) Carbon nanomaterials for biological imaging and nanomedicinal therapy. *Chem. Rev.* 115: 10816.

21. Diao S, Blackburn JL, Hong G, Antaris AL, Chang J, Wu JZ, Zhang B, Cheng K, Kuo CJ, and Dai H (2015) Fluorescence imaging in vivo at wavelengths beyond 1500 nm. *Angew. Chem. Int. Ed.* 54: 14758.

22. Bruchez M, Moronne M, Gin P, Weiss S, and Alivisatos AP (1998) Semiconductor nanocrystals as fluorescent biological labels. *Science* 281: 2013.

23. Chan WCW, and Nie S (1998) Quantum dot bioconjugates for ultrasensitive nonisotopic detection. *Science* 281: 2016.

24. Gao X, Cui Y, Levenson RM, Chung LWK, and Nie S (2004) In vivo cancer targeting and imaging with semiconductor quantum dots. *Nat. Biotechnol.* 22: 969.

25. Michalet X, Pinaud FF, Bentolila LA, Tsay JM, Doose S, Li JJ, Sundaresan G, Wu AM, Gambhir SS, and Weiss S (2005) Quantum dots for live cells, in vivo imaging, and diagnostics. *Science* 307: 538.

26. Zhang Y, Hong G, Zhang Y, Chen G, Li F, Dai H, and Wang Q (2012) Ag_2S quantum dot: a bright and biocompatible fluorescent nanoprobe in the second near-infrared window. *ACS Nano* 6: 3695.

27. Zhang M, Yue J, Cui R, Ma Z, Wan H, Wang F, Zhu S, Zhou Y, Kuang Y, Zhong Y, Pang DW, and Dai H (2018) Bright quantum dots emitting at ~1,600 nm in the NIR-IIb window for deep tissue fluorescence imaging. *Proc. Natl. Acad. Sci. U.S.A.* 115: 6590.

28. Bünzli JCG, and Pecharsky VK (2015) *Handbook on the physics and chemistry of rare earths.* Elsevier.

29. Bünzli JCG, and Piguet C (2005) Taking advantage of luminescent lanthanide ions. *Chem. Soc. Rev.* 34: 1048.

30. Bünzli JCG (2010) Lanthanide luminescence for biomedical analyses and imaging. *Chem. Rev.* 110: 2729.

31. Bünzli JCG (2016) Lanthanide light for biology and medical diagnosis. *J. Lumin.* 170: 866.

32. Brik MG, and Ma CG (2020) *Theoretical spectroscopy of transition metal and rare earth ions.* Taylor & Francis.

33. del Rosal B, Pérez-Delgado A, Misiak M, *et al.* (2015) Neodymium-doped nanoparticles for infrared fluorescence bioimaging: the role of the host. *J. Appl. Phys.* 118: 143104.

34. del Rosal B, Rocha U, Ximendes EC, Rodríguez EM, Jaque D, and Solé JG (2017) Nd^{3+} ions in nanomedicine: perspectives and applications. *Opt. Mater.* 63: 185.

35. Liu B, Li C, Yang P, Hou Z, and Lin J (2017) 808 nm-light-excited lanthanide-doped nanoparticles: rational design, luminescence control and theranostic applications. *Adv. Mater.* 29: 1605434.

36. Zhong Y, Ma Z, Zhu S, Yue J, Zhang M, Antairs AL, Yuan J, Cui R, Wan H, Zhou Y, Wang W, Huang NF, Luo J, Hu Z, and Dai H (2017) Boosting the down-shifting luminescence of rare-earth nanocrystals for biological imaging beyond 1500 nm. *Nat. Commun.* 8: 1.

37. Auzel F (2004) Upconversion and anti-stokes processes with f and d ions in solids. *Chem. Rev.* 104: 139.

38. Chen X, Liu Y, and Tu D (2014) *Lanthanide-doped luminescent nanomaterials: from fundamentals to bioapplications.* Springer.

39. Wang F, and Liu X (2009) Recent advances in the chemistry of lanthanide-doped upconversion nanocrystals. *Chem. Soc. Rev.* 38: 976.

40. Liu Y, Tu D, Zhu H, and Chen X (2013) Lanthanide-doped luminescent nanoprobes: controlled synthesis, optical spectroscopy, and bioapplications. *Chem. Soc. Rev.* 42: 6924.

41. Dong H, Du SR, Zheng XY, Lyu GM, Sun LD, Li LD, Zhang PZ, Zhang C, and Yan CH (2015) Lanthanide nanoparticles: from design toward bioimaging and therapy. *Chem. Rev.* 115: 10725.

42. Zhou J, Leaño JR, Liu Z, Jin D, Wong KL, Liu RS, and Bünzli JCG (2018) Impact of lanthanide nanomaterials on photonic devices and smart applications. *Small* 14: 1801882.

43. Hemmer E, Acosta-Mora P, Méndez-Ramos J, and Fischer S (2017) Optical nanoprobes for biomedical applications: shining a light on upconverting and near-infrared emitting nanoparticles for imaging, thermal sensing, and photodynamic therapy. *J. Mater. Chem. B.* 5: 4365.

44. Hemmer E, Venkatachalam N, Hyodo H, Hattori A, Ebina Y, Kishimoto H, and Soga K (2013) Upconverting and NIR emitting rare earth based nanostructures for NIR-bioimaging. *Nanoscale* 5: 11339.

45. Hemmer E, Benayas A, Légaré F, and Vetrone F (2016) Exploiting the biological windows: current perspectives on fluorescent bioprobes emitting above 1000 nm. *Nanoscale Horiz.* 1: 168.

46. Liu Y, Lu Y, Yang X, Zheng X, Wen S, Wang F, Vidal X, Zhao J, Liu D, Zhou Z, Ma C, Zhou J, Piper JA, Xi P, and Jin D (2017) Amplified stimulated emission in upconversion nanoparticles for super-resolution nanoscopy. *Nature* 543: 229.

47. Liu L, Wang S, Zhao B, Pei P, Fan Y, Li X, and Zhang F (2018) Er^{3+} sensitized 1530 nm to 1180 nm second near-infrared window upconversion nanocrystals for in vivo biosensing. *Angew. Chem. Int. Ed.* 57: 7518.

48. Fan Y, Wang P, Lu Y, Wang R, Zhou L, Zheng X, Li X, Piper JA, and Zhang F (2018) Lifetime-engineered NIR-II nanoparticles unlock multiplexed in vivo imaging. *Nat. Nanotechnol.* 13: 941.

49. Sharifi S, Behzadi S, Laurent S, Forrest ML, Stroeve P, and Mahmoudi M (2012) Toxicity of nanomaterials. *Chem. Soc. Rev.* 41: 2323.

50. Yong KT, Law WC, Hu R, Ye L, Liu L, Swihart MT, and Prasad PN (2013) Nanotoxicity assessment of quantum dots: from cellular to primate studies. *Chem. Soc. Rev.* 42: 1236.

51. Lifante J, Shen Y, Ximendes E, Rodríguez EM, and Ortgies DH (2020) The role of tissue fluorescence in in vivo optical bioimaging. *J. Appl. Phys.* 128: 171101.

52. del Rosal B, and Benayas A (2018) Strategies to overcome autofluorescence in nanoprobe-driven in vivo fluorescence imaging. *Small Methods* 2: 1800075.

53. del Rosal B, Villa I, Jaque D, and Sanz-Rodríguez F (2016) In vivo autofluorescence in the biological windows: the role of pigmentation. *J. Biophotonics* 9: 1059.

54. Eeckhout K, Smet PF, and Poelman D (2010) Persistent luminescence in Eu^{2+}-doped compounds: a review. *Materials* 3: 2536.

55. Eeckhout K, Poelman D, and Smet PF (2013) Persistent luminescence in non-Eu^{2+}-doped compounds: a review. *Materials* 6: 2789.

56. Poelman D, der Heggen D, Du J, Cosaert E, and Smet PF (2020) Persistent phosphors for the future: fit for the right application. *J. Appl. Phys.* 128: 240903.

57. Xu J, and Tanabe S (2019) Persistent luminescence instead of phosphorescence: history, mechanism. *J. Lumin.* 205: 581.

58. Brito HF, Hölsä J, Laamanen T, Lastusaari M, Malkamäki M, and Rodrigues LCV (2012) Persistent luminescence mechanisms: human imagination at work. *Opt. Mater. Express.* 2: 371.

59. Zhuang Y, Katayama Y, Ueda J, and Tanabe S (2014) A brief review on red to near-infrared persistent luminescence in transition-metal-activated phosphors. *Opt. Mater.* 36: 1907.

60. Li Y, Gecevicius M, and Qiu J (2016) Long persistent phosphors-from fundamentals to applications. *Chem. Soc. Rev.* 45: 2090.

61. Viana B, Sharma SK, Gourier D, Maldiney T, Teston E, Scherman D, and Richard C (2016) Long term *in vivo* imaging with Cr^{3+} doped spinel nanoparticles exhibiting persistent luminescence. *J. Lumin.* 170: 879.

62. Jain A, Kumar A, Dhoble SJ, and Peshwe DR (2016) Persistent luminescence: an insight. *Renew. Sust. Energ. Rev.* 65: 135.

63. Arras J, and Bräse S (2018) The world needs new colors: cutting edge mobility focusing on long persistent luminescence materials. *ChemPhotoChem* 2: 55.

64. Smet PF, van den Eeckhout K, de Clercq OQ, and Poelman D (2015) *Persistent phosphors in handbook on the physics and chemistry of rare earths* (Edited by: J-CG Bünzli, VK Pecharsky), Elsevier.

65. Wu S, Pan Z, Chen R, and Liu X (2016) *Long afterglow phosphorescent materials*. Springer.

66. Singh SK (2014) Red and near infrared persistent luminescence nano-probes for bioimaging and targeting applications. *RSC Adv.* 4: 58674.

67. Lécuyer T, Teston E, Ramirez-Garcia G, Maldiney T, Viana B, Seguin J, Mignet N, Scherman D, and Richard C (2016) Chemically engineered persistent luminescence nanoprobes for bioimaging. *Theranostics* 6: 2488.

68. Wang J, Ma Q, Wang Y, Shen H, and Yuan Q (2017) Recent progress in biomedical applications of persistent luminescence nanoparticles. *Nanoscale* 9: 6204.

69. Sun SK, Wang HF, and Yan XP (2018) Engineering persistent luminescence nanoparticles for biological applications: from biosensing/bioimaging to theranostics. *Acc. Chem. Res.* 51: 1131.
70. Wu L, Tang Y, Lu F, Yuan Z (2021) Recent progress of near-infrared persistent phosphors in biorelate and emerging applications. *Chem. Asian J.* (In press) DOI: 10.1002/asia.202100108.
71. Zhou Z, Li Y, and Peng M (2020) Near-infrared persistent phosphors: synthesis, design, and applications. *Chem. Eng. J.* 399: 125688.
72. Le Masne de Chermont Q, Chanéac C, Seguin J, Pellé F, Maitrejean S, Jolivet JP, Gourier D, Bessodes M, and Scherman D (2007) Nanoprobes with near-infrared persistent luminescence for *in vivo* imaging. *Proc. Natl. Acad. Sci. U.S.A.* 104: 9266.
73. Maldiney T, Bessière A, Seguin J, Teston E, Sharma SK, Viana B, Bos AJJ, Dorenbos P, Bessodes M, Gourier D, Scherman D, and Richard C (2014) The *in vivo* activation of persistent nanophosphors for optical imaging of vascularization, tumours and grafted cells. *Nat. Mater.* 13: 418.
74. Maldiney T, Lecointre A, Viana B, Bessière A, Bessodes M, Gourier D, Richard C, and Scherman D (2011) Controlling electron trap depth to enhance optical properties of persistent luminescence nanoparticles for in vivo imaging. *J. Am. Chem. Soc.* 133: 11810.
75. Maldiney T, Ballet B, Bessodes M, Scherman D, and Richard C (2014) Mesoporous persistent nanophosphors for in vivo optical bioimaging and drug-delivery. *Nanoscale* 6: 13970.
76. Maldiney T, Doan BT, Alloyeau D, Bessodes M, Scherman D, and Richard C (2015) Gadolinium-doped persistent nanophosphors as versatile tool for multimodal in vivo imaging. *Adv. Funct. Mater.* 25: 331.
77. Teston E, Richard S, Maldiney T, Lièvre N, Wang GY, Motte L, Richard C, and Lalatonne Y (2015) Non-aqueous sol-gel synthesis of ultra-small persistent luminescence nanoparticles for near-infrared in vivo imaging. *Chem. Eur. J.* 21: 7350.
78. Teston E, Lalatonne Y, Elgrabli D, Autret G, Motte L, Gazeau F, Scherman D, Clément O, Richard C, and Maldiney T (2015) Design, properties, and in vivo behavior of superparamagnetic persistent luminescence nanohybrids. *Small* 11: 2696.
79. Liu F, Yan W, Chuang YJ, Zhen Z, Xie J, and Pan Z (2013) Photostimulated near-infrared persistent luminescence as a new optical read-out from Cr^{3+}-doped $LiGa_5O_8$. *Sci. Rep.* 3: 1554.
80. Chuang YJ, Zhen Z, Zhang F, Liu F, Mishra JP, Tang W, Chen H, Huang X, Wang L, Chen X, Xie J, and Pan Z (2014) Photostimulable near-infrared persistent luminescent nanoprobes for ultrasensitive and longitudinal deep-tissue bio-imaging. *Theranostics* 4: 1112.
81. Liu JM, Liu YY, Zhang DD, Fang GZ, and Wang S (2016) Synthesis of $GdAlO_3:Mn^{4+},Ge^{4+}$@Au core-shell nanoprobes with plasmon-enhanced near-infrared persistent luminescence for in vivo trimodality bioimaging. *ACS Appl. Mater. Interfaces* 8: 29939.
82. Li Z, Zhang Y, Wu X, Huang L, Li D, Fan W, and Han G (2015) Direct aqueous-phase synthesis of sub-10 nm "luminous pearls" with enhanced in vivo renewable near-infrared persistent luminescence. *J. Am. Chem. Soc.* 137: 5304.
83. Li Z, Zhang Y, Wu X, Wu X. Maudgal R, Zhang H, and Han G (2015) In vivo repeatedly charging near-infrared-emitting mesoporous $SiO_2/ZnGa_2O_4:Cr^{3+}$ persistent luminescence nanocomposites. *Adv. Sci.* 2: 1500001.
84. Zou R, Gong S, Shi J, Jiao J, Wong KL, Zhang H, Wang J, and Su Q (2017) Magnetic-NIR persistent luminescent dual-modal ZGOCS@MSNs@Gd_2O_3 core-shell nanoprobes for in vivo imaging. *Chem. Mater.* 29: 3938.
85. Zou R, Huang J, Shi J, Huang L, Zhang X, Wong KL, Zhang H, Jin D, Wang J, and Su Q (2017) Silica shell-assisted synthetic route for mono-disperse persistent nanophosphors with enhanced in vivo recharged near-infrared persistent luminescence. *Nano Res.* 10: 2070.
86. Abdukayum A, Chen JT, Zhao Q, and Yan XP (2013) Functional near infrared-emitting Cr^{3+}/Pr^{3+} co-doped zinc gallogermanate persistent luminescent nanoparticles with superlong afterglow for in vivo targeted bioimaging. *J. Am. Chem. Soc.* 135: 14125.
87. Shi J, Sun X, Zhu J, Li J, and Zhang H (2016) One-step synthesis of amino-functionalized ultrasmall near infrared-emitting persistent luminescent nanoparticles for in vitro and in vivo bioimaging. *Nanoscale* 8: 9798.
88. Shi J, Sun X, Zheng S, Li J, Fu X, and Zhang H (2018) A new near-infrared persistent luminescence nanoparticle as a multifunctional nanoplatform for multimodal imaging and cancer therapy. *Biomaterials* 152: 15.
89. Wang J, Ma Q, Hu XX, Liu H, Zheng W, Chen X, Yuan Q, and Tan W (2017) Autofluorescence-free targeted tumor imaging based on luminous nanoparticles with composition-dependent size and persistent luminescence. *ACS Nano* 11: 8010.

90. Sharma SK, Gourier D, Teston E, Scherman D, Richard C, and Viana B (2017) Persistent luminescence induced by near infra-red photostimulation in chromium-doped zinc gallate for in vivo optical imaging. *Opt. Mater.* 63: 51.
91. Mie G (1908) Beiträge zur optik trüber medien, speziell kolloidaler metallösungen. *Ann. Phys.* 330: 377.
92. Jacques SL (2013) Optical properties of biological tissues: a review. *Phys. Med. Biol.* 58: R37.
93. Richards-Kortum R, and Sevick-Muraca E (1996) Quantitative optical spectroscopy for tissue diagnosis. *Annu. Rev. Phys. Chem.* 47: 555.
94. Marcos-Vidal A, Vaquero JJ, and Ripoll J (2020) *Near infrared-emitting nanoparticles for biomedical applications.* Springer.
95. Weissleder R (2001) A clearer vision for in vivo imaging. *Nat. Biotechnol.* 19: 316.
96. Smith AM, Mancini MC, and Nie S (2009) Second window for in vivo imaging. *Nat. Biotechnol.* 4: 710.
97. Tanabe S (2015) Glass and rare-earth elements: a personal perspective. *Int. J. Appl. Glass Sci.* 6: 305.
98. Liao SK, Yong HL, Liu C *et al.* (2017) Long-distance free-space quantum key distribution in daylight towards inter-satellite communication. *Nat. Photonics* 11: 509.
99. Bünzli JCG, and Eliseeva SV (2010) Lanthanide NIR luminescence for telecommunications, bioanalyses and solar energy conversion. *J. Rare Earths* 28: 824.
100. Rogalski A (2002) Infrared detectors: an overview. *Infrared Phys. Techn.* 43: 187.
101. Yu N, Liu F, Li X, and Pan Z (2009) Near infrared long-persistent phosphorescence in $SrAl_2O_4$:Eu^{2+},Dy^{3+},Er^{3+} phosphors based on persistent energy transfer. *Appl. Phys. Lett.* 95: 231110.
102. Xu J, Murata D, Ueda J, Viana B, and Tanabe S (2018) Toward rechargeable persistent luminescence for the first and third biological windows via persistent energy transfer and electron trap redistribution. *Inorg. Chem.* 57: 5194.
103. Xu J, Murata D, Ueda J, and Tanabe S (2016) Near-infrared long persistent luminescence of Er^{3+} in garnet for the third bio-imaging window. *J. Mater. Chem. C* 5: 11096.
104. Xu J, Murata D, Katayama Y, Ueda J, and Tanabe S (2017) Cr^{3+}/Er^{3+} co-doped $LaAlO_3$ perovskite phosphor: a near-infrared persistent luminescence probe covering the first and third biological windows. *J. Mater. Chem. B.* 5: 6385.
105. Sordillo LA, Pu Y, Pratavieira S, Budansky Y, and Alfano R (2014) Deep optical imaging of tissue using the second and third near-infrared spectral windows. *J. Biomed. Opt.* 19: 056004.
106. Sordillo LA, Pratavieira S, Pu Y, Ramirez KS, Shi L, Zhang L, Budansky Y, and Alfano R (2014) Third therapeutic spectral window for deep tissue imaging. *Proc. SPIE* 8940: 89400V.
107. Shi L, Sordillo LA, Rodríguez-Contreras A, and Alfano R (2016) Transmission in near-infrared optical windows for deep brain imaging. *J. Biophotonics* 9: 38.
108. Strecker HJ (1956) Biochemistry and the central nervous system. *J. Am. Chem. Soc.* 78: 3233.
109. Bhaumik S, DePuy J, and Klimash J (2007) Strategies to minimize background autofluorescence in live mice during noninvasive fluorescence optical imaging. *Lab Anim.* 36: 40.
110. Taroni P, Pifferi A, Torricelli A, Comelli D, and Cubeddu R (2003) In vivo absorption and scattering spectroscopy of biological tissues. *Photochem. Photobiol. Sci.* 2: 124.
111. Frangioni JV (2003) In vivo near-infrared fluorescence imaging. *Curr. Opin. Chem. Biol.* 7: 626.
112. Diao S, Hong G, Antaris AL, Blackburn JL, Cheng K, Cheng Z, and Dai H (2015) Biological imaging without autofluorescence in the second near-infrared region. *Nano Res.* 8: 3027.
113. Viegas MS, Martins TC, Seco F, do Carmo A (2007) An improved and cost-effective methodology for the reduction of autofluorescence in direct immunofluorescence studies on formalin-fixed paraffin-embedded tissues. *Eur. J. Histochem.* 51: 59.
114. Mansfield JR, Gossage KW, Hoyt CC, and Levenson RM (2005) Autofluorescence removal, multiplexing, and automated analysis methods for in-vivo fluorescence imaging. *J. Biomed. Opt.* 10: 041207.
115. Ortgies DH, Tan M, Ximendes EC, del Rosal B, Hu J, Xu L, Wang X, Martín Rodríguez E, Jacinto C, Fernandez N, Chen G, and Jaque D (2018) Lifetime-encoded infrared-emitting nanoparticles for in vivo multiplexed imaging. *ACS Nano* 12: 4362.
116. del Rosal B, Ortgies DH, Fernández N, Sanz-Rodríguez F, Jaque D, and Martín Rodríguez E (2016) Overcoming autofluorescence: long-lifetime infrared nanoparticles for time-gated in vivo imaging. *Adv. Mater.* 28: 10188.
117. Tan M, del Rosal B, Zhang Y, Martín Rodríguez E, Hu J, Zhou Z, Fan R, Ortgies DH, Fernández N, Chaves-Coira I, Núñez Á, Jaque D, and Chen G (2018) Rare-earth-doped fluoride nanoparticles with engineered long luminescence lifetime for time-gated in vivo optical imaging in the second biological window. *Nanoscale* 10: 17771.
118. Gu Y, Guo Z, Yuan W, Kong M, Liu Y, Liu Y, Gao Y, Feng W, Wang F, Zhou J, Jin D, and Li F (2019) High-sensitivity imaging of time-domain near-infrared light transducer. *Nat. Photonics* 13: 525.

119. Cheng S, Shen B, Yuan W, Zhou X, Liu Q, Kong M, Shi Y, Yang P, Feng W, and Li F (2019) Time-gated ratiometric detection with the same working wavelength to minimize the interferences from photon attenuation for accurate in vivo detection. *ACS Cent. Sci.* 5: 299.
120. Li H, Tan M, Wang X, Li F, Zhang Y, Zhao L, Yang C, and Chen G (2020) Temporal multiplexed in vivo upconversion imaging. *J. Am. Chem. Soc.* 142: 2023.
121. Smet PF, Poelman D, and Hehlen MP (2012) Focus issue introduction: persistent phosphors. *Opt. Mater. Express* 2: 452.
122. Blasse G, and Grabmaier BC (1994) *Luminescent materials.* Springer.
123. Rodnyi PA (1997) *Physical processes in inorganic scintillators.* CRC Press.
124. Lecoq P, Annenkov A, Gektin A, Korzhik M, and Pedrini C (2006) *Inorganic scintillators for detector systems: physical principle and crystal engineering.* Springer.
125. Nikl M, and Yoshikawa A (2015) Recent R&D trends in inorganic single-crystal scintillator materials for radiation detection. *Adv. Optical. Mater.* 3: 463.
126. Dujardin C, Auffray E, Bourret E, Dorenbos P, Lecoq P, Nikl M, Vasil'ev AN, Yoshikawa A, and Zhu R (2018) Needs, trends and advances in inorganic scintillators. *IEEE Trans. Nucl. Sci.* 9499: 2.
127. Qin X, Liu X, Huang W, Bettinelli M, and Liu X (2017) Lanthanide-activated phosphors based on 4f-5d optical transitions: theoretical and experimental aspects. *Chem. Rev.* 117: 4488.
128. Weber MJ (2002) Inorganic scintillators: today and tomorrow. *J. Lumin.* 100: 35.
129. Lecoq P (2016) Development of new scintillators for medical applications. *Nucl. Instrum. Methods Phys. Res. A: Accel. Spectrom. Detect. Assoc. Equip.* 809: 130.
130. Norrbo I, Carvalho JM, Laukkanen P, Mäkelä J, Mamedov F, Peurla M, Helminen H, Pihlasalo S, Härmä H, Sinkkonen J, and Lastusaari M (2017) Lanthanide and heavy metal free long white persistent luminescence from Ti doped Li-hackmanite: a versatile, low-cost material. *Adv. Funct. Mater.* 27: 1606547.
131. de Carvalho JM, Pedroso CCS, Machado IP, Hölsä J, Rodrigues LVC, Gluchowski P, Lastusaari M, and Brito HF (2018) Persistent luminescence warm-light LEDs based on Ti-doped RE_2O_2S materials prepared by rapid and energy-saving microwave-assisted synthesis. *J. Mater. Chem. C* 6: 8897.
132. Jablonski A (1933) Efficiency of anti-stokes fluorescence in dyes. *Nature* 131: 839.
133. Xu S, Chen R, Zheng C, and Huang W (2016) Excited state modulation for organic afterglow: materials and applications. *Adv. Mater.* 28: 9920.
134. Hirata S (2017) Recent advances in materials with room-temperature phosphorescence: photophysics for triplet exciton stabilization. *Adv. Optical Mater.* 5: 1700116.
135. Hirata S (2018) Ultralong-lived room temperature triplet excitons: molecular persistent room temperature phosphorescence and nonlinear optical characteristics with continuous irradiation. *J. Mater. Chem. C* 6: 11785.
136. Harvey EN (1957) *A history of luminescence from the earliest times until 1900.* American Philosophical Society.
137. Yen WM, and Weber MJ (2004) *Inorganic phosphors: compositions, preparation and optical properties.* CRC Press.
138. Yen WM, Shionoya S, and Yamamoto H (2007) *Phosphor handbook* (2nd ed.). CRC Press.
139. Hölsä J (2009) Persistent luminescence beats the afterglow: 400 years of persistent luminescence. *Electrochem. Soc. Interface* 4: 42.
140. Matsuzawa T, Aoki Y, Takeuchi N, and Murayama Y (1996) A new long phosphorescent phosphor with high brightness, $SrAl_2O_4$:Eu^{2+}, Dy^{3+}. *J. Electrochem. Soc.* 143: 2670.
141. www.nemoto.co.jp/nlm/.
142. Yamamoto H, and Matsuzawa T (1997) Mechanism of long phosphorescence of $SrAl_2O_4$:Eu^{2+}, Dy^{3+} and $CaAl_2O_4$:Eu^{2+}, Nd^{3+}. *J. Lumin.* 72–74: 287.
143. Lin Y, Tang Z, and Zhang Z (2001) Preparation of long-afterglow $Sr_4Al_{14}O_{25}$-based luminescent material and its optical properties. *Mater. Lett.* 51: 14.
144. Lin Y, Tang Z, Zhang Z, and Nan CW (2002) Anomalous luminescence in $Sr_4Al_{14}O_{25}$:Eu, Dy phosphors. *Appl. Phys. Lett.* 81: 996.
145. Lin Y, Tang Z, Zhang Z, Wang X, and Zhang J (2001) Preparation of a new long afterglow blue-emitting $Sr_2MgSi_2O_7$-based photoluminescent phosphor. *J. Mater. Sci. Lett.* 20: 1505.
146. Botterman J, and Smet PF (2015) Persistent phosphor $SrAl_2O_4$:Eu, Dy in outdoor conditions: saved by the trap distribution. *Opt. Express* 23: A868.
147. Wang X, Zhang Z, Tang Z, and Lin Y (2003) Characterization and properties of a red and orange Y_2O_2S-based long afterglow phosphor. *Mater. Chem. Phys.* 80: 1.

148. Kang CC, Liu RS, Chang JC, and Lee BJ (2003) Synthesis and luminescent properties of a new yellowish-orange afterglow phosphor Y_2O_2S:Ti, Mg. *Chem. Mater.* 15: 3966.

149. Jia D, Jia W, Evans DR, Dennis WM, Liu H, Zhu J, Yen WM (2000) Trapping processes in CaS:Eu^{2+}, Tm^{3+}. *J. Appl. Phys.* 88: 3402.

150. Jia D, Zhu J, and Wu B (2000) Trapping centers in CaS:Bi^{3+} and CaS:Eu^{2+}, Tm^{3+}. *J. Electrochem. Soc.* 147: 386.

151. Jia D (2006) Enhancement of long-persistence by Ce co-doping in CaS:Eu^{2+}, Tm^{3+} red phosphor. *J. Electrochem. Soc.* 153: H198.

152. Poelman D, and Smet PF (2010) Photometry in the dark: time dependent visibility of low intensity light sources. *Opt Express* 18: 26293.

153. McKeever SWS (1985) *Thermoluminescence of Solids*. Cambridge University Press.

154. Bos AJJ (2017) Thermoluminescence as a research tool to investigate luminescence mechanisms. *Materials* 10: 1357.

155. Zhuang Y, Wang L, Lv Y, Zhou TL, and Xie RJ (2017) Optical data storage and multicolor emission readout on flexible films using deep-trap persistent luminescence materials. *Adv. Funct. Mater.* 28: 1705769.

156. Li W, Zhuang Y, Zheng P, Zhou TL, Xu J, Ueda J, Tanabe S, Wang L, and Xie RJ (2018) Tailoring trap depth and emission wavelength in $Y_3Al_{5-x}Ga_xO_{12}$:Ce^{3+},V^{3+} phosphor-in-glass films for optical information storage. *ACS Appl. Mater. Interfaces* 10: 27150.

157. Wang X, Zhao Z, Wu Q, Wang C, Wang Q, Yanyan L, and Wang Y (2016) Structure, photoluminescence and abnormal thermal quenching behavior of Eu^{2+}-doped $Na_3Sc_2(PO4)_3$: a novel blue-emitting phosphor for n-UV LEDs. *J. Mater. Chem. C* 4: 8795.

158. Kim YH, Arunkumar P, Kim BY, Unthrattll S, Kim E, Moon SH, Hyun JY, Kim KH, Lee D, Lee JS, and Im WB (2017) A zero-thermal-quenching phosphor. *Nat. Mater.* 16: 543.

159. Fan X, Chen W, Xin S, Liu Z, Zhou M, Yu X, Zhou D, Xu X, and Qiu J (2018) Achieving long-term zero-thermal-quenching with the assistance of carriers from deep traps. *J. Mater. Chem. C* 6: 2978.

160. Liu Z, Zhao L, Chen W, Xin S, Fan X, Bian W, Yu X, Qiu J, and Xu X (2018) Effects of the deep traps on the thermal-stability property of $CaAl_2O_4$: Eu^{2+} phosphor. *J. Am. Ceram. Soc.* 101: 3480.

161. Ji X, Zhang J, Li Y, Liao S, Zhang X, Yang Z, Wang Z, Qiu Z, Zhou W, Yu L, and Lian S (2018) Improving quantum efficiency and thermal stability in blue-emitting $Ba_{2-x}Sr_xSiO_4$:Ce^{3+} phosphor via solid solution. *Chem. Mater.* 30: 5137.

162. Qiao J, Ning L, Molokeev MS, Chuang YC, Liu Q, and Xia Z (2018) Eu^{2+} site preferences in the mixed cation $K_2BaCa(PO_4)_2$ and thermally stable luminescence. *J. Am. Chem. Soc.* 140: 9730.

163. Aitasalo T, Deren P, Hölsä J, Jungner H, Krupa JC, Lastusaari M, Legendziewicz J, Niittykoski J, and Strek W (2003) Persistent luminescence phenomena in materials doped with rare earth ions. *J. Solid State Chem.* 171: 114.

164. Hölsä J, Laamanen T, Lastusaari M, Malkamäki M, Novák P (2009) Persistent luminescence – Quo vadis? *J. Lumin.* 129: 1606.

165. Ueda J, Kuroishi K, and Tanabe S (2014) Bright persistent ceramic phosphors of Ce^{3+}-Cr^{3+}-codoped garnet able to store by blue light. *Appl. Phys. Lett.* 104: 101904.

166. Ueda J, Dorenbos P, Bos AJJ, Kuroishi K, and Tanabe S (2015) Control of electron transfer between Ce^{3+} and Cr^{3+} in the $Y_3Al_{5-x}Ga_xO_{12}$ host *via* conduction band engineering. *J. Mater. Chem. C* 3: 5642.

167. Hu L, Wang P, Zhao M, Liu L, Zhou L, Li B, Albaqami FH, El-Yoni AM, Li X, Xie Y, Sun X, and Zhang F (2018) Near-infrared rechargeable "optical battery" implant for irradiation-free photodynamic therapy. *Biomaterials* 163: 154.

168. Dorenbos P (2000) The 5d level positions of the trivalent lanthanides in inorganic compounds. *J. Lumin.* 91: 155.

169. Dorenbos P (2003) Relation between Eu^{2+} and Ce^{3+} f↔d-transition energies in inorganic compounds. *J. Phys.: Condens. Matter.* 15: 4797.

170. Dorenbos P (2003) Energy of the first $4f^7 \rightarrow 4f^6 5d$ transition of Eu^{2+} in inorganic compounds. *J. Lumin.* 104: 239.

171. Dorenbos P (2003) Systematic behaviour in trivalent lanthanide charge transfer energies. *J. Phys.: Condens. Matter.* 15: 8417.

172. Dorenbos P (2005) The Eu^{3+} charge transfer energy and the relation with the band gap of compounds. *J. Lumin.* 111: 89.

173. Dorenbos P (2012) Electronic structure engineering of lanthanide activated materials. *J. Mater. Chem.* 22: 22344.

174. Dorenbos P (2012) Modeling the chemical shift of lanthanide 4f electron binding energies. *Phys. Rev. B* 85: 165107.
175. Dorenbos P (2013) Determining binding energies of valence-band electrons in insulators and semiconductors *via* lanthanide spectroscopy. *Phys. Rev. B* 37: 035118.
176. Dorenbos P (2013) A review on how lanthanide impurity levels change with chemistry and structure of inorganic compounds. *ECS J. Solid State Sci. Technol.* 2: R3001.
177. Jia W, Yuan H, Lu L, Liu H, and Yen WM (1998) Phosphorescent dynamics in $SrAl_2O_4$: Eu^{2+}, Dy^{3+} single crystal fibers. *J. Lumin.* 76–77: 424.
178. Tanabe S, and Hanada T (1996) Appearance of light-storage rare earth aluminate phosphors and their optical properties. *New Ceramics* 9: 27.
179. Korthout K, Van den Eeckhout K, Botterman J, Nikitenko S, Poelman D, and Smet PF (2011) Luminescence and x-ray absorption measurements of persistent SrAl₂O₄:Eu,Dy powders: evidence for valence state changes. *Phys. Rev. B* 84: 085140.
180. Ueda J, Nakanishi T, Katayama Y, and Tanabe S (2012) Optical and optoelectronic analysis of persistent luminescence in Eu^{2+}-Dy^{3+} codoped $SrAl_2O_4$ ceramic phosphor. *Phys. Status Solidi. C* 12: 2322.
181. Joos JJ, Korthout K, Amidani L, Glatzel P, Poelman D, and Smet PF (2020) Identification of Dy^{3+}/Dy^{2+} as electron trap in persistent phosphors. *Phys. Rev. Lett.* 125: 033001.
182. Ueda J, Dorenbos P, Bos AJJ, Meijerink A, and Tanabe S (2015) Insight into the thermal quenching mechanism for $Y_3Al_5O_{12}$:Ce^{3+} through thermoluminescence excitation spectroscopy. *J. Phys. Chem. C* 119: 25003.
183. Ueda J, Meijerink A, Dorenbos P, Bos AJJ, and Tanabe S (2017) Thermal ionization and thermally activated crossover quenching processes for $5d$-$4f$ luminescence in $Y_3Al_{5-x}Ga_xO_{12}$:Pr^{3+}. *Phys. Rev. B* 95: 014303.
184. Ueda J, Tanabe S, Takahashi K, Takeda T, and Hirosaki N (2018) Thermal quenching mechanism of $CaAlSiN_3$:Eu^{2+} red phosphor. *Bull. Chem. Soc. Jpn.* 91: 173.
185. Sontakke AD, Ueda J, Xu J, Asami K, Katayama M, Inada Y, and Tanabe S (2016) A comparison on Ce^{3+} luminescence in borate glass and YAG ceramic – understanding the role of host's characteristics. *J. Phys. Chem. C* 120: 17683.
186. Sontakke AD, Ueda J, Katayama Y, Dorenbos P, and Tanabe S (2015) Role of electron transfer in Ce^{3+} sensitized Yb^{3+} luminescence in borate glass. *J. Appl. Phys.* 117: 013105.
187. Sontakke AD, Ueda J, and Tanabe S (2016) Significance of host's intrinsic absorption band tailing on Ce^{3+} luminescence quantum yield in borate glass. *J. Lumin.* 170: 785.
188. Botterman J, Joos JJ, and Smet PF (2014) Trapping and detrapping in SrAl₂O₄:Eu,Dy persistent phosphors: influence of excitation wavelength and temperature. *Phys. Rev. B* 90: 085147.
189. Dobrowolska A, Bos AJJ, and Dorenbos P (2014) Electron tunneling phenomena in YPO₄:Ce,Ln (Ln = Er, Ho, Nd, Dy). *J. Phys. D: Appl. Phys.* 47: 335301.
190. Avouris P, and Morgan TN (1981) A tunneling model for the decay of luminescence in inorganic phosphors: the case of Zn2SiO4:Mn. *J. Chem. Phys.* 74: 4347.
191. Vedda A, and Fasoli M (2018) Tunneling recombinations in scintillators, phosphors, and dosimeters. *Radiat. Meas.* 118: 86.
192. Huntley DJ (2006) An explanation of the power-law decay of luminescence. *J. Phys.: Condens. Mater.* 18: 1359.
193. Xu J, Ueda J, and Tanabe S (2017) Toward tunable and bright deep-red persistent luminescence of Cr^{3+} in garnets. *J. Am. Ceram. Soc.* 100: 4033.
194. Dorenbos P (2005) Mechanism of persistent luminescence in Eu^{2+} and Dy^{3+} codoped aluminate and silicate compounds. *J. Electrochem. Soc.* 152: H107.
195. Dorenbos P (2000) The $4f^n \leftrightarrow 4f^{n-1}5d$ transitions of the trivalent lanthanides in halogenides and chalcogenides. *J. Lumin.* 91: 91.
196. Dorenbos P (2002) Relating the energy of the $[Xe]5d^1$ configuration of Ce^{3+} in inorganic compounds with anion polarizability and cation electronegativity. *Phys. Rev. B* 65: 235110.
197. Dorenbos P (2003) f→d transition energies of divalent lanthanides in inorganic compounds. *J. Phys.: Condens. Matter.* 15: 575.
198. Dorenbos P (2003) Exchange and crystal field effects on the $4f^{n-1}5d$ levels of Tb^{3+}. *J. Phys.: Condens. Matter.* 15: 6249.
199. Dorenbos P (2005) Valence stability of lanthanide ions in inorganic compounds. *Chem. Mater.* 17: 6452.
200. Dorenbos P (2013) Ce^{3+} 5d-centroid shift and vacuum referred 4f-electron binding energies of all lanthanide impurities in 150 different compounds. *J. Lumin.* 135: 93.

201. Dorenbos P (2017) Charge transfer bands in optical materials and related defect level locations. *Opt. Mater.* 69: 8.
202. Lyu T, and Dorenbos P (2018) Charge carrier trapping processes in lanthanide doped $LaPO_4$, $GdPO_4$, YPO_4, and $LuPO_4$. *J. Mater. Chem. C* 6: 369.
203. Xu J, Ueda J, and Tanabe S (2015) Design of deep-red persistent phosphors of $Gd_3Al_{5-x}Ga_xO_{12}:Cr^{3+}$ transparent ceramics sensitized by Eu^{3+} as an electron trap using conduction band engineering. *Opt. Mater. Express* 5: 710.
204. Katayama Y, Kayumi T, Ueda J, Dorenbos P, Viana B, and Tanabe S (2017) The role of Ln^{3+} (Ln = Eu, Yb) in persistent and luminescence in $MgGeO_3:Mn^{2+}$. *J. Mater. Chem. C* 5: 8893.
205. Katayama Y, Kayumi T, Ueda J, and Tanabe S (2018) Enhanced persistent red luminescence in Mn^{2+}-doped $(Mg,Zn)GeO_3$ by electron trap and conduction band engineering. *Opt. Mater.* 79: 147.
206. Back M, Trave E, Ueda J, and Tanabe S (2016) Ratiometric optical thermometer based on dual near-infrared emission in Cr^{3+}-doped bismuth-based gallate host. *Chem. Mater.* 28: 8347.
207. Back M, Ueda J, Brik MG, Lesniewski T, Grinberg M, and Tanabe S (2018) Revisiting Cr^{3+}-doped $Bi_2Ga_4O_9$ spectroscopy: crystal field effect and optical thermometric behavior of near-infrared-emitting singly-activated phosphors. *ACS Appl. Mater. Interfaces* 10: 41512.
208. Back M, Ueda J, Xu J, Asami K, Brik MG, and Tanabe S (2020) Effective ratiometric luminescent thermal sensor by Cr^{3+}-doped mullite $Bi_2Al_4O_9$ with robust and reliable performances. *Adv. Opt. Mater.* 8: 00124.
209. Back M, Ueda J, Nambu H, Fujita M, Yamamoto A, Yoshida H, Tanaka H, Brik MG, and Tanabe S (2021) Boltzmann thermometry in Cr^{3+}-doped Ga_2O_3 polymorphs: the structure matters! *Adv. Opt. Mater.* (in press) DOI: 10.1002/adom.202100033.
210. Yan W, Liu F, Lu Y, Wang X, Yin M, and Pan Z (2010) Near infrared long-persistent phosphorescence in $La_3Ga_5GeO_{14}:Cr^{3+}$ phosphor. *Opt. Express* 18: 20215.
211. Jia D, Lewis LA, and Wang X (2010) Cr^{3+}-doped lanthanum gallogermanate phosphors with long persistent IR emission. *Electrochem. Solid-State Lett.* 13: J32.
212. Tanabe Y, and Sugano S (1954) On the absorption spectra of complex ion I. *J. Phys. Soc. Jpn.* 9: 753.
213. Tanabe Y, and Sugano S (1954) On the absorption spectra of complex ions II. *J. Phys. Soc. Jpn.* 9: 766.
214. Liu F, Liang Y, Chen Y, and Pan Z (2016) Divalent nickel-activated gallate-based persistent phosphors in the short-wave infrared. *Adv. Opt. Mater.* 4: 562.
215. Pellerin M, Castaing V, Gourier D, Chanéac C, and Viana B (2018) Persistent luminescence of transition metal (Co, Ni…)-doped $ZnGa_2O_4$ phosphors for applications in the near-infrared range. *Proc. SPIE* 10533: 1053321.
216. Judd BR (1962) Optical absorption intensities of rare-earth ions. *Phys. Rev.* 127: 750.
217. Ofelt GS (1962) Intensities of crystal spectra of rare-earth ions. *J. Chem. Phys.* 37: 511.
218. Hehlen MP, Brik MG, and Krämer KW (2013) 50th Anniversary of the Judd-Ofelt theory: an experimentalist's view of the formalism and its application. *J. Lumin.* 136: 221.
219. Xu J, Tanabe S, Sontakke AD, and Ueda J (2015) Near-infrared multi-wavelengths long persistent luminescence of nd^{3+} ion through persistent energy transfer in Ce^{3+}, Cr^{3+} Co-doped $Y_3Al_2Ga_3O_{12}$ for the first and second bio-imaging windows. *Appl. Phys. Lett.* 107: 081903.
220. Meng Z, Yoshimura T, Fukue K, Higashihata M, Nakata Y, and Okada T (2000) Large improvement in quantum fluorescence yield of Er^{3+}-doped fluorozirconate and fluoroindate glasses by Ce^{3+} cooping. *J. Appl. Phys.* 88: 2187.
221. Zhou J, Teng Y, Liu X, Ye S, Ma Z, and Qiu J (2010) Broadband spectral modification from visible light to near-infrared radiation using $Ce^{3+}-Er^{3+}$ codoped yttrium aluminium garnet. *Phys. Chem. Chem. Phys.* 12: 13759.
222. Xu J, Murata D, So B, Asami K, Ueda J, Heo J, and Tanabe S (2018) Ho^{3+} perovskite 1.2 μm persistent luminescence of Ho^{3+} in $LaAlO_3$ and $LaGaO_3$ perovskites. *J. Mater. Chem. C* 6: 11374.
223. Katayama Y, Kobayashi H, Ueda J, Viana B, and Tanabe S (2016) Persistent luminescence properties of $Cr^{3+}-Sm^{3+}$ activated $LaAlO_3$ perovskite. *Opt. Mater. Express* 6: 1500.
224. Teng Y, Zhou J, Ma Z, Smedskjaer MM, and Qiu J (2011) Persistent near infrared phosphorescence for rare earth ions co-doped strontium aluminate phosphors. *J. Electrochem. Soc.* 158: K17.
225. Kamimura S, Xu C, Yamada H, Terasaki N, and Fujihala M (2014) Long-persistent luminescence in the near-infrared from Nd^{3+}-doped Sr_2SnO_4 for in vivo optical imaging. *Jpn. J. Appl. Phys.* 53: 092403.
226. Kamimura S, Xu C, Yamada H, Marriott G, Hyodo K, and Ohno T (2017) Near-infrared luminescence from double-perovskite $Sr_3Sn_2O_7:Nd^{3+}$: a new class of probe for in vivo imaging in the second optical window of biological tissue. *Jpn. J. Appl. Phys.* 125: 591.

227. Yuan L, Jin Y, Zhu D, Mou Z, Xie G, and Hu Y (2020) Ni^{2+}-doped yttrium aluminum gallium garnet phosphors: bandgap engineering for broad-band wavelength-tunable shortwave-infrared long-persistent luminescence and photochromism. *ACS Sustain. Chem. Eng.* 8: 6543.

228. Hou D, Zhang Y, Li JY, Li H, Lin H, Lin Z, Dong J, Huang R, and Song J (2020) Discovery of near-infrared persistent phosphorescence and Stokes luminescence in Cr^{3+} and Nd^{3+} doped $GdY_2Al_3Ga_2O_{12}$ dual mode phosphors. *J. Lumin.* 221: 117053.

229. Qin X, Li Y, Zhang R, Ren J, Gecevicius M, Wu Y, Sharafudeen K, Dong G, Zhou S, Ma Z, and Qiu J (2016) Hybrid coordination-network-engineering for bridging cascaded channels to activate long persistent phosphorescence in the second biological window. *Sci. Rep.* 6: 20275.

230. Jiang R, Yang J, Meng Y, Yan D, Liu C, Xu C, and Liu Y (2020) X-ray/red-light excited ZGGO:Cr,Nd nanoprobes for NIR-I/II afterglow imaging. *Dalton Trans.* 49: 6074.

231. Lin H, Yu T, Bai G, Tsang M, Zhang Q, and Hao J (2016) Enhanced energy transfer in Nd^{3+}/Cr^{3+} co-doped $Ca_3Ga_2Ge_3O_{12}$ phosphors with near-infrared and long-lasting luminescence properties. *J. Mater. Chem. C* 4: 3396.

232. Lyu T, and Dorenbos P (2019) Designing thermally stimulated 1.06 μm Nd^{3+} emission for the second bio-imaging window demonstrated by energy transfer from Bi^{3+} in La-, Gd-, Y-, and $LuPO_4$. *Chem. Eng. J.* 372: 978.

233. Nie J, Li Y, Liu S, Chen Q, Xu Q, and Qiu J (2017) Tunable long persistent luminescence in the second near-infrared window via crystal field control. *Sci. Rep.* 7: 12392.

234. Ma C, Liu H, Ren F, Liu Z, Sun Q, Zhao C, and Li Z (2020) The second near-infrared window persistent luminescence for anti-counterfeiting application. *Cryst. Growth Des.* 20: 1859.

235. Castaing V, Sontakke AD, Xu J, Fernández-Carrión AJ, Genevois C, Tanabe S, Allix M, and Viana B (2019) Persistent energy transfer in $ZGO:Cr^{3+},Yb^{3+}$: a new strategy to design nano glass-ceramics featuring deep red and near infrared persistent luminescence. *Phys. Chem. Chem. Phys.* 21: 19458.

236. Caratto V, Locardi F, Costa GA, Masini R, Fasoli M, Panzeri L, Martini M, Bottinelli E, Gianotti E, and Miletto I (2014) NIR persistent luminescence of lanthanide ion-doped rare-earth oxycarbonates: the effect of dopants. *ACS Appl. Mater. Interfaces* 6: 17346.

237. Liang Y, Liu F, Chen Y, Wang X, Sun K, and Pan Z (2016) New function of the Yb^{3+} ion as an efficient emitter of persistent luminescence in the short-wave infrared. *Light Sci. Appl.* 5: e16124.

238. Li Z, Zhang S, Xu Q, Duan H, Lv Y, Lin X, Wang C, Jin Y, and Hu Y (2018) Long persistent phosphor $SrZrO_3:Yb^{3+}$ with dual emission in NUV and NIR region: a combined experimental and first-principles methods. *J. Alloys Compd.* 766: 663.

239. Zou Z, Feng L, Cao C, Zhang J, and Wang Y (2016) Near-infrared quantum cutting long persistent luminescence. *Sci. Rep.* 6: 24884.

240. Zou Z, Wu C, Li X, Zhang J, Li H, Wang D, and Wang Y (2017) Near-infrared persistent luminescence of Yb^{3+} in perovskite phosphor. *Opt. Lett.* 42: 4510.

241. Dai WB, Lei YF, Zhou J, Xu M, Chu LL, Li L, Zhao P, and Zhang ZH (2017) Near-infrared quantum-cutting and long-persistent phosphor $Ca_3Ga_2Ge_3O_{12}$: Pr^{3+}, Yb^{3+} for application in in vivo bioimaging and dye-sensitized solar cells. *J. Alloys Compd.* 726: 230.

242. Liang Y, Liu F, Chen Y, Wang X, Sun K, and Pan Z (2017) Red/near-infrared/short-wave infrared multi-band persistent luminescence in Pr^{3+}-doped persistent phosphors. *Dalton Trans.* 46: 11149.

243. Liang Y, Liu F, Chen Y, Wang X, Sun K, and Pan Z (2017) Extending the applications for lanthanide ions: efficient emitters in short-wave infrared persistent luminescence. *J. Mater. Chem. C* 5: 6488.

244. Duncan R, and Sat YN (1998) Tumour targeting by enhanced permeability and retention (EPR) effect. *Ann. Oncol.* 9: 39.

245. Fan X, Liu Z, Yang X, Chen W, Zeng W, Tian S, Yu X, Qiu J, and Xu X (2019) Recent developments and progress of inorganic photo-stimulated phosphors. *J. Rare Earth* 37: 679.

246. Liu F, Liang Y, and Pan Z (2014) Detection of up-converted persistent luminescence in the near infrared emitted by the $Zn_3Ga_2GeO_8:Cr^{3+}$, Yb^{3+}, Er^{3+} phosphor. *Phys. Rev. Lett.* 113: 177401.

247. Liu F, Chen Y, Liang Y, and Pan Z (2016) Phonon-assisted upconversion charging in $Zn_3Ga_2GeO_8:Cr^{3+}$ near-infrared persistent phosphor. *Opt. Lett.* 41: 954.

248. Chen Y, Liu F, Liang Y, Wang X, Bi J, Wang X, and Pan Z (2018) A new up-conversion charging concept for effectively charging persistent phosphors using low-energy visible-light laser diodes. *J. Mater. Chem. C* 6: 8003.

249. Wegh RT, Donker H, Oskam KD, and Meijerink A (1999) Visible quantum cutting in $LiGdF_4$: Eu^{3+} through downconversion. *Science* 283: 663.

250. Huang X, Han S, Huang W, and Liu X (2013) Enhancing solar cell efficiency: the search for luminescent materials as spectral converters. *Chem. Soc. Rev.* 42: 173.

251. Deng G, Li S, Sun Z, Li W, Zhou L, Zhang J, Gong P, and Cai L (2018) Near-infrared fluorescence imaging in the largely unexplored window of 900-1,000 nm. *Theranostics* 8: 4116.

252. Back M, Ueda J, Xu J, Murata D, Brik MG, and Tanabe S (2019) Ratiometric luminescent thermometers with a customized phase-transition-driven fingerprint in perovskite oxides. *ACS Appl. Mater. Interfaces* 11: 38937.

253. Back M, Casagrande E, Trave E, Cristofori D, Ambrosi E, Dallo F, Roman M, Ueda J, Xu J, Tanabe S, Benedetti A, and Riello P (2020) Confined-melting-assisted synthesis of bismuth silicate glass ceramic nanoparticles: formation and optical thermometry investigation. *ACS Appl. Mater. Interfaces* 12: 55195.

254. Zhuang Y, Ueda J, and Tanabe S (2013) Enhancement of red persistent luminescence in Cr^{3+}-doped $ZnGa_2O_4$ phosphors by Bi_2O_3 codoping. *Appl. Phys. Express* 6: 052602.

255. Van den Eeckhout K, Bos AJJ, Poelman D, and Smet PF (2013) Revealing trap depth distributions in persistent phosphors. *Phys. Rev. B* 87: 045126.

256. Randall JT, and Wilkins MHF (1945) Phosphorescence and electron traps. I. The study of trap distributions. *Proc. R. Soc. London, Ser. A* 184: 365.

257. Zheng B, Bai Y, Chen H, Pan H, Ji W, Gong X, Wu X, Wang H, and Chang J (2018) Near-infrared light excited upconverting persistent nanophosphors in vivo for imaging-guided cell therapy. *ACS Appl. Mater. Interfaces* 10: 19514.

258. Qiu X, Zhu X, Xu M, Yuan W, Feng W, and Li F (2017) Hybrid nanoclusters for near-infrared to near-infrared upconverted persistent luminescence bioimaging. *ACS Appl. Mater. Interfaces* 9: 32583.

259. Tanner PA, Zhou L, Duan C, and Wong KL (2018) Misconceptions in electronic energy transfer: bridging the gap between chemistry and physics. *Chem. Soc. Rev.* 47: 5234.

260. Ramírez-García G, Gutiérrez-Granados S, Gallegos-Corona MA, Palma-Tirado L, d'Orlye F, Varenne A, Mignet N, Richard C, and Martínez-Alfaro M (2017) Long-term toxicological effects of persistent luminescence nanoparticles after intravenous injection in mice. *Int. J. Pharm.* 532: 686.

261. Sun X, Shi J, Fu X, Yang Y, and Zhang H (2018) Long-term in vivo biodistribution and toxicity study of functionalized near-infrared persistent luminescence nanoparticles. *Sci. Rep.* 8: 10595.

262. Rogalski A (2002) Infrared detectors: an overview. *Infrared Phys. Techn.* 43: 187.

263. Rogalski A (2005) HgCdTe infrared detector material: history, status and outlook. *Rep. Prog. Phys.* 68: 2267.

264. Chen Z, Li X, Wang J, Tao L, Long M, Liang SJ, Ang LK, Shu C, Tsang HK, and Xu JB (2017) Synergistic effects of plasmonics and electron trapping in graphene short-wave infrared photodetectors with ultrahigh responsivity. *ACS Nano* 11: 430.

265. Pelayo García de Arquer F, Armin A, Meredith P, and Sargent EH (2017) Solution-processed semiconductors for next-generation photodetectors. *Nat. Rev. Mater.* 2: 1.

12 High-Pressure Study of Phosphors Emission

Sebastian Mahlik

CONTENTS

DOI: 10.1201/9781003098669-12

12.1 INTRODUCTION

The rapid development of technology requires either searching for entirely new materials or adapting existing ones that work with new and more demanding applications. To efficiently design luminescent materials, one must fully understand the relationship between phosphors' optical and structural properties.

Inorganic phosphors are generally considered as a system consisting of the luminescent centers (impurity metal ions) – an environment associated with crystalline solids. The interaction of the impurity metal ions with the crystal lattice mainly involves the group of anions nearest to the impurity metal, called the first coordination zone in which factors such as bond lengths, angles, nature of chemical bonds, and coordination number can be characterized. These factors determine the energy, splitting, and mixing of the electronic states involved in luminescence.

High-pressure spectroscopy identifies the physical and chemical factors that control the optical properties and helps find relationships between luminescence centers and the local crystal environment. Applying high pressure to phosphors reduces the volume of the material. Thus, the distance between the ions that make up the structure directly influences the bonding interactions. The change in the luminescent centers' local environment affects the optical properties such as the wavelength, line shape, efficiency and dynamics of absorption, and luminescence.

In general, understanding the observed effects at high pressures gives a better understanding of systems at atmospheric pressure. A good example could be an observation of a crossover of energy states. High pressure can eliminate the degeneration of the local state with the band states. As a result, an efficient luminescence at high pressure can be obtained from a non-emitting material at atmospheric pressure. Knowledge of the pressure shifts of intersecting states and appropriate extrapolation allows one to receive the energetic structure of the system under atmospheric pressure.

Another important application of high pressure is the ability to create entirely new crystalline structures and energy states that do not exist under normal conditions. Some of the systems modified at high pressures remain altered after pressure is released. Alternatively, the change of the local environment occurring at high pressure can be permanent, thus getting a new material with significantly different spectroscopic properties.

It is essential to use high pressure to distinguish luminescent centers in different local environments. The differences unobservable at atmospheric pressure can become significant at high pressure. Due to varying changes in local surroundings caused by stress, the optical properties of individual centers, indistinguishable or inactive in atmospheric pressure, often change significantly, allowing them to be distinguished.

Finally, high-pressure spectroscopy directly tests physical models such as the crystal field theory or the ligand field theory.

It should be remembered that although we observe fascinating effects at high pressures, the main goal of this type of research is to improve one's knowledge of luminescent materials under normal conditions in which they are applied.

12.2 HIGH-PRESSURE SPECTROSCOPY METHOD

High hydrostatic pressure is an external variable that can alter the system and can be used to tune its physical properties. The magnitude of pressure (P) applied and the compressibility of the material determines how much volume reduction occurs. The compressibility K (i.e., reciprocal bulk modulus B_o) of the material is defined by the relation [1, 2]:

$$K = \frac{1}{B_0} = -\frac{1}{V}\left(\frac{\partial V}{\partial P}\right)_T \tag{12.1}$$

where the volume V of a material decreases upon isothermal compression $\left(\partial V / \partial P\right)_T < 0$.

The above-mentioned relation shows that a more rigid material with higher compressibility (low B_0) exhibits small decreases in volume per unit of applied pressure. In the condensed phase, the intermolecular interactions result in low compressibility (high B_0). The values of bulk modulus B_0 of the most inorganic phosphors are between 500 and 3000 kbar [3, 4]. Therefore, to significantly affect its properties, high pressures on the order of tens of kilobars or higher should be applied. Such high pressures are only possible to obtain with the technique involving diamond anvil cells (DACs). Here, it should be mentioned that spectroscopic measurements discussed in this chapter took place at pressures not exceeding 300 kbar (30 GPa).

By exposing the phosphor material to high pressure, we directly influence the interaction of the crystal environment with luminescent centers and cause significant changes in the energy structures of studied systems. This has been directly observed through changes in the emission and absorption bands position or indirectly through the quenching or enhancement of luminescence.

High-pressure spectroscopy includes luminescence spectra, time-resolved luminescence spectra, luminescence decays, Raman spectra, luminescence excitation spectra, and absorption spectra measurements. However, the most frequently used technique is the measurement of luminescence spectra due to the universality of the apparatus and the ease of carrying out this type of measurement. The sample size is critical factor since none of the dimensions usually exceed 0.1 mm. To study such a small object's optical properties, we must first use a sensitive fluorescence detection and, second, a bright source of excitation. The obvious choice is lasers commonly used in physics laboratories, which are great for measuring luminescence at high pressure. In the research on excitation and absorption spectra, light sources such as Xe lamps are most often used, at the expense of much more demanding experiment conditions and lower signal-to-noise ratios obtained. Focusing the Xe lamp light on such a small sample is quite a challenge and means that only a few research centers in the world perform high-pressure dependence of absorption and excitation measurements.

12.2.1 DIAMOND ANVIL CELL (DAC)

The most frequently used equipment for high-pressure spectroscopic measurements is the DAC. The underlying principle of DAC is that force applied over a tiny area causes high pressure. Several types of force-generating mechanisms were designed for DAC [5, 6]; however, currently, the most popular rely on tightening screws, levers, or pneumatic and hydraulic membranes. The screw mechanism is the simplest and most versatile solution allowing the DAC's miniaturization. Hence, they are commonly used in various experimental systems. An example controlled by screws DAC is shown in Figure 12.1, where the construction of the modified Merrill-Bassett type is presented [7].

The presented DAC does not need any additional pressure generation mechanism; the pressure is generated simply by tightening the screws. Standard dimensions are given in the drawing, and none of the dimensions exceed 40 mm, which allows the DACs to be mounted in, e.g., standard cryostats. The Merrill-Bassett cell (Figure 12.1) consists of two steel elements that can slide relative to each other by the action of the screws. The lower part of the cell has additional metal supporting walls perpendicular to the plates designed to align both parts of the cell. In the center of both parts of the cell, diamond anvils are fixed onto bases with drilled holes about 1 mm in diameter. By using three Allen cap screws with Belleville washers, the diamond anvils are pressed closer together; the force is transmitted to the sample by the pressure transmitting medium (PTM), and the sample pressure increases.

Other types of force-generating mechanisms in DAC use the solution in which the force is transmitted to the diamonds by pneumatic or hydraulic membranes. This type of solution requires adaptation of the apparatus due to the requirement for a constant connection between the DAC and the gas cylinder or the hydraulic mechanism. This solution is somewhat less handy; for example, loading usually requires not only the DAC to be moved but also the whole force-generating system. However, all inconveniences are compensated by the possibility of changing the pressure at a distance without removing the DAC from the measuring system. For this reason, these DACs have substantial advantages in measurements performed at various temperatures.

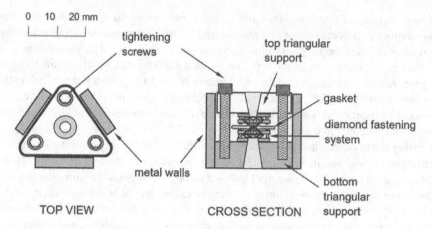

FIGURE 12.1   Schematic depiction of the modified Merrill-Basset DAC.

Regardless of the force-generating mechanism, the force is applied to the diamonds mounted in a mechanical force transmitting system. A sample and pressure sensor along with the PTM are located in a confined space between two parallel polished culets (flat tips) of two opposed diamond anvils and contained on the circumference by the metal gasket (see Figure 12.2).

Forcing the anvils together causes uniaxial pressure that is transformed into hydrostatic pressure by the medium and transmitted to the sample. Four elements must be considered in the high-pressure experiments, i.e., diamond anvils, gasket, PTM, and pressure sensor. These elements are discussed in the subsequent paragraphs.

12.2.1.1   Diamond Anvils

Typical anvils are made of high gem-quality natural or synthetic diamonds. Diamonds are polished according to the (1 0 0)-crystal orientation, usually cut with 16 facets. The larger face of diamond (table) diameter ranges from 2.5 to 4 mm, with typical weight varies between 0.1 and 0.33 carat (20 and 67 mg). Depending on the cell type, the bottom (table) of the diamond can be flat, or it may have bevels.

By definition of pressure $P = F/A$, to achieve ultrahigh pressures P, a relatively large force F should be applied on a small area A. The maximum achievable pressure is mainly determined by the size of the diamond's top face (culet). The relationship between the culet diameter and maximum

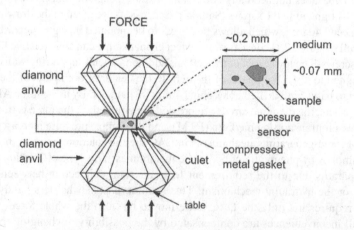

FIGURE 12.2   Diagram of the DACs interior with diamond anvils, the enlargement on the right shows the working area filled with the PTM.

safe pressure limit is not precisely determined. However, a phenomenological rule for the maximum pressure to expect from a culet with a diameter d, $P_{max} = 100/d$ [kbar/mm] is widely accepted [8]. Considering the dependence, for a well-designed DAC, it can usually be expected that for a 0.7-mm diameter culet, a pressure of up to 200 kbar can be achieved, and for a 0.5-mm culet, the maximum pressure is 400 kbar [9].

The widespread use of diamond in high-pressure spectroscopic measurements is associated with its two features: the first, the diamond is the hardest known substance. Second, the diamond is optically transparent and allows for a wide range of spectral studies from UV to IR. The best for spectroscopic measurements are synthetically produced type IIac (nitrogen-free) diamonds characterized by transparency above 240 nm and ultra-weak luminescence [10].

12.2.1.2 Gasket

The metal gasket with a drilled hole placed between diamond culets is designed to hold the transparent PTM and confines the sample and the pressure sensor. Common materials for the gasket in high-pressure spectroscopic measurement are hard, stiff metals such as stainless steel and Inconel alloys [5].

Before the measurement starts, the gasket should be adequately prepared. Initially, the gasket is pre-indented by compression in the DAC to the desired thickness of 40–70 μm from the initial thickness in the 100- to 250-μm range. Usually, it is done in the same DAC in which the measurements are performed. Then, in the center of the indentation, a hole in the gasket is pierced with either a microdrill, an electro-erosion machine, or the laser micromachining system [11]. The hole diameter is usually 1/3 and should not be larger than 2/3 of the diameter of the diamond culet (indentation diameter). The accuracy of the hole drilling and its exact position in the center of indentation is crucial to keeping the volume under load.

12.2.1.3 Pressure Transmitting Medium

The PTM function is to transfer the axial force generated as the diamonds are moved closer together and convert it into a hydrostatic pressure with the homogeneous distribution in the sample's vicinity of the sensor. The PTM can be soft solids, liquids, or gases at ambient pressure. Apart from the possibility of generating hydrostatic pressure in the most extensive possible pressure range, the medium must be transparent in the studied spectral range and should not exhibit any absorption and luminescence. Moreover, it should be nontoxic, chemically stable, not interact with the tested material, and possess compressibility lower than the gasket.

The most used media in spectroscopy are argon, xenon, hydrogen, or helium [8], which are gasses at ambient conditions and require special loading techniques, e.g., remotely, into a cryogenically cooled cell. The more convenient applications that do not require special equipment to load are liquid media like methanol-ethanol mixture [12], methanol-ethanol-water mixture [13], and silicon oils like polydimethylsiloxanes [14, 15].

The pressure range over which the medium provides hydrostatic or near-hydrostatic pressures varies significantly on the medium. For example, argon can be successfully used for pressures up to 300 kbar, helium and polydimethylsiloxanes oil up to 600 kbar. In contrast, the most widely used medium: methanol-ethanol mixture can be used only up to 200 kbar. More information on this subject can be found in extensive studies presented in [5] and [10].

12.2.1.4 Pressure Calibration

Determining the pressure inside the DAC requires a sensor placed together with the sample in the medium. The most frequently used sensor is a ruby crystal ($Al_2O_3:Cr^{3+}$) thanks to significant pressure shift and emission brightness [16–19]. The pressure is determined based on observing the pressure-induced redshift of luminescence lines associated with Cr^{3+} ions. The ruby is characterized by a strong emission related to the $^2E \rightarrow {}^4A_2$ transitions in Cr^{3+} ions. At room temperature, it consists of two narrow emission lines, R_1 at wavelength 694.27 nm and R_2 at 692.8 nm (see Figure 12.2).

The position of R_1 and R_2 lines change linearly (0.365 Å/kbar [−0.759 cm⁻¹/kbar]) with increasing pressure up to 200 kbar [20]. However, above 200 kbar, the dependence is no longer linear, and it was necessary to use the empirical relation found by Mao [21]:

$$P = 3.808 \left[\left(\frac{\lambda(p) - 694.27 \text{ nm}}{694.27 \text{ nm}} + 1 \right)^5 - 1 \right] \text{Mbar} \qquad (12.2)$$

To determine the pressure at temperatures other than the room temperature, one must consider that the position of R_1 and R_2 lines strongly depend on temperature [22].

Other materials proposed for use as spectral pressure sensors include phosphors activated by transition metal (TM) or lanthanide ions, mainly Cr^{3+} or Sm^{2+} [9, 23], but also Ce^{3+} and Eu^{3+} [24]. These materials exhibit visible and near-infrared luminescence, and it is often quite challenging to avoid overlapping sample emission with pressure sensor emission. To avoid such an overlapping problem, a pressure-sensitive material that emits light e.g. only in the ultraviolet range is beneficial. A good example is $KMgF_3:Eu^{2+}$ [25, 26]. Incidentally, $KMgF_3:Eu^{2+}$ is one of few Eu^{2+}-activated materials which show f–f emission in the form of narrow lines instead of typical broadband d–f emission [27]. Figure 12.3 presents emission spectra of $KMgF_3:Eu^{2+}$ at different pressure (left) together with emission spectra of ruby (right).

The $KMgF_3:Eu^{2+}$ luminescence consists of the line attributed to the $^6P_{7/2} \rightarrow {}^8S_{7/2}$ transition, which exhibits linear shifts of a rate equal to −0.815 ± 0.007 cm⁻¹/kbar with pressure up to at least 300 kbar [25]. This value of the pressure shift is similar in value to the pressure shift of ruby (when expressed in cm⁻¹/kbar). However, one must keep in mind that due to the much higher energy of emission of $KMgF_3:Eu^{2+}$ this corresponds to a comparatively much lower pressure shift rate when expressed in nm/kbar. The sensor's exceptional feature is that excitation and emission spectra lie in the UV, making it very convenient to use high-pressure optical measurements in visible and IR

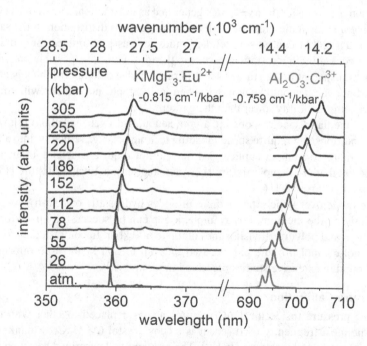

FIGURE 12.3 Emission spectra of $KMgF_3:Eu^{2+}$ at different pressure (left) together with emission spectra of ruby (right). (Modified from Barzowska J, Lesniewski T, Mahlik S, Seo HJ, Grinberg M (2018) $KMgF_3:Eu^{2+}$ as a new fluorescence-based pressure sensor for diamond anvil cell experiments. *Opt. Mater.* 84: 99, with permission).

regions with no interference from the sensor's luminescence. Comprehensive information on spectroscopic measurements with the use of DACs can be found in the work of W.B. Hozapfel and N. S. Isaacs [8]. Additionally, the book by F. Sherman and A. A. Stadtmuller of 1987 [5] provides detailed information on various high-pressure techniques. Especially for scientists entering the field of high-pressure research, papers of D. J. Dunstan and I. L. Spain [28] and [29] are recommended. It should be noted that the information presented earlier relates to the typical spectroscopic measurements at high hydrostatic pressures not exceeding 300 kbar, which is a standard limit achieved in most facilities aside from very advanced laboratories. Measurement at higher pressures and different types of measurements under pressure like XRD, Brillouin scattering, EXAFS, NMR, and electron transport have other requirements and need different materials for gaskets and appropriate designs of the DACs themselves.

12.3 ENERGETIC STRUCTURE AND POSSIBLE ELECTRONIC TRANSITIONS IN INORGANIC PHOSPHORS

In phosphors, considered as systems consisting of solid lattice doped with luminescence center (Lc): usually, TM or lanthanide ions (Ln), we can distinguish three types of energy states. The first group associated with the lattice consists of the delocalized band states represented by valence (VB) and conduction band (CB). The second group comprises the Lc local state, ground $Lc^{\alpha+}$, and the excited states $(Lc^{\alpha+})^*$ (α is the valence of the ion). The third group includes excitonic states related to the electron-hole pairs in which one or both charge carriers are trapped. If charge carriers are trapped at the crystalline ions, we are dealing with the self-trapped exciton (STE) [30], and if charge carriers are trapped on the dopant ions, we are dealing with impurity-trapped exciton (ITE) [31, 32]. In the literature, ITE states are often referred to as either charge transfer (CT) states [33, 34] in case of trapping the hole on the $Ln^{(\alpha+1)+}$ ion or intervalence charge transfer (IVCT) states in the case when an electron is trapped on the ion $Ln^{(\alpha-1)+}$ [35–37]. For the above-described energetic structure of the phosphors, we can distinguish three different groups of transitions illustrated by the vertical arrows in Figure 12.4.

Transitions (absorption and emission) of the first type occur between the Lc localized states $(Lc^{\alpha+} \leftrightarrow (Lc^{\alpha+})^*)$, which, in the case of Ln ions, occur between states belonging to the $4f^n$ (f–f

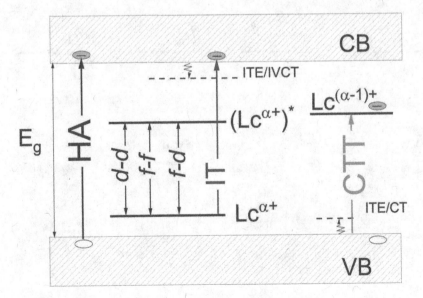

FIGURE 12.4 Energetic structure and possible electronic transitions in inorganic phosphors.

transition) or the $4f^{n-1}5d$ and $4f^n$ electronic configurations (*d–f* transitions). In the case of TM, all electronic states belong to the d^n electronic configuration, and all transitions take place between these states (*d–d* transition) [38].

The second type are the band-to-band transitions between the VB and CB states (host absorption, HA), allowing the determination of the compound's bandgap (E_g). As a result of this transition, the electron is located in the CB and a hole in the VB.

The third type (mainly observed in absorption) includes charge transfer transitions. We can distinguish the ionization transition (IT) in which an electron is excited from the $Lc^{\alpha+}$ to the CB and the charge transfer transition (CTT) in which an electron is excited from the VB to the $Lc^{(\alpha-1)+}$ state.

12.4 HIGH-PRESSURE LUMINESCENCE PHENOMENA

To accurately describe the influence of high pressure on energy states of phosphors, many factors determining the properties system should be considered. The most crucial are changes in the coordinating environment of the immediate vicinity of Lc, such as bond length and angles, covalency, and electron and spin-orbit interactions. However, general changes in the crystal structure are also important. Moreover, the stiffness of the material also largely determines the impact of pressure on the studied electronic states.

Nevertheless, the straightforward and, in most cases, the correct assumption is that the transitions that strongly depend on the crystalline environment react strongly to the pressure changes and accordingly weakly reacting to the crystalline environment react weakly to the pressure changes. Thus, weak pressure-induced changes are expected for the *f–f* and *d(t) – d(t)* transitions in which the d orbital occupancy does not change (transitions within the same crystal field orbital, e.g., t_2–t_2). In contrast, for the remaining transitions, significant pressure-induced changes are expected. Typical behavior of electronic states between which the emission usually takes place is shown schematically in Figure 12.5.

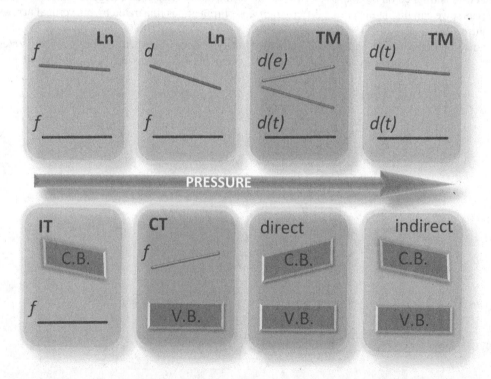

FIGURE 12.5 Typical relative change of the electronic states involved in luminescence under pressure.

One notices that the presented diagram does not describe all the possible transitions (especially absorption transitions to higher excited states) but only shows the expected changes of the luminescence observed at high pressure. For example, in TM systems, significant pressure shifts for transitions in which the d orbital occupancy changes are expected. However, for the most studied dopants, this influence can be observed as a blueshift or redshift of the emission, as in the case of Ti^{3+} and Cr^{3+} (low crystal field case) and Mn^{2+}, respectively.

The next chapter outlines the typical behaviors of electronic transitions and their dependence on pressure and provides a discussion in the context of the nephelauxetic effect, crystal field, and electron-lattice interaction. It also provides several examples in which without high-pressure spectroscopy, it would not be possible to understand the system studies, determine the energy structure, and explain the origin and nature of radiative and nonradiative transitions. Additionally, unusual cases in which atypical behaviors are observed and their explanations are also presented.

12.4.1 Pressure Dependence of the Energy of f–f Transitions

The f–f transitions occur in most luminescent materials in which the luminescence centers are Ln^{3+} ions, except Ce^{3+}. The outer $5s^2 5p^6$ shells effectively shield the electrons occupying the 4f orbitals; therefore, f–f transitions weakly interact with the crystalline environment [39–41]. Also, the influence of pressure on the f–f transitions is feeble. Usually, the observed redshift of luminescence is in the range of 0–3 cm^{-1}/kbar [9, 10] but does not exceed the value of 5 cm^{-1}/kbar.

The f–f transitions are parity-forbidden and result in the narrow excitation and emission lines. According to the Laporte rule, optical transitions within the $4f^n$ configuration occur between levels with the same parity, so they cannot be "pure" electric-dipole transitions [39]. The f–f transitions can result from magnetic dipole transitions and forced electric-dipole transitions. This forcing occurs because of mixing of 4f state with opposite parity 5d states or band states, significantly impacting the changes observed in pressure.

12.4.1.1 Covalency Effect

Due to the nephelauxetic effect, both the spin-orbit coupling constant ζ and the Slater integrals F_k decrease from their free ion values when the ion is placed in a crystal [39]. The increasing pressure causes further nephelauxetic reduction, which can be observed experimentally as a redshift of absorption and luminescence of f–f spectra lines [42]. In the early 1990s, Holzapfel and coworkers deeply examined the changes in the Slater parameters F_k and the spin-orbit coupling constant ζ with pressure for the ions: Pr^{3+}, Nd^{3+}, Eu^{3+}, and Sm^{2+} [43–45]. These works showed that the variation of both the F_k and ζ parameters are generally small in applied pressure. For example, at 80 kbar the parameter values were shifted on average by 2% for F_2 and approx. 0.5% for ζ from their value in atmospheric pressure. Generally, increasing pressure caused a decrease in both the F_k and ζ parameters, and the parameter F_k is several times more sensitive to pressure than ζ.

Two models are commonly used to explain the F_k and ζ parameters' variations, the central field covalency (CFC) and the symmetry restricted covalency (SRC) [46–48]. Combinations of the two models are also often used [9, 49]. Interestingly, both models predict a constant ratio of changes between F_k and ζ parameters: CFC model predicts that the spin-orbit coupling parameter is three times more sensitive to pressure than the Slater parameter. In contrast, the SRC model predicts that the spin-orbit coupling parameter is two times less sensitive to pressure. The experimental data clearly shows that this ratio of pressure changes differed significantly in different materials, as shown by Bray [9], who collected the pressure dependence of the F_k and ζ parameters for a large group of phosphors. The principal reason for this is that the covalency effects are coupled with crystal field effects in the lanthanide systems and cannot be considered separately.

12.4.1.2 Crystal Field Effect

The interaction of the crystal field on the 4f states is considered in the weak field limit. As a first approximation, it is assumed that the hydrostatic pressure causes a proportional decrease of all distances in the lattice without a change in the Ln ion's local symmetry. In such a case, crystal field effect has been calculated using the superposition model [50], linking the parameters B_q^k of the radial crystal field with the length of the bonds between the central ion and its closest neighbors (ligands). The lanthanide binding's local symmetry determines k and q by the charged ligands. In the superposition model, the sum of interactions of each ligand with the lanthanide 4f orbital energies is considered. The ligand-ligand interactions are neglected. The parameters B_q^k are usually obtained empirically from spectroscopic data.

With increasing pressure, the crystal field parameters B_q^k change [43, 51]. However, pressure changes of these parameters are much more difficult to determine than the parameters F_k and ζ. The influence of pressure on parameters B_2^0 B_4^0 B_6^0 is large since these parameters strongly depend on the distance between the ligands and the central ion, while parameters such as: B_3^6 B_6^6 can be completely independent of pressure [52].

The application of the superposition model to describe the effects caused by high pressure is complicated and, in many cases, does not give good results due to local structural distortions. When lanthanide dopants differ in size from the lattice ions they replace, significant distortions of the local environment: bond angles and bond lengths occur. These distortions strongly affect the spectroscopic properties of f–f transitions and are difficult to detect by structural methods. An attempt to solve this problem was made by Gregorian et al. [43], reporting an angular-overlap model that considers the effect of local distortions on the parameters of the crystal field to describe the f–f transition of Pr^{3+} ions in $LaCl_3$ lattice. Although most of the changes in crystal field parameters are described quite well, the model was successfully used to analyze other materials doped with lanthanide ions [24, 44, 53].

In the last 20 years, studies of high-pressure dependence of the covalency and crystal field effects on the 4f states have focused mainly on Yb^{3+} [54–57], Nd^{3+} [58, 59], Pr^{3+} [53, 60, 61], and Eu^{3+} [62–64] ions, and the latest comprehensive study on f–f transitions can be found in [10].

12.4.1.3 Different Ln Ions in the Same Lattice

The magnitude of the pressure shift largely depends on the lattice. However, even within the same host, different lanthanide ions exhibit different pressure shifts. Because the f–f transitions occur between states that are split in the crystal field, it is difficult to indicate the dependence of the pressure-induced redshift on the type of Ln ion. An example of such behavior is shown in Figure 12.6, which presents luminescence spectra of $Gd_2(WO_4)_3$:Ln^{3+} under high hydrostatic pressure [65].

At ambient conditions, the emission consists of sharp lines related to the typical f–f transitions for Dy^{3+}, Eu^{3+}, Tb^{3+}, and Pr^{3+} ions, respectively. As expected, the reduction in bond length causes by pressure enhances both the nephelauxetic and the crystal field effects, causing the redshift of all f–f transitions. However, it is visible that the pressure shifts are much more significant in Tb^{3+} and Pr^{3+} ions than in Eu^{3+} and Dy^{3+} ions. This difference are caused by the impact of $4f^{n-1}5d$ states that lie much closer to the 4f excited states of Tb^{3+} and Pr^{3+} than Eu^{3+} and Dy^{3+}.

One would expect that the pressure effect is more significant for the higher excited states for the selected Ln ion due to a stronger influence of closely lying d states and band states. However, the several studies in which transitions from different excited states were observed do not confirm such a relationship. This can be seen in the example of the Pr^{3+} ion in $Gd_2(WO_4)_3$ (see Figure 12.6(d)), where emission is related to transitions from the higher excited $^3P_0$ state and the lowest excited state $^1D_2$. In this case, the energy of the $^3P_0 \rightarrow {}^3H_4$ transition decrease with increasing pressure with the rate -2.15 cm^{-1}/kbar, very close to the rate obtained for $^1D_2 \rightarrow {}^3H_4$ transitions (2.33 cm^{-1}/kbar). Similar results were obtained for other matrices doped with Pr^{3+} ions and other Ln ions where transitions from higher excited states are observed [10]. This is due to the effect of the crystal field, which split the $^1D_2$ state and does not affect the $^3P_0$ state.

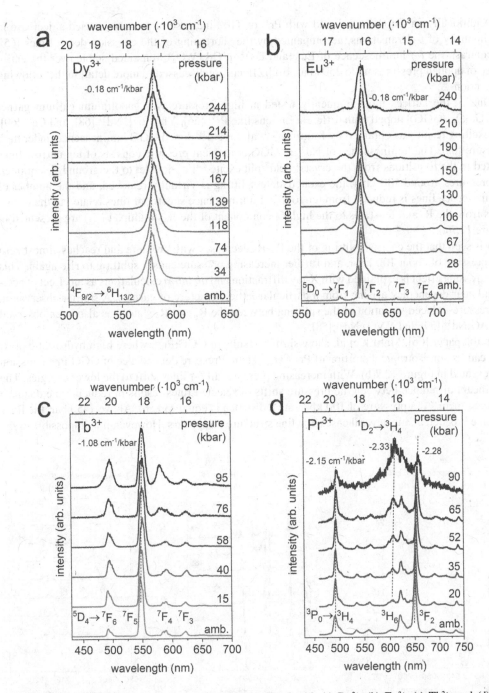

FIGURE 12.6 Luminescence spectra of $Gd_2(WO_4)_3$ doped with (a) Dy^{3+}, (b) Eu^{3+}, (c) Tb^{3+}, and (d) Pr^{3+} obtained at different pressures. The pressure shift rates obtained from the slopes of the linear fits to the experimental data are also presented. (Modified from Mahlik S, Lazarowska A, Grobelna B, Grinberg M (2012) Luminescence of $Gd_2(WO_4)_3$:Ln^{3+} at ambient and high hydrostatic pressure. *J. Phys. Cond.* Matter 24: 485501, with permission).

Additionally, in $Gd_2(WO_4)_3$ doped with Pr^{3+} or Tb^{3+}, increase of pressure caused a decrease in the intensity of $f\text{–}f$ transitions, accompanied by the shortening of the emission decay times [65]. In contrast, the $f\text{–}f$ luminescence of Eu^{3+} and Dy^{3+} in $Gd_2(WO_4)_3$ remained stable over the entire range of applied pressures up to 250 kbar. Both effects are discussed in more detail in the following subsections.

One of the lattices most frequently tested at high pressure was gadolinium gallium garnet [$Gd_3Ga_5O_{12}$ (GGG)] doped with different Ln ions like Pr^{3+} [66], Yb^{3+} [67], Nd^{3+} [68], and Ce^{3+} [69]. An exciting result was obtained by Kamińska et al. for Nd^{3+}-doped GGG luminescence under high pressure [68]. The luminescence of Nd^{3+} in GGG at ambient pressure consists of ten narrow lines related to the transitions from the crystal field split excited $^4F_{3/2}$ doublet to the ground $^4I_{9/2}$ quintet. At pressures around 100 kbar, the excited state splitting is virtually removed, and the number of luminescence lines is reduced from ten to five. Luminescence spectra of lines related to the transitions from the R_1 and R_2 states to the highest component of the $^4I_{9/2}$ multiplet (Z_5) are presented in Figure 12.7(a).

It is seen that the energy splitting of the $R_1\text{–}R_2$ decreases with pressure and reaches almost zero at a pressure of about 100 kbar, and further increasing pressure causes splitting to rise again. This anti-crossing effect is explained with X-ray diffraction and *ab initio* calculations as an effect of accidental near-degeneracy arising from a particular lattice structure of garnets. A similar observation of pressure-induced variation of the splitting between the R_1 and R_2 states has also been observed in YAG:Nd^{3+} [59] and YVO$_4$:Nd^{3+} [58].

In the paper [66], Mahlik et al. show similar results in GGG:Pr^{3+}, where high hydrostatic pressure causes non-isotropic distortion of $Pr^{3+}\text{–}O_8$ system. Pressure dependence of GGG:Pr^{3+} emission is presented in Figure 12.7(b). With increasing pressure, all $f\text{–}f$ lines shift to the lower energies. The significant difference between the pressure shifts of selected lines results in reducing the distance between them. At a pressure of 100 kbar, such distance is equal to zero. Above 145 kbar, the lines become broader, and in several cases, the fine structure disappears. However, it is impossible to say

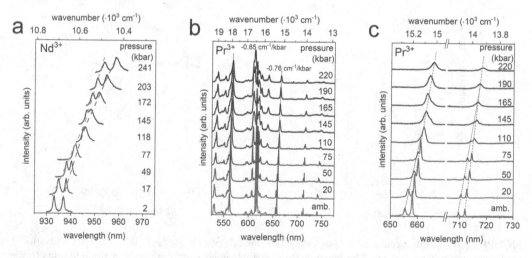

FIGURE 12.7   (a) Pressure dependence of the luminescence spectra of GGG:Nd^{3+} corresponding to the transitions from the R_1 and R_2 states to the highest component of the $^4I_{9/2}$ multiplet. (Modified from Kaminska A, Buczko R, Paszkowicz W, Przybylińska H, Werner-Malento E, Suchocki A, Brik M, Durygin A, Drozd V, Saksena S (2011) Merging of the $^4F_{3/2}$ level states of Nd^{3+} ions in the photoluminescence spectra of gadolinium-gallium garnets under high pressure. Phys Rev B 84: 075483, with permission). (b) (c) Pressure dependence of the luminescence spectra of GGG:Pr^{3+}. (Modified from Mahlik S, Malinowski M, Grinberg M (2011) High pressure and time resolved luminescence spectra of $Gd_3Ga_5O_{12}$:Pr^{3+} crystal. *Opt. Mater.* 33: 1525, with permission).

whether further increasing pressure causes increase splitting again. To better visualize this effect in Figure 12.7(c) selected *f–f* transitions of Pr^{3+} are presented.

12.4.1.4 Single Dopant Ln Ion in Different Lattices

The influence of the hydrostatic pressure on the *f–f* transitions of Yb^{3+} ions-doped different dielectrics and semiconductors lattice was presented by Kamińska et al. in the paper [67]. The results revealed the significant effect of interaction between the band states and the Ln states, as well as the local symmetry of Yb^{3+} on its radiative transition rate. The pressure dependence of the most intense luminescence lines of Yb^{3+} ions in InP and GaN are shown in Figure 12.8(a) and (b), respectively.

In the case of GaN:Yb^{3+}, increasing pressure causes the typical redshift of all the *f–f* transitions induced by the increase of covalency effect and crystal field strength [57]. A similar effect has been observed for numerous dielectric and semiconductor hosts doped with Yb^{3+} ions [54–57]. The only case of the opposite effect was in InP samples doped with Yb^{3+} ions, in which blueshift of all emission lines related to the *f–f* transitions was observed [55]. This very unusual effect has been the subject of intensive research and *ab initio* calculation and was finally explained by increasing the effective spin-orbit coupling parameter of Yb^{3+} relating to the mixing of the Yb wavefunction with those of the surrounding P^{3-} ions. Such mixing was primarily influenced by the pressure-induced approach of the top of the VB of InP crystal host to the local 4f states of Yb^{3+} [56].

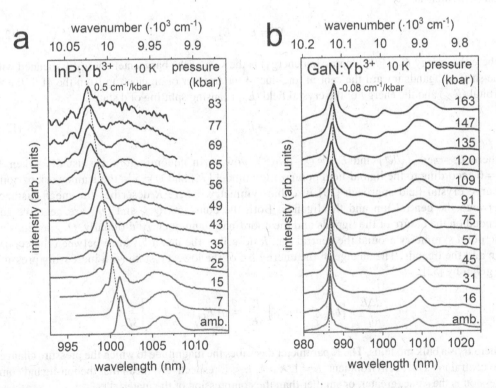

FIGURE 12.8 Pressure dependence of luminescence spectra of (a) InP:Yb^{3+} and (b) GaN:Yb^{3+}, measured at 10 K. (Modified from Kaminska A, Kozanecki A, Ramirez MO, Bausa LE, Boulon G, Bettinelli M, Boćkowski M, Suchocki A (2016) Spectroscopic study of radiative intra-configurational 4f–4f transitions in Yb^{3+}-doped materials using high hydrostatic pressure. *J. Lumin.* 169: 507, with permission.)

12.4.2 Pressure Dependence of the Energy of d–f Transitions

The best known and commonly studied materials in which the d–f luminescence is observed are oxides, halides, fluorides, sulfides, nitrides, oxynitrides, doped with Ce^{3+} and Eu^{2+} ions [27, 70–72]. The high-intensity d–f emission can be observed in the broad spectral range from UV to near-IR region, depending on the lattice. Generally, to describe the strong interaction of 5d electrons with the lattice, the crystal field theory is used [73, 74].

In most lanthanide ions in solids, the d–f luminescence does not occur. The apparent reason may be that these transitions are of high energy, and we do not see them in the available spectral range, especially in trivalent ions [75–77]. In other cases, the reasons for this are varied and complex. The suppression of d–f luminescence through nonradiative transitions to states in $4f^n$ configuration or $4f^{n-1}5d^1$ states degeneration with CB should be mentioned here. Interestingly, in the case of Ce^{3+}-doped garnets using the high pressure can remove the degeneration of states, an example of which is shown in Section 12.4.2.4.

12.4.2.1 Covalency and Crystal Field Effect

For the spectroscopic properties, the location of the lowest state of the $4f^{n-1}5d^1$ configuration to the ground state of the lanthanide ion is the most important issue. A summary of the results of studies on the lowest excited $4f^{n-1}5d^1$ states of Ln ions in inorganic matrices can be found in the extensive Dorenbos reviews [27, 70–72]. In these works, an empirical model was proposed to describe the decrease in the lowest level of the $4f^{n-1}5d^1$ configuration of the lanthanide ion in the lattice in relation to the position of this level for the free E_0 ion. The energy difference ΔE defined in this way can be written as:

$$\Delta E = E_0 - E_{depr} \tag{12.3}$$

where the energy E_{depr} [71] (depression energy) is the sum of the barycenter energy associated with the shift of ligands toward the central ion, due to the transfer of an electron from the 4f to the 5d orbital (E_{cen}) and the energy of the crystal field (E_{CF}) causing splitting of d states:

$$E_{depr} = E_{cen} + E_{CF} \tag{12.4}$$

where $E_{cen} = Q_{cen}(1/R^n)$ and $E_{CF} = Q_{CF}(1/R^m)$. Powers in denominators are usually taken as $n = 6$ according to the ligand-field polarization model [78] and $m = 5$ in the framework of point charge crystal field approach [79] in cubic symmetry, where R describes the mean distance between the central ion and the ligands. Both the constants Q_{cen} and Q_{CF} are negative and depend on the nature of the ligands and the coordination number. Additionally, Q_{CF} depends on the point symmetry around the central ion. R describes the mean distance between the central ion and the ligands. The change of the energy ΔE of the lowest d state with increasing pressure is given by [52]:

$$\frac{d\Delta E}{dp} = \left[nE_{cen} + mE_{CF}\right]\frac{K}{3B_0} = m\left[E_{CF} + \frac{n}{m}E_{cen}\right]\frac{K}{3B_0} \tag{12.5}$$

where B_0 is a bulk modulus. The K parameter describes the magnitude to which the pressure changes the central ion's local environment. $K = 1$, $K > 1$, $K < 1$, respectively, mean that the ion-ligand compression is the same, greater, or smaller than the compression of the material.

To simplify the (12.2) expression, we can assume that $m/n \approx 1$, then we get:

$$\frac{d\Delta E}{dp} \approx n\left[E_{CF} + E_{cen}\right]\frac{K}{3B_0} = nE_{depr}\frac{K}{3B_0} \tag{12.6}$$

Since the energy E_{depr} is negative, the relation (12.7) shows that the increase in pressure causes a decrease in the energy of the d–f luminescence.

12.4.2.2 Electron-Lattice Interaction

The above mentioned considerations do not take into account a crucial element in d–f transitions, namely, the lattice relaxation resulting from the electron-lattice coupling. In the current state of the field, some relationships between the magnitude of lattice relaxation and selected crystal matrix properties have been established. However, no comprehensive model exists that considers all of the host properties. In general, one expects greater lattice relaxation energy for more ionic crystals. The magnitude of lattice relaxation is undoubtedly affected by the ligands. The smallest lattice relaxation is expected when the ligands surrounding dopants have a high symmetry and a high coordination number.

The lattice relaxation energy is described by quantity $S\hbar\omega$ where S is the Huang-Rhys parameter and $\hbar\omega$ is the energy of totally symmetric vibrational mode [38]. The magnitude of the electron-lattice coupling corresponds to the Stokes shift, which describes the energy difference between absorption and emission of $\Delta S = 2S\hbar\omega$.

The relationship describing the change of the electron-lattice coupling $S\hbar\omega$ as a function of pressure can be presented as follow [10]:

$$\frac{dS\hbar\omega}{dp} = -\left[2(n+1)-6\gamma\right]\frac{S\hbar\omega}{R}\frac{dR}{dp} = \left[2(n+1)-6\gamma\right]\frac{S\hbar\omega}{3B_0} \tag{12.7}$$

where n is the exponent that describes the dependence of crystal field strength on the central ion-ligand distance R, and whose experimentally obtained value is usually less than 5. B_0 is the bulk modulus, and γ is the Gruneisen parameter which can be defined as [80]:

$$\gamma = -\frac{V}{\omega}\frac{\partial\omega}{\partial V} \tag{12.8}$$

where V is the volume of the lattice and ω describes phonon frequency. Usually, in the crystal lattice, the energy of optical phonons increases with increasing pressure so that small positive values for γ are normally observed.

It should be mentioned here that the variation of phonon energy under pressure can be measured directly in the Raman or IR spectra. Finding a relationship between the phonon energies obtained from these measurements and effective phonon in the configurational coordinate model is complicated because this model is based on idealized fully symmetric mode, which can be considered the mean of all modes of the center.

Experimental determination of the pressure-induced change in the energy of lattice relaxation at first glance seems to be a simple matter; it is enough to consider the Stokes shift (relative change of excitation and emission spectra) under pressure. However, while emission spectra belong to the standard measurements of high-pressure spectroscopy, excitation or absorption spectra at high pressures appear extremely rarely in the literature. Only a few absorptions results as a function of pressure can be found in the literature. Pressure dependence of electron-phonon coupling in Ce^{3+} has been discussed in GSAG [81] and GGG [69] lattices. In both cases, it was shown that the pressure shift of the absorption band is two times larger than the emission band, which corresponds to the pressure-induced diminishing of the Stokes shift. Additionally, for GSAG:Ce^{3+}, a quantitative analysis in the context of the single configurational coordinate model and the conventional crystal field model allowed to find information about the local stiffness of the Ce^{3+} environment. Both this and subsequent works confirmed that the local bulk modulus is significantly larger than the bulk modulus B_0 of materials [10].

The reduction of the emission energy associated with the transition between the lowest lying excited state $4f^{n-1}5d^1$ and the ground state in the $4f^n$ configuration with increasing pressure was observed many times in fluorides, oxides, sulfides and chlorides. In most of the cases studied, a redshift of the maximum of the d–f emission band ranging from -5 to -40 cm^{-1}/kbar occurred. A review of the results concerning the d–f transitions in Ln ions in high hydrostatic pressures is presented in Refs. [10, 82].

12.4.2.3   High-Pressure Impact on d-f Transition with Extremely Low Lattice Relaxation

In the last decade, very intensively studied luminescent materials are nitrides and oxynitrides doped with Ce^{3+} and Eu^{2+} of which a large group has been tested at high hydrostatic pressures [83–88]. Nitride materials are characterized by very high bulk modulus B_0 (2000–3500 kbar), so the impact of high pressure on the d–f transitions is relatively weak. A prime example is a β-SiAlON material, in which the pressure dependence of luminescence has been described for Pr^{3+} [89] and Eu^{2+} ions [87].

The pressure dependence of emission spectra of β-Si$_{5.82}$Al$_{0.18}$O$_{0.18}$N$_{7.82}$:Eu (denoted as β-SiAlON:Eu (540) and β-Si$_{5.97}$Al$_{0.03}$O$_{0.03}$N$_{7.97}$:Eu (indicated as β-SiAlON:Eu (529)) are presented in Figure 12.9(c and d). The pressure shift of individual lines with numbers 1 to 3 and maximum emission bands of β-SiAlON:Eu (540) (label 4) versus pressure are presented in Figure 12.9(e).

The observed redshift of d–f Eu^{2+} luminescence in the case of β-SiAlON is extremely small, and comparable in magnitude to f–f transitions. This unusual shift is related to the vast bulk modulus value ($\sim$3000 kbar) and the shielding effect of attracted electrons by Eu^{2+}, which occupies the interstitial sites. Namely, when Eu^{2+} occupies the interstitial sites to satisfy the charge neutrality requirement, Eu^{2+} attracts two electrons forming additional bonding-antibonding states [89]. If these two electrons are located between Eu^{2+} and ligand ions, they can shield 5d electrons like the $5s^2 5p^6$ shielding of the $4f^7$ electrons. Consequently, the coupling of the 5d electron to the lattice can be enormously diminished.

The simple configurational coordinate diagrams representing the energetic structure of the Eu^{2+} system for strong (a) and weak (b) electron-lattice coupling of the 5d electron with lattice are presented in Figure 12.9. The diagram of the system of Eu^{2+} describes the ground $4f^7(^8S)$ state and the emitting state $4f^6 5d$. After excitation of the electron from the $4f^n$ state to the $4f^{n-1}5d^1$ state of Ln, the system relaxes. Lattice relaxation is represented by an offset of the minimum of the parabola corresponding to $4f^{n-1}5d^1$ in the configurational space R -proportional to the average distance between the Eu^{2+} ion and ligands.

Usually, the excited electronic configuration $4f^6 5d$ is strongly coupled to the lattice, and the respective situation is presented in the configurational coordinate diagram in Figure 12.10(a). This effect is responsible for diminishing the energy of the system in the excited state and for shifts of excited states in the configurational space to lower values of R, and transitions are seen as broad bands.

When the interaction of 5d electron with lattice ions is sufficiently low, like in β-SiAlON:Eu, the shift of the $4f^6 5d$ states with respect to ground state 4f is slight. Such a situation is presented in Figure 12.10(b). In such a case, the d–f emission band is very narrow. In the extreme case, as in β-SiAlON:Eu, the luminescence is related to purely electronic transitions resulting from the different microenvironment around Eu^{2+} ions.

12.4.2.4   High-Pressure Impact on d–f Luminescence in Oxynitrides

Pressure dependence of the d–f luminescence in oxynitrides (Sr,Ba)Si$_2$O$_2$N$_2$:Eu^{2+} has been presented by Lazarowska et al. [86]. Depending on the crystal structure; the pressure coefficient varies from -40 cm^{-1}/kbar to 0 cm^{-1}/kbar. Figure 12.10 presents pressure dependence of luminescence spectra of (a) (Ba$_{0.98}$Eu$_{0.02}$)Si$_2$O$_2$N$_2$, (b) (Sr$_{0.23}$Ba$_{0.75}$Eu$_{0.02}$)Si$_2$O$_2$N$_2$ and (c) (Sr$_{0.49}$Ba$_{0.49}$Eu$_{0.02}$)Si$_2$O$_2$N$_2$.

In the (Ba$_{0.98}$Eu$_{0.02}$)Si$_2$O$_2$N$_2$ (orthorhombic phase), typical d–f luminesce redshift is observed. The energy of the emission maximum shifts linearly with increasing pressure with a rate of about

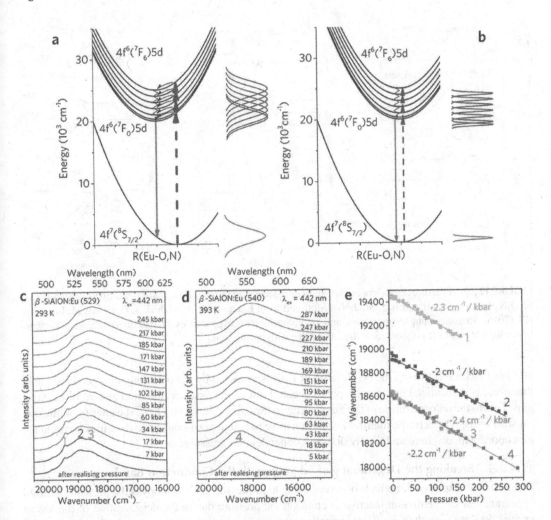

FIGURE 12.9 Configurational coordinate diagrams representing the energetic structure of Eu^{2+} system for strong (a) and weak (b) electron-lattice coupling of the 5d electron, the respective emission and absorption line shapes are presented at the right side of the figures. (c, d) Luminescence spectra of β-SiAlON:Eu (529) and β-SiAlON:Eu (540) obtained for different pressure and (e) pressure shift of the individual lines marked in parts c and d. (From: Zhang X, Fang MH, Tsai YT, Lazarowska A, Mahlik S, Lesniewski T, Grinberg M, Pang WK, Pan F, Liang C, Zhou W, Wang J, Lee JF, Cheng BM, Hung TL, Chen YY, Liu RS (2017) Controlling of Structural Ordering and Rigidity of β-SiAlON:Eu through Chemical Cosubstitution to Approach Narrow-Band-Emission for Light-Emitting Diodes Application. *Chem. Mater.* 29: 6781, with permission.)

$-20\,cm^{-1}$/kbar. The $(Sr_{0.23}Ba_{0.75}Eu_{0.02})Si_2O_2N_2$ (triclinic I) redshift is equal to $-40\,cm^{-1}$/kbar up to 20 kbar. Above 20 kbar, the redshift rate diminishes and is equal to $-20\,cm^{-1}$/kbar. That change in the pressure shift rate results from phase transition from triclinic (I) to orthorhombic phase.

In the case of the $(Sr_{0.49}Ba_{0.49}Eu_{0.02})Si_2O_2N_2$ (triclinic II) under pressure up to 40 kbar, the d–f luminescence spectrum does not change (see Figure 12.10(c)). In the pressure range between 40 and 50 kbar, the pressure-induced phase transition from triclinic (II) to the orthorhombic phase occurs. Further increase of pressure up to 200 kbar causes the shift of emission maximum toward the lower energies with a rate approximately equal to $-20\,cm^{-1}$/kbar.

Lack of pressure shift of d–f luminescence was explained by interactions of the Eu^{2+} 5d electron with the ions forming the second coordination zone. In $(Sr_{0.49}Ba_{0.49}Eu_{0.02})Si_2O_2N_2$ at pressure up to

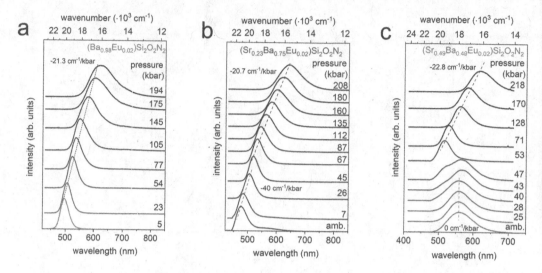

FIGURE 12.10   Pressure dependence of luminescence spectra of (a) $(Ba_{0.98}Eu_{0.02})Si_2O_2N_2$, (b) $(Sr_{0.23}Ba_{0.75}Eu_{0.02})$ $Si_2O_2N_2$ and (c) $(Sr_{0.49}Ba_{0.49}Eu_{0.02})Si_2O_2N_2$. (Modified from Lazarowska A, Mahlik S, Grinberg M, Li G, Liu RS (2016) Structural phase transitions and photoluminescence properties of oxonitridosilicate phosphors under high hydrostatic pressure. *Sci. Rep.* 6: 34010, with permission.)

40 kbar (triclinic II phase), the 5d electron of Eu^{2+} is attracted by the positively charged chain and, therefore, is delocalized and can be partly located out of the first coordination sphere of negative ions. Due to the delocalization of the 5d electron in $(Sr_{0.49}Ba_{0.49}Eu_{0.02})Si_2O_2N_2$, the energy of the lowest state of the 5d electronic manifold is not sensitive to actual positions of negative ions forming the first coordination sphere and finally does not dependent on pressure.

12.4.2.5   Breaking the Degeneracy of 5d State with the Conduction Band

One of the most spectacular effects observed using high-pressure spectroscopy is the pressure-induced appearance of Ce^{3+} emission inactive at atmospheric pressure due to the degeneration of the lowest excited 5d state with the CB. After excitation, the auto-ionization process completely quenches the d–f Ce^{3+} luminescence at ambient pressure. Increasing the pressure eliminates the degeneration of the lowest excited 5d state with the CB, and thus d–f Ce^{3+} luminescence is observed [69, 90, 91].

Suchocki and coworkers first observed the pressure-induced appearance of d–f emission from Ce^{3+} in garnets [69, 91]. This phenomenon had been explained by the decrease of the d–f transition energy under pressure (due to the increase in the crystal field strength resulting from the central ion-ligands distance shortening). Luminescent properties of $Y_3Al_5O_{12}$ (YGG) doped with Ce^{3+} ions at high hydrostatic pressures and variable temperatures were investigated by Mahlik et al. in paper [92]. With increasing pressure, the d–f emission was noticed to appear at the pressure of 20 kbar, although at low temperatures only. Further increase in pressure caused the temperature stabilization of the d–f emission. This process was observed as the increasing of the luminescence decay time (see Figure 12.11(b)) and the possibility of observing the luminescence at ever-higher temperatures. At pressure above 50 kbar, the intense d–f emission was visible even at room temperature, as shown in Figure 12.11(a).

In paper [92], it was shown that in the case of YGG: Ce^{3+}, the d–f transition energy change as a function of pressure is equal to -7.2 cm^{-1}/kbar. The energy of the lowest 5d state with respect to the CB edge decreases with pressure by the quantity of 7 cm^{-1}/kbar, and thus the change of the energy of the ground state $4f(^2F_{5/2})$ of the Ce^{3+} with respect to the CB can be easily determined and is equal to mere 0.2 cm^{-1}/kbar. This means that the energy distance between the ground state of Ce^{3+} and the CB is virtually independent of pressure in YGG.

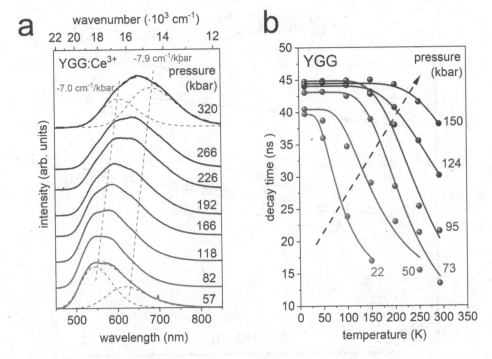

FIGURE 12.11 (a) Luminescence spectra and (b) temperature dependence of the decay time of Ce³⁺ luminescence in YGG:Ce³⁺ obtained for different pressures. (Modified from Mahlik S, Lazarowska A, Ueda J, Tanabe S, Grinberg M (2016) Spectroscopic properties and location of the Ce³⁺ energy levels in $Y_3Al_2Ga_3O_{12}$ and $Y_3Ga_5O_{12}$ at ambient and high hydrostatic pressure. *Phys. Chem. Chem. Phys.* 18: 6683, with permission.)

In 2012 Dorenbos [93] demonstrated a simple model assuming linear position change of the ground states of lanthanide ions relative to the vacuum level with the decreasing distance between the central ion and the ligands. According to the model, the binding energy of the electron at the $4f^n$ electronic configuration of the Ln^{3+} ion $E(Ln^{3+})$ in the lattices is related to the respective binding energy for free ion $E_0(Ln^{3+})$ in a vacuum as follows:

$$E(Ln^{3+}) = E_0(Ln^{3+}) + \frac{3C}{R_Q}$$ (12.9)

The value $C = 1440$ when energies are expressed in eV, and R_Q is given in pm. R_Q is the screening distance, defined in ref. [93], comparable to the Shannon ionic radii [94]. In the case of Ce^{3+}, to calculate pressure-induced changes in the absolute energies of the ground states, one obtains:

$$\frac{dE(Ce^{3+})}{dp} = \frac{dE(^2F_{5/2})}{dp} = \frac{C}{R_Q \cdot B_0} = \frac{E(Ce^{3+}) - E_0(Ce^{3+})}{3 \cdot B_0}$$ (12.10)

where B_0 is the bulk modulus. The value of $E(Ce^{3+})$ in YGG is approximately equal to −5.5 eV (−44,360 cm⁻¹) [95]. Considering that the bulk modulus in YGG is equal to 1707 kbar [96], pressure induces a shift of the ground state of Ce^{3+} equal to 49.2 cm⁻¹/kbar, and CB edge equals 49.4 cm⁻¹/kbar relative to the vacuum level. This result is consistent with the theoretical ab initio calculation presented by Monteseguro et al. [96]. The results obtained in [92] are summarized in Figure 12.12, illustrating the Ce^{3+} and bands states change as a function of pressure.

It is seen that the CB edge, the 5d state, and the 4f ground state of Ce^{3+} increase with increasing pressure. As the result concludes, pressure-induced removing of the degeneration of the 5d state

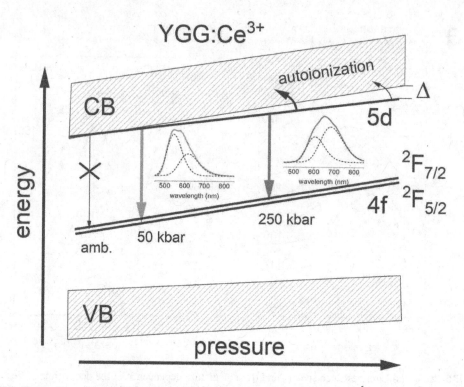

FIGURE 12.12 Pressure dependence of the energetic structure of YGG:Ce^{3+}. (Modified from Mahlik S, Lazarowska A, Ueda J, Tanabe S, Grinberg M (2016) Spectroscopic properties and location of the Ce^{3+} energy levels in Y$_3$Al$_2$Ga$_3$O$_{12}$ and Y$_3$Ga$_5$O$_{12}$ at ambient and high hydrostatic pressure. *Phys. Chem. Chem. Phys.* 18: 6683, with permission.)

with CB in YGG is mainly related to the pressure-induced increase of the bandgap energy. Similar results were obtained for the GGG:Ce^{3+} sample and presented in [97].

One concludes that the CB and local Lc state's degeneration is removed only for the materials where the shift of the CB edges is higher than the Lc states' shift towards the vacuum level. Generally, in the most inorganic lattice, we observe the opposite effect: the increasing pressure causes the degeneration of the local states of lanthanide ions with the CB (extensively described in the next section).

12.4.3 Pressure Dependence of the Energy of Ionization and Charge Transfer Transitions

The information about the position of the ground and excited states of lanthanide ions relative to the edges of the band states is the key to understanding the optical properties of the system studied. It is generally assumed that CTT and IT provide information on the location of lanthanide ions (Ln) levels relative to the VB and CB edges. Dorenbos extensively studied the ground states of all Ln^{3+} and Ln^{2+} ions in different lattices [98–100]. The differences between energies of the ground states of Ln^{2+} as well as Ln^{3+} with the different number of 4f electrons are fixed and independent of the lattice [100]. Characteristic (zig-zag) curves connecting points corresponding to the ground levels of successive lanthanides [100] are consistent with the semi-empirically determined curve obtained by Nakazawa [101, 102]. Locations of the levels of Ln^{2+} and Ln^{3+} with respect to the band edges of a given lattice are schematically presented in Figure 12.13(a). (b) Expected pressure dependence of energies of ground states of Ln^{2+} and Ln^{3+} are presented in Figure 12.13(b).

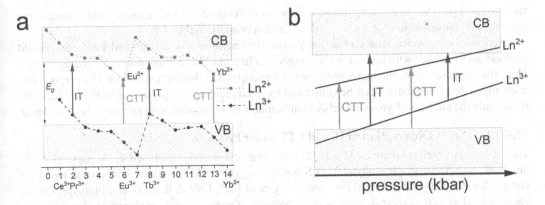

FIGURE 12.13 (a) The diagram shows energies of Ln^{3+} (black points) and Ln^{2+} (red points) in respect to the VB and CB edges. For selected ions, the CTT and IT are indicated by vertical arrows; x axis values correspond to the number of electrons in the 4f shell of Ln^{3+}. (b) Expected pressure dependence of the CTT and IT energies.

The CTT from the top of the VB to the Ln^{2+} state (Ln^{3+} excited state) is observed as a broadband in the excitation spectra of trivalent lanthanide (Ln^{3+}) ions. After the CTT, the system consists of an Ln^{2+} ion and a hole in the CB. The energetic distance between the top of the VB and the ground state of the selected Ln^{2+} is the smallest for Eu^{2+} and Yb^{2+} (see Figure 12.13(a)). Additionally, the Eu^{3+} luminescence spectrum is in the visible range (in contrast to the Yb^{3+} ion that emits light in the IR range); thus, Eu^{3+} is the most widely used ion to determine the CTT energy. In other lanthanide ions, CTT energies are relatively large, usually exceeding the bandgap energy [103]. If the Eu^{3+} CTT is known, the location of all ground state levels of the divalent lanthanides above the CB can be determined unambiguously.

Location below the CB of all trivalent lanthanide (Ln^{3+}) ground levels can be estimated if the IT (blue arrows) is known for a given ion. The IT is a transition from the localized Ln^{3+} state to the CB. The IT energies are most easily determined for Tb^{3+} or Pr^{3+} since they are the lowest for these ions and can be observed as broadbands in the excitation spectra, typically in the UV region.

In both cases, after CTT and IT free charges in the bands are attracted by the lanthanide ions by Coulomb potential and can be trapped, creating the ITE state. The trapping energy, defined as the difference between the energy of the free carriers in the bands and the carriers trapped in the ITE states, depends primarily on the type of lattice and weakly depends on the type of the lanthanide ion on which the exciton state is formed. These states are directly related to the band states [10].

There is no exact procedure for quantitative calculation of the trapping energy of ITE; however, it can be assumed that this energy should not exceed several hundred of cm^{-1}. So far, the only one example where the trapping energy of ITE equal 630 cm^{-1} has been determined from photocurrent excitation measurements was in GGG:Ce^{3+} material [97]. Additionally, when the effect of high pressure on ITE states in Pr^{3+} and Tb^{3+} is considered, it is generally assumed that the trapping energy of ITE is constant, which means that the energy of the ITE state changes exactly like the edge of the CB [104].

Grinberg and Mahlik [31] proposed that the energy of the ITE state can be described by a function proportional to C/R^m. The exponent m depends on the type of interaction. Specifically, $m = 1$ corresponds to electrostatic Coulomb repulsion, $m = 2$ corresponds to the potential of quantum well, $m = 12$ corresponds to short distance Lennard-Jones potential. It appears, however, that within the applied pressure range, where the relative change in distance R does not exceed 30%, it is sufficient to take into account only the effect of classical electrostatic potential, i.e., the $1/R$ relationship. The same dependency also appears in the IVCT model described by Boutinaud et al. [35–37] and in the Dorenbos model assuming linear position change of the ground states of Ln ions relative to the vacuum level with the decreasing distance between the central ion and ligands. This change is the same for all lanthanide ions of a given valency and is proportional to 1/R.

From all these models, the conclusion is that the energies of the Ln^{2+} and Ln^{3+} ions should increase with pressure with respect to the energy of a free ion and ambient pressure binding energy. Thus, the pressure causes the increase of the CTT energies and the decrease of the IT energies. The phenomena described earlier can be illustrated by a simple diagram presented in Figure 12.13(b), where only the changes of ground levels of lanthanide ions with increasing pressure are considered.

12.4.3.1 Direct Observation of IT and CTT under Pressure

The influence of high pressure on IT and CTT has been extensively discussed in the paper [105]. In this work, the spectroscopic study of Y_2O_2S doped with Tb^{3+} and Eu^{3+} allowed the estimation of the energy distance between Ln^{2+} and Ln^{3+} and energies of Ln^{2+}, Ln^{3+}, VB, and CB with respect to the vacuum level as well as their dependences on pressure. The pressure dependence of photoluminescence spectra of $Y_2O_2S:Tb^{3+}$ obtained under IT excitation are presented in Figure 12.14(a).

The emission consists of sharp lines related to the transitions from the Tb^{3+} $^5D_3$ and $^5D_4$ excited states at ambient conditions. The intensity of the luminescence originating from the higher excited state $^5D_3$ decreases with increasing pressure, and the emission is completely quenched at 100 kbar. A similar effect has been observed in various inorganic lattices doped with Tb^{3+} and Pr^{3+} ions. It has been shown that increasing the pressure causing a decrease in f–f transition intensity is accompanied by shortening of the emission decay times. These effects have been explained by the increased energy transfer between the local levels of Pr^{3+} and Tb^{3+} through the ITE states [60, 106–116].

Analysis of the excitation spectra of $Y_2O_2S:Tb^{3+}$ presented in Figure 12.14(b) clearly indicates that a linear shift of IT toward lower energy is observed as the pressure is increased. Such a change has been repeatedly predicted in many papers [60, 107–116]; however, only in [105] was directly observed experimentally.

The photoluminescence spectra of $Y_2O_2S:Eu^{3+}$ obtained under CT excitation are presented in Figure 12.14(c). At ambient pressure, the $Y_2O_2S:Eu^{3+}$ PL spectrum consists of a group of lines corresponding to the transitions from the $^5D_0$, $^5D_1$, and $^5D_2$ excited states. Above 40 kbar, another group of lines appears, which corresponds to the luminescence from the $^5D_3$ excited state of Eu^{3+}. Further increase of pressure causes an increase of the $^5D_3$ luminescence intensity, and this effect is accompanied by an increase of the PL decay time [117]. In the case of the CTT in $Y_2O_2S:Eu^{3+}$, a pressure-induced shift toward higher energies has been observed in the excitation spectra (see Figure 12.14(d)). Such a relation has been predicted for $Y_2O_2S:Eu^{3+}$[117] and $LaAlO_3: Eu^{3+}$ [118] but a direct experimental proof is presented in the paper [105].

The observed pressure-induced changes of the CCT and IT can be explained using the configurational coordinate diagrams shown in Figure 12.15.

Black parabolas correspond to $4f^n$ states while blue and green parabolas correspond to ITE states (solid line corresponds to the pressure of 1 bar, dashed line −100 kbar, dotted line −200 kbar) for Tb^{3+} (ITE_{Tb}) and Eu^{3+}(ITE_{Eu}), respectively.

After IT, there is one less electron in Tb ion, which causes the lattice surrounding to shrink. In contrast, after CTT, the extra electron in Eu ion causes the lattice expansion. Therefore, the lattice relaxation (described by the quantity $S\hbar\omega \sim 3000$–8000 cm^{-1}) causes the shift of the neighboring anions (ΔR), resulting in decreased ITE energy [10].

Figure 12.15(a) is illustrating the situation for $Y_2O_2S:Tb^{3+}$. It is evident that as the pressure increases, the IT (and thus ITE_{Tb}) decreases with respect to the levels of Tb^{3+}, causing quenching of the emission from the $^5D_3$ state. The ITE_{Tb} state transfers energy from the $^5D_3$ to the lower lying $^5D_4$ state of Tb^{3+}.

In the case of $Y_2O_2S:Eu^{3+}$ at ambient pressure, as shown in Figure 12.15(b), the ITE_{Eu} state lies below the excited state $^5D_3$ of Eu^{3+} ion, which causes its quenching. As the pressure increases, the CCT energy (and thus ITE_{Eu}) increases, which causes the exposure of the $^5D_3$ state and the possibility to observe the emission from this state.

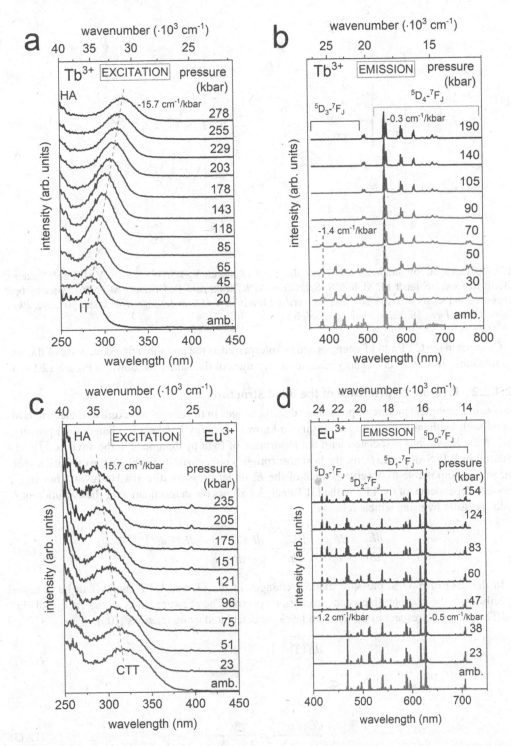

FIGURE 12.14 (a), (c) Pressure dependence of luminescence excitation and (b), (d) luminescence spectra of Y_2O_2S doped with Tb^{3+} and Eu^{3+}, respectively. Luminescence excitation spectra were monitored at maxima of luminescence. The photoluminescence spectra were obtained under IT and CTT excitation. (Modified from Behrendt M, Mahlik S, Szczodrowski K, Kukliński B, Grinberg M (2016) Spectroscopic properties and location of the Tb^{3+} and Eu^{3+} energy levels in Y_2O_2S under high hydrostatic pressure. *Phys. Chem. Chem. Phys.* 18: 22266, with permission.)

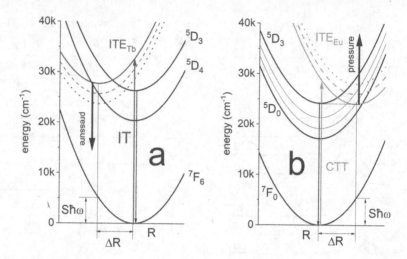

FIGURE 12.15   Configurational coordinate diagrams of the (a) $Y_2O_2S:Tb^{3+}$ and (b) $Y_2O_2S:Eu^{3+}$ systems. (Modified from Behrendt M, Mahlik S, Szczodrowski K, Kukliński B, Grinberg M (2016) Spectroscopic properties and location of the Tb^{3+} and Eu^{3+} energy levels in Y_2O_2S under high hydrostatic pressure. *Phys. Chem. Chem. Phys.* 18: 22266, with permission.)

Changes in the CTT and IT energies can be interpreted as the pressure related increase of the lanthanide ions' ground levels' energy relative to the edges of the bands, as shown in Figure 12.13(b).

12.4.3.2   Pressure Dependence of the Band Structure

To obtain a complete picture of pressure-induced changes in the energy structure of dielectric hosts doped with lanthanide ions, it is necessary to know the energy bandgap E_g change with pressure. Due to the limitations associated with the absorption of light by diamonds in the anvil cell (below 250 nm) in Y_2O_2S-doped Ln ions, the host absorption is only partially visible in the excitation spectra, so it is impossible to observe change of the E_g under pressure directly. However, knowing of pressure dependence of the IT and the CTT energies allows the determination of the evolution of E_g under pressure by using simple relation:

$$\frac{dE_g}{dp} = \frac{dE_{IT}}{dp} + \frac{dE_{CTT}}{dp} + \frac{dE(Tb^{3+})}{dp} - \frac{dE(Eu^{2+})}{dp} \tag{12.11}$$

In the case of Y_2O_2S, pressure induces changes in the IT, and CTT energy can be obtained directly from the photoluminescence excitation spectra. The pressure shifts of the energy of Eu^{2+} and Tb^{3+} ions with respect to the vacuum level are calculated using relations [105]:

$$\frac{dE(Tb^{3+})}{dp} = \frac{2C}{3B_0 \cdot R_{3+}} \tag{12.12}$$

and

$$\frac{dE(Eu^{2+})}{dp} = \frac{3C}{3B_0 \cdot R_{2+}} \tag{12.13}$$

where $C = 64,350$ cm^{-1} Å [93], B_0 is bulk modulus equal to 1250 kbar and the values of R_{2+} and R_{3+} corresponding to the Tb^{3+} and Eu^{2+} ionic radii, respectively. Pressure-induced change of the E_g in Y_2O_2S obtained from Equation (12.8) is equal to 25.08 cm^{-1}/kbar. Pressure dependence of the energetic structure of Y_2O_2S doped with Eu^{3+} and Tb^{3+} is presented in Figure 12.16.

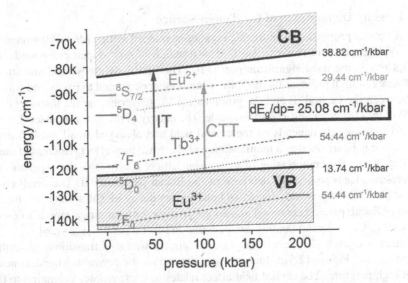

FIGURE 12.16 Pressure dependence of the energetic structure of Y_2O_2S doped with Eu^{3+} and Tb^{3+} (Modified from Behrendt M, Mahlik S, Szczodrowski K, Kukliński B, Grinberg M (2016) Spectroscopic properties and location of the Tb^{3+} and Eu^{3+} energy levels in Y_2O_2S under high hydrostatic pressure. *Phys. Chem. Chem. Phys.* 18: 22266, with permission.)

Similarly, like in the case of garnets doped with Ce^{3+} [92, 97], it was shown that pressure causes the increase of the energy of all states: the localized states related to Ln^{3+} and Ln^{2+}, the CB and VB states in relation to the vacuum level. Moreover, in each of these samples, the largest pressure shift corresponds to Ln^{3+} localized states, whereas the smaller pressure shift characterizes the band states [104]. This means that no matter if the energy gap decreases or increases, in the dielectric materials doped with Ln ions, we always observe a pressure-induced decrease of the IT energy and increase of the CT energy.

12.4.4 PRESSURE DEPENDENCE OF D-D TRANSITION

The d-d transitions are observed in phosphors activated with TM ions. To describe the energetic structure of the TM ions in the lattice, one should consider that the electrons which occupy the d electronic orbitals interact with Coulomb and exchange potential in the ion core and interact with potential created by the crystal [119, 120]. The latter effect is known as the crystal field interaction. When the spin-orbit interaction is neglected, the energetic structure of d electrons of a free ion can be parameterized by two Racah parameters: B and C, combinations of the integrals representing the Coulomb repulsion and exchange interactions between pairs of the electrons, respectively.

The d orbitalsplits into the lower threefold degenerated t_2 state and higher twofold degenerated e state in the octahedral crystal field. In the crystal field model, the splitting energy is equal to $10Dq$, where [38]:

$$Dq = \frac{1}{4\pi\epsilon_0}\frac{Ze^2}{6R^5}\langle r^4 \rangle_{3d} \tag{12.14}$$

R is the average distance between central ion and ligands, e is the charge of electron and Z is the charge of ligand ion represented by the point charge. The parameter $10Dq$ is referred to as the crystal field strength parameter. When considering both crystal and electron-electron interactions, one obtains the ion's full electronic structure. The energy of each multielectron state is parameterized by crystal field strength parameter $10Dq$ and Racah parameters B and C.

12.4.4.1 Pressure Dependence of Cr^{3+} Luminescence

The study of optical properties of inorganic compounds doped with TMs was an important issue in the field of high-pressure studies [9, 10]. The research of Cr^{3+}-doped compounds, especially ruby (Al_2O_3:Cr^{3+}), is the most significant part [121]. The first widely observed and studied effect was the influence of high pressure on the R-lines luminescence related to parity and spin forbidden transition between the excited ^2E and the ground state $^4A_2$ (described as the $d(t)$–$d(t)$ transition in Figure 12.5) of the Cr^{3+} decaying in ms time scale. The energy of the lowest doublet ^2E with respect to the ground state weakly depends on the crystal field and observed small redshift of the $^2E \rightarrow$ $^4A_2$ luminescence is interpreted as a manifestation of the covalency effect – increased metal-ligand orbital overlap due to the compression of the crystal lattice. Analysis of the experimental results show that pressure leads to a small decrease in the Racah parameter B. The small change in B indicates that pressure only weakly increases bonding covalency in TM systems [9, 10]. The pressure changes of Racah parameters based on various (usually semi-empirical) models describing the covalent effects of Cr^{3+} – ligand binding can be found in several papers [122–126].

The situation is entirely different in the case of spin allowed d–d transitions (described as the $d(e)$–$d(t)$ transition in Figure 12.5), which strongly depend on the crystal field and, hence, also the influence of high pressure. The crystal field effect relates to energy states belonging to the crystal field configuration with at least one e orbital occupied. The energy of such state with respect to the ground state $^4A_2$ depends significantly on the crystal field strength parameter Dq. In practice, in phosphors activated by Cr^{3+} ions, the crystal field effect is observed for the $^4T_2 \rightarrow$ $^4A_2$ luminescence and the $^4A_2 \rightarrow$ $^4T_2$,$^4T_1$ absorption transitions.

The pressure-induced diminishing of the distance between the ions in the crystal lattice increases the crystal field strength $10Dq$ [81]:

$$\frac{dDq}{dp} = -Dq\frac{1}{R}\frac{dR}{dp} = \frac{5Dq}{3B_0} \tag{12.15}$$

where B_0 is the bulk modulus and R average distance between ligand and Cr^{3+} ions.

This relationship has been observed in numerous materials by the pressure-induced blueshift of d–d emission between split by crystal field $^4T_2$ and $^4A_2$ states, of the order of tens cm^{-1}/kbar [9, 10].

The pressure-induced blue shift should also be observed in the absorption and excitation spectra of Cr^{3+} containing the $^4A_2 \rightarrow$ $^4T_2$ and $^4T_1$ transitions. However, due to the difficulties associated with measuring this type of spectra as a function of pressure, only a few studies show such results [10, 127].

The combined effect of a large increase in the energy of the $^4T_2$ state and a decrease in the energy of the ^2E state with pressure leads to a possibility of electronic crossover. The $^4T_2$ and ^2E crossover is one of the most spectacular and most studied effects in the field of high-pressure spectroscopy. In 1986, Dolan et al. [128] in K_2NaGaF_6:Cr^{3+} showed that change from the $^4T_2 \rightarrow$ $^4A_2$ to $^2E \rightarrow$ $^4A_2$ emission leads to the observed transformation from broadband to the sharp lines. Since then, such an effect has been widely studied in the literature [9, 10].

One of the most recent examples presenting such phenomena can be found in the paper [129], showing pressure dependence of luminescence spectra of fluoride lattices activated with Cr^{3+} ions. Pressure dependence of luminescence spectra of K_3AlF_6:Cr^{3+} is presented in Figure 12.17(a).

With increasing pressure, the $^4T_2 \rightarrow$ $^4A_2$ emission band shifts toward higher energies due to crystal field strength increasing. Consequently, crossover of the $^4T_2$ and the ^2E state of Cr^{3+} occurs, leading to narrow bands related to the $^2E \rightarrow$ $^4A_2$ electronic transition. As the pressure increases further, the intensity of the $^4T_2 \rightarrow$ $^4A_2$ transition diminishes gradually to zero, and a small redshift of the $^2E \rightarrow$ $^4A_2$ transition is observed.

The most frequently presented model describing the influence of the crystal field on the energy levels of TM ions is the Tanabe-Sugano diagram. The Tanabe-Sugano diagram presents the dependence of the energies of localized d states on the crystal field strength. The x-axis of a

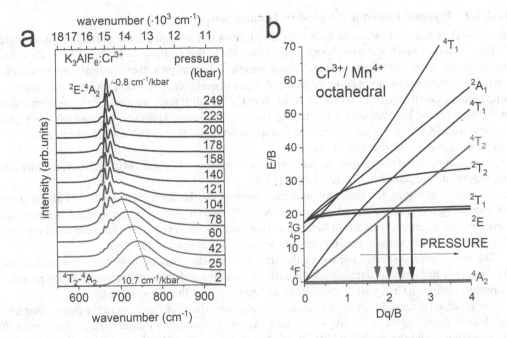

FIGURE 12.17 (a) Pressure dependence of luminescence spectra of $K_3Al_{0.97}F_6{:}0.03Cr^{3+}$. (Modified from Lee C, Bao Z, Fang MH, Lesniewski T, Mahlik S, Grinberg M, Leniec G, Kaczmarek SM, Brik MG, Tsai YT, Tsai TL Liu RS (2020) Chromium(III) doped fluoride phosphors with broadband infrared emission for light-emitting diodes. *Inorg. Chem.* 59: 37. With permission.) (b) Tanabe–Sugano diagram for $3d^3$ system in octahedral crystal field.

Tanabe-Sugano diagram is expressed by crystal field splitting parameter Dq, and the y-axis is in terms of the energy E, both scaled by the Racah parameter B [119]. The d^3 Tanabe–Sugano diagram which represents the selected energetic states of octahedrally coordinated Cr^{3+} (and Mn^{4+}) is presented in Figure 12.17(b). The energy of the $^4T_2$ state in relation to the ground state $^4A_2$ is equal to the crystal field strength $10Dq$. Depending on the crystal field strength, represented in the diagram by the ratio Dq/B, the first excited state is either $^4T_2$ (the low field case for $Dq/B < 2.4$) or ^2E (the high field case for $Dq/B > 2.4$.

Often the results of pressure measurements are also interpreted using Tanabe-Sugano diagrams. For example, crossover observed in Cr^{3+} ions can be explained simply by the pressure-induced increasing the crystal field as shown in Figure 12.17(b). This explains the increase in the energy of the $^4T_2$ state relative to the $^4A_2$ state. However, for the ^2E state in relation to the $^4A_2$ state, Tanabe-Sugano predicts an opposite relationship (slight blueshift) to the observed ones (redshift). It should be remembered that the Tanabe-Sugano diagram is parameterized to the constant Racah B parameter (thus ignores the increase in the covalencyunder pressure), so it does not reflect the pressure dependence, and it can only be qualitatively useful.

One noticed that the pressure-induced crossover allowing observation of narrow lines instead of broadband emission can be used to distinguish between different multiple Cr^{3+} luminescence centers in a material. These are difficult to distinguish of different Cr^{3+} centers if they emit broad and strongly overlapping bands. At the same time, after the crossover, we have well-distinguished lines the proper analysis of which allows us to determine the number and type of centers. An example of such an approach is presented in the refs. [130–132].

The comprehensive reviews on high-pressure spectroscopy of the Cr^{3+} doped materials can be found in [9] and [10].

12.4.4.2   Pressure Dependence of Mn⁴⁺ Luminescence

In the XXI century, an intensive search for a red phosphor to improve the performance of white LEDs resulted in much attention being paid to fluoride lattice activated with Mn^{4+} ions. This also applies to high-pressure measurements of these materials. Importantly, there is significant number of studies reporting information about pressure changes in emission, excitation spectra, and luminescence kinetics in this case [133–136]. In contrast to Cr^{3+} in Mn^{4+} ions, we observe the only emission in the high crystal field region related to the $^2E \rightarrow {}^4A_2$ transition. However, because the electronic configuration Mn^{4+} (3d³) is the same as the electronic configuration of Cr^{3+} models obtained for Cr^{3+} can be used to describe the Mn^{4+} system.

As an example, the photoluminescence excitation spectra of KNaSiF₆:Mn⁴⁺ measured under different pressure are shown in Figure 12.18(a) [133].

At ambient pressure, the excitation spectra consist of two broadbands at about 355 and 460 nm corresponding to the spin-allowed transitions $^4A_2 \rightarrow {}^4T_1$ and $^4A_2 \rightarrow {}^4T_2$, respectively. When pressure increases, the expected blueshift of both excitation bands is observed. The excitation band maxima shift almost linearly with increasing pressure with the rate around 9 cm⁻¹/kbar.

The luminescence spectra consist of the six characteristic lines at 610–650 nm related to the $^2E \rightarrow {}^4A_2$ spin-forbidden d–d transition (the zero-phonon line and phonon repetition). When pressure increases, similar to Cr^{3+}, a slight redshift of all luminescence lines is observed.

The obtained results of the pressure-induced changes qualitatively agree with the theory. Namely, we observe the blueshift of the $^4A_2 \rightarrow {}^4T_2$ and $^4A_2 \rightarrow {}^4T_2$ caused by the increase of the impact of the crystal field and the redshift of the $^2E \rightarrow {}^4A_2$ transition caused by the increasing nephelauxetic effect. However, the spectroscopic data yields pressure rate for crystal field strength that was two times smaller than that calculated using relation (12.15). Such difference was explained by increasing local microscopic compressibility compared to the bulk one. Such difference has been reported earlier for several materials doped with Cr^{3+} [9, 127, 136].

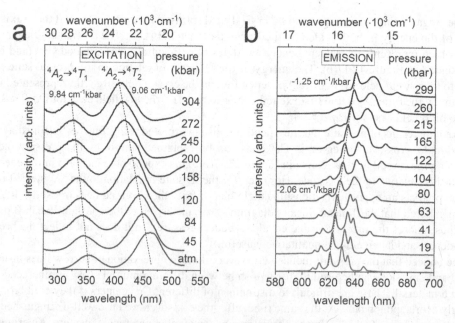

FIGURE 12.18   Pressure dependence of (a) excitation and (b) luminescence spectra of NaKSiF₆:Mn⁴⁺. (Modified from Jin Y, Fang MH, Grinberg M, Mahlik S, Lesniewski T, Brik MG, Luo GY, Lin JG, Liu RS (2016) Narrow Red Emission Band Fluoride Phosphor KNaSiF₆:Mn⁴⁺ for Warm White Light-Emitting Diodes. *ACS Appl. Mater. Interfaces* 8: 11194, with permission.)

12.4.4.3 Pressure Dependence of Mn⁴⁺ Luminescence Kinetic

The decay curves of the $^2E \rightarrow {}^4A_2$ emission in $KNaSiF_6:Mn^{4+}$ observed at different pressure are shown in Figure 12.19(a). Figure 12.19(b) shows the pressure dependence of the decay time of the Mn^{4+} emission in $KNaSiF_6:Mn^{4+}$ fitted by a single exponential equation for all considered pressures.

When pressure increases, the decay time of the $^2E \rightarrow {}^4A_2$ transition increases. Such a situation has also been observed for Cr^{3+} [137, 138], and Mn^{4+} in other lattices [133–136]. The $^2E \rightarrow {}^4A_2$ transition is parity and spin forbidden, and luminescence can appear due to the lowest quartet state $^4T_2$ mixing with the ^2E state due to spin-orbit interaction. In a high field system where the energy the ^2E and $^4T_2$ are separated by a large energy distance, one can express the quartet contribution to the ^2E electronic manifold using the perturbation approach [128]. The $^2E \rightarrow {}^4A_2$ luminescence lifetime can be calculated from the relation:

$$\frac{1}{\tau} \approx \frac{|V_{s-o}|^2}{\tau_T} \frac{1}{(\Delta + S\hbar\omega)^2} \tag{12.16}$$

where $1/\tau_T$ is the probability of the $^4T_2 \rightarrow {}^4A_2$ radiative transition, V_{S-O} is the spin-orbit coupling constant, and $\Delta + S\hbar\omega$ is equal to the energy distance between the emitting level ^2E and the energy of the $^4T_2$ state (corresponding to the excitation band maximum from the ground state to the $^4T_2$ state). The result of calculations for $KNaSiF_6:Mn^{4+}$ is presented by the solid curve in Figure 12.18(a). The fitting has been obtained assuming that the $\left(|V_{s-o}|^2\right)/\tau_T$ is equal to $1.91 \cdot 10^{-1}$ cm^{-2} s^{-1} and is pressure independent. The model used satisfactorily describes the obtained experimental results. However, in many cases, the impact of other influences (strength of local odd parity crystal field, nonradiative relaxation) that are not considered for $KNaSiF_6:Mn^{4+}$ have to be taken into account [133–136]. The complete model that allows calculating the probability of the $^2E \rightarrow {}^4A_2$ transition in high field materials that includes spin-orbit interaction and electron-phonon coupling in the excited states ^2E, $^2T_1$, and $^4T_2$ electronic states with their vibronic structure is discussed in detail in paper [138] and [139].

In contrast to Ln ions, in studies on the behavior of TM luminescence at high pressure, the relationship between local dopant states and band states is rarely considered. Exciting work in this context has been presented by Ueda et al. [140]. In this study, the pressure dependence of the persistent luminescence decay curves of $Y_3AlGa_4O_{12}:Ce^{3+}$ co-doped with Yb^{3+} or Cr^{3+} (electron traps) was

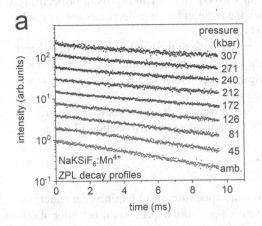
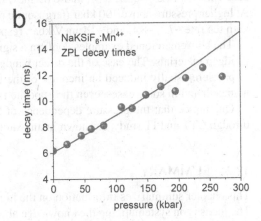

FIGURE 12.19 Decay profiles (a) and (b) decay times of emission in $KNaSiF_6:Mn^{4+}$ versus pressure. The solid curve represents the calculated lifetime by formula (12.17). (Modified from Jin Y, Fang MH, Grinberg M, Mahlik S, Lesniewski T, Brik MG, Luo GY, Lin JG, Liu RS (2016) Narrow Red Emission Band Fluoride Phosphor $KNaSiF_6:Mn^{4+}$ for Warm White Light-Emitting Diodes. *ACS Appl. Mater. Interfaces* 8: 11194, with permission.)

investigated. Obtained results indicate that the trap depth of Yb^{3+} becomes deeper and that of Cr^{3+} becomes shallower with increasing pressure. The electron trap depth increasing for the Yb^{2+} (Yb^{3+} +e) was explained by the pressure-induced increase of the edge of CB. In contrast, the decreasing trend for the Cr^{2+} (Cr^{3+}+e) was described as the increase of the e_g level of 3d orbitals due to increasing crystal field splitting. Under pressure, the splitting progress faster than the rise of the band edge, resulting in the e_g level approaching the CB. Such examples show that in the task related to determining the type and depth of electron and hole traps, in which knowledge is necessary to understand the mechanisms of persistent luminescence, high-pressure spectroscopy is a cognitive tool.

12.4.5 PRESSURE DEPENDENCE OF BAND STATES

The behavior of the band states as a function of pressure in phosphors has been exhaustively studied theoretically and experimentally in semiconductors [141, 142]. Generally, if semiconductors have a direct energy bandgap, the increase in pressure causes an increase in the energy bandgap. In an indirect bandgap semiconductor, increasing the pressure reduces the energy bandgap. Such phenomena usually cause direct bandgap semiconductors to change to an indirect one at high pressure [141, 142].

Similar conclusions can be drawn for dielectrics, for which, however, we have much less experimental and theoretical data in terms of the pressure dependence of band structure. Such a lack of data is caused by the inability to directly measure the dielectric energy bandgap's pressure dependence. The difficulty of measuring pressure-dependent absorption/excitation spectra is caused by difficulties in precise projecting the light of a broadband light source onto a small sample in the DAC. Moreover, numerous wide-gap materials have absorption band located at wavelengths shorter than 240 nm, beyond the measuring range available in high-pressure spectroscopy using the DAC.

High-pressure absorption has been investigated for zircon-type YVO_4, $YbVO_4$, $LuVO_4$, and $NdVO_4$ up to 20 GPa [143]. The pressure-induced zircon-to-scheelite structural phase transition is observed in all the studied materials to cause a collapse of the bandgap energy. The pressure coefficient shows positive values for the zircon phase and negative values for the scheelite phase. The evolution of the bandgap with pressure has also been presented for AWO_4 (A = Ca, Sr, Ba, Pb) [144]. In the pressure range up 60 kbar (scheelite phase), the direct bandgap slightly changes with pressure with the rates −0.21, 0.37, and 0.89 meV/kbar for $CaWO_4$, $SrWO_4$, and $BaWO_4$, respectively. At higher pressure above 100 kbar (fergusonite phase), the indirect bandgap decreases more rapidly with the rate −7.3, −8.1, and −1.1 meV/kbar, respectively.

The abovementioned examples confirm a significant decrease of the bandgap energy in indirect bandgap materials. The case of the direct bandgap dielectrics is not as unequivocal. The influence of pressure usually induced an increase of the bandgap energy. However, the trend is not as pronounced, and in some cases, even the opposite is observed.

One notice that the pressure dependence of band structure can also be investigated indirectly, through CTT and IT study, as shown in an example $Y_2O_2S:Ln^{3+}$ described in Section 12.4.3.2.

12.5   SUMMARY

This chapter summarizes information on the high-pressure spectroscopy of luminescent materials. The focus is on systematizing the knowledge about the high-pressure impact on the behavior of different optical transitions in inorganic phosphors. The effects of pressure on the fundamental properties of *f–f*, *d–f*, *d–d*, and CTTs are considered based on the recent examples from the literature concerning phosphor materials activated with lanthanide and transition metal ions. The observed spectroscopic properties are discussed based on commonly used models containing the nephelauxetic effect, the crystal field theory, and the electron-lattice interactions.

ACKNOWLEDGMENTS

This chapter is dedicated to Professor Marek Grinberg, for whom high-pressure spectroscopy was a scientific adventure in which I had the pleasure to participate. Special thanks go to my teammates N. Majewska, A. Baran, and T. Leśniewski for their assistance in the editing of this chapter.

REFERENCES

1. Murnaghan FD (1944) The compressibility of media under extreme pressures. *Proc. Natl. Acad. Sci. U.S.A.* 30 (9): 244.
2. Callen HB (1985) *Thermodynamics and an introduction to thermostatistics.* New York, NY: Wiley.
3. Batsanov SS (2006) Mechanism of metallization of ionic crystals by pressure. *Russ. J. Phys. Chem.* 80: 135.
4. Ledbetter HM, Kim SA (2000) Bulk moduli systematics in oxides, including superconductors. In *Handbook of elastic properties of solids, liquids, and gases,* ed. Levy M, Bass H, Stern R, 2: 249. San Diego, CA: Academic Press.
5. Sherman WF, Stadtmuller AA (1987) *Experimental techniques in high pressure research.* London: John Wiley and Sons Ltd.
6. Jayaraman A (1983) Diamond anvil cell and high-pressure physical investigations. *Rev. Mod. Phys.* 55: 65.
7. Merrill L, Bassett WA (1974) Miniature diamond anvil pressure cell for single crystal x-ray diffraction studies. *Rev. Sci. Instrum.* 45: 290.
8. Hozapfel WB, Isaacs NS (1997) *High pressure techniques in chemistry and physics: a practical approaches.* Oxford: Oxford University Press.
9. Bray KL (2001) High pressure probes of electronic structure and luminescence properties of transition metal and lanthanide systems. In *Transition metal and rare earth compounds. Topics in current chemistry,* ed. Yersin H, 213: 1. Berlin, Heidelberg: Springer.
10. Grinberg M (2017) Principles of energy transfer based on high pressure measurements. In *Phosphors, up conversion nano particles, quantum dots and their applications,* ed. Liu RS, 1: 67. Berlin, Heidelberg: Springer.
11. Hrubiak R, Sinogeikin S, Rod E, Shen G (2015) The laser micro-machining system for diamond anvil cell experiments and general precision machining applications at the High Pressure Collaborative Access Team. *Rev. Sci. Instrum.* 86: 072202.
12. Fujishiro I, Piermarini GJ, Block S, Munro RG (1982) Viscosities and glass transition pressure in the methanol-ethanol-water system. In *High pressure in research and industry. 8th AIRAPT Conference Proceedings, Arkitektkopia.* Uppsala, Sweden, eds. Backman CM, Johannisson T, Tegner L, II: 608.
13. Piermarini GJ, Block S, Barnnett JD (1973) Hydrostatic limits in liquids and solids to 100 kbar. *J. Appl. Phys.* 44: 5377.
14. Ragan DD, Clark DR, Scheiferl D (1996) Silicone fluid as a high-pressure medium in diamond anvil cells. *Rev. Sci. Instrum.* 67: 494.
15. Shen Y, Kumar RS, Pravica M, Nicol MF (2004) Characteristics of silicone fluid as a pressure transmitting medium in diamond anvil cells. *Rev. Sci. Instrum.* 75: 4450.
16. Forman RA, Piermarini GJ, Barnett JD, Block S (1972) Pressure measurement made by the utilization of ruby sharp-line luminescence. *Science* 176: 284.
17. Piermarini GJ, Block S, Barnett JD, Forman RA (1975) Calibration of the pressure dependence of the R_1 ruby fluorescence line to 195 kbar. *J. Appl. Phys.* 46: 2774.
18. Barnett JD, Block S, Piermarini GJ (1973) An optical fluorescence system for quantitative pressure measurement in the diamond-anvil cell. *Rev. Sci. Instrum.* 44:1.
19. Datchi F, Dewaele A, Loubeyre P, Letoullec R, Le Godec Y, Canny B (2007) Optical pressure sensors for high-pressure-high-temperature studies in a diamond anvil cell. *High Press Res: Int. J.* 27: 447.
20. Mao HK, Xu J, Bell PM (1986) Calibration of the ruby pressure gauge to 800 kbar under quasi-hydrostatic conditions. *J. Geophys. Res.* 91: 4673.
21. Mao HK, Bell PM (1976) High-pressure physics: The 1-megabar mark on the ruby R_1 static pressure scale. *Science* 191: 851.
22. Ragan DD, Gustavsen R, Schiferl D (1992) Calibration of the ruby R_1 and R_2 fluorescence shifts as a function of temperature from 0 to 600 K. *J. Appl. Phys.* 72: 5539.

23. Runowski M, Wozny P, Stopikowska N, Guo O, Lis S (2019) Optical pressure sensor based on the emission and excitation band width (fwhm) and luminescence shift of Ce^{3+}-doped fluorapatite high-pressure sensing. *ACS Appl. Mater. Interfaces* 11: 4131.

24. Shen YR, Gregorian T, Holzapfel WB (1991) Progress in pressure measurements with luminescence sensors. *High Press Res: Int. J.* 7: 73.

25. Barzowska J, Lesniewski T, Mahlik S, Seo HJ, Grinberg M (2018) $KMgF_3:Eu^{2+}$ as a new fluorescence-based pressure sensor for diamond anvil cell experiments. *Opt. Mater.* 84: 99.

26. Wisniewski K, Mahlik S, Grinberg M, Seo HJ (2011). Influence of high hydrostatic pressure on Eu^{2+}-luminescence in $KMgF_3:Eu^{2+}$ crystal. *J. Lumin.* 131: 306.

27. Dorenbos P (2003) Energy of the first $4f^7 \rightarrow 4f^6 5d$ transition of Eu^{2+} in inorganic compounds. *J. Lumin.* 104: 239.

28. Dunstan DJ, Spain IL (1989) Technology of diamond anvil high-pressure cells: I. Principles, design and construction. *J. Phys. E Sci. Instrum.* 22: 913.

29. Spain IL, Dunstan DJ (1989) The technology of diamond anvil high-pressure cells: II. Operation and use. *J. Phys. E Sci. Instrum.* 22: 923.

30. Blasse G, Grabmaier BC (1994) *Luminescent materials*, Berlin: Springer-Verlag.

31. Grinberg M, Mahlik S (2008) Impurity-trapped excitons: experimental evidence and theoretical concept. *J. Non-Cryst. Solids* 354: 4163.

32. Grinberg M, Mahlik S (2013) Impurity trapped exciton states related to rare earth ions in crystals under high hydrostatic pressure. *Crystallogr. Rep.* 58: 139.

33. Struck CW, Fonger WH (1970) Role of the charge-transfer states in feeding and thermally emptying the ^5D states of Eu^{3+} in yttrium and lanthanum oxysulfides. *J. Lumin.* 1–2: 456.

34. Blasse G (1979) Chemistry and physics of R-activated phosphors. In *Handbook on the physics and chemistry of rare earths*, ed. Gschneidner Jr. KA, Eyring L, 4: 237. Amsterdam: North-Holland.

35. Boutinaud P, Mahiou R, Cavalli E, Bettinelli M (2004) Excited state dynamics of Pr^{3+} in YVO_4 crystals. *J. Appl. Phys.* 96: 4923.

36. Boutinaud P, Mahiou R, Cavalli E, Bettinelli M (2006) Luminescence properties of Pr^{3+} in titanates and vanadates: Towards a criterion to predict $^3P_0$ emission quenching. *Chem. Phys. Lett.* 418: 185.

37. Cavalli E, Boutinaud P, Mahiou R, Bettinelli M, Dorenbos P (2010) Luminescence dynamics in Tb^{3+}-doped $CaWO_4$ and $CaMoO_4$ crystals. *Inorg. Chem.* 49: 4916.

38. Avram NM, Brik MG (2013) *Optical properties of 3d-ions in crystals: Spectroscopy and crystal field analysis*. Beijing: Tsinghua University Press, and Berlin, Heidelberg: Springer-Verlag.

39. Henderson B, Imbusch GF (1989) *Optical spectroscopy of inorganic solids*. Oxford: Clarendon Press.

40. Wybourne BG (1965) *Spectroscopic properties of rare earth*. New York, NY: Wiley Interscience.

41. Dieke GH (1968) *Spectra and energy levels of rare earth ions in crystals*. New York, NY: Interscience Publishers.

42. Jorgensen CK (1962) The nephelauxetic series. *Prog. Inorg. Chem.* 4: 73.

43. Gregorian T, d'Amour-Sturm H, Holzapfel WB (1989) Effect of pressure and crystal structure on energy levels of Pr^{3+} in $LaCl_3$. *Phys. Rev. B* 39: 12497.

44. Tröster T, Gregorian T, Holzapfel WB (1993) Energy levels of Nd^{3+} and Pr^{3+} in RCl_3 under pressure. *Phys. Rev. B* 48 (5): 2960.

45. Bungenstock C, Tröster T, Holzapfel WB, Bini R, Ulivi L, Cavalieri S (1998) Study of the energy level scheme of Pr^{3+}: $LaOCl$ under pressure. *J. Phys. Condens. Matter.* 10: 9329.

46. Wang Q, Bulou A (1993) Nephelauxetic effects in europium-doped LnOX (Ln=La, Gd, or Y; X=F, Cl, or Br) crystals: high-pressure effects. *J. Phys. Condens. Matter.* 5: 7657.

47. Cowan RD (1981) *The theory of atomic structure and spectra*. Berkeley, CA: University of California Press.

48. Newman DJ, Ng B, Poon YM (1984) Parametrisation and interpretation of paramagnetic ion spectra. *J. Phys. C Solid State Phys.* 17: 5577.

49. Pal P, Penhouet T, D'Anna V, Hagemann H (2013) Effect of pressure on the free ion and crystal field parameters of Sm^{2+} in $BaFBr$ and $SrFBr$ hosts. *J. Lumin.* 134: 678.

50. Newman DJ, Ng BKC (2000) *Crystal Field Handbook*. Cambridge: Cambridge University Press.

51. Carnall WT, Crosswhite H, Crosswhite HM (1977) *Energy level structure and transition probabilities of the trivalent lanthanides in LaF_3. Argonne national laboratory report*. Argonne: Argonne National Laboratory.

52. Grinberg M (2006) High pressure spectroscopy of rare earth ions doped crystals-new results. *Opt. Mater.* 28: 26.

53. Turos-Matysiak R, Zheng H, Wang JW, Yen WM, Meltzer RS, Lukasiewicz T, Swirkowicz M, Grinberg M (2007) Pressure dependence of the $^3P_0 \rightarrow ^3H_4$ and $^1D_2 \rightarrow ^3H_4$ emission in Pr^{3+}:YAG. *J. Lumin.* 122–123: 322.

54. Kaminska A, Kozanecki A, Ramirez MO, Bausa LE, Boulon G, Bettinelli M, Boćkowski M, Suchocki A (2016) Spectroscopic study of radiative intra-configurational 4f-4f transitions in Yb^{3+}-doped materials using high hydrostatic pressure. *J. Lumin.* 169: 507.

55. Kaminska A, Kozanecki A, Trushkin S, Suchocki A (2010). Spectroscopy of ytterbium-doped InP under high hydrostatic pressure. *Phys. Rev. B* 81: 165209.

56. Brik MG, Kaminska A, Suchocki A (2010) Ab initio calculations of structural, electronic, optical and elastic properties of pure and Yb-doped InP at varying pressure. *J. Appl. Phys.* 108: 103520.

57. Kaminska A, Ma CG, Brik MG, Kozanecki A, Bockowski M, Alves E, Suchocki A (2012) Electronic structure of ytterbium-implanted GaN at ambient and high pressure: experimental and crystal field studies. *J. Phys. Condens. Matter.* 24: 095803.

58. Manjon FJ, Jandl S, Riou G, Ferrand B, Syassen K (2004) Effect of pressure on crystal-field transitions of Nd-doped YVO_4. *Phys. Rev. B* 69: 165121.

59. Kobyakov S, Kaminska A, Suchocki A, Galanciak D, Malinowski M (2006) Nd3+ doped Yttrium Aluminum Garnet crystal as a near-infrared pressure sensor for diamond anvil cells. *Appl. Phys. Lett.* 88 (23): 234102.

60. Gryk W, Dyl D, Ryba-Romanowski W, Grinberg M (2005) Spectral properties of $LiTaO_3$: Pr^{3+} under high hydrostatic pressure. *J. Phys. Condens. Matter.* 17: 5381.

61. Troster T, Lavín V (2003) Crystal fields of Pr^{3+} in $LiYF_4$ under pressure. *J. Lumin.* 101: 243.

62. Changxin G, Bilin L, Yuefen H, Hongbin C (1991). High pressure effect on the luminescence spectra of Eu^{3+} in stoichiometric host microcrystals. *J. Lumin.* 48-49: 489.

63. Yuanbin C, Shensin L, Wufu S, Lizhong W, Guangtian Z (1986) Crystal field analysis for emission spectra of $LaOCl:Eu^{3+}$ under high pressure. *Phys. B+C* 139-140: 555.

64. Shenxin L, Yuanbin C, Xuyi Z, Lizhong W (1997) High pressure luminescence and pressure induced phase transition for $LiYF_4$:Eu, *J. Alloys Compd.* 255: 1.

65. Mahlik S, Lazarowska A, Grobelna B, Grinberg M (2012) Luminescence of $Gd_2(WO_4)_3$: Ln^{3+} at ambient and high hydrostatic pressure. *J. Phys. Condens. Matter.* 24: 485501.

66. Mahlik S, Malinowski M, Grinberg M (2011) High pressure and time resolved luminescence spectra of $Gd_3Ga_5O_{12}$:Pr^{3+} crystal. *Opt. Mater.* 33: 1525.

67. Kaminska A, Biernacki S, Kobyakov S, Suchocki A, Boulon G, Ramirez MO, Bausa L (2007) Probability of Yb^{3+} 4f–4f transitions in gadolinium gallium garnet crystals at high hydrostatic pressures. *Phys. Rev. B* 75: 174111.

68. Kaminska A, Buczko R, Paszkowicz W, Przybylinska H, Werner-Malento E, Suchocki A, Brik M, Durygin A, Drozd V, Saksena S (2011) Merging of the $^4F_{3/2}$ level states of Nd^{3+} ions in the photoluminescence spectra of gadolinium-gallium garnets under high pressure. *Phys. Rev. B* 84: 075483.

69. Kaminska A, Duzynska A, Berkowski M, Trushkin S, Suchocki A (2012) Pressure-induced luminescence of cerium-doped gadolinium gallium garnet crystal. *Phys. Rev. B* 85: 155111.

70. Dorenbos P (2000) The $4f^n \leftrightarrow 4f^{n-1}5d$ transitions of the trivalent lanthanides in halogenides and chalcogenides. *J. Lumin.* 91: 91.

71. Dorenbos P (2000) The 5d level positions of the trivalent lanthanides in inorganic compounds. *J. Lumin.* 91: 155.

72. Dorenbos P (2002) 5d-level energies of Ce^{3+} and the crystalline environment. IV. Aluminates and "simple" oxides. *J. Lumin.* 99: 283.

73. Bethe H (1929) Termaufspaltung in Kristallen. *Annalen der Physik* 395: 133.

74. Van Vleck JH (1937) The puzzle of rare-earth spectra in solids. *J. Phys. Chem.* 41: 67.

75. Dieke GH, Crosswhite HM (1963) The spectra of the doubly and triply ionized rare earths. *Appl. Opt.* 2: 675.

76. Loh E (1966) Lowest 4f→5d transition of trivalent rare-earth Ions in CaF_2 crystals. *Phys. Rev.* 147: 332.

77. Loh E (1967) Ultraviolet absorption spectra of Ce^{3+} in alkaline-earth fluorides. *Phys. Rev.* 154: 270.

78. Morrison CA (1980) Host dependence of the rare-earth ion energy separation $4f^n - 4f^{n-1}$. *J. Chem. Phys.* 72: 1001.

79. Henderson B, Bartram RH (2000) *Crystal – field engineering of solid state laser materials. Cambridge studies in modern optics*. Cambridge: Cambridge University Press.

80. Grüneisen E (1912) Theorie des festen Zustandes einatomiger elemente. *Ann. Phys.* 344: 257.

81. Grinberg M, Barzowska J, Shen YR, Meltzer RS, Bray KL (2004) Pressure dependence of electron-phonon coupling in Ce^{3+}-doped $Gd_3Sc_2Al_3O_{12}$ garnet crystals. *Phys. Rev. B* 69: 205101.

82. Grinberg M (2011) Excited states dynamics under high pressure in lanthanide-doped solids. *J. Lumin.* 131: 433.

83. Lazarowska A, Mahlik S, Grinberg M, Yeh CW, Liu RS (2015) Pressure dependence of the $Sr_2Si_5N_8:Eu^{2+}$ luminescence. *J. Lumin.* 159: 183.

84. Tsai YT, Nguyen HD, Lazarowska A, Mahlik S, Grinberg M, Liu RS (2016). Improvement of the water resistance of a narrow-band red-emitting $SrLiAl_3N_4:Eu^{2+}$ phosphor synthesized under high isostatic pressure through coating with an organosilica layer. *Angew. Chem.* 55: 9652.

85. Leaño Jr. JL, Lin SY, Lazarowska A., Mahlik S, Grinberg M, Liang C, Zhou W, Molokeev MS, Atuchin VV, Tsai YT, Lin CC, Sheu HS, Liu RS (2016) Green light-excitable Ce-doped nitridomagnesoaluminate $Sr[Mg_2Al_2N_4]$ Phosphor for white light-emitting diodes. *Chem. Mater.* 28: 6822.

86. Lazarowska A, Mahlik S, Grinberg M, Li G, Liu RS (2016) Structural phase transitions and photoluminescence properties of oxonitridosilicate phosphors under high hydrostatic pressure. *Sci. Rep.* 6: 34010.

87. Zhang X, Fang MH, Tsai YT, Lazarowska A, Mahlik S, Lesniewski T, Grinberg M, Pang WK, Pan F, Liang C, Zhou W, Wang J, Lee JF, Cheng BM, Hung TL, Chen YY, Liu RS (2017) Controlling of structural ordering and rigidity of β-SiAlON:Eu through chemical cosubstitution to approach narrow-band-emission for light-emitting diodes application. *Chem. Mater.* 29: 6781.

88. Leaño Jr. JL, Lazarowska A, Mahlik S, Grinberg M, Sheu HS, Liu RS (2018) Disentangling red emission and compensatory defects in $Sr[LiAl_3N_4]:Ce^{3+}$ phosphor. *Chem. Mater.* 30: 4493.

89. Mahlik S, Lazarowska A, Grinberg M, Liu TC, Liu RS (2013) Luminescence spectra of β-SiAlON/Pr^{3+} under high hydrostatic pressure. *J. Phys. Chem. C* 117: 13181.

90. Shen YR, Gatch DB, Rodríguez Mendoza UR, Cunningham G, Meltzer RS, Yen WM, Bray KL (2002) Pressure-induced dark-to-bright transition in $Lu_2O_3:Ce^{3+}$. *Phys. Rev. B* 65: 212103.

91. Wittlin A, Przybylińska H, Berkowski M, Kamińska A, Nowakowski P, Sybilski P, Ma CG, Brik MG, Suchocki A (2015). Ambient and high pressure spectroscopy of Ce^{3+} doped yttrium gallium garnet. *Opt. Mater. Express* 5: 1868.

92. Mahlik S, Lazarowska A, Ueda J, Tanabe S, Grinberg M (2016) Spectroscopic properties and location of the Ce^{3+} energy levels in $Y_3Al_2Ga_3O_{12}$ and $Y_3Ga_5O_{12}$ at ambient and high hydrostatic pressure. *Phys. Chem. Chem. Phys.* 18: 6683.

93. Dorenbos P (2012) Modeling the chemical shift of lanthanide 4f electron binding energies. *Phys. Rev. B* 85: 165107.

94. Shannon RD (1976) Revised effective ionic radii and systematic studies of interatomic distances in halides and chalcogenides. *Acta Crystallogr. A* 32: 751.

95. Ueda J, Dorenbos P, Bos AJJ, Kuroishi K, Tanabe S (2015) Control of electron transfer between Ce^{3+} and Cr^{3+} in the $Y_3Al_{5-x}Ga_xO_{12}$ host via conduction band engineering. *J. Mater. Chem. C* 3: 5642.

96. Monteseguro V, Rodriguez-Hernandez P, Lavın V, Manjon FJ, Munoz A (2013) Electronic and elastic properties of yttrium gallium garnet under pressure from ab initio studies. *J. Appl. Phys.* 113: 183505.

97. Lesniewski T, Mahlik S, Asami K, Ueda J, Grinberg M, Tanabe S (2018) Comparison of quenching mechanisms in $Gd_3Al_{5-x}Ga_xO_{12}$: Ce^{3+} (x=3 and 5) garnet phosphors by photocurrent excitation spectroscopy. *Phys. Chem. Chem. Phys.* 20: 18380.

98. Dorenbos P (2005) The Eu^{3+} charge transfer energy and the relation with the band gap of compounds. *J. Lumin.* 111: 89.

99. Dorenbos P (2013) A review on how lanthanide impurity levels change with chemistry and structure of inorganic compounds. *ECS J. Solid State Sci. Technol.* 2: R3001.

100. Dorenbos P, Krumpel AH, van der Kolk E, Boutinaud P, Bettinelli M, Cavalli E (2010) Lanthanide level location in transition metal complex compounds. *Opt. Mater.* 32: 1681.

101. Nakazawa E (2002) The lowest 4f-to-5d and charge-transfer transitions of rare earth ions in YPO_4 hosts. *J. Lumin.* 100: 89.

102. Nakazawa E, Shiga F (2003) Lowest 4f-to-5d and charge-transfer transitions of rare-earth ions in $LaPO_4$ and related host-lattices. *J. Appl. Phys.* 42: 1642.

103. Srivastava AM, Dorenbos P (2009). Charge transfer transitions and location of the rare ion energy levels in Ca-α-SiAlON. *J. Lumin.* 129: 634.

104. Brik MG, Mahlik S, Jankowski D, Strak P, Korona KP, Monroy E, Krukowski S, Kaminska A (2017) Experimental and first-principles studies of high-pressure effects on the structural, electronic, and optical properties of semiconductors and lanthanide doped solids. *J. Appl. Phys.* 56: 05FA02.

105. Behrendt M, Mahlik S, Szczodrowski K, Kukliński B, Grinberg M (2016) Spectroscopic properties and location of the Tb^{3+} and Eu^{3+} energy levels in Y_2O_2S under high hydrostatic pressure. *Phys. Chem. Chem. Phys.* 18: 22266.

106. Gryk W, Dyl D, Grinberg M, Malinowski M (2005) High pressure photoluminescsnce study of Pr^{3+} doped $LiNbO_3$ crystal. *Phys. State Solid C* 2: 188.
107. Gryk W, Dujardin C, Joubert MF, Ryba-Romanowski W, Malinowski M, Grinberg M (2006) Pressure effect on luminescence dynamics in Pr^{3+}-doped $LiNbO_3$ and $LiTaO_3$ crystals. *J. Phys. Condens. Matter.* 18: 117.
108. Mahlik S, Grinberg M, Cavalli E, Bettinelli M, Boutinaud P (2009) High pressure evolution of YVO4:Pr^{3+} luminescence. *J. Phys. Condens. Matter.* 21: 105401.
109. Mahlik S, Grinberg M, Kaminskii AA, Bettinelli M, Boutinaud P (2009) Luminescence of $Ca(NbO_3)_2$:Pr^{3+} at ambient and high hydrostatic pressure. *J. Lumin.* 129: 1219.
110. Mahlik S, Behrendt M, Grinberg M, Cavalli E, Bettinelli M (2012). Pressure effects on the luminescence properties of $CaWO_4$:Pr^{3+}. *Opt. Mater.* 34: 2012.
111. Mahlik S, Grinberg M, Cavalli E, Bettinelli M (2012). High pressure luminescence spectra of $CaMoO_4$:Pr^{3+}. *J. Phys. Condens. Matter.* 24: 215402.
112. Mahlik S, Cavalli E, Bettinelli M, Grinberg M (2013). Luminescence of $CaWO_4$:Pr^{3+} and $CaWO_4$:Tb^{3+} at ambient and high hydrostatic pressures. *Rad. Meas.* 56: 1.
113. Lazarowska A, Mahlik S, Grinberg M, Malinowski M (2011). High pressure luminescence and time resolved spectra of $LiNbO_3$:Pr^{3+}. *Photonics Lett. Pol.* 3: 67.
114. Mahlik S, Behrendt M, Grinberg M, Cavalli E, Bettinelli M (2013) High pressure luminescence spectra of $CaMoO_4$:Ln^{3+} (Ln = Pr, Tb). *J. Phys. Condens. Matter.* 25: 105502.
115. Mahlik S, Lazarowska A, Speghini A, Bettinelli M, Grinberg M (2014) Pressure evolution of luminescence in $Sr_xBa_{1-x}(NbO_2)_3$: Pr^{3+} (x= 1/2 and 1/3). *J. Lumin.* 152: 62.
116. Mahlik S, Cavalli E, Amer M, Boutinaud P (2015) Energy levels in $CaWO_4$: Tb^{3+} at high pressure. *Phys. Chem. Chem. Phys.* 17: 32341.
117. Behrendt M, Szczodrowski K, Mahlik S, Grinberg M (2014) High pressure effect on charge transfer transition in Y_2O_2S:Eu^{3+}. *Opt. Mater.* 36: 1616.
118. Behrendt M, Mahlik S, Grinberg M, Stefańska D, Dereń PJ (2017) Influence of charge transfer state on Eu^{3+} luminescence in $LaAlO_3$, by high pressure spectroscopy. *Opt. Mater.* 63: 158.
119. Sugano S, Tanabe Y, Kamimura H (1970) *Multiplets of transition-metal ions in crystals.* New York and London: Academic Press.
120. Bersuker IB (1996) *Electronic structure and properties of transition metal compounds: Introduction to the theory.* New York, Chichester, Brisbane, Toronto, Singapore: John Wiley & Sons.
121. Syassen K (2008) Ruby under pressure. *High Pres. Res.* 28: 75.
122. Munro RG (1977) A scaling theory of solids under hydrostatic pressure. *J. Chem. Phys.* 67: 3146.
123. Eggert JH, Goettel KA, Silvera IF (1989) Ruby at high pressure. I. Optical line shifts to 156 GPa. *Phys. Rev. B* 40: 5724.
124. Ma D, Zheng X, Xu Y, Zhang Z (1986) Theoretical calculations of the R_1 red shift of ruby under high pressure. *Phys. Lett. A* 115: 245.
125. Ma D, Chen J, Wang Z, Zhang Z (1988) Theoretical calculations of pressure-induced shifts of R, R´, B lines and U, Y bands of ruby. *Phys. Lett. A* 126: 377.
126. Suchocki A, Biernacki SW, Grinberg M (2007) Nephelauxetic effect in high-pressure luminescence of transition-metal ion dopants. *J. Lumin.* 125: 266.
127. Duclos SJ, Vohra YK, Ruoff AL (1990) Pressure dependence of the $^4T_2$ and $^4T_1$ absorption bands of ruby to 35 GPa. *Phys. Rev. B* 41: 5372.
128. Dolan JF, Kappers LA, Bartram RH (1986) Pressure and temperature dependence of chromium photoluminescence in K_2NaGaF_6: Cr^{3+}. *Phys. Rev. B* 33: 7339.
129. Lee C, Bao Z, Fang MH, Lesniewski T, Mahlik S, Grinberg M, Leniec G, Kaczmarek SM, Brik MG, Tsai YT, Tsai TL Liu RS (2020) Chromium(III) doped fluoride phosphors with broadband infrared emission for light-emitting diodes. *Inorg. Chem.* 59: 376.
130. Kaminska A, Kaczor P, Durygin A, Suchocki A, Grinberg M (2002) Low-temperature high-pressure spectroscopy of lanthanum lutetium gallium garnet crystals doped with Cr^{3+} and Nd^{3+}. *Phys. Rev. B* 65: 104106.
131. Kaminska A, Suchocki A, Arizmendi L, Callejo D, Jaque F, Grinberg M (2000) Spectroscopy of near-stoichiometric $LiNbO_3$: MgO, Cr crystals under high pressure. *Phys. Rev. B* 62: 10802.
132. Shen YR, Bray KL, Grinberg M, Barzowska J, Sokolska I (2000) Identification of multisite behavior in a broadly emitting transition-metal system using pressure. *Phys. Rev. B* 61: 14263.
133. Jin Y, Fang MH, Grinberg M, Mahlik S, Lesniewski T, Brik MG, Luo GY, Lin JG, Liu RS (2016) Narrow red emission band fluoride phosphor $KNaSiF_6$:Mn^{4+} for warm white light-emitting diodes. *ACS Appl. Mater. Interfaces* 8: 11194.

134. Wu WL, Fang MH, Zhou W, Lesniewski T, Mahlik S, Grinberg M, Brik MG, Sheu HS, Cheng BM, Wang J, Liu RS (2017) High color rendering index of Rb_2GeF_6:Mn^{4+} for light-emitting diodes. *Chem. Mater.* 29: 935.
135. Fang MH, Wu WL, Jin Y, Lesniewski T, Mahlik S, Grinberg M, Brik MG, Srivastava AM, Chiang CY, Zhou W, Jeong D, Kim SH, Leniec G, Kaczmarek SM, Sheu HS, Liu RS (2018). Control of luminescence by tuning of crystal symmetry and local structure in Mn^{4+}-activated narrow band fluoride phosphors. *Angew. Chem.* 57: 1797.
136. Fang MH, Yang TH, Lesniewski T, Mahlik S, Grinberg M, Peterson VK, Didier C, Pang WK, Su C, Liu RS (2019) Hydrogen-containing $Na_3HTi_{1-x}Mn_xF_8$ narrow-band phosphor for light-emitting diodes. *ACS Energy Lett.* 4: 527.
137. Galanciak D, Perlin P, Grinberg M, Suchocki A (1994) High pressure spectroscopy of LLGG doped with Cr^{3+}. *J. Lumin.* 60-61: 223.
138. Grinberg M, Suchocki A (2007) Pressure-induced changes in the energetic structure of the $3d^3$ ions in solid matrices. *J. Lumin.* 125: 97.
139. Lesniewski T, Mahlik S, Grinberg M, Liu RS (2017) Temperature effect on the emission spectra of narrow band Mn^{4+} phosphors for application in LEDs. *Phys. Chem. Chem. Phys.* 19: 32505.
140. Ueda J, Harada M, Miyano S, Yamada A, Tanabe S (2020) Pressure-induced variation of persistent luminescence characteristics in $Y_3Al_{5-x}Ga_xO_{12}$:Ce^{3+}–M^{3+} (M = Yb, and Cr) phosphors: opposite trend of trap depth for 4f and 3d metal ions. *Phys. Chem. Chem. Phys.* 22: 19502.
141. Goñi AR, Syassen K (1989) Chapter 4 Optical properties of semiconductors under pressure. *Semiconduct. Semim.* 54: 247.
142. Suski T (1998) *High pressure in Semiconductor Physics I.* ed. Suski T, Paul W, San Diego, CA: Academic Press.
143. Panchal V, Errandonea D, Segura A, Rdríguez-Hernández P, Muñoz A, Lopez-Moreno S, Bettinelli M (2011) The electronic structure of zircon-type orthovanadates: effects of high-pressure and cation substitution. *J. Appl. Phys.* 110: 043723.
144. Lacomba-Paralles R, Errandonea D, Segura A, Ruiz-Fuertes J, Rodríguez-Hernández P, Radescu S, López- Solano J, Mujica A, Muñoz A (2011) A combined high-pressure experimental and theoretical study of the electronic band-structure of scheelite-type AWO_4 (A = Ca, Sr, Ba, Pb) compounds. *J. Appl. Phys.* 110: 043703.

13 Absolute Photoluminescence Quantum Yield of Phosphors

Kengo Suzuki

CONTENTS

DOI: 10.1201/9781003098669-13

13.1   DEFINITION AND MEASUREMENT METHODS

The photoluminescence (PL) quantum yield Φ_{PL} for a molecule or material is defined as the ratio of the number of emitted photons $PN(Em)$ to the number of absorbed photons $PN(Abs)$. This is one of the most fundamental and important photophysical parameters for luminescent materials.

$$\Phi_{PL} = \frac{PN(Em)}{PN(Abs)} \tag{13.1}$$

The methods for measuring Φ_{PL} can be classified into absolute and relative methods (Table 13.1) [1]. In the absolute method, except for the calorimetric method, Φ_{PL} is determined by measuring the number of absorbed photons and the number of emitted photons within a unit time. In the calorimetric method, Φ_{PL} is determined by measuring the ratio of the heat energy released from a sample by nonradiative processes to the excitation energy absorbed. In the relative methods, Φ_{PL} is determined by comparing the integrated PL intensity for the sample with that for a standard material under the same excitation and emission measurement conditions.

13.2   VAVILOV METHOD AND ITS APPLICATION TO PHOSPHORS

13.2.1   VAVILOV METHOD

The first reliable absolute Φ_{PL} measurement method was introduced by Vavilov [2]. In the Vavilov method, the absolute quantum yield is determined by comparing the PL intensity with excitation light scattered from a solid scatterer. Melhuish measured the Φ_{PL} of several sample solutions using this method [3].

The measurement setup used by Melhuish consists of an excitation light source, a sample holder, and a photodetection system as pictured in Figure 13.1. A mercury lamp is used as the excitation light source and the excitation wavelength is selected using a filter. A solid scatterer (magnesium oxide) or a cuvette for sample solutions is mounted on the sample holder. The photodetection system includes a rhodamine B quantum counter, a red filter, and a photomultiplier tube. The rhodamine B quantum counter is a dense solution of rhodamine B, which outputs light, the intensity of which is proportional to the incident light intensity and independent of the wavelength of the incident light.

The protocol of the method used by Melhuish can be summarized as follows:

1. The solid scatterer is mounted on the sample holder and is irradiated with the excitation light at 45 degrees. The excitation light scattered from the solid scatterer enters the rhodamine B quantum counter positioned in the direction normal to the surface of the solid scatterer. The scattered excitation light is converted into the emission of the rhodamine B quantum counter and the emitted light is detected by the photomultiplier tube after passing through a red filter.

TABLE 13.1

Representative Measurement Methods for Determining the Φ_{PL}

	Method
I. Absolute method	Vavilov method (using magnesium oxide as a standard)
	Weber and Teale method (using solution scatterer as a standard)
	Calorimetric method
	Integrating sphere method
II. Relative method	Optically dilute/dense method

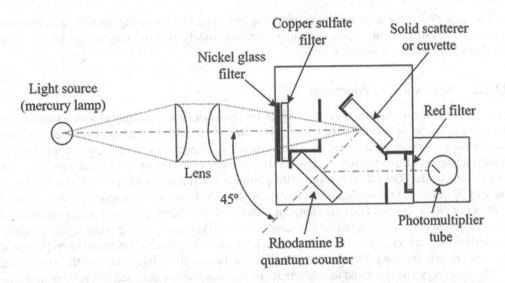

FIGURE 13.1 Schematic diagram of a setup for the Vavilov method. (Adapted with permission from Melhuish, W. H. 1955. *N. Z. J. Sci. Technol.* 37: 142. © The Optical Society.)

2. The solid scatterer is replaced by a cuvette filled with a highly concentrated sample solution. The excitation light is completely absorbed by the sample solution. The PL emitted from the sample solution is detected by the photodetection system.

Φ_{PL} is calculated from the ratio of the output intensities of the photodetection system for the solid scatterer and the sample solution, S and F, respectively, and correction terms using the following equation.

$$\Phi_{PL} = 4\left(\frac{F}{S}\right) n^2 \left(\frac{I_0}{I_\theta}\right)_{AV} \frac{R_1 R_2}{1 - R_e - R_f} \tag{13.2}$$

where, n is the refractive index of the solvent, R_1 is the relative reflectance of the solid scatterer at 45 degrees relative to the normal direction, R_2 is the absolute reflectance of the solid scatterer at the excitation wavelength, R_e is the fraction of the excitation light reflected off the cuvette when the excitation light is irradiated at 45 degrees, R_f is the fraction of the PL reflected off the cuvette when the PL is output vertically from inside the cuvette, $(I_0/I_\theta)_{AV}$ is the distribution of the PL intensity averaged over the aperture angle, $R_1 R_2$ is a correction term related to the measurement of the excitation light when the solid scatterer is mounted on the sample holder, and n^2, $(I_0/I_\theta)_{AV}$, and $(1 - R_e - R_f)$ are correction terms related to the PL measurement of the sample solution.

In the Vavilov method, part of the PL output from the sample solution is reduced by reabsorption because a highly concentrated solution is used for the measurement. The effect of the reabsorption becomes more pronounced as the overlap between the absorption and PL spectra increases. Melhuish corrected the Φ_{PL} value for the reabsorption by comparing the PL spectrum of the diluted solution with that of the dense solution used for the Φ_{PL} measurement.

Melhuish determined the Φ_{PL} value for 0.5-M sulfuric acid solution of quinine bisulfate (QBS) at a concentration of 5.0×10^{-3} M to be 0.508 [3]. The following constants were used to calculate the Φ_{PL}.

$$\theta = 18°, \; (I_0/I_\theta)_{AV} = 1.023, \; R_1 = 0.92, \; R_2 = 0.96, \; 1 - R_e - R_f = 0.90$$

Then, the Φ_{PL} value at an infinite dilution was estimated to 0.546 using the self-quenching rate constant. The estimated Φ_{PL} value at infinite dilution is widely used by many researchers as one of the standards for relative methods.

13.2.2   APPLICATION TO PHOSPHORS

Bril and Jager-Veenis measured the Φ_{PL} values for phosphors using equipment with a similar configuration to that in the Vavilov method [4]. As shown in Figure 13.2, a high-pressure mercury lamp is used as an excitation light source and the excitation wavelength is selected by interference filters. A solid scatterer (barium sulfate) or a layer of the phosphor with a thickness of about 2 mm is placed on a sample holder. The solid scatterer or the phosphor is irradiated with the excitation light at an angle of 50 degrees to the surface of the sample holder. The excitation light reflected from the solid scatterer or the PL emitted from the phosphor is detected by a thermoelement detector (thermopile) fixed perpendicular to the surface of the sample holder. Thermopiles are thermoelectric detectors that utilize the Seebeck effect, in which a thermal electromotive force is generated in proportion to the incident light intensity. The spectral response of a thermopile is independent of the wavelength.

The PL energy efficiency and the reflectance of the phosphor are obtained by the three measurement steps described below.

1. Measurement of the excitation light reflected from the solid scatterer
2. Measurement of the excitation light reflected from the phosphor together with its PL with no filters mounted at the photodetector side
3. Measurement of the PL with filters mounted between the phosphor and the thermoelectric detector

Expressing the thermal electromotive forces obtained from the measurements 1–3 as V_R, V_P, and $V_{P,F}$, respectively, the following relationships apply.

$$CV_R = IR \tag{13.3}$$

$$CV_P = Ir_p + L \tag{13.4}$$

$$CV_{P,F} = \tau L \tag{13.5}$$

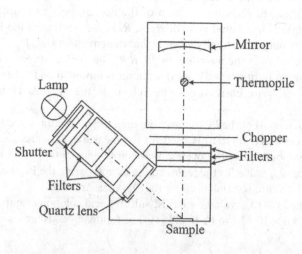

FIGURE 13.2   Schematic diagram of the measurement setup used by Bril and de Jager-Veenis. (From Bril, A. and de Jager-Veenis, A. W. 1976. *J. Res. Natl. Bur. Stand. A* 80A: 401. Used with permission.)

where I is the excitation light intensity irradiated onto the solid scatterer or the phosphor, R is the reflectance of the solid scatterer, r_p is the reflectance of the phosphor, L is the PL intensity of the phosphor, τ is the transmittance of the filters, and C is a constant.

From Equations (13.3)–(13.5), the PL energy efficiency η_p and r_p are derived as follows.

$$\eta_p = \frac{L}{I(1-r_p)} = \frac{RV_{P,F}}{\tau(1-r_p)V_R} \tag{13.6}$$

$$r_p = \frac{R(V_P - V_{P,F}/\tau)}{V_R} \tag{13.7}$$

When the PL spectrum of the phosphor is known, Φ_{PL} can be obtained from η_p.

$$\Phi_{PL} = \eta_p \frac{\int \lambda p(\lambda)\,d\lambda}{\lambda_{ex}\int p(\lambda)\,d\lambda} \tag{13.8}$$

where $p(\lambda)$ is the PL intensity and λ_{ex} is the excitation wavelength.

To determine the PL energy efficiency accurately, η_p needs to be corrected by considering the loss of the PL absorbed by the phosphor itself. The corrected PL energy efficiency η_i can be approximated using the reflectance of an infinite thick layer r_∞ using the following equation.

$$\eta_i = \frac{2\eta_p}{1 + r_\infty} \tag{13.9}$$

Using this method, Bril and Jager-Veenis determined the Φ_{PL} values for sodium salicylate to be 0.60 at $\lambda_{ex} = 260$ nm and 0.41 at $\lambda_{ex} = 365$ nm [4]. Sodium salicylate is a commonly used standard for phosphors.

13.3 CALORIMETRIC METHODS

In calorimetric methods, Φ_{PL} is determined by measuring the ratio of the heat energy released from a sample by nonradiative processes to the excitation energy absorbed. The methods can be classified into the conventional calorimetric method and photothermal spectroscopy [1, 5].

The conventional calorimetric method measures the temperature rise of a luminescent sample and nonluminescent reference solutions by photoexcitation [6]. The Φ_{PL} value for the sample solution is obtained by comparing the temperature change relative to a reference solution. The sample solution is adjusted to the same absorbance (optical density) as the reference solution.

Photothermal spectroscopy measures thermal or acoustic waves generated from a sample by nonradiative processes following the absorption of the excitation light [7, 8]. A representative photothermal method is photoacoustic spectroscopy. Adams et al. measured the Φ_{PL} for sample solutions and thin films by photoacoustic spectroscopy [8]. The measurement setup they used consists of a continuous light source, a rotating chopper, a monochromator, an optoacoustic cell, a lock-in amplifier, and a recorder. The optoacoustic cell includes an input window, a sample room, and a microphone. The continuous light is modulated using the rotating chopper and focused into the entrance slit of the monochromator to generate monochromatic light. The sample is mounted in the optoacoustic cell and irradiated with the monochromatic light. When the sample excited by the monochromatic light releases heat energy in nonradiative relaxation processes, a periodic pressure wave (photoacoustic wave) is generated in the medium or gas surrounding the solution or solid sample. The photoacoustic signal is detected by the microphone and extracted by the lock-in amplifier directly connected to the microphone using a reference signal generated by the rotating chopper.

The amplitude of the photoacoustic signal PA at the excitation wavelength λ_{ex} is expressed as follows [9].

$$PA = KP^{abs}\gamma \tag{13.10}$$

where K is a constant related to the surface area and thermal properties of the sample and the instrumental configuration, P^{abs} is the excitation light energy absorbed by the sample, and γ is the fraction of the absorbed energy released as thermal energy by nonradiative relaxation processes.

The amplitude of the photoacoustic signal for a luminescent sample PA_{sample} is given as follows.

$$PA_{sample} = K_{sample} P_{sample}^{abs} \left[1 - \Phi_{PL} + \Phi_{PL} \left(\frac{v_0 - \bar{v}_{PL}}{v_0} \right) \right] \tag{13.11}$$

where v_0 and $\bar{v}_{PL}$ are the frequency of the excitation light and the mean frequency of the PL, respectively.

For a nonluminescent sample ($\gamma = 1$) used as a reference, the photoacoustic signal amplitude $PA_{reference}$ is given as follows.

$$PA_{reference} = K_{reference} P_{reference}^{abs} \tag{13.12}$$

The following equation is derived from a combination of Equations (13.11) and (13.12),

$$\Phi_{PL} = \frac{\bar{\lambda}_{PL}}{\lambda_{ex}} \left(1 - \frac{PA_{sample}}{PA_{reference}} \cdot \frac{K_{reference}}{K_{sample}} \cdot \frac{P_{reference}^{abs}}{P_{sample}^{abs}} \right) \tag{13.13}$$

where $\bar{\lambda}_{PL}$ is the mean PL wavelength.

For solution samples, the thermal properties of the luminescent sample and nonluminescent reference can be considered to be equal ($K_{sample} = K_{reference}$). Assuming the absorptance of the sample is identical to that of the reference, Equation (13.13) can be simplified to the following equation.

$$\Phi_{PL} = \frac{\bar{\lambda}_{PL}}{\lambda_{ex}} \left(1 - \frac{PA_{sample}}{PA_{reference}} \right) \tag{13.14}$$

Adams *et al.* [10] determined the Φ_{PL} value for 0.05-M sulfuric acid solution of QBS to be 0.53 ± 0.02 at $\lambda_{ex} = 366$ nm. A 0.1-M hydrochloric acid solution of QBS was used as the nonluminescent reference.

For solid samples, K_{sample} and $K_{reference}$ are generally not equal since the sample and the reference have different surface areas and thermal diffusivities. Hence, it is necessary to match K_{sample} and $K_{reference}$ by a careful preparative technique in which a very thin sample or reference layer is deposited on a substrate material with a thermal depth L [9]. The thermal depth L is defined as the sample thickness that contributes to the generation of the photoacoustic signal at the sample–gas interface,

$$L = 2\pi \sqrt{\frac{\alpha}{\pi f}} \tag{13.15}$$

where α is the thermal diffusivity (cm^2 s^{-1}) and f is the modulation frequency (s^{-1}). For the typical thermal diffusivity of organic substances and a modulation frequency of 10^{-3} cm^2 s^{-1} and 30 Hz, respectively, L is about 200 μm.

The relationship between the sample thickness l and absorbance A follows Lambert–Beer's law and l is given by the following.

$$l = \frac{A}{\varepsilon C} \qquad (13.16)$$

The sample should have an absorbance large enough to absorb all the excitation light. Assuming a sample absorbance of three, a typical absorption coefficient for organic substances ε of $10{,}000\ \mathrm{Lmol^{-1}\ cm^{-1}}$, and a sample concentration C of 10 M, l is approximately $0.3\ \mu m$.

Adams *et al.* determined the Φ_{PL} value for sodium salicylate to be 0.58 ± 0.01 at $\lambda_{ex} = 310$ nm by photoacoustic spectroscopy using Congo red as the nonluminescent reference [9]. The sodium salicylate was thinly and uniformly sprayed on a glass substrate. In the determination of Φ_{PL}, they considered the background signal from the optoacoustic cell and the reflectivity of the sample.

13.4 RELATIVE METHOD

13.4.1 SYSTEM CONFIGURATION AND MEASUREMENT PROCEDURE

The absolute methods described in Sections 13.2 and 13.3 require sophisticated techniques and various complex corrections to accurately determine Φ_{PL}, making them generally difficult for nonexperts to use. As an alternative, relative methods are widely used by many researchers.

In a relative method, the Φ_{PL} value for a sample is estimated by a comparison with a standard substance whose Φ_{PL} is well known [1, 11]. A spectrophotometer and a calibrated spectrofluorometer are used to measure the absorbance and the PL spectrum of a sample, respectively.

In a typical spectrofluorometer, a high-pressure xenon lamp coupled with an excitation monochromator is used as the excitation light source (Figure 13.3). The monochromatic light is focused on the sample mounted on a sample holder. The PL emitted from the sample is detected by a photodetector after passing through focusing lenses and an emission monochromator at 90 degrees to the optical axis of the excitation light. A photomultiplier tube is commonly used as the photodetector.

A photomultiplier tube is a highly sensitive photodetector that utilizes the external photoelectric effect [12]. It consists of a vacuum tube sealed in a glass tube, with an input window, a grid, a photocathode, focusing electrodes, dynodes, and an anode (Figure 13.4). The incident light hits

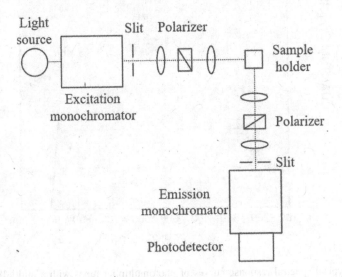

FIGURE 13.3 Schematic diagram of a spectrofluorometer.

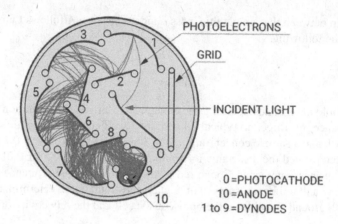

FIGURE 13.4   Cross section of a photomultiplier tube (circular-cage type).

the photocathode after passing through the input window. Photoelectrons generated in the photocathode by the external photoelectric effect are multiplied 10^6–10^7 times by collisions with the dynodes and collected by the anode. The spectral response of a photomultiplier tube depends on the transmittance of the input window and the photocathode material. Typical spectral response curves for photomultiplier tubes with a multialkali photocathode and different window materials (Hamamatsu, R928 and R955) are shown in Figure 13.5.

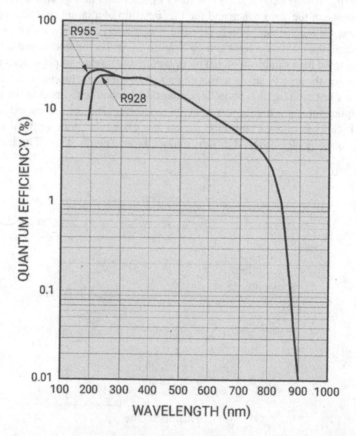

FIGURE 13.5   Typical spectral response curves of photomultiplier tubes with a multialkali photocathode material and different window materials (Hamamatsu, R928 and R955).

The PL spectrum of a sample is acquired by scanning the wavelength with the emission mono-chromator. The acquired PL spectrum is distorted from the true spectrum because the spectral response of the photodetection system in the spectrofluorometer is dependent on the detection wave-length range. The acquired PL spectrum needs to be corrected for the spectral response of the photodetection system to obtain the true spectrum. The spectral response of the detection system is determined by the product of instrumental functions, such as the transmittance of the focusing lenses, the diffraction efficiency of the grating mounted on the emission monochromator, the reflec-tance of the mirrors, and the spectral response of the photomultiplier tube.

When the PL spectrum of a standard substance and sample are measured under the identical experimental conditions (excitation wavelength, slit widths of the excitation and emission mono-chromators, gain of the photodetector, etc.), the Φ_{PL} for the sample can be obtained by a comparison between the sample and the standard using the following equation [1, 11],

$$\Phi_{PL} = \Phi_{PL,std} \left(\frac{I_{em}}{I_{em,\,std}} \right) \left(\frac{A_{std}}{A} \right) \left(\frac{n^2}{n_{std}^2} \right) \tag{13.17}$$

where I_{em}, A, and n are the corrected PL intensity (photons per time), the absorbance at the excita-tion wavelength, and the refractive index of the solvent used for the sample or standard, respectively. The subscript "std" represents the standard.

In a relative method, it is generally recommended to select a standard with an emission wave-length range close to that for the sample to minimize the measurement error caused by uncertainties due to the sensitivity corrections of the photodetection system of the spectrofluorometer [13–15]. The excitation wavelength used for the standard should be chosen by referring to the literature, because the Φ_{PL} value for the standard possibly depends on the excitation wavelength. For solution samples, the absorbance at the excitation wavelength should be less than 0.05 to minimize measure-ment errors due to inner filter effects and the formation of dye aggregates and/or excimers at high concentration.

The relative method is widely used because Φ_{PL} can be obtained with a relatively simple mea-surement procedure using commercially available spectrofluorometers. The biggest issue is that reliable standard substances are very limited. Especially, standards for the near-infrared range and solid standards have not yet been fully established [16]. Moreover, the relative method cannot be applied to highly scattering solutions since in this case the absorbance is overestimated due to scat-tering of the reference beam of the spectrophotometer [17].

13.4.2 STANDARD SOLUTIONS

In the IUPAC technical report on standard solutions reviewed by Brouwer [15], QBS in sulfuric acid or perchloric acid, rhodamine 6G in ethanol, and fluorescein in 0.1-M NaOH are recommended as well-established standards. In addition, rhodamine 101 in ethanol is recommended as a gold standard in a critical review of the determination of Φ_{PL} by Würth et al. [5]. These standards are commercially available and the variation in the Φ_{PL} values in numerous literature reports is small. Table 13.2 lists the reported Φ_{PL} values for standard solutions for various excitation and emission wavelength ranges.

The QBS solutions exhibit PL in the visible range (380–650 nm) for excitation in the ultraviolet range (315–365 nm). QBS solutions are widely used as standards for two reasons: (1) the Φ_{PL} is less affected by the self-absorption effect due to the large Stokes shift (the large energy gap between the absorption and emission maxima) and (2) it is not quenched by oxygen, making it unnecessary to purge the solution with nitrogen or argon gas prior to the measurement. However, the Φ_{PL} for a QBS solution depends on the QBS and acid concentrations, so the solutions should be prepared at the same concentrations as described in the literature. The most popular solution for the QBS standard

TABLE 13.2

Excitation Wavelength (λ_{exc}), Emission Wavelength Range (λ_{em}), and PL Quantum Yields (Φ_{PL}) of Standard Solutions

Material	Solvent	λ_{exc} (nm)	λ_{em} (nm)	Φ_{PL}	Method	Note	Reference
Quinine bisulfate	0.5-M H_2SO_4	366	380–650	0.508	Vavilov	5×10^{-3} M	[3]
		350		0.52 ± 0.02	Integrating sphere	5×10^{-3} M	[18]
	0.5-M H_2SO_4	366		0.546	Vavilov	infinite dilution	[3]
		365		0.54 ± 0.02	Weber and Teale	infinite dilution	[19]
		350		0.60 ± 0.02	Integrating sphere		[18]
	0.1-M $HClO_4$	347.5	375–650	0.60 ± 0.02	NIST SRM 936		[20]
		315, 348, 365		0.64 ± 0.04	Integrating sphere	1×10^{-5} M	[21]
Rhodamine 6G	EtOH	488	525–700	0.94	Thermal lens		[32]
		465–530		0.90 ± 0.05	Integrating sphere		[23]
		500		0.92 ± 0.05	Integrating sphere		[24]
		532		1.00 ± 0.06	Photoacoustic		[24]
Fluorescein	0.1-M NaOH	313.1, 365.5, 435.8	500–650	0.87	Weber and Teale		[29]
		488		0.92 ± 0.03	Thermal lens		[25]
		470		0.91 ± 0.05	Integrating sphere		[26]
		460		0.88 ± 0.03	Integrating sphere		[18]
Rhodamine 101	EtOH	525, 565	550–675	0.90 ± 0.05	Integrating sphere		[23]
		525		0.92 ± 0.03	Integrating sphere		[27]

is a 0.5-M H_2SO_4 solution whose Φ_{PL} values are reported to be 0.51–0.52 at a QBS concentration of 5×10^{-3} M and 0.55–0.60 at 1×10^{-5} M. The 0.1-M $HClO_4$ solution of QBS is certified by the National Institute of Standards and Technology (NIST) as Standard Reference Material (SRM) 936. The reported Φ_{PL} values are 0.60–0.64. Rhodamine 6G in ethanol, fluorescein in 0.1-M NaOH, and rhodamine 101 in ethanol exhibit absorption and PL in the visible range. The PL spectra of these solutions are observed in the range between 500 and 700 nm for excitation between 488 and 565 nm. These solutions do not need to be purged with nitrogen or argon gas due to the PL insensitivity to dissolved oxygen. For fluorescein in 0.1-M NaOH, a fresh solution should be used just after preparation, as it is not stable.

9,10-Diphenylanthracene in cyclohexane, anthracene in ethanol, and tris(2,2′-bipyridyl) ruthenium ($[Ru(bpy)_3]^{2+}$) in acetonitrile or H_2O are used as popular standards [15, 18]. Φ_{PL} values have been reported for 9,10-diphenylanthracene in cyclohexane ($\Phi_{PL} = 0.90$–1.06 at λ_{ex} between 308 and 366 nm) [6, 18, 28], anthracene in ethanol ($\Phi_{PL} = 0.27$–0.28, $\lambda_{ex} = 340$ nm) [13, 18], $[Ru(bpy)_3]^{2+}$ in acetonitrile ($\Phi_{PL} = 0.090$–0.095, $\lambda_{ex} = 450$ nm) [18, 29], and $[Ru(bpy)_3]^{2+}$ in H_2O ($\Phi_{PL} = 0042$–0.063, $\lambda_{ex} = 450$ nm) [18, 30]. The PL of these solutions is quenched by dissolved oxygen and the Φ_{PL} values for the aerated solutions are notably smaller. For example, the reported Φ_{PL} value for $[Ru(bpy)_3]^{2+}$ in aerated acetonitrile (0.018) is about five times smaller than that in argon-saturated acetonitrile (0.095) [18]. To obtain accurate Φ_{PL} values, the standard solutions should be used immediately after removing the dissolved oxygen by purging with nitrogen or argon gas. For $[Ru(bpy)_3]^{2+}$, careful attention should be paid to the measurement temperature as well as oxygen quenching since Φ_{PL} shows a significant temperature dependence especially in acetonitrile [31].

Certain standards can be excited below 300 nm, namely naphthalene in cyclohexane ($\Phi_{PL} = 0.23$, $\lambda_{ex} = 270$ nm) [13, 18] and tryptophan in H_2O ($\Phi_{PL} = 0.14$–0.15, $\lambda_{ex} = 270$ nm) [13, 18]. The reported Φ_{PL} values were obtained for deaerated or argon-saturated solutions.

There are only a few standards available for the wavelengths longer than 600 nm because the measurement error in Φ_{PL} tends to be larger in the near-infrared region due to the instability and low purity of near-infrared fluorophores [5], absorption of PL by solvents [32, 33], and the low sensitivity of the photodetection system in spectrofluorometers. The representative standards are indocyanine green (ICG), IR26, and IR1061. The most popular standard in the near-infrared region is ICG in dimethyl sulfoxide (DMSO). The absorption and emission maxima are observed at 790 and 830 nm, respectively. The frequently used Φ_{PL} value is 0.13, which was determined for a highly concentrated solution (2.7×10^{-4} M) by the relative method using sodium fluorescein solution as a standard [34]. Two research groups have reported the Φ_{PL} values for IR125 (with a chemical structure identical to that of ICG) and ICG in DMSO to be 0.23 [27] and 0.26 [32], respectively, employing the absolute method using an integrating sphere (IS). According to Würth *et al.*, the low purity of the dye used for the Φ_{PL} measurement accounts for the smaller value (0.13) [5]. IR26 and IR1061 emit PL at wavelengths longer than 1000 nm. The absorption and emission maxima are observed at 1080 and 1130 nm for IR26 in 1,2-dichloroethane (DCE) [35] and at 1030 and 1050 nm for IR1061 in dichloromethane (DCM) [32], respectively. The Φ_{PL} values for IR26 in DCE and IR1061 in DCM have been reported to be 0.0011 [33] and 0.0059 [32], respectively. For wavelengths longer than 1000 nm, careful attention should be paid to the effect of solvent absorption in the PL measurements because many solvents show absorption bands attributed to overtones of vibration modes in the wavelength range (see 13.7).

Standard solutions should be prepared and measured under conditions identical to those described in the literature. For the preparation of a standard solution, a solute and solvent with high purity should be prepared under the conditions used in the literature (solvent, concentration, acid or basic concentration, with or without dissolved oxygen, etc.). The standards should be measured under the same conditions (temperature, excitation wavelength, etc.) as in the literature [13, 15, 36].

13.4.3 SOLID STANDARDS

Solid-state standards have not yet been established. One reason is the difficulty in validating solid samples. Φ_{PL} is likely to be affected by various factors inherent in solid samples. Katoh *et al.* identified three main factors that can influence Φ_{PL} for crystalline materials: chemical impurities, structural defects, and reabsorption of the PL [37]. They demonstrated the influence of the three factors on Φ_{PL} for aromatic hydrocarbon crystals by the absolute method using an IS.

First, they compared PL spectra and Φ_{PL} for a highly purified single crystal of anthracene and several unpurified commercial products, to study the effect of impurities. For the unpurified commercial products, additional PL bands were observed. The determined Φ_{PL} values of 0.12–0.49 were smaller than that for the highly purified single crystal (0.64). This result indicates that excitons are rapidly captured during migration by impurities or structural defects in the unpurified product, leading to PL quenching.

In a second experiment, the dependence of the PL spectra and Φ_{PL} on the crystal size in *p*-terphenyl and anthracene were measured. Highly purified single crystals were milled to different crystal sizes. Their PL were also studied to investigate the reabsorption effect. Generally, crystal materials have high refractive indices and the PL is efficiently reflected on the surface back into the crystal. This is known as the waveguide effect and results in a strong reabsorption effect in the wavelength region where the PL spectrum overlaps with the absorption spectrum. In large crystals this propagation is more pronounced, resulting in a stronger decrease in PL intensity. Thus, we can expect that the reabsorption is suppressed in small crystals.

For *p*-terphenyl, the PL intensity in the shorter wavelength band (357 nm) increased and the Φ_{PL} value increased from 0.67 to 0.80 by milling of a single crystal (Figure 13.6). Anthracene showed different results. While the Φ_{PL} value for anthracene decreased from 0.64 to 0.27 after the milling process, as expected, the PL spectrum showed an increase not only in the shorter wavelength region

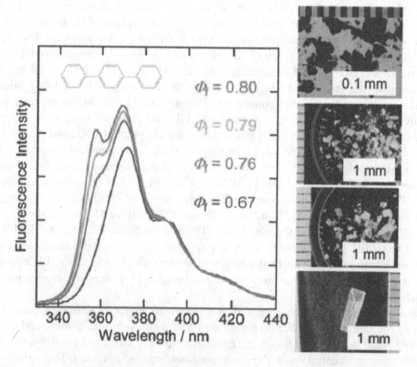

FIGURE 13.6 PL spectra, showing changes in Φ_{PL} during the mechanical milling process, and microscopy images for *p*-terphenyl. (From Katoh, R., *et al.* 2009. *J. Phys. Chem. C* 113: 2961. Used with permission.)

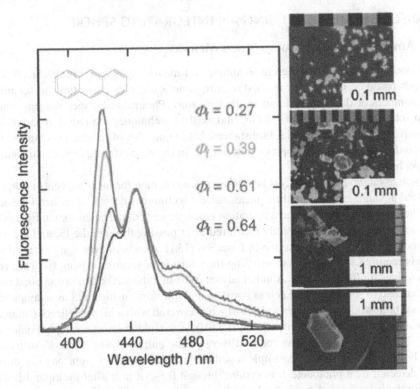

FIGURE 13.7 PL spectra, showing changes in Φ_{PL} during the mechanical milling process, and microscopy images for anthracene. (From Katoh, R., *et al.* 2009. *J. Phys. Chem. C* 113: 2961. Used with permission.)

(425 nm), but also in the wavelength region longer than 450 nm (Figure 13.7). By milling the single crystal of anthracene, the reabsorption is suppressed; however, the PL is quenched due to the introduction of luminescent structural defects.

Sodium salicylate powder is a commonly used standard phosphor for Φ_{PL} measurements by the relative method in the ultraviolet region because the Φ_{PL} value is constant over an excitation wavelength range up to 350 nm [4]. Several research groups have reported absolute Φ_{PL} values for sodium salicylate obtained using optical and photoacoustic methods at excitation wavelengths between 253.7 and 360 nm. However, there is a large discrepancy between the reported values. Φ_{PL} values of 0.41–0.64 for thin films, sprayed on a glass substrate or on the light-receiving surface of a photodetector [4, 9, 38], and 0.56 for a 1-mm thick pellet [39] have been reported. Rohwer and Martin posed the question as to whether the same Φ_{PL} values are obtained for sodium salicylate in thin film (sprayed form) and powder form [40].

The Φ_{PL} values for the conventional phosphor powders, BaMgAl$_{10}$O$_{17}$:Eu^{2+} (BAM) and Y$_3$Al$_5$O$_{12}$:Ce^{3+} (YAG:Ce), have been evaluated. BAM is a blue phosphor exhibiting PL under excitation in the ultraviolet region. Φ_{PL} values of 0.93 at λ_{ex} = 365 nm [41], 0.72 at λ_{ex} = 395 nm [41], 0.88 at λ_{ex} = 254 nm [42], 0.924 at λ_{ex} = 325 nm [43], and 0.93 at λ_{ex} = 390 nm [40] have been reported. YAG:Ce is a yellow phosphor exhibiting PL under excitation from the ultraviolet to visible region. Φ_{PL} values of 0.95 and 0.87 at λ_{ex} = 451 nm [44], and 0.95 at λ_{ex} = 340 and 460 nm [39], have been reported. Typically, the Φ_{PL} values for conventional phosphors vary due to sample-intrinsic factors such as activator concentration and particle size. For example, the Φ_{PL} value for BAM dramatically decreases from 0.924 to 0.497 with increasing activator concentration from 11.6 to 42.5 mol% [43], and YAG:Ce shows a decrease in Φ_{PL} from 0.89 to 0.81 with increasing activator concentration from 1% to 3% [45]. Establishing standards based on conventional phosphors is challenging because it is not easy to precisely control the sample-intrinsic factors that influence Φ_{PL}.

13.5   ABSOLUTE METHOD USING AN INTEGRATING SPHERE

13.5.1   ADVANTAGES OF INTEGRATING SPHERE METHOD

In the previous sections, three different techniques for measuring Φ_{PL} were introduced. The absolute methods based on the Vavilov method require various complex corrections to accurately estimate the numbers of the absorbed and emitted photons. Photoacoustic spectroscopy can only be applied to solution samples and thin films, and skillful techniques are required to fabricate thin films. Relative methods need standard substances, but the number of reliable standards is very limited and standards for solution samples exhibiting PL in the near-infrared region and solid samples have not yet been established.

The absolute method using an IS is becoming more common for measurements of Φ_{PL} because it has several advantages over the three measurement techniques described earlier. One advantage of the method is that the measurement operation is simple and easy. The excitation light profile and PL spectrum of the sample are directly measured after passing through the IS, and the Φ_{PL} value is calculated according to the definition in Equation (13.1). Another advantage is that no reference substance is required in this method, avoiding the measurement errors arising from the reliability of the Φ_{PL} values for the standards. Another advantage is that the method can be applied not only to solutions but also to solid samples such as powders and thin films using an IS as a sample chamber.

An IS is a hollow globe the inner wall of which is covered with a highly reflective material such as barium sulfate or polytetrafluoroethylene (PTFE). A sample is set in a sample holder mounted on the IS. The excitation light is introduced through the entrance port of the IS to irradiate the sample. The PL emitted from the sample together with the excitation light passing through the sample is detected by a photodetection system through the exit port after multiple diffuse reflections within the inner wall. A baffle is placed between the sample and the exit port to avoid direct detection of the PL from the sample. Luminescent materials, especially solid samples such as thin films and powders, often display significant optical anisotropy in the PL. The IS can eliminate the optical anisotropy of the PL by multiple diffuse reflections inside the inner wall. In this way, the IS method allows Φ_{PL} to be determined without the complicated corrections required for other absolute methods.

13.5.2   SYSTEM CONFIGURATION

Absolute PL quantum yield measurement systems based on the IS method are generally composed of an excitation light source, an IS, a photodetection system, and a data analyzer. The measurement systems can be classified into two groups: (1) custom-made setups [24, 40, 46–55] or commercial products [18, 21, 32] dedicated to measuring Φ_{PL} and (2) spectrofluorometers with a mounted IS accessory instead of a sample holder [26, 41, 43, 54, 56, 57].

As the excitation light source, a continuous-wave xenon lamp coupled with a monochromator (monochromatic light source) [21, 24, 40, 52, 54], a laser diode [43, 44, 46–51, 53, 55], or a light-emitting device (LED) [44, 53, 54] can be used. A white light source coupled to a monochromator is widely used because the excitation wavelength is adjustable to the absorption wavelength of the measured samples. Laser diodes can minimize the spectral overlap with the PL spectrum of the sample due to the narrow spectral bandwidth. However, the radiant power needs to be attenuated using filters to avoid signal saturation of the photodetector. LEDs are often used for evaluating phosphors. However, the spectral overlap with the emission spectrum of a sample should be taken into account in selecting the output wavelength of the LED due to the broad bandwidth.

For the photodetection system, a photodiode [40, 46–48], a photomultiplier tube combined with a monochromator [43, 44, 54, 55], or an image sensor attached to a spectrograph [21, 24, 32, 49–52, 54] can be employed. When a photodiode is used for the measurement, band-pass filters appropriate to the wavelength region and bandwidth of the excitation profile or PL spectrum need to be used to separate the PL from the excitation light. For a photomultiplier-tube based photodetection system,

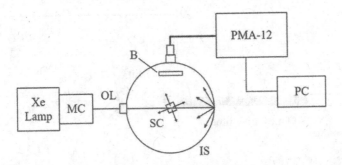

FIGURE 13.8 Absolute PL quantum yield measurement system (Hamamatsu, C9920-02). MC: monochroma-tor, OL: optical light guide, SC: sample cuvette, B: baffle, PC: personal computer. (From Suzuki, K., *et al.* 2009. *Phys. Chem. Chem. Phys.*, 11: 9850. Used with permission.)

the excitation profile and PL spectrum are measured by scanning the detection wavelength with a monochromator. An image sensor based system can measure both the excitation profile and the PL spectrum simultaneously in the measurement wavelength range of the photodetection system. As an image sensor, linear image sensors based on charge coupled devices (CCD) and/or indium gallium arsenide (InGaAs) are used depending on the measurement wavelength range.

Figure 13.8 shows an example of a system for measuring Φ_{PL} (Hamamatsu, C9920-02) [18]. A monochromatic light source is used as the excitation light source, consisting of a monochromator and a xenon arc lamp with a lamp rating of 150 W. The xenon arc lamp, with a high output stability of 1.0% (peak to peak), reduces the measurement error in Φ_{PL} arising from fluctuations of the output intensity. The excitation light is introduced into an IS through an optical lightguide. The IS has an inner diameter of about 84 mm, with Spectralon® (Labsphere) on the inner surface as a high reflectance material (99% reflectance for wavelengths from 350 to 1650 nm and over 96% for wavelengths from 250 to 350 nm). A solution sample in a quartz cuvette or a solid sample in a Petri dish is attached in the corresponding sample holder mounted on the IS. For solution samples, a 10-mm path length quartz cuvette is positioned in the center of the IS. For solid samples, a Petri dish with an inner diameter of 15 mm is placed on the sample holder on the bottom surface of the inner wall. The solution and solid sample holders are designed so that the excitation light does not escape from the entrance port after reflecting from the solution or solid sample container. A baffle between the sample and the photodetector prevents direct detection of the PL from the sample. A photonic multichannel analyzer PMA-12 (Hamamatsu, C10027-01) is used as the photodetection system.

PMA-12 is composed of an optical fiber probe, a spectrograph, and a back-thinned CCD (BT-CCD) linear image sensor (Figure 13.9). The optical fiber probe is made of optical fiber bundles. Measured light is introduced through the optical fiber probe to the input slit of a Czerny–Turner type spectrograph. In the spectrograph, the measured light is dispersed into various wavelengths by the grating and the spectrum is projected onto the BT-CCD with 1024×128 pixels and a pixel size of 24×24 μm^2. The BT-CCD transmits electric charge, and using the device the wavelength dispersed light is converted from photons to electric charges. First, the electric charges are transmitted perpendicularly to a horizontal shift resister, as shown in Figure 13.10. Then, the electric charges on each horizontal line are summed by the horizontal shift register to generate a 1-dimensional data. Finally, spectral profile data are obtained by reading the horizontal shift register. BT-CCD has high sensitivity over a wide spectral range from 200 to 1100 nm, as shown in Figure 13.11.

The sensitivity of the measurement system is fully calibrated for the spectral region from 250 to 950 nm using deuterium and halogen standard light sources. These standard light sources are calibrated in accordance with measurement standards traceable to primary standards (national standards) located at the National Metrology Institute of Japan. The primary measurement standards are based on the physical units of measurement according to the International System of Units (SI).

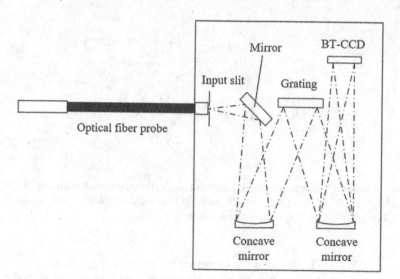

FIGURE 13.9   Schematic diagram of PMA-12/C10027-01.

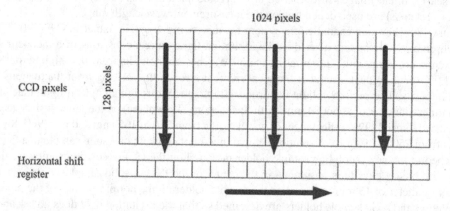

FIGURE 13.10   Diagrammatic illustration of a BT-CCD.

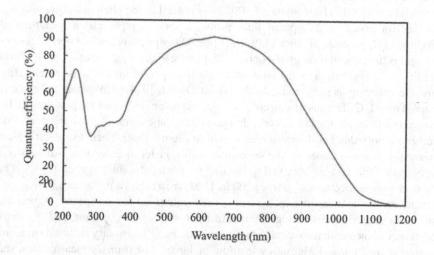

FIGURE 13.11   Spectral response curve of BT-CCD.

13.5.3 Spectral Sensitivity Correction of Photodetection System

13.5.3.1 Spectral Response of Photodetection System and Spectral Sensitivity Correction

It is essential to obtain the true spectral profiles of the excitation light and the PL to determine Φ_{PL} accurately. However, raw spectral profiles acquired from uncalibrated measurement systems are distorted compared to the true spectral profiles due to the wavelength-dependent spectral response of photodetection systems $R(\lambda)$. In the case of spectrofluorometers (Figure 13.3) with an IS accessory, $R(\lambda)$ can be expressed by the product of the reflectance of the IS, the transmittance of the focusing lenses at the photodetector side, the diffraction efficiency of the grating, the reflectance of the mirrors mounted on the monochromator, and the spectral response of the photomultiplier tube. For an IS system based on a photonic multichannel analyzer (Figure 13.8), $R(\lambda)$ can be obtained as the product of the reflectance of the IS, the transmittance of the optical fiber probe, the diffraction efficiency of the grating, the reflectance of the mirrors mounted on the spectrograph, and the spectral response of the BT-CCD. A true spectral profile $I_{corr}(\lambda)$ is obtained as the product of a raw spectral profile acquired from a photodetection system $I_{obs}(\lambda)$ and the reciprocal of $R(\lambda)$,

$$I_{corr}(\lambda) = \frac{I_{obs}(\lambda)}{R(\lambda)} = I_{obs}(\lambda) \cdot S(\lambda) \tag{13.18}$$

where $S(\lambda)$ is the spectral sensitivity correction function of the photodetection system.

$S(\lambda)$ for a photodetection system is obtained using reference light sources the light intensity of which is well characterized as a function of wavelength. In general, physical source-based or fluorophore-based transfer standards are used as reference light sources [58].

13.5.3.2 Spectral Sensitivity Correction Using Physical Source-Based Transfer Standards

As physical source-based transfer standards, standard lamps (e.g., halogen and deuterium lamps) are used for the spectral sensitivity correction of photodetection systems. Radiation from the standard lamp is introduced to the entrance port of the IS, and the radiant spectrum is acquired. The spectral sensitivity correction function $S'(\lambda)$ is obtained by comparing the radiant quantity of the standard lamp $E_{in}(\lambda)$ with the raw spectral intensity acquired from the measurement system $I_{acq}(\lambda)$.

$$S'(\lambda) = \frac{E_{in}(\lambda)}{I_{acq}(\lambda)} \tag{13.19}$$

Figure 13.12 shows a typical $S'(\lambda)$ of an IS system based on a photonic multichannel analyzer. An observed spectrum $I_{obs}(\lambda)$ is corrected using $S'(\lambda)$ to obtain the true spectrum $I'_{corr}(\lambda)$.

$$I'_{corr}(\lambda) = I_{obs}(\lambda) \cdot S'(\lambda) \tag{13.20}$$

The units for the true spectrum are energy per unit time. In the calculation of Φ_{PL}, the units for the true spectrum intensity have to be converted to the number of photons per unit time.

$$I_{corr}(\lambda) = \frac{\lambda}{hc} \cdot I'_{corr}(\lambda) \tag{13.21}$$

For the correction method using physical source-based transfer standards, the reliability of the spectral radiant quantity of the standard lamps and the transfer accuracy of the radiant quantity from standard lamps to the photodetection system are important. Standard lamps with radiant quantities that are well characterized or traceable to the national standard are highly recommended. A correction over the whole measurement range of the photodetection system requires a standard lamp with a broadband radiation spectrum or several standard lamps providing radiation over different wavelength ranges (e.g., halogen and deuterium lamps). Standard lamps exhibiting unstructured

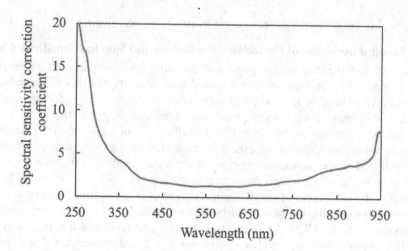

FIGURE 13.12   Spectral sensitivity correction curve for an IS system based on a multichannel analyzer. The correction curve is normalized to one at the minimum value.

radiation spectra are recommended to increase the transfer accuracy. The error in the spectral sensitivity correction can increase in wavelength ranges where $R(\lambda)$ is small due to the low sensitivity of the photodetector and/or the low diffraction efficiency of the grating.

13.5.3.3   Fluorophore-Based Transfer Standards

Corrected emission spectra reported for some standard fluorophores (fluorescence standards) can be used for spectral sensitivity correction, similar to the physical source-based transfer standards [59]. When the PL is given in terms of the number of photons per unit time, the spectral sensitivity correction function of the photodetection system $S(\lambda)$ is obtained by comparing the PL intensity of the standard $F(\lambda)$ with the raw spectral intensity acquired from the measurement system $I_{acq}(\lambda)$.

$$S(\lambda) = \frac{F(\lambda)}{I_{acq}(\lambda)} \qquad (13.22)$$

The true spectrum $I_{corr}(\lambda)$ is obtained from the observed spectrum $I_{obs}(\lambda)$ and $S(\lambda)$ using Equation (13.18).

If the PL spectrum intensity for the standard solution is in units of energy per unit time, the spectral sensitivity correction function $S'(\lambda)$ is obtained by the following equation.

$$S'(\lambda) = \frac{F'(\lambda)}{I_{acq}(\lambda)} \qquad (13.23)$$

where $F'(\lambda)$ is the PL spectrum in energy per unit time.

The observed spectrum $I_{obs}(\lambda)$ is converted to the true spectrum with units of photons per unit time $I_{corr}(\lambda)$ using Equations (13.20) and (13.21).

Figure 13.13 shows the corrected (true) PL spectra of representative standard solutions [18]. The PL spectra of the standard solutions should be well characterized using a carefully calibrated photodetection system. Because the emission bandwidth for each standard solution is not broad enough to cover the whole measurement wavelength range of the photodetection system, several standard solutions the emission bands of which overlap need to be used for spectral sensitivity correction. In the wavelength range between 300 and 770 nm, commercial kits of certified standard solutions are available, for example, from the Federal Institute of Materials Research and Testing (BAM) in

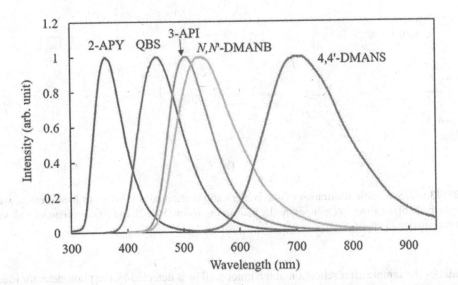

FIGURE 13.13 Corrected PL spectra for 2-aminopyridine (2-APY; 10^{-5} M in 0.1-N H_2SO_4), quinine bisulfate (QBS; 10^{-5} M in 0.1-N H_2SO_4), 3-aminophthalimide (3-API; 5×10^{-4} M in 0.1-N H_2SO_4), N,N'-dimethylaminonitrobenzene (N,N'-DMANB; 10^{-4} M in benzene-hexane (3:7, v/v)) and 4-dimethylamino-4'-nitrostilbene (4,4'-DMANS; 10^{-3} M in o-dichlorobenzene).

Germany [60]. Standard solutions and the procedures to acquire the PL spectra and calculate the spectral sensitivity correction coefficients are discussed in the literature [16, 58].

13.5.4 MEASUREMENT APPROACHES

For the determination of Φ_{PL} by the IS method, two different measurement approaches are proposed: an approach based on a three-step procedure [26, 43, 46–48, 50, 56] and an approach with a two-step procedure [18, 21, 40, 49, 55, 57]. This section provides an overview of the principle and the measurement procedures for the two measurement approaches and a comparison between them. At the end of this section, an example of Φ_{PL} measurement by the two-step procedure is provided.

13.5.4.1 Three-Measurement Approach (3MA)

The three-measurement approach (3MA) was proposed by de Mello *et al.* [47]. This approach uses a measurement procedure with three steps.

Step 1. Measurement of the excitation light intensity without a sample (Figure 13.14(a))
Step 2. Measurement of the excitation light intensity and sample PL intensity under indirect excitation (Figure 13.14(b))
Step 3. Measurement of excitation light intensity and sample PL intensity under direct excitation (Figure 13.14(c))

Figure 13.15 shows a typical example of a measurement result obtained by 3MA. According to the notation proposed by de Mello *et al.*, the number of photons for the excitation light and the PL are represented as L and P, respectively. At Step 1, the excitation light introduced to the IS is detected after multiple reflections inside the inner wall. L_a is the number of photons for the excitation light acquired at Step 1. At Step 2, the excitation light indirectly irradiates the sample after being reflected at the inner wall of the IS. The excitation light that is not absorbed by the sample

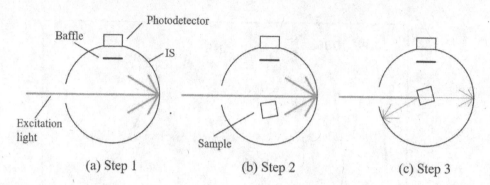

FIGURE 13.14 Schematic illustration of the configurations required for 3MA: (a) measurement without sample (Step 1), (b) measurement with sample by indirect excitation (Step 2), and (c) measurement with sample by direct excitation (Step 3).

re-irradiates the sample after reflection at the inner wall or is detected by the photodetector together with the PL emitted from the sample. L_b and P_b are the numbers of photons for the excitation light that are not absorbed by the sample and the PL obtained at Step 2, respectively. At Step 3, the incident excitation light directly irradiates the sample. The excitation light reflected from the inner wall of the IS or transmitted through the sample then re-irradiates the sample or is detected by the photodetector. L_c and P_c are the numbers of photons for the excitation light that are not absorbed by the sample and the PL obtained at Step 3, respectively.

In 3MA, Φ_{PL} is calculated using the following equation.

$$\Phi_{PL} = \frac{PN(Em)}{PN(Abs)} = \frac{P_c - (1-A)P_b}{L_a A}$$

(13.24)

where A is given by the following.

$$A = 1 - \frac{L_c}{L_b}$$

(13.25)

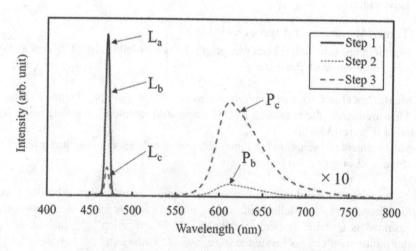

FIGURE 13.15 Typical measurement example of the excitation light profiles and the PL spectra of a sample by 3MA.

3MA is based on an experimental method proposed by Greenham *et al.* [46]. They used the following equation to determine Φ_{PL} of thin films,

$$\Phi_{PL} = \frac{P_c - (R+T)P_b}{(1-R-T)L_a} \tag{13.26}$$

where R and T are, respectively, the normal reflectance and transmittance of a thin film at the excitation wavelength. In the measurement of R and T, the contribution of the excitation light component diffusely scattered from the thin film is neglected.

de Mello *et al.* denoted the fraction of the excitation light that is absorbed by a sample as A and substituted the term $(R+T)$ in Equation (13.26) with $(1-A)$. Using this notation, the denominator of Equation (13.24) represents the number of photons directly absorbed by the sample except for the contribution from indirect excitation. The numerator of Equation (13.24) corresponds to the number of emitted photons produced only by direct excitation. According to de Mello *et al.*, the proposed method can accurately determine Φ_{PL} for highly scattering samples because this method can measure the number of absorbed photons with high accuracy by using the IS without additional measurements of R and T.

13.5.4.2 Two-Measurement Approach (2MA)

In the two-measurement approach (2MA), the Φ_{PL} measurement is performed with a two-step procedure, and thus sample measurement by indirect excitation is not required [18, 40, 49, 53, 54].

Step 1. Measurement of the excitation light intensity without a sample (Figure 13.14(a))
Step 2. Measurement of the excitation light intensity and sample PL intensity under direct excitation (Figure 13.14(c))

In 2MA, Φ_{PL} is given as follows.

$$\Phi_{PL} = \frac{PN(Em)}{PN(Abs)} = \frac{P_c}{L_a - L_c} \tag{13.27}$$

The denominator of Equation (13.27) represents the total number of photons absorbed by a sample in the processes of direct and indirect excitation. The numerator of Equation (13.27) corresponds to the total number of emitted photons produced by direct and indirect excitation.

13.5.4.3 Comparison between 3MA and 2MA

These two different approaches have been theoretically and experimentally compared by several research groups.

Würth *et al.* compared Φ_{PL} for an aqueous solution of rhodamine 6G at different concentrations obtained by three different measurement methods, 2MA, 3MA, and a two-step measurement approach with indirect illumination in Step 1 and Step 2, depicted in Figure 13.14 (2MA$_{indir}$) [24]. Rhodamine 6G shows a significant overlap between the absorption and PL spectra (small Stokes shift), and Φ_{PL} is likely to be influenced by the self-absorption effect. Dimers and higher aggregates are formed in water at a higher concentration. Due to the self-absorption effects and the presence of aggregates, the observed Φ_{PL} decreases with increasing concentration of rhodamine 6G. However, the three different methods gave identical Φ_{PL} values at each concentration except for the lowest concentration of 9.27×10^{-7} M. For example, the observed Φ_{PL} values were 0.77 (2MA), 0.77 (2MA$_{indir}$), and 0.77 (3MA) at 2.69×10^{-6} M, and 0.59 (2MA), 0.58 (2MA$_{indir}$), and 0.59 (3MA) at 2.32×10^{-5} M. At the lowest concentration, 9.27×10^{-7} M, the Φ_{PL} values obtained by 2MA$_{indir}$ (0.91) deviated from those by 2MA (0.84) and 3MA (0.81). The deviation of the measured value (2MA$_{indir}$) was attributed to measurement error resulting from the small absorption. At the lowest concentration, the fraction of the excitation light absorbed in 2MA was 7.7%, while only 2.7% of the excitation light was absorbed in $2MA_{indir}$. Würth *et al.* concluded that for transparent dye solutions, only 2MA (the measurement procedure with Step 1 and Step 3 shown in Figure 13.14) is required for Φ_{PL} measurements.

Faulkner *et al.* compared 3MA and 2MA by measuring Φ_{PL} for an ethanol solution of rhodamine 6G, thin films (polymer samples spin-coated on glass substrates), and a powder of $NaYF_4$ [53]. They concluded that it is unlikely that there any significant differences between the two measurement approaches for typical samples with single-photon processes. However, they predicted that the measurement results might differ for samples with multiphoton processes that show a dependence of Φ_{PL} on the excitation light intensity, such as second harmonic generation and upconversion.

Leyre *et al.* analyzed 3MA and 2MA theoretically and compared experimental results for Φ_{PL} obtained by the two different measurement approaches [54]. The theoretical analysis was based on rigorous IS theory and indicated that both approaches gave identical Φ_{PL} values under the assumption that the excitation light profile does not overlap the PL spectrum. In an experimental comparison of the two approaches using two series of luminescent powder samples (commercial blue, green, yellow, and red phosphor powders used in lighting applications and some silver-exchanged zeolites), the Φ_{PL} values obtained by 3MA and 2MA agreed well. Additionally, the obtained Φ_{PL} values were compared with those measured with different excitation light sources and photodetection systems at two independent laboratories. Regardless of the measurement approach and setup, acceptable agreement of the measured values was observed for all powder samples used in the study. Leyre *et al.* concluded that the difference in the measured values between 3MA and 2MA is mainly due to the error in the sensitivity correction for the measurement setups rather than any differences in the measurement approaches employed, and 2MA is sufficient for determining Φ_{PL}.

The results of the comparative studies of the two measurement approaches can be summarized as follows:

1. The two measurement approaches give identical Φ_{PL} values.
2. The difference in the measured values using the two approaches is mainly due to the error in the sensitivity correction for the photodetection systems in the measurement setups. The measurement error for the number of absorbed photons can also contribute to a difference in the obtained Φ_{PL} values between the two measurement approaches when the sample has low absorption.
3. For typical samples with a single-photon process, 2MA with a two-step measurement procedure is sufficient to determine Φ_{PL}.

13.5.4.4   General Measurement Procedure Based on 2MA

This section details an example of measurement using QBS in 0.5-M H_2SO_4 by 2MA using a commercial product dedicated to Φ_{PL} measurement (see 13.5.2). The following equation is used to determine Φ_{PL},

$$\Phi_{PL} = \frac{PN(Em)}{PN(Abs)} = \frac{\int \left[I_{em}^{sample}(\lambda) - I_{em}^{reference}(\lambda) \right] d\lambda}{\int \left[I_{ex}^{reference}(\lambda) - I_{ex}^{sample}(\lambda) \right] d\lambda} \tag{13.28}$$

where $I_{ex}^{reference}$ and I_{ex}^{sample} are the excitation light intensities without and with a sample, respectively, and I_{em}^{sample} and $I_{em}^{reference}$ are the PL intensity with a sample and the baseline, respectively. The units for the intensity for the excitation light and the PL are the number of photons per unit time.

Figure 13.16 shows the excitation profile and the PL spectrum of quartz cuvettes without and with QBS in 0.5-M H_2SO_4. The measurement procedure for Φ_{PL} is described as follows. First, an empty quartz cuvette is mounted in the IS and the cuvette is irradiated with excitation light with a

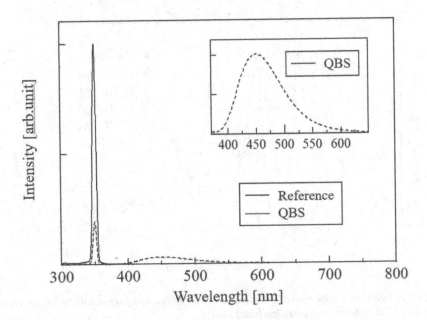

FIGURE 13.16 Excitation light profiles and PL spectrum obtained by 350 nm excitation of a reference and quinine bisulfate in 0.5-M H_2SO_4. The inset shows an expanded PL spectrum of QBS. (From Suzuki, K., et al. 2009. *Phys. Chem. Chem. Phys.*, 11: 9850. With permission.)

peak wavelength at 350 nm to obtain the excitation profile, shown by the solid line in Figure 13.16 (reference measurement). Then, the empty quartz cuvette is replaced by a cuvette containing QBS in 0.5-M H_2SO_4, and the sample solution is irradiated with excitation light (sample measurement). The excitation profile at around 350 nm is reduced by absorption due to QBS and the PL spectrum is observed in the wavelength range between 380 and 650 nm, as shown by the dotted line in Figure 13.16. The excitation light profile and the PL spectrum in Figure 13.16 are carefully corrected for the spectral sensitivity of the photodetection system. The number of photons absorbed by the QBS solution is proportional to the difference in the integrated excitation light intensities obtained from the reference and sample measurements, while the number of photons emitted from the sample solution is proportional to the integrated PL intensity. Thus, Φ_{PL} can be calculated from the ratio of the difference of the integrated excitation light intensity to the integrated PL intensity, according to Equation (13.28).

Suzuki *et al.* determined the Φ_{PL} value for QBS in 0.5-M H_2SO_4 at a concentration of 5.0×10^{-3} M to be 0.52 ± 0.02 [18], which agrees well with the literature value (0.508) reported by Melhuish [3].

13.6   SELF-ABSORPTION EFFECT

13.6.1   SELF-ABSORPTION EFFECT AND CORRECTION METHOD

Figure 13.17 shows the absorption and PL spectra of an ethanol solution of anthracene purged with argon gas [18]. The PL spectra were measured in the concentration range between 1×10^{-6} and 1×10^{-3} M using an IS system. Anthracene shows a small Stokes shift and the 0–0 absorption band significantly overlaps the PL spectrum. As the dye concentration of the solution increases from 1×10^{-6} to 1×10^{-3} M, the PL intensity in the 0–0 band region is progressively attenuated and Φ_{PL} decreases from 0.278 to 0.220 (Figure 13.17 and Table 13.3).

The distortion of the PL spectra and the decrease in Φ_{PL} result from reabsorption of the PL by the sample solution itself (the self-absorption effect) [1]. Generally, the magnitude of the self-absorption

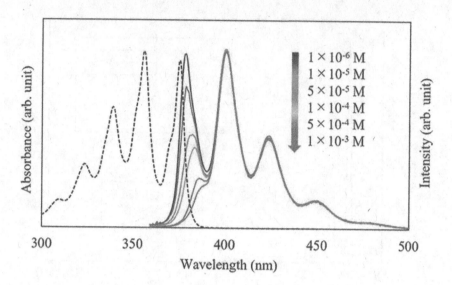

FIGURE 13.17 Absorption spectrum of ethanol solution of anthracene purged with argon gas (broken line) and the PL spectra at different concentration (solid lines).

effect depends on the amount of overlap between the absorption and PL spectra of the measured sample, the absorption cross section, and the concentration of fluorophore in solution (or the activator concentration in the host material in the case of phosphors [44]). Using an IS system, the self-absorption effect is dramatically enhanced compared to measurement systems without an IS because the effective optical path length increases due to multiple reflections from the inner wall. The sample size against the inner diameter of the IS also influences the self-absorption effect [27].

Ahn *et al.* suggested a correction method for the self-absorption effect in Φ_{PL} measurements using the IS method [51]. When a sample in the IS emits PL with Φ_{PL}, the emitted photons are reflected from the inner surface of the IS and then reabsorbed by the sample itself. The probability of reabsorption (the self-absorption probability) is represented as a. The PL is reemitted from the sample after reabsorption with a probability given by Φ_{PL}. The emitted photons that are not reabsorbed by the sample are detected by the photodetector with a photon escape probability $1 - a$. The observed PL quantum yield Φ_{PL}^{obs} after the reabsorption/reemission cycle is given by a geometric series as in the following equation.

$$\Phi_{PL}^{obs} = \Phi_{PL}(1-a)(1+a\Phi_{PL}+a^2\Phi_{PL}^2+\cdots) = \frac{\Phi_{PL}(1-a)}{1-a\Phi_{PL}} \tag{13.29}$$

TABLE 13.3

Observed and Corrected PL Quantum Yields (Φ_{PL}^{obs} and Φ_{PL}) and Self-Absorption Probability (a) for Ethanol Solution of Anthracene Purged with Argon Gas

Concentration (M)	Φ_{PL}^{obs}	a	Φ_{PL}
1×10^{-5}	0.278	0.066	0.290
5×10^{-5}	0.262	0.142	0.294
1×10^{-4}	0.252	0.179	0.291
5×10^{-4}	0.235	0.251	0.289
1×10^{-3}	0.220	0.271	0.280

The true PL quantum yield Φ_{PL} is derived from Equation (13.29).

$$\Phi_{PL} = \frac{\Phi_{PL}^{obs}}{1 - a + a\Phi_{PL}^{obs}} \qquad (13.30)$$

The self-absorption probability a can be estimated by comparing the observed PL spectrum $F_{obs}(\lambda)$ with the true spectrum, which is not influenced by the self-absorption effect,

$$a = 1 - \frac{\int F_{obs}(\lambda)\,d\lambda}{\int F'(\lambda)\,d\lambda} \qquad (13.31)$$

where $F'(\lambda)$ is the PL spectrum rescaled by normalizing the true spectrum $F(\lambda)$ to the observed spectrum at the long wavelength edge. The true spectrum can be estimated from the spectrum measured for a sufficiently low-concentration sample using the same detection system or by measuring the spectrum without using the IS.

Table 13.3 summarizes a, Φ_{PL}^{obs}, and Φ_{PL} for anthracene in ethanol with different concentrations. As the concentration increases, a increases and Φ_{PL}^{obs} decreases from 0.278 to 0.220 due to the self-absorption effect. After correction using Equations (13.30) and (13.31), Φ_{PL} shows an almost constant value of 0.29 over the entire concentration range. This result demonstrates that this correction method is useful for samples with high concentration and is recommended when using an IS for samples with a small Stokes shift.

13.6.2 Influence of Self-Absorption Effect on Phosphor Powders

The correction method proposed by Ahn *et al.* has been used not only for solutions and thin films but also for phosphor powders. This section introduces measurement examples for garnet and oxynitride phosphor powders [61].

13.6.2.1 Garnet Phosphor

Figure 13.18(a) shows images of different amounts of garnet phosphor prepared in a Petri dish. G1 and G2 are thinly spread garnet phosphors with diameters of about 5 and 15 mm in Petri dishes, respectively. G3 and G4 are phosphors spread over the inner diameter of the Petri dishes with thicknesses of about 1 and 2 mm, respectively. Figure 13.18(b) shows plots of the absorption ratio as a function of wavelength for G1–G4 measured with an IS system. The absorption ratio is given by the following equation,

$$\text{Absorption ratio}(\lambda) = \frac{\int \left[I_{ex}^{reference}(\lambda) - I_{ex}^{sample}(\lambda) \right] d\lambda}{\int I_{ex}^{reference}(\lambda)\,d\lambda} \qquad (13.32)$$

where $I_{ex}^{reference}(\lambda)$ and $I_{ex}^{sample}(\lambda)$ are the intensities of the excitation light without and with a sample, respectively. Figure 13.18(c) shows the PL spectra normalized at 645 nm. With increasing amount of garnet phosphor from G1 to G4, the absorption ratio increases. On the other hand, the PL intensity is attenuated over the wavelength range where the PL spectrum overlaps with the absorption spectrum (480–650 nm). Despite the distortion of the PL spectra and increase in the self-absorption probability with increasing sample amount, Φ_{PL}^{obs} shows an almost constant value between 0.92 and 0.93, which is very close to the corrected values of Φ_{PL} (0.93–0.94, see Figure 13.18(c) and Table 13.4). This fact can be explained by Equation (13.30). The samples with Φ_{PL} close to 1 can reemit the PL with a probability close to 1 after reabsorption of the emitted photons. Thus, Φ_{PL}^{obs} approaches Φ_{PL}.

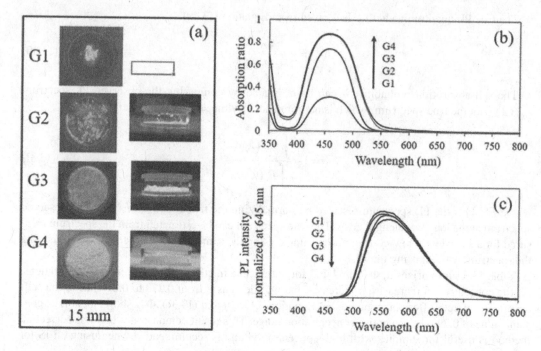

FIGURE 13.18   (a) Garnet phosphor of different amounts in Petri dishes (G1–G4), (b) plots of the absorption ratio and (c) PL spectra normalized at 645 nm.

In the measurement example, the PL spectrum of the garnet phosphor with the smallest amount (G1) was used to obtain the true spectrum because it is nearly unaffected by the self-absorption effect. Gorrotxategi *et al.* used a garnet phosphor with low activator concentration (3%) as the true spectrum to correct Φ_{PL}^{obs} for the phosphors with higher concentrations (20% and 70%) [44].

13.6.2.2   Oxynitride Phosphor

Figure 13.19(b) and (c) depict the absorption ratios and PL spectra, respectively, of oxynitride phosphors O1–O4 measured by a similar procedure to that of the garnet phosphors. Images of O1–O4 are shown in Figure 13.19(a). The absorption features a broadband in the ultraviolet to visible range and levels off at about 550 nm. With increasing sample amount from O1 to O4, the absorption ratio increases as expected. However, for O3 and O4, the absorption appears to stretch toward longer wavelengths. As this absorption overlaps with the PL spectra (see inset figure in Figure 13.19(c)), increased reabsorption can be expected.

TABLE 13.4

Self-Absorption Probability (*a*) and Observed and True PL Quantum Yields (Φ_{PL}^{obs} and Φ_{PL}) of Garnet Phosphor of Different Amounts (G1–G4)

Sample	*a*	Φ_{PL}^{obs}	Φ_{PL}
G1	0.10	0.93	0.94
G2	0.11	0.93	0.93
G3	0.15	0.92	0.93
G4	0.19	0.92	0.93

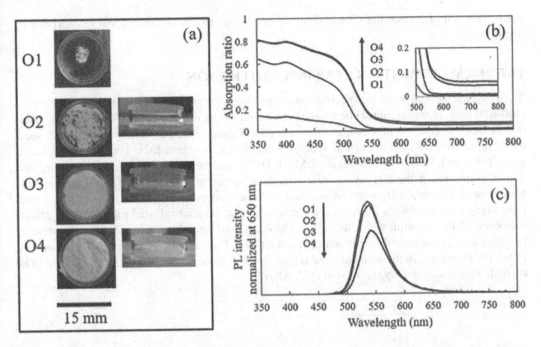

FIGURE 13.19 (a) Oxynitride phosphor of different amounts in Petri dishes (O1–O4), (b) plots of the absorption ratio, and (c) PL spectra normalized at 650 nm.

Fukuda investigated the relationship between the photophysical properties and the carbon content for a series of green-emitting oxynitride phosphors [62]. It was found that the absorption tail component extended to the near-infrared region and might be due to residual carbon in the phosphor.

Table 13.5 lists a, Φ_{PL}^{obs}, and Φ_{PL} for the oxynitride phosphors O1–O4. The Φ_{PL}^{obs} of O1 and O2 are 0.87 and 0.86, respectively, which are almost the same as the corrected values ($\Phi_{PL} = 0.87$ for both O1 and O2). On the other hand, the Φ_{PL}^{obs} values for O3 (0.79) and O4 (0.77) are about 10% smaller than those for O1 and O2. Even after the self-absorption correction, the Φ_{PL} values for O3 (0.85) and O4 (0.84) are still slightly smaller than those for O1 (0.87) and O2 (0.87). In the correction method proposed by Ahn *et al.*, the true PL spectrum needs to be rescaled to the observed spectrum at longer wavelengths where there is no overlap between the absorption and PL spectra, so that the self-absorption probability a can be estimated. For oxynitride phosphors with larger sample volumes (O3 and O4), the Φ_{PL} values are less than those with small sample volumes (O1 and O2) because the absorption tail component prevents the estimation of a accurately. In the case of the oxynitride,

TABLE 13.5

Self-Absorption Probability (a) and Observed and True PL Quantum Yields (Φ_{PL}^{obs} and Φ_{PL}) of Oxynitride Phosphor of Different Amounts (O1–O4)

Sample	a	Φ_{PL}^{obs}	Φ_{PL}
O1	0.07	0.87	0.87
O2	0.11	0.86	0.87
O3	0.34	0.79	0.85
O4	0.34	0.77	0.84

a small amount of the powder should be used for the Φ_{PL} measurement to avoid reabsorption by impurities.

13.7 MEASUREMENTS IN NEAR-INFRARED REGION

Typically, the solvents used for spectroscopic measurements are transparent in the visible region (400–700 nm). However, absorption bands attributed to the overtones of C–H and O–H vibration modes are observed in the near-infrared region. Figure 13.20 shows the transmittance spectra of representative solvents in the visible to near-infrared region [32]. The absorption bands around 900 and 1200 nm for toluene, DMSO, DCM, and ethanol correspond to the third and second overtones of the CH_3 symmetric stretching vibration, respectively, and the absorption band around 980 nm for H_2O can be ascribed to the second overtone of the OH stretching vibration. Many solvents absorb PL from sample solutions in the near-infrared region, which affects the observed PL quantum yields Φ_{PL}^{obs}. When an IS is used, the absorption effect by the solvents becomes more pronounced due to multiple reflections of the sample emission on the inner wall of the IS. Therefore, in the near-infrared region the Φ_{PL}^{obs} values measured using an IS need to be corrected for absorption by the solvents [32, 33].

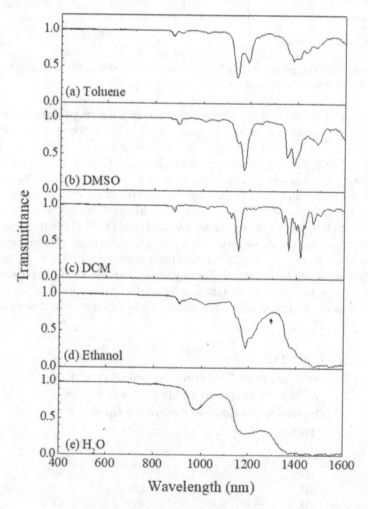

FIGURE 13.20 Transmittance spectra of representative solvents in the visible to near-infrared region (400–1600 nm).

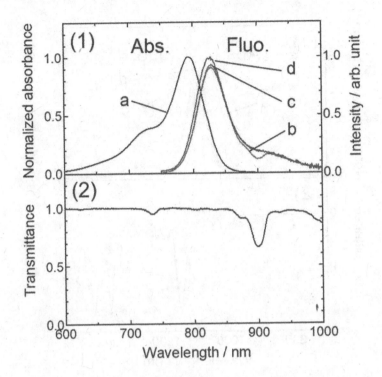

FIGURE 13.21 (1) Observed absorption (a) and PL (b) spectra, PL spectrum corrected for solvent absorption (c), and true PL spectrum (d) of ICG in DMSO. (2) Transmittance spectrum of DMSO measured with an IS instrument. (From Hoshi R., *et al.* 2020. *Anal. Chem.* 92: 607–11. Used with permission.)

Hoshi *et al.* have proposed a method to correct for the effect of solvent absorption on Φ_{PL}^{obs} values in the near-infrared region [32]. ICG and IR1061 solutions were used in the experiment, as these fluorophores emit PL in the first near-infrared (NIR-I, 700–950 nm) and the second near-infrared (NIR-II, 1000–1700 nm) windows, respectively. The solid line (b) in Figure 13.21(1) shows the PL spectrum of ICG in DMSO measured using an IS. The observed PL spectrum has a dip at around 900 nm. The dip is attributed to absorption by DMSO, as the wavelength of the dip matches that of the absorption band in the transmittance spectrum of DMSO (Figure 13.21(2)).

The PL quantum yield can be corrected for solvent absorption using the transmittance of the solvent $T(\lambda)$,

$$\Phi_{PL}^{Tcorr} = \frac{\int \left[\left(\left(I_{em}^{sample}(\lambda) \right) / T(\lambda) \right) - \left(\left(I_{em}^{reference}(\lambda) \right) / T(\lambda) \right) \right] d\lambda}{\int \left[\left(\left(I_{ex}^{reference}(\lambda) \right) / T(\lambda) \right) - \left(\left(I_{ex}^{sample}(\lambda) \right) / T(\lambda) \right) \right] d\lambda} \quad (13.33)$$

where Φ_{PL}^{Tcorr} is the corrected PL quantum yield, $I_{ex}^{reference}$ and I_{ex}^{sample} are the excitation light intensities without and with a sample, respectively, and I_{em}^{sample} and $I_{em}^{reference}$ are the PL intensity and the baseline, respectively. The excitation light profile and the PL intensity have units of number of photons per unit time. Equation (13.33) is derived from the relationship $I_0(\lambda) = I(\lambda) / T(\lambda)$ based on the definition of the transmittance of a sample, where $I_0(\lambda)$ and $I(\lambda)$ are the light intensities before and after passing through the sample.

The solid line (c) in Figure 13.21(1) is the PL spectrum of the ICG solution corrected for the absorption by DMSO using Equation (13.33). No dip is seen in the corrected PL spectrum, and the profile is close to that of the true spectrum (solid line (d) in Figure 13.21(1)). The slight difference in

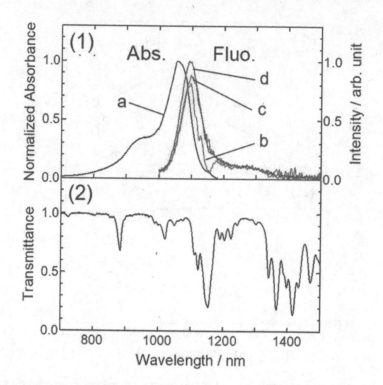

FIGURE 13.22   (1) Observed absorption (a) and PL (b) spectra, PL spectrum corrected for solvent absorption (c), and true PL spectrum (d) of IR1061 in DCM. (2) Transmittance spectrum of DCM measured with an IS instrument. (From Hoshi R., *et al.* 2020. *Anal. Chem.* 92: 607–11. Used with permission.)

the shorter wavelength region between the corrected and true spectra is due to the self-absorption effect caused by spectral overlap between the absorption and PL spectra. The observed and corrected quantum yields, Φ_{PL}^{obs} and Φ_{PL}^{Tcorr}, are 0.235 and 0.247, respectively, and the correction factor $\Phi_{PL}^{Tcorr}/\Phi_{PL}^{obs}$ is 1.05. The influence of absorption by the solvent on the Φ_{PL}^{obs} value is fairly small in the NIR-I region.

They also measured the PL spectrum of the NIR-II fluorophore IR1061 in DCM using an IS instrument (solid line (b) in Figure 13.22(1)). The observed spectrum is significantly distorted due to absorption by DCM. The distortion was corrected using Equation (13.33) to obtain the corrected PL spectrum (solid line (c)). Φ_{PL}^{obs} and Φ_{PL}^{Tcorr} are 0.0041 and 0.0050, respectively. The correction factor (1.22) is significantly larger than that for the ICG solution (1.05). The fact that most solvents tend to exhibit stronger absorption in the NIR-II region leads to a larger effect on the Φ_{PL}^{obs} values compared to the NIR-I region.

13.8   VALIDATION OF SYSTEM PERFORMANCE BY MEASURING STANDARD SUBSTANCES

For absolute PL quantum yield measurement systems based on the IS method, it is essential to certify that the measurement systems can provide accurate values of Φ_{PL} for samples. The easiest and most reliable way to verify measurement systems is to check the performance using standard materials whose Φ_{PL} values are well known or certificated. In the verification, the measured Φ_{PL} values are compared with the literature values. For the standards used for the verification, the excitation and emission wavelength regions should be similar to those for a sample being measured, because the reliability of the measured Φ_{PL} values mainly depends on the accuracy of the spectral sensitivity correction coefficient for the measurement system. Multiple standards need to be used to validate the system performance for the case when samples with different excitation and emission wavelength ranges are to be used.

The standard solutions introduced in 13.4.2 are highly recommended for use as standards for verification. One of the advantages of these solutions is that the fluorophores and solvents used can be purified to remove impurities by purification techniques such as high-performance liquid chromatography. Standards with high purity exhibit a high measurement reproducibility of the Φ_{PL} values.

. In the IS method, the self-absorption effect is more pronounced due to multiple reflections of sample PL inside the IS. When a standard solution with a small Stokes shift is measured, the observed PL quantum yield Φ_{PL}^{obs} is affected more significantly by the self-absorption effect. To minimize the measurement error arising from the self-absorption effect, dilute solutions should be used for the measurement, or Φ_{PL}^{obs} should be corrected for self-absorption (see 13.6).

In Φ_{PL} measurements in the near-infrared region, Φ_{PL}^{obs} may be affected by solvent absorption because many solvents have absorption bands attributed to the overtones of C–H and O–H vibration modes. To obtain accurate Φ_{PL} values, the Φ_{PL}^{obs} should be corrected for absorption by the solvent (see 5.5.3).

13.9 CONCLUSION

In this chapter, three measurement techniques for determining the Φ_{PL} for phosphors have been reviewed: the absolute methods based on the Vavilov method, photoacoustic spectroscopy, and relative methods. Then, the absolute method using an IS method has been introduced and detailed, including descriptions of the principle of the method, system configuration, and measurement procedure.

The IS method is becoming more common for Φ_{PL} measurements because it has several advantages over the three measurement techniques. The IS method does not require standard substances. The IS method can be applied for not only to solutions but also solid samples. In addition, the operation of the IS method is simple and easy. On the other hand, there are limitations arising from the measurement principle. In the IS method, the self-absorption effect is more pronounced due to multiple reflections of sample PL inside the IS. When a sample with a small Stokes shift is measured, the observed PL quantum yield Φ_{PL}^{obs} is affected more significantly by the self-absorption effect. In Φ_{PL} measurements in the near-infrared region, Φ_{PL}^{obs} may be affected by solvent absorption. Correction methods for the self-absorption and the absorption by the solvent have been introduced, respectively.

For absolute PL quantum yield measurement systems based on the IS method, it is essential to calibrate the spectral sensitivity of the photodetection system over the detection wavelength region. Standard solutions are highly recommended for use as standards to certify that the measurement systems can provide accurate values of Φ_{PL} for samples.

REFERENCES

1. Demas, J. N., and Crosby, G. A. 1971. "The Measurement of Photoluminescence Quantum Yields. A Review." *J. Phys. Chem.* 75: 991.
2. Vavilov, S. I. 1924. "Die Fluoreszenzausbeute von Farbstofflösungen." *Z. Phys.* 22: 266.
3. (a) Melhuish, W. H. 1955. "The Measurement of Absolute Quantum Efficiencies of Fluorescence." *N. Z. J. Sci. Technol.* 37: 142. (b) Melhuish, W. H. 1961. "Quantum Efficiencies of Fluorescence of Organic Substances: Effect of Solvent and Concentration of the Fluorescent Solute." *J. Phys. Chem.* 65: 229.
4. Bril, A., and de Jager-Veenis, A. W. 1976. "Some Methods of Luminescence Efficiency Measurements." *J. Res. Natl. Bur. Stand. A* 80A: 401.
5. Würth, C., Geißler, D., Behnke, T., Kaiser, M., and Resch-Genger, U. 2015. "Critical Review of The Determination of Photoluminescence Quantum Yields of Luminescent Reporters." *Anal. Bioanal. Chem.* 407: 59.
6. Mardelli, M., and Olmsted III, J. 1977. "Calorimetric Determination of the 9,10-Diphenylanthracene Fluorescence Quantum Yield." *J. Photochem.* 7: 277.

7. Braslavski, S. E., and Heihoff, K. 1989. "Photothermal Methods." In *Handbook of Organic Photochemistry*, ed. Scaiano, J. C., 327. Boca Raton, FL: CRC Press.

8. Adams, M. J., King, A. A., and Kirkbright, G. F. 1976. "Analytical Optoacoustic Spectrometry. Part I. Instrument Assembly and Performance Characteristics." *Analyst* 101: 73.

9. (a) Adams, M. J., Highfield, J. G., and Kirkbright, G. F. 1980. "Determination of the Absolute Quantum Efficiency of Luminescence of Solid Materials Employing Photoacoustic Spectroscopy." *Anal. Chem.* 52: 1260. (b) Adams, M. J., Highfield, J. G., and Kirkbright, G. F. 1981. "Determination of the Absolute Quantum Efficiency of Sodium Salicylate Using Photoacoustic Spectroscopy." *Analyst* 106: 850.

10. Adams, M. J., Highfield, J. G., and Kirkbright, G. F. 1977. "Determination of Absolute Fluorescence Quantum Efficiency of Quinine Bisulfate in Aqueous Medium by Optoacoustic Spectrometry." *Anal. Chem.* 49: 1850.

11. Parker, C. A., and Rees, W. T. 1960. "Correction of Fluorescence Spectra and Measurement of Fluorescence Quantum Efficiency." *Analyst* 85: 587.

12. Hamamatsu Photonics K.K. 2017. "Basic Principles of Photomultiplier Tubes." In *Photomultiplier Tubes – Basics and Applications*, ed. Inproject, Inc., 13–21. Hamamatsu, Japan: Hamamatsu Photonics K.K., Electron Tube Division.

13. Eaton, D. F. 1988. "Reference Materials for Fluorescence Measurement." *Pure Appl. Chem.* 60: 1107.

14. Resch-Genger, U., and Rurack, K. 2013. "Determination of the Photoluminescence Quantum Yield of Dilute Dye Solutions (IUPAC Technical Report)." *Pure Appl. Chem.* 85: 2005.

15. Brouwer, A. M. 2011. "Standards for Photoluminescence Quantum Yield Measurements in Solution (IUPAC Technical Report)." *Pure Appl. Chem.* 83: 2213.

16. Grabolle, M., Spieles, M., Lesnyak, V., Gaponik, N., Eychmüller, A., and Resch-Genger, U. 2009. "Determination of the Fluorescence Quantum Yield of Quantum Dots: Suitable Procedures and Achievable Uncertainties." *Anal. Chem.* 81: 6285.

17. Martini, M., Montagna, M., Ou, M., Tillement, O., Roux, S., and Perriat, P. 2009. "How to Measure Quantum Yields in Scattering Media: Application to the Quantum Yield Measurement of Fluorescein Molecules Encapsulated in Sub-100 nm Silica Particles." *J. Appl. Phys.* 106: 094304–1.

18. Suzuki, K., Kobayashi, A., Kaneko, S., *et al.* 2009. "Reevaluation of Absolute Luminescence Quantum Yields of Standard Solutions Using a Spectrometer with an Integrating Sphere and a Back-thinned CCD Detector." *Phys. Chem. Chem. Phys.* 11: 9850.

19. Dawson, W. R., and Windsor, M. W. 1968. "Fluorescence Yields of Aromatic Compounds." *J. Phys. Chem.* 72: 3251.

20. Velapoldi, R. A., and Mielenz, K. D. 1980. "Standard Reference Materials: A Fluorescence Standard Reference Material: Quinine Sulfate Dihydrate." In *National Bureau of Standards Special Publication*, 260–4. Washington, DC: U.S. Government Print Office.

21. Würth, C., Lochmann, C., Spieles, M., *et al.* 2010. "Evaluation of a Commercial Integrating Sphere Setup for the Determination of Absolute Photoluminescence Quantum Yields of Dilute Dye Solutions." *Appl. Spectrosc.* 64: 733.

22. Fischer, M., and Georges, J. 1996. "Fluorescence Quantum Yield of Rhodamine 6G in Ethanol as a Function of Concentration Using Thermal Lens Spectrometry." *Chem. Phys. Lett.* 260:115.

23. Würth, C., Grabolle, M., Pauli, J., Spieles, M., and Resch-Genger, U. 2011. "Comparison of Methods and Achievable Uncertainties for the Relative and Absolute Measurement of Photoluminescence Quantum Yields." *Anal. Chem.* 83: 3431.

24. Würth, C., González, M. G., Niessner, R., Panne, U., Haisch, C., and Resch-Genger, U. 2012. "Determination of the Absolute Fluorescence Quantum Yield of Rhodamine 6G with Optical and Photoacoustic Methods – Providing the Basis for Fluorescence Quantum Yield Standards." *Talanta* 90: 30.

25. Shen, J., and Snook, R. D. 1989. "Thermal Lens Measurement of Absolute Quantum Yields Using Quenched Fluorescent Samples as References." *Chem. Phys. Lett.* 155: 583.

26. Porrès, L., Holland, A., Pålsson, L.-O., Monkman, A. P., Kemp, C., and Beeby, A. 2006. "Absolute Measurements of Photoluminescence Quantum Yields of Solutions Using an Integrating Sphere." *J. Fluoresc.* 16 (2): 267.

27. Würth, C., Pauli, J., Lochmann, C., Spieles, M., and Resch-Genger, U. 2012. "Integrating Sphere Setup for the Traceable Measurement of Absolute Photoluminescence Quantum Yields in the Near Infrared." *Anal. Chem.* 84: 1345.

28. Ware, W. R., and Rothman, W. 1976. "Relative Fluorescence Quantum Yields Using an Integrating Sphere. The Quantum Yield of 9,10-Diphenylanthracene in Cyclohexane." *Chem. Phys. Lett.* 39: 449–53.

29. Abedin-Siddique, Z., Ohno, T., Nozaki, K., and Tsumomura, T. 2004. "Intense Fluorescence of Metal-to-Ligand Charge Transfer in [Pt(0)(binap)$_2$] [binap = 2,2'-Bis(diphenylphosphino)-1,1'-binaphthyl]." *Inorg. Chem.* 43: 663.

30. Van Houten, J., and Watts, R. J. 1976. "Temperature Dependence of the Photophysical and Photochemical Properties of the Tris(2,2'-bipyridyl)ruthenium(II) Ion in Aqueous Solution." *J. Am. Chem. Soc.* 98, 4853.

31. Ishida, H., Tobita, S., Hasegawa, Y., Katoh, R., and Nozaki. K. 2010. "Recent Advances in Instrumentation for Absolute Emission Quantum Yield Measurements." *Coord. Chem. Rev.* 254: 2449.

32. Hoshi R., Suzuki, K., Hasebe, N., *et al.* 2020. "Absolute Quantum Yield Measurements of Near-Infrared Emission with Correction for Solvent Absorption." *Anal. Chem.* 92: 607.

33. Hatami, S., Würth, C., Kaiser, M., *et al.* 2015. "Absolute Photoluminescence Quantum Yields of IR26 and IR-Emissive Cd$_{1-x}$Hg$_x$Te and PbS Quantum Dots – Method- and Material-Inherent Challenges." *Nanoscale* 7: 133.

34. Benson, R. C., and Kues, H. A. 1977. "Absorption and Fluorescence Properties of Cyanine Dyes." *J. Chem. Eng. Data* 22: 379.

35. Semonin, O. E., Johnson, J. C., Luther, J. M., Midgett, A. G., Nozik, A. J., and Beard, M. C. 2010. "Absolute Photoluminescence Quantum Yields of IR-26 Dye, PbS, and PbSe Quantum Dots." *J. Phys. Chem. Lett.* 1: 2445.

36. Würth, C., Grabolle, M., Pauli, J., Spieles, M., and Resch-Genger, U. 2013. "Relative and Absolute Determination of Fluorescence Quantum Yields of Transparent Samples." *Nat. Protoc.* 8: 1535.

37. Katoh, R., Suzuki, K., Furube, A., Kotani, M., and Tokumaru, K. 2009. "Fluorescence Quantum Yield of Aromatic Hydrocarbon Crystals." *J. Phys. Chem. C* 113: 2961.

38. Kristianpoller, N. 1964. "Absolute Quantum Yield of Sodium Salicylate." *J. Opt. Soc. Am.* 54: 1285.

39. Bai, X., Caputo, G., Hao, Z., *et al.* 2014. "Efficient and Tunable Photoluminescent Boehmite Hybrid Nanoplates Lacking Metal Activator Centres for Single-Phase White LEDs." *Nat. Commun.* 5: 5702. DOI: 10.1038/ncomms6702.

40. Rohwer, L. S., and Martin, J. E. 2005. "Measuring the Absolute Quantum Efficiency of Luminescent Materials." *J. Lumin.* 115: 77.

41. Do, Y. R, and Bae, J. W. 2000. "Application of Photoluminescence Phosphors to a Phosphor-Liquid Crystal Display." *J. Appl. Phys.* 88: 4660.

42. Jüstel, T., Krupa, J.-C., and Wiechert, D. U. 2001. "VUV Spectroscopy of Luminescent Materials for Plasma Display Panels and Xe Discharge Lamps." *J. Lumin.* 93: 179.

43. Kim, K.-B., Kim, Y.-I., Chun, H.-G., Cho, T.-Y., Jung, J.-S., and Kang, J.-G. 2002. "Structural and Optical Properties of BaMgAl$_{10}$O$_{17}$:Eu^{2+} Phosphor." *Chem. Mater.* 14: 5045.

44. Gorrotxategi, P., Consonni, M., and Gasse, A. 2015. "Optical Efficiency Characterization of LED Phosphors Using a Double Integrating Sphere System." *J. Solid State Light.* 2:1.

45. Bachmann, V., Ronda, C., and Meijerink, A. 2009. "Temperature Quenching of Yellow Ce^{3+} Luminescence in YAG:Ce Phosphor." *Chem. Mater.* 21:2077.

46. Greenham, N. C., Samuel, I. D. W., Hayes, G. R., *et al.* 1995. "Measurement of Absolute Photoluminescence Quantum Efficiencies in Conjugated Polymers." *Chem. Phys. Lett.* 241: 89.

47. de Mello, J. C., Wittmann, H. F., and Friend, R. 1997. "An Improved Experimental Determination of External Photoluminescence Quantum Efficiency." *Adv. Mater.* 9: 230.

48. Mattoussi, H., Murata, H., Merritt, C. D., Iizumi, Y., Kido, J., and Kafafi, Z. H. 1999. "Photoluminescence Quantum Yield of Pure and Molecularly Doped Organic Solid Films." *J. Appl. Phys.* 86: 2642.

49. Kawamura, Y., Sasabe, H., and Adachi, C. 2004. "Simple Accurate System for Measuring Absolute Photoluminescence Quantum Efficiency in Organic Solid-State Thin Films." *Jpn. J. Appl. Phys.* 43: 7729.

50. Zhang, Y., Russo, R. E., and Mao, S. S. 2005. "Quantum Efficiency of ZnO Nanowire Nanolasers." *Appl. Phys. Lett.* 87: 043106.

51. Ahn, T.-S., Al-Kaysi, R. O., Müller, A. M., Wentz, K. M., and Bardeen, C. J. 2007. "Self-absorption Correction for Solid-State Photoluminescence Quantum Yields Obtained from Integrating Sphere Measurements." *Rev. Sci. Instrum.* 78: 086105.

52. Johnson, A. R., Lee, S.-J., Klein, J., and Kanicki, J. 2007. "Absolute Photoluminescence Quantum Efficiency Measurement of Light-Emitting Thin Films." *Rev. Sci. Instrum.* 78: 096101.

53. Faulkner, D. O., McDowell, J. J., Price, A. J., Perovic, D. D., Kherani, N. P., and Ozin, G. A. 2012. "Measurement of Absolute Photoluminescence Quantum Yields Using Integrating Spheres – Which Way to Go?" *Laser Photonics Rev.* 6: 802.

54. Leyre, S., Coutino-Gonzalez, E., Joos, J. J., *et al.* 2014. "Absolute Determination of Photoluminescence Quantum Efficiency Using an Integrating Sphere Setup." *Rev. Sci. Instrum.* 85: 123115.

55. Makowiecki, J., and Martynski, T. 2014. "Absolute Photoluminescence Quantum Yield of Perylene Dye Ultra-Thin Films." *Org. Electron.* 15: 2395.

56. Pålsson, L.-O., and Monkman, A. P. 2002. "Measurements of Solid-State Photoluminescence Quantum Yields of Films Using a Fluorimeter." *Adv. Mater.* 14: 757.

57. Wilson, L. R., and Richards, B. S. 2009. "Measurement Method for Photoluminescent Quantum Yields of Fluorescent organic Dyes in Polymethyl Methacrylate for Luminescent Solar Concentrators." *Appl. Opt.* 48: 212.

58. Gardecki, J. A., and Maroncelli, M. 1998. "Set of Secondary Emission Standards for Calibration of the Spectral Responsivity in Emission Spectroscopy." *Appl. Spectrosc.* 52: 1179.

59. Lakowicz, J. R. 2006. "Appendix I, Corrected Emission Spectra." In *Principles of Fluorescence Spectroscopy*, 3rd ed., 873–82. New York, NY: Springer Science+Business Media.

60. Pfeifer, D., Hoffmann, K., Hoffmann, A., Monte, C., and Resch-Genger, U. 2006. "The Calibration Kit Spectral Fluorescence Standards – A Simple and Certified Tool for the Standardization of the Spectral Characteristics of Fluorescence Instruments." *J. Fluoresc.* 16: 581.

61. (a) Suzuki, K. 2013. "Absolute Photoluminescence Quantum Yield Measurement by Using an Integrating Sphere." Paper presented at Phosphor Global Summit 2013, New Orleans. (b) Suzuki, K. 2013. "Absolute Photoluminescence Quantum Yield Measurement by Using an Integrating Sphere, -The Reliability of Measured Quantum Yield and Self-Absorption Effect-." Paper presented at Phosphor Safari 2013, Jeju, Republic of Korea.

62. Fukuda, Y. 2013. "Evaluation of Luminescence Properties of Eu^{2+}-Doped Green-Emitting Sr-Containing Sialon Phosphors by Electron Spin Resonance." *ESC J. Solid State Sci. Technol.* 2: R56.

14 Photoionization Analysis on Phosphors

Jumpei Ueda

CONTENTS

14.1 INTRODUCTION

In this chapter, the history of the photoionization process for lanthanide ions (*Ln*) in compounds and the luminescence quenching is introduced briefly and then several analysis methods to demonstrate the photoionization process are summarized. Photoionization for *Ln* ions is the process that an electron of *Ln* ion in a compound moves into the conduction band (CB) by photon illumination. The research on the photoionization phenomenon for *Ln* ions began in several research areas such as photochromic glasses, excited-state absorption for laser application, electroluminescence (EL), anomalous luminescence (impurity trapped exciton) and luminescence quenching. The observations of photoionization phenomena had attracted the energy-level location of lanthanide ions with respect to the host electronic structure such as valence band (VB) and CB. After that, to determine its energy level, photoionization measurements have been utilized. Today, the energy levels of *Ln* ions with respect to the host band structure become clear, and it is accepted that the photoionization process is critical to understand the quenching process for the 5d–4f and 4f–4f luminescence of *Ln* ions and the persistent luminescence mechanism in the phosphor area.

14.2 HISTORY OF PHOTOIONIZATION FOR LANTHANIDE IONS

The photoionization of *Ln* ions was first reported in 1947. Stookey in Corning company reported the photosensitive glass, whose color can be changed after exposure to UV light (300–350 nm) [1]. The glass contains gold, silver or copper ions, and these metal ions are reduced to metal colloid during the exposure, which can be an efficient color center due to the localized surface plasmon absorption. The Ce^{3+} ion acts as an excellent photosensitization for the Au^+-doped photosensitive glass, and its photochemical reaction in the glass is written as follows:

$$Ce^{3+} + Au^+ + h\nu \rightarrow Ce^{4+} + Au^0 \tag{14.1}$$

DOI: 10.1201/9781003098669-14

The Ce^{3+} was photoionized by UV light and the photoelectron was captured by Au^+. After that, the photoreduced Au metal can form the Au colloid and show strong coloration.

Thanks to the great work by Stookey, it was widely accepted in the glass area that the Ce^{3+} can be photoionized by UV light in the 1960s. For instance, Stroud investigated the photoionization process from Ce^{3+} to Ce^{4+} in glasses by absorption and electron paramagnetic resonance (EPR) spectroscopy [2, 3].

On the other hand, in the process of exploring a new solid laser material doped with *Ln* ions after the invention of the first ruby laser in 1960, the photoionization process becomes important to understand the excited-state absorption and the energy level of *Ln* ions. For instance, the *Ln* ions–doped calcium fluoride crystal was considered to be an excellent candidate for the laser material because of high bandgap energy, the easy substitution of *Ln* ions and ease growing the single crystal. In actuality, the $CaF_2:Sm^{2+}$ was known to be the third successful laser developed by Sorokin and Steven in 1961 [4]. In that time, many trivalent and divalent *Ln* ions–doped CaF_2 crystals were prepared. As one of the preparation methods of $CaF_2:Ln^{2+}$, the photochemical reaction using γ-ray irradiation was used as follows:

$$Eu^{2+} + Ln^{3+} + h\nu \rightarrow Eu^{3+} + Ln^{2+} \tag{14.2}$$

Thus, the photoionization process of Eu^{2+} in fluoride crystals was also known in the 1960s, although the excitation is the high-energy photon [5]. In 1964, Anderson and Kiss in RCA (Radio Corporation of America) laboratory discussed the photoionization process of Ln^{2+} in CaF_2 by the photocurrent per incident photon (photocurrent excitation [PCE] spectrum) [5, 6]. The absorption spectrum of $CaF_2:Tm^{2+}$ corresponds to the PCE spectrum, indicating the autoionization of 5d electron into the CB [5, 6]. In 1969, Heyman also reported the PCE spectrum of $CaF_2:Gd^{2+}$ [7]. At the same time, Abbruscato reported the PCE band which is attributed to the 4f–5d transition of Eu^{2+} in $SrAl_2O_4$ to discuss the persistent luminescence in 1971 [8].

Through the photoconductivity measurement for the *Ln* ions–doped compounds, the energy-level location with respect to the CB became clearer. In 1979, Pedrini, McClure and Anderson determined the threshold energy of photoionization for divalent *Ln* ions (Ho^{2+}, Dy^{2+}, Tm^{2+}) in various fluoride compounds (CaF_2, SrF_2, BaF_2) by photoconductivity measurement [9]. In the late 1970s, to investigate the mechanism that YAG:Ce^{3+} is not lasing, the excited-state absorption from the lowest 5d energy level was studied by some groups [10–12]. To investigate the ESA process, Pedrini and McClure et al. measured the threshold energy of photoionization of Ce^{3+} by photoconductivity and also suggested that the strong ESA is caused by the transition from 5d to CB in 1986 [13].

From the 1980s to 1990s, in addition to the research group of Pedrini and McClure, the group of Takahashi and Shionoya, the group of Kaplyanskii and Basun, the group of Evans and Dennis with Basun, the group of Yen with Basun and some other groups (the short history of photoconductivity measurement in the early period is reviewed in the paper by Pédrini, Joubert and McClure [14]) started to measure the photoconductivity spectra in the transition metals or *Ln* ions–doped compounds to discuss the influence for the luminescence quenching process of phosphors by photoionization [15–25]. In the same period, Świątek and Godlewski demonstrated the photoionization process of *Ln* ions using an EPR under light illumination [26–28]. Due to these great contributions to understanding the photoionization process, the photoionization quenching for phosphors had been accepted in the 1990s.

14.3 HISTORY OF LUMINESCENCE QUENCHING BY PHOTOIONIZATION

In the past, the luminescence quenching and thermal quenching processes were mainly explained to be caused by the cross-over process between the excited state and the ground state [29, 30]. Thermally activated cross-over, as shown by the red arrow in the CC diagram of Figure 14.1 is the nonradiative relaxation process from the excited 5d potential curve to the lower 4f potential curve

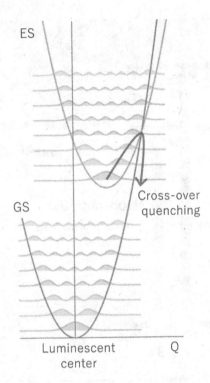

FIGURE 14.1 Configuration coordinate diagram for thermally activated cross-over quenching. GS: 4f ground state, ES: 5d excited state

through the crossing point. At the crossing point, the energy of the 5d vibronic state coupling with a few phonons matches that of a higher 4f vibronic state, and the resonant transition to the 4f state is followed by rapid nonradiative relaxation to lower vibronic 4f states. The high 4f vibrational levels involved have their amplitude concentrated almost exclusively at the extremes of the parabola. For this reason, the thermally activated cross-over is often depicted as a thermally activated process with the energy difference between the lowest vibrational level of the excited state and the crossing point of the parabola as activation energy. This activation barrier decreases for a larger configurational offset and a smaller energy difference between the states.

On the other hand, it was suggested by Blasse in 1978 that there is another type of quenching process related to electron transfer (charge transfer) [31]. With the history of the demonstration of the photoionization process as introduced in Section 14.1, the luminescence quenching related to the electron transfer has been regarded one of the main quenching processes, especially for the 5d–4f luminescence. The schematic energy diagram of photoionization quenching using Ce^{3+}:5d–4f transition is shown in Figure 14.2. For photoionization, there are mainly two processes with or without the thermally activated pathway. When we focus on the excited 5d electron, the electron transfer process without thermal assistance from the 5d excited state to the CB is called autoionization as shown in the dark green arrow of Figure 14.2, while thermal ionization is the thermally activated electron transfer process as shown by the red arrow of Figure 14.2. In brief, the processes of photon excitation into 5d state and auto-ionization are regarded photoionization and the processes of photon excitation and thermal ionization is thermally assisted photoionization.

The discussion of the (thermally assisted) photoionization quenching for the 5d–4f luminescence was stimulated by various experimental results obtained in $Y_3Al_5O_{12}$ (YAG) doped with Ce^{3+}, which is a famous and excellent yellow phosphor. The thermal quenching of the Ce^{3+} luminescence in the YAG was studied in 1973, when Weber measured the temperature dependence of the 5d–4f luminescence in YAG:Ce^{3+} and YAG:Pr^{3+} and reported that the lifetime of the $5d_1$ excited state rapidly

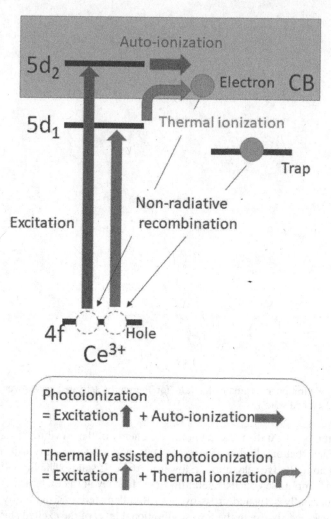

FIGURE 14.2 Schematic energy diagram for photoionization and thermally assisted photoionization. (Reproduced with permission from Ref. [36].)

decreases above 600K in YAG:Ce^{3+} [32]. He tried to explain the quenching process by multiphonon emission from the 5d state to the 4f ground state. From the late 1970s, the excited-state absorption from the lowest 5d energy level was studied by some groups [10–13] as introduced earlier. In 1991, Lyu and Hamilton reported a similar quenching curve ($T_{50\%}$ ~630K) based on lifetime measurements in YAG:Ce^{3+} to that reported by Weber et al. [33] From the rough agreement between the quenching activation energy (6500 cm^{-1}, 0.81 eV) and the $5d_1$-CB energy gap estimated from the excited-state absorption spectra (10,000 cm^{-1}, 1.24 eV) [33, 34], Lyu and Hamilton suggested that the quenching process is caused by the thermal ionization. In the same year, Blasse et al. also suggested the thermal ionization quenching process in YAG:Ce^{3+} on the basis of the results of ESA by Hamilton [34] and photoconductivity by Pedrini [13], also predicting the nonluminescent Ce^{3+}-doped compounds such as La_2O_3 and La_2O_2S have a strong photoionization process [35]. However, there was no direct evidence that the luminescence quenching was caused by the ionization process.

In 1996, the direct evidence of photoionization quenching for *Ln* ions was first reported by Yen et al. in Ce^{3+}-doped Y_2O_3, Lu_2O_3 and Lu_2SiO_5 from the PCE spectrum. In the PCE spectrum of Lu_2O_3:Ce^{3+}, a strong PCE band was observed in the visible range at both room temperature and liquid nitrogen temperature, as shown in Figure 14.3. This PCE band is caused by the Ce^{3+}:4f–5d

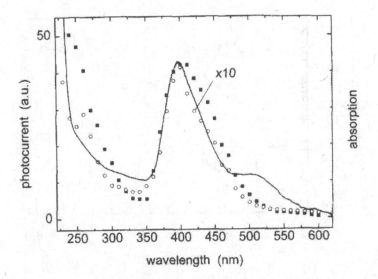

FIGURE 14.3 PCE spectrum of $Lu_2O_3:Ce^{3+}$ at room temperature (solid squares) and at liquid nitrogen temperature (open circles), normalized to incident photon flux. The solid line is the absorption spectrum at room temperature. (Reproduced with permission from Ref. [22].)

absorption and the consequent electron transfer to the CB. Yen et al. suggested the schematic energy diagram in two situations; totally quenched luminescence and efficient luminescence for Ce^{3+} as shown in Figure 14.4. The energy-level location of $Lu_2O_3:Ce^{3+}$ as well as $Y_2O_3:Ce^{3+}$ follows the situation in Figure 14.4(a), and as a result, there is no luminescence. On the other hand, the PCE spectra of $Lu_2SiO_5:Ce^{3+}$ were changed dramatically at room temperature and at liquid nitrogen temperature (Figure 14.5). The lowest $5d_1$ PCE band disappears at low temperature, indicating the $5d_1$ level is located below the bottom of CB, which follows the situation in Figure 14.4(b). In this situation, the Ce^{3+} luminescence can be observed up to a certain temperature. After the reports by Yen et al.,

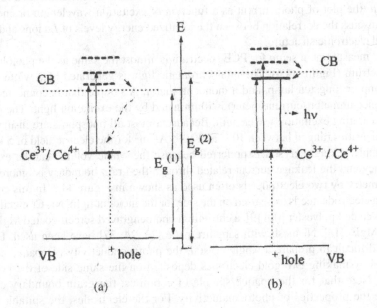

FIGURE 14.4 Schematic energy diagrams of Ce^{3+}-doped compound for (a) nonluminescent material and (b) luminescent material. (Reproduced with permission from Ref. [22].)

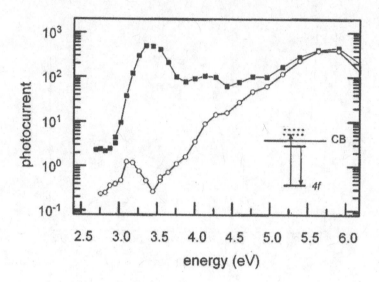

FIGURE 14.5 PCE of $Lu_2SiO_5:Ce^{3+}$ at room temperature (solid squares) and at liquid nitrogen temperature (open circles), normalized to incident photon flux. (Reproduced with permission from Ref. [21].)

the photoionization process has been observed in many compounds doped with Ce^{3+}, Pr^{3+}, Tb^{3+} and Eu^{2+}. By the apparent evidence of photoionization and thermally assisted photoionization, the quenching process of *Ln* ions became clear, which have helped to develop many LED phosphors with excellent thermal stability.

14.4 PHOTOCONDUCTIVITY TECHNIQUE

The most widely used technique to detect the photoionization process is photoconductivity measurement. As introduced in the brief history, the energy-level location of *Ln* ions and the luminescent quenching process had been proved by using photoconductivity measurement. Especially, the PCE spectrum (the plot of photocurrent as a function of excitation wavelength or energy) has the advantage to discuss the correlation between the localized energy levels of *Ln* ions and the delocalized host band electronic structure.

The optical measurement setup for PCE spectrum is almost the same as the photoluminescence excitation spectrum (Figure 14.6). The monochromatic light is generated by a white light source, usually Xe lamp or tungsten lamp, and a monochromator. A sample that is connected to a power supply and a picoammeter (current meter) is illuminated by the excitation light. The photocurrent is plotted by scanning excitation wavelength. Because almost all phosphors are insulators, typical photocurrent densities ranged between 10^{-12} and 10^{-14} A/cm² for the electric field of 5 kV/cm [9].

For the sample, a single crystal is preferred because the whole volume can be excited due to the transparency and the leakage current related through the grain boundary is ignored. Thus, the sandwich geometry by two electrodes is often used as shown in Figure 14.7. In this configuration, the transparent electrode needs to be used on the side for the incident light. Ni–Cr metal-coated sapphire [9], gold-coated polyester film [9], a chromium photoengraved screen-coated MgF_2 plate [13], In_2O_3-coated MgF_2 [13], Ni mesh with sapphire plate [23, 24, 37] have been used. For the samples that it is difficult to prepare a single crystal, the photoconductivity of opaque ceramics has been measured by making two gold electrodes deposited on the same side of the pellet [38, 39] (Figure 14.8). Note that for the opaque sample of ceramics, the grain boundary considerably may influence the properties of photoconductivity. For the electrodes, the suitable metal must be selected to make ohmic contact. Also, the saturation of photocurrent should be avoided. Before the photoconductivity measurement, the photocurrent as a function of excitation light

Measurement system of photocurrent excitation (PCE) spectrum

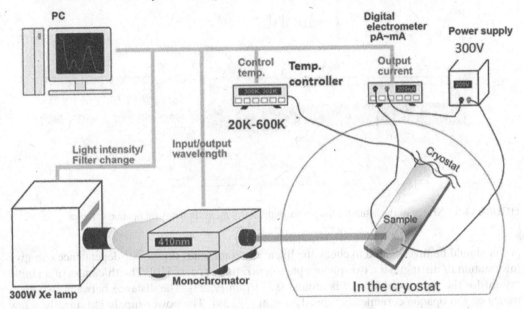

FIGURE 14.6 Schematic diagram of general photoconductivity measurement system. (Reproduced with permission from Ref. [36].)

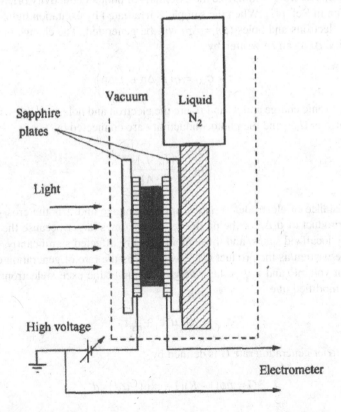

FIGURE 14.7 Sketch of the setup for photoconductivity measurements for the single crystals. (Reproduced with permission from Ref. [22].)

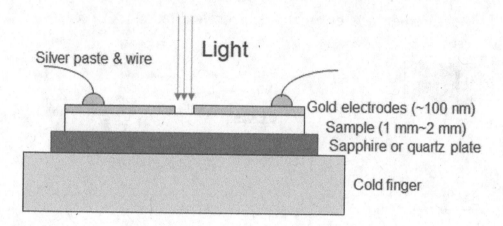

FIGURE 14.8   Sketch of the setup for the photoconductivity measurements for opaque samples.

power should be investigated to check the linearity. Sometimes this power dependence can give information to distinguish a two-photon photoionization process [40]. The thickness of a single crystal for the photoconductivity is around 0.2–3 mm [9, 23]. The distance between two electrodes on the opaque ceramics is typical ~1 mm [38, 39]. The power supply is typically a few tenths of V to 1 kV.

The principles of photoconductivity for semiconductors were explained comprehensively in some books [41, 42]. Here, we summarize the essentials of photoconductivity briefly here by following the explanation in Ref. [42]. When the sample is irradiated by excitation light, the equal excess densities of free electrons and holes ($\Delta n = \Delta p$) will be generated. The change of the conductivity (photoconductivity, σ_{PC}) can be written by

$$\Delta\sigma = \sigma_{PC} = q\left(\mu_n \Delta n + \mu_p \Delta p\right) \tag{14.3}$$

where q is the electronic charge and μ_n and μ_p are the electron and hole mobilities, respectively. The observed photocurrent (I_{PC}) and the photoconductivity are connected by

$$\sigma_{PC} = \frac{\left(I_{PC} \cdot L\right)}{\left(V \cdot A\right)} \tag{14.4}$$

where L is the distance of electrodes, V is the applied voltage and A is the cross-section area. In many cases, the product of $\mu_n \Delta n$ or the product $\mu_p \Delta p$ becomes larger because the electron or hole can be trapped by localized states, and those mobilities are affected significantly. Also, the excess density Δn can be written as the product $G\tau_{n,p}$, where G is the rate of generation of free electrons and holes per unit volume, and $\tau_{n,p}$ is the average lifetime of the excess electrons or holes. Thus, Eq. (14.3) can be modified into

$$\sigma_{PC} = qG\left(\mu_n \tau_n + \mu_p \tau_p\right) \tag{14.5}$$

The average carrier generation rate G is defined by

$$G = \eta\phi(1 - R)\left(1 - \exp(-\alpha d)\right)/d \tag{14.6}$$

where η is the quantum efficiency of the carrier generation process, φ is the photon flux, R is the reflection coefficient, α is the optical absorption coefficient and d is the thickness of the sample.

When the value of d is small with respect to the optical absorption depth $1/\alpha$, Eq. (14.6) can then be simplified to

$$G \cong \eta\phi(1-R)\alpha \tag{14.7}$$

In this case, the G value can be assumed to be a constant. On the other hand, for the sample with a thick thickness or/and a high absorption coefficient, the photoionization process will only be caused for the side of the incident light. We should avoid this saturation for the photoconductivity measurement. When we assume that the free electrons are the main carriers for the photoconductivity, the free electron lifetimes, τ_n in Eq. (14.5), can be affected by the holes. Thus, the recombination rate can be written as $\tau^{-1}_n = b(p_0 + \Delta p)$, where b is recombination constant, and p_0 and Δp are the equilibrium and excess minority carrier densities, respectively. Thus, eq. 14.5 can be changed into

$$\sigma_{PC} \propto \Delta n = G\tau_n = G/b\left(p_0 + \Delta p\right) \tag{14.8}$$

This equation is valid for a semiconductor with free electrons in the majority. However, for the insulator compounds, the concentration of the free hole, p_0, can be very small. Also, for the phosphors doped with Ln ions, when the 4f electron was photoionized to the CB, the hole is strongly trapped by Ln ions such as Ce^{3+}, Pr^{3+}, Tb^{3+}, Eu^{2+} and Yb^{2+}, because these ions can act as deep hole traps. Thus, in this situation, the recombination rate between free electrons and hole-trapped Ln ions is dominant. For instance, this recombination rate was assumed to be $\sim 10^{10}$ s^{-1} in Ce^{4+} and free electron by van der Kolk et al. [43].

For the PCE spectrum, the photocurrent intensity is calibrated by the incident photon flux. Assuming that R and τ_n are not changed largely as a function of excitation wavelength, the PCE spectrum is produced by the product of the absorption coefficient spectrum and the ionization efficiency. Thus, the onset energy of the PCE spectrum can be regarded the threshold energy for photoionization, which is the energy gap between the ground state of the Ln ion and the bottom of the CB. However, the PCE spectrum at room temperature includes the thermally assisted photoionization process, leading to the possibility of underestimating the energy-level locations.

Figures 14.9 and 14.10 show the temperature dependence of PCE spectra for Ce^{3+}-doped $Y_3Al_2Ga_3O_{12}$ (YAGG) and $Y_3Ga_5O_{12}$ (YGG), respectively. YAGG:Ce^{3+} shows green luminescence, which is attributed to the transition from the lowest $5d_1$ to 4f state and YGG:Ce^{3+} is a nonluminescent material. When the PCE spectra of these two phosphors at 300K are compared, similar results can be obtained. Two PCE bands attributed to the 4f–$5d_1$ and $5d_2$ (second-lowest 5d) are observed around 430 and 350 nm, respectively. On the basis of the PCE spectra at room temperature, the photoionization threshold energy is located below approximately 2.48 eV (500 nm) in both phosphors. However, the PCE band attributed to 4f-$5d_1$ of the YAGG:Ce^{3+} disappears at low temperature, indicating that the excited electrons at $5d_1$ state are assisted by thermal energy to transfer to the CB. Thus, the YAGG:Ce^{3+} follows the energy diagram as shown in Figure 14.4(b), whereas the 4f-$5d_1$ PCE band of YGG:Ce^{3+} is still observed at low temperature, indicating the $5d_1$ band is located within the CB (Figure 14.4(a)). Thus, to determine the precise energy-level location between 4f, 5d and the CB, low-temperature photoconductivity measurements are necessary.

The temperature dependence of PCE spectra enables to estimate the activation energy from the $5d_1$ to the CB under the situation of thermally assisted photoionization. We can assume that η is only a temperature-dependent parameter in Eq. (14.7), the Arrhenius plot of the integrated area of PCE intensity leads into the thermal activation energy. However, the linearity in the Arrhenius plot becomes worse at low temperatures, which can be attributed to the tunneling process without a thermally activated process [44]. Thus, the fitting equation becomes

$$\text{PCE intensity} = C_1 + C_2 \times \exp(-E/kT) \tag{14.9}$$

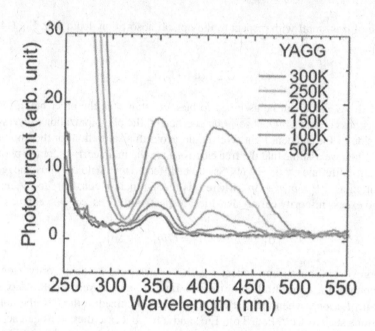

FIGURE 14.9   Temperature variations of PCE spectra of Ce^{3+}-doped $Y_3Al_2Ga_3O_{12}$ (YAGG). (Reproduced with permission from Ref. [39].)

Figure 14.11 shows the Arrhenius plot of photocurrent intensity for YAGG:Ce^{3+}. The curves were fitted well by Eq. (14.9), and the activation energy for the thermal ionization process was estimated. Figure 14.12 shows the temperature dependence of photoluminescence and photocurrent in YAGG:Ce^{3+}. The opposite tendency for photocurrent and photoluminescence intensity as a function of temperature can be observed. Thus, the thermal ionization process is determined to be the main thermal quenching process.

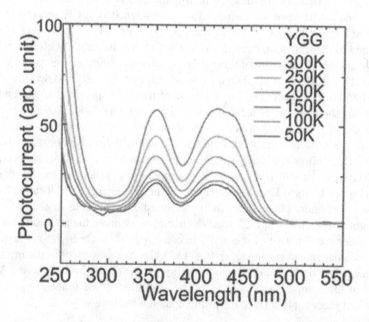

FIGURE 14.10   Temperature variations of PCE spectra of Ce^{3+}-doped $Y_3Ga_5O_{12}$ (YGG). (Reproduced with permission from Ref. [39].)

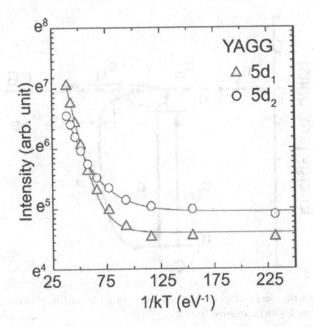

FIGURE 14.11 Arrhenius plot of PCE intensity Ce^{3+}-doped $Y_3Al_2Ga_3O_{12}$ (YAGG). (Reproduced with permission from Ref. [39].)

On the other hand, Van der Kolk et al. discussed the quenching process by thermally assisted photoionization process quantitatively [43]. The rate equation was considered in Ce^{3+}-doped $GdAlO_3$:Ce^{3+}, in which the thermal ionization quenching exists (Figure 14.13) [43]. The population of each excited level under illumination is written as the following equation,

$$n_1' = -c_{12} \cdot n_1 + c_{21} \cdot n_2 + c_{31} \cdot n_3 \tag{14.10}$$

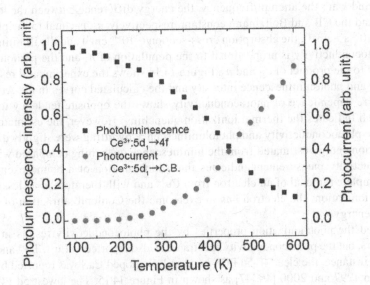

FIGURE 14.12 Temperature dependence of photoluminescence and photocurrent in $Y_3Al_2Ga_3O_{12}$:Ce^{3+} [45]. (Reproduced with permission from Ref. [45].)

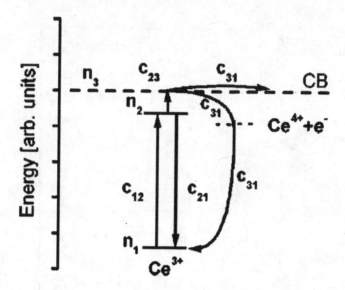

FIGURE 14.13 Schematic energy diagram for the rate equation related to photoionization quenching in GdAlO$_3$:Ce^{3+}. (Reproduced with permission from Ref. [43].)

$$n_2' = c_{12} \cdot n_1 - (c_{21} + c_{23}) \cdot n_2 \tag{14.11}$$

$$n_3' = c_{23} \cdot n_2 - c_{31} \cdot n_3 \tag{14.12}$$

in which n_1 and n_2 are the population of Ce^{3+} ions in the ground state and in the excited state, respectively. n_3 is equal to the number of the electron in the CB and at the same time the number of Ce^{4+} ions. n' is dn/dt. The constant coefficients c_{ij}(s^{-1}) are given by $c_{12} = 10^{-5}$ as the excitation rate of Ce^{3+}, $c_{21} = 5 \times 10^7$ as the decay rate of Ce^{3+}, $c_{23} = f_0 \times \exp(-\Delta E/kT)$ as thermally ionization rate and $c_{31} = 10^{10}$ as recombination rate [43].

The f_0, ΔE and k are the attempt frequency, the energy difference between the lowest energy Ce^{3+} 5d state and the CB and Boltzmann constant, respectively. c_{12} is equal to the product of the photon flux (~10^{12} s^{-1}·cm^{-2}), the absorption cross-section (~10^{-16} cm^2) and the illuminated surface area. The photoconductivity is proportional to the population of n_3 and the photoluminescence is proportional to the product of c_{21} and n_2. Figure 14.14 shows the experimental results for photoconductivity and photoluminescence intensity and the calculated curves in the Arrhenius plot. The temperature dependence of photoconductivity shows the opposite tendency to photoluminescence, which indicates the thermal ionization quenching. However, the estimated activation energies for the photoconductivity and photoluminescence by fitting were slightly different. The smaller activation energy estimated from the luminescence quenching compared with that from the photoconductivity measurement indicates that the formation of a bound exciton does not involve the complete removal of an electron from Ce^{3+} and will, therefore, cost less energy [43]. In the case of ionization, the electron has to overcome the Coulomb attraction of Ce^{4+}, which requires more energy.

We explained the photoionization properties by the photoconductivity results in some Ce^{3+}-doped phosphors, but the photoconductivity spectra were also reported in Eu^{2+}, Tb^{3+} and Pr^{3+}-doped phosphors. For instance, the clear 4f–5d PCE band in Eu^{2+}-doped CaS was reported by Basun and Happek et al. in 1997 and 2000 [46, 47] as shown in Figure 14.15. The lowest 5d PCE band can be observed at 300K, but it was not observed at 150K. Thus, this photoionization process can be activated by thermal energy.

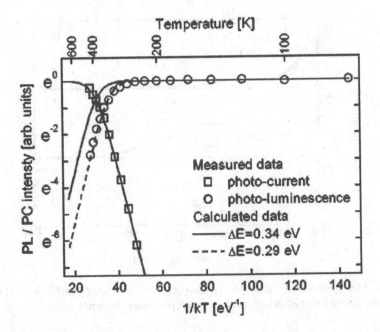

FIGURE 14.14 Calculated and measured temperature dependence of photocurrent and photoluminescence intensity. (Reproduced with permission from Ref. [43].)

In 2002, Jia, Wang and Yen reported the PCE spectrum of Tb^{3+} in $CaAl_2O_4$ host for the first time [48]. Then, the same group reported the photoconductivity spectra of $CaAl_4O_7:Tb^{3+}$ and $Y_2O_3:Tb^{3+}$ in 2004 [49]. Figure 14.16 shows the PCE spectrum of $CaAl_2O_4:Tb^{3+}$, and the 4f–5d band is observed only at room temperature, which indicates the thermal ionization [49]. In 2005, van der Kolk first reported the photoconductivity spectra of Pr^{3+} (Figure 14.17) [50]. According to the temperature-dependent photoconductivity, it is confirmed that the Pr^{3+} has the thermally assisted photoionization process in Y_2SiO_5.

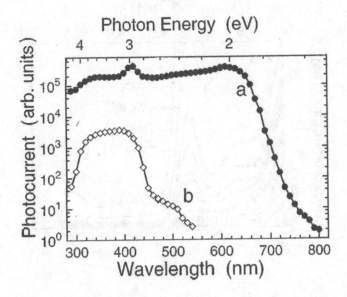

FIGURE 14.15 PCE spectra of $CaS:Eu^{2+}$. Curve (a): $T = 300K$ and curve (b): 150K. (Reproduced with permission from Ref. [46].)

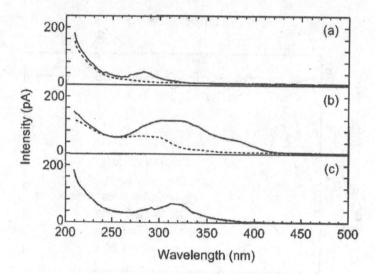

FIGURE 14.16   PCE spectra of the $CaAl_4O_7$: (a) $CaAl_4O_7$:Tb^{3+}, (b) $CaAl_4O_7$:Ce^{3+}, (c) $CaAl_4O_7$:Tb^{3+}, Ce^{3+} (solid lines: 290K, dashed lines: 140K). (Reproduced with permission from Ref. [49].)

14.5   THERMOLUMINESCENCE SPECTROSCOPY

There is no doubt that the photoconductivity experimental technique is very convincing to demonstrate photoionization and thermally assisted photoionization. However, some disadvantages still exist, including the constraint on the sample form (only single crystal or ceramics), the need to make electrodes and to apply high voltage. If it is possible to detect the photoionization process by a contactless method for powder samples, it will become a good alternative experimental way. One of the alternative ways is to measure the thermoluminescence (TL) excitation (TLE) spectrum. TL is caused by detrapping of charges that were previously trapped followed by recombination on a

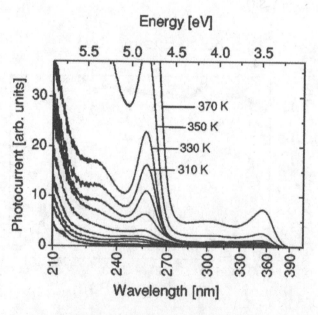

FIGURE 14.17   PCE spectra of $Y_2Si_2O_5$:Pr^{3+} between 110 and 370K with incremental steps of 20K. (Reproduced with permission from Ref. [50].)

luminescent center. Charge trapping occurs when electrons in the excited state of luminescence centers are transferred to the CB (e.g., through autoionization or thermal ionization) and then captured by traps in the host. By measuring the TL intensity as a function of the charging wavelength, a TLE spectrum is obtained that provides information on the threshold energy of the photoionization. The attempt to determine the energy-level location with respect to the host CB using TL technique was first reported in $Y_2O_2S:Tb^{3+}$ by Amiryan in 1977 [51] and then have been studied actively by some research groups [52–54] from the 2000s.

The measurement system of TLE spectroscopy is constructed by the combination of the TL and PLE measurement systems (Figure 14.18). For the monotonic excitation light, the white light source (Xe lamp or tungsten lamp) and monochromatic apparatus (monochromator or bandpass filter) are

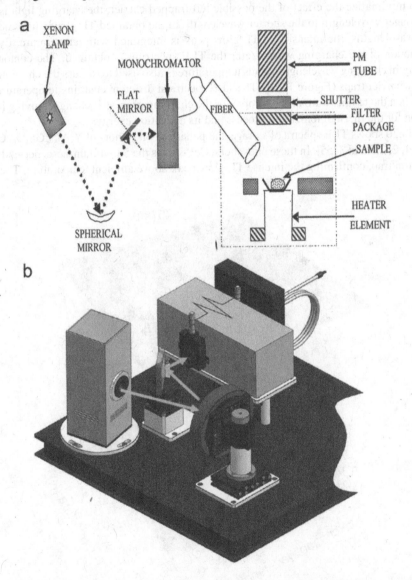

FIGURE 14.18 (a) Schematic of the setup to measure a thermoluminescence excitation (TLE) spectrum. Left image illustrates how the emission from a Xe-lamp is focused on the entrance slit of the monochromator. Monochromatic light is focused by an optical fiber on the sample. Right part of the figure shows the TL readout part of the setup. (b) 3D drawing of the TL excitation part of the setup with the Xe-lamp, focusing mirrors, monochromator, shutter and optical fiber. (Reproduced with permission from Ref. [54].)

used. The TL setup is a typical system, which includes a heater, optical filters, shutter and high-sensitivity detector (generally photomultiplier). Before the TL measurement, the sample should be bleached thermally at a high temperature. The TL glow curves are usually measured twice after charging of the monochromatic light at a certain wavelength for a certain time. The TL glow curves at first read-out become the primary data, and at the second read-out, one has two purposes: one is to subtract the background caused by black body radiation and another is to bleach all the trapped carriers. The traps can be regarded as completely empty after the second read-out. Sometimes, the second reading out process is omitted when there is almost no practical background with respect to the TL glow peak, and no filled trap is confirmed after the first read-out. After that, the charging wavelength is changed to another wavelength and repeats the TL glow curve measurements. In general, to minimalize the effect of the possible left trapped carrier, the charging light is scanned from the longer wavelength to the shorter wavelength. In the obtained TL glow curves at different charging wavelengths, the intensity of TL glow peak is integrated with temperature. By plotting it as a function of the charging wavelength, the TLE spectrum is obtained. The contour plot of TL intensity in charging wavelength versus temperature is also used to discuss the charging process for multiple carrier traps (Figure 14.19). The TLE spectra at different charging temperatures can be obtained when the temperature of the sample for the charging process is changed, giving important information for the thermal ionization process and its activation energy.

Figure 14.20 shows TLE spectra of storage and persistent phosphors of $Y_3Al_{5-x}Ga_xO_{12}$:Ce^{3+}–Cr^{3+} with $x = 0, 1, 2, 2.5$ and 3 [55]. In these phosphors, Ce^{3+} shows the $5d_1$–$4f$ luminescence and Cr^{3+} acts as an electron trap, confirming the intense TL glow peak above ambient temperature. The samples

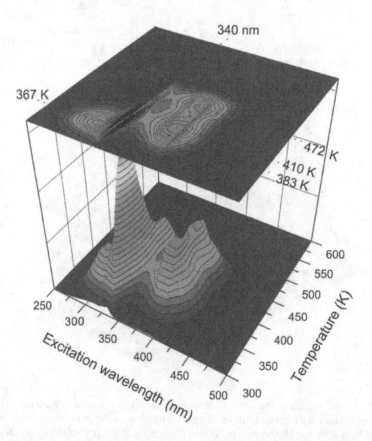

FIGURE 14.19 TL intensity of ZnS:Cu$^+$ as a function of excitation wavelength and temperature as contour plot (top) and in 3D representation (bottom). (Reproduced with permission from Ref. [54].)

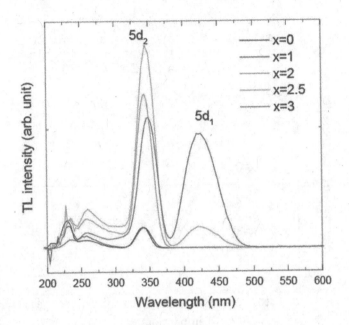

FIGURE 14.20 TLE spectra of $Y_3Al_{5-x}Ga_xO_{12}$:Ce^{3+}–Cr^{3+} with different x charged at ambient temperature. (Reproduced from Ref. [55] with permission from the Royal Society of Chemistry.)

were charged at ambient temperature by monochromatic light at a different wavelength. The TL glow curves were measured, and then the obtained TL glow peak intensity was integrated. In the TLE spectrum of $Y_3Al_2Ga_3O_{12}$:Ce^{3+}–Cr^{3+}, two TLE bands attributed to $4f$–$5d_1$ and $5d_2$ are observed at 430 and 350 nm, showing the (thermally assisted) photoionization process. The TLE spectrum of $Y_3Al_2Ga_3O_{12}$:Ce^{3+}–Cr^{3+} is very similar to the PCE spectrum of $Y_3Al_2Ga_3O_{12}$:Ce^{3+} shown in Figure 14.9. From this accordance, it is regarded that the effect by Cr^{3+} co-doping is limited for the charging process, and the charging process is mainly determined by the energy location between Ce^{3+}:$5d$ energy levels and the bottom of CB. Ideally, to discuss the photoionization process of the phosphors, the codopant should not be included. Based on the TLE spectra at ambient temperature, it is found that the thermal ionization process from the $5d_1$ excited level to the bottom of CB occurs only in $x = 3$ and 2.5 samples, but not in $x = 0$, 1 and 2. These results are in good agreement with the results of PL intensity as a function of temperature as shown in Figure 14.21. The samples with $x = 0$, 1 and 2 do not show luminescence quenching at 300K, while the samples with $x = 2.5$ and 3 show clear quenching. Thus, the Ce^{3+}:$5d_1$–$4f$ luminescence in $x = 2.5$ and 3 samples is quenched by thermal ionization at ambient temperature.

The TLE spectrum analysis was also applied to the discussion of the thermal quenching for YAG:Ce^{3+}[56]. Figure 14.22 shows the TLE spectra of YAG:Ce^{3+} charged at 303 and 573K, which are the temperatures that YAG:Ce^{3+} does not show quenching and undergoes quenching, respectively (Figure 14.23). At 303K, a TLE band is observed at around 340 nm, while at 573K, an additional TLE band is observed at around 450 nm. The TLE bands at 450 and 340 nm are attributed to the transition from the $4f$ ground level to the $5d_1$ and the $5d_2$, respectively. Because the trap filling proceeds by the electron transport through the CB, these results provide evidence that the electrons in the $5d_2$ level are thermally ionized to the CB already at 303K, but the electrons at the $5d_1$ level are only thermally ionized efficiently at higher temperatures around 573K. At 573K, a temperature corresponding to the onset of thermal quenching of the Ce^{3+} luminescence, the excitation in the lowest $5d_1$ band at 450 nm gives rise to a TL signal with temperature, indicating that thermal ionization is responsible for thermal quenching of the Ce^{3+} luminescence (Figure 14.23) [56].

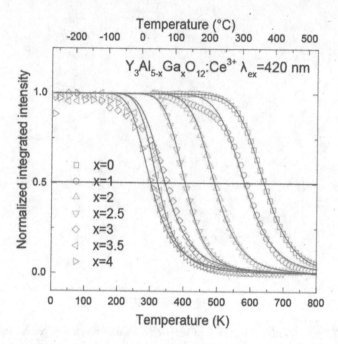

FIGURE 14.21 Temperature dependence of photoluminescence intensity excited by 420 nm for $Y_3Al_{5-x}Ga_x$ $O_{12}:Ce^{3+}$ with different x. (Reproduced from Ref. [55] with permission from the Royal Society of Chemistry.)

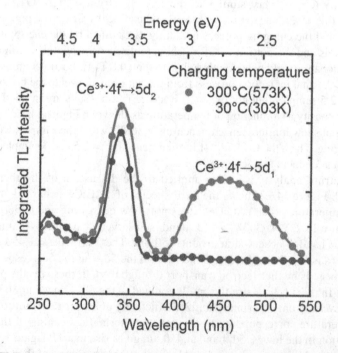

FIGURE 14.22 TLE spectra of YAG:Ce^{3+} charged at 303 and 573K. (Reproduced with permission from Ref. [56].)

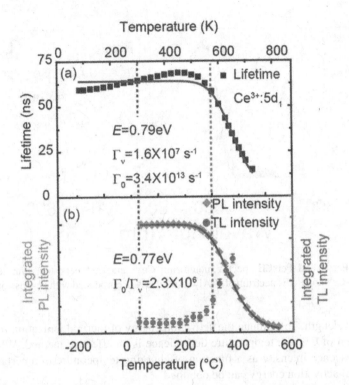

FIGURE 14.23 Temperature dependence of (a) Ce^{3+} luminescence lifetime ($\lambda_{em} = 540$ nm, for 340 nm excitation with a ps LED) and (b) integrated PL intensity and TL intensity for YAG:Ce^{3+}(0.5%). TL intensity was integrated between 623 (350°C) and 773K (500°C) in each TL glow curve after 450 nm excitation at different temperatures from 303 (30°C) to 648K (375°C). Blue and red dashed lines express the temperatures; room temperature (no quenching) and 300°C (onset of quenching), where the TLE experiments were conducted. (Reproduced with permission from Ref. [56].)

14.6 PERSISTENT LUMINESCENCE (DELAYED RECOMBINATION LUMINESCENCE) SPECTROSCOPY

The idea to use persistent luminescence (delayed recombination luminescence) for detecting the photoionization process is almost the same as that using TL. The method using persistent luminescence to investigate the thermal ionization from the excited state of *Ln* ion was seen in the paper by Aguirre de Cárcer et al. in 1988 [25]. Also, Nikl et al. have actively investigated the thermal ionization process based on the delayed recombination luminescence [57–61]. Similar to TL, the charging process is necessary to cause persistent luminescence or delayed recombination luminescence. In the TL glow curves, the sample is heated at an individual heating rate, but for persistent luminescence, the temperature of the sample is kept at constant. Instead of temperature, the persistent luminescence intensity is plotted by time (persistent luminescence decay curve). When we plot the persistent luminescence intensity at a certain time after ceasing excitation as a function of charging wavelength, the persistent luminescence excitation (PersLE) spectrum can be obtained, which is almost the same as the TLE method. For the persistent luminescence excitation (PersLE) spectroscopy, the measurement system of TLE or PLE can be used for PersLE measurement.

For the PersLE, the persistent luminescence intensity at a particular time after ceasing the excitation is plotted as a function of charging wavelength. Different from the TLE method, the trapped carriers are not released at high temperatures during the measurement of persistent luminescence decay. Thus, to move the next charging process, it is necessary to wait for enough long time or to detrap forcibly by heating. Same as the TLE, the wavelength is scanned from the longer wavelength

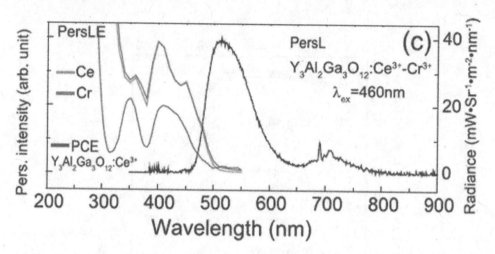

FIGURE 14.24 PersL and PersLE spectra monitoring Ce^{3+} and Cr^{3+} persistent luminescence band in Y$_3$Al$_2$Ga$_3$O$_{12}$:Ce^{3+}–Cr^{3+} and PCE spectrum of Y$_3$Al$_2$Ga$_3$O$_{12}$:Ce^{3+}. (Reproduced with permission from Ref. [62].)

to the shorter wavelength. To estimate the activation energy of thermal ionization from the excited state to the bottom of CB, the temperature dependence is an effective method. When we plot the persistent luminescence intensity as a function of charging temperature in a certain temperature range, the thermal activation energy can be obtained.

Figure 14.24 shows PersL and PersLE spectra of Y$_3$Al$_2$Ga$_3$O$_{12}$:Ce^{3+}–Cr^{3+} persistent phosphors, which is also introduced in Section 14.5. In the PersLE spectrum monitoring Ce^{3+}:5d–4f persistent luminescence at around 510 nm, the PersLE bands attributed to 5d$_1$ and 5d$_2$ of Ce^{3+} are observed at 420 and 350 nm. Thus, the electron traps by Cr^{3+} are filled after the excitation upon Ce^{3+}:5d$_1$ and 5d$_2$ levels, indicating the photoionization of Ce^{3+}. The result of PersLE spectrum is almost the same as that of the PCE spectrum in YAGG:Ce^{3+}, which means that the PersLE is also one of the effective methods. However, only one PersLE spectrum at ambient temperature cannot give a detailed energy location of Ln ions with respect to the CB; there are two possibilities that the Ce^{3+}:5d$_1$ is located below the CB or within the CB. We may be able to determine the 5d$_1$ level to be below the CB from the fact the Ce^{3+}:5d$_1$–4f luminescence can be observed at ambient temperature. To get the direct evidence, the PerLE spectrum should be measured at low temperatures without the thermal activation process. However, the persistent luminescence intensity depends on the carrier trap depth and temperature. If there is no persistent luminescence at low temperatures, the PersLE spectrum technique cannot be used.

Another possibility is to measure the charging temperature dependence of persistent luminescence (delayed recombination luminescence) intensity in the appropriate temperature range. Figure 14.25 shows the temperature dependence of delayed recombination luminescence intensity of Lu$_2$SiO$_5$:Pr^{3+} by the excitation in 247 nm upon the Pr^{3+}:5d$_1$ state [58]. When the charging temperature increases, the delayed recombination luminescence intensity becomes stronger. This result shows the charging process is caused by thermally assisted photoionization. On the basis of the fitting by $I_{DR} = C_0 + C_1 \times \exp(-E_1/kT)$, the activation energy of the thermal ionization process from the Pr^{3+}:5d$_1$ to the bottom of the CB is estimated to be 0.29 eV. For the Pr^{3+} ion, it is reported that the relaxation process from the exited 5d state to the next lower 4f excited state can affect the photoconductivity intensity [50]. Also, the thermal ionization from the excited state can occur after lattice relaxation. Thus, the activation energy can include some energetical deviation compared with the energy gap between the 5d$_1$ and the bottom of CB based on the energy diagram before lattice relaxation. This deviation is observed not only for the activation energy estimated from the persistent luminescence spectroscopy but also for other measurement methods (photoconductivity, TL and so on).

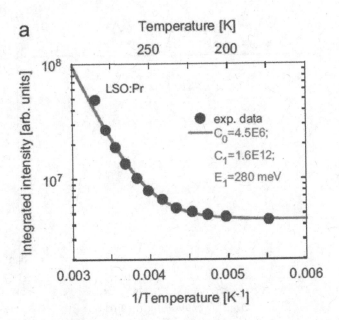

FIGURE 14.25 Temperature dependence of delayed recombination luminescence intensity of $Lu_2SiO_5:Pr^{3+}$ by the excitation in 247 nm upon the $Pr^{3+}:5d_1$ state. (Reproduced with permission from Ref. [58].)

14.7 PHOTOCONDUCTIVITY USING THE MICROWAVE RESONATOR TECHNIQUE

Another method to detect the ionization process of *Ln* ions is the photoconductivity microwave resonant cavity technique. When the *Ln* ions are photoionized, the change of the dielectric loss factor of the dielectric permittivity is caused. This change can be detected based on the microwave absorption by the dielectric crystal doped with *Ln* ions inserted in a resonant microwave cavity with illumination [63–65]. This contactless technique is also valid for powder samples [63]. To detect the small change of dielectric loss factor by photoionization process, the highly sensitive resonant microwave cavity which is developed for EPR is used.

Figure 14.26 shows the photoconductivity detected by microwave resonant cavity technique as a function of excitation wavelength for $CaF_2:Eu^{2+}$ [65]. Almost the same information as the conventional photoconductivity method is obtained. In this PCE spectrum, the 4f–5d PCE bands of Eu^{2+} were observed, indicating the photoionization process of Eu^{2+}. However, the effect of excited-state absorption cannot be excluded because high power density laser excitation is usually used.

14.8 EPR (ELECTRON PARAMAGNETIC RESONANCE)

EPR, electron spin resonance (ESR), is one of the methods to detect unpaired electrons. The fundamental of EPR is found in some articles and books, for instance Ref. [66]. The family of *Ln* ions possesses electrons in the 4f subshell which can accommodate 14 electrons. Due to the 4f electrons, the *Ln* ions show paramagnetism. The magnetic moment of *Ln* ions is affected by spin-orbit coupling. The states of *Ln* ions are described by term symbols $^{2S+1}L_J$. S is the total spin angular momentum, L is the total orbital angular momentum and J is the total angular momentum. The magnetic moment, μ can be expressed by

$$\mu = -g_{Land\acute{e}}\beta J \qquad (14.13)$$

$$g_{Land\acute{e}} = 1 + (g_e - 1)\{J(J+1) + S(S+1) - L(L+1)\}/\{2J(J+1)\} \qquad (14.14)$$

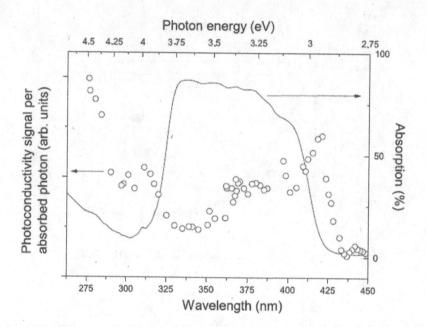

FIGURE 14.26 Photoconductivity intensity of CaF$_2$:Eu^{2+} as a function of the excitation wavelength (open circles) and absorption spectrum (solid line) at room temperature. (Reproduced with permission from Ref. [65].)

where β is the electronic Bohr magneton and $g_e = 2.0023$. Based on these equations, only the ground state ($^7F_0$) of Eu^{3+} and Sm^{2+} with 4f^6 electron configuration takes $J = 0$ among the *Ln* ions with unpaired electrons, resulting in non-activity for EPR. Practically, because the next higher state ($^7F_1$) of Eu^{3+} an Sm^{2+} is thermally populated at around ambient temperature, the EPR signal is detected sometimes. The La^{3+} and Ce^{4+} ions with 4f^0 electron configuration and the Lu^{3+} and Yb^{2+} with 4f^{14} do not show EPR signals due to the non-unpaired electrons. Other *Ln* ions can be detected by EPR spectroscopy. In actuality, the observed EPR spectrum can also be affected by hyperfine interaction with a nucleus and by ligand field. The general EPR apparatus enables the measurement under light illumination in situ, detecting the photoionization directly.

In the research area of EL, photoionization using EPR techniques had been discussed often. In 1986, Godlewski and Hommel determined the threshold energy for the photoionization of Eu^{2+} in the ZnS host by using ESR under light illumination [67]. After that, the photoionization in ZnS:Yb^{2+} and ZnSe:Eu^{2+} was also confirmed by the EPR measurement under light illumination [26, 28].

Figure 14.27 shows the EPR intensity of Eu^{2+} in ZnSe as a function of time [28]. During the EPR measurement, the light at 2.2 eV illuminated to the sample for 50 s and stopped, and then another light at 1 eV illuminated again to the sample and stopped. Under the light illumination at 2.2 eV, the EPR signal of Eu^{2+} clearly decreased, which indicates that the Eu^{2+} was photoionized by light at 2.2 eV and changed into the Eu^{3+} state. After stopping excitation light, it seems that some electrons captured by shallow traps are thermally recombined with the Eu^{3+}. After the equilibrium, the deep traps were released by another excitation light at 1 eV, changing into Eu^{2+} by recombination with the Eu^{3+}.

When the deference of Eu^{2+} ESR intensities before and after light illumination with different photon energy is plotted, the threshold of photoionization can be obtained (Figure 14.28). Above 1.9 eV illumination, the Eu^{2+} ions are effectively photoionized. The EPR spectroscopy is one of the effective methods to investigate the photoionization process.

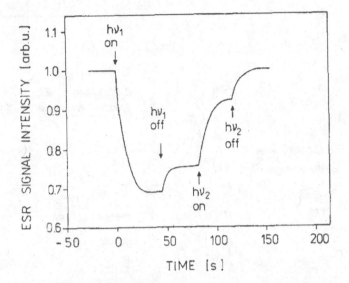

FIGURE 14.27 Time dependence of Eu²⁺ EPR signal in ZnSe under the $h\nu_1 = 2.2$ eV and $h\nu_2 = 1$ eV illumination. (Reproduced with permission from Ref. [28].)

14.9   XANES (X-RAY ABSORPTION NEAR-EDGE STRUCTURE) SPECTROSCOPY

X-ray absorption spectrum (XAS) is a plot of the absorption coefficient of an atom or ion in the X-ray energy range from 0.1 to 100 keV. The XAS shows the fine structure, which reflects the electronic state and the local geometry around the specific atom. Thus, the valence state of Ln ions can be determined from the X-ray absorption spectra in the near edge region, called XANES (X-ray Absorption Near-Edge Structure). When the XANES spectrum is measured during light

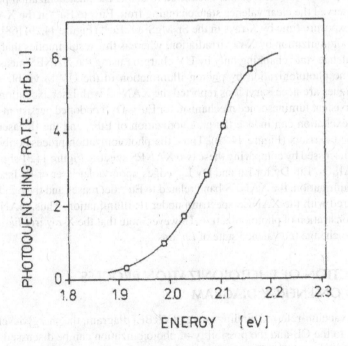

FIGURE 14.28 Decline of EPR intensity by illumination with different photon energy (photoquenching) for ZnSe:Eu²⁺ at 4.2K. (Reproduced with permission from Ref. [28].)

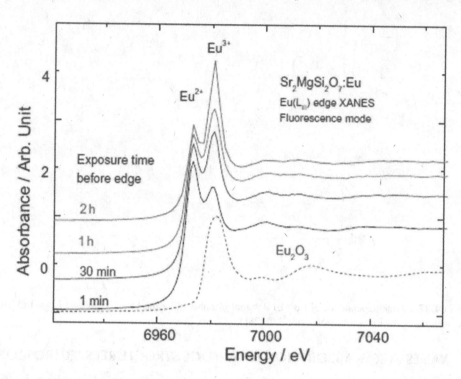

FIGURE 14.29 Effect of exposure time on the Eu LIIIedge XANES spectra in $Sr_2Mg_2Si_2O_7:Eu^{2+}$. (Reproduced with permission from Ref. [68].)

illumination, the valence state changing caused by photoionization can be detected in principle. So far, the XANES spectrum of *Ln* ions has been tried to detect the photoionization process. In 2009, Carlson et al. observed the clear valence state changing from Eu^{2+} to Eu^{3+} in the XANES spectrum with increasing exposure time by X-rays in the $Sr_2MgSi_2O_7:Eu^{2+}$ (Figure 14.29) [68]. The result was a good sign of photoionization by X-ray irradiation, whereas this result implies that it is not easy to determine the valence state changing only by UV charging using the XANES measurement. Thus, to demonstrate the photoionization by photon illumination in the UV to visible using XANES, additional techniques are necessary. Joos reported the XANES with laser excitation (violet and IR) to explain the persistent luminescence mechanism for Eu^{2+}–Dy^{3+} codoped persistent phosphors [69]. The violet laser excitation can induce the photoionization of Eu^{2+}, and the IR laser excitation can release the trapped carriers (Figure 14.30). Thus, the photoionization process only by violet light charging can be discussed by comparing these two XANES spectra. Figure 14.31 shows the XANES spectrum of $Sr_4Al_{14}O_{25}:Eu$–Dy for Eu and Dy L_{III} edges upon violet laser or IR laser illumination. By violet laser illumination, the XANES band related to Eu^{2+} decreases and the band of Eu^{3+} clearly increased compared with the XANES spectrum under IR illumination. Thus, XANES can also be used for the demonstration of photoionization. However, note that the X-ray irradiation during measurement can also change the valence state of *Ln* ions.

14.10 PREDICTION OF PHOTOIONIZATION PROCESS BASED ON ENERGY DIAGRAM

According to the vacuum-referred binding energy (VRBE) diagram, the energy-level location of *Ln* ions with respect to the CB and the possibility of photoionization can be discussed (Figure 14.32). For instance, Ce^{3+}, Pr^{3+} and Tb^{3+} have higher VRBE of -6.0, -7.8 and -7.6 eV, respectively, among the Ln^{3+} ions as shown in Figure 14.32. Also, all the Ln^{2+} ions have higher VRBE than Ln^{3+}. The

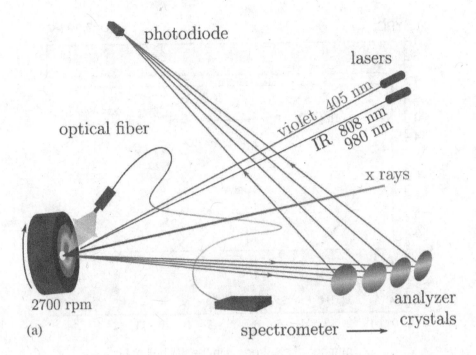

FIGURE 14.30 Schematic diagram of XANES measurement with laser excitation. (Reproduced with permission from Ref. [69] by Joos et al., DOI: 10.1103/PhysRevLett.125.033001.)

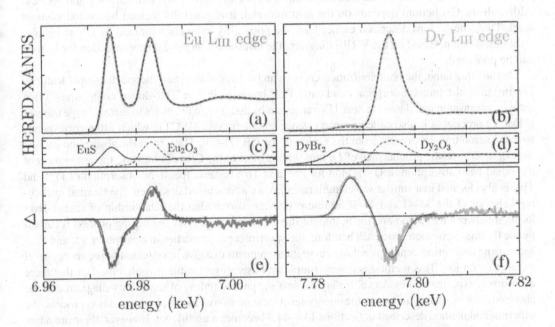

FIGURE 14.31 High energy resolution fluorescence detected XANES spectra for the (a) Eu and (b) Dy LIII edges upon violet (solid line) or IR (dotted line), compared to conventional XANES of reference spectra for (c) EuS and Eu$_2$O$_3$ and (d)Dy$_2$O$_3$ and DyBr$_2$, and (e), (f) the associated difference spectra Δ. (Reproduced with permission from Ref. [69] by Joos et al., DOI: 10.1103/PhysRevLett.125.033001.)

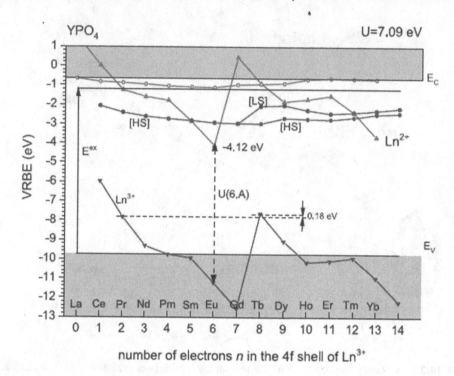

FIGURE 14.32   VRBE diagram for the divalent and trivalent $4f^q$ lanthanide ground state levels and $4f^{q-1}5d$ excited-state levels in YPO_4. (Reproduced with permission from Ref. [70].)

Eu^{2+} and Yb^{2+}, which can be stabilized in many compounds, have the VRBE of -4.1 and -3.7 eV. Although the CB bottom depends on the host material, it is generally located below the vacuum level. Thus, the *Ln* ions described earlier have the possibility to be photoionized by UV to visible-light illumination. Based on the VRBE diagram, the threshold energy of photoionization for *Ln* ions can be predicted.

On the other hand, the photoionization energy can be determined from the spectroscopic features. For instance, the broad absorption bands and PLE bands in Pr^{3+} or Tb^{3+}-doped compounds (often including transition metal ions) in near UV range can be assigned to the metal to metal charge transfer (MMCT) process (it is also called as intervalence charge transfer [IVCT]), which often corresponds to the electron transfer process from the *Ln* ion to the CB [71–73]. Thus, from the absorption energy, we can determine the ground state of *Ln* ions with respect to the CB bottom. Note the assignment of the broad band absorption in UV region for Pr^{3+} and Tb^{3+} because the 4f–5d absorption of Pr^{3+} and Tb^{3+} is also located in a similar wavelength range. Also, as discussed described, the thermal quenching behavior of the 5d–4f and 4f–4f luminescence can also predict the relationship of energy-level location between the *Ln* ion and the bottom of the CB, conversely, if the quenching process is caused by the thermal ionization process. Therefore, the spectroscopic properties in absorption, PL and PLE spectra and the luminescence quenching curve measurements can give information of the energy-level location of *Ln* ion. This method is a very simple and useful way, while it is also the fact that there is no direct experimental evidence. It may cause a misunderstanding of the energy diagram and the photoionization process. To ensure the energy-level location as well as the photoionization process, the experimental method described in Sections 14.4–14.9 becomes a useful tool. However, there are a few phosphors whose photoionization process was confirmed experimentally.

The phosphors with photoionization process which are confirmed experimentally are summa-rized in Table 14.1. As expected from the energy diagram, the photoionization process was reported often in Ce^{3+}, Pr^{3+} and Eu^{2+}-doped compounds as shown in Table 14.1.

TABLE 14.1
List of Phosphors Whose Photoionization Process Was Confirmed Experimentally

Host	Dopant	Absorption Transition for Ionization	Threshold Energy (eV)	Method	Ref.
CaF_2	Sm^{2+}	4f–5d or MMCT@RT	1.7	PC	[13]
	Eu^{2+}	4f–5d	3.8	PC, PCM	[13, 65]
	Tm^{2+}	4f–5d or MMCT@RT	2.75	PC	[9]
	Dy^{2+}	4f–5d or MMCT@RT	1.76	PC	[9]
	Ho^{2+}	4f–5d or MMCT@RT	1.75	PC	[9]
	Yb^{2+}	4f–5d @RT	~4	PC	[74]
SrF_2	Tm^{2+}	4f–5d @RT	2.08	PC	[74]
	Yb^{2+}	MMCT@RT	3.0	PC	[74]
BaF_2	Tm^{2+}	4f–5d or MMCT@RT	1.16	PC	[9]
$LiYF_4$	Ce^{3+}	4f–$5d_{1,2}$ @RT (two photon process)	~4.6	PCM	[75]
$LiLuF_4$	Ce^{3+}	4f–$5d_{1,2}$ @RT (two photon process)	~4.6	PCM	[75]
Cs_3LuCl_6	Ce^{3+}	4f–5d (e_g) @140K	5.6–5.9	TL	[76]
$Cs_2LiLuCl_6$	Ce^{3+}	4f–5d (e_g) @140K	5.6–5.9	TL	[76]
Lu_2O_3	Ce^{3+}	4f–$5d_1$ @RT 4f–$5d_1$ @LNT	2.5	PC	[20–22]
	Tb^{3+}	4f–$5d_1$ (C_2)@100K 4f–$5d_1$ (C_2, C_{3i})@300K	~4.8	PC	[77]
Y_2O_3	Ce^{3+}	4f–$5d_1$@LNT	~2.5	PC	[20, 22]
	Pr^{3+}	4f–$5d_1$@RT	3.9	PC	[78]
	Tb^{3+}	4f–$5d_1$@RT	~4.3	PC	[49]
$Lu_3Al_5O_{12}$	Ce^{3+}	4f–$5d_3$@RT	4.5	PersL	[61]
$Y_3Al_5O_{12}$	Ce^{3+}	4f–$5d_{3,2}$ @RT 4f–$5d_{1,2}$ @ 573K	3.8	PC, TL	[13, 39, 56]
$Y_3Al_4Ga_1O_{12}$	Ce^{3+}	4f–$5d_2$@RT		TL	[55]
$Y_3Al_3Ga_2O_{12}$	Ce^{3+}	4f–$5d_2$@RT		TL	[55]
$Y_3Al_2Ga_3O_{12}$	Ce^{3+}	4f–$5d_2$ @50K 4f–$5d_{1,2}$@RT	3.2	PC, TL, PersL	[39, 55, 62]
	Pr^{3+}	4f–$5d_{1,2}$ @RT		TL	[79]
$Y_3Ga_5O_{12}$	Ce^{3+}	4f–$5d_{1,2}$@RT, 50K	2.8	PC	[39]
$Gd_3Al_2Ga_3O_{12}$	Ce^{3+}	4f–$5d_{1,2}$, MMCT@RT 4f–$5d_2$, MMCT @140K	~2.9	PC, TL	[80, 81]
$Gd_3Ga_5O_{12}$	Ce^{3+}	4f–$5d_{1,2}$, MMCT@ RT,140K	~2.8	PC	[80]
$Gd_3Sc_2Al_3O_{12}$	Ce^{3+}	4f–$5d_2$, Gd^{3+}:4f–4f (cross-ionization) @LNT	3.4	PC	[82]
Lu_2SiO_5	Ce^{3+}	4f–$5d_{1,2,3}$ @RT 4f–$5d_{2,3}$ @LNT	~3.8	PC, TL, PersL	[21, 22, 44, 50, 53, 58]
	Pr^{3+}	4f–$5d_{1,2}$ @RT 4f–$5d_2$ @110K	~5.3	PC, PersL	[50, 58]
Y_2SiO_5	Ce^{3+}	4f–$5d_{1,2,3}$ @RT 4f–$5d_{2,3}$ @LNT	~3.8	PC, TL	[50, 52, 83]
	Pr^{3+}	4f–$5d_{1,2}$ @RT 4f–$5d_2$ @110K	~5.3	PC, PersL	[50, 58]

(Continued)

TABLE 14.1 *(Continued)*

List of Phosphors Whose Photoionization Process Was Confirmed Experimentally

Host	Dopant	Absorption Transition for Ionization	Threshold Energy (eV)	Method	Ref.
$(Sr_{1-x}Ca_x)TiO_3$	Pr^{3+}	MMCT, 4f–4f(^3P_J)@RT (x = 0, 0.3, 0.5, 0.8, 1) MMCT@100K (x = 1)	3.2~3.5	PC, PersL	[84]
$SrAl_2O_4$	Eu^{2+}	4f–5d (weak CFS site) @100K 4f–5d (weak, strong CFS site) @RT	~3.1	PC	[8, 85, 86]
	Ce^{3+}	4f–5d$_1$@290, 140K	~3.8	PC	[87]
$Sr_4Al_{14}O_{25}$	Eu^{2+}	4f–5d@RT		TL,	[54]
$SrHfO_3$	Ce^{3+}	4f–5d$_1$	3.88 + 0.25	TL	[57, 88]
$CaAl_2O_4$	Eu^{2+}	4f–5d @20K, 300K	~3.0	PC	[89]
	Ce^{3+}	4f–5d$_{2,3}$, 5d$_1$ (two photons process) @RT	4.55	PC	[40]
	Tb^{3+}	4f–5d$_{1,2}$ @RT	4.77	PC	[48]
$CaAl_4O_7$	Ce^{3+}	4f–5d$_{1,2}$ @290K 4f–5d$_2$ @140K	3.79	PC	[90]
	Tb^{3+}	4f–5d$_{1,2}$ @290K 4f–5d$_2$@140K	4.78	PC	[49, 90]
$Ca_3Si_2O_7$	Eu^{2+}	4f–5d$_{low, high}$ @473K 4f–5d$_{high}$ @RT	~2.9	PersL	[91]
NaCl	Eu^{2+}	4f–5d$_{high}$ @RT	~4.8	PC, PersL	[25]
KCl	Eu^{2+}	4f–5d$_{high}$ @RT	~5.0	PC, PersL	[25]
KBr	Eu^{2+}	4f–5d$_{high}$ @RT	~4.7	PC, PersL	[25]
KI	Eu^{2+}	4f–5d$_{high}$ @RT	~4.7	PC, PersL	[25]
CaS	Ce^{3+}	4f–5d$_2$ @300K	4.5	PC	[37]
	Eu^{2+}	4f–5d$_{low, high}$ @300K 4f–5d$_{high}$ @80K	2.3	PC	[46, 47]
ZnS	Eu^{2+}	4f–5d@RT	2.2	ESR	[27]
	Yb^{2+}	4f–5d@RT	1.65	ESR	[26]
$CaGa_2S_4$	Ce^{3+}	4f–5d$_2$@26K 4f–5d$_{1,2}$@300K	3.4	PC	[92, 93]
	Eu^{2+}	4f–5d@300K	2.35	PC	[92, 93]
$CaAlSiN_3$	Eu^{2+}	4f–5d$_{low, high}$ @450K 4f–5d$_{high}$ @300K	~2.9	TL, PersL	[94]
$Ca_2Si_5N_8$	Eu^{2+}	4f–5d@RT	~3	TLE	[95]
$Ba_2Si_5N_8$	Eu^{2+}	4f–5d@RT	~3	TLE	[95]
$Sr(SCN)_2$	Eu^{2+}	4f–5d@~20K	~2.7	TL	[96]
ZnSe	Eu^{2+}	4f–5d@RT	1.9	ESR	[28]

Note: Host, luminescence center, absorption transition which causes the ionization with temperature (RT, room temperature, LNT: liquid nitrogen temperature), threshold energy for photoionization and experimental methods (PC: conventional photoconductivity, PCM: photoconductivity using the microwave resonator technique, TL: thermoluminescence, PersL: persistent luminescence) For Ce^{3+}, Pr^{3+} and Tb^{3+}, the split 5d levels by crystal filed splitting is distinguished (the lowest level: 5d$_1$, the second level:5d$_2$ and so on.) For Eu^{2+}:5d excited state, the higher and lower 5d excited levels are roughly distinguished. If there are several cation sites that show the detectable difference for the photoionization, the site was also shown.

14.11   CONCLUSIONS

In summary, the history, experimental techniques and examples for the photoionization process of lanthanide ions are introduced in this chapter. On the basis of experimental facts and the energy diagram, the compounds doped with lanthanide ions such as Ce^{3+}, Eu^{2+}, Pr^{3+}, Tb^{3+} and Yb^{2+} tends to show the photoionization process. These ions are also excellent luminescence centers for practical applications such as scintillator and LED and LD phosphors. To design the high luminescence quantum efficiency and high quenching temperature, we must understand and control the photoionization process. Also, the phosphors in the current high power laser excitation application undergoes the multiphoton photoionization, indicating more severe photoionization quenching. For the luminescence quenching process, it is important to detect the photoionization process directly using the experimental method explained in this chapter in addition to the consideration based on the energy diagram.

REFERENCES

1. Stookey SD (1949) Photosensitive glass. *Ind. Eng. Chem.* 41: 856.
2. Stroud JS (1961) Photoionization of Ce^{3+} in glass. *J. Phys. Chem.* 35: 844.
3. Stroud JS (1962) Color centers in a cerium-containing silicate glass. *J. Phys. Chem.* 37: 836.
4. Sorokin PP, Stevenson MJ (1961) Solid-state optical maser using divalent samarium in calcium fluoride. *IBM J. Res. Dev.* 5: 56.
5. Hayes W, *Crystals with the Fluorite Structure: Electronic, Vibrational, and Defect Properties* (Clarendon Press, Oxford, 1974).
6. Anderson C, Kiss Z (1964) Photoconductivity of divalent rare-earth-doped calcium fluoride. *Bull. Am. Phys. Soc.* 9: 87.
7. Heyman PM (1969) photoconductivity of photochromic $Gd:CaF_2$. *Appl. Phys. Lett.* 14: 81.
8. Abbruscato V (1971) Optical and electrical properties of $SrAl_2O_4:Eu^{2+}$ *J. Electrochem. Soc.* 118: 930.
9. Pedrini C, McClure DS, Anderson CH (1979) Photoionization thresholds of divalent rare earth ions in alkaline earth fluorides. *J. Phys. Chem.* 70: 4959.
10. Jacobs RR, Krupke WF, Weber MJ (1978) Measurement of excited-state-absorption loss for Ce^{3+} in $Y_3Al_5O_{12}$ and implications for tunable 5d $\rightarrow$ 4f rare-earth lasers. *Appl. Phys. Lett.* 33: 410.
11. Miniscalco WJ, Pellegrino JM, Yen WM (1978) Measurements of excited-state absorption in Ce^{3+}: YAG. *J. Appl. Phys.* 49: 6109.
12. Owen JF, Dorain PB, Kobayasi T (1981) Excited-state absorption in Eu^{+2}: CaF2 and Ce^{+3}: YAG single crystals at 298 and 77 K. *J. Appl. Phys.* 52: 1216.
13. Pedrini C, Rogemond F, McClure DS (1986) Photoionization thresholds of rare-earth impurity ions. $Eu^{2+}:CaF_2$, $Ce^{3+}:YAG$, and $Sm^{2+}:CaF_2$. *J. Appl. Phys.* 59: 1196.
14. Pédrini C, Joubert MF, McClure DS (2007) Photoionization processes of rare-earth dopant ions in ionic crystals. *J. Lumin.* 125: 230.
15. Takahashi K, Kohda K, Miyahara J, Kanemitsu Y, Amitani K, Shionoya S (1984) Mechanism of photostimulated luminescence in $BaFX:Eu^{2+}$ (X=Cl,Br) phosphors. *J. Lumin.* 31–32: 266.
16. Basun SA, Feofilov SP, Kaplyanskii AA, Sevastyanov BK, Sharonov MY, Starostina LS (1992) Photoelectric studies of two-step photoionization of Ti^{3+} ions in oxide crystals. *J. Lumin.* 53: 28.
17. Wong WC, McClure DS, Basun SA, Kokta MR (1995) Charge-exchange processes in titanium-doped sapphire crystals. I. Charge-exchange energies and titanium-bound excitons. *Phys. Rev. B* 51: 5682.
18. Wong WC, McClure DS, Basun SA, Kokta MR (1995) Charge-exchange processes in titanium-doped sapphire crystals. II. Charge-transfer transition states, carrier trapping, and detrapping. *Phys. Rev. B* 51: 5693.
19. Basun SA, Danger T, Kaplyanskii AA, McClure DS, Petermann K, Wong WC (1996) Optical and photoelectrical studies of charge-transfer processes in $YAlO_3:Ti$ crystals. *Phys. Rev. B* 54: 6141.
20. Raukas M, Basun S, Dennis WM, Evans DR, Happek U, Van Schaik W, Yen WM (1996) Optical properties of Ce^{3+}-doped Lu_2O_3 and Y_2O_3 single crystals. *J. Soc. Inf. Disp.* 4: 189.
21. Raukas M, Basun SA, Van Schaik W, Yen WM, Happek U (1996) Luminescence efficiency of cerium doped insulators: The role of electron transfer processes. *Appl. Phys. Lett.* 69: 3300.
22. Yen WM, Raukas M, Basun SA, van Schaik W, Happek U (1996) Optical and photoconductive properties of cerium-doped crystalline solids. *J. Lumin.* 69: 287.

23. Dong Y, Su M-Z (1995) Luminescence and electro-conductance of BaFBr:Eu^{2+} crystals during X-irradiation and photostimulation. *J. Lumin.* 65: 263.
24. Evans DR, Dennis WM, Basun S, Kane J, Yocum PN (1996) Fluorescent transient and photoconductivity measurements in Ce^{3+}-doped thiogallate crystals. *J. Soc. Inf. Disp.* 4: 271.
25. Aguirre de Cárcer I, Cussó F, Jaque F (1988) Afterglow and photoconductivity in europium-doped alkali halides. *Phys. Rev. B* 38: 10812.
26. Przybylińska H, Świątek K, Stąpor A, Suchocki A, Godlewski M (1989) Recombination processes in Yb-activated ZnS. *Phys. Rev. B* 40: 1748.
27. Świątek K, Godlewski M, Hommel D (1990) Deep europium-bound exciton in a ZnS lattice. *Phys. Rev. B* 42: 3628.
28. Świątek K, Godlowski M (1992) Europium impurity as a recombination center in ZnSe lattice. *J. Lumin.* 53: 406.
29. Blasse G (1969) Thermal quenching of characteristic fluorescence. *J. Chem. Phys.* 51: 3529.
30. Blasse G (1974) Thermal quenching of characteristic luminescence. II. *J. Solid State Chem.* 9: 147.
31. Blasse G (1978) Quenching of luminescence by electron transfer. *Chem. Phys. Lett.* 56: 409.
32. Weber MJ (1973) Nonradiative Decay from 5d States of Rare Earths in Crystals. *Sol. State Commun.* 12: 741.
33. Lyu LJ, Hamilton DS (1991) Radiative and nonradiative relaxation measurements in Ce^{3+} doped crystals. *J. Lumin.* 48–49: 251.
34. Hamilton DS, Gayen SK, Pogatshnik GJ, Ghen RD, Miniscalco WJ (1989) Optical-absorption and photoionization measurements from the excited states of Ce^{3+}:Y$_3$Al$_5$O$_{12}$. *Phys. Rev. B* 39: 8807.
35. Blasse G, Schipper W, Hamelink JJ (1991) On the quenching of the luminescence of the trivalent cerium ion. *Inorg. Chim. Acta* 189: 77.
36. Ueda J, Tanabe S (2019) Review of luminescent properties of Ce^{3+}-doped garnet phosphors: new insight into the effect of crystal and electronic structure. *Opt. Mater.: X* 1: 100018.
37. Jia D, Meltzer RS, Yen WM (2002) Ce^{3+} energy levels relative to the band structure in CaS: evidence from photoionization and electron trapping. *J. Lumin.* 99: 1.
38. Nakanishi T, Tanabe S (2010) Construction of photoconductivity measurement system as functions of excitation wavelength and temperature: application to Eu^{2+}-activated phosphors. *Proc. SPIE* 7598: 759809.
39. Ueda J, Tanabe S, Nakanishi T (2011) Analysis of Ce^{3+} luminescence quenching in solid solutions between Y$_3$Al$_5$O$_{12}$ and Y$_3$Ga$_5$O$_{12}$ by temperature dependence of photoconductivity measurement. *J. Appl. Phys.* 110: 053102.
40. Jia D, Yen WM (2003) Trapping mechanism associated with electron delocalization and tunneling of CaAl$_2$O$_4$:Ce^{3+}, a persistent phosphor. *J. Electrochem. Soc.* 150: H61.
41. Bube RH, *Photoelectronic Properties of Semiconductors* (Cambridge University Press, Cambridge, 1992).
42. Reynolds S, Brinza M, Benkhedir ML, Adriaenssens GJ, in *Springer Handbook of Electronic and Photonic Materials*, edited by S. Kasap, P. Capper (Springer International Publishing, Cham, 2017).
43. van der Kolk E, Dorenbos P, de Haas JTM, van Eijk CWE (2005) Thermally stimulated electron delocalization and luminescence quenching of Ce impurities in GdAlO$_3$. *Phys. Rev. B* 71: 045121.
44. Van der Kolk E, Basun SA, Imbusch GF, Yen WM (2003) Temperature dependent/spectroscopic studies of the electron delocalization dynamics of excited Ce ions in the wide band gap insulator, Lu$_2$SiO$_5$. *Appl. Phys. Lett.* 83: 1740.
45. Ueda J (2015) Analysis of optoelectronic properties and development of new persistent phosphor in Ce^{3+}-doped garnet ceramics. *J. Ceram. Soc. Jpn.* 123: 1059.
46. Basun SA, Raukas M, Happek U, Kaplyanskii AA, Vial JC, Rennie J, Yen WM, Meltzer RS (1997) Off-resonant spectral hole burning in CaS:Eu by time-varying Coulomb fields. *Phys. Rev. B* 56: 12992.
47. Happek U, Basun SA, Choi J, Krebs JK, Raukas M (2000) Electron transfer processes in rare earth doped insulators. *J. Alloys Compd.* 303–304: 198.
48. Jia D, Wang X-j, Yen WM (2002) Electron traps in Tb^{3+}-doped CaAl$_2$O$_4$. *Chem. Phys. Lett.* 363: 241.
49. Jia D, Wang X-j, Yen WM (2004) Delocalization, thermal ionization, and energy transfer in singly doped and codoped CaAl$_4$O$_7$ and Y$_2$O$_3$. *Phys. Rev. B* 69: 235113.
50. van der Kolk E, Dorenbos P, van Eijk CWE, Basun SA, Imbusch GF, Yen WM (2005) 5d electron delocalization of Ce^{3+} and Pr^{3+} in Y$_2$SiO$_5$ and Lu$_2$SiO$_5$. *Phys. Rev. B* 71: 165120.
51. Amiryan AM, Gurvieh AM, Katomina RV, Petrova IY, Soshchin NP, Tombak MI (1977) Recombination processes and emission spectrum of terbium in oxysulfides. *J. Appl. Spectrosc.* 27: 1159.

52. Fleniken J, Wang J, Grimm J, Weber MJ, Happek U (2001) Thermally stimulated luminescence excitation spectroscopy (TSLES): a versatile technique to study electron transfer processes in solids. *J. Lumin.* 94: 465.

53. Dorenbos P, Bos AJJ, Eijk CWEv (2002) Photostimulated trap filling in Lu_2SiO_5:Ce^{3+}. *J. Phys.: Condens. Matter* 14: L99.

54. Bos AJJ, van Duijvenvoorde RM, van der Kolk E, Drozdowski W, Dorenbos P (2011) Thermoluminescence excitation spectroscopy: a versatile technique to study persistent luminescence phosphors. *J. Lumin.* 131: 1465.

55. Ueda J, Dorenbos P, Bos AJJ, Kuroishi K, Tanabe S (2015) Control of electron transfer between Ce^{3+} and Cr^{3+} in the $Y_3Al_{5-x}Ga_xO_{12}$ host via conduction band engineering. *J. Mater. Chem. C* 3: 5642.

56. Ueda J, Dorenbos P, Bos AJJ, Meijerink A, Tanabe S (2015) Insight into the thermal quenching mechanism for $Y_3Al_5O_{12}$:Ce^{3+} through thermoluminescence excitation spectroscopy. *J. Phys. Chem. C* 119: 25003.

57. Jary V, Mihokova E, Nikl M, Bohacek P, Lauria A, Vedda A (2010) Thermally-induced ionization of the Ce^{3+} excited state in $SrHfO_3$ microcrystalline phosphor. *Opt. Mater.* 33: 149.

58. Pejchal J, Nikl M, Mihokova E, Novoselov A, Yoshikawa A, Williams RT (2009) Temperature dependence of the Pr^{3+} luminescence in LSO and YSO hosts. *J. Lumin.* 129: 1857.

59. Fasoli M, Vedda A, Miháková E, Nikl M (2012) Optical methods for the evaluation of the thermal ionization barrier of lanthanide excited states in luminescent materials. *Phys. Rev. B* 85: 085127.

60. Nikl M, Kamada K, Kurosawa S, Yokota Y, Yoshikawa A, Pejchal J, Babin V (2013) Luminescence and scintillation mechanism in Ce^{3+} and Pr^{3+} doped $(Lu,Y,Gd)_3(Ga,Al)_5O_{12}$ single crystal scintillators. *Phys. Stat. Sol. C* 10: 172.

61. Blazek K, Krasnikov A, Nejezchleb K, Nikl M, Savikhina T, Zazubovich S (2004) Luminescence and defects creation in Ce^{3+}-doped $Lu_3Al_5O_{12}$ crystals. *Phys. Stat. Sol. B* 241: 1134.

62. Ueda J, Kuroishi K, Tanabe S (2014) Bright persistent ceramic phosphors of Ce^{3+}-Cr^{3+}-codoped garnet able to store by blue light. *Appl. Phys. Lett.* 104: 101904.

63. Loudyi H, Guyot Y, Kazanskii SA, Gâcon J-C, Pédrini C, Joubert M-F (2007) What may be expected from the "microwave resonant cavity technique" applied to rare-earth-doped insulating materials? *Phys. Stat. Sol. C* 4: 784.

64. Joubert MF, Kazanskii SA, Guyot Y, Gâcon JC, Pédrini C (2004) Microwave study of photoconductivity induced by laser pulses in rare-earth-doped dielectric crystals. *Phys. Rev. B* 69: 165217.

65. Loudyi H, Guyot Y, Kazanskii SA, Gâcon JC, Moine B, Pédrini C, Joubert MF (2008) Analysis of the photoconduction in CaF_2:Eu^{2+} crystals using the microwave resonant cavity technique. *Phys. Rev. B* 78: 045111.

66. Bertrand P, *Electron Paramagnetic Resonance Spectroscopy: Fundamentals* (Springer International Publishing, Switzerland, 2020).

67. Godlewski M, Hommel D (1986) Eu^{2+} photocharge transfer processes in ZnS crystals determined by photo-ESR measurements. *Phys. Stat. Sol. A* 95: 261.

68. Carlson S, Hölsä J, Laamanen T, Lastusaari M, Malkamäki M, Niittykoski J, Valtonen R (2009) X-ray absorption study of rare earth ions in $Sr_2MgSi_2O_7$:Eu^{2+},R^{3+} persistent luminescence materials. *Opt. Mater.* 31: 1877.

69. Joos JJ, Korthout K, Amidani L, Glatzel P, Poelman D, Smet PF (2020) Identification of Dy^{3+}/Dy^{2+} as electron trap in persistent phosphors. *Phys. Rev. Lett.* 125: 033001.

70. Dorenbos P (2020) Improved parameters for the lanthanide $4f^q$ and $4f^{q-1}5d$ curves in HRBE and VRBE schemes that takes the nephelauxetic effect into account. *J. Lumin.* 222: 117164.

71. Boutinaud P, Pinel E, Oubaha M, Mahiou R, Cavalli E, Bettinelli M (2006) Making red emitting phosphors with Pr^{3+}. *Opt. Mater.* 28: 9.

72. Boutinaud P, Mahiou R, Cavalli E, Bettinelli M (2007) Red luminescence induced by intervalence charge transfer in Pr^{3+}-doped compounds. *J. Lumin.* 122–123: 430.

73. Boutinaud P, Putaj P, Mahiou R, Cavalli E, Speghini A, Bettinelli M (2007) Quenching of lanthanide emission by intervalence charge transfer in crystals containing closed shell transition metal ions. *Spectrosc. Lett.* 40: 209.

74. McClure DS, Pedrini C (1985) Excitons trapped at impurity centers in highly ionic crystals. *Phys. Rev. B* 32: 8465.

75. Nurtdinova L, Semashko V, Guyot Y, Korableva S, Joubert MF, Nizamutdinov A (2011) Application of photoconductivity measurements to photodynamic processes investigation in $LiYF_4$:Ce^{3+} and $LiLuF_4$:Ce^{3+} crystals. *Opt. Mater.* 33: 1530.

76. Grimm J, Fleniken J, Krämer KW, Biner D, Happek U, Güdel HU (2007) On the determination of photoionization thresholds of Ce^{3+} doped Cs_3LuCl_6, $Cs_2LiLuCl_6$ and Cs_2LiYCl_6 by thermoluminescence. *J. Lumin.* 122–123: 325.

77. Majewska N, Lesniewski T, Mahlik S, Grinberg M, Kulesza D, Ueda J, Zych E (2020) Properties of charge carrier traps in Lu_2O_3:Tb,Hf ceramic storage phosphors observed by high-pressure spectroscopy and photoconductivity. *J. Phys. Chem. C* 124: 20340.

78. Jia D, Wang X-J, Yen W (2008) Ground-state measurement of Pr^{3+} in Y_2O_3 by photoconductivity. *Phys. Sol. State* 50: 1674.

79. Ueda J, Meijerink A, Dorenbos P, Bos AJJ, Tanabe S (2017) Thermal ionization and thermally activated crossover quenching processes for 5d-4f luminescence in $Y_3Al_{5-x}Ga_xO_{12}$:Pr^{3+}. *Phys. Rev. B* 95: 014303.

80. Lesniewski T, Mahlik S, Asami K, Ueda J, Grinberg M, Tanabe S (2018) Comparison of quenching mechanisms in $Gd_3Al_{5-x}Ga_xO_{12}$:Ce^{3+} (x = 3 and 5) garnet phosphors by photocurrent excitation spectroscopy. *PCCP* 20: 18380.

81. Babin V, Boháček P, Jurek K, Kučera M, Nikl M, Zazubovich S (2018) Dependence of Ce^{3+} - related photo- and thermally stimulated luminescence characteristics on Mg^{2+} content in single crystals and epitaxial films of $Gd_3(Ga,Al)_5O_{12}$:Ce,Mg. *Opt. Mater.* 83: 290.

82. Happek U, Choi J, Srivastava AM (2001) Observation of cross-ionization is $Gd_3Sc_2Al_3O_{12}$:Ce^{3+}. *J. Lumin.* 94–95: 7.

83. Choi J, Basun SA, Lu L, Yen WM, Happek U (1999) Excited state impurity band conductivity in $Y_2(SiO_4)O$:Ce^{3+}. *J. Lumin.* 83–84: 461.

84. Katayama Y, Ueda J, Tanabe S (2014) Photo-electronic properties and persistent luminescence in Pr^{3+} doped (Ca,Sr)TiO_3 ceramics. *J. Lumin.* 148: 290.

85. Yuan HB, Jia W, Basun SA, Lu L, Meltzer RS, Yen WM (2000) The long-persistent photoconductivity of $SrAl_2O_4$:Eu^{2+}, Dy^{3+} single crystals. *J. Electrochem. Soc.* 147: 3154.

86. Ueda J, Nakanishi T, Katayama Y, Tanabe S (2012) Optical and optoelectronic analysis of persistent luminescence in Eu^{2+}-Dy^{3+} codoped $SrAl_2O_4$ ceramic phosphor. *Phys. Stat. Sol. C* 9: 2322.

87. Jia D, Wang XJ, Jia W, Yen WM (2006) Temperature-dependent photoconductivity of Ce^{3+}-doped $SrAl_2O_4$. *J. Lumin.* 119–120: 55.

88. Mihóková E, Jarý V, Fasoli M, Lauria A, Moretti F, Nikl M, Vedda A (2013) Trapping states and excited state ionization of the Ce^{3+} activator in the $SrHfO_3$ host. *Chem. Phys. Lett.* 556: 89.

89. Ueda J, Shinoda T, Tanabe S (2015) Evidence of three different Eu^{2+} sites and their luminescence quenching processes in $CaAl_2O_4$:Eu^{2+}. *Opt. Mater.* 41: 84.

90. Jia D, Jia W, Wang XJ, Yen WM (2004) Quenching of thermo-stimulated photo-ionization by energy transfer in $CaAl_4O_7$: Tb^{3+}, Ce^{3+}. *Sol. State Commun.* 129: 1.

91. Ueda J, Maki R, Tanabe S (2017) Vacuum referred binding energy (VRBE)-guided design of orange persistent $Ca_3Si_2O_7$:Eu^{2+} phosphors. *Inorg. Chem.* 56: 10353.

92. Hiraguri K, Hidaka C, Takizawa T (2006) Photoconductivity of $CaGa_2S_4$ single crystals doped with Eu^{2+} and Ce^{3+}. *Phys. Stat. Sol. C* 3: 2730.

93. Hidaka C, Hiraguri K, Takizawa T (2008) Photoconductivity and thermally stimulated current in $CaGa_2S_4$ single crystals doped with Eu^{2+} and Ce^{3+}. *Phys. Stat. Sol. A* 205: 2903.

94. Ueda J, Tanabe S, Takahashi K, Takeda T, Hirosaki N (2018) Thermal quenching mechanism of $CaAlSiN_3$:Eu^{2+} red phosphor. *Bull. Chem. Soc. Jpn.* 91: 173.

95. Smet PF, Van den Eeckhout K, Bos AJJ, van der Kolk E, Dorenbos P (2012) Temperature and wavelength dependent trap filling in $M_2Si_5N_8$:Eu (M=Ca, Sr, Ba) persistent phosphors. *J. Lumin.* 132: 682.

96. Wen B (2004) Thesis, The University of Georgia.

15 Synchrotron Radiation Analysis on Phosphors

Chun Che Lin and Sarmad Ahmad Qamar

CONTENTS

15.1 INTRODUCTION TO PHOSPHORS AND SYNCHROTRON RADIATION

The term 'phosphor' is used for the solid materials that exhibit luminescence upon absorption of energy by the external source. The emission of light by a solid material when exposed to energy source (exciting radiations) is known as fluorescence. However, the afterglow emission, which is detectable after the excitation period, is known as phosphorescence. It is essential to highlight these terminologies with regard to inorganic materials; however, these definitions are somewhat different for describing organic materials. In organic materials, single-state emission of light is known as fluorescence; however, triplet-state emission of light is referred as phosphorescence. To understand the synchrotron radiation analysis on phosphors, it is essential to have a look on what synchrotron radiations actually are. In late nineteenth century, it was understood that any charge subjected to an acceleration must radiate energy in the form of electromagnetic radiations. When the field is electric, such radiations are called bremsstrahlung; however, under the influence of magnetic field, these radiations are called synchrotron radiations. The loss of energy is significant for high-energy particles *via* synchrotron radiations, and scale proportionate to $B^2\gamma^2$, where 'B' indicates the magnetic field. Protons require much higher amount of energy (γmc^2) to produce same radiation compared to electrons, because of much higher weight of protons. Therefore, negligible synchrotron radiations are produced by the proton accelerators while high-energy electron accelerators produce significant synchrotron radiations.

The characteristic properties of synchrotron radiations have been well researched by both theoretical and experimental point-of-views. The topic is somewhat of a nuisance from an accelerator engineering point-of-view for the further exploration of synchrotron radiation characteristics. By growing electron energy, total power of synchrotron radiations produced grows quickly. Therefore, synchrotron radiation facilities are designed about a specified fixed-energy machine known as storage ring, designed for optimum performance to recirculate the beam for as much as turns are possible. If electronic beam-producing chambers are adequately evacuated, the lifetime of electron beam is about 5–100 hours. The huge amount of synchrotron radiation is intricate in an electron accelerator working above 100 GeV. The applications of synchrotron radiation in synchrotron X-ray diffraction (SXRD) for the refinement of structures, synchrotron vacuum ultraviolet-photoluminescence (VUV-PL), X-ray absorption near-edge structure (XANES), and X-ray absorption fine structure (EXAFS) have been comprehensively discussed below.

DOI: 10.1201/9781003098669-15

15.2 SYNCHROTRON X-RAY DIFFRACTION (SXRD) REFINEMENT FOR STRUCTURES

X-ray diffraction (XRD) has long been practiced for structure determination of materials. The XRD relies on the principle of short wavelength of X-rays equivalent to the atomic distances in a condensed matter. When a material exhibits periodic atomic order i.e., crystal lattice exposed to X-rays, it acts as well defined, extended grating, and produced various diffraction spots known as Bragg's peaks. The determination of intensities and position of these diffraction peaks enable materials researchers to analyze spatial properties of grating, such as 3D structural determination of material under investigation. By the passage of time, XRD system has been successfully employed for a variety of different materials. Moreover, for the determination of structure of micrometer-sized or nanometer-sized materials, XRD has been successfully used i.e., glasses, liquids, nanocrystals, and phosphors. These materials, while exposed to X-rays, act as imperfect gratings and form patterns in the form of peaks. Therefore, XRD has been demonstrated to be a useful tool for the determination of both regular and/or specific structures. The modern science is rapidly moving toward smaller materials production. All kinds of different microscale materials are being rapidly synthesized day by day. Microcrystalline materials fall between the microcrystals and regular crystals depending on their structural point-of-view. These microcrystalline materials produce XRD pattern in the form of diffuse components and Bragg's peaks. The diffraction data should be collected and analyzed with high accuracy and with improved magnitude than classical analysis. To achieve such level of accuracy, longer time than usual, XRD data should be used [1]. Larger area detection and SXRD can minimize the collection interval to seconds [2]. Recent studies describing phosphors synthesis and characterization using variety of different dopants with key findings have been summarized in Table 15.1.

TABLE 15.1

Few Recent Studies Describing Synchrotron X-Ray Diffraction (SXRD) Refinement for Microcrystal Structures with Different Dopant Ions and Key Conditions

Host Matrix	Dopant	Conditions	Reference
$Sr_{1.98}Si_5N_8:Eu_{0.02}$	$Sr_{1.98-x}(Ca_{0.55}Ba_{0.45})_x$	1. M1–N distance indicate less variation 2. M2–N decrease with x 3. The diffraction results indicate significant distortion on both sites with no cationic segregation as x increases.	[3]
$Cs_{0.96}MgPO_4:0.04Eu^{2+}$	$P_{0.96}O_4:0.04[Eu^{2+}–Si^{4+}]$	1. Si^{4+} co-doping resulted in improved crystallinity. 2. Steric configuration was managed by cationic co-substitution of $[Cs^{1+}–P^{5+}]$ by $[Eu^{2+}–Si^{4+}]$ occurred.	[4]
$Na_2F_6:Mn^{4+}$	(Si_xGe_{1-x}) and (Ge_yTi_{1-y})	1. Cations Ti^{4+}, Ge^{4+}, and Si^{4+} were substituted by Mn^{4+} forming octahedral MnF_6^{2-} moieties. 2. Structural characteristics of solid solution were in the range of cationic content. 3. Electronic/structural characteristics were in range of cationic content.	[5]
$Sr[LiAl_3N_4]$	Ce^{3+}	1. Refinement parameters were in agreement to SLA host, synthesized by radiofrequency and solid-state approach.	[6]
K_2GeF_6	Mn^{4+}	1. By increasing ethanol concentration phase transfer from $P\bar{3}m1$ to $P6_3mc$ was observed. 2. High stability of $P\bar{3}m1$ than $P6_3mc$ was observed in terms of energy.	[7]

(Continued)

TABLE 15.1 (*Continued*)
Few Recent Studies Describing Synchrotron X-Ray Diffraction (SXRD) Refinement for Microcrystal Structures with Different Dopant Ions and Key Conditions

Host Matrix	Dopant	Conditions	Reference
$M_5(PO_4)_3Cl$ (M = Sr, Ba)	Eu^{2+}	1. Unequal distribution of Ba^{2+} among Sr^{2+} sites. 2. The occupancy of Ba^{2+} at Sr(I) site increased at $x > 1.5$. 3. The Ba^{2+} could be preferred occupation in Sr^{2+} sites.	[8]
$Sr_3Ce(PO_4)_3$	Eu^{2+}	1. Eu^{2+}, Ce^{3+}, and Sr^{2+} cations occupy single 16c sites in the lattice. 2. Lattice shrinkage (1.474 Å) was observed compared to host matrix (1.529 Å).	[9]
$KNaSiF_6$	Mn^{4+}	1. Mn^{4+} ions substituted the Si^{4+} sites to make MnF_6^{-2} group (rMn^{4+}, 0.053 nm as compared to rSi^{4+}, 0.040 nm). 2. Si^{4+} was bounded by six F^- ions, with two angles i.e., $\angle F_3SiF = 179.29°$ and $\angle F_1SiF_2 = 174.11°$.	[10]
$BaLa_2Si_2S_8$	$(Ba_{0.98}Eu_{0.02})$	1. Crystal lattice was formed by trigonal prisms of $[(1/3Ba + 2/3La)S_8]$ and SiS_4 tetrahedral, linked on edges and corners. 2. The Ba/LaS_8 polyhedron was consisted of eight S and single Ba/La atom with 8 coordination no.	[11]
K_2GeF_6	Mn^{4+}	1. Crystallographic expansion of volume by linearly increased parameters with temperature. 2. Bond length F and Ge^{4+} for 52°C prepared samples (1.810 Å) was longer than prepared at 0°C (1.776 Å), confirming expansion.	[12]
K_2MF_6 (M = Ge, Si)	(Ge/Si):Mn^{4+}	1. Si-doped phosphors showed light color as compared with Ge doped, emitting bright red emission after 460 nm of excitation light. 2. Si^{4+} and Ge^{4+} owning coordination as SiF_6^{2-} and GeF_6^{2-} formed octahedron.	[13]
$MSi_2O_2N_2$ (M = Sr, Ba)	$(Sr_{0.98-x}Ba_xEu_{0.02})$	1. SX-D indicated 3-phases, phase I ($x = 0–0.63$) phase II ($x = 0.68–0.77$), phase III ($x = 0.78–0.98$). 2. Increased lattice parameters by the increase in x, indicate that solid solution was formed. At $x > 0.20$.	[14]
$BaLa_2Si_2S_8$	$Ba(La_{0.94}Ce_{0.06})_2Si_2S_8$	1. The decrease in bond distances and lattice parameters was noticed by increasing dopant (Ce^{3+}) concentration.	[15]
$Ca_4Si_2O_7F_2$	$(Ca_{0.89}Eu_{0.01}Mn_{0.1})_4$	1. Decrease in lattice volume was observed by increasing the dopant (Mn^{2+}) ions concentration.	[16]

Wei et al. [4] have reported the synthesis of $CsMgPO_4$:Eu^{2+} phosphors, which belonged to *Pnma*-(62) space group with orthorhombic structure. The crystal structure of $CsMgPO_4$ microcrystal showed the existence of single P, Mg, and Cs site with three O sites, correspondingly. Tetrahedral $[PO_4]$ and tetrahedra $[MgO_4]$ association, by sharing oxygen atom at the vertex, making 3D structure was reported. The degree of condensation shows convenient management of the steric configuration by cationic co-substitution method. Two centrally antisymmetric $[CsO_6]$ polyhedra, on the direction of b-axis, share an edge, making a Cs_2O_{10} group, which is located with alternating four $[MgO_4]$ and four $[PO_4]$. At the direction of a-axis, $[CsO_6]$ polyhedra contributes one $[MgO_4]$ tetrahedron, and five $[PO_4]$ tetrahedra. Moreover, $[PO_4]$ tetrahedron, connects with $[CsO_6]$ polyhedron by sharing vertex 'O' atom; however, $[MgO_4]$ tetrahedron connects with $[CsO_6]$ polyhedron by sharing edge as well as vertex. Considering the lattice size-mismatch, and the imbalanced charge caused by Eu^{2+} (CN = 6, r = 1.17 Å, here 'r' represents the ionic radii, and 'CN' shows coordination number), replacing Cs^{1+} monovalent ions with CN = 6 and r = 1.67 Å, few faults may be produced. The probable Cs or O vacant defects was solved by cationic co-substitution of $[Eu^{2+}–Si^{4+}]$ for $[Cs^{1+}–P^{5+}]$ was designed. XRD results of $Cs_{1-x}MgP_{1-x}O_4{:}x[Eu^{2+}–Si^{4+}]$ could beresembled with $CsMgPO_4$ (ICSD 260423) orthorhombic from x = 0.005 to 0.05, representing that the newly designed microcrystals are pure phases. The co-substitution of $[Eu^{2+}–Si^{4+}]$ did not affect the overall crystal structure, and the phase purity of $CsMgPO_4$. For further refinement of crystal structure, High quality synchrotron XRD patterns of $Cs_{0.96}MgPO_4{:}0.04Eu^{2+}$ and $Cs_{0.96}MgP_{0.96}O_4{:}0.04[Eu^{2+}–Si^{4+}]$ were recorded and analyzed. The R-factor (R_p = 0.05, R_{wp} = 0.09, and χ^2 = 3.46 for $Cs_{0.96}MgPO_4{:}0.04Eu^{2+}$; and R_p = 0.05, R_{wp} = 0.07, and χ^2 = 2.27 for $Cs_{0.96}MgP_{0.96}O_4{:}0.04[Eu^{2+}–Si^{4+}]$ values also indicated the synthesis of solid solution. The degree of crystallization was further improved by co-doping with Si^{4+} ions for charge compensation. The crystallographic data of as-prepare samples showed no Eu^{2+} or Si^{4+} trace because of low dopant concentration. $Cs_{0.7}MgP_{0.7}O_4{:}0.30[Eu^{2+}–Si^{4+}]$ specimen showed replacement of Cs^{1+} with Eu^{2+} ions at x = 0.4952, y = 0.2500, z = 0.7033, and P^{5+} were replaced by Si^{4+} at x = 0.1955, y = 0.2500, z = 0.4257. Doping level for purified Eu^{2+} was 0.29, and for Si^{4+} was 0.32 [4].

Similarly, Fang et al. [5] studied structural refinement of $Na_2(Si_xGe_{1-x})F_6{:}Mn^{4+}$ and $Na_2(Ge_yTi_{1-y})F_6{:}Mn^{4+}$ phosphors *via* SXRD (λ = 0.774910 Å) analysis. Both structures i.e., Na_2SiF_6 (NaSF) and Na_2GeF_6 (NaGF) are crystallized in trigonal ($P321$ space group); however, Na_2TiF_6 (NaTF) exhibited triclinic structure with $P1$ space group. The cations Ti^{4+}, Ge^{4+}, and Si^{4+} was substituted by Mn^{4+} making octahedral structure MnF_6^{2-} moieties. Leaño Jr et al. [6] studied Ce^{3+}-doped $Sr[LiAl_3N_4]$ (SLA) preparation using all-nitride in a gas pressure sintering furnace (GPS). The synthesized powder was analyzed using SXRD, where the patterns exhibit wide range of Ce^{3+} content in agreement with CSD-427067. The Rietveld refinement of SXRD showed triclinic crystal structure belonging to $P1$ space group. The a = 5.85241, b = 7.48525, and c = 9.92537 Å, were in agreement with SLA host synthesized by solid-state and radiofrequency approaches using Eu doping. Leaño Jr et al. [17] also studied the preparation of $Sr[Mg_2Al_2N_4]$ phosphor Eu-doping using all-nitride reaction mixture in GPS furnace. Light orange-colored powder was obtained exploring the wide spectrum of Eu^{2+} still exhibiting well similarity with CSD-425321. Rietveld refinement of SXRD showed that the crystal with tetragonal structure was made relating to $I4/m$ (No. 87) space group. Comparative analysis indicated that the $Sr[Mg_2Al_2N_4]$ matrix was formed by radiofrequency method with activator (Ce^{3+}). Highly comparable ionic radii of Eu^{2+} and Sr^{2+} remained pure phases showing comparatively wider x-values. Similarly, to study the phases of thiogallate phosphors doped with Eu^{2+} and determine the crystal structure, Lee et al. [18] performed Rietveld refinement using $Ba_2Ga_8SiS_{16}$ (ICSD 194875) and $BaGa_2SiS_6$ (ICSD 184747) crystal structures, as references to determine approximation of newly designed crystal structure. The R-factors of SXRD results of Rietveld refinement were $(Ba_{0.90}Eu_{0.10})_2Ga_8SiS_{16}$ (R_p = 1.59%, R_{wp} = 2.72%) and $(Ba_{0.95}Eu_{0.05})Ga_2SiS_6$ (R_p = 3.04%, R_{wp} = 4.28%).

Zhou et al. [7] studied the morphological and optical characteristics of $K_2GeF_6{:}Mn^{4+}$ phosphor microcrystals, which were influenced by the volume ratio between ethanol and hydrogen fluoride. The volume of HF was fixed to 5 mL; however, ethanol was varied to 5, 10, and 25 mL, and synthesized samples were denoted as $R1$, $R2$, and $R5$. Diffraction outcomes indicated the phase of phosphors transferred from $P\bar{3}m1$ to $P6_3mc$ by the increase in ethyl alcohol concentration (Figure 15.1(a)). $R1$ is pure phase of $P\bar{3}m1$ that was studied *via* SXRD analysis (Figure 15.1(b)), However, $R5$ was pure $P6_3mc$ that was partially transferred to $P\bar{3}m1$ using SXRD, which indicated comparatively high stability of $P\bar{3}m1$ than $P6_3mc$ in terms of energy. $R2$ was found as the combination of both. By scanning electron microscope (SEM), $R1$ appears around 20 μm with octahedral

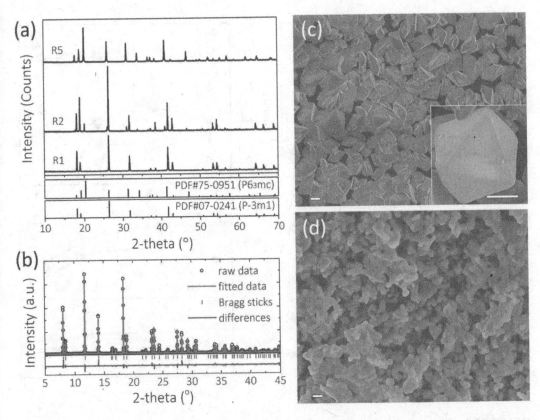

FIGURE 15.1 (a) XRD pattern of K_2GeF_6:Mn^{4+} phosphors microcrystals; (b) SXRD refinement of sample $R1$; and the SEM micrographs of (c) $R1$ (10 μm), and (d) $R5$ (1 μm). Reprinted from Zhou et al. [7] with permission.

shape with a smooth and clear surface (Figure 15.1(c)). $R5$ showed particle size of about 0.6 μm, with small and big particles, which were $P6_3mc$ and $P\bar{3}m1$ KGFM according to SXRD analysis (Figure 15.1(d)).

Lin et al. [3] studied $Sr_{1.98-x}(Ca_{0.55}Ba_{0.45})_xSi_5N_8$:$Eu_{0.02}$ phosphors ($x = 0, 0.5, 1.0$, and 1.5) by SXRD analysis for structural determination. Sample with $x = 0.5$ data has been shown in Figure 15.2(a). The diffraction results indicated that materials with $x = 0$–1.5 adopted $M_2Si_5N_8$ orthorhombic structure with $Pmn2_1$ space group. Lattice parameters 298K (Figure 15.2(b)) represent irregular modifications of less than 0.1% as x enhances from 0% to 1.5%, which could be explained by the average size of unit cell and the constant lattice parameters to excellent estimation. Neutron diffraction was studied at 25–400°C temperature from all samples to determine thermostability of newly designed materials across 150–200°C (working temperature). All samples showed similar thermal characteristics across the described range, with less modification and high temperature expansion coefficients $(15$–$20) \times 10^{-6}$/K. Two M-cationic inequivalent regions existed in $M_2Si_5N_8$ orthorhombic structure, and simultaneous refinement against neutron data and 298K X-rays were used to determine cationic residences using element scattering contrasts for neutrons and X-rays, as shown in Figure 15.2(c). By fixing dopant content to 1%, results indicated more disordered 10-coordinate M1 site, with nearly equal amount of Ba and Ca completely replacing Sr at $x = 1.5$, where M1 is $\sigma_1^2 = 0.020$ Å^2. M2 was slightly smaller with eight coordinates and was occupied predominantly by Sr and Ca with little amount of Ba, causing smaller $\sigma_2^2 = 0.011$ Å^2 at $x = 1.5$. Trend was seen by M–N bond; however, M1–N distance indicates less difference along the series. The M2–N decline with x is as indicated in Figure 15.2(d). Though, the σ^2 value at both places varies, the outcome indicates that regions are significantly disordered with no cationic separation as x increases.

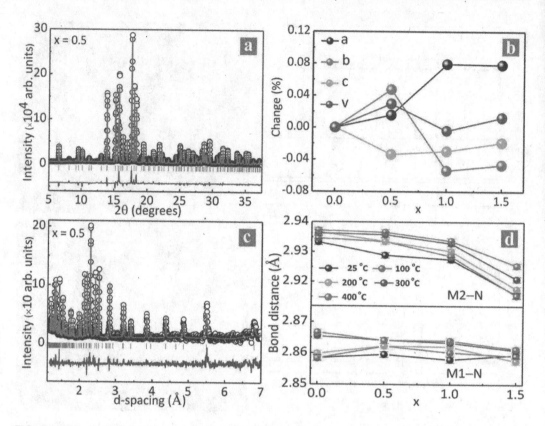

FIGURE 15.2 Crystal description of phosphors ($Sr_{1.98-x}(Ca_{0.55}Ba_{0.45})_xSi_5N_8:Eu_{0.02}$). (a) SXRD refinement for samples ($x = 0.5$); (b) relative alterations in lattice parameters at 298K using SXRD as function of x; (c) neutron diffraction data for samples ($x = 0.5$); (d) x and thermal necessity of M–N bond distance using neutron data and Rietveld refinement data. Reprinted form Lin et al. [3] with permission.

Fang et al. [8] studied $(Sr,Ba)_5(PO_4)_3Cl:Eu^{2+}$ phosphors for their structure *via* SXRD and Rietveld refinement. Lattice parameters showed linear increase due to the increase in ionic size of Ba^{2+}. Remarkably, the distribution of Ba^{2+} was not equal within Sr^{2+} sites. The occupancy of Ba^{2+} at Sr(I) sites was about zero, which slightly increased while x was higher than that of 1.5. The results indicated that Ba^{2+} could be a highly preferred occupation in various Sr^{2+} regions. Similarly, Lee et al. [19] performed Rietveld analysis for the evaluation of phase purity and to study detailed structural information of $(La_{0.9}Ce_{0.1})_3Br(SiS_4)_2$ microcrystals. The isotype single structure of $La_3Br(SiS_4)_2$ (ICSD 411996) crystallographic data as reference for Ce^{3+}-doping $La_3Br(SiS_4)_2$ reaches an approximation of microcrystal structure. The refined SXRD patterns were obtained for $(La_{0.9}Ce_{0.1})_3Br(SiS_4)$, which revealed the R_{wp} as 10.05%. Microstructures $(La_{0.9}Ce_{0.1})_3Br(SiS_4)_2$ designed in this study crystallized in $C2/c$ (No. 15) space group with monoclinic system (Z = 4), iso-typical to A-type $La_3Cl(SiO_4)_2$. Zhang et al. [20] studied $SrLiAl_3N_4:Eu^{2+}$ phosphors for their structural evaluation and impurity content using SXRD and Rietveld refinement. The initial parameters were analyzed by $SrLiAl_3N_4$ single crystal results of Pust et al. [21]. The calculation, experimental, and various Rietveld refined XRD patterns at ambient temperature have been indicated in Figure 15.3. Almost every XRD spectra could correspond to data giving goodness of fit of $R_p = 2.36\%$, $R_{wp} = 3.65\%$, and $\chi^2 = 1.08$. The AlN second phase was analyzed to be ~5.48 wt% by Rietveld refinement.

Dai et al. [9] examine the probability occupancy of cations and the crystal structure of $Sr_3Ce(PO_4)_3:Eu$ (SCPO) phosphors using SXRD and Rietveld refinement. The structural model was generated by crystallographic reference data $Sr_3La(PO_4)_3$ (ICSD 69431) that belonged to

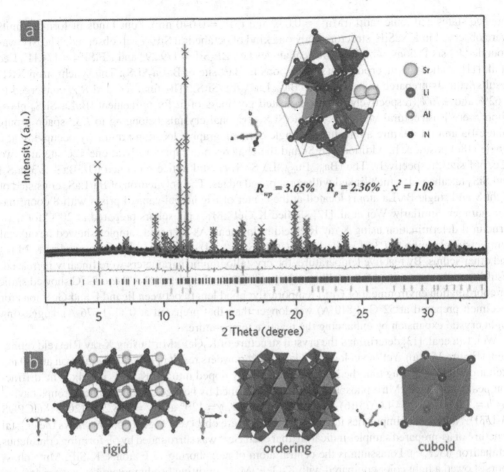

FIGURE 15.3 (a) SXRD Rietveld refinement of $SrLiAl_3N_4$:Eu^{2+} phosphors, x – (observed), red lines (calculated), green and blue lines indicate Bragg's peaks of AlN and $SrLiAl_3N_4$, respectively. Grey line represents observed-calculated difference, crystal structure of unit-set (inset); (b) crystal structure indicating AlN_4 (blue tetrahedral), LiN_4 (orange tetrahedral) cuboid SrN_8 (pink polyhedral), coordinated by LiN_4, and AlN_4. Reprinted from Zhang et al. [20] with permission.

$M^{II}_3Ln^{III}(PO_4)_3$ (143d space group) and isostructural to SCPO compound. Study reported that Eu^{2+}, Ce^{3+}, and Sr^{2+} ions occupy single 16c sites in unit cells. In Rietveld refinement method, it was assumed that 7% of 16c sites were filled by Eu atoms and 93% by Ce and Sr without temperature factor and site-mixing for Eu, Ce, and Sr in 16c site. The R-factors for the SXRD patterns were $R_p = 5.14\%$ and $R_{wp} = 8.72\%$. The refined lattice parameters for microcrystals were $a = 10.169(7)$ Å and $V = 1051.791(1)$ Å^3, indicating the deceptive lattice shrinkage compared with $Sr_3La(PO_4)_3$. Jin et al. [10] studied crystal structure, composition, and phase purity of microcrystals $KNaSiF_6$:Mn^{4+} by synchrotron XRD and Rietveld refinement analysis. The $KNaSiF_6$ crystallographic data was used as a structural model during refinement, followed by the construction of an actual structure of $KNaSiF_6$:Mn^{4+}. The results indicated the orthorhombic crystal structure with *Pnma* structure (ICSD 07–1334). The lattice structure included $\alpha = \beta = c = 90°$ where $a = 9.33208(21)$, $b = 5.50742(11)$, $c = 9.80288(22)$ Å, with a cellular volume of 503.825(19) Å^3. R-Bragg and χ^2 values were 6.01 and 1.44, respectively, which exhibited rationality of Rietveld refinement method. The bond angle from central atom was analyzed to investigate the association between PL spectra and phosphor structure. The axial bond pair was nearly 179.29°; however, the angle between two other paired atoms in equatorial was 174.11°. Mn^{4+} ions substituted the Si^{4+} sites to make MnF_6^{-2} group, comparing with valence states and ionic radii (rMn^{4+} = 0.053 and rSi^{4+} = 0.040 nm). Four kinds of formula units were observed in $KNaSiF_6$ structure, only one kind of octahedral SiF_6^{-2} was observed. The Si^{4+} was bounded by six F^- ions, with two different angles i.e., $\angle F_3SiF$ = 179.29° and $\angle F_1SiF_2$ = 174.11°. Lee et al. [11] studied the incorporation of Eu^{2+} ions in Ba^{2+} site of $BaLa_2Si_2S_8$. The synchrotron XRD results from as-prepared microcrystals $(Ba_{0.98}Eu_{0.02})La_2Si_2S_8$. The final R_{wp} and R_{exp} converged to 6.62% and 9.76%, respectively, which indicated goodness-of-fit for refinement. $BaLa_2Si_2S_8$ phosphors were isostructural with that of $PbR_2Si_2S_8$, trigonal crystals belonging to $R\bar{3}c$ space group, where Ba and La mixture atoms have a single crystallographic location arbitrarily occupying at a single (18e) position. In addition, the Si and the chalcogen occupied fully at one 12c site and two 12c, 36f sites, respectively. The $(Ba_{0.98}Eu_{0.02})La_2Si_2S_8$ crystal lattice remained [(1/3Ba + 2/3La)S_8] and SiS_4 tetrahedral, which linked with corners and edges. The polyhedron of Ba/LaS_8 consisted of eight S and single Ba/La atoms located at the center of trigonal bi-capped prism with 8 coordination number. Similarly, Wei et al. [12] studied K_2GeF_6:Mn^{4+} phosphors prepared at 20°C for their structural determination using X-ray Rietveld refinement. As-prepared samples showed hexagonal symmetry belonging to $P\bar{3}m1$ space group (JCPDS 73-1531). No trances were observed of K_2MnF_6 and other scums. By reactive temperature, the crystallographic parameters were linearly increased, resulting in the expansion of crystalline volume. Two samples prepared at 0 and 52°C showed similar coordination environment of Ge^{4+}, temporarily; bond length between ligand F and Ge^{4+} ions for specimen prepared at 52°C (1.810 Å) was longer than that prepared at 0°C (1.776 Å), suggesting slight crystal expansion by enhancing the reactive temperature.

Wei, Lin et al. [13] determined the crystal structure of K_2GeF_6:Mn^{4+} using X-ray Rietveld refinement (Figure 15.4(a)). Yellow-colored K_2GeF_6:Mn^{4+} powders result in bright red emission at 460 nm excitation light, showing that the Mn^{4+} was effectively doped into the K_2GeF_6 matrix. The diffraction peaks of K_2GeF_6:Mn^{4+} phosphors could be described by hexagonal $P\bar{3}m1$ space groups having $a = b = 5.63171(6)$, $c = 4.66751(6)$ Å, with $\alpha = \beta = 90°$, $\gamma = 120°$ and $V = 128.2027(20)$ Å^3 (JCPDS 73-1531). No traces of impurities including K_2MnF_6 were observed. Figure 15.4(b) shows the crystal structure of as-prepared samples indicating that each Ge^{4+} was surrounded by 6F forming continuous octahedron (GeF_6^{-2}). Potassium is the central atom of neighboring 12 F atoms. K_2SiF_6:Mn^{4+} phosphors showed a light color compared with K_2GeF_6:Mn^{4+}, emitting bright red emission after 460 nm of excitation light. The K_2SiF_6:Mn^{4+} phosphors showed high purity and association with $Fm3m$ with $a = b = c = 8.13107(7)$ Å, where $\alpha = \beta = \gamma = 90°$ and $V = 537.579(8)$ Å^3 (JCPDS 37-1155). Si^{4+} was residing face-centered and vertex position of cubit cells, and 4 K^+ ions were consistently dispersed inside cubic structure. Each Si^{4+} was bounded by 6 F to synthesize regular SiF_6^{2-} octahedron.

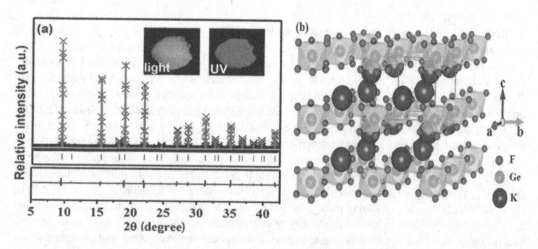

FIGURE 15.4 (a) SXRD Rietveld refinement and (b) schematic microcrystal structural diagram of K_2SiF_6:Mn^{4+} phosphors. Reprinted from Wei et al. [13] with permission.

Moreover, central Si^{4+} and Ge^{4+} own different coordination environments as SiF_6^{2-} and GeF_6^{2-} octahedron lying in different crystal lattices. Huang et al. [22] studied $SrSiAl_2O_3N_2$ phosphors for their crystal structure using SXRD Rietveld refinement. The $SrSiAl_2O_3N_2$ crystal structure represented the yellow balls (Sr atoms), pink tetrahedron (three types of AlO_2N_2 and AlO_3N), and blue tetrahedron ($SiON_3$) corner-sharing structural units. In the structure of $SrSiAl_2O_3N_2$, Sr is found coordinated by three nitrides and six oxides. The results demonstrated that the crystals of both $BaSiAl_2O_3N_2$ and $SrSiAl_2O_3N_2$ belong to same space group and symmetry (orthorhombic, $P2_12_12_1$), and due to the larger size, lattice parameters also became larger.

Li et al. [14] studied the phase purity and structure of $(Sr_{0.98-x}Ba_xEu_{0.02})Si_2O_2N_2$ microcrystals using SXRD analysis (Figure 15.5). The SXRD results showed that prepared microcrystals could be divided into three categories i.e., phase I where $x = 0–0.63$, phase II where $x = 0.68–0.77$, and phase III where $x = 0.78–0.98$, based on the relative intensities difference of diffraction peaks. In addition, XRD high-definition patters of $(Sr_{0.98-x}Ba_xEu_{0.02})Si_2O_2N_2$ were recorded on $11°–18°$ angle range, which indicated the presence of three phases due to significant difference in diffraction peaks. Rietveld refinement results indicated that the phase I and phase III are isostructural single-phase crystal structures of $SrSi_2O_2N_2$ and $BaSi_2N_2O_2$, respectively. The isotropic (Eu) displacement for Ba/Sr in microcrystals $(Sr_{0.98-x}Ba_xEu_{0.02})Si_2O_2N_2$ were seen to be equal. The $(Sr_{0.98-x}Ba_xEu_{0.02})Si_2O_2N_2$ diffraction peaks were in agreement with the (ICSD 172877) reported values, which can

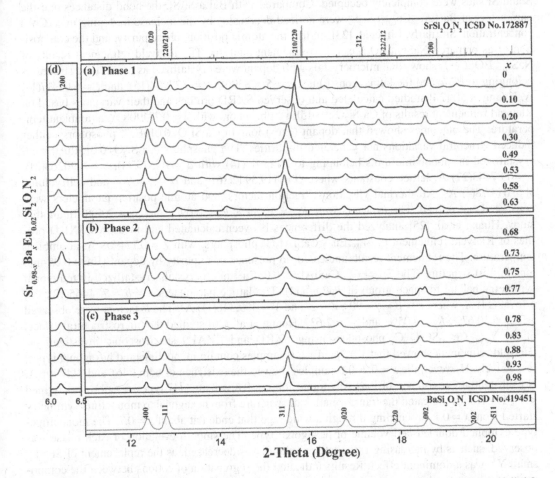

FIGURE 15.5 XRD patterns of $(Sr_{0.98-x}Ba_xEu_{0.02})Si_2O_2N_2$ phosphors showing (a) $x = 0–0.63$ [P1 – triclinic]; (b) $x = 0.68–0.77$ [P2 – triclinic]; (c) $x = 0.78–0.98$ [P3 – orthorhombic]. Reprinted from Li, Lin et al. [14] with permission.

be designated as the triclinic phase of $SrSi_2O_2N_2$ (P1). The ionic radius of Sr^{2+} was 1.21 Å, CN = 7, significantly smaller than that of Ba^{2+} as 1.38 Å, CN = 7, in $SrSi_2O_2N_2$ isostructural; the directional shift (2θ) of all the diffraction peaks in phase I could be observed by increasing x from 0 to 0.63, which confirms the successful Ba-incorporation into the crystal lattice. Smoothly increasing lattice parameters by the increase in x, indicates that $(Sr_{0.98-x}Ba_xEu_{0.02})Si_2O_2N_2$ solid solution was formed. At $x > 0.20$, the broadening of anisotropic peaks was observed. The phase III XRD patterns were in agreement with $BaSi_2O_2N_2$ standard (ICSD 419450). The SXRD refinement indicated that among $(Sr_{0.98-x}Ba_xEu_{0.02})Si_2O_2N_2$, all the crystals were orthorhombic (Pbcn) at $x = 0.78$–0.98. The Vegard type dependence was difficult to observe on Ba content, indicating phase II could be solid-solution with less structure differences. Phase II series indicated substantial alterations in position and the relative intensity of XRD results compared with phase I and phase III, which could be described the structural differences.

Lee et al. [15] studied $Ba(La_{0.94}Ce_{0.06})_2Si_2S_8$ and $BaLa_2Si_2S_8$ phosphors for structural determination; the SXRD spectra showed converged weight profile of R_{wp} as 8.77% and 9.18%, respectively. Results revealed the presence of single phase without the identified diffraction peaks by impurity. Microcrystals were identified hexagonal crystallization belonging to the space group $R\bar{3}c$ (No. 167) with $Z = 6$. Among unit cell and the crystal lattice of $BaLa_2Si_2S_8$, there was 18e single site occupied randomly by the mixture of La and Ba atoms, replacing Si atoms, and sites of chalcogen atoms. The S and Si sites were completely occupied. Compared with $BaLa_2Si_2S_8$, the bond distances and the parameters of $Ba(La_{0.94}Ce_{0.06})_2Si_2S_8$ were noticed decreasing by the increase in doping ions (Ce^{3+}) concentration. Similarly, Lin et al. [23] studied the atomic position, phase purity, and the composition of as-prepared $NaSr_{0.99}BO_3:Ce_{0.01}$ by refinement analysis. The Rietveld refinement results of $NaSr_{0.99}BO_3:Ce_{0.01}$ shows that microcrystals with dopant were crystallized as monoclinic structures belonging to $P2_1/c$ and the lattice constraints, $a = 5.32839(9)$, $b = 9.27904(15)$, and $c = 6.07417(14)$ Å. Huang et al. [24] studied calculated and observed SXRD profiles and their variances based on Rietveld refinement results of $Ca_3Si_2O_7:0.015Eu^{2+}$ phosphors with $\lambda = 0.774908$ Å, at ambient temperature. The outcomes showed that dopant (Eu^{2+}) ions in $Ca_3Si_2O_7:0.015Eu^{2+}$ phosphors neither induced structural variations nor generated impurities. The phosphors ($Ca_3Si_2O_7:0.015Eu^{2+}$) were crystallized in monoclinic units belonging to $P12_1/a1$ (14) with $a = 10.5775(4)$, $b = 8.90263(31)$, $c = 7.87443(27)$ Å, lattice constants where $\beta = 119.5942(16)°$ and $\alpha = \gamma = 90°$ and cell volume $V = 644.78(4)$ Å^3. All thermal vibrations, fraction factors, and atomic position parameters were converged and refined. The R-factors were $R_{wp} = 10.38\%$, $R_p = 6.09\%$, and $\chi^2 = 2.608$. In another study, Huang et al. [25] analyzed the differences between calculated and measured SXRD profiles of Rietveld refinement of $Sr_{0.95}Eu_{0.05}CaSiAl_2O_7$ phosphors, with $\lambda = 0.774908$ Å, at ambient temperature. The refinement results indicated no presence of impurity by dopant (Eu^{2+}) 0.05-mol ions in host structure. The $Sr_{0.95}Eu_{0.05}CaSiAl_2O_7:0.05Eu^{2+}$ microcrystals crystallized in tetragonal symmetry belong to space group $P42_1m$ (14). The lattice parameters $a = b = 7.74465(12)$, $c = 5.15752(14)$ Å, where $a = \beta = \gamma = 90°$, $Z = 2$, and $V = 309.346(11)$ Å^3. The R-factors were observed as $R_{wp} = 10.67\%$, $R_p = 7.05\%$, and $\chi^2 = 2.637$. Huang et al. [26] studied the microstructural effects of $Sr_{1-x}Y_{0.98+x}Ce_{0.02}Si_4N_{7-x}C_x$ phosphors using SXRD and EXAFS spectroscopic investigations. Rietveld refinement showed that $x = 1$ and $x = 0$ samples contain monoclinic and hexagonal types, respectively. However, the $x = 0.2$–0.6 samples showed the existence of both 2461 and 1147 types. No substantial coexistence of ions was found for $x = 0.8$, where 1147-types fraction was refined <1%. The results indicated the transformation of structure from hexagonal to monoclinic symmetry started from $x = 0.1$, proceeding through a wide rage that ended at about $x = 0.7$. The monoclinic-type exhibited doubled cell volume of hexagonal type. The smooth evolution of each phase was observed, such as by increasing x, the lattice volume was decreased, as the replacement of Sr^{2+} by small Y^{3+} was a dominant effect. Results indicated the segregation of cations between the components of microcrystals.

Huang et al. [16] studied the calculated and observed SXRD profiles and their differences by Rietveld refinement of $(Ca_{0.89}Eu_{0.01}Mn_{0.1})_4Si_2O_7F_2$ and $(Ca_{0.99}Eu_{0.01})_4Si_2O_7F_2$ phosphors with

$\lambda = 0.774908$ Å. The results indicated Eu^{2+} or Eu^{2+}/Mn^{2+} dopant ions in host ($Ca_4Si_2O_7F_2$) matrix, remain single-phased. $Ca_4Si_2O_7F_2$:Eu^{2+}, Mn^{2+} microcrystals were crystallized into monoclinic structures belonging to $P2_1/c$ (14) space group. The lattice refinement parameters were $a = 7.5624(1)$, $b = 10.5722(2)$, $c = 10.9451(2)$ Å, and $\beta = 109.5984(11)°$ and $V = 824.37(9)$ Å^3. The R-factors were determined to be $R_{wp} = 7.49\%$, $R_p = 5.23\%$, and $\chi^2 = 1.96$. Among as-prepared phosphors, Ca^{2+} ions were replaced by Mn^{2+} ions and lattice parameters were $a = 7.5463(2)$, $b = 10.5532(4)$, $c = 10.9381(4)$ Å, $\beta = 109.5782(20)°$, and $V = 820.72(5)$ Å^3, while R-factors become $R_{wp} = 10.30\%$, $R_p = 7.05\%$, and $\chi^2 = 3.09$. Microcrystals were doped by high content of Mn^{2+} (above 5 mol%); minor impurities were noted. In another study Huang et al. [27] performed the alteration of $Sr_8ZnSc(PO_4)_7$:$0.05Eu^{2+}$ and $Sr_8ZnSc(PO_4)_7$ phosphor structure keeping $\lambda = 0.7749$ Å and 298K. Refinement results indicated no impurity or secondary phase formation in host lattice of undoped and doped microcrystals. The $Sr_8ZnSc(PO_4)_7$:$0.05Eu^{2+}$ crystals were crystallized in monoclinic cell structure belonging to $I2/a$ space group. For undoped $Sr_8ZnSc(PO_4)_7$ microcrystals, with $a = 18.0681(9)$, $b = 10.6813(5)$, $c = 18.4116(9)$ Å, and $V = 2600.74(23)$ Å^3, R-factors were noticed to be $R_{wp} = 13.05\%$, $R_p = 9.06\%$, and $\chi^2 = 5.31$. In the host lattice, the Sr^{2+} ions were replaced by smaller Eu^{2+} ions, the lattice parameters of microcrystals after dopant addition $a = 18.0597(10)$, $b = 10.6744(6)$, $c = 18.4036(9)$ Å, and $V = 3510.8$ Å^3, and the R-factors became $R_{wp} = 13.79\%$, $R_p = 9.49\%$, and $\chi^2 = 5.81$. Wang et al. [28] studied $CaAlSiN_3$:Eu micromaterial by SXRD Rietveld refinement, also of Li and La series at $x = 0.5$ (Figure 15.6). Ionic radii of parental $CaAlSiN_3$:Eu micromaterial [$^{[6]}r(Ca^{2+}) + ^{[4]}r(Al^{3+}) + ^{[4]}r(Si^{4+}) = 1.00 + 0.39 + 0.26 = 1.65$ Å] was less compared with La-lattice [$^{[6]}r(La^{3+}) + ^{[4]}r(Al^{3+}) + ^{[4]}r(Al^{3+}) = 1.032 + 0.39 + 0.39 = 1.812$ Å]; however, it was bigger compared with Li-lattice [$^{[6]}r(Li^+) + ^{[4]}r(Si^{4+}) + ^{[4]}r(Si^{4+}) = 0.76 +$

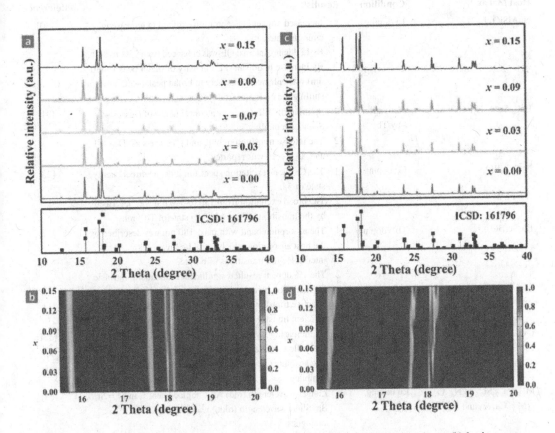

FIGURE 15.6 (a) SXRD patterns of $(Ca_{1-x}La_x)(Al_{1+x}Si_{1-x})N_3$:Eu ($x = 0–0.15$) phosphors; (b) lattice contraction and expansion in La series; (c) SXRD patterns of $(Ca_{1-x}Li_x)(Al_{1-x}Si_{1+x})N_3$:Eu ($x = 0–0.15$) phosphors; and (d) lattice contraction and expansion in Li series. Reprinted from Wang et al., [28] with permission.

$0.26 + 0.26 = 1.28$ Å] [29]. Through, lattice expansion with x in La material could be described by the cationic replacement among phosphors. The contraction in Li-material series could be seen. In both series, variations in the sum of ion radius was closely in agreement with the relative charge in volume. The overall results indicated that Li^+/Si^{4+} and La^{3+}/Al^{3+} have been successfully introduced into the crystal lattice.

15.3 PHOTOLUMINESCENCE (PL)

PL is stated to the emission of light after the irradiation of materials by the light from another source. Few related examples have been discussed below in Table 15.2. Zhao et al. [30] studied the Li-doping effect into $NaAlSiO_4$:Eu system and reported that the substitution at Al site besides Na, in the corresponding ($NaAlSiO_4$:Eu) system, leads toward enhancement in emission efficiency/intensity, converting Eu^{3+} to Eu^{2+} ions. PL spectrum of N$\square$ASO:yLi, Eu ($y = 0$, where $\square$ represents Na vacancy, V_{Na}) phosphors, exhibited a broadband with relatively minor peak at ~450 nm and dominant peak at ~535 nm. The consistent PL excitation (PLE) results were recorded at 535 nm, which exhibited

TABLE 15.2
Few Recent Studies Describing PL Characteristics of Microcrystal Phosphors with Change Conditions and Key Results Exhibited by PL Spectra

Host Matrix	Change Condition	Results	Reference
$NaAlSiO_4$:Eu	Li doping	1. Increased dopant concentration resulted in increase in emission intensity. 2. By Li doping, internal QE was enhanced from 75% to 86%. 3. By increase in dopant, PL intensity of minor peak (~450 nm) was enhanced relatively to broad peak (~535 nm), shifting the emission from yellow to green.	[30]
$Ba_{2.89}Si_6O_{12}N_2$	Activation by Tb	1. Splitting of ^{2S+1}LJ state as a consequence of Russell–Saunders coupling. 2. The intense transition in blue and green sets as $^5D_3 \rightarrow {}^7F_6$ and $^5D_4 \rightarrow {}^7F_5$ was reported.	[31]
$NaSr_{0.99}BO_3$	Ce doping	1. The Ce^{3+} ions excitation spectrum indicate direct energy state of 5d. 2. The broad excitation from 180 to 300 nm was described by the transition ($4f^8 \rightarrow h4f^75d^1$) state of Tb^{3+} ions.	[23]
$Ba_3Sc(BO_3)_3$	Tb^{3+} doping	1. The absorption band with max. 180 nm was described by the host absorption which could be linked with intermolecular transitions of borate. 2. The 1% dopant resulted significant reduction in intrinsic UV emission by enhancing Tb^{3+} emission, indicating effective transfer of energy from STE to Tb^{3+}. 3. Efficient transfer of energy from host to dopant (Tb^{3+}).	[32]
$(Na_{1-x}Ca_x)(Sc_{1-x}Mg_x)Si_2O_6$	Eu doping	1. By increasing x values, transfer in the emission spectra i.e., yellow, while, and blue can be observed. 2. The absolute QE (12.2%) by these phosphors was exhibited.	[33]
$M_{1.95}Eu_{0.05}Si_{5-x}Al_xN_{8-x}O_x$ (M = Ca, Sr, and Ba)	Eu doping	1. Longer wavelength (red) AlO^+ replacement x, in M = Sr, Ba. Short wavelength (blue) was observed in Ca analogues. 2. Results indicated Eu dopant was slightly smaller than that of Sr^{2+} ion.	[34]

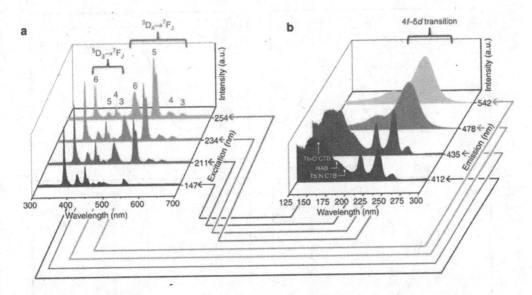

FIGURE 15.7 Synchrotron PL of $Ba_{2.89}Si_6O_{12}N_2:Tb_{0.11}$ microcrystal phosphors. (a) Representation of emission spectra by the excitation of varying wavelength synchrotron radiation, and (b) excitation spectra determined at transition ($^5D_4$ and $^5D_3$) at room temperature. Reprinted from Lin et al. [31] with permission.

a broadband with dual peaks at 275, and 325 nm, due to the transition ($4f \rightarrow 5d$) state of Europium (Eu^{2+}). By increase in the concentration of Li, the PL emission intensity was improved. Therefore, by Li-incorporation, the internal quantum efficiency (QE) was enhanced from 75% to 86%. The broadband with relatively minor peak at ~450 nm and a dominant peak at ~535 nm can be described with Eu^{2+} locations at Na1 and Na2. By enhancing the dopant content, the PL intensity of minor peak (~450 nm) was enhanced relatively to broad peak (~535 nm), shifting the emission (yellow to green) [30]. Lin et al. [31] studied $Ba_{2.89}Si_6O_{12}N_2:Tb_{0.11}$ phosphors microcrystal by PL spectroscopy by exiting the phosphors under specified synchrotron radiation $\lambda = 147, 211, 234$, and 254 according to excitation peak in PLE (Figure 15.7(a)). Fine structural characteristics in PL spectra were caused by splitting of ^{2S+1}LJ state as a consequence of Russell–Saunders coupling. Each precise transition $^5D_{J=3, 4}$ to $^7F_{J=3-6}$ was labeled in inset. The intense most transition in blue and green sets are $^5D_3 \rightarrow ^7F_6$ and $^5D_4 \rightarrow ^7F_5$. PLE spectral investigations were performed at 412 nm ($^5D_3 \rightarrow ^7F_5$), 435 nm ($^5D_3 \rightarrow ^7F_4$), 478 nm ($^5D_4 \rightarrow ^7F_6$), and 542 nm ($^5D_4 \rightarrow ^7F_5$) wavelength as given in Figure 15.7(b).

Lin et al. [23] studied $NaSr_{0.99}BO_3:Ce_{0.01}$ microcrystals by ultraviolet (UV)/Vis PLE and PL analysis as shown in Figure 15.8. The Ce^{3+} ions excitation spectrum indicate direct energy state of 5d, which could be applied to study transition (f–d) state in similar host lattice due to the resemblance of lattice influences on 5d state. Moreover, $[Xe]4f^n–5d^1$ configuration can be studied using empirical equations:

$$E_{abs}(n, Q, A) = E_{Afree}(n, Q) - D(Q, A)$$

$$E_{em}(n, Q, A) = E_{Afree}(n, Q) - D(Q, A) - \Delta S(Q, A)$$

which is the difference in energy between Ce^{3+} (5d) ions and $NaSrBO_3$ compounds (10.745×10^3 cm^{-1}). This semiquantitative centroid shift could be determined based on average electronegativity value of matrix using equation:

$$\varepsilon_c = 1.79 \times 10^{13} \sum_{i=1}^{N} \frac{a_{sp}^i}{(R_i - 0.6\Delta R)^6}$$

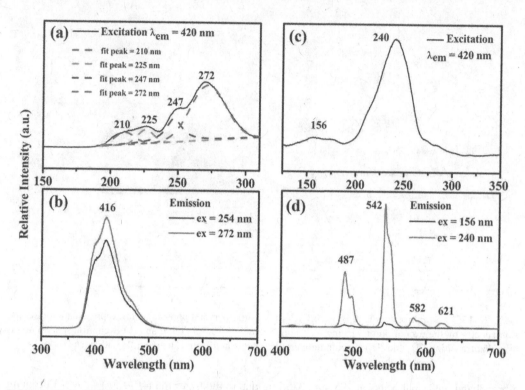

FIGURE 15.8 VUV-PL and PLE spectra of $NaSr_{0.99}BO_3:Ce_{0.01}^{3+}$ at (a) PLE determined at 420 nm; (b) PL excitation at 272 and 254 nm; and that of $NaSr_{0.93}BO_3:Tb_{0.07}^{3+}$ phosphors at (c) PLE determined at 542 nm; and (d) PL excited at 240 and 156 nm. Reprinted from Lin et al. [23] with permission.

where a_{sp}^i ($a_{sp}^o = 0.33 + 4.8/\chi_{av}^2$) shows polarization of ionic ligand, χ_{av} is the average weight electronegativity of positive ions in oxide, and 'R_i' is distance between anionic ligand and Ce^{3+} ions, ΔR shows the difference in ionic radii of substituted cations and Ce^{3+} ions, i.e., $\chi_{NaSrBO3} = (1\chi_{Na} + 2\chi_{Sr} + 3\chi_B)/6 = 1.49$, $\alpha_{sp}^o = 2.49$, $\Delta R = 11.4$ pm, and $R_i = 257.9$–283.4 pm. The ε_c value of few borate compounds have been showed between 8.730×10^3 and $12.730 \times 10_3$ cm^{-1}, as reported by previous studies, which indicate that the calculated findings were in agreement to that of experimental findings. The broad excitation could be observed (Figure 15.8(c)) from 180 to 300 nm which could be described by transition ($4f^8 \rightarrow 4f^75d^1$) state of Tb^{3+} ions.

Wang et al. [32] studied $Ba_3Sc(BO_3)_3:Tb^{3+}$ phosphors for the interpretation of luminescence spectra using PL and PLE spectral investigations. The PL/PLE spectra recorded by undoped $Ba_3Sc(BO_3)_3$ phosphors at room temperature have been indicated in Figure 15.9. At 172 nm excitation, the $Ba_3Sc(BO_3)_3$ emit intrinsic broad-spectrum UV luminescence band with highest emission at 336 nm. The emission may have resulted from self-trapped excitation, which could have been associated with molecular transitions or bandgap excitations with BO_3^{3-}. In PLE spectrum of $Ba_3Sc(BO_3)_3$, the broad spectrum peaking at 180 nm was analyzed by determination of intrinsic UV emission. The intrinsic matrix emission was extended to $\lambda = 450$ nm, which overlapped partially (4f–4f) Tb^{3+} such as the transition state $^7F_6 \rightarrow {}^5D_3$. Tb^{3+} ions introduces among host $[Ba_3Sc(BO_3)_3]$ lattice as a result of spectral overlap; the efficient transfer of energy from host to Tb^{3+} could be expected. The addition of dopant (Tb^{3+}, 1%) ions resulted in significant reduction in intrinsic UV emission and escorted increased Tb^{3+} emission, indicating effective transfer of energy from STE to Tb^{3+} (Figure 15.9). The PL of $Ba_3Sc(BO_3)_3$ comprising 1%Tb^{3+} dopant content could be divided into two regions, a green-emission region resulted by dopant (Tb^{3+}) transition $^5D_4-^7F_j$ (where, $j = 3$–6) and an UV emission region denoted by the emission overlap by the host and dopant (Tb^{3+}) transition $^5D_3-^7F_6$. The weak transition ($^5D_3-^7F_6$) of Tb^{3+} doping ions could be observed in PL spectra around

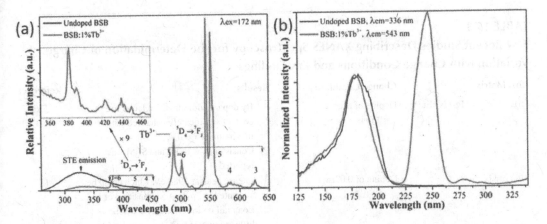

FIGURE 15.9 (a) PL of undoped $Ba_3Sc(BO_3)_3$ and doped $Ba_3Sc(BO_3)_3$:1%Tb^{3+} phosphors under 172 nm excitation. Enlarged portion is PL spectrum of $Ba_3Sc(BO_3)_3$:1%Tb^{3+} phosphors at 350–470 nm (inside), (b) PLE of $Ba_3Sc(BO_3)_3$:1%Tb^{3+} phosphors determining emission at 543 nm and that of undoped $Ba_3Sc(BO_3)_3$ at 336 nm. Reprinted from Wang et al. [32] with permission.

380 nm in $Ba_3Sc(BO_3)_3$:1%Tb^{3+}, indicating host energy transfer to dopant (Tb^{3+}) involving $^5D_3$ state. PLE spectra of both undoped and doped samples were in agreement showing broad absorption band below 210 nm, indicating similarity in origin i.e., host absorption. The obtained results indicated the transfer of energy from host to dopant (Tb^{3+}). Moreover, the PLE spectrum of $Ba_3Sc(BO_3)_3$:1%Tb^{3+} indicated strong excitation around 220–270 nm (max. around 246 nm) and could be designated to spin-allowed transition ($4f^8–4f^75d^1$) of dopant. Relatively weak (around 278 nm) excitation band could be assigned to spin-forbidden transition ($4f^8–4f^75d^1$) of dopant [32].

Xia et al. [33] studied $(Na_{1-x}Ca_x)(Sc_{1-x}Mg_x)Si_2O_6$:0.03$Eu^{2+}$ microcrystals. Among microcrystals, the dopant concentration was fixed at optimum value (0.03), and the 'x' was varied over range 0–1. At $x = 0$, microcrystals exhibited yellow emission band from 410 to 700 nm, peaking at 532 nm. By increasing $CaMgSi_2O_6$ content, blue emission band appeared, which exhibited intense emission by increasing 'x' value. The phosphors exhibited tunable emission with UV excitation from yellow to blue. By increasing x values, transfer in the emission spectra i.e., yellow, while, and blue can be observed. The PLE spectra of $(Na_{0.7}Ca_{0.3})(Sc_{0.7}Mg_{0.3})Si_2O_6$:0.03$Eu^{2+}$ showed broad excitation bands by two different emission centers i.e., 449 nm (yellow emission) and 532 nm (blue emission) which could be described by the transition Eu^{2+} between $4f^7$ and $4f^65d$. The absolute QE (12.2%) by these phosphors was shown by the study. It was reported that as-prepared solid solution showed low luminescence efficiency, which could be tuned by controlling the crystalline defects, morphologies, size distribution, and particle size of phosphors. Chen et al. [34] studied luminescence properties of $M_{1.95}Eu_{0.05}Si_{5-x}Al_xN_{8-x}O_x$ phosphors. The local coordination effects and irregular sequence reflected alterations in crystal structure types in Ca and Sr. The PL shifted to longer red (longer wavelength) by enhancing AlO^+ replacement x, in M = Sr, Ba, while short wavelength (blue emission) was determined in calcium equivalents. The switching and variation of energy shift was not observed in other microcrystal types. The shift in energy for phosphor samples showed size-mismatch. The host M = Ca, Sr, and Ba showed significant size distribution 1.12, 1.26, and 1.42 Å, respectively. However, for Ca^{2+} smaller ions, this switched toward blue shifting.

15.4 X-RAY ABSORPTION NEAR-EDGE STRUCTURE (XANES) SPECTROSCOPY

XANES spectroscopy enables the determination of local symmetry of elements and chemical analysis and is frequently being used for the characterization of phosphor microcrystals. Few recent studies describing XANES spectroscopy for the determination of charge variation with change conditions and key findings have been summarized in Table 15.3 and discussed next. For instance,

TABLE 15.3

Few Recent Studies Describing XANES Spectroscopy for the Determination of Charge Variation with Change Conditions and Key Findings

Host Matrix	Change Condition	Results	Reference
$Sr(LiAl_3)_{1-x}(SiMg_3)_xN_4:Eu^{2+}$	Doping of Mg^{2+} and Si^{4+} ions	1. By dopant addition, the relative Eu^{2+} content significantly enhanced with maximum at $x = 0.4$. 2. Consistency with PL and SEM was reported.	[36]
$Rb_3YSi_2O_7$	Doping of 0.02Eu	1. Intense peak at 6982 eV, which could be attributed to $2p_{3/2} \rightarrow 5d$ transition of Eu^{3+}. 2. Long tail on lower energy state shows little amount of Eu^{2+}.	[37]
N7ng tyLi, Eu ($y = 0$)	Li doping	1. Results indicate optimistic influence of Li-incorporation in enhancing Eu^{2+} compared with Eu^{3+} ions. 2. Valence electrons at V_{Na1} position transfer to Eu^{3+} to convert them to Eu^{2+}, enhancing I_{Eu2+}/I_{Eu3+} ratio.	[30]
$Cs_{0.96}MgPO_4:0.04Eu^{2+}$	Cationic co-substitution i.e., $[Eu^{2+}–Si^{4+}]$ to $[Cs^{+1}–P^{5+}]$	1. By cationic substitution i.e., $[Eu^{2+}–Si^{4+}]$ to $[Cs^{+1}–P^{5+}]$, the Eu^{3+} ions significantly reduced in the host. 2. The intensity of emission by nanocrystals was much improved in $Cs_{0.96}MgP_{0.96}O_4:0.04[Eu^{2+}–Si^{4+}]$ phosphors.	[4]
Mg_2TiO_4	Nb doping	1. The Nb-absorption edge near to Nb_2O_5 (Nb^{5+}) rather NbO_2 (Nb^{4+}) indicated the pentavalent (Nb), shifting the Fermi toward conduction band, and giving additional electrons for donor–acceptor recombination.	[38]
$SrLiAl_3N_4:Eu^{2+}$	Increased applied pressure, which could be explained by charge compensation of Eu and Li ions Eu^{3+} ions incorporate into Sr^{2+} sites, closer to Li^+ to maintain local coordination	1. The presence of Eu^{3+} was much higher than Eu^{2+} ions, however, no luminescence was noticed to be produced by Eu^{3+} ions. 2. Slightly increased in Eu^{2+} at higher pressure contribute to final luminescence yield.	[39]
$Sr_4Al_{14}O_{25}$	Mn doping	1. Coexistence of Mn^{4+}, Mn^{2+}, and Mn^{3+} ions among $Sr_4Al_{14}O_{25}:Mn$ nanocrystals was observed.	[40]
CLPSO-x ($x = 0, 1, 2, 4, 6$)	Eu doping	1. Almost all Eu existed in Eu^{2+} form except minute amount of Eu^{3+} ions in CLSPO-0. 2. The cationic co-substituent indicated systematically increased intensity of Eu^{2+}, by decreasing Eu^{3+} ions. 3. Co-substitution could result adjustment and coexistence of Eu^{2+}/Eu^{3+} in apatite structure.	[41]

(Continued)

TABLE 15.3 *(Continued)*

Few Recent Studies Describing XANES Spectroscopy for the Determination of Charge Variation with Change Conditions and Key Findings

Host Matrix	Change Condition	Results	Reference
Mo_2C	P doping	1. The pre-edge broad feature suggested the development of Mo-P species. 2. Co-doping of P and N into C-matrix was observed.	[42]
$Sr[Mg_2Al_2N_4]$	Ce doping	1. Coexistence of Ce^{3+} and Ce^{4+} was observed, with high Ce^{3+}/Ce^{4+} at lower x values.	[43]
$(Sr,Ba)_5(PO_4)_3Cl$	Eu doping	1. The addition of dopant (Ba^{2+}) slightly shifted the signal toward Eu^{3+} by reducing of Eu^{2+}. 2. When Ba^{2+} was doped into Sr(I), the forceful shrinkage on Eu^{2+} was resulted and, therefore, oxidized to smaller Eu^{3+} ions.	[8]
CLSPO-1250/1350	Cationic co-substitution of $[Ca^{2+}-P^{5+}]$ for $[La^{3+}-Si^{4+}]$	1. Electric-dipole transition $^5D_0-^7F_2$ resulted in the highest emission at 613 nm, higher than magnetic-dipole transition. 2. Cationic co-substitution indicated the amount of Eu^{2+}/Eu^{3+} increased.	[44]
$Sr_xCa_{0.993-x}AlSiN_3$	Eu doping	1. Coexistence of Eu^{3+} and Eu^{2+} in nanocrystals. 2. Increase in dopant (Sr^{2+}) concentration resulted in the increased intensity of Eu^{2+} ions.	[45]

the influence of graphite on effective phosphorus usage and the influence of polyaniline on the improvement of stability were investigated by Jin et al. [35], using X-ray absorption spectroscopic (XAS) analysis. The XAS spectra were recorded to determine the state of P in black phosphorus (BP) and BP–graphite (BP–G). The absorption-edge of BP–G was higher (2145.7 eV) than that of plain BP (2145.5 eV), which can be attributed to the strong transfer of charge from P to C, due to high electronic affinity of C (1.262 eV) compared with P (0.746 eV). Moreover, *ex situ* XAS analysis of as-prepared samples was performed at different charge states. After sodiation, the P K-edge of BP–G was transferred upward (2146.0 eV), exhibiting a pre-edge, which indicates the synthesis of Na_xP ($1 < x < 3$). The P K-edge XAS spectrum of BP indicates no modification, which shows no alteration in electronic density of BP during the discharging process. Remarkably, during the recharging of BP–G back to 2.0 V, the shifting of P K-edge back to previous stage was observed. However, significant change was observed by BP–G after discharging step because of the existence of graphite, enabling excessive use of BP during de-sodiation/sodiation mechanism [35].

Fang and coworkers in 2019, studied the development of $Sr(LiAl_3)_{1-x}(SiMg_3)_xN_4:Eu^{2+}$ microcrystals *via* solid-solution method. The XANES spectra of phosphors were recorded to analyze the relative Eu^{2+}/Eu^{3+} ratio in newly designed samples. Mostly, among oxynitride or nitride phosphors, the Eu^{2+} concentration was much higher than Eu^{3+} ions content (Table 15.3). The XANES spectra of $Sr(LiAl_3)_{1-x}(SiMg_3)_xN_4:Eu^{2+}$ showed much lower content of Eu^{2+} ions; however, by incorporation of dopant (Mg^{2+} and Si^{4+}) ions, the relative Eu^{2+} content was significantly enhanced and reached the highest at $x = 0.4$. This study demonstrated such higher content of Eu^{2+} in UCr_4C_4 families for

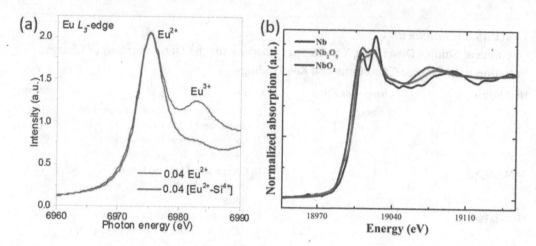

FIGURE 15.10   (a) XANES spectra of $Cs_{0.96}MgPO_4{:}0.04Eu^{2+}$ and $Cs_{0.96}MgP_{0.96}O_4{:}0.04[Eu^{2+}{-}Si^{4+}]$. Reprinted from Wei et al. [4]; and (b) XANES spectra of MTO:Mn,$Nb_{0.5}$ (black), NbO_2 (blue), and Nb_2O_5 (red) powders. Reprinted from Huang et al. [38] with permission.

the first time. In addition, the consistency of XANES spectrum with PL and SEM spectra has been reported [36]. Similar to that of Eu_2O_3 peak, $Rb_3YSi_2O_7{:}0.02Eu$ phosphors showed an intense peak at 6982 eV, which could be attributed to $2p_{3/2} \rightarrow 5d$ transition of Eu^{3+}. In addition, the indication of long tail on lower energy state shows the existence of little Eu^{2+} content in $Rb_3YSi_2O_7{:}0.02Eu$ phosphors [37]. Similarly, XANES spectra of N☐ASO:yLi,Eu ($y = 0$, where ☐ represents Na vacancy, V_{Na}) phosphors indicated two peaks at ~6975 and ~6982 eV, which could be explained by the transition ($2p_{3/2} \rightarrow 5d$) state of Eu^{2+} and Eu^{3+}, and their increased and decreased strengths, respectively, by enhancing Li concentration (Table 15.3). The sum of peaks area is nearly constant; therefore, (I_{Eu2+}/I_{Eu3+}) ratio could be used as a ratio of Eu^{2+} and Eu^{3+} ions. The results indicate optimistic influence of Li-incorporation in enhancing Eu^{2+} as compared with Eu^{3+} ions [30].

Li, Pei, & He et al. [46] investigated the synergistic effect of BP nanosheets and PtRu nanoclusters by XANES measurements. The shifting of Ru-absorption edge to high energy ~22,120 eV exhibits synergistic effect of BP and PtRu nanoclusters and slight transfer of electrons from Ru to BP, though the PtRu NC/BP peak at ~11,568 eV seems analogous to PtRu nanoclusters. In $CsMgPO_4$, the valence of Eu was studied by Eu L_3-edge XANES spectroscopy (Figure 15.10(a)). Two evident peaks at 6983.2 and 6975.2 eV in $Cs_{0.96}MgPO_4{:}0.04Eu^{2+}$ structure could be designated to $2p_{3/2} \rightarrow 5d$ transition of Eu^{3+}/Eu^{2+}. Though the synthesis of $Cs_{0.96}MgPO_4{:}0.04Eu^{2+}$ was carried out under reducing environment, and the major emission was caused by Eu^{2+} ions, still there were many Eu^{3+} ions present. This could also be attributed to the low efficacy of red emission by Eu^{2+}. By cationic substitution i.e., $[Eu^{2+}{-}Si^{4+}]$ to $[Cs^{1+}{-}P^{5+}]$, the Eu^{3+} significantly reduced. Therefore, the emission intensities by microcrystals were much improved in $Cs_{0.96}MgP_{0.96}O_4{:}0.04[Eu^{2+}{-}Si^{4+}]$. The charge compensation *via* co-substitution demonstrated by Wei et al. [4] provided an efficient system that could significantly enhance the luminescence. Huang et al. [38] determined the atomistic site of niobium (Nb) in Mg_2TiO_4 (MTO) matrix and its valence state using XAS spectroscopy. The Nb K-edge XANES spectra exhibited the Nb-absorption edge near to Nb_2O_5 (Nb^{5+}) rather NbO_2 (Nb^{4+}), suggesting the pentavalent state of Nb, shifting the Fermi toward the conduction band and giving additional electrons for donor–acceptor recombination (Figure 15.10(b)). Ce-$L3$ edge XANES analysis of $Sr[LiAl_3N_4]$ was conducted, which represented Ce^{4+} and Ce^{3+} co-existed. The existence of Ce^{4+} can be stabilized by the discrepancies. Therefore, compensation of charges defects became crucial because of dual ionic existence of Ce i.e., Ce^{4+} and Ce^{3+}. In $Sr[LiAl_3N_4]$ matrix, compensation of Eu^{2+} vacancies were performed by Sr^{2+}. However, Li or Li^+ vacancy in Sr^{2+} produced compensation flaws in $Sr[LiAl_3N_4]$. The similar flaws were needed to compensate Ce^{3+} in Sr positions [6]. The Eu L_3-edge XANES spectra of $Sr[Mg_2Al_2N_4]$ microcrystals, doped with Eu^{2+}, were recorded, which exhibited the oxidation state of Eu. Eu_2O_3 and $BaMgAl_{10}O_{17}:Eu^{2+}$ were selected as standard for Eu^{3+} and Eu^{2+} determination. Two peaks at 6984 and 6974 eV were recorded representing the relative concentrations of Eu^{3+} and Eu^{2+} ions, respectively.

Fang et al. [39] studied the formation of $SrLiAl_3N_4:Eu^{2+}$ phosphors. The oxidation of Eu was determined by recording XANES spectra, which showed two absorption peaks at 6.984 and 6.974 eV, indicating $2p_{3/2} \rightarrow 5d$ transition of Eu^{3+} and Eu^{2+}, respectively (Figure 15.11). The presence of Eu^{3+} was noticed much higher than Eu^{2+} ions; however, no luminescence was noticed to be produced by Eu^{3+} ions. The shift in oxidation state to Eu^{2+} could be observed by XANES spectrum by enhancing the applied pressure, which could be explained by charge compensation perspective of Eu and Li ions. By the evaporation of one Li^+ ion, one positive ion must be incorporated at that site to manage the electronic neutrality. Compared with Eu^{2+}, oxidation of Li^+ and Sr^{2+} is tougher due to higher ionization energies. Thus, the oxidation of Eu^{2+} to Eu^{3+} takes place, resulting in less QE and luminescence intensity. Moreover, Eu^{3+} ions preferably incorporate into the Sr^{2+} sites, closer to Li^+ defects to maintain local coordination and electronically neutral environment. A slight increase in Eu^{2+} concentration in samples formed at higher pressure contribute to final luminescence intensity, although Eu^{2+} content is relatively less in XANES spectrum, as shown in Figure 15.11(b).

Li, Cui et al. [47] studied the phosphorus extraction from wastewater by co-fermentation and Fe-dosing fermenter design. The samples of activated sludge were collected from aerobic membrane

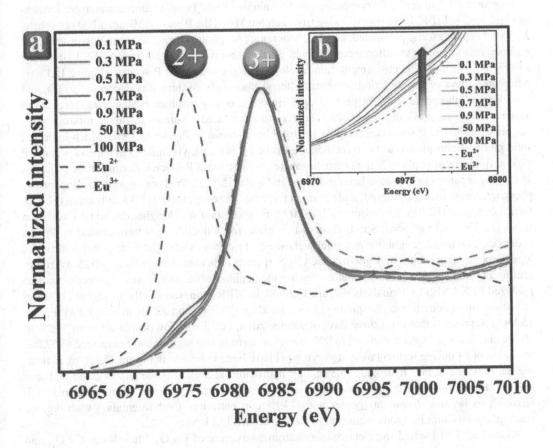

FIGURE 15.11 Eu L_3-edge XANES spectra of $SrLiAl_3N_4:Eu^{2+}$ phosphors prepared under varying pressure; (b) XANES Eu^{2+} spectral signal. Reprinted from Fang et al. [39] with permission.

bioreactor (MBR) following 90 days of reaction. The sludge was removed and fed into fermentation bioreactors for operation. To determine the influence of different parameters such as pH, incubation time, and organic addition, on the transition of P and Fe throughout the bioprocess, operational conditions was varied, and the sludge specimens were collected at different stages i.e., time interval 0, 24, 48, and 72 hours, and pH of 3, 4, 5 and 6, with/without organic loading. First bioreactor was adjusted with organic waste-loading deprived of pH adjustment; the sludge specimens were obtained after various incubation times from 0 to 72 hours. The 2nd to 5th bioreactors were adjusted with organic loading and pH adjustment from 3 to 6, and specimens were obtained after 72 hours of fermentation. The 6th reactor was kept without organic loading and pH adjustment. Samples were processed by centrifugation at 6000 rpm for 15 minutes, and the pallets were kept at −20°C for 4 hours and lyophilized for 24 hours. The sludge pallets were pressed into 1-to 2-mm thickness and 5-mm diameter. The oxidation state of both P and Fe was determined by Fe K-edge (7112.0 eV) and P K-edge (2145.5 eV) XANES. The X-ray energy calibration was performed by Fe-metal foil for Fe with 7124.6 eV and L_{III}-edge for Nb for P with at 2153.4 eV. The measurement was performed using ionization chamber in fluorescence mode using a Lytle detector under ambient conditions. The P K-edge XANES spectra obtained have been shown in Figure 15.12. The adsorbed P and Ferric phosphate exhibited pre-edge peak around 2150 eV, which could be attributed to the hybridization of P 3p, O 2p, and Fe 3d. In majority of samples pre-edge peaks were obtained, indicating adsorbed P and/or ferric phosphate in the sludge. The control and activated sludge at pH 6 represented three peaks at post-edge region, representing the existence of vivianite. The $FePO_4$ was apparently found as amorphous.

Moreover, Li, Cui et al. [47] comprehensively evaluated P and Fe speciation in anaerobic fermenters and aerobic MBR. In activated sludge from aerobic MBR, the P was 60.5% adsorbed phosphate, 24.3% amorphous ferric phosphate, and 15.2% organic phosphate. During fermentation process, the P transformation and speciation were difficult. P-speciation was analyzed by XANES LCF analysis (Figure 15.13(a)). The co-fermentation resulted from the change in P species (Figure 15.13(a)). After three days of fermentation incubation, ferric phosphate content was decreased to 0% and the adsorbed phosphate from 59.4% to 37.9%. However, orthophosphate content was increased to 49.4%, which was previously just 0.4%, and vivianite was hardly noticed in sludge mixture without pH control. The P species at varying pH have been shown in Figure 15.13(b), which indicated solid phase of phosphorus at pH 6 including adsorbed-P (46.5%), vivianite (39.7%), and organic-P (7.6%). By decreasing pH to 3, the orthophosphates became main P species. Fermentative P species at varying organic loading have been indicated in Figure 15.13(c). Without organic loading, orthophosphates were barely observed, and the main P species were organic-P (15.3%), vivianite (14.6%), ferric phosphate (12.1%), and adsorbed-P (57.0%). Fe speciation was tougher due to two oxidation states i.e., Fe^{2+} and Fe^{3+} ions. Solid phase of Fe^{2+} after iron reduction may have existed in ferrous hydroxides, vivianite, magnetite, and siderite, however, Fe^{3+} may exist in the form of iron hydroxide, lepidocrocite, akageneite, and goethite. XANES spectra indicated first peak at ~2125 eV representing Fe^{2+} ions, and another peak at ron hydroxide, lepidocrocite, $akag^{3+}$ ions. Fe-speciation was analyzed by XANES LCF analysis (Figure 15.13(c)). In MBR, iron was mostly present in Fe^{3+} form including amorphous ferric phosphate (18.1%), goethite (28.7%), and ferrihydrite (52.4%). Figure 15.13(*a) represents that after three days of fermentation, Fe^{3+} ions were decreased from 59.0% to 16.5% and ferric phosphate reduced to 0%. However, ferrous ions significantly enhanced to 57.2%. Variation of Fe during fermentation at varying pH have been indicated in Figure 15.13(*b), indicating the presence of Fe^{2+} hydroxide (6.3%), Fe^{3+} hydroxide (11.8%), goethite (10.6%), and vivianite (54.7%). By decreasing the pH, vivianite was largely dissolved. By further decrease in pH to 3, ferrous ions became dominant species with 77.9% concentration. Co-fermentation with organic loading, the resultant Fe^{2+} ions increased to 69.6% (Figure 15.13(*c)).

Xiao et al. [48] studied the electronic and atomic structure of Co_3O_4, Phosphorus-Co_3O_4, and oxygen vacant-Co_3O_4 *via* synchrotron XANES spectroscopy. The absorption peak at 7730 eV indicates energy position and spectral profile of Co XANES of unmodified, oxygen vacant, and

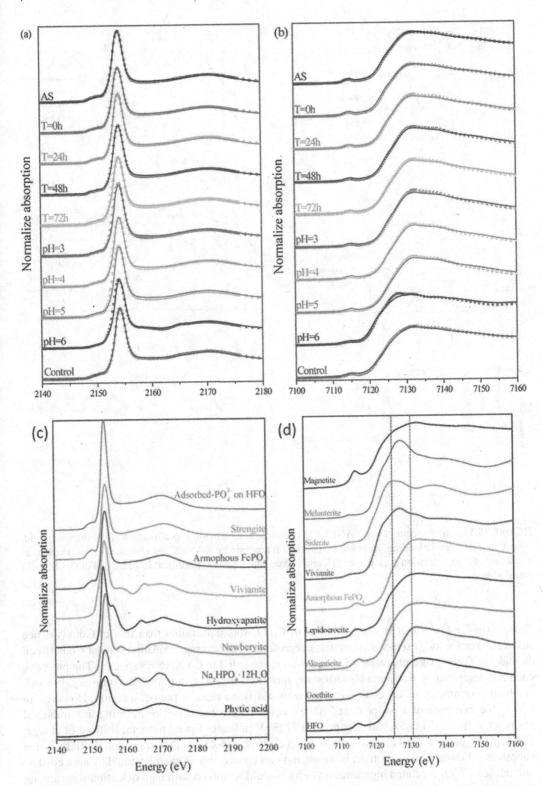

FIGURE 15.12 (a) P, and (b) Fe XANES of specimens indicating spectral data lines from linear combination fitting; (c) and (d) reference spectra for P and Fe XANES, respectively. Reprinted from Li, Cui et al. [47] with permission.

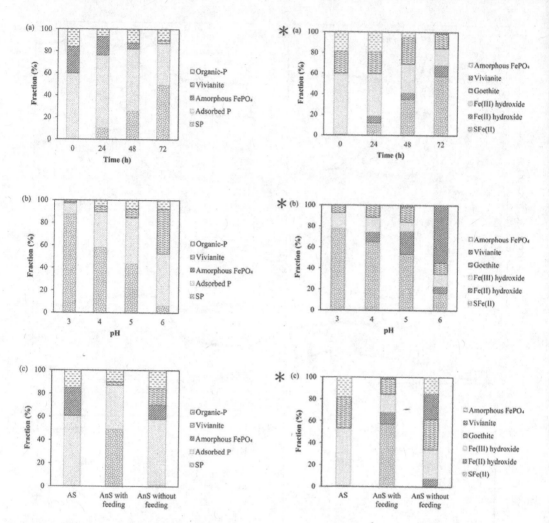

FIGURE 15.13 (a) P speciation by XANES and chemical determination, as a function of incubation period, (b) pH, and (c) organic loading, however, Fe speciation analysis by XANES and chemical determination, as a function of (*a) incubation period, (*b) pH, and (*c) organic loading. Reprinted from Li, Cui et al. [47] with permission.

Phosphorus-Co_3O_4, which is similar to that of Co_3O_4 standard, rather than that of CoO (valence state of 2+) or Co foil (valance state of zero), representing the average oxidation state of Co between 2+ and 3+. Three major characteristics of Co_3O_4 represented in Co XANES spectra. The pre-peak could be described by transition (1s–3d) state, however, indicating tetrahedral symmetry. In CoO, octahedral symmetry of Co could be seen, with nearly no peak at pre-peak region. However, in Co_3O_4, the existence of a pre-peak at 7709 eV could be attributed to Co^{2+} residing in tetrahedral positions of spinel Co_3O_4. A shoulder peak at 7723 eV indicates ligand-to-metal shifting of charge. The primary peak at 7730 eV indicates 1s to 4p electron-dipole–resulted transition. Initially, three nanosheets showed similar profiles; however, indirect charge among samples could be noticed. The unmodified Co_3O_4 exhibited high intensity, which could be linked with high oxidation state among three studied samples. In the case of oxygen vacant-Co_3O_4, shift toward lower energy could be seen as compared to unmodified sample, exhibiting lower average oxidation state of Co. However, after the occupancy of vacancies by P, the shifting of energy back to higher value could be observed, which shows higher oxidation state in Phosphorus-Co_3O_4 compared with oxygen vacant-Co_3O_4. However, the reduced intensity was observed compared to pristine, which could be explained by the fact that P have smaller electronegativity than that of O; therefore, capability of P to attract electrons is seemingly less than that of O. As P residing in O sites, the electrons may transfer to P from Co sites. Hence, the peak intensity in Phosphorus-Co_3O_4 was high compared with oxygen vacant-Co_3O_4, however, still less than that of unmodified Co_3O_4. In right bottom, pre-edge of Phosphorous-Co_3O_4/pristine Co_3O_4 exhibits lowest/highest energy, showing the convenient transfer of electrons to Co 3d orbital of P-Co_3O_4 compared with oxygen vacant-Co_3O_4. Second, the pre-peak region was reduced in oxygen vacant-Co_3O_4, which indicated the entry of charges to Co 3d orbital while oxygen vacancies were generated. In addition to Co^{2+} in tetrahedral, there was another peak at high energy, which originated from octahedral symmetry. The indication of this peak was a result of fact that nanosheets had high Co-atoms at the surface and more reduced states than pristine Co_3O_4. Lower coordination, surface atoms produce distortion in atomic structure and may break octahedral symmetry, which allows the existence of Co^{3+} in pre-peak region. Similarly, Zhang et al. [40] studied the $Sr_4Al_{14}O_{25}$:Mn phosphors for charge varieties of Mn (Figure 15.14). For Mn valence state 2+,

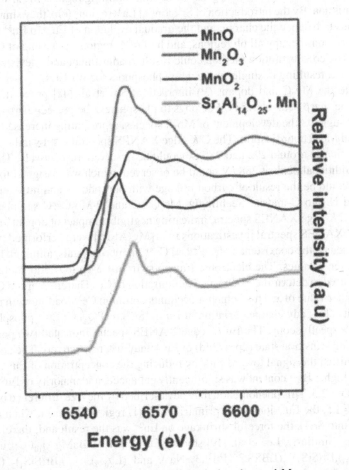

FIGURE 15.14 XANES spectra of the $Sr_4Al_{14}O_{25}$:Mn red phosphor and Mn precursors. Reprinted form Zhang et al. [40] with permission.

3+, and 4+, reference materials MnO, Mn_2O_3, and MnO_2 were used, respectively. The Mn K-edge XANES spectra exhibited two peaks at 6547 and 6539 eV, indicating the presence of Mn^{2+} ions, one peak at 6561 eV indicating the presence of Mn^{4+} ions, and other at 6570 eV indicating the presence of Mn^{3+} ions. Therefore, spectral results indicated the coexistence of Mn^{2+}, Mn^{3+}, and Mn^{4+} among $Sr_4Al_{14}O_{25}$:Mn microcrystals.

Wei et al. [41] studied the valence transformation of Eu^{2+}/Eu^{3+} in CLSPO-x, using Eu L_3-edge XANES spectroscopy, to tune desired single phase white light-emitting diodes (WLEDs). Two peaks were observed at 6985 and 6975 eV, which could be attributed to the $2p_{3/2} \rightarrow 5d$ transition state in Eu^{2+} and Eu^{3+}, respectively. The increased shielding effect is caused by the addition of 4f electron in Eu^{2+} rather than in Eu^{3+}. Almost all Eu exist in the form of Eu^{2+} except a minute amount of Eu^{3+} ions in CLSPO-0. The significantly higher emission of at 613 nm by $^5D_0/^7F_2$ compared with 588 nm emission by $^5D_0/^7F_1$ exhibits that the Eu^{3+} ions primarily enter crystalline inverse symmetry. The substituent $[La^{3+}-Si^{4+}]$ of $[Ca^{2+}-P^{5+}]$ indicated systematically increased intensity of Eu^{2+}; however, for Eu^{3+}, the corresponding intensity was decreased. The results indicated the coexistence of Eu^{2+} and Eu^{3+}, which mutually transform by cationic co-substitution. The possible co-substitution transform mechanism of Eu^{2+} and Eu^{3+} has been indicated clearly. Initially, among the lattices $Ca_8La_2(PO_4)_6O_2$, Eu ions preferably occupy Ca^{2+} ions sites, showing broadband emission by Eu^{2+} ions. Slight sintering temperature cause the microphase mixing of $Ca_2La_8(SiO_4)_6O_2$ and $Ca_8La_2(PO_4)_6O_2$ in domain regions, generating multi-emitting regions in Eu^{2+} ions producing a yellow-green emission. By the introduction of Si^{4+} ions at La^{3+} regions (6h), they simultaneously enter Ca^{2+}/Eu^{2+} regions to balance the charge, and the gradual evolution of Eu^{2+} to Eu^{3+} ions takes place. At beginning, La^{3+} ions occupy all 6h regions, and half of 4f regions at the end of $Ca_2La_8(SiO_4)_6O_2$. In general, cationic co-substitution method could result in adjustment and coexistence of Eu^{2+}/Eu^{3+} in apatite structure, resulting in single-phase white phosphors for WLEDs.

To investigate the Mo_2C and dopant (P) interaction, Shi et al. [42] proceeded with XANES spectroscopy. The sharp peak at 2153 eV belongs to phosphates, the pre-edge broad feature around 2145 eV further suggests the development of Mo-P species, representing increased correspondence toward increased Mo-P coordination. The C K-edge XANES spectral results indicated that C–C σ^* and C=C π^* resonance around 292 and 285 eV could be observed, respectively. The shoulder peak at 284 eV and additional peak at 288 eV could be observed, which was assigned to Mo_2C. Another peak at 287 eV could be the result of carbon linkage with phosphorus and nitrogen, indicating co-doping of P and N into C-matrix. Seemingly, $Mo_2C@C$ and $P-Mo_2C@C$ samples indicate slight modification in C K-edge XANES spectra, indicating negligible impact of dopant on carbon matrix. The Ce $L3$-edge XANES spectral investigations in $Sr[Mg_2Al_2N_4]$ were performed by Leaño Jr et al. [43]. Spectra indicated the coexistence of Ce^{3+} and Ce^{4+} in microcrystals, although higher Ce^{3+}/Ce^{4+} was noted at lower x values. The electronic transition from Ce-2p to 4f5d6s outermost shell has frequently been used to determine electronic configuration of Ce. Therefore, more Ce^{3+} occur while $x = 0.02$ than higher value of x. This behavior accounts optimum Ce^{3+} loading with relatively higher emission intensity. To study charge variation of Eu in $(Sr,Ba)_5(PO_4)_3Cl$:Eu^{2+} phosphors, Fang et al. [8] used XANES spectroscopy. The Eu L_3-edge XANES spectra indicated two peaks at 6984 and 6974 eV, indicating transition state ($2p_{3/2}/2dd$) of Eu^{3+} and Eu^{2+}, respectively. The addition of dopant (Ba^{2+}) slightly shifted the signal toward Eu^{3+} by reducing the concentration of Eu^{2+}. In samples, the Eu^{3+} signal with higher Ba^{2+} content was significantly enhanced, and majority of Eu^{2+} ions were converted to Eu^{3+} as $x = 2.5$. This phenomenon may be described by the size difference between Sr^{2+} and Ba^{2+} ions. At $x = 1.5$, the Eu^{2+} ions were primarily at Sr(I) region. However, when relatively larger Ba^{2+} was doped into Sr(I), the forceful shrinkage on Eu^{2+} was the result and, therefore, oxidized to smaller Eu^{3+} ions. Similarly, Lee et al. [19] studied Eu_2O_3 (Eu^{3+}), $BaMgAl_{10}O_{17}$:Eu^{2+} (BAM:Eu^{2+}), $(La_{0.97}Eu_{0.03}Na_{0.03})_3Br(SiS_4)_2$ (LBSS,3%Eu^{2+},3%Na^+), and $(La_{0.97}Eu_{0.03})_3Br(SiS_4)_2$ (LBSS, 3%Na^+) microcrystals for their charge variation using XANES spectroscopy. Every spectrum showed two absorption peaks at 6983 and 6975 eV, which could be described by the transition state ($2p_{3/2} \rightarrow 5d$) of Eu^{3+} and Eu^{3+} ions, respectively. Results indicated the coexistence of Eu^{2+} and Eu^{3+} ions among tested specimens. Sodium ions were used as charge compensators among LBSS,3%Eu^{2+},3%Na^+ microcrystals. The absorption intensities increased by that of Eu^{2+} (6975 eV), however, decreased by Eu^{3+} ions (peak at 6983 eV) compared to that of $(La_{0.97}Eu_{0.03})_3Br(SiS_4)_2$. To study the valence state of dopant Eu in microcrystals, XANES analysis was performed by Zhang et al. [20]. Using the references Eu_2O_3 and $BaMgAl_{10}O_{17}$:Eu^{2+}, the coexistence of Eu^{3+} and Eu^{2+} was noticed. Eu^{3+} absence was checked by choosing varied characteristic excitation wavelengths of 230, 393, and 464 nm to excite respective phosphors. However, the sharp red emission was caused by f–f transition of Eu^{3+} in emission spectra. This could be due to the electronic existence at excited state of Eu^{3+} undergoing a non-radiative transition to ground state due to lattice relaxation diminishing the transfer of charge below $^5D_0$ state regardless of lower or higher wavelength excitation energy. Li et al. [44] used XANES spectroscopy to determine the valence state of Eu, owing to difference in energy threshold of Eu^{3+} (4f6) and Eu^{2+} (4f7) from 4f to unoccupied 5d state. Due to shielding effect of nuclear potential by the excessive 4f electrons in Eu^{2+}, these ions represent low binding energy than Eu^{3+} ions, which are the corresponding core electrons. To further analyze the Eu valence state and explore the transformative mechanism in 2+ and 3+ system by the substitution of $[Ca^{2+}–P^{5+}]$ for $[La^{3+}–Si^{4+}]$, Eu L_3-edge XANES spectra was recorded by CLSPO-1350 and CLSPO-1250 (Figure 15.15(a)). Two evident peaks at 6984 and 6975 eV were recorded which could be described by the electronic

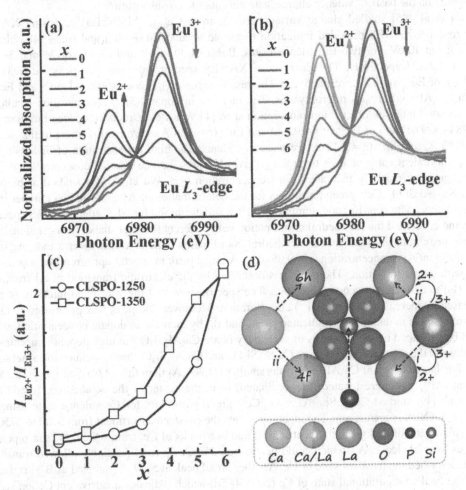

FIGURE 15.15 XANES spectra of $Ca_{0.98(2+x)}La_{0.98(8x)}Eu_{0.2}(SiO_4)_{6-x}(PO_4)_xO_2$ phosphors at (a) and (b) 1250 and 1350°C; (c) Eu^{2+}, Eu^{3+} emission dependence incorporation into phosphors; (d) schematic diagram of possible transformation mechanisms of cationic co-substitution. Reprinted from Li et al. [44] with permission.

($2p_{3/2} \rightarrow 5d$) transition of Eu^{3+} and Eu^{2+}, respectively. For samples, CLSPO-1250/1350-0, majority of ions exist in the form of Eu^{3+}. In addition, electric-dipole transition $^5D_0$–$^7F_2$ resulted in the highest emission at 613 nm (Figure 15.15(b)), which was significantly higher than magnetic-dipole transition $^5D_0$–$^7F_1$. Results indicated that Eu^{3+} likely entered into site deviating from inversion symmetric sites. The substitution for [La^{3+}–Si^{4+}] by [Ca^{2+}–P^{5+}], the relative absorption intensity at 6975 eV by Eu^{2+}, and at 6983 eV by Eu^{3+} ions, gradually increased, indicating the amount of Eu^{2+}/Eu^{3+} gradually increased. The Eu^{3+} to Eu^{2+} transformation in two series differs significantly, which could be described by the enhanced sintering temperature, which induces the La^{3+} occupancy by the Eu and the decrease in Eu^{3+} ions in CLSPO-1350 specimens compared with CLSPO-1250 specimens. From abovementioned results, study described possible transformation mechanism of Eu^{3+} and Eu^{2+} (Figure 15.15(d)). In host matrix, La^{3+} occupies all 6h regions and half of 4f, however, other half of the 4f occupied by the Ca^{2+} ions. La^{3+} primarily occupied by Eu in two series, thus synthesizing Eu^{3+} and emitting red light of transition $^5D_0$–^7F_j. When Si^{4+} gradually substituted by P^{5+}, the gradual replacement of Ca^{2+} for Eu^{3+}/La^{3+} occurs to maintain charge balance (Figure 15.15(c)). Hence, transformation of Eu^{3+} to Eu^{2+} occurs with the generation of 5d–4f emission. By managing the sintering temperature, crystalline strength at Eu^{2+} is increased due to increase in crystallinity, therefore, preventing tunable blue to green emission. This process attained optical characteristics adjustment of phosphors on the basis of valence charge state *via* cationic co-substitution.

Tsai et al. [45] studied charge variation of Eu in $Sr_xCa_{0.993-x}AlSiN_3:Eu^{2+}_{0.007}$ using XANES spectroscopy. XANES provided ionization threshold around Eu3+ ($4f^6$) and Eu2+ ($4f^7$) absorption edge at 10 eV. For two Eu valence states, $BaMgAl_{10}O_{17}:Eu^{2+}$ and Eu_2O_3 for 2+ and 3+ ionization states, respectively. The Eu L_3-edge XANES spectra indicated two peaks at 6984 and 6974 eV of Eu^{3+} and Eu^{2+}, respectively. The results revealed the coexistence of Eu^{3+} and Eu^{2+} in $Sr_xCa_{0.993-x}AlSiN_3:Eu^{2+}_{0.007}$ microcrystals. The increase in dopant (Sr^{2+}) concentration resulted in the increased intensity of Eu^{2+} ions absorption at 6974 eV, however, reduced concentration of Eu^{3+} at 6984 eV. The radii of Eu^{3+} ($r = 0.95$ Å) and Eu^{2+} ($r = 1.14$ Å) were in agreement to that of Ca^{2+} ($r = 0.99$ Å) and Sr^{2+} ($r = 1.12$ Å), respectively. Hence, the high concentration of Sr ions results in the convenient entry of Eu^{2+} to $Sr_xCa_{0.993-x}AlSiN_3:Eu^{2+}_{0.007}$. The relative concentration of Eu^{3+} is decreased by partially increasing Eu^{2+} concentration [45]. Wu et al. [49] studied as-prepared $Ca_{2.955}Sc_{2-x}Al_xSi_3O_{12}:Ce^{3+}$ green phosphors for the determination of Al coordination number in the crystal lattice. The coordination number of Al^{3+} was determined by Al K-edge XANES spectroscopy and concluded the tetrahedral coordination with different spectral shape, stronger signal, and reverse peak intensity than that in octahedral coordination. The XANES spectra indicate series with different Al^{3+} concentration ($x = 0.05$–0.4). A broad peak in spectra appeared that was associated with dipole-transition. The peak was described by the electronic transition of Al from core state (1s) to derived (unoccupied 3p) state. The spectra show that the width and intensity of peak increased by increasing dopant (Al^{3+}) concentration. Moreover, the peak was progressively modified from single to dual peaks (indicated by A_1 and B_1) by increase in dopant concentration, which could be described by the synthesis of secondary phase $Ca_3Al_2Si_3O_{12}$ in high dopant-concentration samples. By the previous reports, both $Ca_3Sc_2Si_3O_{12}$ and $Ca_3Al_2Si_3O_{12}$ are associated to garnet structure. The lattice constant $Ca_3Al_2Si_3O_{12}$ was smaller (11.864 Å) than that of $Ca_3Sc_2Si_3O_{12}$ (12.25 Å). Dopant (Al^{3+}) ions were located (6-coordinated) in in the center of the octahedron. In addition, Wu et al. [49] studied $Ca_{2.955}Sc_{2-x}Al_xSi_3O_{12}:Ce^{3+}$ green phosphors for Ce valence state using Ce L_3-edge XANES spectroscopy. The inset represents the enlarged assortment from 5700 to 5750 eV after subtraction of background. Spectra indicated two peaks of Ce, from which, the sharp peak at 5726 eV was labeled as A_2, which could be attributed to Ce 2p to $4f^15d$ dipole-allowed transition, which described the trivalent state of Ce. Another broad peak at 5737 eV (labeled as B_2) originated from the final configurational state of Ce (2p to $4f^05d$) which indicates quadrivalent Ce ion signal. The abovementioned findings indicated the coexistence of Ce^{3+} and Ce^{4+} ions in newly prepared phosphors. The material used for Ce was $Ce_2(C_2O_4)_3 \cdot 9H_2O$ (Ce^{3+}); CeO_2 was produced in trace amount in sintering process at high temperature. The spectral investigations indicated that the peak

A_2 intensity was much higher than that of B_2, which indicated higher concentration of Ce^{3+} ions compared with Ce^{4+} ions. Moreover, by dopant addition, A_2 peak intensity was increased directly proportional to that of Al^{3+} concentration. No significant modifications were observed in B_2 peak intensity. The emission and excitation wavelengths were 503 and 450 nm, respectively. The wide emission peaks and absorption band of Ce^{3+} could be described by the transition (4f–5d) state of Ce. The inset represents the integral emission intensity of $Ca_{2.955}Sc_{2-x}Al_xSi_3O_{12}$:$Ce^{3+}$ series compared to that of $x = 0$ and the variation of peak position as Al^{3+} function. By the emission intensity comparison, the intensity can significantly be improved by co-doping Al^{3+}. After co-doping, significant increase in the concentration of Ce^{3+} was observed. However, at $x = 0.4$, A_2 peak intensity turned toward weaker and B_2 became stronger, which resulted in significantly decreased emission intensity. This could be described by the sudden change caused by appearance of secondary phase $Ca_3Al_2Si_3O_{12}$ destroying the PL characteristic [49].

The Ca $L_{3,2}$-edge XANES spectra of $Ca_{2.955}Sc_{2-x}Al_xSi_3O_{12}$:$Ce^{3+}$ green phosphors showed multiple peaks that were generated by the dipole-transition of Ca $L_{3,2}$-edge from 2p to 3d. The dissociation of a_1/a_2 and b_1/b_2 could be due to non-liner value of crystal parameter (10 Dq) field. Previous researches have designated the discrepancies of Ca local environment indicating various coordination states such as six-coordinated octahedron and eight-coordinated cube. However, this study [49] reported as-prepared $Ca_{2.955}Sc_{2-x}Al_xSi_3O_{12}$:$Ce^{3+}$ green phosphors as eight-coordinated distorted-cubic environment. Moreover, dopant (Al^{3+}) addition had not influenced the crystal lattice of the host. The increased dopant ions (Al^{3+}) concentration resulted in higher intensity and the energy as indicated by Ca $L_{3,2}$-edge signal. At $x = 0.4$, the $Ca_3Al_2Si_3O_{12}$ showed the secondary phase; the trend was reversed. The systematic shift in energy of Ca $L_{3,2}$-edge was related to the symmetric change around Ca sites. By abovementioned indications, the Ce^{3+} ions were described to occupy Ca^{2+} sites. The declined symmetry of occupying the environment enhances the splitting and shifts the emission spectra toward red light. By the increase in dopant ions concentration, the redshifted peak from 503 to 507 nm was observed, however, shifted back to 505 nm. The valence state of O was also determined by using normalized O K-edge XANES spectroscopic investigations. The difference in intensity has been presented at the bottom, relative to the sample without dopant (Al^{3+}) addition. The clear change in peak was observed before and after dopant (Al^{3+}) ions addition into the host, which synthesized at 527.8 eV (designated as peak A_3). The new peak intensity was enhanced by the enhancement in dopant concentration. This could be described by the excitation of O electrons (1s to 2p) highly hybridized state with 3d of Al. Because the diversity in components of as-synthesized $Ca_{2.955}Sc_{2-x}Al_xSi_3O_{12}$:$Ce^{3+}$ phosphors, the spectra were recorded by all characteristic combination peaks from interactional modes of O with Ce, Si, Al, Sc, and Ca atoms. From XANES spectra of raw materials, peak observed could be attributed to Ce^{3+}. These indications were in agreement with Ce L_3-edge XANES results, which showed that dopant ions (Al^{3+}) can maintain Ce^{3+} and inhibit the synthesis of CeO_2 impurity phase. Peaks observed at 529.8, 532.0, 534.2, 536.6, 538.4, and 539.8 eV (denoted by B_3 to G_3) were obtained by the host $Ca_3Sc_2–Si_3O_{12}$. Moreover, the signal variation was decreased indicating Sc^{3+}, which was replaced by Al^{3+} dopant ions. The photon energy was shifted slightly toward lower energy as presented in O K-edge XANES spectra of $Ca_{2.955}Sc_{2-x}Al_xSi_3O_{12}$:$Ce^{3+}$, representing that the host gained electrons. The crystal field influenced energy splitting of Ce^{3+} and the covalency effect, competing with each other. It could be clearly observed that the addition of Al^{3+} induced a compressive stress at Ce^{3+}, shifting the excited state energy toward lower energies, which induced redshift emission. The XANES spectra revealed that after dopant addition, electronic migration occurred toward the oxygen side. In as-synthesized phosphors, the electronegativities of Sc^{3+} (1.36), Al^{3+} (1.61), and O^{2-} (3.44) were observed. The difference in electronegativities of O^{2-} and cations is called electronegativity mismatch. The addition of co-doping Al^{3+} ions reduced the electronegativity mismatch, which caused high covalency in host causing splitting in energy of Ce^{3+} to alter to lower state, ultimately causing blue emission. The covalency and the crystal field effects induced blue and redshift emission, respectively.

Vij, Gautam et al. [50] studied the electronic structure and the existence of Sm in SrS using NEXAFS spectroscopy in both total fluorescence yield (TFY) and total electron yield (TEY) mode at Sm $M_{5,4}$-edges SrS:Sm (1.0 mol%). The $M_{5,4}$-edges spectra were dominated by excitation transition ($3d^{10}4f^n$–$3d^94f^{n+1}$). Due to spin orbital association of $3d^9$ host was larger than that of exchange ($3d^94f^{n+1}$), the NEXAFS spectra consisted of two well-defined line groups near $3d_{3/2}$ (M4) and $3d_{5/2}$ (M5). At 1104 and 1079 eV, the spectral properties for 0.1, 0.5, and 1.0 mol% showed spectral similarities. Moreover, the pre-edge characteristic feature at 1077 eV was also identified for 1.0 mol% of Sm in SrS, which could not be observed in other spectra. The spectral analysis was performed by Sm $M_{5,4}$-edges for Sm^{2+} and Sm^{3+} ions by atomic multiple correlation calculations using experimental data. The results indicated the spectral similarities with Sm^{3+}, indicating the existence of Sm^{3+} valence state of ions in SrS. In addition, spectral features at 1 mol% were in agreement with that of simulated Sm^{2+} spectrum, exhibiting the coexistence of Sm^{2+} and Sm^{3+} ions. For further clarifications, NEXAFS of 1.0 mol% SrS:Sm was performed in TFY mode, which indicated high spectral similarities in TFY mode analysis as in TEY mode except 'feature A', which was absent in TFY. The results indicated that the Sm^{2+} ions were present on the surface on SrS:Sm. In view of abovementioned results, it could be concluded that the exceptional feature was due to the decrease of some Sm^{3+} ions to Sm^{2+} ions. Vij, Gautam et al. [50] also studied the co-doping effect of Ce on the PL and the structure of SrS:Sm phosphors by fixing the individual concentration of both ions to 0.5 mol%. Ce and Sm $M_{4,5}$-edges NEXAFS of SrS:Ce,Sm was performed to determine the valence state of both doping ions in host matrix. The Ce $M_{5,4}$-edges NEXAFS spectra indicated spectral similarities with trivalent CeF_3 standard. Atomic multiplet calculations were used to study Ce^{3+} Ce $M_{5,4}$-edges, which showed excellent similarities with experimental data, indicating the valence state of Ce as 3+ in the phosphors. Results indicated that both the dopants were present in trivalent forms in the SrS:Ce,Sm phosphors.

Yeh et al. [51] used XANES spectral investigation to study the valence state of Eu ions in $M_{2-x}Si_5N_8:Eu_x$ (M = Sr, Ba) phosphors (Figure 15.16). Herein, Eu_2O_3 and commercial $BaMgAl_{10}O_{17}$:Eu phosphor were used as standard for Eu^{3+} and Eu^{2+} valence state. The coexistence of Eu^{3+}/Eu^{2+} ions was observed by both XANES and XPS investigations. The ESCA analysis showed the conversion of Eu^{3+} more than that of Eu^{2+} ions (without thermal treatment), showing material oxidization at the surface resulting conversion of Eu^{2+} to Eu^{3+}, which was in agreement with the XANES investigations, which showed a high amount of Eu^{2+} compared to Eu^{3+} ions. Large amount of Eu^{3+}

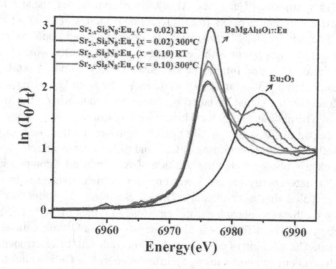

FIGURE 15.16   Eu L_3-edge XANES of $M_{2-x}Si_5N_8:Eu_x$ (M = Sr, Ba) phosphors. Reprinted from Yeh et al. [51] with permission.

was determined before thermal treatment for $Sr_2Si_5N_8$:Eu ($x = 0.02$). The increased Eu^{3+} ions were observed in $x = 0.10$ phosphors. The microcrystal contact with air caused the oxidization of Eu^{2+} to Eu^{3+} ion from the surface. Therefore, $Sr_2Si_5N_8$:Eu thermal degradation could be associated with increased Eu^{3+} content after heat treatment.

Huang et al. [52] studied valence state of Eu in $Ca_{12}Al_{14-z}Si_zO_{32+z}F_{2-z}$:Eu phosphors using XANES spectroscopy, which could easily differentiate valence states of Eu^{3+} ($4f^6$) and Eu^{2+} ($4f^7$) because of the threshold energy difference around 8 eV of their white-light resonance, which could be attributed to the transition (4f–5d) state. The difference of energy caused by the shielding effect caused by additional 4f electrons in Eu^{2+} significantly reduced the binding energy of core electrons. To study Eu valence state in $Ca_{12}Al_{14-z}Si_zO_{32+z}F_{2-z}$:Eu phosphors, Eu L_3-edge XANES was performed keeping Eu_2O_3 and $BaMgAl_{10}O_{17}$:Eu^{2+} references for Eu^{3+} and Eu^{2+} valence states, respectively. The normalized XANES spectra revealed two peaks at 6983 and 6975 eV, which could be described by the transition ($2p_{3/2}$/2dd) in Eu^{3+} and Eu^{2+}, respectively. The results indicated the coexistence of both 2+ and 3+ valence states among these phosphors. The systemic decrease and increase in relative absorption intensities by Eu^{3+} at 6983 eV and Eu^{2+} at 6975 eV could be correlated with the amount of Si^{4+}–O^{2-} incorporation. The increasing emission intensity and the ratios of Eu^{2+}/Eu^{3+} shows similar trend for the enhancement of Si^{4+}–O^{2-} incorporation, indicating that the luminescence basically correlates with the concentration of Eu^{2+} ions [52].

15.5 EXTENDED X-RAY ABSORPTION FINE STRUCTURE (EXAFS) SPECTROSCOPY

EXAFS determine the number and bond length of surrounding atoms to that molecule or materials. This spectroscopic technique has widely been used for the determination of local environmental variation in phosphors (Table 15.4). Pt L_3-edge EXAFS spectra of Pt NCs/BP indicated main peak at 2.5 Å, showing the near coordination of Pt-atoms (Figure 15.17). The lower primary peak in PtRu NCs/BP represents the synthesis of nanoalloy, and the reduced primary peak intensities in PtRu NCs/BP and Pt NCs/BP represent the synthesis of nanoclusters, which was confirmed by the redshift of PtRu NCs/BP peak position, because of the formation of relatively short Pt–Ru bonds. Furthermore, at 1.5 Å, no significant peak of Ru–O coordination can be noticed indicating that the nanoclusters are not apparently oxidized on the surface [46]. The coordination between Nb and Mg_2TiO_4 matrix was investigated using EXAFS analysis by Huang et al. [38]. The details structural analysis showed that the bond lengths between Nb and neighboring Mg (8a) and O (32e) were closer to those of Mn and Ti in same Mg_2TiO_4 matrix. This shows that the crystal structure of Nb has high

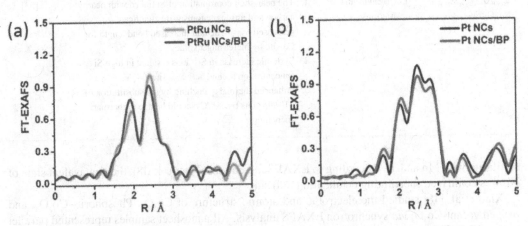

FIGURE 15.17 (a) Ru K-edge EXAFS spectra for Pt NCs/BP and PtRu nanocrystals, and (b) Pt L_3-edge EXAFS spectra for Pt NCs/BP and PtRu nanocrystals. Reprinted from Li, Pei, & He et al. [46] with permission.

TABLE 15.4

Few Recent Studies Describing EXAFS Spectroscopic Investigation for the Determination of Local Environment and Bond Lengths of Phosphors with Change Conditions and Key Findings

Host Matrix	Change Condition	Results	Reference
Pt nanoclusters/black phosphorus hybrids	Ru doping	1. Lower primary peak in PtRu NCs/BP represents synthesis of nanoalloy. 2. Reduced primary peak intensities in PtRu NCs/BP and Pt NCs/BP represent the synthesis of nanoclusters. 3. No surface oxidation was observed due to the absence of Ru–O peak.	[46]
Mg_2TiO_4	Nb doping	1. High crystal similarity of Nb with Mn and Ti was reported. 2. Pentavalent state of Nb was confirmed substituting octahedral (16d) site.	[38]
$CaAlSiN_3$	Eu doping	1. The average La–N bond distances of La-series materials were shorter compared with Ca–N distance. 2. The binding strength of Ca–N was lower than that of La–N.	[28]
Mo_2C	P doping	1. By the increasing dopant concentration, increased intensity of Mo-P was observed. 2. Dopant addition results in Mo O/C shifting to lower radical distance. 3. P-Mo linkages in P-Mo_2C@C, show direct link between C and dopant.	[42]
K_2MF_6:Mn^{4+} (M = Si, Ge)	Si^{4+} and Ge^{4+} coordination	1. The bond length between ligand and Ge^{4+} (1.81 Å) was longer than that of ligand with Si^{4+} (1.66 Å). 2. The coordination environment strongly influenced the luminescence properties.	[13]
$Sr_{0.92-x}Ba_xSiAl_2O_3N_2$: $Ce^{3+}_{0.04}$,$Eu^{2+}_{0.04}$	Activator ions addition (Ce^{3+} or Eu^{2+})	1. Progressive replacement of Sr^{2+} by larger Ba^{2+} ions indicated decreased enlargement of unit cells. 2. Site occupation by activator ions resulted in structural relaxation with redshifting.	[22]
$Ca_{0.99}Al_{1-48/3-x}Si_{1+\delta+x}$ $N_{3-x}C_x$:$Eu^{2+}_{0.01}$	Local environment variations	1. The near-shell coordination of Si linked with nearby carbon and nitrogen atoms were significantly decreased for Si-site and increased bond length for Ca-site by increasing x values. 2. Systemic increase in Si^{4+} ions resulted in high Si concentration around activator sites. 3. Enhanced thermal quenching by the substitution of Al^{3+}/N^{3-} ions by Si^{4+}/C^{-4} could be due to thermal quenching.	[53]

similarity with Mn and Ti. According to EXAFS indications (Figure 15.18), the pentavalent state of Nb was confirmed substituting octahedral (16d) site.

Xiao et al. [48] studied the electronic and atomic structure of Co_3O_4, Phosphorus-Co_3O_4, and oxygen vacant-Co_3O_4 *via* synchrotron EXAFS analysis. All nanosheet samples represented parallel peaks at EXAFS spectra as reference Co_3O_4 spinel, which indicated similar spinal structure of all nanosheets. The weak power of oscillation from nanosheets (Co_3O_4) was observed rather than that

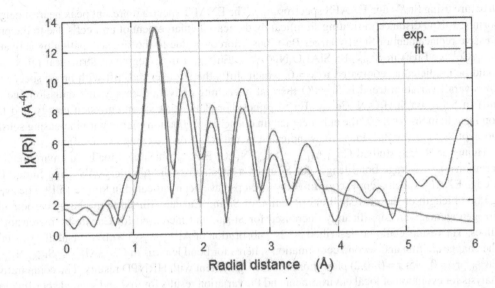

FIGURE 15.18 Nb EXAFS spectra of Mg_2TiO_4:Mn,Nb phosphors. Reprinted from Huang et al., [38] with permission.

of bulk Co_3O_4, which could be described by the presence of more atoms on the surface and thus exhibiting more disorder in Co_3O_4 nanostructure. In addition, three Fourier-transformed intensities indicate the scattering of various coordination shells. Initial peak was observed from two Co–O bonds, consisting of both kinds of Co^{3+}–O octahedral and Co^{2+}–O tetrahedral shells. The next peak could be attributed to Co–Co bonds for [Co] octahedral ions. Remarkably, the major alteration in Fourier-transformed intensities could only be observed in first coordination shell. Oxygen vacant-Co_3O_4 in first shell showed relatively low coordination number compared with pristine Co_3O_4, which indicated the formation of oxygen vacancies after plasma etching. Comparatively high coordination number was observed in first shell in P-Co_3O_4 than in oxygen vacant-Co_3O_4, and nearly similar in pristine-Co_3O_4, which could be described by Ar-plasma treatment in the presence of sodium biphosphate, resulting in efficient P-filling among vacancies. However, interbonds distance of tetrahedral Co^{2+}–O and octahedral Co^{3+}–O were too near to be distinguished; therefore, the creation of oxygen vacancies was not clear by EXAFS results.

Wang et al. [28] studied $CaAlSiN_3$:Eu microcrystals for the determination of interactive effect of dopant ions and host matrix material using EXAFS spectroscopy. The nearest bond shows that distances of N^{3-} surrounding anions and La^{3+} cations. The average La–N bond distances of La-series materials were shorter compared with Ca–N distance, determined by SXRD Rietveld refinement of as-synthesized samples, which could be due to the large cationic size of La^{3+} than that of Ca^{2+}. Therefore, the combination of La^{3+} with nitride cation would result in high covalence compared with Ca^{2+} ions, indicating the binding strength of Ca–N is lower than that of La–N. On the other hand, the L–N theoretic binding strength was weaker than that of Ca–N due to relative weaker covalent bonds between nitride anions and Li^+ cations. Shi et al. [42] investigated the Mo_2C and dopant (P) interaction by EXAFS and XANES analyses.

Wei, Lin et al. [13] used EXAFS spectroscopy associated with Fourier transformation to analyze the K_2GeF_6:Mn^{4+} coordination environment with Si^{4+} and Ge^{4+}. The bond length between ligand (F^- ions) and Ge^{4+} ions (1.81 Å) was longer than that of F^- ions and Si^{4+} ions (1.66 Å). The difference between K_2SiF_6:Mn^{4+} and K_2GeF_6:Mn^{4+} phosphors in a lattice symmetry and coordinated environment with host structure strongly affect the optical properties of phosphors. Similarly, Huang et al. [22] studied $Sr_{0.92-x}Ba_xSiAl_2O_3N_2$:$Ce^{3+}_{0.04}$,$Eu^{2+}_{0.04}$ phosphors for the determination of Sr ionic structure using Sr K-edge EXAFS spectroscopy. The EXAFT spectra represent peak nearest neighboring Sr–N/O distance indicating significantly decreased enlargement of unit cells due to the progressive replacement of Sr^{2+} by larger Ba^{2+} ions. Moreover, due to the site occupation by activator ions (Ce^{3+} or Eu^{2+}) in $Sr_{0.92-x}Ba_xSiAl_2O_3N_2:Ce^{3+}_{0.04},Eu^{2+}_{0.04}$ a high degree of structural relaxation could be predicted as compared to $x = 0$, which shifts the stronger redshift with broad spectrum. The overall variation trend is Sr–N/O thermal quenching activation energy and emission energy shift in $Sr_{0.92-x}Ba_xSiAl_2O_3N_2:Ce^{3+}_{0.04},Eu^{2+}_{0.04}$ phosphor. The decrease in emission energy and the bond length in Sr–N/O, and the enhancement in quenching activation energy with increasing substitution of Sr^{2+} ions by Ba^{2+} ions are as indicated by the activator site compression effect.

Huang et al. [53] studied $Ca_{0.99}Al_{1-4\delta/3-x}Si_{1+\delta+x}N_{3-x}C_x:Eu^{2+}_{0.01}$ phosphors for local environmental variations, i.e., average bond length for Si and Ca sites across the microcrystals using Si and Ca K-edge EXAFS spectroscopy; coordination envelops have been indicated in Figure 15.19. The average bond length or the near-shell coordination of Si and Ca sites, linked with nearby carbon and nitrogen atoms, were significantly decreased for Si-site and increased for Ca-site by increasing x values. The results could be described by the substitution of Al^{3+} (larger) with Si^{4+} (smaller) sphere. The changes in first and second coordination spheres for bond lengths in $Ca_{0.99}Al_{1-4\delta/3-x}Si_{1+\delta+x}N_{3-x}C_x:Eu^{2+}_{0.01}$ ($\delta = 0.345$; $x = 0$–0.2) phosphors were in agreement with HRNPD results. The comparative analysis for evolution of local environment and PL variation results for first and second coordination sphere induced by the replacement of Al^{3+}/N^{3-} ions by Si^{4+}/C^{4-} in $Ca_{0.99}Al_{1-4\delta/3-x}Si_{1+\delta+x}N_{3-x}C_x:Eu^{2+}_{0.01}$ phosphors' lattice. Systemic enhancement in Si^{4+} ions in second coordination sphere resulted in high Si concentration around activator sites in first sphere, resulting in loose activator sites, weaker crystal field splitting, longer Ca–N/C bond lengths, and higher emission shifting toward blue light. The overall trend in Si–N/C bond length based on thermal quenching and EXAFS analysis has been shown in Figure 15.19. In first coordination sphere, central complex $[EuN_5]^{-13}$ ions were expanded

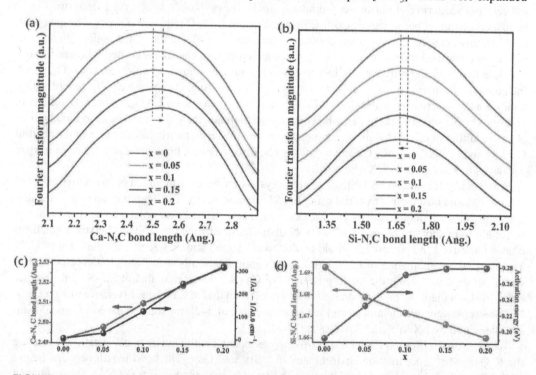

FIGURE 15.19 (a) Ca K-edge Fourier-transformed EXAFS showing Ca–C,C envelope for $Ca_{0.99}Al_{1-4\delta/3-x}Si_{1+\delta+x}N_{3-x}C_x:Eu^{2+}_{0.01}$ phosphors and (b) Si K-edge Fourier-transformed EXAFS showing Si–C,N envelope; (c) Trend of average Ca–C,N bond length from EXAFS spectra and (d) trend of average Si–C,N bond length from EXAFS spectra. Reprinted from Huang et al. [53] with permission.

by increase in temperature or excitation. Meanwhile, the second coordination sphere shrinkage resulted by Si^{4+} ions for charge compensation enhanced the capability to respond to the expansion of central complex in initial coordination sphere. The reduced second coordination sphere in $Ca_{0.99}Al_{1-4\delta/3-x}Si_{1+\delta+x}N_{3-x}C_x$:$Eu^{2+}_{0.01}$ phosphors causes high TQ activation energy. The gradual enlargement trend in first sphere and increased emission energy, with second sphere shrinkage, and enhanced thermal quenching by the substitution of Al^{3+}/N^{3-} ions by Si^{4+}/C^{-4} in $Ca_{0.99}Al_{1-4\delta/3-x}Si_{1+\delta+x}N_{3-x}C_x$:$Eu^{2+}_{0.01}$ phosphors could be described by the shrinkage effect of second sphere on the activation site.

15.6 CONCLUSIONS

In this chapter, we have summarized analysis methods of synchrotron radiation for many phosphors, including SXRD, VUV-PL, XANES, and EXAFS, and have been comprehensively discussed here. Synchrotron radiation is a powerful tool for the accurate structure, oxidation state, coordination environment, VUV excitation, and emission of fluorescent materials. Moreover, the sample preparation, operation technology, and data analysis are also important if you want to get the perfect results.

ACKNOWLEDGMENTS

This work was supported by the Ministry of Science and Technology of Taiwan (Contract Nos. MOST 109-2113-M-027-004, MOST 109-2622-M-027-001-CC2, MOST 110-2113-M-027-012, and MOST 110-2622-M-027-001) and Joint Research Program Funding Sponsorship by University System of Taipei (Contract Nos. USTP-NTUT-TMU-109-02 and USTP-NTUT-TMU-110-02).

REFERENCES

1. Toby, B. H., Billinge, S. J. L. (2004) Determination of standard uncertainties in fits to pair distribution functions. *Acta Cryst.* A60: 315.
2. Chupas, P. J., Chapman, K. W., Lee, P. L. (2007) Applications of an amorphous silicon-based area detector for high-resolution, high-sensitivity and fast time-resolved pair distribution function measurements. *J. Appl. Crystallogr.* 40: 463.
3. Lin, C. C., Tsai, Y. T., Johnston, H. E., Fang, M. H., Yu, F. J., Zhou, W. Z., Whitfield, P., Li, Y., Wang, J., Liu, R. S., Attfield, J. P. (2017) Enhanced photoluminescence emission and thermal stability from introduced cation disorder in phosphors. *J. Am. Chem. Soc.* 139: 11766.
4. Wei, Y., Gao, J. S., Xing, G. C., Li, G. G., Dang, P. P., Liang, S. S., Huang, Y. S., Lin, C. C., Chan, T. S., Lin, J. (2019) Controllable Eu^{2+}-doped orthophosphate blue-/red-emitting phosphors: charge compensation and lattice-strain control. *Inorg. Chem.* 58: 6376.
5. Fang, M. H., Wu, W. L., Jin, Y., Lesniewski, T., Mahlik, S., Grinberg, M., Brik, M. G., Srivastava, A. M., Chiang, C. Y., Zhou, W. Z., Jeong, D., Kim, S. H., Leniec, G., Kaczmarek, S. M., Sheu, H. S., Liu, R. S. (2018) Control of luminescence by tuning of crystal symmetry and local structure in Mn^{4+}-activated narrow band fluoride phosphors. *Angew. Chem. Int. Ed.* 57: 1797.
6. Leaño Jr, J. L., Lazarowska, A., Mahlik, S., Grinberg, M., Sheu, H. S., Liu, R. S. (2018) Disentangling red emission and compensatory defects in $Sr[LiAl_3N_4]$:Ce^{3+} phosphor. *Chem. Mater.* 30: 4493.
7. Zhou, W., Fang, M. H., Lian, S., Liu, R. S. (2018) Ultrafast self-crystallization of high-external-quantum-efficient fluoride phosphors for warm white light-emitting diodes. *ACS Appl. Mater. Interfaces* 10: 17508.
8. Fang, M. H., Ni, C. C., Zhang, X. J., Tsai, Y. T., Mahlik, S., Lazarowska, A., Grinberg, M., Sheu, H. S., Lee, J. F., Cheng, B. M., Liu, R. S. (2016) Enhance color rendering index via full spectrum employing the important key of cyan phosphor. *ACS Appl. Mater. Interfaces* 8: 30677.
9. Dai, P., Lee, S. P., Chan, T. S., Huang, C. H., Chiang, Y. W., Chen, T. M. (2016) $Sr_3Ce(PO_4)_3$:Eu^{2+}: a broadband yellow-emitting phosphor for near ultraviolet-pumped white light-emitting devices. *J. Mater. Chem. C* 4: 1170.
10. Jin, Y., Fang, M. H., Grinberg, M., Mahlik, S., Lesniewski, T., Brik, M. G., Luo, G. Y., Lin, J. G., Liu, R. S. (2016) Narrow red emission band fluoride phosphor $KNaSiF_6$:Mn^{4+} for warm white light-emitting diodes. *ACS Appl. Mater. Interfaces* 8: 11194.

11. Lee, S. P., Chan, T. S., Chen, T. M. (2015) Novel reddish-orange-emitting $BaLa_2Si_2S_8:Eu^{2+}$ thiosilicate phosphor for LED lighting. *ACS Appl. Mater. Interfaces* 7: 40.

12. Wei, L. L., Lin, C. C., Wang, Y. Y., Fang, M. H., Jiao, H., Liu, R. S. (2015) Photoluminescent evolution induced by structural transformation through thermal treating in the red narrow-band phosphor $K_2GeF_6:Mn^{4+}$. *ACS Appl. Mater. Interfaces* 7: 10656.

13. Wei, L. L., Lin, C. C., Fang, M. H., Brik, M. G., Hu, S. F., Jiao, H., Liu, R. S. (2015) A low-temperature co-precipitation approach to synthesize fluoride phosphors $K_2MF_6:Mn^{4+}$ (M = Ge, Si) for white LED applications. *J. Mater. Chem. C* 3: 1655.

14. Li, G. G., Lin, C. C., Chen, W. T., Molokeev, M. S., Atuchin, V. V., Chiang, C. Y., Zhou, W. Z., Wang, C. W., Li, W. H., Sheu, H. S., Chan, T. S., Ma, C. G., Liu, R. S. (2014) Photoluminescence tuning via cation substitution in oxonitridosilicate phosphors: DFT calculations, different site occupations, and luminescence mechanisms. *Chem. Mater.* 26: 2991.

15. Lee, S. P., Huang, C. H., Chan, T. S., Chen, T. M. (2014) New Ce^{3+}-activated thiosilicate phosphor for LED lighting-synthesis, luminescence studies, and applications. *ACS Appl. Mater. Interfaces* 6: 7260.

16. Huang, C. H., Chan, T. S., Liu, W. R., Wang, D. Y., Chiu, Y. C., Yeh, Y. T., Chen, T. M. (2012) Crystal structure of blue–white–yellow color-tunable $Ca_4Si_2O_7F_2:Eu^{2+},Mn^{2+}$ phosphor and investigation of color tunability through energy transfer for single-phase white-light near-ultraviolet LEDs. *J. Mater. Chem.* 22: 20210.

17. Leaño Jr, J. L., Lesniewski, T., Lazarowska, A., Mahlik, S., Grinberg, M., Sheu, H. S., Liu, R. S. (2018) Thermal stabilization and energy transfer in narrow-band red-emitting $Sr[(Mg_2Al_2)_{1-y}(Li_2Si_2)_yN_4]:Eu^{2+}$ phosphors. *J. Mater. Chem. C* 6: 5975.

18. Lee, S. P., Chan, T. S., Dutta, S., Chen, T. M. (2018) Novel Eu^{2+}-activated thiogallate phosphors for white LED applications: structural and spectroscopic analysis. *RSC adv.* 8: 11725.

19. Lee, S. P., Liu, S. D., Chan, T. S., Chen, T. M. (2016) Synthesis and luminescence properties of novel Ce^{3+}-and Eu^{2+}-doped lanthanum bromothiosilicate $La_3Br(SiS_4)_2$ phosphors for white LEDs. *ACS Appl. Mater. Interfaces* 8: 9218.

20. Zhang, X., Tsai, Y. T., Wu, S. M., Lin, Y. C., Lee, J. F., Sheu, H. S., Cheng, B. M., Liu, R. S. (2016) Facile atmospheric pressure synthesis of high thermal stability and narrow-band red-emitting $SrLiAl3N_4:Eu^{2+}$ phosphor for high color rendering index white light-emitting diodes. *ACS Appl. Mater. Interfaces* 8: 19612.

21. Pust, P., Weiler, V., Hecht, C., Tücks, A., Wochnik, A. S., Henß, A. K., Wiechert, D., Scheu, C., Schmidt, P. J. Schnick, W. (2014). Narrow-band red-emitting $Sr[LiAl_3N_4]:Eu^{2+}$ as a next-generation LED-phosphor material. *Nat. Mater.* 13: 891.

22. Huang, W. Y., Yoshimura, F., Ueda, K., Shimomura, Y., Sheu, H. S., Chan, T. S., Chiang, C. Y., Zhou, W. Z., Liu, R. S. (2014) Chemical pressure control for photoluminescence of $MSiAl_2O_3N_2:Ce^{3+}/Eu^{2+}$ (M = Sr, Ba) oxynitride phosphors. *Chem. Mater.* 26: 2075.

23. Lin, C. C., Liu, Y. P., Xiao, Z. R., Wang, Y. K., Cheng, B. M., Liu, R. S. (2014) All-in-one light-tunable borated phosphors with chemical and luminescence dynamical control resolution. *ACS Appl. Mater. Interfaces* 6: 9160.

24. Huang, C. H., Liu, W. R., Chan, T. S., Lai, Y. T. (2014) Orangish-yellow-emitting $Ca_3Si_2O_7:Eu^{2+}$ phosphor for application in blue-light based warm-white LEDs. *Dalton Trans.* 43: 7917.

25. Huang, C. H., Lai, Y. T., Chan, T. S., Yeh, Y. T., Liu, W. R. (2014) A novel green-emitting $SrCaSiAl_2O_7:Eu^{2+}$ phosphor for white LEDs. *RSC Adv.* 4: 7811.

26. Huang, W. Y., Yoshimura, F., Ueda, K., Shimomura, Y., Sheu, H. S., Chan, T. S., Greer, H. F., Zhou, W. Z., Hu, S. F., Liu, R. S., Attfield, J. P. (2013) Nanosegregation and neighbor-cation control of photoluminescence in carbidonitridosilicate phosphors. *Angew. Chem. Int. Ed.* 125: 8260.

27. Huang, C. H., Chiu, Y. C., Yeh, Y. T., Chan, T. S., Chen, T. M. (2012) Eu^{2+}-activated $Sr_8ZnSc(PO_4)_7$: a novel near-ultraviolet converting yellow-emitting phosphor for white light-emitting diodes. *ACS Appl. Mater. Interfaces* 4: 6661.

28. Wang, S. S., Chen, W. T., Li, Y., Wang, J., Sheu, H. S., Liu, R. S. (2013) Neighboring-cation substitution tuning of photoluminescence by remote-controlled activator in phosphor lattice. *J. Am. Chem. Soc.* 135: 12504.

29. Shannon, R. D. (1976) Revised effective ionic radii and systematic studies of interatomic distances in halides and chalcogenides. *Acta Cryst.* 32A: 751.

30. Zhao, M., Xia, Z., Huang, X., Ning, L., Gautier, R., Molokeev, M. S., Zhou, Y. Y., Chuang, Y. C., Zhang, Q. Y., Liu, Q. L., Poeppelmeier, K. R. (2019) Li substituent tuning of LED phosphors with enhanced efficiency, tunable photoluminescence, and improved thermal stability. *Sci. Adv.* 5: eaav0363.

31. Lin, C. C., Chen, W. T., Chu, C. I., Huang, K. W., Yeh, C. W., Cheng, B. M., Liu, R. S. (2016) UV/VUV switch-driven color-reversal effect for Tb-activated phosphors. *Light Sci. Appl.* 5: e16066.

32. Wang, D. Y., Chen, Y. C., Huang, C. H., Cheng, B. M., Chen, T. M. (2012) Photoluminescence investigations on a novel green-emitting phosphor $Ba_3Sc(BO_3)_3$:Tb^{3+} using synchrotron vacuum ultraviolet radiation. *J. Mater. Chem.* 22: 9957.

33. Xia, Z., Zhang, Y., Molokeev, M. S., Atuchin, V. V., Luo, Y. (2013) Linear structural evolution induced tunable photoluminescence in clinopyroxene solid-solution phosphors. *Sci. Rep.* 3: 1.

34. Chen, W. T., Sheu, H. S., Liu, R. S., Attfield, J. P. (2012) Cation-size-mismatch tuning of photoluminescence in oxynitride phosphors. *J. Am. Chem. Soc.* 134: 8022.

35. Jin, H. C., Zhang, T. M., Chuang, C. H., Lu, Y. R., Chan, T. S., Du, Z. Z, Ji, H. X., Wan, L. J. (2019) Synergy of black Phosphorus–Graphite–Polyaniline-based ternary composites for stable high reversible capacity Na-ion battery anodes. *ACS Appl. Mater. Interfaces* 11: 16656.

36. Fang, M. H., Mahlik, S., Lazarowska, A., Grinberg, M., Molokeev, M. S., Sheu, H. S., Lee, J. F., Liu, R. S. (2019) Structural evolution and effect of the neighboring cation on the photoluminescence of $Sr(LiAl_3)_{1-x}(SiMg_3)_xN_4$:$Eu^{2+}$ phosphors. *Angew. Chem. Int. Ed.* 58: 7767.

37. Qiao, J., Ning, L., Molokeev, M. S., Chuang, Y. C., Zhang, Q., Poeppelmeier, K. R., Xia, Z. (2019) Site-selective occupancy of Eu^{2+} toward blue-light-excited red emission in a $Rb_3YSi_2O_7$:Eu phosphor. *Angew. Chem. Int. Ed.* 58: 11521.

38. Huang, C. S., Huang, C. L., Liu, Y. C., Lin, S. K., Chan, T. S., Tu, H. W. (2018) Ab initio-aided sensitizer design for Mn^{4+}-activated Mg_2TiO_4 as an ultrabright fluoride-free red-emitting phosphor. *Chem. Mater.* 30: 1769.

39. Fang, M. H., Tsai, Y. T., Sheu, H. S., Lee, J. F., Liu, R. S. (2018) Pressure-controlled synthesis of high-performance $SrLiAl_3N_4$:Eu^{2+} narrow-band red phosphors. *J. Mater. Chem. C* 6: 10174.

40. Zhang, N. M., Tsai, Y. T., Fang, M. H., Ma, C. G., Lazarowska, A., Mahlik, S., Grinberg. M., Chiang, C. Y., Zhou, W. Z., Lin, J. G., Lee, J. F., Zheng, J., Guo, C., Liu, R. S. (2017) Aluminate red phosphor in light-emitting diodes: theoretical calculations, charge varieties, and high-pressure luminescence analysis. *ACS Appl. Mater. Interfaces* 9: 23995.

41. Wei, Y., Jia, H., Xiao, H., Shang, M. M., Lin, C. C., Su, C. C., Chan, T. S., Li, G. G., Lin, J. (2017) Emitting-tunable $Eu^{(2+/3+)}$-doped $Ca_{(8-x)}La_{(2+x)}(PO_4)_{6-x}(SiO_4)_xO_2$ apatite phosphor for n-UV WLEDs with high-color-rendering. *RSC Adv.* 7: 1899.

42. Shi, Z., Nie, K., Shao, Z. J., Gao, B., Lin, H., Zhang, H., Liu, B., Wang, Y., Zhang, Y., Sun, X., Cao, X. M., Hu, P., Gao, Q. Tang, Y. (2017) Phosphorus-Mo_2C@carbon nanowires toward efficient electrochemical hydrogen evolution: composition, structural and electronic regulation. *Energy Environ. Sci.* 10: 1262.

43. Leaño Jr, J. L., Lin, S. Y., Lazarowska, A., Mahlik, S., Grinberg, M., Liang, C., Zhou, W. Z., Molokeev, M. S., Atuchin, V. V., Tsai, Y. T., Lin, C. C., Sheu, H. S., Liu, R. S. (2016) Green light-excitable Ce-doped nitridomagnesoaluminate $Sr[Mg_2Al_2N_4]$ phosphor for white light-emitting diodes. *Chem. Mater.* 28: 6822.

44. Li, G., Lin, C. C., Wei, Y., Quan, Z., Tian, Y., Zhao, Y., Chan, T. S., Lin, J. (2016) Controllable Eu valence for photoluminescence tuning in apatite-typed phosphors by the cation cosubstitution effect. *Chem. Commun.* 52: 7376.

45. Tsai, Y. T., Chiang, C. Y., Zhou, W., Lee, J. F., Sheu, H. S., Liu, R. S. (2015). Structural ordering and charge variation induced by cation substitution in $(Sr,Ca)AlSiN_3$:Eu phosphor. *J. Am. Chem. Soc.* 137: 8936.

46. Li, Y., Pei, W., He, J., Liu, K., Qi, W., Gao, X., Zhou, S., Xie, H., Yin, K., Gao, Y., He, J., Zhao, J., Hu, J., Chan, T. S., Li, Z., Zhang, G., Liu, M. (2019). Hybrids of PtRu nanoclusters and black phosphorus nanosheets for highly efficient alkaline hydrogen evolution reaction. *ACS Catal.* 9: 10870.

47. Li, R. H., Cui, J. L., Li, X. D., Li, X. Y. (2018). Phosphorus removal and recovery from wastewater using Fe-dosing bioreactor and cofermentation: investigation by X-ray absorption near-edge structure spectroscopy. *Environ. Sci. Technol.* 52: 14119.

48. Xiao, Z., Wang, Y., Huang, Y. C., Wei, Z., Dong, C. L., Ma, J., Shen, S., Li, Y., Wang, S. (2017) Filling the oxygen vacancies in Co_3O_4 with phosphorus: an ultra-efficient electrocatalyst for overall water splitting. *Energy Environ. Sci.* 10: 2563.

49. Wu, Y. F., Chiou, J. W., Nien, Y. T., Yeh, C. L., Chen, I. G., Lee, J. F. (2015). X-ray absorption near-edge spectroscopy study of photoluminescent $Ca_{2.955}Sc_{2-x}Al_xSi_3O_{12}$:$Ce^{3+}$ green phosphors. *Ceram. Int.* 41: 4538.

50. Vij, A., Gautam, S., Kumar, V., Brajpuriya, R., Kumar, R., Singh, N., Chae, K. H. (2013) X-ray absorption spectroscopy and photoluminescence study of rare earth ions doped strontium sulphide phosphors. *Appl. Surf. Sci.* 264: 237.

51. Yeh, C. W., Chen, W. T., Liu, R. S., Hu, S. F., Sheu, H. S., Chen, J. M., Hintzen, H. T. (2012) Origin of thermal degradation of $Sr_{2-x}Si_5N_8$:Eu$_x$ phosphors in air for light-emitting diodes. *J. Am. Chem. Soc.* 134: 14108.

52. Huang, K. W., Chen, W. T., Chu, C. I., Hu, S. F., Sheu, H. S., Cheng, B. M., Chen, J. M., Liu, R. S. (2012) Controlling the activator site to tune europium valence in oxyfluoride phosphors. *Chem. Mater.* 24: 2220.

53. Huang, W. Y., Yoshimura, F., Ueda, K., Pang, W. K., Su, B. J., Jang, L. Y., Chiang, C. Y., Zhou, W. Z., Duy, N. H., Liu, R. S. (2014) Domination of second-sphere shrinkage effect to improve photoluminescence of red nitride phosphors. *Inorg. Chem.* 53: 12822.

Index

Note: Locators in *italics* represent figures and **bold** indicate tables in the text.

Printed in the United States
by Baker & Taylor Publisher Services

Printed in the United States
by Baker & Taylor Publisher Services